Lecture Notes in Networks and Systems

Volume 514

The series "Lecture Notes in Networks and Systems" publishes the latest developments in Networks and Systems—quickly, informally and with high quality. Original research reported in proceedings and post-proceedings represents the core of LNNS.

Volumes published in LNNS embrace all aspects and subfields of, as well as new challenges in, Networks and Systems.

The series contains proceedings and edited volumes in systems and networks, spanning the areas of Cyber-Physical Systems, Autonomous Systems, Sensor Networks, Control Systems, Energy Systems, Automotive Systems, Biological Systems, Vehicular Networking and Connected Vehicles, Aerospace Systems, Automation, Manufacturing, Smart Grids, Nonlinear Systems, Power Systems, Robotics, Social Systems, Economic Systems and other. Of particular value to both the contributors and the readership are the short publication timeframe and the world-wide distribution and exposure which enable both a wide and rapid dissemination of research output.

The series covers the theory, applications, and perspectives on the state of the art and future developments relevant to systems and networks, decision making, control, complex processes and related areas, as embedded in the fields of interdisciplinary and applied sciences, engineering, computer science, physics, economics, social, and life sciences, as well as the paradigms and methodologies behind them.

Indexed by SCOPUS, INSPEC, WTI Frankfurt eG, zbMATH, SCImago.

All books published in the series are submitted for consideration in Web of Science.

For proposals from Asia please contact Aninda Bose (aninda.bose@springer.com).

More information about this series at https://link.springer.com/bookseries/15179

Joy Iong-Zong Chen · João Manuel R. S. Tavares · Fuqian Shi
Editors

Third International Conference on Image Processing and Capsule Networks

ICIPCN 2022

Springer

Editors
Joy Iong-Zong Chen
Department of Electrical Engineering
Da-Yeh University
Changhua, Taiwan

João Manuel R. S. Tavares
Departamento de Engenharia Mecânica,
Faculdade de Engenharia
Universidade do Porto
Porto, Portugal

Fuqian Shi
Rutgers Cancer Institute of New Jersey
New Brunswick, NJ, USA

ISSN 2367-3370 ISSN 2367-3389 (electronic)
Lecture Notes in Networks and Systems
ISBN 978-3-031-12412-9 ISBN 978-3-031-12413-6 (eBook)
https://doi.org/10.1007/978-3-031-12413-6

This Springer imprint is published by the registered company Springer Nature Switzerland AG
The registered company address is: Gewerbestrasse 11, 6330 Cham, Switzerland

We would like to dedicate this proceeding to all members of advisory committee, program committee for providing their excellent guidance. We also dedicate this proceeding to the members of the review committee for their excellent cooperation throughout the conference. We also record our sincere thanks to all the authors and participants.

Preface

It is our pleasure to welcome you to the 3rd International Conference on Image Processing and Capsule Networks [ICIPCN 2022] organized on May 20–21, 2022. A major goal and feature of the conference is to bring academia and industries together to share and exchange their significant research experiences and results in the field of imaging science, with a particular interest on the capsule network algorithms and models by discussing the practical challenges encountered and solutions adopted to it.

This conference will deliver a technically productive experience to the budding researchers in the field of image processing and capsule networks by stimulating a good awareness on this emerging research field. ICIPCN promises to provide a bright landscape for image processing research works, while the response and research eagerness witnessed have far exceeded our expectations. We are overwhelmed with gratitude at the conclusion of the conference event.

The response for the conference event has been increasing at an unprecedented rate both from Thailand and overseas. Due to the professional expertise of both internal and external reviewers, the papers have been selectively accepted based on their extensive research and publication quality. We have received a total submission of 298 out of which only 65 papers were accepted for publication based on its research effectiveness and applicability. We acknowledge our gratitude to the keynote speaker Dr. R. Kanthavel, Department of Computer Engineering, King Khalid University, Kingdom of Saudi Arabia, for valuable research insights of capsule networks. We would also like to extend our thanks to the members of the organizing committee for their hard work in delivering spontaneous response to all the conference participants.

We are now enthusiastic to get the proceedings of the ICIPCN conference event covered by Springer. We also appreciate all the authors of ICIPCN 2022 for their timely response to all the queries raised from the conference.

Finally, we would like to thank the Springer publications for producing this volume.

Joy Iong-Zong Chen
João Manuel R. S. Tavares
Fuqian Shi
Guest Editors

Contents

Brain-Inspired Spatiotemporal Feature Extraction Using Convolutional Legendre Memory Unit

G. Sudharshan[1(✉)], V. Khoshall[1], and M. Saravanan[2]

[1] Vellore Institute of Technology, Chennai, India
sudarchakra2000@gmail.com
[2] Ericsson India Global Services Pvt. Ltd., Chennai, India
m.saravanan@ericsson.com

Abstract. Feature extraction techniques in image processing are used to obtain useful numerical features from image data which can be further used in classification or regression tasks. Extracting features from single images only provides spatial information and extending them to sequential images provides both the temporal and spatial features. However, such networks require more time and power to run. SNN are efficient and consume less memory which makes them power efficient. In this paper, we extract spatiotemporal features relevant for SNN that can classify sequential images. We train our network to recognize the images even they are transformed or occluded over time. In this paper, we have proposed a new type of SNN called Convolutional Legendre Memory Unit (ConvLMU) inspired by Convolutional Long Short-Term Memory and LMU. Our model outperforms some of the state-of-the-art spatiotemporal feature extraction models run on a synthetic dataset generated from the benchmark MNIST dataset.

Keywords: Deep learning · Spiking Neural Networks · Spatiotemporal models · Convolutional Long Short-Term Memory · Convolutional Legendre Memory Unit

1 Introduction

Feature Extraction is a process of extracting useful features/information from the input data, often reducing the dimension of the data in the process. This process of reducing the dimensions is also called dimensionality reduction [1]. Feature extraction helps in building better and faster models [2, 3]. It is especially an important step while dealing with image data. Extracting features such as edges, contours, shapes, etc. can help improve the performance of the model [4]. Spatial feature extraction involves extracting features related to space, location, or geometry of the data. The spatial features of an image could include aspect ratio, shape, color, coordinate location, etc. Temporal features, as the name suggests, refer to time-domain features. Such features may change over time. Examples of temporal features for a sequence of images are transformation or change in resolution over time [5]. In this paper, we focus on extracting both spatial and

J. I.-Z. Chen et al. (Eds.): ICIPCN 2022, LNNS 514, pp. 1–12, 2022.
https://doi.org/10.1007/978-3-031-12413-6_1

temporal features related to sequential image data and using those features to perform classification tasks on sequential image data.

Feature extraction is a pre-processing step where the input data is transformed into values that can be forwarded while conserving the actual information from the dataset. This step was carried out before building a Machine Learning model to perform the classification tasks. However, with the advent of Neural Networks, feature extraction has become an implicit part of the neural network architecture, carried out by convolutional and pooling layer pairs. Traditional Convolutional Neural Networks (CNNs) are more capable of extracting spatial features related to images, however, It does not capture the temporal features in tandem [5]. The use of recurrent units such as LSTMs (Long Short-Term Memory) can help in extracting the temporal information of the data, and a combination of LSTMs and CNNs can perform the task of spatiotemporal feature extraction exceptionally well in most situations [1, 4]. However, such large neural networks require a considerable amount of time and power to perform such tasks. So, we explore an alternative to such traditional Artificial Neural Network (ANN) architectures, called Spiking Neural Networks (SNNs) [6] in this paper.

1.1 Neuromorphic Computing

Neuromorphic computing [7] is a crucial research area for the future of Artificial Intelligence (AI). Neuromorphic systems that are inspired by the working mechanism of human brains possess a massively parallel architecture with closely coupled memory and computing. The main challenge in neuromorphic research is to match a human's capability to learn from unstructured stimuli along with the efficiency of our complex brain system. SNNs are a novel network under neuromorphic computing to mimic some of the natural neural networks present in our brain [6]. SNNs are temporal which makes them suitable for spatiotemporal data. The emerging area of neuromorphic computing has great potential in improving hardware efficiency, upgrading system scalability, as well as accelerating the capability of intelligence systems [8].

In this paper, we have proposed a new type of SNN called Convolutional Legendre Memory Units (ConvLMU) for the feature extraction related to image processing. Here we have taken inspiration from Convolutional LSTMs [9] and LMUs [10] to introduce a new Convolutional LMU model which exhibits a simpler architecture that provides faster execution and processing based on the effect of SNNs. We have designed and implemented ConvLMU model and tested it on the MNIST dataset [11]. Our model outperforms some state-of-the-art spatiotemporal feature extracting models on a synthetic dataset generated from the MNIST dataset.

2 Related Works

Spatiotemporal Feature Extraction has been used in the telecommunication industry to a great extent [12–14]. One of the major fields in which spatiotemporal feature extraction is used is to estimate network traffic and then using the network details, various other processes can be achieved. In [12], the spatiotemporal feature of network traffic is proposed to implement network traffic estimation at first, and on this basis, an anomaly detection algorithm is put forward.

In Idle Time Window (ITW) prediction in Cellular Networks with Deep Spatiotemporal Modeling [13], is studied in mobile networks based on network subscribers' demand and mobility behaviors were observed by network operators. The ITW prediction is formulated into a regression problem with an ITW presence confidence index that facilitates direct ITW detection estimation. Feature extraction on the demand and mobility history are then proposed to capture the current trends of subscribers' demand.

In [14], a deep learning-based end-to-end network traffic classification framework, called TEST is presented. Here, CNN and LSTM are combined and implemented to help the network intrusion detection systems which automatically extract features from both spatial and time-related features of the raw traffic. In [6], insight is provided into spatiotemporal feature extraction of convolutional SNNs where the experiments are designed to exploit this property. The shallow convolutional SNN outperforms state-of-the-art spatiotemporal feature extractor methods such as C3D, ConvLSTM, and similar networks. Furthermore, a new deep spiking architecture was presented to tackle real-world problems (classification tasks) which achieved superior performance compared to other SNN methods on NMNIST, DVS- CIFAR10, and DVSGesture datasets [12] where ANN methods performed well on UCF-101 and HMDB-51 datasets [9]. It is also worth noting that the training process is implemented based on the variation of spatiotemporal backpropagation methodology.

A new type of recurrent neural network (RNN) known as the Legendre Memory Unit (LMU) was proposed in [10] and it achieves state-of-the-art performance compared with several benchmark datasets. In [15], a linear time-invariant (LTI) memory component of the LMU was proposed to construct a simplified variant that can be parallelized during training. Recently, the performance of ConvLSTM methods is comparatively better than the SNN methods [9]. Hence, in this paper, we have proposed a new method known as Convolutional LMU which combines the features of ConvLSTM and LMU's to train on the image data where most of the models were tested.

Earlier works have used spatiotemporal feature extraction as a means to a bigger solution for specific applications [12, 14]. In this paper, we have focused on brain-inspired spatiotemporal feature extraction by using ConvLMUs, a new model added to the extended ones. Previous models have used bulky architecture to perform the task at hand which makes it slightly difficult to complete the execution of the processes. The proposed model has a simpler architecture that provides faster execution and processing. One significant advantage that the proposed model possesses is faster training speed, which is achieved using SNNs, whilst not compromising on accuracy, the results of which we have provided in Sect. 5.

3 Proposed Convolutional LMU Model

The proposed system is a three-stage architecture that consists of pre-processing of the dataset which is then fed into the network which consists of an input layer, ConvLMU cell and an output layer. The values in the output layer are then applied to the argmax function to find the predicted output. Sparse categorical cross-entropy is chosen as the evaluation metric to classify the sequences into 10 different classes. Figure 1 shows the complete system proposed in this paper.

Fig. 1. Block diagram of the proposed model using ConvLMU cell

Our proposed ConvLMU cell (as shown in Fig. 2) is built like a Legendre Memory Unit [10, 15]. ConvLMUs also orthogonalize the continuous-time history of their input signal which is u(t) across a sliding window of size. The structure and internal mechanics of the ConvLMU are derived from an LMU similar to how ConvLSTMs were derived from LSTMs. The difference between ConvLSTM and an LSTM is that ConvLSTMs have convolution operation replacing matrix multiplications inside the LSTM structure. Analogous to that ConvLMUs which replace the internal multiplications into convolution operation. ConvLMUs are more suitable for spatiotemporal predictions for images or videos as it is designed for 3D inputs because of this property.

The ConvLMU cell gets input vector x_t and it generates a hidden state h_t. Both these vectors are convolved before they are fed into the network. The memory state m_t interacts with the hidden state to calculate non-linear functions across time and simultaneously write to memory. The hidden state is calculated similarly to an LMU calculation where the hidden state is a function of the input (convolved), previously hidden state, and current memory as shown below:

$$h_t = f(W_x \circledast x_t + W_h \circledast h_{t-1} + W_m m_t) \tag{1}$$

where W_x, W_h, and W_m are weights that were trained and it shows some non-linearity. The input signal u_t is calculated as:

$$u_t = {e_x}^T \circledast x_t + {e_h}^T \circledast h_{t-1} + {e_m}^T m_{t-1} \tag{2}$$

where $\mathrm{e_x}$, $\mathrm{e_h}$ and $\mathrm{e_m}$ are learned encoding vectors.

Fig. 2. Proposed ConvLMU cell

4 Synthetic Dataset and Evaluation Measures

We have generated a synthetic dataset from the standard MNIST dataset [11]. This dataset was generated with the idea of challenging the spatiotemporal feature extraction property of the model. Hence, it contains 5 different sequences of MNIST images, and each sequence contains 10 images.

The MNIST dataset contains 60000 images. We split the MNIST dataset into 5 categories which contain 10000 images each and the last 10000 images are used to generate the test set. The last 10000 images are again split into 5 categories which contain 2000 images each to generate the sequences for the test set.

The first sequence contains images that are zoomed out from 110% to 10% in a time frame of 10. Each image in the sequence is 10% smaller than the previous one. The main challenge for the model is to take the whole sequence as input and classify it as a number between 0 and 9.

The second sequence contains images that are rotated from 0 to 360° within a time frame of 10. Hence after looking at this sequence the model will classify if the sequence belongs to a number between 0 and 9.

The third sequence is like the first sequence, but instead of zooming out from 110% to 10%, the dataset is zoomed out from 90% to 40%. Also, each image is zoomed out 5% from the previous image instead of 10%. This sequence is to challenges the model to figure out the subtle difference between images in the sequence.

The fourth sequence is the occlusion of digits. This sequence is completely randomized where 14 * 14 square is removed randomly from the 28 * 28 sized images. This mask is moved randomly among the images in the sequence. Only the first image in the sequence is presented. The rest of the images are occluded.

The last sequence is also randomized. The images in the sequence are randomly rotated between 0 and 360°. Here also the first image in the sequence is presented as it is, i.e., without any random rotation. The rest of the 9 images are rotated randomly. The sequences are given in Fig. 3.

The model is trained with a sequence and then tested with a test set of the same sequence. For evaluation, the network is trained 10 different times, and top accuracy is captured and noted. Both the training and testing accuracies are noted with the 5 different sequences that were generated.

Fig. 3. Image sequences

5 Results and Analysis

We will compare the performance of several spatiotemporal models and spiking neural networks in this section. We compare the training and testing accuracies of the different sequences that were generated before with these networks. The synthetic dataset was created to demonstrate the key characteristics of these models. We trained a network using a single sequence and then put it to the test with images that followed the same pattern. We can use this method to figure out which sequences the network has trouble with respect to classifying tasks.

Gradient descent is used by all common neural networks, such as RNNs and CNNs, to change the weights between the layers. However, the Leaky Integrate-And-Fire model, which is used to describe neurons spiking and firing, is non-differentiable, which means that we can't calculate gradients or update weights [16]. During the training process, the spiking neurons are simulated in a differentiable rate-based manner so that the gradient and weights could be determined. The learned weights might then be used in the actual SNN.

Fig. 4. Layers present in our spiking convolutional neural network

In this section, we describe the setup of the experiments implemented in this paper. All models were run on a local Intel CPU Processor. The configuration of the CPU is as follows: 8 cores, with 2 threads per core. Each processor is a 2.30 GHz Intel(R) Core(TM) i7-8550U CPU and has a 64 KB first-level cache, 12 GPU Memory, and a 256 KB second-level cache. We also tried with neuromorphic hardware implementation (Loihi Cloud [17]), but due to the complexity of LMU codes, it is not materialized.

With the available setup, we ran the spiking convolutional network to train the synthetic dataset. As illustrated in Fig. 4, this network was created with Nengo [18] and it has six layers of spiking convolutional layers.

We begin with a simple core network, which is then repeated two times in parallel, with the results totaled. It's possible to think of this as a variation of ensemble learning. Instead of stacking all of the images into a ten-image sequence, the network was shown each image individually. The neuron type used for this test is spiking ReLU (Rectified linear unit) [19]. These sequences were all trained for a total of 50 epochs. Without any synaptic filters, the model is formed and trained. By adding synaptic filters to the trained model, the model's performance can be expected to be improved.

After training, we test the model using the test set that was generated for different sequences. Figure 5 shows output activity from the network while we present the test images one by one. Each image is presented for 0.1 s. We can see clearly that the prediction still stays the same even when the image is modified.

For the first sequence, the predicted value stays on 7 while the image is zoomed out. At the end when an image with 7 is zoomed out by a factor of 0.1, the model gets confused and outputs a noise. But when this sequence ends and the next sequence starts, the probability of the next predicted value goes up while other values go down. 5 has the highest probability value for sequence 2 which stays on the top even when the image is rotated after 0.1 s. Sequence 3 was an easy one for this model, as we can see a clear gap between the top predicted value and probability values for other numbers.

Fig. 5. Spiking CNN predictions on test set

Sequence 4 was a hard sequence for this model, as throughout the entire time frame the model confuses the images between 9 and 7. Sequence 5 was once again hard because it confuses to differentiate 6 and 9 and the probability of 9 goes slightly down while the probability of 6 goes slightly up, but it still predicts 9 most of the time frame.

Figure 6 shows an LSTM network that was built on a par with the LMU network for easy comparison of how these networks will work on the synthetic dataset. This LSTM model was trained with a batch size of 128 and has 212 units. Additionally, we have added a dropout of 20%. This model was trained for 50 epochs for all the sequences. It performs well for sequences 2 and 3 compared to other sequences. Sequence 4 and sequence 5 have average training and testing accuracies while the model completely fails to classify sequence 1. It only has 87% testing accuracy on sequence 1.

On the spiking side, we have LMUs. We built an LMU model having 212 units with an order of 256. This model accepts a sequence and processes each image in the sequence in order. Using LMUs instead of LSTMS improved the training accuracy by a huge margin. Sequences 2 and 4 have a 100% training accuracy after training for just 50 epochs. This model performs very well on sequences 2, 3, and 4 and performs above par on sequences 1 and 5.

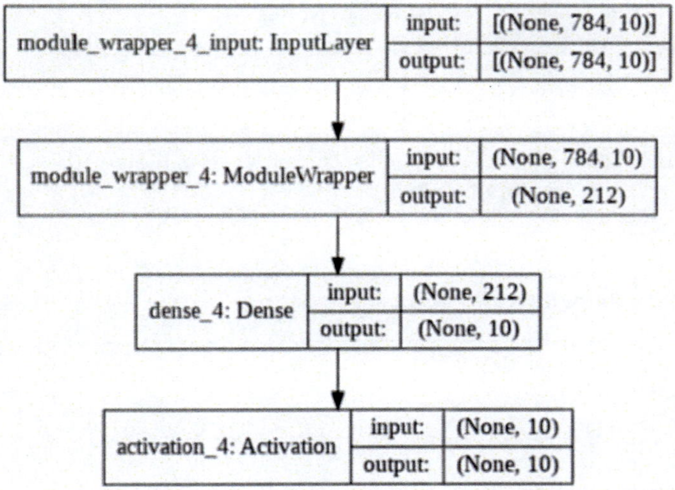

Fig. 6. LSTM network

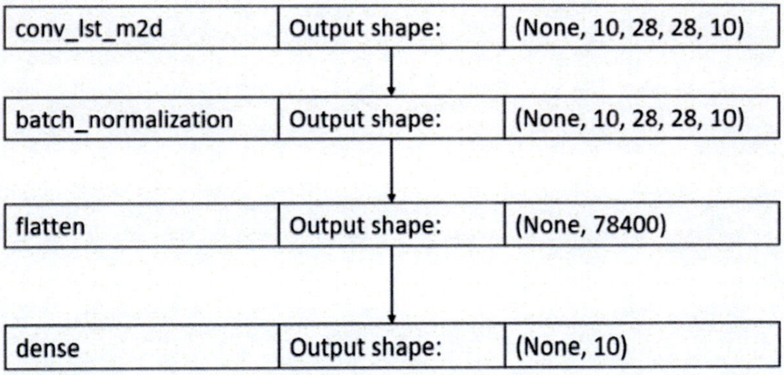

Fig. 7. Conv LSTM network

The artificial neural network which is used for spatiotemporal feature extraction in most problems will be Conv LSTM. As expected, ConvLSTM shown in Fig. 7 performs better than all the networks that are presented in this study. It has a 100% training accuracy for all the sequences and has 96+% accuracy on the test set.

At last, we present our novel ConvLMU model in Fig. 8. This model has 212 hidden units and 256 dimensions for memory. We set the window size as 10 s. It accepts the sequence and processes the images in the sequence one by one. The first step is convolving the input image into another 3-dimensional vector which is then passed into the network. Similarly, it has a recurrent convolution that performs convolution on the state vector h_t before it is fed back to itself. This model also has a 100% training accuracy on all the sequences and has a 94+% accuracy on the test set.

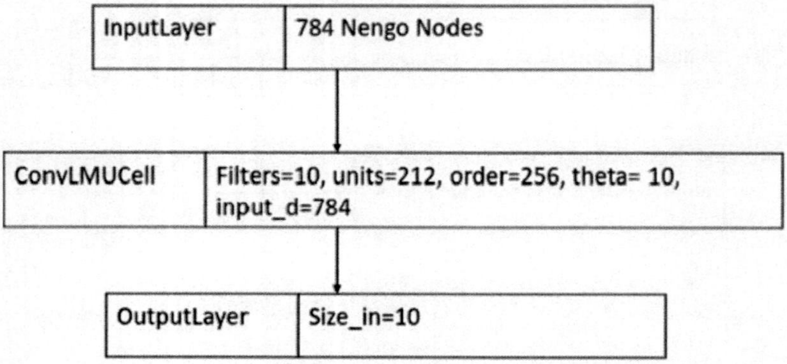
InputLayer
784 Nengo Nodes
ConvLMUCell
Filters=10, units=212, order=256, theta= 10,
input_d=784
OutputLayer
Size_in=10

Table 3. Training summary for different models

Model	Total training time	Epochs
Conv CNN	250 s	50
LSTM	300 s	50
LMU	200 s	50
ConvLSTM	574 s	30
ConvLMU	120 s	30

6 Conclusion

We have shown how different models perform on different sequences generated from the MNIST dataset. In this paper, we present a novel brain-inspired model ConvLMU, to perform spatiotemporal feature extraction. We also demonstrated how, because of its inherent convolutional structure, ConvLMU keeps the benefits of LMU while also being better suitable for spatiotemporal sequences. It exhibits a faster convergent comparison to other models. For future work, we can attempt creating a more complex network that has been tested on a variety of various sequences, such as the moving MNIST dataset to show the spatiotemporal dynamics of the model. We further present a comparison of our model with other existing traditional and state-of-the-art models in a sequential image classification task run on a synthetic dataset. We would like to run both convolutional models on neuromorphic hardware to understand the adeptness of the proposed ConvLMU model.

References

1. Zhao, W., Du, S.: Spectral–spatial feature extraction for hyperspectral image classification: a dimension reduction and deep learning approach. IEEE Trans. Geosci. Remote Sens. **54**(8), 4544–4554 (2016)
2. Alpaydin, E.: Introduction to Machine Learning. The MIT Press, London (2010)
3. Vijayakumar, T., Vinothkanna, R., Duraipandian, M.: Fusion based feature extraction analysis of ECG signal interpretation–a systematic approach. J. Artif. Intell. **3**(01), 1–16 (2021)
4. Tripathi, M.: Analysis of convolutional neural network based image classification techniques. J. Innov. Image Process. (JIIP) **3**(02), 100–117 (2021)
5. Rao, S., Gopalapillai, R.: Effective spam image classification using CNN and transfer learning. In: Smys, S., Tavares, J., Balas, V., Iliyasu, A. (eds.) Computational Vision and Bio-inspired Computing, ICCVBIC 2019. Advances in Intelligent Systems and Computing, vol. 1108, pp.1378–1385. Springer, Cham (2020). https://doi.org/10.1007/978-3-030-37218-7_145
6. Samadzadeh, A., Far, F.S.T., Javadi, A., Nickabadi, A., Chehreghani, M.H.: Convolutional Spiking Neural Networks for Spatio-temporal feature extraction. arXiv preprint arXiv:2003.12346 (2020)
7. Li, H., Qiu, Q., Wang, Y.: Guest editorial: design and applications of neuromorphic computing system. IEEE Trans. Multi-Scale Comput. Syst. **2**(04), 223–224 (2016)
8. Schuman, C.D., et al.: A Survey of Neuromorphic Computing and Neural Networks in Hardware, arXiv preprint arXiv:1705.06963 (2017)

9. Simonyan, K., Zisserman, A.: Two-Stream Convolutional Networks for Action Recognition in Videos. Computer Vision and Pattern Recognition, arXiv:1406.2199 (2014)
10. Voelker, A., Kajic, I., Eliasmith, C.: Legendre memory units: continuous-time representation in recurrent neural networks. In: Advances in Neural Information Processing Systems 32 (NeurIPS 2019) (2019)
11. Deng, L.: The MNIST database of handwritten digit images for machine learning research. IEEE Signal Process. Mag. **29**(6), 141–142 (2012)
12. Nie, L., Li, Y., Kong, X.: Spatio-temporal network traffic estimation and anomaly detection based on convolutional neural network in vehicular ad-hoc networks. IEEE Access **6**, 40168–40176 (2018)
13. Fang, L., Cheng, X., Wang, H., Yang, L.: Idle time window prediction in cellular networks with deep spatiotemporal modeling. IEEE J. Sel. Areas Commun. **37**(6), 1441–1454 (2019)
14. Zeng, Y., Qi, Z., Chen, W., Huang, Y.: EST: an end-to-end network traffic classification system with spatio-temporal features extraction. In: 2019 IEEE International Conference on Smart Cloud (Smart- Cloud), pp. 131–136 (2019). https://doi.org/10.1109/SmartCloud.2019.00032
15. Chilkuri, N.R., Eliasmith, C.: Parallelizing legendre memory unit training. In: Proceedings of the 38th International Conference on Machine Learning, PMLR, vol. 139, pp. 1898–1907 (2021)
16. Vaila, R., Chiasson, J., Saxena, V.: Deep Convolutional Spiking Neural Networks for Image Classification, arXiv:1903.12272 (2019)
17. Davis, M., et al.: advancing neuromorphic computing with Loihi: a survey of results and outlook. Proc. IEEE **109**(5), 911–934 (2021)
18. Bekolay, T.: Nengo: a Python tool for building large-scale functional brain models. Front. Neuroinform. **7** (2014)
19. Nair, V., Geoffrey, E., Hinton, E.: Rectified linear units improve restricted Boltzmann machines. In: Proceedings of the 27th International Conference on Machine Learning (ICML), pp. 807–814, Haifa, Israel (2010)

Underwater Image Enhancement Using Image Processing

V. Nagamma(✉) and S. V. Halse

Department of Electronics, Karnataka State Akkamahadevi Women's University, Vijayapur, Karnataka, India
nagini125@gmail.com

Abstract. The field of underwater image processing has evolved to become more thought about subject in recent decades, and it has made significant progress. In this study, we look at some of the most modern underwater approaches that have been created specifically for the environment. These strategies can improve image contrast and resolution while also expanding the range of underwater imaging. We focus on the numerous methodologies available in the literature after discussing the basic physics of wave scattering in the water medium. The circumstances under which they were created, as well as the quality evaluation procedures that were utilised to measure their performance, are highlighted. The HWD transform is then used to sharpen the picture. A highpass filter is used to eliminate the low-frequency background. Image histograms are mapped based on the intermediate colour channel to narrow the gap between the inferior and dominant colour channels. Wavelet fusion and an adaptive local histogram definition technique are utilised after that. It presents an improved underwater picture improvement system based on an image reconstruction method that can accurately restore underwater photos. The given approach takes a single image as input and performs a series of operations on it, including gamma evaluations, white balancing, sharpening. To produce the desired output, multiscale picture fusion of the inputs is performed. The colour imprecise input image is white balanced in the first stage to remove colour casts and retain a realistic undersea image. The output photographs from the proposed approach might then be used for detection and identification to extract more useful data.

Keywords: Underwater ımage · Bi-spectrum · HWD · Wavelength

1 Introduction

It must first grasp the underlying physics of light propagation in the water medium before it can deal with underwater image processing. Degradation effects are caused by the medium's physical qualities, which are not present in regular photographs captured in air. Low visibility is characteristic of underwater images because light is dimmed as it goes through water. As a result, the scenes are hazy and poorly contrasted. Due to light attenuation, visual distance is restricted to around twenty metres in clear water and five metres or less in muddy water. Absorption (the removal of light energy) and dispersion

J. I.-Z. Chen et al. (Eds.): ICIPCN 2022, LNNS 514, pp. 13–22, 2022.
https://doi.org/10.1007/978-3-031-12413-6_2

are the two factors that create light attenuation (which alters the light path's orientation). The total efficacy of underwater imaging equipment is influenced by light absorption and scattering in water. Image enhancement is a technique for transforming an underwater image into a correct, clearly visible image that can be used in a variety of research applications. This image enhancement technique is useful for increasing the amount of information in an image. The mage's visuality is affected by this image, which is advantageous to the observer's image information. Because the enhancement technique would eliminate the information already there in the image, image improvement in underwater images would be challenging. The characteristic in the image is identified through image enhancement.

The enhancement method improves picture properties such as edge and contrast in order to improve the image. pictures for the purpose of analysis and study Qualitative An objective technique is used to enhance the process.

Display the stunning photographs. Enhancement of images Many operations are included, such as contrast stretch, Noise extract, pseudo-coloring, and noise filtering are all examples of noise clipping technique. The active range of image characteristics has the many recognised features have magnified the signal. The photos have low quality due to the nature of the light, according to many known improvement algorithms. Because water is a denser medium than air, light entering it is refracted, absorbed, and dispersed. When light reaches the water, it is dispersed in all directions, resulting in the formation of these light droplets. Light scattering is caused by the blur result of light and the loss of colour distinction. These variations in water in underwater images are generated by creatures and other things in the water, as well as the composition of the water.

1.1 Problem Statement

It is necessary to explain the physical basis of the situation before creating an algorithm. There will not be any refraction if the sea surface is calm and light is directed straight up from the flat land underneath it. The spectator at point a will see the object O in this case. The normal to the sea surface N, on the other hand, is inclined at an angle when there are ocean waves. As a result, at point b, the observer sees O instead of O. The ratio c and d is used to compute the reflectivity of a material according to Snell's law.

The average-based strategy is the most typical solution to such a difficulty. The average-based technique involves computing the picture ensemble's temporal average[1]. It performs effectively in settings that are quite calm. When the target is too small and contains a lot of information, however, it fails. Several scholars have recommended that the target image be formed by identifying and merging the least distorted regions across the sequence of raw images. This approach produces a considerably crisper result than the average-based strategy.

2 Literature Survey

Sungheetha, Akey, and Rajesh Sharma [1] discussed Human-robot interaction provides a variety of aided services in a variety of real-time applications (HRI). In robotic systems, the notion of convergence of a three-dimensional (3D) picture into a plane-based projection is employed for object recognition via digital visualisation. During the convergence process, projections in multiple planes are mistaken, resulting in recognition mistakes. The use of an input processing strategy that is based on the projection technique helps reduce these misidentifications in object recognition. By projecting and examining the input image in all possible dimensions, the conjoining indices are determined. To increase recognition processing speed and accuracy, a machine learning approach is used. Intersections with no conjoined indices are separated via labelled analysis.

Darney, P. Ebby, and I. Jeena Jacob [2] discussed Many public outdoor crimes have been recorded on film during the wet season, and they are unable to recognise picture features due to a lack of detailed feature information. For indexing and extracting additional information from photographs with rain streaks, rain streak reduction methods are ideal. Rain also reduces the overall picture quality of vision systems in outdoor recording circumstances by drastically altering the intensity of images and recordings. The elimination of rain streaks in photographs will need a complex trial and error technique in order to be effective. Using statistics on photon counts, chromaticity, and the frequency of rain streaks, several algorithms have been used to identify and eliminate rainy effects in digital images.

Kumar, T. Senthil [3] discussed VR and AR are fast evolving technologies that have considerably improved the internet buying experience and retail selling environment. Disintegration in VR and AR adds to the technology's interdisciplinary origins in terms of applications and academic study in terms of practical application and academic research. The retail application and study activities that leverage VR and AR technologies are contrasted and examined in this study. The phrases "implementation," "customer acceptability," "applications," "problems," and others are compared. This research lays the groundwork for future research in the realm of retail applications.

Jacob, I. Jeena, and P. Ebby Darney [4] discussed The Internet of Things (IoT) is a multi-device, multi-connection ecosystem with a high number of users and a tremendous volume of data. Due to its suitability for "big data" challenges and future concerns, deep learning is particularly well suited for these circumstances. Nonetheless, ensuring security and privacy has become a major concern for IoT management. Deep learning algorithms have shown to be increasingly efficient in completing security evaluations for IoT devices without the need of manual rules in a number of recent situations. This study combines principal component analysis (PCA) with higher performance for feature extraction.

Shakya, Subarna [5] discussed Because of its effectiveness in determining the outside temperature and the capacity to estimate crop water levels, thermal imaging is used in agricultural crop water management. With the lowered weight and better resolution of thermal imaging systems, for agricultural and civil engineering purposes, thermal imaging was considered a suitable integration in Unmanned Aerial Vehicles. When used on-site, this approach was able to solve a variety of problems, such as estimating the amount of water in a plant in farms or fields while taking into account formally caused variations or naturally occurring water levels. Using high-resolution thermal imaging, the proposed project tries to quantify the quantity of water content in a vineyard.

PhilominaSimon [6] gives Color contrast is reduced due to light dispersion. Because of dispersion and the existence of undersea animals, water has an impact on underwater photographs. It introduce an enhanced underwater picture enhancement system based on the fusion approach that can recover underwater photographs accurately. The given work takes a single image and performs several operations on it, including many processing of image scheme. The final output is created via multi-scale picture fusion of the inputs. To eliminate colour casts and retain a realistic underwater view, the color-distorted input image is white-balanced in the first step.

Peng Liu; Guoyu Wang [7] First Explanatory synthetic underwater pictures are developed as training data for CNN models using CycleGAN. Second, VD-SR and the Underwater Res-net model, a residual learning network for underwater picture improvement tasks, are presented to the VD-SR. The training mode and loss function have also been upgraded. Using edge difference loss and mean squared error loss suggested, a multi-term loss function is created. To increase the presentation of the multi-term loss purpose, an asynchronous training strategy is also given. Finally, the effects of batch normalisation are examined.

Chengyi Cai; Yiheng Zhang [8] discussed a high-quality photograph with no interfering objects is ideal for studying the deep ocean globe map. The image quality is harmed by light attenuation, water thickness, and scattering impact, just as it is in the water. Furthermore, dynamic interference may have an impact on the actual undersea map. Frmo this research [5, 6, 8–12], it offer a multi-step and all-around underwater image dispensation system that improves picture quality, removes dynamic interference, and reconstructs the image for underwater photographs collected in succession. First and foremost, picture contrast and colour are increased by using a murky channel prior and a better grey world algorithm.

3 Methodology

Picture processing may be approached from two perspectives: as a means of image restoration or as a way of image enhancement:

(I) Image recovery is an inverse problem in which a damaged image is reconstructed using a degradation model and original image production. These methods are exact, but they need a large number of model parameters (such as attenuation and diffusion coefficients, which reflect water turbidity) that are seldom found in tables and can vary greatly. The depth estimate of a single item in the shot is another crucial issue to consider.

(II) Image enhancement, rather than relying on a physical model, employs qualitative subjective criteria to create a more visually appealing image. These techniques are frequently simpler and quicker than de-convolution techniques.

In arithmetic, nearby factual examination, the bispectrum is a measurement used to look for nonlinear communications.

A. Picture obtaining Input picture is caught from the underwater through the camera. This caught picture is as RGB picture which comprises of red, green and blue area. The caught picture goes through the pre-preparing stage the greater the pixel esteems the more in main focus the photo. As an outcome this calculation picks the infocus territories from every single enter photo through picking the best an incentive for every single pixel, bringing about especially focused on yield. The value of the pixel P (p, q) of each image is taken and when contrasted with one another.
B. Preparation The number of cycles in the pre-handling phases of photo handling varies depending on the information picture. Preprocessing processes are applied to information images obtained from various sources. The RGB image is the information picture. The RGB image is transformed into a dim scale image with a pixel range of 0 to 255. Under the pre-handling step, there is a filtering cycle. This filtering cycle reduces the cacophony that can be heard in the image.
C. Detection The discovery interaction which incorporates the upgrade of picture which is taken as an info picture. In the proposed work upgrade measure is done utilizing HWD change method. Utilizing this method commotion can be eliminated and the upgrade of picture will be higher.

4 Architecture

Dark scale is essentially lessening intricacy: from a 3-D pixel esteem (R, G, B) to a 1D worth. Dim scale is a scope of shades of dim without clear tone. Dark scale Images are more appropriate for some applications, for example, edge discovery. The pre-preparing stage comprises of subsequent stages (Fig. 1).

The picture will be initial changed over into regular logarithm space before the Fast-Fourier change is practical. The high pass channel is then used utilizing the Butterworth channel. Butterworth high-pass channel is a powerful high-pass channel to wipe out low recurrence signal which is typically utilized for foundation expulsion.

Mapping of the Histogram. The red shading channel is usually the least effective, whereas the green and blue shading channels are the most popular. The red shading channel has the lowly shading channel rate, whereas the green or blue shading channel has the highest shading channel rate.

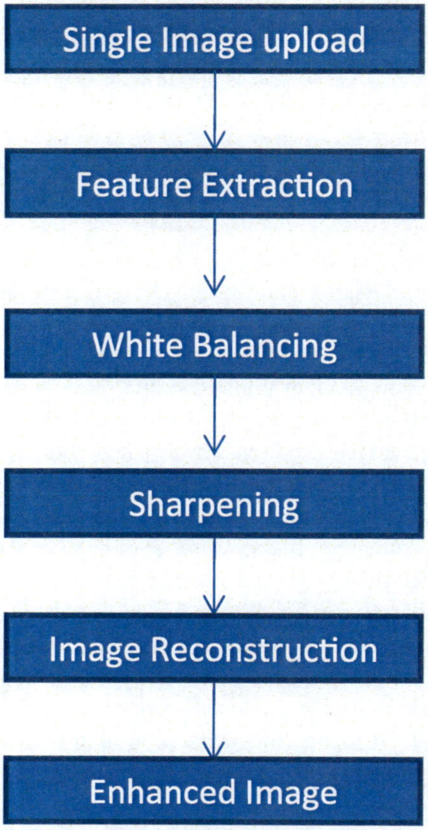

Fig. 1. Flow diagram of proposed system

The above image has been processed by using the image prcoessing method and one can observe that the image has been enhanced and a clear image can be seen.

Different image has been used for the processing purpose, as one can observe that the image quality has been improved which can be seen by comparing with the adjustant image.

(a)

(b)

(c)

(d)

Fig. 2. Image reconstruction

The above image is a underater image which is used for processing purposes, white balancing and image sharpening method has been used for this and one can see that the image clearity is enhanced (Fig. 2).

Improved IMAGE These histograms will deliver two autonomous pictures which have under-and over-upgraded impacts. In this way, for three fundamental histogram channels of red, green, and blue will deliver six free histograms.

Working

This white balancing approach [25] focuses on restoring colours that have been damaged as a result of white light absorption as it travels through water. The biggest issue with underwater photographs is the greenish-blue colour caused by wave scattering as depth increases. Waves with longer wavelengths are absorbed first. As a result, red will be absorbed first, followed by blue, and so on. The amount of colour that is lost is proportional to the distance between the observer and the plane. The first step is to compensate the red channel, and the second is to use the Gray-World Algorithm to produce the white balanced picture.

Mathematically, the adjusted red channel Irc at each pixel point () should be expressed as follows:

Irc() is equal to Ir() + (g r).

(1) Ir().

Ig(), (1).

Here Ir and Ig are the red and green colour channels of image I, respectively, and r and g signify the mean value of Ir and Ig, respectively, at the interval [0, 1].

The second input comes from the image's sharpened version, which is produced by the white balance portion. So, to blurred or unsharp, employ the Gussian filter, which is based on an unsharp masking concept. The sharpened picture.

S = I + (I G I),

Here I is the image to be sharpened, GI is the "Gaussian" ltered version of I, and is a parameter, is not a simple equation to solve. A small does not sharpen picture I, but a very big does, resulting in oversaturated regions with deeper shadows and brighter highlights. To solve this problem, we define the sharpened image S as S = (I + NI GI)/2, (6) N. representing the linear normalising operator, commonly known as histogram stretch.

Despite the importance of white balance, it is necessary to recover the colours that are lost when light flows through water. This is insufficient when dealing with edges and resolving the dehazing problem using scattering effects. As a result, an effective fusion based on contrast histogram equalization, gamma correction, and sharpening was introduced to minimise the fogginess of the white balanced image.

Histogram Equalization Histogram adjustment strategy for the most part used to expand the differentiation of numerous pictures, particularly when the significant information of that picture is addressed by close difference esteems. Through this change, the power can be better circulated across the histogram. Histogram evening out adequately spreading out the most incessant force esteems. This strategy is helpful for both splendid and dull districts in clinical imaging. This technique is straightforward, quick and it get satisfactory outcomes for some applications, these are the fundamental focal points of this strategy. Weakness of this strategy is, it might build the differentiation of foundation clamor, while it diminishes the usable sign. Wavelet Transform WT is notable for its denoising capacity. it is superior to other changes like Fourier change, Hilbert change and so on because of following reasons It gives the recurrence portrayal of crude sign at some random time stretch.

- Wavelet change can catch the restricted element which is the recurrence range of a modest segment.
- The calculation time which is N for the WT and N(log N) for the DFT.
- It has the properties of multiresolution, sparsity and edge recognition.
- Use of both recurrence area field and spatial field in the wavelet change gives better upgraded picture and it can likewise lessen the commotion in these high recurrence sub-groups. K-L Transform Karhunen-Loeve Transform (KLT) which was based on factual based properties. The exceptional preferred position of KLT is a decent de-relationship. In the MSE (Mean Square Error) sense, it is the best change, and it has a significant situation in the information pressure innovation. KLT has 4 qualities: De-relationship: After change the weight if vector signal Y irrelevant.
- Energy focus: After change of N-dimensional vector signal, the most extreme difference is in the previous of M lower sub-vector. Under estimating of the MSE: The twisting is not exactly other change. It is the amount of the sub-vectors which were precluded. No speedy calculation and the diverse sign example assortment has distinctive change framework. (it is the inadequacy of KLT)

5 Conclusion

It propose a unique plan to remake a lowered article twisted by affecting water surface. It expect the normals of the water surface are Gaussian conveyed. The bi-spectrum strategy is utilized to recuperate the period of the genuine item. In spite of the fact that tests show that the methodology is promising, there exist a few cutoff points. One breaking point is that the calculation needs a huge PC memory and weighty calculation since that the bi-spectrum of a picture is four dimensional. Another cutoff is the recursive stage recuperation technique with just a subset of the stage data of the arrived at the midpoint of bi-spectrum being utilized. This may diminish the goal of the yield. To defeat such cutoff points is the following stage in the examination.

References

1. Sungheetha, A., Sharma, R.: 3D image processing using machine learning based input processing for man-machine interaction. J. Innov. Image Process. (JIIP) **3**(01), 1–6 (2021)
2. Darney, P.E., Jacob, I.J.: Rain streaks removal in digital images by dictionary based sparsity process with MCA estimation. J. Innov. Image Process. **3**(3), 174–189 (2021)
3. Kumar, C., Senthil, T.: Study of retail applications with virtual and augmented reality technologies. J. Innov. Image Process. (JIIP) **3**(02), 144–156 (2021)
4. Jacob, I.J., Darney, P.E.: Design of deep learning algorithm for IoT application by ımage based recognition. J. ISMAC **3**(03), 276–290 (2021)
5. Ceballos, A., Bolano, I.D., Sanchez-Torres, G.: Analyzing preprocessing filters sequences for underwater ımage enhancement. Contemp. Eng. Sci. **10**(16), 751–771 (2017)

6. Ancuti, C.O., Ancuti, C., Vleeschouwer, C.D., Bekaert, P.: Color balance and fusion for underwater ımage enhancement. IEEE Trans. Image Process. **27**(1), 379–393 (2018)
7. Bora, D.J.: Importance of Image Enhancement Techniques in Color Image Segmentation: A Comprehensive and Comparitive Study (2017)
8. Lu, H., Li, Y., Zhang, Y., Chen, M., Serikawa, S., Kim, H.: Underwater Optical Image Processing: Comprehensive Review. Kyushi Institute of Technology, 1-1 Sensui, Tobata, Kita Kyushi 804-8550, Japan
9. Bora, D.J.: Importance of ımage enhancement techniques in color ımage segmentation: a comprehensive and comparitive study. Indian, J. Sci. Res. **15**(1), 115–131 (2017)
10. Setiawan, A.W., Mengko, T.R., Santoso, O.S., Suksmono, A.B.: Color retinal ımage enhancement using CLAHE. In: International Conference on ICT for Smart Society (ICISS) (2013)
11. Sowmyashree, M.S., Bekal, S.K., Sneha, R., Priyanka, N.: A survey on the various underwater image enhancement techniques. Int. J. Eng. Sci. Invent. **87**, 19–23 (2014)
12. Çelebi, A., Erturk, S.: Visual enhancement of underwater images using empirical mode

Fake News Detection on Indian Sources

Navyadhara Gogineni(✉), Yashashvini Rachamallu, Ruchitha Mekala, and H. R. Mamatha

Department of Computer Science and Engineering, PES University, Bangalore, India
navyadhara79@gmail.com, mamathahr@pes.edu

Abstract. Everyone relies on many online resources for news in today's world, where the internet is pervasive. As the use of social media platforms such as Twitter, Facebook, and others has grown, news has traveled quickly among millions of users in a short amount of time. Fake news has far-reaching effects, ranging from the formation of biased opinions to manipulating election outcomes in favor of specific politicians. Furthermore, spammers profit from click-bait ads by utilizing appealing news headlines. Sometimes, humans find it more difficult to determine the article's authenticity without additional verification. In this paper, we developed a classification model with the help of Deep Learning and Natural Language Processing Techniques, that classifies the article as real or fake news. The model has been tested on Indian news source - Times Of India along with other sources such as Politifact and was able to give decent accuracy in verifying the authenticity of the news.

Keywords: NLP · Word2Vec · LSTM · Artificial neural networks · N-gram analysis

1 Introduction

Information sharing has become a simple undertaking in today's world of continuously developing technology. We link ourselves so much to news but sometimes fail to verify its authenticity. As we spend more time interacting online through social media platforms, more people are seeking out and consuming news from social media rather than traditional news organizations. It was also discovered that, as a key news source, social media currently exceeds television. Despite the advantages of social media, the quality of social media stories is lower than that of traditional news organizations.

Large volumes of fake news, i.e. news pieces with purposely incorrect material, are created online for a variety of reasons, including financial and political gain because it is inexpensive to supply news online and significantly faster and easier to distribute through social media. Many people were harmed by this, and their trust in online social networks was eroded (a platform where news is easily spread worldwide). This deception creates importance and necessity for a system that verifies the authenticity of a news article and detects fake news. Many scientists believe that artificial intelligence and machine learning can assist solve the problem of fake news.

J. I.-Z. Chen et al. (Eds.): ICIPCN 2022, LNNS 514, pp. 23–35, 2022.
https://doi.org/10.1007/978-3-031-12413-6_3

Various models are employed to achieve a 60–75% accuracy range. Naive bayes classifier, Linguistic characteristics based, PCA and Chi-Square in feature extraction followed by CNN-LSTM model, Support Vector Machines, and others are included. The parameters that are taken into account do not provide a high level of precision.

This project tries to solve this real-world problem, which is challenging since the bots are improving and deceiving humans. It is not simple to check the article's authenticity; hence we need better systems that help us understand fake news patterns and prevent confusion in the world. We have utilized an available dataset for training our model and performed analysis on the text data along with sentiment analysis to understand the polarity of data. Feature extraction techniques like Word2Vec are employed to generate sentence features and a model built with LSTM and dropout layers for training and validation of data. To test the data on real time data, web scraping has been done on websites like Politifact and Times of India to test systems accuracy.

This paper is prepared as follows. Section 2 highlights the different models, unique techniques and also different types of datasets used for training. Section 3 gives a clear view on methodology and makes up for shortcomings. Section 4 helps you in visualizing the model architecture. Sections 5 and 6 talks about usage of our model and the future works that can be done respectively. The final section consists of the results obtained on testing our model with real time data.

2 Related Works

Following a wave of widespread fake news in recent years, several measures have been developed to detect it. Social bots, trolls, and cyborg users are the three sorts of contributors to fake news. According to the definition of a social bot, a social media account that is managed by a computer algorithm. Content can be generated by the social bot automatically. Second trolls are genuine people who seek to disrupt online communities in order to elicit an emotional response from social media users. Cyborg is the other one. Automated actions with human input are what cyborg users are. To engage in social media activities, people create accounts and use programs.

Some of them propose a general natural language way of performing fake news detection using traditional NLP techniques, some propose novel techniques using deep learning methods.

We began our investigation by learning the basic methodology for developing a fake news detector, which includes news aggregation (gathering of news), authentication (training a model to detect the authenticity of news), and recommendation (assistance in determining the authenticity of news).

We started our research by comparing different word to numeric transformation techniques starting from Traditional methods like Tf-Idf, followed by word2vec and doc2vec embedding algorithms and deep learning approaches. We were also introduced to new concepts namely LSTM and GRU, which helps in overcoming the shortcomings of RNN i.e. long-term dependence [1] and also we were able to figure out the metrics used for finding the accuracy of models.

We went through different approaches of feature extraction [2–4]. These papers used PCA and Chi-Square for dimensionality reduction. And in some papers they used N-gram and sequence model for the text transformations along with the conversion they used regularization techniques to make sure data is not being overfitted. Some papers used a traditional method tf-idf we need to threshold the vocabulary so that we won't get a sparse matrix. From the research we concluded that Word2Vec is a suitable method, since it converts human readable language to system readable language and can also maintain semantics.

A brief research on classification methods have made clear the significance of SVM, KNN in fake news detection. Based on the research done we were able to figure out SVM is one of the better approaches for classification [5]. On further exploring we were able to figure out some of the neural networks consisting of LSTM layers [6] gave better results for the classification. By this, we concluded to use simple neural networks for the classification.

For data scraping from web sources, this paper [12] helped us to understand the process of scraping data from web sources.

3 Proposed Solution

Figure 1 depicts the Pipeline/Workflow for the implementation of the fake news detector.

Fig. 1. Model pipeline

3.1 Dataset

The training of our fake news detection model has been done on ISOT fake news dataset [9, 10]. The dataset consists of Two Comma separated files, Fake.csv, Real.csv. Refer to Fig. 2 for a sample of the dataset. Each of the files has the title of the news, text data, topic/category, date published and whether the news is Fake or Not.

Fig. 2. Sample dataset

3.2 Data Cleaning

The data we choose consists of two CSV's, each for real and fake data. While the data was mostly clean, In order to increase the usability of data for training the following steps have been used:

1. Cleaned the empty (NA) values in the data.
2. The titles without a publisher name have been removed in both the data files.
3. A new column, "*isfake*", is created which represents whether the data is fake or real, i.e. token 0 for real data and 1 for fake data.
4. Both the data files have been merged to create a single data file for easy processing.

3.3 Data Analysis

We have performed data analysis with the goal of getting a deeper understanding of the data. The bar graph (Fig. 3) is displayed to understand the categorical distribution from which the majority of fake news is generated. We were able to deduce from charting that authentic data has only two categories, whereas bogus news has seven. Following the basic text cleaning, a word cloud (Fig. 4) is generated to determine the most commonly used words. The final N-gram analysis was completed. In computational linguistics, an N-gram is explained as an adjoining sequence of n elements in a sample of text or speech. N-gram analysis can be a major asset in deciding the best possible tool for text analysis (Table 1).

Fig. 3. Plot depicting categorical distribution of data

Fig. 4. Word cloud with most famous words

3.4 Text Preprocessing and Text Transformation

As a part of text preprocessing, the title and text parts of the data have been merged for faster processing and better understanding. With the use of the NLTK library various preprocessing steps were implemented. They include:

1. Removing stop words
2. Removing punctuations
3. Utilizing regex to remove single characters.
4. Tokenization of sentences.

Table 1. Plots of ngram analysis.

Bi gram - Fake news	
Tri gram- Fake news	

(*continued*)

Table 1. (*continued*)

Bi gram - real news	
Tri gram - real news	

For text transformation, Word2Vec is used to convert text in human readable form to embeddings that can be understood by the model. It is a two-layered neural network which is used to reconstruct the linguistics of words. It has weights, much like all neural networks, and the purpose of training is to lower the loss function by adjusting the weights. However, we will not utilize Word2Vec for the job for which it was trained; instead, we will use its hidden weights as our word embeddings and discard the rest of the model.

It takes a large corpus of words as its input and produces a vector space of several hundred dimensions, with each unique word being assigned a corresponding vector in the multi dimensional vector space. Word vectors are positioned in the vector space in such a way that words with similar contexts in the corpus are close together in the space. For learning word embeddings from raw text, Word2Vec is a computationally efficient predictive model. The Continuous Bag-of-Words (CBOW) model and the Skip-gram model are also available.

Word2vec builds the vocabulary and also extracts the sentence features which are later used to train the system. The size of the input is capped at 700 words as most of the sentences fall in this category thus removing any outliers. Refer to Fig. 5 for the same.

Fig. 5. Input outlier detection

3.5 Model

To train the data we relied on a custom neural network. Figure 6 displays the architecture of the Neural Network. The Neural network consists of many layers and the most important of them is the LSTM layer.

3.5.1 Long Short Term Memory

Long short term memory, i.e., LSTM [11] is an improved variant of recurrent neural networks(RNN). The issue with RNN is that they take a long time to back-propagate, majorly caused due to the decay of error and its backflow which makes it futile in situations where we need to store the memory. LSTM, which can hold short term memories longer than RNN can be used in these situations and they are very capable. LSTM maintains different cell states and gates (input gate, output gate and forget gate) (Fig. 7).

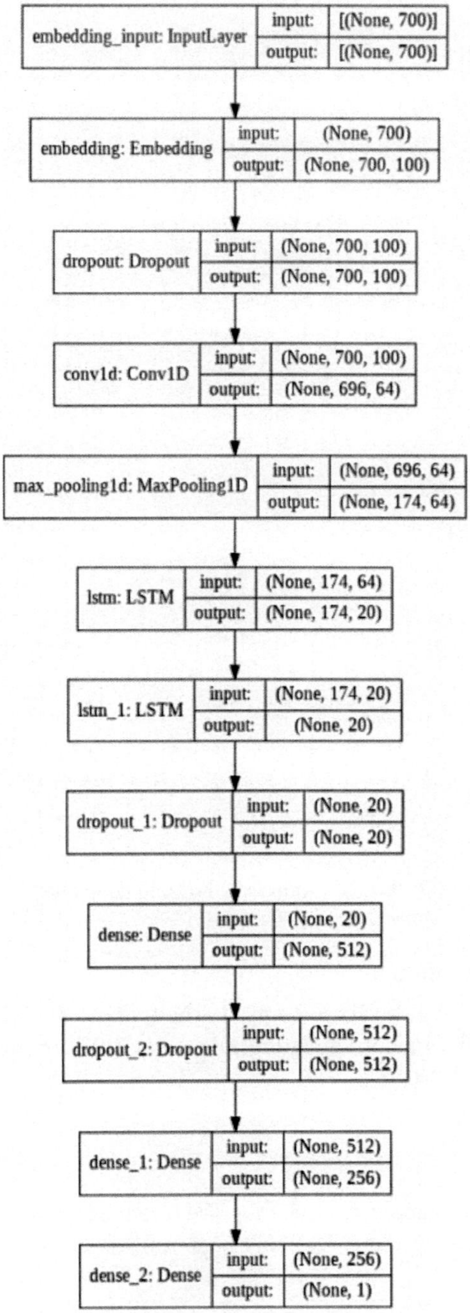

Fig. 6. Model architecture

Fig. 7. LSTM at a timestamp t and equations

3.5.2 Training and Validation

To avoid overfitting, dropout layers were used. The training and validation accuracies were impressive with the model validating fake news 99% on test data (Fig. 8 and Table 2).

3.6 Testing

The testing was done on Indian news websites. We tested on real time news by scraping the website named 'Times of India' by using beautiful soap (To read the html file) and

Fig. 8. Training procedure

Table 2. Confusion matrix of training

	Precision	Recall	f1-score	support
Fake (1)	0.99	1.00	0.99	5713
Real (0)	1.00	0.99	0.99	5354
Accuracy			0.99	11067
Macro avg	0.99	0.99	0.99	11067
Weighted avg	0.99	0.99	0.99	11067

request libraries. After scraping the text from the site, it is pre processed and passed to our model. And our model is able to give an accuracy around 67%, 75% in the detection of fake news respectively.

4 Results

We have scraped data from websites like "Times Of India" and "Politifcat". After pre-processing, the text and extracted text features are passed to the Model. The results in terms of accuracy are represented in Table 3.

Table 3. Results in terms of accuracy of predictions

	Accuracy when tested on		
Data	Title + Description	Only description	Only title
Training data	98.96	98.5	86.14
Times Of India	75.01	68.18	71.3
Politifact (American)	66.67	61.66	63.34

5 Use Cases

When people try to impersonate someone else or a trusted source to spread false information, this can also be considered as false news. In most of the cases, the people who spread false news have an agenda that can be political, economic, or to change behavior or thinking about a topic. And there are also countless scenarios where fake news is also spread from automated bots. Our application can be used to detect and filter these fake news sources, which can help to report them, thus helping to provide reliable news.

6 Future Works

1. The model can be incorporated in social media websites which can check the content of the message and refrain them from sharing or posting fake news itself by avoiding the spreading of fake news initially.
2. It can also be integrated with UI which lets users give their piece of text and check for its authenticity.
3. This approach can be further extended to testing on various other news of other languages and nations.
4. In real time, the availability of data in a clear format is minimal, hence more concentration can be given to work with unformatted data in an unsupervised manner to make users life's simpler.

7 Conclusion

In this paper, we focused on detecting the fake news using neural networks on Indian news. We used word2vec for the text transformations followed by neural networks for the classification of fake and real news. Testing the model against real data and attaining an acceptable accuracy of 75% has further reinforced our research work. In conclusion, this is a simple yet yielding deep learning approach for the classification of real and fake news using traditional NLP methods.

References

1. Palacio Marín, I., Arroyo, D.: Fake news detection. In: Herrero, Á., Cambra, C., Urda, D., Sedano, J., Quintián, H., Corchado, E. (eds.) CISIS 2019. AISC, vol. 1267, pp. 229–238. Springer, Cham (2021). https://doi.org/10.1007/978-3-030-57805-3_22
2. Hamid, A., et al.: Fake News Detection in Social Media using Graph Neural Networks and NLP Techniques: A COVID-19 Use-case (2020). https://doi.org/10.13140/RG.2.2.26073.34407
3. Traylor, T., Straub, J., Gurmeet, Snell, N.: Classifying fake news articles using natural language processing to identify in-article attribution as a supervised learning estimator. In: 2019 IEEE 13th International Conference on Semantic Computing (ICSC), pp. 445–449 (2019). https://doi.org/10.1109/ICOSC.2019.8665593
4. Oshikawa, R., Qian, J., Wang, W.Y.: A Survey on Natural Language Processing for Fake News Detection. LREC (2020)
5. Hirlekar, V.V., Kumar, A.: Natural language processing based online fake news detection challenges – a detailed review. In: 2020 5th International Conference on Communication and Electronics Systems (ICCES), pp. 748–754 (2020). https://doi.org/10.1109/ICCES48766.2020.9137915

6. Jain, A., Shakya, A., Khatter, H., Gupta, A.K.: A smart system for fake news detection using machine learning. In: International Conference on Issues and Challenges in Intelligent Computing Techniques (ICICT) 2019, pp. 1–4 (2019). https://doi.org/10.1109/ICICT46931.2019.8977659
7. Kaliyar, R.K.: Fake news detection using a deep neural network. In: 2018 4th International Conference on Computing Communication and Automation (ICCCA), pp. 1–7 (2018). https://doi.org/10.1109/CCAA.2018.8777343
8. Hossain, Md., Rahman, Md., Islam, Md.S., Kar, S.: BanFakeNews: A Dataset for Detecting Fake News in Bangla (2020)
9. Ahmed, H., Traore, I., Saad, S.: Detecting opinion spams and fake news using text classification. Secur. Priv. **1**, e9 (2017). https://doi.org/10.1002/spy2.9
10. Ahmed, H., Traore, I., Saad, S.: Detection of online fake news using n-gram analysis and machine learning techniques. In: Traore, I., Woungang, I., Awad, A. (eds.) Intelligent, Secure, and Dependable Systems in Distributed and Cloud Environments, ISDDC 2017. LNCS, vol. 10618, pp. 127–138. Springer, Cham (2017). https://doi.org/10.1007/978-3-319-69155-8_9
11. Hochreiter, S., Schmidhuber, J.: Long short-term memory. Neural Comput. **9**(8), 1735–1780 (1997). https://doi.org/10.1162/neco.1997.9.8.1735
12. Shu, K., Sliva, A., Wang, S., Tang, J., Liu, H.: Fake news detection on social media: a data mining perspective. ACM SIGKDD Explor. Newsl **19** (2017). https://doi.org/10.1145/3137597.3137600
13. Detecting Fake News in Social Media Networks (2018)
14. Zhang, J., Dong, B., Yu, P.S.: FakeDetector: effective fake news detection with deep diffusive neural network. In: 2020 IEEE 36th International Conference on Data Engineering (ICDE) (2020). https://doi.org/10.1109/icde48307.2020.00180

Exploring Self-supervised Capsule Networks for Improved Classification with Data Scarcity

Ladyna Wittscher(✉) and Christian Pigorsch

Friedrich-Schiller-University, Carl-Zeiß-Str. 3, 07743 Jena, Germany
{ladyna.wittscher,christian.pigorsch}@uni-jena.de

Abstract. While Capsule Networks overcome many shortcomings of Convolutional Neural Networks by taking spatial hierarchies between features into consideration and providing a new approach to routing between layers, they are sensible towards overfitting. Self-supervised training improves the semantic understanding of a network significantly and improves generalization without requiring additional images. Therefore, self-supervised Capsule Networks are a promising approach to improve learning under data scarcity, which makes the combination interesting for a huge variety of applications. Our approach improves test accuracy by up to 11.7% for small data sets and by up to 11.5% for small and imbalanced data sets. Furthermore, we explore the synergies and characteristics of this innovative method combination and give a general overview on the possibilities of self-supervised Capsule Networks.

Keywords: Capsule Network · Self-supervised learning · Data scarcity · Data imbalance · Robust neural networks

1 Introduction

The Capsule Network (CapsNet) architecture, which has been praised as the new sensation in Deep Learning, was developed to overcome some of the shortcomings of Convolutional Neural Networks (CNN), like their incapability to recognize pose and deformation of objects [17,52,62,66–68]. The structure of CapsNets and their functionality was adapted from the human visual cortex [9] and makes them most suitable for computer vision [52]. As in traditional CNN the max pooling layers only feed the information of the most active neurons into the subsequent layer, spatial hierarchies between features are lost [80]. Consequently, false positive classifications can occur if an image is classified into a certain category as soon as a specific key feature is detectable, even if it is in the wrong position [52]. While features are transmitted via a scalar output in CNN, CapsNets use routing-by-agreement and have an output vector specifying the features of the object as well as the likelihood of its existence [56,63]. CapsNets can learn directionality of image objects on different detail levels and their interrelationships as spatial information is passed from one layer to another [8,23].

J. I.-Z. Chen et al. (Eds.): ICIPCN 2022, LNNS 514, pp. 36–50, 2022.
https://doi.org/10.1007/978-3-031-12413-6_4

Therefore, CapsNets can especially advance fields of application where the data are spatio-temporal [52], e.g. autonomous driving and traffic recognition [27,32], protein classification [13,22] or text recognition [26,43].

A number of other typical data set characteristics which pose a challenge to CNN can be handled more successfully using CapsNets. They are in general characterized by good pixel level classification capabilities [52] and can deal well with rotated data [17,31,45,66]. Given class-imbalance, CapsNets also perform better [24]. They deal well with deficient labels [19], affine input transformations [44,49] and deformations [52]. In general, they show a high degree of robustness. Sabour et al. [56] under-trained their CapsNet on MNIST and then tested it on affNIST, where it achieved an improvement of 13 percentage points compared to a CNN with the same accuracy on MNIST. CapsNets perform well on small data sets [31] and retain more information [65]. Being able to learn and store the spatial relationship information of images, CapsNets need fewer samples for training in contrast to CNN [24,56,69,77]. Their generalization ability results in successful training with small data sets, enabling good results without data augmentation [1,2,63,76,77]. In contrast, CNN often rely on immense, high-quality data sets, as models trained with very small data sets are often characterized by a lack of generalization capability and overfitting [51]. Still, small data sets are inevitable if data collection requires extensive experimental operations, high detection costs or lengthy analysis, such is the case in toxicity analysis [5,11,40], material science [20], flexible manufacturing [4] or targeted niche marketing [29]. Therefore, it is important to optimize learning under conditions of data scarcity. This work does not focus on one specific area of application with data scarcity, but rather investigates the combination of CapsNets with an optimized self-supervised training approach in regard to its capability to perform better than the original CapsNet approach given small data sets.

Self-supervised learning, which is a sub-type of unsupervised learning [25, 30,48], boosts the network's ability to cope with small data sets. During an auxiliary prediction tasks (pretext tasks) performed on the same data set using self-generated labels, the weights are pretrained to improve the performance of the downstream task [36]. Rotation, prediction of a missing image patch, colorization or the order of a jigsaw puzzle are common pretext tasks [25]. Thereby, the semantics of the data are learned from the data itself, so that one part of the data works as a self-supervisory signal used to predict another part [21,25]. Self-supervised learning can also be understood as a form of pretraining that makes use of the same data set twice or an adapted transfer learning that does not change the data, but the task of the neural network. As the benefits from pretraining are reduced when no images from a similar domain are available, self-supervision is superior in these cases [48,55]. Self-supervised training can help to focus on key objects and disregard irrelevant background information [58,73]. In contrast, the original CapsNet implementation does not perform well on images with complex backgrounds, as its feature extraction ability is not accurate enough and a large number of training parameters are needed [23,71]. The absence of a pooling function results in CapsNets being more sensible to background noises [33], which increases the runtime [46] and creates models with a large computation complexity [38]. While the improved inter-layer communication due to dynamic-routing can boost time and space complexity, too many iterations can also cause overfitting and decrease performance [60]. Likewise, too many layers can lead to overfitting [53,60], which limits the complexity of the model [66]. Self-supervised learning is a successful strategy to reduce overfitting [30] and enables fast convergence [15]. Self-supervision both enables efficient learning with small data sets [14] and increases the model's ability to generalize [3]. Even when the number of training data is substantially smaller than the number of model parameters, self-supervision can significantly improve model performance [54]. If approximate conditional independence can be assumed between the random variables of input features and downstream labels, solving the pretext tasks based on known information can improve learning useful representations and enable nearly linear separation of the downstream targets with a small estimation error even with very few labeled training samples [36]. Self-supervision can furthermore generate significant improvements using large-scale imbalanced data sets due to the fact that its pretext task is based on the balanced pseudo-labels, resulting in a better downstream classification [35,72].

The combination of self-supervised training and CapsNet architecture is very promising not only because it can mitigate some of the disadvantages of CapsNets, but also because they both deal well with data set properties that most CNN struggle with. As these two methods have not been yet studied in combination in detail (see Sect. 2.2), we are primarily interested in the performance and characteristics of the new approach under conditions of data scarcity and imbalance using the benchmarking data set MNIST (see Sect. 3.1). As a reference scenario, we use the non-pretrained CapsNet. We analyze the differences in performance in different scenarios and their dependence on the difficulty of the learning conditions, study to what extent pretext task accuracy and downstream task accuracy correlate and examine the potential and synergies of self-supervised CapsNets.

2 Related Work

2.1 Functionality of Capsule Networks

CapsNets were first introduced in 2011 by Hinton et al. [17] as autoencoders being able to recognize pose and had their break-through in 2017 due to the introduction of vector CapsNets with dynamic routing by Sabour et al. [56]. The eponymous capsules are organized in layers and include groups of related neurons [67]. CapsNets compute the transformation matrix between sub-objects and higher-level objects and are therefore capable of representing part-whole relationships [56,78]. Lower-level positional information is "place-coded" while the higher-level with more degrees of freedom is "rate-coded", so capsule dimensionality increases with hierarchy [56]. The output vector of a capsule is condensed using a non-linear function (squashing), which makes it possible to choose which parent in the layer above the capsule output has to be sent to [66]. Furthermore, squashing increases the convergence speed of the routing algorithm and compresses information [12]. Each capsule output should be directed to a successive capsule that is fed with the most similar other inputs which works due to "coincidence filtering" as in a high-dimensional space the probability that agreements are only coincidence is very low [9]. The current capsule transfers its output to the correct parent capsule by choosing the one where the scalar product between prediction vector and output of the higher level capsule is largest, which is also known as "dynamic routing by agreement" [52,59]. This results in a top-down feedback and a part-whole relationship, as relevant paths are determined by the agreement of lower- and higher-level capsules [52]. CapsNets achieve regularization using a reconstruction autoencoder [41,60,64]. This methodology enabled the network to learn more general visual representation of the input data and makes traditional strategies to deal with small data input like data augmentation unnecessary [66].

2.2 Self-supervision and Capsule Networks

Given that both self-supervision and CapsNets are relatively new concepts within machine learning, the combination has not been thoroughly researched. There are only very few papers that combine these two strategies or aspects of them. As CapsNets learn transformation equivariant representations, Du et al. classify all publications based on Hinton's original work on learning transformation capsules [17] as a "family of self-supervised approaches" [10], although this definition is not in accordance with standard approaches towards self-supervised learning. One publication that combines self-supervision and the consideration of contextual information is "Dual-stream Multiple Instance Learning Network" by Li et al. [37] using self-supervised contrastive learning. But their method differs significantly from CapsNets as the weights between the nodes are no learned parameters. Sabour et al. [57] proposed a self-supervised method to learn visual part descriptors for CapsNets, thereby doing pioneering work on combining those two methods, although they do not analyze in detail the effects the two methods have on each other. In their publication, primary capsule encoders are used to identify the elementary parts of an image. A single input image is used for the encoder, but the part discovery makes use of motion-based self-supervision [6,42,57]. The resulting model is evaluated on unsupervised part segmentation and unsupervised image classification, demonstrating robustness towards viewpoint changes and adversarial attacks [57]. They do not execute a detailed analysis on the synergies of self-supervised training and CapsNet or evaluate the possibilities for pure image data using the given combination. Their model is effective on real and synthetic data and can complete shapes that have partially been occluded [57]. Sun et al. [61] use a self-supervised CapsNet for 3D point cloud reconstruction, autoencoding, canonicalization and unsupervised classification which achieves state-of-the art performance. Their downstream training is unsupervised [61]. They use self-supervision for object decomposition with pairs of randomly rotated objects [61]. Thereby they strive "to aggregate the attention masks into semantic keypoints, and use these to supervise a decomposition that satisfies the capsule invariance/equivariance properties" [61].

2.3 Pretrained Capsule Networks

There are also several authors that use pretrained approaches with CapsNets. These differ from self-supervised learning by using a different data set and no pseudo-labels. There is either the possibility of pretraining the encoder [79], the decoder [79], using pretrained weights (transfer learning strategy) [1,50] or pretrained word embedding in text classification [70,74]. Pretraining can decrease the train loss, improve low-level image representation and significantly reduce the number of trainable parameters [50]. Furthermore, pretraining with a data set of similar nature can improve both accuracy (increased by 2.6 percentage points) and specificity (increased by 2.8% points) significantly in the context of Covid19 detection in X-ray images [1].

3 Methods

3.1 Data Set

We use MNIST data set including 60,000 training images and 10,000 test images with handwritten numbers, each of these consisting of 28×28 pixels in grayscale values for all computations in this paper [34]. Being one of the most common benchmarking data sets to train machine vision algorithms, training with MNIST allows comparability of different methods. Instead of using a per se small data set, MNIST is artificially reduced by different factors to simulate different degrees of data scarcity. This has the advantage of simulating the model's behaviour given different data set sizes. Data imbalance is simulated by reducing the number of samples within the classes 0, 3, 5, 6 and 8 by a certain percentage, e.g. for the 10% example, 90% of the original samples in those categories are excluded. At the same time, the amount of samples in the other categories remains unchanged. As this also reduces the total amount of data available, we focus on analyzing the combination of small and imbalanced data sets, which is also very common in practical applications (e.g. engineering [75] and data mining [18]). All data are normalized, but no augmentation or deformation is executed. In contrast to Sabour et al. [56], the individual images are not changed by shifting.

3.2 Capsule Network Model

To analyze the synergies of self-supervised learning and CapsNet, we adapt the original dynamic-routing algorithm by Sabour et al. [56] and add a pretext step which pretrains the weights for the actual classification (see Sect. 3.3). We compare the self-supervised version to a reference scenario without pretraining, which otherwise uses the same model and hyperparameters. An overview of the model can bee seen in Fig. 1. In contrast to CNN, hyperparameters like learning rate and its decay or batch size have a reduced influence on the performance of CapsNets, while the routing iterations and optimizer have a significant impact [7,39]. We use Adam optimizer [28], 3 routing iterations, a learning rate of 0.001 and batch size 10. Training epochs vary with data set size to avoid overfitting, but reference and downstream task are always trained for the same number of epochs to ensure comparability. In general, CapsNets have way less parameters compared to CNN, as instead of links between individual neurons, only connections between capsules are of importance [52]. Also, CapsNet can be more shallow than CNN achieving comparable or better performance than CNN [56]. Too many layers can even lead to overfitting [53,60]. The CapsNet encoder is made up of three different parts, followed by a decoder. Our CapsNet includes a convolutional layer with 1 input channel, 256 output channels and kernel size 9. The primary capsules have 256 input channels, 32 output channels and kernel size 9. The digit capsules have 1152 routes, 8 input channels and 16 output channels. The decoder consists of three linear layers with an output size of 784. Margin loss and reconstruction loss are summed up for training and test loss, as recommended by [56] to achieve a good regularization. ReLu is used as activation function, reconstruction loss calculation is based on mean squared error. To evaluate the results, we use test loss and test accuracy, as the size of the training set is not changed for the different scenarios. In case of the imbalanced scenarios, we use weighted margin loss for training and additionally calculate the accuracy of the underrepresented classes for evaluation to make sure these have been successfully learned.

Fig. 1. Schematic overview of the self-supervised model with pretext and downstream task and the reference model.

3.3 Self-supervision

In general, it is important to note that the specific combination of pretext and downstream task has a significant influence on the model's behaviour. Even when using the same pretext task, different but similar models produce significantly different results [30]. As CapsNets are sensible to spatial orientation, we make use of rotation as pretext task. We are convinced that this is an interesting combination as it emphasises the semantic nature of the pretext task and therefore could improve the spatial knowledge of the network further. Using rotation in the pretext task avoids the generation of easily detectable low-level visual artifacts which would result in the network learning trivial features with no practical use for the downstream task [15]. An image is randomly either rotated by 0°, 90°, 180° or 270° to generate the pseudo-labels. The number of epochs for training is selected individually for the different scenarios so that the precision can be maximized. A detailed discussion on the influence of the pretext accuracy on the downstream classification can be found in Sect. 4.4. Like most self-supervised learning applications, we use the same network architecture for both pretext and downstream task [47]. Reference and downstream task are always trained for the same number of epochs (between 3 and 10 epochs depending on the data set size), the pretext task training is mostly shorter (between 5 and 1 epoch).

4 Results and Discussion

4.1 Data Scarcity

Self-supervised learning can boost CapsNet performance regarding small data sets significantly. If the amount of samples available for training is decreased, the difference in performance between the pretrained and the reference CapsNet model increases. Table 1 shows the percentage of data used from the original MNIST data set, the resulting classification accuracy and loss of the self-supervised as well as the reference model. The applied method for the generation of the given results is described in Sect. 3.2. If only 60 training images are available (0.1%), self-supervision enables a test accuracy improvement of 11.7%, while the loss is decreased by 13.5% compared to the reference model. If 1% of the original data set is available, the self-supervised approach has an accuracy of 94.78% instead of 93.73% (+1.1%) while the loss is decreased by 55.0%. Loss and accuracy are obviously not always improved to a similar extent, but the loss is in general very small. As can be seen by the 10% scenario, where an accuracy improvement of only 0.1% is achieved and loss is reduced by 3.8%, the difference between the two models decreases with increased data set size. When the full data set is available for classification, the difference between the two approaches is only marginal: the self-supervised version has a test accuracy of 98.87%, while the reference version has 98.84%. The test loss in this case is 0.0024 and 0.0027. Still, self-supervision can offer improvements in all tested scenarios and in general increases robustness and learning quality. In scenarios where the reference model

Table 1. Test accuracy and test loss for self-supervised (Self) and reference model (Reference) with different amounts of data from the original MNIST data set.

Data [%]	Self accuracy [%]	Self loss	Reference accuracy [%]	Reference loss
0.1	75.81	0.0576	67.84	0.0666
1	94.78	0.0190	93.73	0.0422
10	97.97	0.0050	97.88	0.0052
25	98.59	0.0028	98.57	0.0031
50	98.70	0.0024	98.50	0.0037
100	98.87	0.0024	98.84	0.0027

already performs good, the pretraining can contribute less to the performance than in more challenging scenarios.

4.2 Learning Behaviour of the Self-supervised CapsNet

As we aim to understand the synergies between CapsNets and self-supervised learning better and why the combination has positive effects concerning robustness towards data scarcity, we analyzed the confusion matrix for the 0.1% scenario. Therefore, we calculated the confusion matrix for both the self-supervised and the reference scenario and then computed the difference to make the two scenarios easily comparable in one figure. The results can be seen in Fig. 2, showing the difference in accuracy of the two models in percentage points. The horizontal digits are the target predictions, the vertical ones are the actual predictions, meaning that the diagonal numbers signify all correctly classified digits, while the rest are false classifications. Consequently, positive values on the diagonal signify that the self-supervised model is better than the reference model while positive values in other fields mean that the self-supervised model has more misclassifications. Fields are marked in green if the self-supervised model is at least 5% points better and red if it is at least 5% points worse. All digits have been classified more accurately by the self-supervised approach besides 3, where the accuracy is 12.6% points lower. For digit 1, the classifications accuracy is only increased by 4% points due to self-supervision. For every other digit the improvement is 5% points or more. The most significant improvements can be seen concerning digits 4 (+23.1), 5 (+12.9) and 2 (+11.8). In these three cases, misclassifications as 9, 3 and 8 have been predominantly reduced. Furthermore, there is a relationship between the improvement by pretraining and the classification accuracy of the reference model. The classification of digit 4 was the one with the worst classification accuracy of the reference model (21.3%), while the improvement due to pretraining was the most significant one (23.1% points). The second and third best improvements (+12.9 and +11.8) were achieved for the two digits with the second and third worst classification accuracy of the reference model (47.3% and 58.5%). In contrast to that, digit 1 had the highest accuracy (92.5%) and self-supervision could only generate and improvement by

4.0% points. Still, this is only a general tendency and there are outliers to it (e.g. digit 3). Self-supervised learning improves the overall classification and not only the one of few classes, so it is a holistic approach. The additional information that can be drawn from the same training set due to the pretext task leads to an overall enhanced learning process and makes the model more robust, but the benefits are most pronounced for classes which are more difficult to learn.

Fig. 2. Difference in test accuracy of the two models in percentage points per digit using 0.1% of the MNIST data set. The figure was attained by subtracting the confusion matrix of the reference model from the one of the self-supervised model. Positive values on the diagonal indicate a better performance of the self-supervised model while positive values in other fields mean that the reference model performed better. If the self-supervised model is 5 or more percentage points better, a field is marked in green. The opposite is marked in red.

4.3 Data Scarcity and Imbalance

Both models significantly decrease in performance with increased imbalance and decreased data set size, but the self-supervised version decreases considerably less. The method for the generation of the results is explained in Sect. 3.2. Looking at the accuracy of the underrepresented classes, the self-supervised CapsNet achieves a better classification accuracy in all analyzed scenarios, though the overall accuracy is slightly lower or the same in some cases (see Table 2). We take both metrics into consideration to make sure that the underrepresented classes are sufficiently learned, but not at the expense of overall accuracy. In case only 1% of the samples from the underrepresented classes are available and the data set is reduced to 10% of its original size, self-supervision improves the accuracy of the underrepresented classes by 3.6%, while overall test accuracy is the same. If only 1% of the data set is available using the same imbalancing, the performance of both models is low, demonstrating that this combination poses a big challenge. But self-supervision improves the accuracy of underrepresented classes by 11.5% and the overall accuracy by 2.4%. If only 1% of all data are available and underrepresented classes are reduced to 10%, self-supervision is 0.7% better regarding the accuracy of the underrepresented classes but also 3.6% worse regarding overall accuracy. With 10% of the data set and the same amount of imbalance, the self-supervised version is 0.3% better and the overall accuracy is improved by 0.06%. Comparing scenario 2 and 3 that have the same data set size but different imbalancing factors, the additional challenge imbalance adds becomes obvious. Again, self-supervision poses most advantage in the most difficult scenarios. As balanced pseudo-labels are used for the pretext task, the bias introduced by imbalanced labels can be mitigated [35,72]. Furthermore, the combination of two different tasks allows to extract more diverse information from the same data set, which seems to be especially beneficial if there are only few training samples available. For sole imbalancing, the self-supervised version still does not provide significant improvements, as CapsNets are already dealing very well with data imbalance [24]. In general, the accuracies of imbalanced and small data sets are considerably lower than the ones with data scarcity only. Both challenges reinforce each other and jointly pose a huge challenge as only very few examples of the underrepresented classes are available. Self-supervised learning especially improves the results for the underrepresented classes, while the overall accuracy is not always improved. Even if there are no accuracy improvements, the combination of a pretext task and a supervised downstream task can provide a strong regularization of the network and thereby boost robustness [16]. Thus, the appropriateness of using self-supervision depends heavily on the intended use and the significance of the underrepresented classes.

Table 2. Accuracies of the underrepresented classes and overall accuracies using small and imbalanced versions of MNIST with self-supervised CapsNet and a non-pretrained reference version.

Data characteristics		Self-supervised		Reference	
Imbalance [%]	Data [%]	Underrepresented [%]	All [%]	Underrepresented [%]	All [%]
1	10	84.61	81.21	81.70	81.21
1	1	22.73	47.66	20.38	46.53
10	1	57.17	62.24	56.77	64.58
10	10	96.76	92.44	96.47	92.38

4.4 Correlation of Pretext and Downstream Accuracy

Higher pretext accuracies generally result in better downstream classification results if the same training conditions are applied [15]. Still, pretext task accuracy can vary significantly with model architecture and parameters [30]. In our case, we do not change the model, but significantly modify the used data set. Consequently, the linear correlation between pretext task and downstream test accuracy for the different scenarios is very low ($R^2 = 0.16$), meaning pretext accuracy cannot act as an indicator for the overall performance. This indicates that a higher accuracy of the learning in the pretext task does not necessarily guarantee that the learned visual representations are also improving the downstream training. We assume that a sort of overfitting can also occur in this case. Consequently, pretext task epoch number is an important hyperparameter which has to be chosen carefully.

5 Conclusion

Our research demonstrates that the combination of self-supervised learning and CapsNet architecture has a lot of potential. The improvement due to pretraining is more significant the more difficult the task is as for less demanding scenarios, CapsNets are already very performative. Also looking at the individual classes, those the reference network classifies with the lowest accuracy are improved most significantly. Self-supervision's unique way of learning helps the model to deal better with data scarcity and imbalance because it allows to extract more knowledge from the same amount of samples. Given the combination of data scarcity and imbalance, the classification accuracy for the underrepresented classes is also considerably improved, although the overall accuracy can be slightly reduced in less demanding scenarios. A certain trade-off of robustness and performance is a typical problem in machine learning and could contribute to the tendency that the self-supervised model shows most advantage under difficult learning conditions. Further research using different pretext tasks that do not change spatial orientation would be interesting in combination with rotation invariant data sets to analyze the role of the pretext more. The fact that self-supervision can have tailored pretext tasks allows very specific optimization for individual datasets and fields of application, which make it interesting for diverse use cases. Consequently, self-supervision in combination with CapsNets can in general be a useful tool to improve the robustness of models, making it possible to decrease less in performance when the quality of the available training data are decreased. Furthermore, self-supervision can improve classification under data scarcity without requiring data augmentation or large models.

References

1. Afshar, P., Heidarian, S., Naderkhani, F., Oikonomou, A., Plataniotis, K.N., Mohammadi, A.: COVID-caps: a capsule network-based framework for identification of COVID-19 cases from X-ray images. Pattern Recogn. Lett. **138**, 638–643 (2020)
2. Afshar, P., Naderkhani, F., Oikonomou, A., Rafiee, M.J., Mohammadi, A., Plataniotis, K.N.: MIXCAPS: a capsule network-based mixture of experts for lung nodule malignancy prediction. Pattern Recogn. **116**, 107942 (2021)
3. Albuquerque, I., Naik, N., Li, J., Keskar, N., Socher, R.: Improving out-of-distribution generalization via multi-task self-supervised pretraining. arXiv preprint arXiv:2003.13525 (2020)

4. Andonie, R.: Extreme data mining: inference from small datasets. Int. J. Comput. Commun. Control **5**(3), 280–291 (2010)
5. Basak, S.C., Grunwald, G.D., Gute, B.D., Balasubramanian, K., Opitz, D.: Use of statistical and neural net approaches in predicting toxicity of chemicals. J. Chem. Inf. Comput. Sci. **40**(4), 885–890 (2000)
6. Bear, D.M., et al.: Learning physical graph representations from visual scenes. arXiv preprint arXiv:2006.12373 (2020)
7. Chauhan, A., Babu, M., Kandru, N., Lokegaonkar, S.: Empirical study on convergence of capsule networks with various hyperparameters (2018)
8. Ding, X., Wang, N., Gao, X., Li, J., Wang, X.: Group reconstruction and max-pooling residual capsule network. In: IJCAI, pp. 2237–2243 (2019)
9. Dombetzki, L.A.: An overview over capsule networks. Network architectures and services (2018)
10. Du, B., Gao, X., Hu, W., Li, X.: Self-contrastive learning with hard negative sampling for self-supervised point cloud learning. In: Proceedings of the 29th ACM International Conference on Multimedia, pp. 3133–3142 (2021)
11. Duan, Q., Lee, J.: Fast-developing machine learning support complex system research in environmental chemistry. New J. Chem. **44**(4), 1179–1184 (2020)
12. Edstedt, J.: Towards Understanding Capsule Networks (2020)
13. Fang, C., Shang, Y., Xu, D.: Improving protein gamma-turn inception capsule networks prediction using inception capsule networks. Sci. Rep. **8**, 15741 (2018)
14. Gidaris, S., Bursuc, A., Komodakis, N., Pérez, P., Cord, M.: Boosting few-shot visual learning with self-supervision. In: Proceedings of the IEEE/CVF International Conference on Computer Vision, pp. 8059–8068 (2019)
15. Gidaris, S., Singh, P., Komodakis, N.: Unsupervised representation learning by predicting image rotations. arXiv preprint arXiv:1803.07728 (2018)
16. Hendrycks, D., Mazeika, M., Kadavath, S., Song, D.: Using self-supervised learning can improve model robustness and uncertainty. arXiv preprint arXiv:1906.12340 (2019)
17. Hinton, G.E., Krizhevsky, A., Wang, S.D.: Transforming auto-encoders. In: Honkela, T., Duch, W., Girolami, M., Kaski, S. (eds.) ICANN 2011. LNCS, vol. 6791, pp. 44–51. Springer, Heidelberg (2011). https://doi.org/10.1007/978-3-642-21735-7_6
18. Hu, Y., et al.: An improved algorithm for imbalanced data and small sample size classification. J. Data Anal. Inf. Process. **3**(03), 27 (2015)
19. Huang, R., Li, J., Wang, S., Li, G., Li, W.: A robust weight-shared capsule network for intelligent machinery fault diagnosis. IEEE Trans. Industr. Inf. **16**(10), 6466–6475 (2020)
20. Hutchinson, M.L., Antono, E., Gibbons, B.M., Paradiso, S., Ling, J., Meredig, B.: Overcoming data scarcity with transfer learning. arXiv preprint arXiv:1711.05099 (2017)
21. Jayaraman, D., Grauman, K.: Learning image representations tied to ego-motion. In: Proceedings of the IEEE International Conference on Computer Vision, pp. 1413–1421 (2015)
22. de Jesus, D.R., Cuevas, J., Rivera, W., Crivelli, S.: Capsule networks for protein structure classification and prediction. arXiv preprint arXiv:1808.07475 (2018)
23. Jia, B., Huang, Q.: DE-CapsNet: a diverse enhanced capsule network with disperse dynamic routing. Appl. Sci. **10**(3), 884 (2020)

24. Jiménez-Sánchez, A., Albarqouni, S., Mateus, D.: Capsule networks against medical imaging data challenges. In: Stoyanov, D., et al. (eds.) LABELS/CVII/STENT -2018. LNCS, vol. 11043, pp. 150–160. Springer, Cham (2018). https://doi.org/10.1007/978-3-030-01364-6_17
25. Jing, L., Tian, Y.: Self-supervised visual feature learning with deep neural networks: a survey. IEEE Trans. Pattern Anal. Mach. Intell. **43**, 4037–4058 (2020)
26. Kim, J., Jang, S., Park, E., Choi, S.: Text classification using capsules. Neurocomputing **376**, 214–221 (2020)
27. Kim, M., Chi, S.: Detection of centerline crossing in abnormal driving using CapsNet. J. Supercomput. **75**(1), 189–196 (2019)
28. Kingma, D.P., Ba, J.: Adam: a method for stochastic optimization. arXiv preprint arXiv:1412.6980 (2014)
29. Kitchin, R., Lauriault, T.P.: Small data in the era of big data. GeoJournal **80**(4), 463–475 (2014). https://doi.org/10.1007/s10708-014-9601-7
30. Kolesnikov, A., Zhai, X., Beyer, L.: Revisiting self-supervised visual representation learning. In: Proceedings of the IEEE/CVF Conference on Computer Vision and Pattern Recognition, pp. 1920–1929 (2019)
31. Kruthika, K., Maheshappa, H., Initiative, A.D.N., et al.: CBIR system using capsule networks and 3D CNN for Alzheimer's disease diagnosis. Inform. Med. Unlocked **14**, 59–68 (2019)
32. Kumar, A.D.: Novel deep learning model for traffic sign detection using capsule networks. arXiv preprint arXiv:1805.04424 (2018)
33. LaLonde, R., Bagci, U.: Capsules for object segmentation. arXiv preprint arXiv:1804.04241 (2018)
34. LeCun, Y., Bottou, L., Bengio, Y., Haffner, P.: Gradient-based learning applied to document recognition. Proc. IEEE **86**(11), 2278–2324 (1998)
35. Lee, H., Hwang, S.J., Shin, J.: Rethinking data augmentation: self-supervision and self-distillation. arXiv preprint arXiv:1910.05872 (2019)
36. Lee, J.D., Lei, Q., Saunshi, N., Zhuo, J.: Predicting what you already know helps: provable self-supervised learning. arXiv preprint arXiv:2008.01064 (2020)
37. Li, B., Li, Y., Eliceiri, K.W.: Dual-stream multiple instance learning network for whole slide image classification with self-supervised contrastive learning. In: Proceedings of the IEEE/CVF Conference on Computer Vision and Pattern Recognition, pp. 14318–14328 (2021)
38. Li, H., Guo, X., Dai, B., Ouyang, W., Wang, X.: Neural network encapsulation. In: Ferrari, V., Hebert, M., Sminchisescu, C., Weiss, Y. (eds.) ECCV 2018. LNCS, vol. 11215, pp. 266–282. Springer, Cham (2018). https://doi.org/10.1007/978-3-030-01252-6_16
39. Lin, A., Li, J., Ma, Z.: On learning and learned representation with dynamic routing in capsule networks. arXiv preprint arXiv:1810.04041 **2**(7) (2018)
40. Liu, L., Liu, S.S., Yu, M., Zhang, J., Chen, F.: Concentration addition prediction for a multiple-component mixture containing no effect chemicals. Anal. Meth. **7**(23), 9912–9917 (2015)
41. Lu, C., Duan, S., Wang, L.: An improved capsule network based on newly reconstructed network and the method of sharing parameters. In: Lu, H., Tang, H., Wang, Z. (eds.) ISNN 2019. LNCS, vol. 11554, pp. 116–123. Springer, Cham (2019). https://doi.org/10.1007/978-3-030-22796-8_13
42. Mahendran, A., Thewlis, J., Vedaldi, A.: Self-supervised segmentation by grouping optical-flow. In: Leal-Taixé, L., Roth, S. (eds.) ECCV 2018. LNCS, vol. 11133, pp. 528–534. Springer, Cham (2019). https://doi.org/10.1007/978-3-030-11021-5_31

43. Mandal, B., Dubey, S., Ghosh, S., Sarkhel, R., Das, N.: Handwritten indic character recognition using capsule networks. In: 2018 IEEE Applied Signal Processing Conference (ASPCON), pp. 304–308. IEEE (2018)
44. Marchisio, A., Nanfa, G., Khalid, F., Hanif, M.A., Martina, M., Shafique, M.: CapsAttacks: robust and imperceptible adversarial attacks on capsule networks. arXiv preprint arXiv:1901.09878 (2019)
45. Marcos, D., Volpi, M., Tuia, D.: Learning rotation invariant convolutional filters for texture classification. In: 2016 23rd International Conference on Pattern Recognition (ICPR), pp. 2012–2017. IEEE (2016)
46. Nair, P., Doshi, R., Keselj, S.: Pushing the limits of capsule networks. arXiv preprint arXiv:2103.08074 (2021)
47. Noroozi, M., Vinjimoor, A., Favaro, P., Pirsiavash, H.: Boosting self-supervised learning via knowledge transfer. In: Proceedings of the IEEE Conference on Computer Vision and Pattern Recognition, pp. 9359–9367 (2018)
48. Ohri, K., Kumar, M.: Review on self-supervised image recognition using deep neural networks. Knowl. Based Syst. **224**, 107090 (2021)
49. Ohta, N., Kawai, S., Nobuhara, H.: Analysis and learning of capsule networks robust for small image deformation. In: 2020 International Joint Conference on Neural Networks (IJCNN), pp. 1–7. IEEE (2020)
50. Ozbulak, G.: Image colorization by capsule networks. In: Proceedings of the IEEE/CVF Conference on Computer Vision and Pattern Recognition Workshops, pp. 2150–2158 (2019)
51. Pasupa, K., Sunhem, W.: A comparison between shallow and deep architecture classifiers on small dataset. In: 2016 8th International Conference on Information Technology and Electrical Engineering (ICITEE), pp. 1–6. IEEE (2016)
52. Patrick, M.K., Adekoya, A.F., Mighty, A.A., Edward, B.Y.: Capsule networks-a survey. J. King Saud Univ. Comput. Inform. Sci. **34**, 1295–1310 (2019)
53. Peer, D., Stabinger, S., Rodriguez-Sanchez, A.: Limitation of capsule networks. Pattern Recogn. Lett. **144**, 68–74 (2021)
54. Pinto, L., Gupta, A.: Supersizing self-supervision: learning to grasp from 50k tries and 700 robot hours. In: 2016 IEEE International Conference on Robotics and Automation (ICRA), pp. 3406–3413. IEEE (2016)
55. Raghu, M., Zhang, C., Kleinberg, J., Bengio, S.: Transfusion: understanding transfer learning for medical imaging. arXiv preprint arXiv:1902.07208 (2019)
56. Sabour, S., Frosst, N., Hinton, G.E.: Dynamic routing between capsules. arXiv preprint arXiv:1710.09829 (2017)
57. Sabour, S., Tagliasacchi, A., Yazdani, S., Hinton, G.E., Fleet, D.J.: Unsupervised part representation by flow capsules. arXiv preprint arXiv:2011.13920 (2020)
58. Sayed, N., Brattoli, B., Ommer, B.: Cross and learn: cross-modal self-supervision. In: Brox, T., Bruhn, A., Fritz, M. (eds.) GCPR 2018. LNCS, vol. 11269, pp. 228–243. Springer, Cham (2019). https://doi.org/10.1007/978-3-030-12939-2_17
59. Shahroudnejad, A., Afshar, P., Plataniotis, K.N., Mohammadi, A.: Improved explainability of capsule networks: relevance path by agreement. In: 2018 IEEE Global Conference on Signal and Information Processing (GlobalSIP), pp. 549–553. IEEE (2018)
60. Sun, K., Yuan, L., Xu, H., Wen, X.: Deep tensor capsule network. IEEE Access **8**, 96920–96933 (2020)
61. Sun, W., et al.: Canonical capsules: self-supervised capsules in canonical pose. In: 35th Conference on Neural Information Processing Systems (2021)
62. Vijayakumar, T.: Comparative study of capsule neural network in various applications. J. Artif. Intell. **1**(01), 19–27 (2019)

63. Wang, D., Liang, Y., Xu, D.: Capsule network for protein post-translational modification site prediction. Bioinformatics **35**(14), 2386–2394 (2019)
64. Wang, Z., et al.: A novel method for intelligent fault diagnosis of bearing based on capsule neural network. Complexity **2019**, 1–17 (2019)
65. Wu, F., Smith, J.S., Lu, W., Pang, C., Zhang, B.: Attentive prototype few-shot learning with capsule network-based embedding. In: Vedaldi, A., Bischof, H., Brox, T., Frahm, J.-M. (eds.) ECCV 2020. LNCS, vol. 12373, pp. 237–253. Springer, Cham (2020). https://doi.org/10.1007/978-3-030-58604-1_15
66. Xi, E., Bing, S., Jin, Y.: Capsule network performance on complex data. arXiv preprint arXiv:1712.03480 (2017)
67. Xiang, C., Zhang, L., Tang, Y., Zou, W., Xu, C.: MS-CapsNet: a novel multi-scale capsule network. IEEE Signal Process. Lett. **25**(12), 1850–1854 (2018)
68. Xinyi, Z., Chen, L.: Capsule graph neural network. In: International Conference on Learning Representations (2018)
69. Xu, Q., Chen, K., Zhou, G., Sun, X.: Change capsule network for optical remote sensing image change detection. Remote Sens. **13**(14), 2646 (2021)
70. Yang, M., Zhao, W., Ye, J., Lei, Z., Zhao, Z., Zhang, S.: Investigating capsule networks with dynamic routing for text classification. In: Proceedings of the 2018 Conference on Empirical Methods in Natural Language Processing, pp. 3110–3119 (2018)
71. Yang, S., et al.: RS-CapsNet: an advanced capsule network. IEEE Access **8**, 85007–85018 (2020)
72. Yang, Y., Xu, Z.: Rethinking the value of labels for improving class-imbalanced learning. arXiv preprint arXiv:2006.07529 (2020)
73. Yuan, D., Chang, X., Huang, P.Y., Liu, Q., He, Z.: Self-supervised deep correlation tracking. IEEE Trans. Image Process. **30**, 976–985 (2020)
74. Zhang, N., Deng, S., Sun, Z., Chen, X., Zhang, W., Chen, H.: Attention-based capsule networks with dynamic routing for relation extraction. arXiv preprint arXiv:1812.11321 (2018)
75. Zhang, T., et al.: Intelligent fault diagnosis of machines with small & imbalanced data: a state-of-the-art review and possible extensions. ISA Trans. **119**, 152–171 (2022)
76. Zhang, X., Luo, P., Hu, X., Wang, J., Zhou, J.: Research on classification performance of small-scale dataset based on capsule network. In: Proceedings of the 2018 4th International Conference on Robotics and Artificial Intelligence, pp. 24–28 (2018)
77. Zhao, T., Liu, Y., Huo, G., Zhu, X.: A deep learning IRIS recognition method based on capsule network architecture. IEEE Access **7**, 49691–49701 (2019)
78. Zhao, W., Ye, J., Yang, M., Lei, Z., Zhang, S., Zhao, Z.: Investigating capsule networks with dynamic routing for text classification. arXiv preprint arXiv:1804.00538 (2018)
79. Zhao, Y., Birdal, T., Deng, H., Tombari, F.: 3D point capsule networks. In: Proceedings of the IEEE/CVF Conference on Computer Vision and Pattern Recognition, pp. 1009–1018 (2019)
80. Zhu, Z., Peng, G., Chen, Y., Gao, H.: A convolutional neural network based on a capsule network with strong generalization for bearing fault diagnosis. Neurocomputing **323**, 62–75 (2019)

A Novel Architecture for Improving Tuberculosis Detection from Microscopic Sputum Smear Images

S. Pitchumani Angayarkanni, V. Vanitha(✉), V. Karan, and M. Sivant

Sri Ramachandra Faculty of Engineering and Technology, Sri Ramachandra Institute of Higher Education and Research, Porur, Chennai, India
{angayarkanni,vanithav,karan.e0119039,
sivant.e0119004}@sret.edu.in

Abstract. Tuberculosis (TB) is a communicable disease that mainly affects lungs by a bacteria Mycobacterium Tuberculosis. It remains a major global health challenge in both developed and developing countries and identified as the leading reason for global death after Covid-19. In several countries in Africa and Asia, screening for pulmonary tuberculosis is done through Ziehl-Neelsen sputum smear images due to cost-effectiveness. Manual detection is time-consuming and strenuous work leading to misdiagnosis and low detection rate. It takes several hours to analyze a single slide to screen the patient for tuberculosis. This study focusses on proposing a novel architecture based on deep learning for mask generation and segmentation. Also, to determine suitable pre-processing techniques for sputum images. The quality of the sputum smear images captured under microscope depends on various factors, mainly the staining procedure and nature of microscope. The pre-processing techniques were analyzed in detail and an effective preprocessed image was determined using image quality metrics analysis. A novel pixel-based mask generation architecture segZnet is proposed. The preprocessed images and corresponding masks are fed to Unet architecture for segmentation. The accuracy of the proposed method is 98.5%.

Keywords: Sputum images · Tuberculosis · Mycobacterium · Segmentation · UNet

1 Introduction

World health organization estimated that 4.1 million people currently suffer from Tuberculosis (TB). In 2020, 10 million people fell sick and around 1.5 million people lost their life [1]. The emergence of COVID-19 has upturned the global progress in preventing tuberculosis and deaths have raised [2]. India is one of the top listed countries and contributes most to the global cases. The underdeveloped countries in Africa and Asia rely on sputum smear microscopy for the detection of TB. Most National TB control programs in developing countries are implementing bright-field microscopic examination of a stained sputum smear using a bright-field microscope owing to its ease to perform,

J. I.-Z. Chen et al. (Eds.): ICIPCN 2022, LNNS 514, pp. 51–62, 2022.
https://doi.org/10.1007/978-3-031-12413-6_5

simplicity, less expensive, and rapid [3]. It has been also highly recommended as the first line of treatment by the world health organization (WHO).

However, TB screening using bright-field microscopy has its limitations. The accuracy of the result is influenced by two important factors; (i) quality of smear reading, (ii) skills and expertise of microscopists. The smears need to be examined for at least 100 fields of view (FOV) to improve the accuracy of detection [4]. The detection and quantification of bacilli are prone to error even by experienced technicians due to physical strain and mental concentration. This labour-intensive task takes several minutes to few hours, depending on the number of bacilli present and also on the stage of infection as the early infections have fewer number of bacilli) [5]. If the sputum image has less than 5000 bacilli/ml, it is highly unlikely to be diagnosed. Also, many slides analyzed vary from 40-to 100 images [6] depending patient's level of infection. Due to WHO quality assurance guidelines on sputum microscopy [7], the technician examines very few slides a day [8] to avoid misdiagnosis and ensure the reliability of the report. Several studies from literature claimed that manual detection misses around 30–55% of active cases. Detection of TB bacilli from the slides by lab technicians is associated with a high misdiagnosis rate due to (i) prolonged and focussed visual acuity (ii) the need to inspect several fields of view. Moreover, they can inspect only a few numbers of samples due to the above constraints which may result in a delay of diagnosis and treatment. The automated diagnosis helps to reduce misdiagnosis, thereby increasing the accuracy of detection. The main advantage of an automated screening system is that many smear slides can be inspected, with several fields of views (FOV) in each slide. In addition to that, a smear slide with few numbers of bacilli indicating the initial stage can be also detected which can be easily missed by humans [9].

To overcome the aforementioned problems, automated analysis of sputum smear images has been proposed aiming to provide quality health care, patient safety efficiency in clinical outcomes. Being said that, automated detection for the presence of bacilli from microscopic smear sputum images poses greater challenges due to uneven light intensity, the complex nature of background, and overlapping of bacilli.

In this paper, we analyzed and evaluated various pre-processing and use the one that produced a better result for slides with different image quality. For comparison of pre-processing techniques, we utilized various retinex methods and their variations, traditional filtering methods including median and Gaussian, and Contrast limited adaptive histogram equalization (CLAHE). A mask is generated for pre-processed images and we proposed a novel pixel-based architecture called segZnet. The pre-processed image and mask are trained with UNet architecture.

The contributions of this paper are:

1. Automated detection and quantification to overcome the disadvantages of low accuracy in manual screening.
2. A novel pixel-based architecture to generate a mask for pre-processed images.
3. Segmentation of overlapping bacteria in the sputum smear images.

2 Literature Review

Various image processing techniques proposed in the literature for the detection of M. tuberculosis using bright-field microscopy based on the review study [10] is presented. Costa et al. [11] were the pioneers in the field of automatic detection of TB bacilli from bright field microscopy. They extracted an R-G image to isolate it from the background in which it appears as light-colored regions on a dark background. Threshold value based on the histogram R-G image was computed to segment the bacilli. The artifacts present in the segment are removed using morphological operations. Though the accuracy is good, specificity is high.

Bayesian segmentation was proposed in [12] to segment TB bacilli using prior knowledge of colour images. This approach is based on predicting the pixels based red green blue values that were significantly different from the non-TB background. The shape descriptors like axis ratio and eccentricity are applied to distinguish true TB bacilli from other TB bacilli artifacts. The accuracy, sensitivity, and specificity were not discussed. Khutlang et al. [13] combined various two-pixel classifiers for segmentation. The necessary features are extracted from segmented bacilli and several classifier methods was compared.

Osman et al. [14] utilized the hybrid multi-layered perceptron network (HMLP) for the identification of Tuberculosis bacilli using HSV images. In [15], k-means clustering followed by feature extraction using Zernike moments was employed to segment the bacteria. In [16], the authors extended the previous work on Neural network techniques to improve the accuracy of the detection. Zhai et al. [17] proposed a two-step segmentation method and a decision tree-based approach for the classification of segmented TB bacilli. Ghosh et al. [18] developed a fuzzy-based decision-making approach to segment the bacilli using shape, color, and granularity features. The contour boundary was traced using the region growing technique.

Lopez et al. [19] designed a convolutional neural network (CNN) model for detecting bacilli. Panicker et al. [20] developed a two-stage approach for the segmentation of TB bacilli. The foreground and background are separated through Otsu's method followed by morphological opening and closing operations to fine-tune them.

Even though the above discussed methods have produced satisfactory results, there are a few drawbacks. Firstly, the efficacy of the proposed techniques was not tested on diverse quality microscopic images. Secondly, non-overlapping bacteria are not considered. Thirdly, the bacteria count is not obtained which is essential to decide the infection level.

3 Methodology

In this section, the proposed approach towards Segmentation of TB is discussed. The method is composed of four major steps: (1) Pre-processing to enhance the quality of image, (2) Preparation of mask using SegZNet CNN Architecture, (3) Augmentation of the images and masks using augmentations library to increase the size of the sputum smear dataset and further increasing the accuracy of the model by making it less biased due to rotations, flips, blurs, uneven light illuminations and etc., and finally (4) Adopt UNET model to train with pre-processed sputum spear images and masks generated in the previous step.

Two data sets containing Ziehl-Neelsen sputum smear microscopic images were used in this study. Open-Source dataset available in Kaggle and dataset from [21] are used for this study. It has 1265 positive images and fewer negative images. dataset 2 has 1735 images. As few numbers of negative images are available, various augmentation methods including rotation, flipping, and resizing were performed on infected and normal images, and eventually obtained a set of 4000 samples. The images are split into two sets with a ratio of 80:20. The training set has 3040 images and a validation set with 800 images. The testing set is limited to 100 images from dataset 2 as it has to be annotated manually by a field expert.

All the experiments are conducted using a 16 GB Graphics Processing Unit, and PyTorch framework.

3.1 Preprocessing

With use of modern imaging technologies to capture images, it is very crucial to uphold and preserve image details. The digital images are prone to noise and other artifacts, hence it is vital to minimize the amount of noise in the images before moving on to the subsequent mask creation process.

There are two things to be taken into account while handling medical sputum smear images 1) un uneven distribution of light and 2) uneven distribution of the stain in the slide. To segregate the bacilli from the stained microscopic images, deconvolution technique can be used to detach the red intensities (since the color of the bacilli is red after the sputum slide immersed with Carbol Fuchsin for 7 to 8 min) from the sputum images. Along with that several pre-processing techniques were used to find the suitable technique.

As the smear images are taken under non-linear illumination circumstances, we tried with and without Multi-Scale Retinex with Color Restoration (MSRCR) of the sputum smear image and then sharpened the images and finally, deconvolution of the image. Then using the ground truth bounding boxes from the Kaggle dataset, we can verify to what extent does our pre-processing technique is good enough to segregate the red intensities of the bacilli. The combination of pre-processing techniques experimented on is presented in Fig. 1.

Fig. 1. Various combination of pre-processing techniques used.

a. MSRCR + Dilate + Deconvolution:

We perform MSRCR technique to enhance images captured under non-linear illumination conditions to the level at which the image is perceived by the naked eyes. Then by further sharpening the image to overcome the blurring introduced by the camera and further increase its legibility.

Then finally we apply deconvolution operation on the image to segregate the red intensities of the bacilli from the TB image. To locate the bacilli, we further annotate the image with bounding boxes using the ground truth XML file provided with the dataset. Then for better visualization, we convert the RGB image to Gray-scale and apply binary otsu and invert it to equalize all the intensities of the bacilli present in the image.

b. Automated MSRCR + Dilate + Deconvolution:

The colour intensity of the microscopic images can be enhanced using MSRCR. However, appropriate hyper parameters has to be set which depends on the image and tend to vary from one image to another. Hence, we can use Automated MSRCR to overcome this problem.

Using variance of histograms and frequency of pixels in the images as a control measure, the automated msrcr algorithm finds the optimal parameter values for the individual image. Then as mentioned in the previous method we sharpen the image, then apply deconvolution operation to segregate the red intensities and further annotate the image with the ground truth XML file and convert it to grayscale, invert it and apply binary otsu for better visualization.

c. Dilate + Deconvolution:

If the image is already captured under proper illumination, there is no need to apply MSRCR image restoration. Hence, we now directly sharpen the image to overcome the blurring introduced by the camera and proceed with deconvolution operation to segregate the intensities of the bacilli in the TB image, annotate it with the ground truth xml file and further covert it to gray scale, invert it and apply binary otsu for better visualization.

Out of the above mentions pre-processing combinations, direct sharpening and deconvolution has performed better than the other combination for the isolation of red intensities from the bacilli and avoid the regions with uneven stains to the maximum.

3.2 Mask Generation Using SegZNet Architecture

The output of the pre-processed images from the above steps consists of segregated red intensities of the bacilli from the sputum smear image. It consists of dominant red intensities which are visible to the naked eyes, as well as a non-dominant quantity of fragile red intensities at the other parts of the images, hence it is crucial to eliminate these weak intensities from the image and make the red intensities of the bacilli clearer. The idea behind the SegZNet Architecture is to find those dominant red intensities using the convolution operations bypassing the pre-processed image through the layers of the CNN architecture, isolate the regions of the dominant intensities, and further cluster those regions by reducing the number of classes of the intensities in the subsequent iterations. Finally, we get a mask generated for the pre-processed image, which overlaps the regions of the bacilli with dominant red intensities.

Fig. 2. SegZNet architecture.

3.3 Data Augmentation

To develop a good model based on deep learning techniques require a good number of data and balanced samples. However, it is challenging to collect an abundant number of medical images. More to that, collected data has to be labelled by expertise which is time consuming process. In image classification task, correct annotation of data is crucial for better accuracy. An accurate box drawn around the region of interest is needed for an object detection task. Hence it requires manual effort and could be a tedious job to label the entire training data without error. In the case of labelling medical images, we need domain experts to verify the annotations. Moreover, several policies on privacy has be preserved while working with healthcare data, and hence getting the data might be feasible but they are very expensive in most cases.

Using image augmentation process we can increase the training samples. To generate a sample, we need to alter the original image, for example, we change to make the image brighter, rotate it, blur it, and use a random combination of multiple image augmentation techniques to create an sufficient quantity of distinct images. For training the UNet model with the sputum smear images, we need to pass positive and negative images with marks generated on the above step as shown in Fig. 3a and b.

Fig. 3. (a) Image and mask for a negative image. (b) Image and mask for a positive image

Using albumentations library in PyTorch we can generate as many numbers training samples as required for training the UNet by creating an albumentation pipeline consisting of the image augmentation techniques that can be applied to the images as well the image mask. The techniques used are tabulated in Table 1.

Table 1. Image augmentation techniques

Methods	Parameters/Range
Flip	Left, left_right, top_bottom
Transpose	0.6
Blur (only one)	Blur, probability = 0.4, blur_limit = 3 Median blur, probability = 0.4, blur_limit = 3 Gaussian blur, probability = 0.4
Distortion (only one)	Optical Distortion, probability = 0.4 Grid Distortion
rotate90	Probability = 0.5
Brightness/Contrast	0.5
Hue Saturation	0.5

Few samples of augmented images are presented below in Fig. 4.

Fig. 4. (a) Image and mask for an augmented negative image. (b) Image and mask for an augmented positive image

3.4 UNet Segmentation

Convolutional Neural Networks are mostly used for image classification like identifying cats and dogs in an image or classifying handwritten letters and digits. They can also perform well on the mainstream tasks such as image segmentation and signal processing. UNet is a CNN based architecture designed for medical images to identify and segment the region of interest. The UNet architecture is presented in Fig. 5.

Fig. 5. UNet architecture

The UNet architecture has an encoder and a decoder. The encoder is used to extract the feature values in the image. The decoder uses transposed convolution for image localization. Moreover, the encoder reduces the spatial dims in every layer and increases the channels. Whereas, the decoder increases the spatial dimensions while reducing the channels. Thus, it provides precise and quick segmentation of biomedical images.

The Augmented TB images and masks created on the previous step are passed as an input for the UNet model after resizing the images to 256×256 dimension. The hyperparameters are: batch size = 16, learning rate = 0.001 with the Adam optimizer and the number of epochs is set to 40. Moreover 90% of the samples are used to train the model whereas the remaining data is used for validating the model.

The plot of the training accuracy vs Val accuracy is depicted in Fig. 6a. The plot of the training loss vs Val-accuracy is depicted in Fig. 6b.

Fig. 6. (a) Training vs val accuracy. (b) Training vs val loss

After the model the trained, the model is tested with sample positive and negative sputum image. The raw image, mask given as input and the output segmented image for both positive and negative sputum smear images are depicted in Fig. 7.

Fig. 7. (a) Actual and predicted mask for a positive TB image. (b) Actual and predicted mask for a negative TB image

4 Result and Discussion

We have evaluated the proposed model on various metrics, namely Intersection over union (IOC), precision, recall and model accuracy. The analysis of the model using sample positive and negative sputum image is presented in the Table 2:

Table 2. Evaluation for negative and positive images

Image	IOU	Model accuracy	Adapted Rand Precision	Adapted Rand Recall
Positive	0.8728	98.5	0.9153	1.0
Negative	1.0	98.5	1.0	1.0

Discussion:

A. Medical images are prone to noises and artifacts during to acquisition and transmission. This interference leads to lower the image quality, which not only impact the effectiveness of the medical image but also seriously impact on diagnosis. There are two challenges need to tackled with Sputum smear microscopic images: (i) uneven/poor distribution of light intensity as shown in Fig. 2a. (ii) uneven distribution of stain in the slide as shown in Fig. 2b. We performed pre-processing step to enhance the quality of microscopic image. A pipeline of Automated MSRCR followed by deconvolution and dilation produced better preprocessed image.
B. General data augmentation techniques performed on the dataset to increase the training set. Augmented techniques are chosen in such a way that information in the image is retained. Zoom and crop techniques were opted out for this reason.
C. The novel pixel based segZnet architecture for automated generation is proposed. Mask can be generated easily for simple geometric images or for images that need a single mask. In smear images of TB, there are several images that need to be masked and has complex differing shapes and sizes. The proposed architecture has stack of 10 convolutional layers followed by 10 ReLu layers and finally 10 batch normalization layers.

5 Conclusion

The proposed technique has an advantage of detecting Tuberculosis bacteria and robust in preserving the image details and effectively reduces the computational cost. This can automate the detection of bacilli with good accuracy to start the treatment on time and prevent the spread of bacteria.

This research is a step forward in the eradication of TB and offers two major advantages to treat the TB infected humans.

1) Manual error is excluded as the lab technician or pathologist does not need to visually detect Mycobacterium tuberculosis in the stained sputum smear images;
2) Rapid diagnosis is feasible with good accuracy.

References

1. WHO: Global Tuberculosis Report 2020 (2020)
2. Chakaya, J., Khan, M., Ntoumi, F., et al.: Global Tuberculosis Report 2020 – reflections on the Global TB burden, treatment and prevention efforts. Int. J. Infect. Dis. **113**, S7–S12 (2021). https://doi.org/10.1016/j.ijid.2021.02.107
3. Desikan, P.: Sputum smear microscopy in tuberculosis: is it still relevant? Indian J. Med. Res. **137**(3), 442 (2013)
4. Xiong, Y., Ba, X., Hou, A., Zhang, K., Chen, L., Li, T.: Automatic detection of mycobacterium tuberculosis using artificial intelligence. J. Thorac. Dis. **10**(3), 1936 (2018)
5. Mithra, K.S., Emmanuel, W.S.: Automatic methods for Mycobacterium detection on stained sputum smear images: a survey. Pattern Recogn. Image Anal. **28**(2), 310–320 (2018)
6. Sotaquirá, M., Rueda, L., Narvaez, R.: Detection and quantification of bacilli and clusters present in sputum smear samples: a novel algorithm for pulmonary tuberculosis diagnosis. In: Proceedings of International Conference on Digital Image Processing, pp. 117–121 (2009)

7. World Health Organization: Quality assurance of sputum microscopy in DOTS programmes: Regional guidelines in countries in the Western Pacific. WHO Regional Office for the Western Pacific, Manila (2003)
8. Desalegn, D.M., et al.: Misdiagnosis of pulmonary tuberculosis and associated factors in peripheral laboratories: a retrospective study, Addis Ababa, Ethiopia. BMC Res. Notes **11**(1), 1–7 (2018)
9. Mhimbira, F.A., Cuevas, L.E., Dacombe, R., Mkopi, A., Sinclair, D.: Interventions to increase tuberculosis case detection at primary healthcare or community-level services. Cochrane Database Syst. Rev. **11** (2017). https://doi.org/10.1002/14651858.CD011432.pub2. ISSN = 1465-1858
10. Panicker, R.O., Soman, B., Saini, G., Rajan, J.: A review of automatic methods based on image processing techniques for tuberculosis detection from microscopic sputum smear images. J. Med. Syst. **40**(1), 1–13 (2016)
11. Forero, M.G., Sroubek, F., Cristóbal, G.: Identification of tuberculosis bacteria based on shape and color. Real Time Imaging **10**, 251–262 (2004)
12. Sadaphal, P., Rao, J., Comstock, G.W., Beg, M.F.: Image processing techniques for identifying Mycobacterium tuberculosis in Ziehl-Neelsen stains. Int. J. Tuberc. Lung Dis. **12**(5), 579–582 (2008)
13. Khutlang, R., et al.: Classification of Mycobacterium tuberculosis in images of ZN-stained sputum smears. IEEE Trans. Inf. Technol. Biomed. **14**(4), 949–957 (2009)
14. Osman, M.K., Mashor, M.Y., Jaafar, H.: Detection of Mycobacterium tuberculosis in Ziehl-Neelsen stained tissue images using Zernike moments and hybrid multilayered perceptron network. In: Proceedings of IEEE International Conference on Systems Man and Cybernetics (SMC), Istanbul, pp. 4049–4055 (2010). https://doi.org/10.1109/ICSMC.2010.5642191
15. Osman, M.K., Mashor, M.Y., Jaafar, H.: Segmentation of tuberculosis bacilli in Ziehl-Neelsen tissue slide images using Hibrid Multilayered Perceptron network. In: Proceedings of 10th International Conference on Information Sciences Signal Processing and their Applications (ISSPA), Kuala Lumpur, pp. 365–368 (2010). https://doi.org/10.1109/ISSPA.2010.5605524
16. Osman, M.K., Noor, M., Mashor, M.Y., Jaafar, H.: Compact single hidden layer feed forward network for Mycobacterium tuberculosis detection. In: 2011 IEEE International Conference on Control System, Computing and Engineering, pp. 432–436 (2011). https://doi.org/10.1109/ICCSCE.2011.6190565
17. Zhai, Y., Liu, Y., Zhou, D., Liu, S.: Automatic Identification of Mycobacterium tuberculosis from ZN-stained sputum smear: algorithm and system design. In: Proceedings of IEEE International Conference on Robotics and Biomimetics (ROBIO), Tianjin, pp. 41–46 (2010). https://doi.org/10.1109/ROBIO.2010.5723300
18. Ghosh, P., Bhattacharjee, D., Nasipuri, M.: A hybrid approach to diagnosis of tuberculosis from sputum. In: 2016 International Conference on Electrical, Electronics, and Optimization Techniques (ICEEOT), pp. 771–776. IEEE (2016)
19. López, Y.P., Costa Filho, C.F.F., Aguilera, L.M.R., Costa, M.G.F.: Automatic classification of light field smear microscopy patches using Convolutional Neural Networks for identifying Mycobacterium Tuberculosis. In: 2017 CHILEAN Conference on Electrical, Electronics Engineering, Information and Communication Technologies (CHILECON), pp. 1–5. IEEE (2017)
20. Panicker, R.O., Kalmady, K.S., Rajan, J., Sabu, M.K.: Automatic detection of tuberculosis bacilli from microscopic sputum smear images using deep learning methods. Biocybern. Biomed. Eng. **38**(3), 691–699 (2018)
21. Ma, J., Fan, X., Ni, J., Zhu, X., Xiong, C.: Multi-scale retinex with color restoration image enhancement based on Gaussian filtering and guided filtering. Int. J. Mod. Phys. B **31**, 1744077 (2017). https://doi.org/10.1142/S0217979217440775

TapasQA - Question Answering on Statistical Plots Using Google TAPAS

Himanshu Jain(✉), Sneha Jayaraman, I. T. Sooryanath, and H. R. Mamatha

PES University, Bengaluru, India
nhimanshujain@gmail.com, mamathahr@pes.edu

Abstract. The proposed research aims to build a question answering system for statistical plots that will help analysts question and understand plots on a large scale and automate decision-making capabilities. The built model aids in answering open-ended questions and Yes/No binary questions on images of statistical plots that include a vertical bar, horizontal bar, line, and dot plots. Google TAble PArSing (TAPAS) supported with custom operations is employed, which has proved to be better than other state-of-the-art models.

Keywords: Visual question answering · Statistical plots · PlotQA · Deep learning · Natural language processing · Google TAPAS · Table parsing

1 Purpose

Statistical charts are an intuitive and simple way to represent data. They form a way of representing structured data in the form of graphical visualizations. Such graphical visualizations aid people in better interpreting features of data. Visual plots are commonly found in research papers, scientific journals, business records e.t.c. Automation of plot analysis through the means of question-answering will aid an individual to draw statistical inferences quickly from them. The most important benefit of visual question answering models on statistical plots is that it will help data analysts question and understand plots on a large scale, and thereby automate the decision-making capabilities in several sectors such as the financial sector. Given this motivation, the research aims to build a Visual Question Answering system that accepts statistical plots along with questions on the plot with respect to the elements of the plot and provides answers to the questions posed. The system should discover relationships between the elements of a plot and provide relational reasoning to answer questions on the plot [11]. Therefore, it involves an understanding of localized image elements and the query language to be able to provide visual reasoning. This work however restricts its scope to a certain amount of selective plots and their inner variants that are frequently occurring in most common data representations. Figure 1 shows four

J. I.-Z. Chen et al. (Eds.): ICIPCN 2022, LNNS 514, pp. 63–77, 2022.
https://doi.org/10.1007/978-3-031-12413-6_6

sample plots that have been taken from the dataset used and have been considered to be within the scope. As shown, a grouped vertical and horizontal bar plot, line plot, and dot plot have been plotted with different color combinations. This is used to illustrate that there are no restrictions on colors or the number of elements in the plot.

Fig. 1. Types of plots

We propose a system, TapasQA, for answering questions on statistical plots. An input plot is analyzed by our model to obtain the bounding box representations of all visual and textual elements in the plot. Textual elements are processed by an OCR module and the visual elements are processed by our system to obtain all the information represented in the statistical plot. This information obtained is then populated into a semi-structured table from which the questions posed are answered. All of these steps are visualized in the form of a 4-staged pipeline. The pipeline consists of stages, namely, Plot Element Detection (PED), Optical Character Recognition (OCR), Semi-Structured Table Generation and Table Question-Answering (QA) stage. A binary classifier is introduced in the pipeline to distinguish between open-ended and boolean (Yes/No) questions. The use of a binary classifier allows for only a single pipeline for our model and eliminates the use of two different pipelines for open-ended questions and binary questions. Our key area of contribution is in the table QA stage where additional helper functions have been added to the Google Tapas model, which improves the overall accuracy.

TapasQA model eradicates the need for two separate pipelines to handle boolean and descriptive questions, as mentioned earlier. Both, boolean and open-ended queries can be handled within a single pipeline due to the presence of binary classifier and the new approach for Table-QA Stage. It also generates a tabulated format of plot data, which can be made available to the user if needed.

2 Previous Work

There have been attempts in the recent past to improve machine reasoning capabilities through visual question answering systems on graphical plots. PlotQA [11] model is a step towards developing a holistic plot-based visual question answering model, which can handle both in-vocabulary and open-ended queries using a hybrid approach on the PlotQA dataset. Existing solutions for Visual Question Answering (VQA) fall under two categories: (i) extricate the response from the graphical data input (like in LoRRA [16]) or (ii) respond with an answer to the query posed based on the existing vocabulary. Such approaches seem to work well for data compilations as portrayed in DVQA [6,10], but under fit for PlotQA with a considerable majority of out of vocabulary queries.

Chart question answering was performed in the paper - Answering Questions about Charts and Generating Visual Explanations [9]. The paper under consideration proposes a question answering system that generates chart-specific answers along with an explanation of how the answer was obtained. The paper uses the Sempre model [1,8] for question answering. The drawback here is that the model is not generic and works only for vertical bar charts and pie charts. The model proposed cannot answer questions that require numerical operations. Chartnet [15] proposes to solve the problem of reasoning over statistical charts (only bar charts and pie charts) using MAC-Network (Memory, Attention, and Composition). The model is capable of answering open-ended questions and gives chart-specific answers. Open-ended QA is addressed by using supervised domain adaptation [18] to learn multiple features across different modalities. This research is tested using the benchmark datasets VQA 2.0 [3] and VizWiz [2] for realistic VQA tasks. The medical VQA model [12,14] is built using a classification and generative model for higher accuracy.

The RNN architecture and the CNN-LSTM architecture form the baseline comparison models with an accuracy of 75% and 60% respectively on the open-source dataset called FigureQA [7]. This in comparison to human accuracy falls short by a large margin. Moreover, the FigureQA dataset supports only Yes/No answers. A recent paper publication introduced the FigureNet [13] architecture that was able to achieve an accuracy of approximately 85% on the FigureQA dataset. This, however, only gives a yes/no binary output to a question posed, and is limited to only bar and pie charts.

The recent development of models to address issues presented in VQA is growing rapidly. The findings have proved that machine intelligent models are capable of achieving near-human accuracy. Although FigureNet can handle only binary questions, the dataset presented is of high relevance to training better models. ChartNet has addressed open-ended questions and uses compositional models. Google Table Parser (TAPAS) [5] has advanced functions that enable easy integration with deep learning pipelines to train QA models. This can address both open-ended and binary questions.

The domain of visual question answering on statistical plots has a lot of scope of improvement in terms of future enhancements of these models. Most of the existing models are limited to bar and pie charts, binary or fixed-vocabulary questions or cannot address numerical questions. Thus, there is a possibility of expanding the types of charts to those beyond bar and pie charts and even improving on accuracy through model adaptation. In our work, we attempted to fine-tune the existing work and also tried out the alternatives for the table question answering stage, focusing on the PlotQA dataset (capable of answering Yes/No as well as open-ended questions).

3 Methodology

3.1 Dataset

Statistical plots are used to learn about data and their important features. Building a visual question answering system requires statistical plots and well-defined questions and answers. PlotQA dataset [11] is used for testing the TapasQA model. It consists of images of statistical plots (bar, dot, and line) with corresponding annotations (bounding boxes of elements of the plot) and question-answer pairs. The set of annotation files is well-formatted using a JSON structure. These annotations are used to test the Plot Element Detection (PED) Stage, and the question-answer pairs are used to test the Table Question Answering (QA) Stage.

There are 3 types of questions in this dataset, namely, structural questions, data retrieval questions, and questions based on reasoning. Structural questions do not require any reasoning as they question the overall structure of a plot. Data retrieval questions are those whose answers can be obtained from a single element in the plot. Reasoning questions require detailed reasoning of more than one element in the plot. There are 3 types of answers supported in this dataset, namely, Yes/No answer, textual answers, and numeric answers.

The dataset has a total of 224,377 images and 28,952,641 Question-Answer pairs. The dataset split up is shown in Table 1.

3.2 Pipeline

In this subsection, we describe the different stages of our model to generate answers for the given input plots and questions. There are four main stages,

Table 1. PlotQA dataset statistics

Dataset split	#Images	#QA pairs
Train	157,070	20,249,479
Validation	33,650	4,360,648
Test	33,657	4,342,514
Total	224,377	28,952,64

viz., (i) Plot Element Detection, (ii) Optical Character Recognition, (iii) Semi-Structured Table Generation, and (iv) Table Question Answering stage. Each stage contributes towards identifying different plot elements and question structures to generate the final answer. Figure 2 shows the entire pipeline.

Fig. 2. Tapas-QA pipeline

Plot Element Detection (PED) Stage: The purpose of this stage is to detect all the elements in the input statistical plot and produce bounding box representations of all detected elements. We have used Detectron-2 as the object (element) detection and bounding box generation tool. Bounding boxes are a vector representation of the plot elements in a graph which refer to the coordinates of the element object (top_x , top_y , bottom_x , bottom_y).

The components of this stage consist of a deep learning FASTER R-CNN module for object localization because we aim to localize and produce bounding boxes around the plot elements and extricate them rather than classifying an entire image. The reason behind choosing Faster-RCNN as our object detection model is due to the presence of the RPN (The regional proposal network is known for locating feature targets accurately) and the inference time. Resnet-101 [4] is used as our feature extractor due to skipping connections and residual blocks which can reduce the problem of vanishing gradients in a very deep network. Resnet is a pre-trained CNN. The output of this model will be a JSON file that maps all the detected objects (plot elements) to the class to which it belongs, with an appropriate confidence value and its bounding box tensor. There are a total of 11 classes that have been defined by us. These are divided into textual and visual elements. The title of the plot, x-label, y-label, x-tick label, y-tick label, and the legend label form the textual elements. Bar, line, dot-line, and legend preview form the visual elements. Additionally, the background of the plot forms an element. Using the JSON file, we generate more readable formatted data that can be used for further processing in the pipeline.

Optical Character Recognition (OCR) Stage: To read the textual and numeric data of the textual components, we make use of the formatted textual output from the previous stage and an Optical Character Recognition module. With the help of bounding box coordinates extracted, we can accurately and locally capture the text information within the bounding boxes rather than passing an entire image into the OCR module. The captured text is then read using the OCR module and classified into its category. The OCR module used is pyocr which is a wrapper for the Tesseract OCR engine. Detected textual elements are cropped to the bounding box size (which is obtained from the previous stage), then converted to gray-scale and passed onto the pyocr module. The output of this stage is textual data corresponding to the detected textual elements.

Semi Structured Table Generation Stage: The previous stage dealt with reading textual and numeric data of textual elements. This stage deals with extracting information of visual elements. As mentioned earlier, bar, line, dot-line, and legend preview form the visual elements.

This stage is responsible for mapping legend values to the legend color, x-ticks to the x-axis label, and the y-ticks to the y-axis label. This is done by associating the legend/x-tick/y-tick value bounding box to the closest legend color/x-axis/y-axis boundary respectively.

For the visual elements, each element is associated with an axis and a corresponding legend. The color of the visual element is matched with the legend colors, and the legend of the closest match is associated with the element. To find the value associated with the bar, the information of height is taken from the bounding box representation, and the closest y-tick is mapped. Doing this for all visual elements will fill the table and result in a table that is stored as a comma-separated file. This file is then passed to the Table QA stage.

Table Question Answering (QA) Stage: Given a semi-structured table and a relevant natural language question as input, this stage is responsible for producing an answer to the question from the table as an output. The questions can be classified into two types. The first type corresponds to open-ended questions that have an unrestricted answer domain. The second type corresponds to questions that require a Yes/No (binary) answer. To handle these questions, we have made use of the existing TaPas (Table Parsing) model [5]. This model is based on the BERT's encoder with certain modifications - positional embeddings are used to encode tabular data and two additional classification layers are introduced to select cells of the table and the aggregation operation to be performed.

Our work makes use of a pre-trained TaPas model that has been trained on the WikiTables Questions dataset with intermediate pre-training. This model can handle three types of aggregation operations - SUM, COUNT, AVERAGE. To add to the capabilities of this model, we have added other operations such as RATIO, DIFFERENCE, MEDIAN, TREND, RANGE, and QUARTILES. To handle questions that require a Yes/No answer, we have used a TaPas model trained on the TabFact dataset [17]. This is a dataset used for table entailment and fact verification. We have extended its capabilities by adding other operations like in the earlier mentioned model. An important aspect here is that given an input question, we would need to know what type of question it is (whether it is an open-domain question or a yes/no question). For this, we have implemented a binary question classifier.

Binary Classifier: There are two categories of questions that can be addressed: Open-ended and Yes/No. Each of these is addressed by an independent model and hence to integrate it into a single pipeline, a binary classifier is used.

The input question is passed onto the Binary Classifier which then classifies the question into either the Yes/No class (class 0) or the Open-ended class (class 1). If the model outputs class 0, the question is passed to the TABFACT model and if the model outputs class 1, the question is passed to the TAPAS model to obtain answers via the Table Question Answering (QA) Stage.

The classifier model is trained on questions of all types of plots (i,e, vbar_categ-orical, hbar_categorical, dot_line, line) and the answers are converted into the categories of 0 or 1. The Deep Learning model is trained to classify the input questions into correct classes. Figure 3 shows the architecture of the binary classifier. The input image is converted into vector embeddings and passed through max-pooling layers and convolution layers of different dimensions to obtain a confidence value for one of the classes. The model is trained using the GloVe embeddings and binary cross-entropy with adam optimizer. The model was trained for 10 Epochs in a batch size of 10. The trained model was saved and loaded to classify test images.

4 Major Research Findings

4.1 Questions Handled by Our Model

TapasQA model can handle open-ended questions and boolean (Yes/No) type questions. Tapas model [5] handles only open-ended questions. These questions can be simple aggregation, complex queries, or structural operations. Count, Sum, Average, Minimum and Maximum operations are the basic operations provided by the pre-trained model of Tapas on the WikiTables dataset. We have added additional functions that can perform complex operations on the statistical plots. Statistical operations such as range, median, quartiles, inter-quartile range, difference, ratio, and other complex operations can be calculated for a particular class.

Fig. 3. Binary classifier

Tapas model trained on Tabfact dataset [17] supports boolean questions that are related to fact verification using the operations mentioned above. Structural questions corresponding to the naming of the x-axis, y-axis, and title of the plot are also supported.

The complete description of the questions handled can be found in Sect. 6.

4.2 Training Details

Plot Element Detection Stage: For PED, we have trained a Faster R-CNN object detection model on the Resnet-101 feature extractor. The model has been trained on the Train split of the dataset with a batch size of 512 for 200,000 iterations. The initial learning rate has been kept to 0.004.

Binary Classifier: As mentioned earlier, the binary classifier is trained using the GloVe embeddings and binary cross-entropy with the adam optimizer. The model was trained for 10 Epochs with a batch size of 10.

4.3 Evaluation Metric

Plot Element Detection Stage: Detectron-2 is an object detection tool and hence we make use of Average Precision (AP) as the evaluation metric. AP@(alpha) is the area under the Precision-Recall (PR) Curve for an Intersection over Union (IoU) set at threshold alpha. IoU determines the degree of overlap between the predicted bounding box and the expected bounding box. Since we are interested in true positives (the correct predictions made by the model), we need to set a threshold alpha such that if the IoU for detection is greater than or equal to alpha, we treat it as a true positive or a correct detection.

The area of overlap and union between the predicted bounding box and the expected bounding box is calculated and the AP for an IoU of alpha 0.5 is determined.

$$IOU = \frac{AreaofOverlap}{AreaofUnion} \tag{1}$$

$$AP@(alpha) = \int_0^1 p(r)dr \tag{2}$$

The higher the value of AP, the higher the overlap and the better is the prediction made by the model. This again requires the right selection of models, an appropriate number of training iterations, and a learning rate. We have chosen to evaluate the model using AP50 (AP calculated at IoU threshold of 0.5) and AP75 (AP calculated at IoU threshold of 0.75). Additionally, AP for the average of AP50, AP75, and AP90 is also calculated.

Below, in Table 2, we display the results for various parameter settings and the incremental growth achieved by the model.

Table 2. Average Precision at different number of iterations

Number of iterations	AP	AP50	AP75
100,000	79.621	91.956	90.397
130,000	86.584	92.819	92.076
150,000	87.014	92.825	92.086
170,000	87.053	92.823	92.083
200,000	87.179	92.823	92.088

It can conclusively be seen that the AP increases with the number of iterations. Better training produces better results. The total number of iterations we have trained for is 200,000. Per category bounding box AP value also tends to increase along with more training. Moreover, we observe that the accuracy falls as we increase the threshold from 0.5 to 0.75. A higher value of AP helps us to capture the textual and pictorial elements better in the further stages of the pipeline.

Table Question Answering (QA) Stage: This stage answers the question using the generated semi-structured table. The predicted answer is evaluated against the expected answer of the PlotQA dataset. We have used accuracy as the evaluation metric. For textual answers, the answer would contribute to the accuracy only if an exact match was found between the expected and the predicted answers. However, in the case of numeric answers, we have allowed for an error window of 5%. Answer values within this range will be considered correct. Not allowing for this error window would seem too strict for evaluation considering that numerical answers could be floating-point values.

Every image of the statistical plot in the dataset consists of multiple question-answer pairs. For every image, we calculate the Per Image Accuracy by assessing the number of correctly answered questions out of all the questions on that image. Using this, overall accuracy per plot type is determined.

$$PerImageAccuracy = \frac{\sum CorrectlyAnsweredQuestions}{\sum TotalQuestions} \tag{3}$$

$$OverallAccuracy = \frac{\sum PerImageAccuracy}{\sum TotalImages} \tag{4}$$

5 Result Implications

5.1 Plot Element Detection Stage

The aim in this stage was to ensure that there was maximal overlap between the bounding boxes proposed by our object detection trained model and the actual bounding box locations of the test image which are hidden from the model. AP aka Average precision is the score to look out for. Below, in Table 3, is the class-wise split-up of Average Precision obtained on the model trained on 200,000 iterations. We observe that most textual elements are being accurately captured, which is essential for question answering.

The accuracy of the Optical Character Recognition(OCR) stage is dependent on the accuracy of the open-source Tessaract pyocr module.

Table 3. PED evaluation

Class	AP
Bar	88.819
Line	61.466
Dot-Line	77.439
X-Label	97.840
Y-Label	98.438
Title	90.056
X-tick Label	96.500
Y-tick Label	71.094
Legend Label	95.550
Preview	94.589

5.2 Table Question Answering (QA) Stage

The Table QA Stage outputs an answer for the question using the CSV generated from the semi-structured table generation stage. The accuracy of this stage depends on the accuracy of previous stages. The answers are categorized into two classes to arrive at the final accuracy. One type is the floating-point answers. Since the answer cannot be exact, we have allowed an error window of 5%. Values that are within this 5% range will be accepted. The other type is the String answers. Here, we consider an Exact Match metric. Answers that exactly match the ground truth values are considered towards accuracy.

We have evaluated the TapasQA model on a total of 186,504 questions from 8000 images. Out of those questions, 79207 were correctly answered which gives the average overall accuracy of 41.52%.

To compare the results on different plot types, we have calculated a per-plot-type accuracy. We have tested on 2000 images of plots for each category.

Table 4 shows the spilt-up based on different plot types.

Table 4. Table QA results

Plot type	Total number of questions	Number of correct answers	Average accuracy (in %)
Dot	53970	25104	46.965499
Vertical	47940	19898	41.474200
Horizontal	49241	20128	40.990114
Line	35353	14077	36.669402

We have compared our results with the PlotQA model. The PlotQA model was tested on 5860 questions from 160 images to obtain human accuracy. The PlotQA paper reports the Human accuracy as 80.47%, an accuracy of 58% on the DVQA dataset, and accuracy of 22.52% on the PlotQA dataset. Our model has an average accuracy of 41.52% across all types of plots and structural, data-retrieval, and reasoning type questions on the PlotQA dataset which is a significant improvement over the accuracy of 22.52% reported in the PlotQA paper.

6 End-To-End Example

Figure 4 shows a sample input image of a grouped horizontal bar plot, and the corresponding CSV generated by the model pipeline. Using this example we demonstrate the types of questions the model can address.

Types of questions addressed:

1. Count
   Q: what are the total number of countries
   A: 6

Fig. 4. End-to-end example

2. Average
   Q: what is the average number of male workers in the year 1980
   A: 49.36953171828586
3. Minimum
   Q: across all countries, what is the minimum percentage of male workers employed in service sector in 1980
   A: 41.56915064054187
4. Difference
   Q: what is the difference between the average number of male workers employed for the year 1980 and 1981
   A: DIFFERENCE = 0.10125573610311278
5. Median
   Q: what is the median number of male workers employed in the year 1980
   A: MEDIAN = 50.22946206732409
6. Ratio
   Q: what is the ratio of male workers employed in 1981 to 1980 for the country hong kong
   A: RATIO = 1.0093722657385955
7. Trend (Increasing or Decreasing)
   Q: what is the trend of male workers employed for the countries finland, france, hong kong in 1980
   A: TREND = INCREASING
   Q: what is the trend of male workers employed for the countries israel, italy in 1980
   A: TREND = DECREASING
8. Selection operation on cell
   Q: what is the number of male workers employed for country france in the year 1981
   A: 48.40140328049439

9. Select operation on cell after applying aggregation operation
   Q: which country has the minimum number of male workers employed in the year 1981
   A: Finland
10. Selection and Aggregation operation on subset of rows
   Q: what is the average number of male workers employed for the countries france, finland in the year 1980
   A: 45.26962827644094
11. Project operation on column
   Q: list out the countries
   A: Finland, France, Hong Kong, Israel, Italy, Norway
12. Range
   Q: what is the range of % of male employment for the year 1980
   A: RANGE = 12.622419158216474
13. Quartiles (Q1 and Q3)
   Q: find the quartiles for the year 1980
   A: FIRST QUARTILE (Q1) = 43.97133559037385
   SECOND QUARTILE (Q2) = 50.22946206732409
   THIRD QUARTILE (Q3) = 53.90779749715967
14. IQR
   Q: find the interquartile range for the year 1980
   A: INTER-QUARTILE RANGE = 9.936461906785823
15. Structural Query
   Q: what is the title of the graph ?
   A: TITLE OF THE GRAPH = Percentage of male workers employed in Service sector
16. YES / NO
   Q: in the year 1981, hong kong has the highest employment percentage.
   A: YES
   Q: the average employment rate of hong kong is greater than the average employment rate of finland.
   A: YES

7 Value and Limitations

Statistical plots are used widely by academicians and business employees because they are a simple way to represent data. Our visual question answering system will help analysts question and understand plots on a large scale. It will aid in analyzing and discovering relationships between elements of a plot in addition to the complex task of reasoning. It is specifically useful for data analysts to understand plots that require complex reasoning and will help them navigate through the information represented in the form of lots of visualizations in real-time. Our system will aid people to quickly get inferences from statistical plots and will make a significant impact particularly when one has to go over a large number of plots, each with a large amount of information to decode. This work is therefore a step towards machine reasoning capabilities.

In real-time, a stream of queries is posed, the system processes the answer at an average of 30 s per query. This research does have limitations in that it can only reason on specific plot types - horizontal bar, vertical bar (grouped), dot, and line plots. Further, the model is restricted to a limited type of questions. Future research aims to address this limitation.

8 Conclusion and Future Work

We have proposed an alternative solution to perform question answering on statistical charts. The input chart goes through the PED stage, OCR stage, and Semi-Structures table generation stage to produce a CSV table. This table combined with the questions passes through the Table QA stage to generate the answer. Additional helper functions help us achieve higher accuracy. The PED model produced an Average Precision of 87.014 % after training for 2 lakh iterations. Further, the Table QA model produced significantly better results compared to the PlotQA model.

TapasQA model eliminates the need for two separate pipelines to answer open-ended questions and boolean questions. Using the binary classifier in the Table QA stage helps us achieve a single pipeline. The accuracy obtained by the PlotQA model and our model tested on TAPAS are significantly lower than the human performance. Hence, there is a wide scope of improvement in the Table QA stage. Further research in this area is required to match human performance.

References

1. Berant, J., Chou, A., Frostig, R., Liang, P.: Semantic parsing on Freebase from question-answer pairs. In: Empirical Methods in Natural Language Processing (EMNLP) (2013)
2. Davis, N., Xie, B., Gurari, D.: Quality of images showing medication packaging from individuals with vision impairments: implications for the design of visual question answering applications. Proc. Assoc. Inf. Sci. Technol. **57**(1), e251 (2020)
3. Goyal, Y., Khot, T., Summers-Stay, D., Batra, D., Parikh, D.: Making the v in VQA matter: elevating the role of image understanding in visual question answering. In: Proceedings of the IEEE Conference on Computer Vision and Pattern Recognition, pp. 6904–6913 (2017)
4. He, K., Zhang, X., Ren, S., Sun, J.: Deep residual learning for image recognition. In: Proceedings of the IEEE Conference on Computer Vision and Pattern Recognition, pp. 770–778 (2016)
5. Herzig, J., Nowak, P.K., Müller, T., Piccinno, F., Eisenschlos, J.M.: Tapas: weakly supervised table parsing via pre-training. arXiv preprint arXiv:2004.02349 (2020)
6. Kafle, K., Price, B., Cohen, S., Kanan, C.: DVQA: understanding data visualizations via question answering. In: Proceedings of the IEEE Conference on Computer Vision and Pattern Recognition, pp. 5648–5656 (2018)
7. Kahou, S.E., Michalski, V., Atkinson, A., Kádár, Á., Trischler, A., Bengio, Y.: Figureqa: an annotated figure dataset for visual reasoning. arXiv preprint arXiv:1710.07300 (2017)

8. Karthigaikumar, P.: Industrial quality prediction system through data mining algorithm. J. Electron. Inf. **3**(2), 126–137 (2021)
9. Kim, D.H., Hoque, E., Agrawala, M.: Answering questions about charts and generating visual explanations. In: Proceedings of the 2020 CHI Conference on Human Factors in Computing Systems, pp. 1–13 (2020)
10. Manoharan, J.S.: Capsule network algorithm for performance optimization of text classification. J. Soft Comput. Paradigm (JSCP) **3**(01), 1–9 (2021)
11. Methani, N., Ganguly, P., Khapra, M.M., Kumar, P.: Plotqa: reasoning over scientific plots. In: Proceedings of the IEEE/CVF Winter Conference on Applications of Computer Vision, pp. 1527–1536 (2020)
12. Moholkar, K.P., Patil, S.H.: Deep ensemble approach for question answer system. In: Pandian, A.P., Fernando, X., Islam, S.M.S. (eds.) Computer Networks, Big Data and IoT. LNDECT, vol. 66, pp. 15–24. Springer, Singapore (2021). https://doi.org/10.1007/978-981-16-0965-7_2
13. Reddy, R., Ramesh, R., Deshpande, A., Khapra, M.M.: Figurenet: a deep learning model for question-answering on scientific plots. In: 2019 International Joint Conference on Neural Networks (IJCNN), pp. 1–8. IEEE (2019)
14. Ren, F., Zhou, Y.: Cgmvqa: a new classification and generative model for medical visual question answering. IEEE Access **8**, 50626–50636 (2020)
15. Sharma, M., Gupta, S., Chowdhury, A., Vig, L.: Chartnet: visual reasoning over statistical charts using mac-networks. In: 2019 International Joint Conference on Neural Networks (IJCNN), pp. 1–7. IEEE (2019)
16. Singh, A., et al.: Towards VQA models that can read. In: Proceedings of the IEEE/CVF Conference on Computer Vision and Pattern Recognition, pp. 8317–8326 (2019)
17. Chen, W., et al.: Tabfact: a large-scale dataset for table-based fact verification. In: International Conference on Learning Representations (ICLR), Addis Ababa, Ethiopia (2020)
18. Xu, Y., Chen, L., Cheng, Z., Duan, L., Luo, J.: Open-ended visual question answering by multi-modal domain adaptation. arXiv preprint arXiv:1911.04058 (2019)

Face Sketch-Photo Synthesis and Recognition

K. M. Mitravinda(✉), M. Chandana, Monisha Chandra, Shaazin Sheikh Shukoor, and H. R. Mamatha

Department of Computer Science and Engineering, PES University, Bangalore, Karnataka, India
mitravinda462@gmail.com, mamathahr@pes.edu

Abstract. This paper presents a simple Face Sketch-Photo Synthesis and Recognition system. Face Sketch Synthesis provides a way to compare and match the faces present in two different modalities (i.e. face-sketches and face-photos). The aim is to significantly reduce the differences between face-sketches and face-photos and also decrease the texture irregularity between them by converting photos to sketches and vice-versa. This results in effective matching between the two thus simplifying the process of facial recognition. This system is modeled using three major components: (i) For a given input face-photo, obtaining an output face-sketch. It is designed using image processing techniques like 2 scale image decomposition and color dodging. (ii) For a given input face-sketch, obtaining an output face-photo. Convolutional Neural Networks are used to model this component. (iii) For a given query face-sketch or face-photo, recognition of face-photo or face-sketch in the database. It is implemented using Fisherface Linear Discriminant Analysis.

Keywords: Face-sketch synthesis · Face-photo synthesis · Face-sketch recognition · Face-photo recognition · Convolutional Neural Networks · Fisherface Linear Discriminant Analysis

1 Introduction

Face sketching is an artistic portrayal of a person's face. The most crucial perceptual information is captured in a face-sketch [1]. Synthesis of face-sketch from face-photo has multiple applications which are mainly in the fields of law enforcement and digital entertainment. There are often situations where the police do not have any photo of the suspect except for one or more eye witnesses. In such circumstances, forensic artists are requested to sketch the suspect's face based on descriptions provided by eye-witnesses, victims, or low-resolution surveillance footage. Alternatively, software tools that generate facial sketches based on a facial description can be used to create composite sketches. These forensic sketches or composite sketches are often used to search through the entire database which contains mugshot photos of criminals and suspects. But, the forensic sketches drawn might not suffice accurate identification of criminals or suspects. Similarly, composite sketches are made up of separated facial parts. Integrating these parts in order to form a complete face image adds even more distortion than

Mitravinda, K.M., Chandana M., Monisha Chandra, Shaazin Sheikh Shukoor — Contributed equally to this work.

J. I.-Z. Chen et al. (Eds.): ICIPCN 2022, LNNS 514, pp. 78–95, 2022.
https://doi.org/10.1007/978-3-031-12413-6_7

drawing a forensic sketch continuously with a pencil. This increases the complexity of facial recognition of composite sketches [2]. The complexity of the same is also due to the reason that the sketches and the photos belong to two different modalities with completely different structural and appearance features [3]. A possible solution could be to either convert the photo to a sketch or vice versa followed by facial recognition of the generated image.

The problem of computer-synthesizing facial sketches from face-photos is interesting. The implicit human process of generating sketches is very complex to accurately express in rules and grammar [4]. Though face-sketches aren't much different from face-photos, sketch artists have the ability to portray the most distinctive and unique features of the human face. Thus, the synthesized sketch must be able to capture such intricacies too. Similarly, synthesizing face-photos from face-sketches is challenging as the face-sketches contain different textures when compared to face-photos. Moreover, the face-sketches are grayscale while the face-photos are colored which further increases the complexity of face-photo synthesis. This paper aims to develop a simple face-sketch and face-photo synthesis and recognition system which works accurately for different types of skin tones, hair colors and other facial features.

2 Related Work

Various models have been used for the purpose of converting a face photograph to a facial sketch [5]. Zhang et al. [6] described a Markov Random Fields model and Wang et al. [4] described an a multiscale Markov Random Fields model for sketch synthesis where the face is divided into patches. The main idea is that if 2 patches are analogous, then the corresponding sketches are bound to be similar as well. Wan et al. [7] worked on a GAN approach where a face photo image is input into the generator, which gives the desired sketch. Three images are received by the discriminator to configure the triplet sample input to learn the ability to extract facial features. Li et al. [8] designed a simplified framework for converting a face photograph to a face-sketch. 2 scale image decomposition with bilateral filtering is adopted to obtain prominent facial features and shadow details of the input face. Then, a color similarity map is computed to select the hair and skin regions of the input face photograph separately, followed by hair and unapparent facial feature creation. Sharma et al. [9] designed a framework which uses Principal Component Analysis followed by Feed Forward Neural Network for face-sketch synthesis and recognition. The background detail is removed from the face images and are converted into sketches using PCA. PCA output is used to obtain feature vectors which are then used to train the neural network for facial recognition. Yu et al. [10] used Composition-Aided GANs for face sketch synthesis. Realism of the synthesized sketch was improved by using face composition as supplemental input and training with compositional loss and the quality was improved by stacking Composition aided GANs.

Xiao et al. [11] used Embedded Hidden Markov Model for sketch-to-photo conversion. Liang et al. [12] used BiCE descriptors to encode local patches. A multi MRF which estimates the photo patches is used. Finally, the output face photograph is generated based on the photo patches selected by MRF. Lu et al. [13] described a methodology for image generation from sketch using Contextual Generative Adversarial Networks. Sketch-photo synthesis is modeled as image completion rather than conversion. Tejomay et al. [14] used Cycle consistent adversarial networks for face photo-sketch synthesis. The network consisted of 2 generators made of U-Net architecture and 2 discriminators made of PatchGANs.

Galea et al. [15] designed a framework which tunes a state-of-the-art pre-trained face-photo recognition model for face photo-sketch recognition using transfer learning. Han et al. [16] represented face-photos with the help of local feature descriptors like multiscale local binary patterns. Galoogahi et al. [17] used the Histogram of Averaged Oriented Gradients to perform face-sketch recognition. Tang et al. [18] proposed a methodology where first the shape and texture information of a face photo is separated and transformation is conducted on them separately. Recognition of the probing sketch from the synthesized pseudo-sketches is done using Bayesian classifier. Jacob et al. [19] proposed a framework with deep learning and PCA to build an IoT image-based identification system. PCA was used to perform feature extraction and to gather the image features for the recognition process. CNN was used for the image identification process.

3 Research Gap

On completion of literature survey, it was found that the methodologies in [9] and [18] approximate the conversion process from face-photo to face-sketch as linear. This can lead to inaccuracies in the conversion and recognition processes due to the non linear nature of the relationship between face-photo and face-sketch. Further in [18], removal of the hair region from the face image leads to loss of discriminative information. The facial recognition process considers only the frontal facial position, thus failing to perform accurate recognition on face images with different facial positions other than the frontal position. The MRF model proposed in [4] and [6] is quite complex, requires a lot of parameter setting and is computationally expensive.

The framework proposed in [8] involves setting different benchmark colors for computing the skin color similarity and hair color similarity maps. The need for setting benchmark colors each time can be disadvantageous as the algorithm can't be generalized for input photos with different skin colors and hair colors. It can also lead to issues and inaccuracies if the skin color and hair color of the input face-photo are similar or if there is absence of hair in the input face-photo. In these cases, it becomes difficult to accurately select the hair and skin regions separately.

The methodology used in [13] fails to preserve all the facial features of the face-sketch in the generated face-photo. It also poses limitations in identification of facial attributes such as eye glasses and beard. In [14] there is limited number of performance metrics included to evaluate the results obtained.

Thus, the methodology to be discussed in this paper is designed to address the research gaps found. The sketch and photo synthesis models are implemented to work accurately for variations such as, absence of hair, multi colored hair, presence of beard and/or moustache, similar or same hair color and skin color and presence of spectacles. This paper also includes a thorough analysis of the results obtained by computing the degree of similarity between face photos and face sketches and the accuracy of the facial recognition process. The results of photo and sketch synthesis are specified using numerical figures that measure the structural and featural similarity of the original and synthesized images. The similarity is further evaluated using a human visual perception experiment which solidifies the accuracy of the obtained results.

4 Data

4.1 CUFS Dataset

The database of CUHK Face Sketch (consisting of CUHK student dataset, AR dataset and XM2VTS dataset) is used in the interest of face-sketch generation. It consists 606 face sketch-photo pairs of dimensions 200 × 250 in JPG format. Every face-photo has a corresponding artist-drawn face-sketch. The face photos are colored and are captured in a frontal position, in good lighting conditions, and with minimal facial expressions. The images are cropped to centralize faces.

4.2 CelebA Dataset

CelebFaces Attributes dataset is a dataset with more than 2 lakh pictures of famous and diverse people, each with many attribute annotations. The pictures cover many variations in the pose and background noise like people wearing glasses, with brown hair and expressions like smiling. The dataset consists of 10,177 numbers of identities and 202,599 numbers of face images with 40 binary attribute annotations per image in 5 landmark locations. The images are available in JPG format, where each image is of 178 × 218 dimensions.

4.3 ORL Dataset

The ORL Database is a set of face images taken at Olivetti Research Laboratory. It consists of grayscale face-photos of 40 identities, each of them having 10 different images with different poses. There are 400 images in the database stored in PGM format and each image has a size of 92 × 112 pixels. These images vary with respect to light, time of photos taken and facial expressions and are taken against a dark homogenous background.

5 Tools and Experimental Settings

Operating System – Ubuntu 20.04, Graphics card - NVIDIA 106 512 MB, Processor – Intel i7, Hard disk – 512 GB, RAM – 6 GB. Programming Language used was Python 3.8. The major libraries used were: OpenCV, NumPy, PIL, tflearn, Pandas, image_similarity_measures, Matplotlib and Scikit-image.

6 Proposed Methodology

6.1 Face-Sketch Synthesis

Two Scale Image Decomposition with Bilateral Filtering

2 scale image decomposition involves splitting the image into 2 parts, detailed-image and base-image. The base-images are obtained by applying bilateral filtering on the input image with varying values of σ_r (size of the range kernel) (Fig. 1). For an input RGB face-photo I, p being the coordinates of the current pixel to be filtered and S being the window centered in p, bilateral filter can be formulated as:

$$BF[I]_p = \frac{1}{W_p} \sum\nolimits_{q \epsilon S} G_{\sigma_s}(\|p - q\|) G_{\sigma_r}\left(\left|I_p - I_q\right|\right) I_q \tag{1}$$

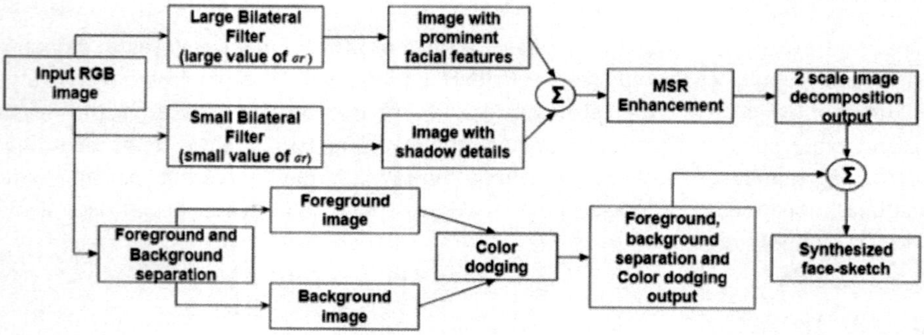

Fig. 1. Block diagram of face-sketch synthesis module

$BF[I]_p$ is the filtered result at the pixel p obtained after normalizing the sum of the intensities I_q at all pixels q in the window S weighted by spatial weight $G_{\sigma_s}(\|p - q\|)$ and range weight $G_{\sigma_r}\left(\left|I_p - I_q\right|\right)$. G_{σ_s} is the Gaussian function for the spatial kernel that smooths the differences in coordinates. Similarly, G_{σ_r} is the Gaussian function for the range kernel that smoothes the differences in intensities. W_p is the normalization factor which can be given by:

$$W_p = \sum_{q \epsilon S} G_{\sigma_s} \|p - q\| G_{\sigma_r}(|Ip - Iq|) \tag{2}$$

The large σ_r bilateral filter smooths the edges in the input image directly. Similarly, the small σ_r bilateral filter smooths the pixels in the regions close to the edges of facial features, producing heavier shadow details without changes in edges. Thus, the base-images are subtracted from their inputs to obtain the detailed-images. The image with prominent facial features and the image with shadow details are obtained by subtracting from the input image, the large and small σ_r bilateral filter outputs respectively and are added using weighted sum. This image is further enhanced using Multiscale Retinex

(MSR) technique which is a perception-based image enhancement algorithm (Fig. 2). For an input image I, MSR is

$$MSR = \sum_{n=1}^{N} w_n\left[\log(I_i(x, y)) - \log(F_n(x, y) \cdot I_i(x, y))\right] \tag{3}$$

where $I_i(x, y)$ is the intensity of the x and y coordinates for the ith color channel of the RGB model, N is the total count of scales, w_n is the weight of each scale and $F_n(x, y)$ refers to the Gaussian function in the nth scale given by,

$$F_n(x, y) = C_n exp\left(\frac{-(x^2 + y^2)}{2\sigma_n^2}\right) \tag{4}$$

Fig. 2. (a) Input image (b) Large σ_r bilateral filter output (c) Small σ_r bilateral filter output (d) Image with prominent facial features (e) Image with shadow details (f) Sketch resulting from applying enhancement on the weighted sum of detailed images

Image Foreground and Background Separation and Blending

The input RGB face-photo image's background and foreground are separated. Foreground image is obtained by converting the input image from RGB to grayscale color space. Similarly, background image BI at a pixel p is obtained after applying Gaussian blur on the inverted foreground image.

$$BI[IG]_p = \sum_{q\epsilon S} G_\sigma(\|p - q\|)IG_q \tag{5}$$

The Gaussian blur calculates the sum of the intensities IG_q at all the pixels q in the window S weighted by the normalized Gaussian function $G_\sigma(\|p - q\|)$.

This is followed by color dodging to blend the foreground and background images. In color dodging, the bottom layer of the image is divided by the inverted top layer. The bottom layer is lightened depending on the top layer's value. For an input image I and foreground image FI, blended image BL can be obtained as:

$$BL = \frac{(FI \cdot 256)}{(255 - BI)} \tag{6}$$

At the end, the synthesized face-sketch SK is obtained by weighted addition of the outputs of 2 scale image decomposition and color dodging (Fig. 3).

$$SK = (0.3(I_1) + (1 - 0.3)(BL)) \tag{7}$$

Fig. 3. (a) Input image (b) Background part (c) Foreground part (d) Blended image obtained after color dodging is applied (e) Final synthesized face-sketch

Fig. 4. Block diagram of face-photo synthesis module

6.2 Face-Photo Synthesis

Preprocessing

Preprocessing is done on the images of CelebA database to detect the faces in the photos using the Haar Cascade classifier in the OpenCV library. This method detects if the picture contains a face with frontal posture. This information is then used to crop the image so that it contains only the face of the identity (Fig. 4).

Generating the Database of Face-Sketches

The database containing face-sketches is created by converting the preprocessed face photographs to face-sketches using the algorithm described in Sect. 6.1 (Fig. 5).

Fig. 5. Left: Input image; Centre: Preprocessed face-photo; Right: Generated face-sketch

Training the Model and Generating the Face-Photo
The generated face-sketch database is split into training and testing datasets. The Convolutional neural network model on training on the sketch database, generates a face photograph for a given facial sketch. The CNN is modeled using 5 convolutional layers followed by 2 deconvolutional layers and finally 2 convolutional layers (Fig. 4). ReLU activation function is adopted to introduce nonlinearity into the model. It also has its advantage over sparsity and in reducing the likelihood of vanishing gradients. For an input value x, the ReLU activation function $r(x)$ is formulated as:

$$r(x) = \max\{0, x\} \tag{8}$$

The loss is calculated using the Mean Squared Error (MSE), which works well for computer vision and image-related deep learning applications. For a given number of samples N, predicted value $\overline{y}$ and true value y, the loss function is described as:

$$L(y, \overline{y}) = \frac{1}{N}\sum_{i=0}^{N}(y - \overline{y_i})^2 \tag{9}$$

Adam Optimizer is used to efficiently calculate gradients and to obtain the least possible mean square loss between pixel values of the output and the target images. It is formulated as follows:

$$m_t = \beta v_{t-1} + (1 - \beta)\left[\frac{\delta L}{\delta w_t}\right]^2 \tag{10}$$

where β is the rate of decay of the gradients' mean, v_t is the summation of squares of previous gradients, L is the loss function and w is the weights (Fig. 6).

Fig. 6. Left: Input face-sketch; Centre: Synthesized face-photo; Right: Ground truth photo

6.3 Facial Recognition

The design approach for facial recognition is based on Fisherface Linear Discriminant Analysis (LDA) whose main aim is to do dimensionality reduction along with preserving the discriminative information. The dataset containing M images which belong to C classes is given as the input to the model. The input dataset is separated into the testing set and the training set, where the testing set consists of M images which belong to j classes and the training set consists of M images which belong to (C-j) classes. Next, Principal Component Analysis is applied upon the images in the training set. This projects the high dimensioned training images onto an (M-1) PCA subspace. This helps in reducing the dimensionality of the training images. In this process, training images data is normalized and covariance matrix, eigenvalues and eigenvectors are computed. For set of input images x in the training dataset, covariance matrix is described as

$$C = \frac{1}{N}\sum_{i=1}^{N}(x_i - \overline{x})(x_i - \overline{x})^T \tag{11}$$

where N is to the total count of data samples, x_i refers to every image in the training set and $\overline{x}$ refers to the sample set mean which is given by,

$$\overline{x} = \frac{1}{N}\sum_{i=1}^{N} x_i \tag{12}$$

The computed pairs of eigenvalues and eigenvectors are sorted in decreasing order. The K largest eigenvectors, $U = [\overline{u_1}, \overline{u_2}, \ldots\ldots, \overline{u_k}]$ are then selected to span the PCA subspace. The projection of the high dimensioned image data to low dimensioned PCA subspace can be formulated as follows:

$$y = U^T x \tag{13}$$

The mean of each class of projected images $\overline{\mu_i}$ and the mean of all projected images $\overline{\mu}$ are computed as follows:

$$\overline{\mu_i} = \frac{1}{N_i}\sum\nolimits_{y\epsilon\omega_i} y \tag{14}$$

$$\overline{\mu} = \frac{1}{N}\sum\nolimits_{for\ all\ y} y \tag{15}$$

where N_i is the count of data samples in every class i, N is the total count of data samples and y refers to the projected training images.

Fisherface LDA is then used on the projected training images for facial recognition. The goal is to maximize a function that measures the difference between projected means normalized by a measure of within-class variability called 'scatter.' Hence, for the projected training images, the between class scatter matrix S_b and the within class scatter matrix S_w are generated. S_w and S_b are:

$$S_w = \sum\nolimits_{i=1}^{C} S_i = \sum\nolimits_{i=1}^{C}\sum\nolimits_{y\varepsilon\omega_i}(y - \overline{\mu_i})(y - \overline{\mu_i})^T \tag{16}$$

$$S_b = \sum\nolimits_{i=1}^{C} N_i(\overline{\mu_i} - \overline{\mu})(\overline{\mu_i} - \overline{\mu})^T \tag{17}$$

S_w measures the variability within each of the C classes after projection, while S_b measures the variability between all classes after projection.

The criterion function for projection matrix W is defined as,

$$J(W) = \frac{|S_b|}{|S_w|} \tag{18}$$

Using the criterion function $J(W)$ eigenvalues and eigenvectors are computed and are sorted in descending order. The data from the training images is then projected onto the eigenspace to provide projected eigenface data. Fisherface LDA is then used for projecting the testing face-photo or face-sketch images which results in an array of projected test images. The difference between the projected train and test images is obtained and the eigenvector which minimizes that difference is obtained. For each image in the testing dataset, the obtained eigenvector is used to decide if it matches with any of the images in the training dataset (Fig. 7).

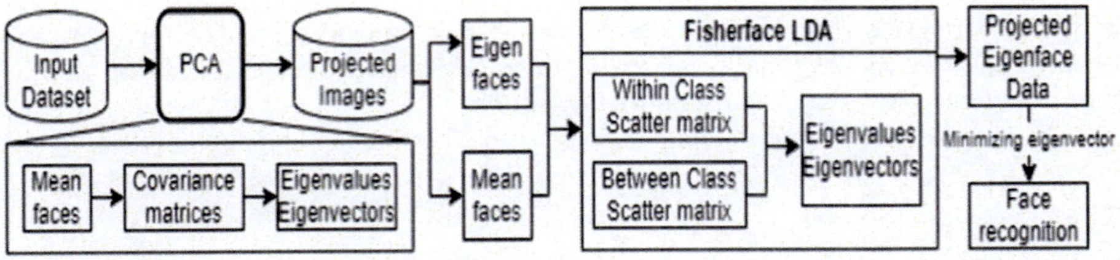

Fig. 7. Framework for facial recognition

7 Results

7.1 Face-Sketch Synthesis

The algorithm for conversion of face-photos to face-sketches described in Sect. 6.1 was implemented. Face-photo images from the CUHK dataset as well as face-photo images obtained from the internet were used to test the algorithm.

In Fig. 8, one can see that the synthesized face-sketches capture all the facial features. The shadow details represented in the synthesized face-sketches closely match the shadow details represented in the artist-drawn face-sketches.

Figure 9(a) and (d) represent dark skin and hair colors. Figure 9(b) and (f) represent light skin and hair colors, while Fig. 9(c) represents dark skin color and red hair color. Despite inputting face-photos with different combinations of skin and hair colors, the framework accurately represents the skin region, hair region and all the facial features in the synthesized sketch without the need to exclusively provide any threshold hair and skin colors. Figure 9(d) represents minimal hair and beard and dark colored beard. Figure 9(e) represents no hair, wheatish skin tone, heavy and dark colored beard, while Fig. 9(f) represents presence of spectacles, minimal and red colored beard. This shows that the exceptional cases of absence of hair and presence of beard and spectacles are handled effectively by the framework.

Fig. 8. Comparing the generated face-sketch and the artist drawn face-sketch for light and wheatish skin tones and dark hair color. Row1: Input face-photo from CUHK dataset; Row2: Synthesized face-sketch; Row3: Artist-drawn face-sketch

Fig. 9. Comparing face-photos to synthesized face-sketches Row1: Input face-photo from the internet; Row2: Synthesized face-sketch.

Figure 10 shows the sketches synthesized using different methods on CUHK dataset. Synthesized sketches obtained using methods proposed in [4] and [6] suffer from distortions and failure to capture all the facial features. Thus, the proposed methodology is seen obtain better synthesized face-sketches. As an extension, the algorithm was applied to convert photos of natural scenery to corresponding sketches. Accurate sketches of natural scenery were synthesized (Fig. 11).

a)Input b)Artist-drawn c)Synthesized d)Wang et al.[4] e)Zhnag et al.[6]

Fig. 10. Comparing face-sketches synthesized by different methods

Fig. 11. Conversion of natural scenery photos to sketches

7.2 Face-Photo Synthesis

The face-photo synthesis algorithm described in Sect. 6.2 was implemented. The algorithm uses a convolutional neural network model which was trained using 95,000 training face-sketches and 5,000 validation face-sketches. The model was then tested on 25,000 face-sketches from the CelebA database's test dataset.

In Fig. 12(a) it can be seen that the skin and hair colors of the synthesized photos match that of the ground truth photos to a great degree. Features like earrings, beard and spectacles are accurately captured in the output face-photos. In Fig. 12(b) it can be seen that the facial features and shadow details are rightly captured in the output face-photos along with predicting appropriate skin, hair and beard colors.

Fig. 12. Comparing the synthesized face-photos with the original face-photos. Col1: Input face-sketch from test dataset; Col2 & Col5: Synthesized face-photo; Col3: Ground truth face-photo; Col4: Input face-sketch from internet; Row1: face with light skin color and dark hair color; Row2: face with wheatish or dark skin color and dark hair color; Row3: face with wheatish or dark skin color, dark hair color and features like spectacles and beard

7.3 Facial Recognition

The recognition algorithm described in Sect. 6.3 was implemented to recognize faces in both photos and sketches. The model was trained on ORL database's face-photos (and face-sketches) of 40 identities belonging to 6 classes and was tested on face-photos (and face-sketches) of 40 identities belonging to 4 classes.

Face-Photo Recognition
Figure 13, shows that face-photos in the test dataset and the synthesized face-photos are accurately matched with the face-photos in the database, despite variations in their facial positions. The unknown face-photos are also accurately identified.

Face-Sketch Recognition
The face-photos of ORL database were converted to face-sketches using the proposed face-sketch synthesis algorithm, thus creating a face-sketch database.

Figure 14 shows that the face-sketches in the test dataset are accurately matched with the face-sketches in the database, despite variations in their facial positions. The unknown face-sketches are also accurately identified.

Fig. 13. Col1, Col3 & Col5: Input face-photos; Col2, Col4 & Col6: Matched and unmatched face-photos; Row1: Testing for face-photos in the test dataset; Row2: Testing for the synthesized face-photos (the face-photos in the test dataset were converted to face-sketches which were then converted to face-photos); Row3: Testing for unknown photos

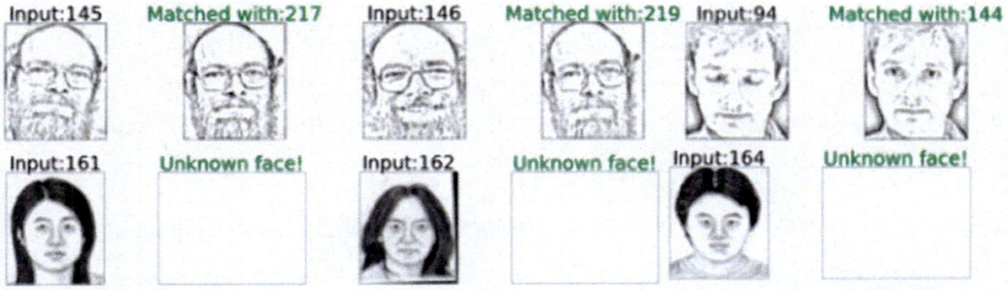

Fig. 14. Col1, Col3 & Col5: Input; Col2, Col4 & Col6: Matched and unmatched sketches

8 Evaluation

8.1 Face-Sketch Synthesis

188 CUHK face-photo and face-sketch pairs were considered to evaluate the results of the face-sketch synthesis algorithm, using the performance metrics, FSIM-Feature Similarity Index Measure, PSNR-Peak Signal to Noise Ratio and SSIM-Structural Similarity Index Measure. SSIM metric is used to calculate the structural similarity between synthesized and artist-drawn face-sketches. FSIM metric is used to calculate the featural similarity between synthesized face-sketches and grayscale face-photos. PSNR metric is used to depict the quality of synthesized face-sketches in comparison to grayscale face-photos. FSIM and PSNR measures are formulated as:

$$FSIM = \frac{\sum_{x \epsilon \Omega} S_L(x).PC_m(x)}{\sum_{x \epsilon \Omega} PC_m(x)} \tag{19}$$

where PC is a feature map, S_l is the similarity score refering to the spatial domain

$$PSNR = 10.\log_{10}\left(\frac{MAX_I^2}{MSE}\right) \tag{20}$$

where MSE is Mean Squared Error, MAX_I is image's maximal pixel value feasible.

Lastly, Human Visual Perception experiment was done to capture human's visual perception on the similarity of input face photos to the generated face sketches. Over 100 people were asked to judge how similar the input photos and the output face sketches were on a scale of 1 to 10 and the average rating was calculated (Fig. 15).

Fig. 15. Snapshot of the google form rolled out for Human Visual Perception experiment

Table 1. Average values of evaluation metrics for face sketch synthesis and photo synthesis

Sl. no	Evaluation metric	Average value for sketch synthesis	Average value for photo synthesis
1	SSIM	0.876 ± 0.028	0.751 ± 0.04
2	FSIM	0.616 ± 0.016	NA
3	PSNR	33.827 ± 1.552	17.78 ± 1.95
4	UIQ	NA	0.657 ± 0.05
5	Human visual perception similarity index	8.571 ± 1.14	7.652 ± 1.34

For sketch synthesis, values of the 4 evaluation metrics obtained in Table 1 are quite high. Approximately 87% structural similarity and 61% featural similarity can be seen. Thus the face-sketch synthesis algorithm is seen to work accurately for a variety of inputs with different skin tones, hair colors and facial features.

8.2 Face-Photo Synthesis

25,000 face-sketches of the CelebA's test dataset were considered to test and evaluate the results of the face-photo synthesis algorithm, using the performance metrics, Peak

Signal to Noise Ratio (PSNR), Universal image Quality Index (UIQ) and Structural Similarity Index Measure (SSIM). SSIM, PSNR and UIQ metrics are applied between the generated and the original face-photos.

Lastly, the Human Visual Perception experiment was done to capture the human's visual perception on similarity between the original face-photos and the synthesized face-photos. Over 100 participants were asked to rate the similarity between the original face-photos and their corresponding synthesized face-photos on a scale of 1 to 10 and the average rating was calculated. In Table 1. 75% structural similarity can be seen between the original and generated face-photos. High UIQ value and Human visual perception similarity index show that there was a high degree of match between the ground truth face-photos and synthesized face-photos. PSNR value shows that there isn't much noise in the synthesized face-photos. This indicates the effective working of the algorithm for a variety of inputs with different skin tones, hair colors and facial features.

8.3 Facial Recognition

The model was evaluated using recall, precision, F1 score and accuracy (Table 2).

Table 2. Evaluation metrics of face-photo and face-sketch recognition

Metric	Face-photo recognition	Face-sketch recognition
Accuracy	91.875%	70.731%
Recall	1.0	1.0
Precision	0.9	0.7
F1 score	0.958	0.828

In face-photo recognition, high accuracy of 91.875% was obtained after testing the facial recognition model. Similarly, high values of precision, recall and F1 score were obtained, suggesting that the facial recognition model worked quite well for recognition of face-photos. In face-sketch recognition, an accuracy of 70.731% was obtained after testing recognition model. A comparative dip in the accuracy seen here can be attributed to the loss of higher degree of discriminative information in the face-sketches than that in the face-photos after application of PCA. High values of precision, recall and F1 score were obtained suggesting that the facial recognition model worked considerably well for face-sketch recognition.

9 Conclusion

This paper presented a face sketch-photo synthesis and facial recognition system. For conversion of face photo to sketch, a simple framework with 2 scale image decomposition and color dodging was designed. The framework was seen to accurately work on scenery photos as well. CNN model was used for face-photo synthesis. The model was trained on the dataset of face sketches generated using sketch synthesis algorithm. Face recognition was implemented using LDA. The entire system was tested, evaluated and seen to work accurately for faces with variations in skin color, hair color and other facial features.

10 Future Work

Future enhancements of this study include improving the accuracy of the face-sketch recognition model and focusing on the conversion of natural scene sketches to photos and colorizing grayscale images.

References

1. Uhl, R.G., Lobo, N.D.V.: A framework for recognizing a facial image from a police sketch. In: Proceedings of the IEEE International Conference on Computer Vision and Pattern Recognition, pp. 586–593 (1996)
2. Peng, C., Wang, N., Gao, X., Li, J.: Face recognition from multiple stylistic sketches: scenarios, datasets, and evaluation. Pattern Recogn. **84**, 262–272 (2018)
3. Farid, N.M., Fard, M.S., Nickabadi, A.: Face sketch to photo translation using generative adversarial networks. arXiv preprint arXiv:2110.12290 (2021)
4. Wang, X., Tang, X.: Face photo-sketch synthesis and recognition. IEEE Trans. Pattern Anal. Mach. Intell. **31**(11), 1955–1967 (2008)
5. Balayesu, N., Kalluri, H.K.: An extensive survey on traditional and deep learning-based face sketch synthesis models. Int. J. Inf. Technol. **12**, 995–1004 (2020)
6. Zhang, W., Wang, X., Tang, X.: Lighting and pose robust face sketch synthesis. In: Daniilidis, K., Maragos, P., Paragios, N. (eds.) ECCV 2010. LNCS, vol. 6316, pp. 420–433. Springer, Heidelberg (2010). https://doi.org/10.1007/978-3-642-15567-3_31
7. Wan, W., Lee, H.J.: Generative adversarial multi-task learning for face sketch synthesis and recognition. In: 2019 IEEE International Conference on Image Processing (ICIP), pp. 4065–4069 (2019). https://doi.org/10.1109/ICIP.2019.8803617
8. Li, X., Cao, X.: A simple framework for face photo-sketch synthesis. Math. Probl. Eng. **2012**, Article ID 910719 (2012)
9. Sharma, A.R., Devale, P.R.: An application to human face photo-sketch synthesis and recognition. Int. J. Adv. Eng. Technol. **3**(2), 395 (2012)
10. Yu, J., et al.: Toward realistic face photo–sketch synthesis via composition-aided GANs. IEEE Trans. Cybern. **51**(9), 4350–4362 (2020)
11. Xiao, B., Gao, X., Tao, D., Li, X.: A new approach for face recognition by sketches in photos. Signal Process **89**(8), 1576–1588 (2009)
12. Liang, Y., Song, M., Xie, L., Bu, J., Chen, C.: Face sketch-to-photo synthesis from simple line drawing. In: Proceedings of the 2012 Asia Pacific Signal and Information Processing Association Annual Summit and Conference, pp. 1–5 (2012)
13. Lu, Y., Wu, S., Tai, Y.-W., Tang, C.-K.: Image generation from sketch constraint using contextual GAN. In: Ferrari, V., Hebert, M., Sminchisescu, C., Weiss, Y. (eds.) ECCV 2018. LNCS, vol. 11220, pp. 213–228. Springer, Cham (2018). https://doi.org/10.1007/978-3-030-01270-0_13
14. Tejomay, K.A., Kamarajugadda, K.K.: Sketch to photo conversion using cycle-consistent adversarial networks (2019)
15. Galea, C., Farrugia, R.A.: Forensic face photo-sketch recognition using a deep learning-based architecture. IEEE Signal Process. Lett. **24**(11), 1586–1590 (2017)

16. Han, H., Klare, B., Bonnen, K., Jain, A.: Matching composite sketches to face photos: a component-based approach. IEEE Trans. Inf. Forensic Secur. **88**(1), 191–204 (2013)
17. Galoogahi, H., Sim, T.: Inter-modality face sketch recognition. In: Proceedings of International Conference on Multimedia and Expo, pp. 224–229 (2012)
18. Tang, X., Wang, X.: Face sketch synthesis and recognition. In: Proceedings Ninth IEEE International Conference on Computer Vision. IEEE (2003)
19. Jacob, I.J., Darney, P.E.: Design of deep learning algorithm for IoT application by image based recognition. J. ISMAC **3**(3), 276–290 (2021)

Toward Robust Image Pre-processing Steps for Vehicle Plate Recognition

Mohamed Alkalai(✉)

IT Faculty, Computer Science Department, University of Benghazi, Benghazi, Libya
Mohamed.alkalai@uob.edu.ly

Abstract. Generally, improving the quality of vehicle plate images has proved to be the key of obtaining a promising recognition accuracy rate. Therefore, researchers tend to revisit related works with the intention to add some new image pre-processing steps in attempting to improve their current results. In this paper, a new version of previous recognition approach is introduced where more attention to image pre-processing steps is given. Firstly, an illustration of a new skew detection and fixing technique is provided. Then, a coherent overview of the previous approach, after adding the new steps, is firmly defined using pseudocode. For evaluation, experiments conduct using standard vehicle plates datasets. Promising results is obtained.

Keywords: Image pre-processing · Image deskew methods · Hough transform · Projection profile · Cross-correlation · Nearest-neighbor clustering · Vehicle plates (HDR) dataset

1 Introduction

Automated Number Plate Recognition (ANPR), is a framework that built to locate and interpret characters of vehicle plates. The demand of such systems are gradually expanded since the COVID-19 outbreak [4], to assist in imposing social distancing guidelines. The report in [3] states that authorities of several countries utilise this technology to monitor people's vehicle movements especially those whom may be close to the pandemic restricted zones. Nevertheless, there are still unsolved accuracy issues encounter ANPR systems which mainly due to the quality degradation of input images. Skewed-content, background colors of vehicle plate image and different illumination positions during capturing images are among certain factors that affect the capturing of high-quality images [5]. In attempt to contribute in tackling such problems, our previous work is revisited [2]. The main focus here is on adding new method in the image preprocessing stage which is developed to deal with a well-known cause of low-quality images named skewing.

J. I.-Z. Chen et al. (Eds.): ICIPCN 2022, LNNS 514, pp. 96–104, 2022.
https://doi.org/10.1007/978-3-031-12413-6_8

Certain circumstances prior the image capturing of vehicle plates are mainly the reason of having skewed images. These situations encompass: having the camera in improper angle during the capturing process or having dislocated vehicle plate ....etc. [10]. The methods that are used widely for skew detection which include Hough transform, projection profile, cross-correlation and nearest-neighbor clustering are described in [7]. These techniques have drawbacks which affect their performance.

For the Hough transform algorithm, a set of aligned pixels are selected to utilise in calculating the skew angle of an image [13]. Remarkable errors usually occur when dealing with images contain figure. Therefore, the input pixels for this algorithm should selected from text-only areas. However, this approach also suffers from its complexity. So, if these number of input pixels are selected from the text-only areas, the precision is, as a consequence, severely affected to the extend that the results might become humble [9].

Although, the projection profile approach relies on calculating and comparing histograms of projection profiles using each time a different skew angles (It is extremely intensive to the CPU), errors still might be occurred as a result of applying this technique on non-text areas, as this develops distortions in these histograms. In other words, good precision can be achieved, if this method runs over a text-only image [6].

Cross-correlation methods start by extracting from the image some thin vertical segments and then apply projection profiling on them to utilise its results in computing the skew angle [8]. Such methods work well on images with a homogeneous horizontal structure. However, their performance are sharply deteriorated when they run over images contain heterogeneous structure components (for instance, two-column images that contain pictures, equations and tables [11]). One can overcome this issue by firstly executing image segmentation phase then extract text areas from such images. Again, this way has its own issues as segmentation methods require skew-free images as an input to yield good results [1].

The main advantage of techniques that are based on nearest neighbor clustering is their ability to produce reasonable results even if they are applied on images with non-text areas. Having said that, there are still several cases where the performance are degraded. For example, clustering of glyphs that contain mixture of small and capital characters like (P, p, J, j, G, g, Y, y), even from skew-free images usually create lines which are not straight that cause premature commitment to errors prior computing the skew angle (if any). Also, Images that are distortion by noise, have an essential impact on the results [8].

To contribute in resolving issues associated with current skew detection and fix methods [12], a new approach is introduced. It is applied on the outcome of segmentation phase where the non text regions within vehicle plate images are eliminated which, as mentioned earlier, have a negative impact on the performance of current deskew methods. In this paper, the proposed method is firstly illustrated. Then, a description of how to insert it into our previous work in [2] is given. For evaluation, firstly, the proposed deskew method is separately run over HDR Dataset [15] to observe its ability to lay out fruitful output. Then, the new version of the previous work (after inserting this method) executes on the same dataset to compare its results with the results from previous one. The results expose satisfied improvement of performance.

2 The Proposed Deskew Approach

This approach is developed taking into an account the issues encounter current deskew methods. Therefore, it starts by utilising the candidate glyphs that are supposedly corresponding to the vehicle plate numbers which can be extracted using our previous work [2]. This resolves the issue of computing the skew angle as follows:

- Firstly, find the bottom-left *(BL)* and bottom-right *(BR)* coordinates of first and last plate numbers respectively. To accomplish this task, JASON information is utilised which is also introduced in our previous work [2]. Among these information are the top-left coordinates, height and width of plate number glyphs. The next definition 1 firmly describes the calculation of *(BL)* and *(BR)* using such information:

Definition 1 (BL and BR Coordinates). *Let* g_1 *and* g_n *be two glyphs represent first and last plate numbers such that predefined* $TL_{x,y}(g_1|g_n)$, $H(g_1|g_n)$ *and* $W(g_1|g_n)$ *are their top-left* $(x - y)$ *coordinates, height and width respectively. Then, we can say,* $BL_{x,y}(g_1)$ *and* $BR_{x,y}(g_n)$ *are their the bottom-left and bottom-right* $(x - y)$ *coordinates which are computed as such:*

- ***For Bottom-Left's x-coordinate:*** $BL_x(g_1) = TL_x(g_1)$
- ***For Bottom-Left's y-coordinate:*** $BL_y(g_1) = TL_y(g_1) + H(g_1)$
- ***For Bottom-Right's x-coordinate:*** $BR_x(g_n) = TL_x(g_n) + W(g_n)$
- ***For Bottom-Right's y-coordinate:*** $BR_y(g_n) = TL_y(g_n) + H(g_n)$

- Then, calculate the difference between $BL_x(g_1)$ and $BR_x(g_n)$ coordinates and also between $BL_y(g_1)$ and $BR_y(g_n)$ coordinates to obtain the length of *Adjacent* and *Opposite* sides of the skew angle respectively (see Fig. 1). Definition 2 expresses this process mathematically.

Definition 2 (Adjacent and Opposite Sides). *Let Adj and Opp be the Adjacent and Opposite sides of a skew angle* θ *respectively. Such that, the absolute length value of them are computed using the following equations:*

Fig. 1. Examples of adjacent and opposite sides of a skew angle

- ***For Adjacent side:*** $|Adj| = BR_x(g_n)$ - $BL_x(g_1)$
- ***For Opposite side:*** $|Opp| = BR_y(g_n)$ - $BL_y(g_1)$

– Now, a skew angle θ can be determined by dividing $|Opp|$ over $|Adj|$. Then, find the inverse of tangent of this division result which corresponds to the skew angle degree (See Definition 3).

Definition 3 (Skew Angle Determination). *Let tan^{-1} be an inverse tangent of angle θ and θ^o be a skew angle degree. Therefore, we can say that such θ^o is computed as follows:*

$$\theta^o = tan^{-1}(Opp\,/\,Adj).$$

– Once skew angle degree θ^o is defined, the deskewing process is started on vehicle plate images Im in which anti-clockwise rotation on these images is applied based on the θ^o and *Opp* be a positive value. However, when *Opp* has a negative value, clockwise rotation is utilised instead. Definition 4 expresses this process firmly.

Definition 4 (Deskew Process). *Let $ROT_c(Im, \theta^o)$ and $ROT_{c^{-1}}(Im, \theta^o)$ be image clockwise and anti-clockwise rotation functions using a defined θ^o respectively. One of these functions runs over a particular vehicle plate image when its associated condition is satisfied:*

1. $ROT_c(Im, \theta^o)$ *if* $(Opp < 0)$
2. $ROT_{c^{-1}}(Im, \theta^o)$ *if* $(Opp > 0)$

Now, the algorithm that is introduced in [2], is revisited to show how the proposed deskew steps insert into the end-to-end recognition approach. The new version which is expressed in Algorithms 1 and 2 encompasses the pseudocode of these steps. The consequence modifications are highlighted.

Algorithm 1: An overview of how the proposed deskew steps insert into the end-to-end approach

```
input  : A vehicle plate image
output: Characters or Numbers of this vehicle plate, which are editable
begin
    let Img be a colored vehicle-plate Image
    let Thresh be a threshold which is given an initial value
                                                        ▷ (ex: Thresh = 100)
    let Inc − Value be a value that added to Thresh each iteration
                                                        ▷ (ex: Inc − Value = 1)
    let Maxi − Char be a maximum no. of Glyphs          ▷ (ex: Maxi − Char = 0)
    while Thresh ≥ 0 do
        convert Img to binary image (B-W) based on Thresh.
        perform Connected Component method on B-W image.
                          ▷ (Output: The boundaries info of glyphs G that
                            extracted from B-W image. such as G =
                            {g1, g2, ...., gn})
        while G ≠ ∅ do
            If Segment(g1, Img) then add g1 to G' ∈ G.
                          ▷ Segment is a function for checking the likelihood
                      of g1 to be a vehicle-plate number
            eliminate g1 from G
                          ▷ Len is a function for counting glyphs in G'
        if Maxi − Char < Len(G') then
            assign Len(G') to Maxi − Char
            assign Thresh to Thresh'
            copy G' into MaxChNo
                          ▷ (MaxChNo would always equal to the G'
                      list which has maximum no. of glyphs)
        let G' = ∅
        let Thresh = Thresh - Inc − Value
    call Algorithm2 with input parameters MaxChNo and Img
                          ▷ (Output: a deskewed image (Img'))
    convert Img' to binary image (B-W) based on Thresh'
    perform Connected Component method on B-W
    while G ≠ ∅ do
        If Segment(g1, Img') then add g1 to G'            ▷ G' ∈ G
        eliminate g1 from G
    while G' ≠ ∅ do
        append Sub − Img(g1, Img') to Char − Img
                          ▷ Sub − Img is a function cuts a piece of Img'
                      using g1 boundaries info. and Char − Img
                      contains a list of these pieces
        eliminate g1 from G'
    recognise the characters that each piece of image has, in Char − Img
                          ▷ (Output: numbers and characters that can be
                            edited (editable-numbers))
    return editable-numbers
```

Fig. 2. The two vehicle plate images in Fig. 1 after deskewing them

---

Algorithm 2: An overview of the proposed deskew steps

---

input : Glyphs of vehicle plate number candidates $MaxChNo$
input : A vehicle plate image Img
output: An deskewed image Img' of Img

begin
  compute $BL_{(x,y)}(g_1 \in MaxChNo)$ **and** $BR_{(x,y)}(g_n \in MaxChNo)$
    ▷ ($BL_{(x,y)}$ and $BR_{(x,y)}$ represent the bottom-left and bottom-right coordinates of a glyph respectively)
  assign ($BR_x(g_n)$ - $BL_x(g_1)$) **to** Adj
  assign ($BR_y(g_n)$ - $BL_y(g_1)$) **to** Opp
    ▷ (Adj and Opp represent *Adjacent* and *Opposite* sides of Img skew angle θ)
  find θ^o **by calculating** $tan^{-1}(|Opp| \ / \ |Adj|)$
    ▷ (tan^{-1} is the inverse of tangent and θ^o is a skew angle degree)
  If $Opp < 0$ **then**
  execute $ROT_c(Img, \theta^o)$
  else
  execute $ROT_{c^{-1}}(Img, \theta^o)$
    ▷ (ROT_c and $ROT_{c^{-1}}$ are functions that rotate Img using θ^o in which c and c^{-1} indicate clockwise and anticlockwise) (Output: a deskewed image (Img'))
  return Img'

Fig. 3. Sample of skewed vehicle plate images that our deskew method fails

3 Performance Evaluation

The lines of Algorithm 1 up to line 24 are executed on HDR dataset [14] to assess the performance of our deskew approach. The dataset contains 326 vehicle plate images (first part) and another 326 of the same images (second part) but taken by different picture capturing circumstances.

Table 1. Experimental results of applying the proposed deskew approach on (a) first part of HDR dataset, (b) second part of HDR dataset

	First part of HDR dataset	Second part of HDR dataset
No. of images correctly deskewed	320	322
Precision percentage	320/326 98.2%	322/326 98.8%

Table 1 shows in numbers the results of this experiment which are obtained manually. As can be inferred from the promising accuracy rates (98.2%, 98.8%) that one can rely on our deskew technique to avoid premature commitment to errors as proceeding toward robust end-to-end vehicle plate recognition approach. Figure 2 shows the two vehicle plate images that are in Fig. 1 after deskewing them. Almost all of deskew failure cases occur due to images with severely low-quality. Figure 3 illustrates a sample of these cases.

To observe the effectiveness of integrating the proposed deskew steps with the end-to-end vehicle plate recognition approach described in Algorithm 1 in [2], an experiment is conducted on the first and second HDR dataset.

Table 2 represents the results of this experiment. The accuracy rates have increased compared with the one in previous version of end-to-end vehicle plate recognition approach which were 76.4% and 79.8% [2] for first and second parts of HDR dataset respectively.

Table 2. Experimental results of applying the new version of end-to-end approach on (a) first part of HDR dataset, (b) second Part of HDR dataset

	First part of HDR dataset	Second part of HDR dataset
No. of char correctly recognized	1850	1870
Precision percentage	1850/2277 81.25%	1870/2277 82.13%

4 Conclusion

In the response of the need of a reliable vehicle plate recognition tools, a development of such applications has previously conducted. The image pre-processing phase plays an essential role in having these methods produced accurate output. Therefore, in this paper, a deskewed approach is introduced to tackle such skewed plates. This method takes into an account the issues which encounter current methods. Both experiments which are separately conducted on this proposed technique and on its integration with the end to end approach, demonstrate their robustness and consistency.

References

1. Ali, A., Ali, A., Suresha, M.: A novel approach to correction of a skew at document level using an Arabic script. Int. J. Comput. Sci. Inf. Technol. **8**(5), 569–573 (2017)
2. Alkalai, M., Lawgali, A.: Image-preprocessing and segmentation techniques for vehicle-plate recognition. In: 2020 IEEE 4th International Conference on Image Processing, Applications and Systems (IPAS), pp. 40–45. IEEE (2020)
3. Brown, J.: Automated number plate recognition (ANPR) surveillance during covid-19. Technical report, Melbourne Activist Legal Support Group (2020)
4. Confederation, A.T.U.: Impact of the covid 19 on the transport industry. Technical report, Arab Trade Union, International Transport Workers, Federation and Danish Trade Union (2020)
5. Markets and Markets Research Group. Anpr system market with covid-19 impact analysis by type (fixed, mobile, portable), application (traffic management, law enforcement, electronic toll collection, parking management, access control), component, and geography - global forecast to 2025. Technical report, Marketsand-Markets, copyrights 2019–2020
6. Huang, K., Chen, Z., Min, Yu., Yan, X., Yin, A.: An efficient document skew detection method using probability model and q test. Electronics **9**(1), 55 (2020)
7. Hull, J.: Document image skew detection: survey and annotated bibliography (1998)
8. Jundale, T., Hegadi, R.: Research survey on skew detection of Devanagari script. Int. J. Comput. Appl. **975**, 8887 (2015)
9. Kumar, D., Singh, D.: Modified approach of Hough transform for skew detection and correction in documented images. Int. J. Res. **2**, 37–40 (2012)
10. Ondrej Martinsky, P.: Recognition of vehicle number plates. In: ACM CZ, p. 33 (2007)

11. Papandreou, A., Gatos, B., Perantonis, S.J., Gerardis, I.: Efficient skew detection of printed document images based on novel combination of enhanced profiles. Int. J. Doc. Anal. Recogn. **17**(4), 433–454 (2014). https://doi.org/10.1007/s10032-014-0228-5
12. Rehman, A., Saba, T.: Document skew estimation and correction: analysis of techniques, common problems and possible solutions. Appl. Artif. Intell. **25**(9), 769–787 (2011)
13. Kumar Shukla, B., Kumar, G., Kumar, A.: An approach for skew detection using Hough transform. Int. J. Comput. Appl. **136**(9), 20–23 (2016)
14. Špaňhel, J., et al.: HDR dataset. https://academictorrents.com/details/8ed33d02_d6b36c389dd077ea2478cc83ad117ef3. Accessed 28 Feb 2022
15. Špaňhel, J., et al.: Holistic recognition of low quality license plates by CNN using track annotated data. In: 2017 14th IEEE International Conference on Advanced Video and Signal Based Surveillance (AVSS), pp. 1–6. IEEE (2017)

Multi-focus Image Fusion Using Morphological Toggle-Gradient and Guided Filter

Manali Roy(✉) and Susanta Mukhopadhyay

Indian Institute of Technology (Indian School of Mines), Dhanbad, Jharkhand 826004, India
manalir66@gmail.com

Abstract. Digital image acquisition devices suffer from a narrow depth of field (DoF) due to optical lenses installed in them. As a result, the generated images have varying focus, thereby losing essential details. Multi-focus image fusion aims to synthesize an all-in-focus image for better scene perception and processing. This paper proposes an effective region-based focus fusion approach based on a novel focus measure derived from multi-scale half gradients extracted from the morphological toggle-contrast operator. The energy of the same focus measure is combined with spatial frequency to design a composite focusing criterion (CFC) to roughly differentiate between the focussed and defocussed regions. The high-frequency information obtained is further enhanced using a guided filter, taking focus guidance from the source images. The best focus region is selected using a pixel-wise maximum rule which is further converted into a refined binarized decision map using a small region removal technique. At this point, the guided filter is re-utilized to verify the spatial correlation concerning the initial fused image to obtain the final decision map for final fusion. Experimental results exhibit the discussed algorithm's efficacy over current fusion approaches in visual perception and quantitative metrics on registered and un-registered multi-focus datasets.

Keywords: Focus fusion · Guided filter · Toggle gradients · Spatial frequency · Synthetic images

1 Introduction

Multi-focus image fusion is an active area of research in the field of computer vision as it solves the problem of the limited depth of field exhibited by imaging devices [13,22]. The purpose behind focus fusion is to colligate the focussed content distributed in multiple source images to produce a synthetic image with improved perceptual quality for further applications (Fig. 1). Generally, fusion algorithms are carried out in spatial, transform or hybrid (combination of spatial and transform) domains [10,23]. Transform domain fusion methods (i.e., DWT, SWT NSST, NSCT, MWT) has the overhead of spatial-to-transform conversion and usually suffer from illumination defects, colour distortion and compromised

J. I.-Z. Chen et al. (Eds.): ICIPCN 2022, LNNS 514, pp. 105–118, 2022.
https://doi.org/10.1007/978-3-031-12413-6_9

resolution while reconstruction of the fused output [12,25,26]. On the contrary, spatial domain methods directly evaluate the focus information at individual pixel locations and retain it in the fused result [9,29]. A primary characteristic of fusion methods in the spatial domain is the absence of inverse transform for the reconstruction of the fused image. Pixel-based methods in the spatial domain are susceptible to noise sensitivity, blurring effects and mis-registration errors. Region-based spatial methods overcome the problem by segmenting the semantic regions within an image [14,20,21]. In this context, the semantic regions which are high in focus or sharpness are usually identified using various sharpness measures [7,8]. Being a cost-effective way to widen the depth-of-field of optical lenses, multi-focus image fusion approach has successful applications in the areas of agriculture, biology, medicine, etc.

Fig. 1. Multi-focus fusion: (a) Focus on background; (b) Focus on foreground; (c) All-in-focus fused image

In this paper, a gradient-based focus measure exploiting the morphological toggle-contrast operator brings out initial salient regions which are confirmed by a composite focusing degree criterion. The rough regions are converted into an optimized edge-smoothed decision map using a guided filter in two steps. The outline of the paper is as follows:

- A novel focus measure is devised from multi-scale weighted gradients extracted from morphological erosion-dilation based toggle contrast operators.
- The focussed pixels are aggregated into focussed regions using a composite focus criterion.
- The final optimized decision maps for image fusion are formed using guided filtering in two steps.
- The method performs identically well for unregistered, registered and synthetic multi-focus image pairs.

The remainder of the article is structured as follows: Sect. 2 briefly discusses the pre-requisite concepts applied for the work. The algorithm is presented in Sect. 3 followed by discussions on objective and subjective results in Sect. 4. Section 5 has the conclusion of the paper.

2 Preliminaries

2.1 Toggle Contrast Operator

Morphological toggle mappings were first introduced in [15] as an alternative to top-hat operators for contrast enhancement in images. The two-state toggle mapping (TM) based on morphological dilation ($\oplus$) and erosion ($\ominus$) operations is expressed as below:

$$TM_S(x,y) = \begin{cases} f \oplus S(x,y), \text{ if} \\ (f \oplus S - f)(x,y) < (f - f \ominus S)(x,y) \\ f \ominus S(x,y) \text{ if} \\ (f \oplus S - f)(x,y) > (f - f \ominus S)(x,y) \end{cases} \tag{1}$$

where f is the image and S denotes the structuring element. Each pixel value in a toggle-mapped output image switches between the eroded and dilated versions of the source image depending on which is closer to the input pixel value. This operator along with its multi-scale variants are widely explored in image fusion [3,4].

Fig. 2. (a) Focus on background; (b) Toggle-mapped output; (c) Beucher's gradient at one scale; (d) Beucher's gradient at three scales; (e) Toggle external gradient at one scale; (f) Toggle external gradient at three scales, i.e., MTGFM output

2.2 Spatial Frequency

Spatial frequency is prominently used as a clarity measure in image fusion tasks. It simulates the human visual system in detecting the amount of frequency content within an image. For a $M \times N$ block, it is defined as,

$$SF = \sqrt{RF^2 + CF^2} \tag{2}$$

where

$$RF = \sqrt{\frac{1}{XY} \sum_{x=0}^{X-1} \sum_{y=1}^{Y-1} [F(x,y) - F(x,y-1)]^2}$$

and

$$CF = \sqrt{\frac{1}{XY}\sum_{x=1}^{X-1}\sum_{y=0}^{Y-1}[F(x,y) - F(x-1,y)]^2}$$

represents the row and column frequency respectively.

2.3 Guided Filter

The guided filter is a widely popular edge-preserving filter expressed as a local linear model between an input image (p), a sharp guidance image (I) and filter output (q) [1,6]. The filtered output is expressed as a linear transform given by,

$$q_i = a_k I_i + b_k, \forall i \in w_k$$

where w_k is a window centered at pixel k and (a_k, b_k) are linear coefficients which is determined by solving a cost function (Eq. 3) to minimise the difference between q_i and p_i while maintaining the linearity in the model [1].

$$E(a_k, b_k) = \sum_{i \in w_k} (a_k I_i + b_k - p_i)^2 + \epsilon a_k^2 \quad (3)$$

where ϵ is the regularization parameter. The linear coefficients are obtained as follows,

$$a_k = \frac{\frac{1}{|w|}\sum_{i \in w_k} I_i p_i - \mu_k \bar{p}_k}{\sigma_k^2 + \epsilon}, \quad b_k = \bar{p}_k - a_k \mu_k$$

where μ_k and σ_k^2 are the mean and variance of I in w_k. $\bar{w}$ is the number of pixels in w_k and $\bar{p} = \frac{1}{|w|}\sum_{i \in w_k} p_i$ is the mean of p in w_k. Averaging the same throughout the whole image, we get,

$$q_i = \frac{1}{|w|}\sum_{i \in w_k} (a_k I_i + b_k) = \bar{a_i} I_i + \bar{b_i}$$

where

$$\bar{a_i} = \frac{1}{|w|}\sum_{k \in w_i} a_k, \quad \bar{b_i} = \frac{1}{|w|}\sum_{k \in w_i} b_k$$

3 Proposed Method

3.1 Morphological Toggle-Gradient Based Focus Measure

The basic toggle mapping (TM) operator computes two transformations for an input image I, i.e., (a) dilation and (b) erosion which denotes the maximum and minimum value within a structuring element (S) respectively. Thus each pixel in the filtered output is modified either by (a) or (b) choosing the one closest to the original pixel value (Fig. 2b). The conventional morphological gradient operator

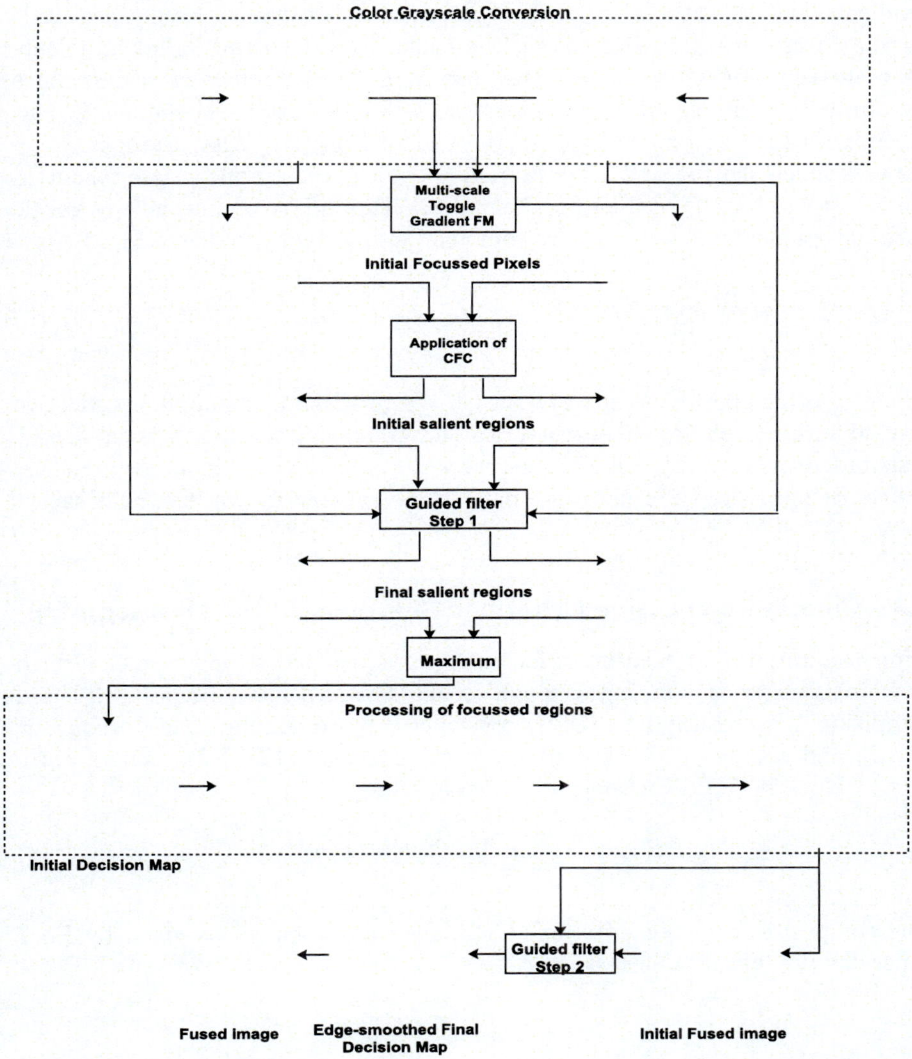

Fig. 3. Process flowchart for the proposed approach

(Beucher's gradient) highlights variations within a neighbourhood determined by a structuring element. Alternately, half gradients (internal and external) extract the outer and inner boundaries of an edge [19]. Likewise, a set of half-gradients (g_i, g_e) can be designed from toggle-contrast operator (Eq. 1) as follows,

$$\begin{aligned} g_e(x,y) &= TM_S(f(x,y)) - f(x,y) \\ g_i(x,y) &= f(x,y) - TM_S(f(x,y)) \end{aligned} \tag{4}$$

where g_i and g_e stands for internal and external gradient respectively. Now Beucher's gradient used at multiple scales (practically, $S \geq 1$) yields thick gradients and is suitable for smooth transitions between objects. Also, it is unable to enhance the edges along the concavities. Contrarily, half gradients generate thin edges for sharp transitions at a lesser number of scales. Since we aim at combining focussed regions, we select the external gradient (g_e) as it captures the outer boundaries of the focussed objects. Also, the focus map appears much sharper and more prominent which is just appropriate for forthcoming steps (Fig. 2f). Exploiting the multi-scale property of morphology, toggle external gradients at all scales are combined into a single gradient (Fig. 3).

$$MTGFM(x,y) = \sum_{j=1}^{n} \alpha_j \cdot 2g_e(x,y), \quad \alpha_j = \frac{1}{2 \times j + 1} \tag{5}$$

where α_j is the gradient weight at scale 'j'. Weighted sum is a simple yet effective way to fuse morphological features, so the gradient information is uniformly distributed by assigning smaller (larger) weights to features at larger (smaller) scales. Being isotropic in nature, a flat disk is chosen with its initial radius set to 2.

3.2 Composite Focusing Criterion (CFC) and Initial Decision Map

After the initial identification of focussed pixels, the focus maps are confirmed and binarized to form an initial rough region. This is achieved by a block-based composite focusing criterion (CFC) constructed by combining spatial frequency (Eq. 2) and energy of toggle gradient focus measure ($MTGFM$). Energy of a $m \times m$ block in $MTGFM$ output is computed as,

$$E_{FM} = \sum_{m}\sum_{m}[MTGFM(x,y)]^2 \tag{6}$$

The composite focussing criterion (CFC) for source images as given by Eq. 7 generates the initial salient regions.

$$CFC = R_{i=i_1,i_2,\ldots i_n}(x,y) = \begin{cases} 1, \text{if} \quad SF_{i=i_1} \geq SF_{i \neq i_1} \\ \wedge \quad E_{FM}^{i=i_1}(x,y) \geq E_{FM}^{i \neq i_1}(x,y) \\ 0, \quad \text{otherwise} \end{cases} \tag{7}$$

The high-frequency regions is further enhanced by guided filter using the source grayscale images as the guidance image (Eq. 8).

$$R_i^g = GF_{w_g,\epsilon}[I_i(x,y), R_i] \tag{8}$$

The best of the guided focus regions are selected using pixel-wise maximum rule (Eq. 9) to finalize the initial decision map.

$$D_i = \max[R_{i=i_1}^g, R_{i=i_2}^g, \ldots, R_{i=i_n}^g] \tag{9}$$

Here, $i = i_1, i_2, \ldots i_n$ denotes the number of source images.

3.3 Final Decision Map and Fusion

Practically, it is impossible to get source images free from noise and spurious information (imaging sensor defects produce electrical noise), which may not get well detected. To resolve this, the initial map is processed using a small region removal strategy [11] based on 8-connectivity with a ratio factor set to 0.01. Thus, the map is further utilized to obtain the initial fused image using Eq. 10.

$$F_i(x,y) = D_i I_n(x,y) + (1 - D_i) I_{n-1,n}(x,y) \tag{10}$$

where n = 1, 2, ... N−1. Due to the presence of sharp edges in D_i, the fused image, $F_i(x,y)$ results in artifacts along the boundaries of definite focussed and definite defocussed regions. So to introduce smooth transitions for the boundaries, the guided filter is re-utilized to fine-tune the initial decision map (D_i) using $F_i(x,y)$ as the guidance image. This gives us the edge-smoothed final decision map (D_f) for the final image fusion following Eq. 11,

$$D_f(x,y) = GF_{w_g,\epsilon}[D_i, F_i]$$

$$F_f(x,y) = D_f I_n(x,y) + (1 - D_f) I_{n-1,n}(x,y) \tag{11}$$

The primary role of guided filter in our work is to perform the spatial consistency verification with respect to source images. As, guided filter collects high-frequency structural information from the guidance image (i.e., input image), it corrects the misalignment of decision map with the object boundaries thereby perfecting the edges of the focussed maps.

4 Experimental Results and Discussion

This section presents a qualitative and quantitative comparison of the approach with other recently developed fusion algorithms.

4.1 Execution Setup

Table 1. Parameters

Name	Value
Number of scales in MTGFM	3
Sliding window for CFC, (m)	4 × 4
Local window (Guided filter, w_g)	5
Regularization Parameter (Guided filter, ϵ)	0.1

The above method is tested on a pair of registered multi-focus datasets, a) Lytro [16], b)MFI-WHU [30] and an unregistered dataset c) Pxleyes [2]. The experiments along with the comparison methods are executed on MATLAB R2017b, on

64 bit windows operating system, Intel 2.60 Hz Core i7 CPU and 16 GB RAM. Four competing fusion methods, GCF [24], IFCNN [32], PMGI [31] and U2 [27] are chosen for comparison with respect to following five fusion metrics: Piella's metric (Q_o) [17], mutual information ($MI_{\mathrm{AB}}^{\mathrm{F}}$) [18], Xydeas's metric ($Q_{\mathrm{AB}}^{\mathrm{F}}$) [28], feature mutual information (FMI) [5] and Zhao's metric (P'_{blind}) [33]. The parameters utilized in the algorithm and average value of metrics corresponding to the datasets are presented in Table 1 and Table 2 respectively.

Table 2. Average objective evaluation on multifocus image sets

Methods						
Images	Metrics	GCF [24]	IFCNN [32]	PMGI [31]	U2 [27]	Proposed
Lytro [16]	$MI_{\mathrm{AB}}^{\mathrm{F}}$	**1.1115**	0.9421	0.8332	0.7710	1.0827
(512 × 512)	Q_o	0.9399	**0.9456**	0.8734	0.8606	0.9451
	FMI	0.9062	0.8999	0.8882	0.8889	**0.9035**
	$Q_{\mathrm{AB}}^{\mathrm{F}}$	0.7309	0.7064	0.5666	0.5812	**0.7374**
	P'_{blind}	0.8441	0.8015	0.6369	0.5909	**0.8481**
Pxyeles [2]	$MI_{\mathrm{AB}}^{\mathrm{F}}$	**0.6431**	0.5186	0.5457	0.5247	0.6341
(512 × 512)	Q_o	0.8333	0.8040	0.7963	0.7900	**0.8521**
	FMI	0.8846	0.8649	0.8782	0.8783	**0.8891**
	$Q_{\mathrm{AB}}^{\mathrm{F}}$	0.5869	0.5420	0.5408	0.5175	**0.6083**
	P'_{blind}	0.4633	0.3330	0.3781	0.4068	**0.5295**
MFI-WHU [30]	$MI_{\mathrm{AB}}^{\mathrm{F}}$	0.8635	0.8308	0.8492	0.7372	**0.8638**
(512 × 512)	Q_o	0.9326	0.9317	0.8878	0.8764	**0.9329**
	FMI	0.8758	0.8754	0.8690	0.8689	**0.8762**
	$Q_{\mathrm{AB}}^{\mathrm{F}}$	0.6172	0.6057	0.5787	0.5304	**0.6191**
	P'_{blind}	0.6771	0.6779	0.6755	0.6461	**0.6828**

4.2 Subjective Evaluation

A proper fusion approach (a) must not generate image artifacts/distortions, (b) should not enhance existing features and (c) should be consistent with the source images. This subsection studies the relative performances of the proposed algorithm with other fusion algorithms. Results from registered multi-focus datasets, i.e., Lytro and MFI-WHU is presented in Fig. 4, Fig. 5, Fig. 6 and Fig. 7 respectively while Fig. 8 and Fig. 9 exhibits the same for Pxleyes dataset [2]. For all the datasets, PMGI and U2 method has highly intensified the colour contrast of the fused images as depicted by Fig. 4(g, h), Fig. 8(g, h), Fig. 6(g, h) and Fig. 7(g, h). In addition, results from PMGI lacks prominence, sharpness and edge clarity (Fig. 9g). For the same source pair from the Pxleyes dataset, results from GCF, IFCNN, PMGI and U2 have introduced a white vertical distortion in the lamp (marked by a red box within a yellow box). This can be attributed to variations in the location of pixels (also known as volumetric differences) in the source images due to mis-registration. In Fig. 5e, GCF based results perform poorly in capturing the foreground information (checkered floor) with respect to other methods. IFCNN gives comparable visual results but compromises over

Fig. 4. Registered Lytro dataset [16] and fused results: (a) Background in focus; (b) Foreground in focus; (c, d) Focus Maps; (e) GCF result; (f) IFCNN result; (g) PMGI result; (h) U2 result; (i) Generated Decision Map; (j) Result using proposed method

Fig. 5. Source pairs and fused results: same order as in Fig. 4

Fig. 6. MFI-WHU dataset [30] and fused results: (a) Background in focus; (b) Foreground in focus; (c, d) Focus Maps; (e) GCF result; (f) IFCNN result; (g) PMGI result; (h) U2 result; (i) Generated Decision Map; (j) Result using proposed method

Fig. 7. Source pairs and fused results: same order as in Fig. 6

Fig. 8. Unregistered Pxyeles dataset [2] and fused results: (a) Background in focus; (b) Foreground in focus; (c, d) Focus Maps; (e) GCF result; (f) IFCNN result; (g) PMGI result; (h) U2 result; (i) Generated Decision Map; (j) Result using proposed method

Fig. 9. Source pairs and fused results: same order as in Fig. 8

the values of fusion quantitative metrics. Our method surpasses other methods in perceptual clarity, preservation of focus information and robustness to misregistration errors. It is further confirmed by the highest (bolded) average value of fusion metrics tabulated in Table 2 for all the datasets.

(a) (b) (c) (d) SSIM = 0.9977

(a1) (b1) (c1) (d1) SSIM = 0.9995

Fig. 10. Performance on synthetic multi-focus images: (a), (a1) Focus on Background; (b), (b1) Focus on Foreground; (c), (c1) Ground-truth image; (d), (d1) Fused by the proposed algorithm

4.3 Fusion on Synthetic Source Pairs

Figure 10(a, b: collected; a1, b1: manually generated using photo-editing tools) illustrates the performance of the algorithm on a pair of synthetic multi-focus images. The availability of ground-truth all-in-focus images aids to judge the quality of fusion exhibited by the algorithm using reference-based fusion quality metrics. Here, the authors have evaluated SSIM (structural similarity index) which establishes the perceptual quality of the results. Regardless of the source image being real or synthetic, the focus measure along with the composite focus criterion largely separates the focussed pixels from the defocused ones, thus fastening the region conversion process.

5 Conclusion

This article discusses a multi-focus image fusion approach that uses a novel focus measure designed utilizing gradients from morphological erosion/dilation based toggle mappings at multiple scales. The energy of the focus measure in conjunction with spatial frequency is used as a composite criterion to extract the high-frequency regions. Due to edge-preserving property and low computational complexity, a guided filter is applied over the initial rough regions at two stages to obtain the confirmed decision map for final fusion. The superior objective and subjective efficacy of the proposed approach over other competing fusion algorithms are experimentally validated on three multi-focus image datasets using suitable fusion quality metrics.

References

1. https://analyticsindiamag.com
2. www.pxleyes.com
3. Bai, X.: Morphological image fusion using the extracted image regions and details based on multi-scale top-hat transform and toggle contrast operator. Digit. Sig. Process. **23**(2), 542–554 (2013)
4. Bai, X., Zhou, F., Xue, B.: Edge preserved image fusion based on multiscale toggle contrast operator. Image Vis. Comput. **29**(12), 829–839 (2011)
5. Haghighat, M.B.A., Aghagolzadeh, A., Seyedarabi, H.: A non-reference image fusion metric based on mutual information of image features. Comput. Electr. Eng. **37**(5), 744–756 (2011)
6. He, K., Sun, J., Tang, X.: Guided image filtering. IEEE Trans. Pattern Anal. Mach. Intell. **35**(6), 1397–1409 (2012)
7. He, K., Gong, J., Xu, D.: Focus-pixel estimation and optimization for multi-focus image fusion. Multimedia Tools Appl. **81**(6), 7711–7731 (2022)
8. Jing, Z., Pan, H., Li, Y., Dong, P.: Evaluation of focus measures in multi-focus image fusion. In: Non-Cooperative Target Tracking, Fusion and Control. IFDS, pp. 269–281. Springer, Cham (2018). https://doi.org/10.1007/978-3-319-90716-1_15
9. Kahol, A., Bhatnagar, G.: A new multi-focus image fusion framework based on focus measures. In: 2021 IEEE International Conference on Systems, Man, and Cybernetics (SMC), pp. 2083–2088. IEEE (2021)
10. Kaur, H., Koundal, D., Kadyan, V.: Image fusion techniques: a survey. Arch. Comput. Meth. Eng. **28**(7), 4425–4447 (2021)
11. Li, S., Kang, X., Hu, J.: Image fusion with guided filtering. IEEE Trans. Image Process. **22**(7), 2864–2875 (2013)
12. Liu, S., et al.: A multi-focus color image fusion algorithm based on low vision image reconstruction and focused feature extraction. Sig. Process. Image Commun. **100**, 116533 (2022)
13. Liu, Yu., Wang, L., Cheng, J., Li, C., Chen, X.: Multi-focus image fusion: a survey of the state of the art. Inf. Fusion **64**, 71–91 (2020)
14. Meher, B., Agrawal, S., Panda, R., Abraham, A.: A survey on region based image fusion methods. Inf. Fusion **48**, 119–132 (2019)
15. Meyer, F., Serra, J.: Contrasts and activity lattice. Sig. Process. **16**(4), 303–317 (1989)
16. Nejati, M., Samavi, S., Shirani, S.: Multi-focus image fusion using dictionary-based sparse representation. Inf. Fusion **25**, 72–84 (2015)
17. Piella, G., Heijmans, H.: A new quality metric for image fusion. In: Proceedings 2003 International Conference on Image Processing (Cat. No. 03CH37429), vol. 3, p. III-173. IEEE (2003)
18. Qu, G., Zhang, D., Yan, P.: Information measure for performance of image fusion. Electron. Lett. **38**(7), 313–315 (2002)
19. Rivest, J.F., Soille, P., Beucher, S.: Morphological gradients. J. Electron. Imaging **2**(4), 326–336 (1993)

20. Roy, M., Mukhopadhyay, S.: Multi-focus fusion using image matting and geometric mean of DCT-variance. In: Singh, S.K., Roy, P., Raman, B., Nagabhushan, P. (eds.) CVIP 2020. CCIS, vol. 1376, pp. 212–223. Springer, Singapore (2021). https://doi.org/10.1007/978-981-16-1086-8_19
21. Roy, M., Mukhopadhyay, S.: A scheme for edge-based multi-focus color image fusion. Multimedia Tools Appl. **79**(33), 24089–24117 (2020)
22. Singh, P., Diwakar, M., Chakraborty, A., Jindal, M., Tripathi, A., Bajal, E.: A non-conventional review on image fusion techniques. In: 2021 IEEE 8th Uttar Pradesh Section International Conference on Electrical, Electronics and Computer Engineering (UPCON), pp. 1–7. IEEE (2021)
23. Singh, V., Kaushik, V.D.: A study of multi-focus image fusion: state-of-the-art techniques. In: Tiwari, S., Trivedi, M.C., Kolhe, M.L., Mishra, K.K., Singh, B.K. (eds.) Advances in Data and Information Sciences: Proceedings of ICDIS 2021, pp. 563–572. Springer, Singapore (2022). https://doi.org/10.1007/978-981-16-5689-7_49
24. Tan, W., Zhou, H., Rong, S., Qian, K., Yu, Y.: Fusion of multi-focus images via a gaussian curvature filter and synthetic focusing degree criterion. Appl. Opt. **57**(35), 10092–10101 (2018)
25. Tan, Y., Yang, B.: Multi-focus image fusion with cooperative image multiscale decomposition. In: Ma, H., et al. (eds.) PRCV 2021. LNCS, vol. 13021, pp. 177–188. Springer, Cham (2021). https://doi.org/10.1007/978-3-030-88010-1_15
26. Wan, H., Tang, X., Zhu, Z., Li, W.: Multi-focus image fusion method based on multi-scale decomposition of information complementary. Entropy **23**(10), 1362 (2021)
27. Xu, H., Ma, J., Jiang, J., Guo, X., Ling, H.: U2Fusion: a unified unsupervised image fusion network. IEEE Trans. Pattern Anal. Mach. Intell. **44**(1), 502–518 (2020)
28. Xydeas, C., Petrovic, V.: Objective image fusion performance measure. Electron. Lett. **36**(4), 308–309 (2000)
29. You, C.-S., Yang, S.-Y.: A simple and effective multi-focus image fusion method based on local standard deviations enhanced by the guided filter. Displays **72**, 102146 (2022)
30. Zhang, H., Le, Z., Shao, Z., Xu, H., Ma, J.: MFF-GAN: an unsupervised generative adversarial network with adaptive and gradient joint constraints for multi-focus image fusion. Inf. Fusion **66**, 40–53 (2021)
31. Zhang, H., Xu, H., Xiao, Y., Guo, X., Ma, J.: Rethinking the image fusion: a fast unified image fusion network based on proportional maintenance of gradient and intensity. In: Proceedings of the AAAI Conference on Artificial Intelligence, vol. 34, pp. 12797–12804 (2020)
32. Zhang, Yu., Liu, Yu., Sun, P., Yan, H., Zhao, X., Zhang, L.: IFCNN: a general image fusion framework based on convolutional neural network. Inf. Fusion **54**, 99–118 (2020)
33. Zhao, J., Laganiere, R., Liu, Z.: Performance assessment of combinative pixel-level image fusion based on an absolute feature measurement. Int. J. Innov. Comput. Inf. Control **3**(6), 1433–1447 (2007)

Security Enhancement of Fog Nodes in IoT Networks Using the IBF Scheme

N. A. Natraj[1(✉)], V. Kamatchi Sundari[2], K. Ananthi[3], S. Rathika[4], G. Indira[4], and C. R. Rathish[5]

[1] Symbiosis Institute of Digital and Telecom Management, Symbiosis International (Deemed University), Pune, India
natraj@sidtm.edu.in
[2] Department of ECE, SRM Institute of Science and Technology, Chennai, India
[3] Department of Mechatronics Engineering, Sri Krishna College of Engineering and Technology, Coimbatore, India
[4] Department of ECE, Prince Shri Venkateshwara Padmavathy Engineering College, Chennai, India
[5] Department of CE, New Horizon College of Engineering, Bengaluru, India

Abstract. Internet of Things provides autonomous control to the internet based physical devices, by reducing the human efforts. Data exchange occurs between multiple devices through internet. IoT There are two forms of system architecture. Architectures based on cloud and fog computing. The fog computing design decreases cloud load and enables effective computation and data collecting operations. But this method of computing faces major challenges like privacy, security, delay, etc. Multiple research works are being carried out to overcome these challenges. In the existing method, DTCH (Double Trapdoor Chameleon Hash Function) based one way hash function was used to enhance the privacy and security. This method proved to be effective in the expense of processing time. Here we propose an IBE based Boneh Frankin scheme (IBF) for Secured Fog computing enhance the security and to overcome the processing time delay of the existing method. The simulation was conducted in Ns-2 and the results compared to those from the existing method proved to be successful.

Keywords: IoT · Fog computing · Security · Privacy · IBE · Encryption

1 Introduction

Internet of Things are generally defined as a network, embedded with hardware and software, where these physical objects, gather and exchange data between them. So the main aim is to extend the internet connectivity from traditional devices to real time applications. The physical world comes in to the meet with the digital world through network. In simple IoT cannot be related as a single technology. It is a cognizance of multiple technologies interwork together to meet the need of real world applications. The IoT architecture has been defined differently by multiple researchers. The 3 layer protocol architecture is the most basic form. They are perception layer, network layer and

J. I.-Z. Chen et al. (Eds.): ICIPCN 2022, LNNS 514, pp. 119–129, 2022.
https://doi.org/10.1007/978-3-031-12413-6_10

application layer. They are more similar to the OSI layer models defined. The perception layer here describes the physical layer, which is responsible for sensing and analyzing the information gathered from the environment through available sensors. The network layer's main role is to process the data and also to connect with the intended network devices. The application layer in the top, provides application oriented services to the user. The usage of IoT is envisioned through this layer. The Basic IoT Architecture is shown in Fig. 1.

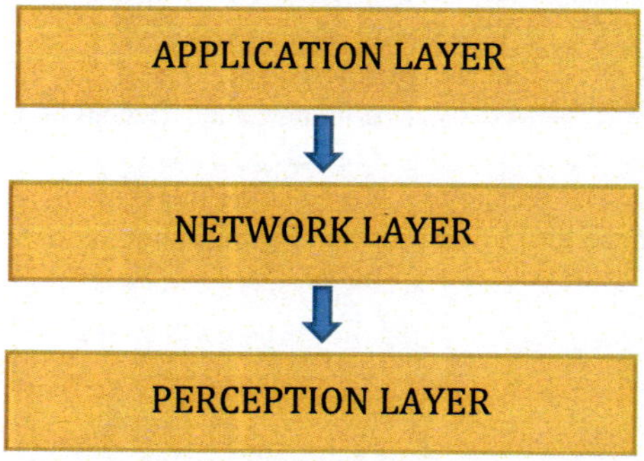

Fig. 1. Basic IoT architecture

The system architecture is generally classified of 2 types. They are fog computing based architecture and cloud computing based architecture. We are unaware about the data handled by IoT devices. This data handling is done by large centralized computers. These computers are cloud based and they are in the center to the applications and the network. Cloud architectures placed at the center should handle data from multiple sensors for different applications. The other type of computing based architecture is fog computing, which delivers a layered kind of approach. The fog architecture layers performs monitoring, preprocessing, temporary storing and security functions. Cisco has termed fog computing as smart gateways and smart sensors. The fog computing based architectures perform multiple functions to coordinate the data processing effectively. They are also placed in the center to balance the application and IoT server data operation.

A cloud provider, one or more fog nodes at the network's edge interacting with the cloud, and a potentially complex network of IoT devices are all common components of fog computing-based IoT applications. Sensor nodes are Internet of Things sensors that can detect a certain quantity (e.g., temperature, humidity, motion, light, and so on) and are located in specific locations known to the nodes. They only respond to requests from Fog nodes. A fog computing IoT application consists of a cloud service provider, one or more fog nodes at the network's edge that are connected to the cloud, and a potentially complicated network of IoT devices beneath them. There are several concerns that must be addressed while using Fog Computing. Privacy and security should be considered as fundamental concerns. The distributed architecture of fog computing poses many challenges to security, including authentication, access control, nefarious node detection, and revocation.

Following is an outline of the remaining sections of this paper. In Sect. 2, we have briefly reviewed the related works. In the Sect. 3, we explain about the IBF based fog computing method. Section 4, deals with the simulation results, and concluding remarks are presented in Sect. 5.

2 Related Works

In article [1], the authors proposed a Double Trap Door Chameleon Hash Function in conjunction with the paillier homomorphic encryption. This method was proposed to increase the security and decrease the delay. By generating two hash functions, double layer security mechanism was provided in preserving the credentials.

In this article [2], the authors revisited a variant of Paillier's scheme. They showed that, when a different subgroup is considered, a different scheme can be obtained, which can be used for interesting applications. In the proposed scheme, the homomorphic cryptosystem security is assumed to be not based on a residuosity-related assumption. In the proposed new trap door scheme, security relied on factorization problem and it proved to be better than many methods.

In article [3], the authors proposed a Data Security Encryption Scheme Based on Identity in Fog Computing The goal of this IBE scheme was to provide safe data delivery to authorised users. They proposed a scheme named as Hierarchical Identity-Based Architecture for Fog Computing (HIBAF). The authors compared the scheme with other cryptography mechanisms and found HIBAF as better than the existing methods in terms of security and privacy.

Heng Chuan Tan et al. [4] presented a methodology for end-to-end integrity. To improve data integrity and Hashing, they employed elliptic curve-based chameleon hashing. They used Meter Data Management System (MDMS) for data verification. It may check the concentrator's data's integrity and validity.

The writers of the article [5] addressed the topic of data outsourcing in the Internet of Things (IoT) to authorised data centres. They demonstrated a blockchain-based distributed cloud architecture with SDN that allows controller fog nodes to be located at the network's edge. This proposal provided low-cost, secure, and on-demand access to the most competitive computing infrastructures in an IoT network.

Attribute-Based Encryption (ABE) was explained in article [6]. ABE allows for a more adaptable approach of implementing access control. It features cryptographic access control, which means that sensitive resources are encrypted and can only be decrypted by organisations with the proper credentials (attributes). As a result, ABE-based systems do not require a centralised infrastructure that matches access policies with entity credentials. ABE has been recognised as the most often utilised way to adding security to untrustworthy, honest-but-curious data stream management systems due to these inherent benefits.

The authors devised an efficient key exchange technique in this study [7] to enable legitimate and secret interactions between groups of fog nodes. The approach used was cypher text-policy attribute-based encryption (CP-ABE). The writers wanted to make sure that the communications of the participants were secure.

In article [8], the authors explained about the process of Data trimming during the integration of Cloud Computing with IoT. They have used Smart Gateway for the implementation. The authors aimed to reduce the burden of cloud computing in through their research.

The Autors provided a general transformation over prime order pairing groups from any affine message authentication code (MAC) to an identity-based encryption (IBE) scheme in this Article [9]. This method aimed to introduce conventional model for tight security through HIBE.

The authors devised an approach to enhance the security in Data Theft during the fog computing process. They proposed an Elliptic Curve Cryptography method along with Decoy technology for enhancing the security and efficiency in this article [10].

3 IBE Based Fog Computing

Fig. 2. Fog structure

Network nodes (IoT devices and fog nodes (Fig. 2)) that are strictly necessary to deliver a service can be recognised and segregated using fog orchestration, protecting the resources of nodes that are not participating in the service and lowering bandwidth consumed to broadcast data. The usage of a network to organise a network monitoring request necessitates that the request be multicast across network nodes. Because the fog nodes and sensors are untrustworthy, security and privacy are jeopardised. To address the security and privacy challenges in fog nodes, we use Enhanced Boneh Franklin-based IBE.

3.1 Identity Based Encryption Using BF Scheme

Adi Samir introduced and proposed the concept of IBC in 1984. This method enables, creating an encryption key without using certificates. In IBE, the unique identity of the user is utilized as a public key. This could be any of the user's public parameters. It might be anything that identifies the user uniquely. These identifications must be converted into 0/1 binary strings. Any string that has been translated to binary can be used as the user's identification. The IBE plan consists of four phases. They are

Setup Phase

The Public Key Generator should set the parameters. The fog nodes are grouped as Bilinear pairing map Groups G1, G2 of prime order r in which $G1 \times G1 \rightarrow G_2$. A trusted Certificate authenticator Public Key Generator is P which is available for each fog group G1.

An encoding function, K_1, which maps the public identities of the user to the elements of the group G.

$$K_2 \colon G_2 \rightarrow \{0, 1\}^n \quad \rightarrow \tag{1}$$

Here n is the message length.

The Public Key Generator's (PKG) master secret key is marked as $S \in Usr$.

The PKG's public key is given by

$$P_{PKG} = sP \quad \rightarrow \tag{2}$$

Registration (Key-Extraction) Phase

A user packet Usr meets the PKG securely. The public identity of user is encoded. This is indicated as

$$P_{usr} = K_1(ID_{usr}) \in G1 \quad \rightarrow \tag{3}$$

where $ID_{usr} = sP_{usr} \in G1$ to Usr.

Encryption Phase

Consider a node-a, which is willing to a packet to the server S. The Node-a message is given by

$$M \in \{0, 1\}^n \quad \rightarrow \tag{4}$$

The fog node computes the message including the encoded data as

$$P_S = K_1(ID_S) \quad \rightarrow \tag{5}$$

The Fog node is identified as group G and

$$G1 = e(P_S, P_{PKG}) \in G_2 \quad \rightarrow \tag{6}$$

A random master key is chosen by the fog node and it is given by

$$a \in Usr \rightarrow \tag{7}$$

Here from Eq. (7)

$$X = aP \text{ and } Y = M \oplus K_2\left(g^a\right) \rightarrow \tag{8}$$

Fig. 3. IBE based fog computing

The cipher text for the message M is the pair

$$(X, Y) \in G1 \times \{0, 1\}^{n} \rightarrow \tag{9}$$

The $K_2(g^a)$ acts as mask to the hide the message M.

Decryption Phase

The decryption phase takes place in the server end. The server S, recovers the message M in the following way from the (X, Y).

$$M = Y + K_2(e(D_S, X)) \rightarrow \tag{10}$$

The Fog nodes, perform encryption of the messages from the sensor nodes in the physical region. This encryption is done with the help of user attributes and the certified authenticator PKG, and the decryption of the cipher text takes place in the server. The flow of the scheme is depiceted in Fig. 3. The IBE based Boneh Franklin scheme is integrated among the fog nodes for effective security and privacy measures.

Identifications in BF Scheme

1. In the setup phase, the user parameters are publicly available.
2. The extraction phase takes as inputs parameters, master-key and this must be done over a secure channel. The public availability of user parameters, enables decryption of cipher texts.
3. The BF schemes makes use of Chosen Cipher texts, proves to reduce the processing time.

4 Results and Discussion

The Network Simulator tool is used to simulate IBF Scheme. It is a free and open source simulator. NS2 involves more updates than other tools. C++ is the base language, while tool command language is the front end (TCL). The wireless network, as well as the ad hoc sensor network, are simple to develop and run. The simulation parameters are depicted in the Table 1 below.

The Simulation

Table 1. Simulation setup

Nodes used	50 nodes
Area size	1200 × 1200 m^2
Transmisson range	200 m
Total simulation period	100 s
Size of the packet	80 bytes
Mac	802.11
Package rate	5 pkt/s

The sensor nodes in the IoT network, forwards the requests to the fog nodes. Their role is to gather the requests, authenticate and encrypt the information using IBF Scheme, then to forward the encrypted information to the server. It decrypts the received data with the assistance from the PKG in the fog node. The performance of IBF scheme is analyzed using various metrics. The Experimental Setup is provided in the Fig. 4 below.

Fig. 4. Data transfer from fog nodes to sensor nodes

Nodes Versus Processing Time

This measure depicts how processing time differs depending on the size of the nodes utilized in both DTCH and IBF. The DTCH scheme uses trap door mechanism with ABE Mechanism. The Processing time is more in the mentioned scheme and it shown in Fig. 5.

Fig. 5. No. of nodes versus processing time

Delay

The graph in Fig. 6 below indicates the amount of time it takes for the user to receive a response to his or her data request. The IBF scheme reduced end to end delay compared with DTCH scheme. The encryption and decryption mechanisms of IBF scheme consumes less time than the Double trap door mechanism of DTCH method.

Fig. 6. No. of nodes versus delay in ms

Packet Delivery Ratio (%)

The data from the user is sent across the system in little packets that are segmented. Throughput is the number of successful packet deliveries made in a given time period to the proper intended destination. The Fig. 7 shows that the DTCH and IBF scheme is analyzed in terms of PDR and IBF has offered better packet delivery ratio than the existing DTCH method.

Fig. 7. No. of nodes versus packet delivery (%)

Energy Consumption

The energy consumption of both the schemes are analyzed in Fig. 8 below. Due to the reduction in processing time, the energy consumed by the number of nodes has considerably reduced in the IBF scheme than the DTCH scheme.

Fig. 8. No. of nodes versus energy consumption (Joules)

5 Conclusion

In fog computing, data security is a key concern because many IoT devices are connected via fog nodes. Our Proposed IBF based fog computing scheme, involves Certified Authenticator based encryption mechanism with the help PKG. The encrypted cipher text is communicated to the server and the decryption mechanism is effectively carried out with the assistance from PKG in Fog network. The proposed scheme's performance was evaluated using necessary metrics, which includes packet delivery ratio, energy consumption, and processing time and end to end delay. The comparison results of IBF is found to be better than the existing method DTCH. The IBF scheme utilizes the public parameters of the User node for the key generation purpose. Though this scheme proves to be effective in multiple aspects, the certified authenticator PKG must be completely trusted and it must be secured. Future research can enhance the trust and security of the public key generator.

References

1. Rhupini, S.K., Preneetha, J., Natraj, N.A.: Double trap door Chameleon hash function based security for fog IoT network. Int. J. Adv. Res. Innov. Ideas Educ. **5**(2) (2019)
2. Bresson, E., Catalano, D., Pointcheval, D.: A simple public-key cryptosystem with a double trapdoor decryption mechanism and its applications. In: Laih, C.-S. (ed.) ASIACRYPT 2003. LNCS, vol. 2894, pp. 37–54. Springer, Heidelberg (2003). https://doi.org/10.1007/978-3-540-40061-5_3
3. Farjana, N., Roy, S., Mahi, M.J.N., Whaiduzzaman, M.: An identity-based encryption scheme for data security in fog computing. In: Uddin, M.S., Bansal, J.C. (eds.) Proceedings of International Joint Conference on Computational Intelligence. AIS, pp. 215–226. Springer, Singapore (2020). https://doi.org/10.1007/978-981-13-7564-4_19
4. Tan, H.C., Lim, K., Keoh, S.L., Tang, Z., Leong, D., Sum, C.S.: Chameleon: a blind double trapdoor hash function for securing AMI data aggregation. In: IEEE 4th World Forum on Internet of Things (WF-IoT), pp. 225–230 (2018)
5. Sharma, P.K., Chen, M.Y., Park, J.H.: A software defined fog node based distributed block chain cloud architecture for IoT. IEEE Access **6**, 115–124 (2018)
6. Sahai, A., Waters, B.: Fuzzy identity-based encryption. In: Cramer, R. (ed.) EUROCRYPT 2005. LNCS, vol. 3494, pp. 457–473. Springer, Heidelberg (2005). https://doi.org/10.1007/11426639_27
7. Alrawais, A., et al.: An attribute-based encryption scheme to secure fog communications. IEEE Access **5**, 9131–9138 (2017)
8. Aazam, M., Huh, E.N.: Fog computing and smart gateway based communication for cloud of things. In: 2014 International Conference on Future Internet of Things And Cloud (Ficloud), pp. 464–470. IEEE (2014)
9. Blazy, O., Kiltz, E., Pan, J.: Identity-based encryption from affine message authentication. In: Garay, J.A., Gennaro, R. (eds.) CRYPTO 2014. LNCS, vol. 8616, pp. 408–425. Springer, Heidelberg (2014). https://doi.org/10.1007/978-3-662-44371-2_23
10. Dong, M.T., Zhou, X.: Fog computing: comprehensive approach for security data theft attack using elliptic curve cryptography and decoy technology. Open Access Libr. J. **3**(09), 1 (2016)
11. Jiang, Y., Susilo, W., Mu, Y., Guo, F.: Ciphertext-policy attribute-based encryption against key-delegation abuse in fog computing. Future Gener. Comput. Syst. **78**, 720–729 (2018)
12. Kahvazadeh, S., Souza, V.B., Masip-Bruin, X., Marn-Tordera, E., Garcia, J., Diaz, R.: Securing combined fog-to-cloud system through SDN approach. In: Proceedings of the 4th Workshop on Crosscloud Infrastructures & Platforms, p. 2. ACM (2017)
13. Li, H., Dai, Y., Yang, B.: Identity-based cryptography for cloud security. IACR Cryptol. Eprint Arch. **2011**, 169 (2011)
14. Rathish, C.R., Natraj, N.A., Sindhuja, P., Manikandan, G.: An energy efficient distributed route selection approach for minimizing delay in WSNs. Solid State Technol. **63**(6), 19714–19723 (2020)
15. Natraj, N.A., Bhavani, S.: SSMDGS: scheduling based stable multicast data gathering scheme for WSN. ARPN J. Eng. Appl. Sci. **12**(24), 7378–7385 (2017)

16. Schridde, C., Dörnemann, T., Juhnke, E., Freisleben, B., Smith, M.: An identity-based security infrastructure for cloud environments. In: 2010 IEEE International Conference on Wireless Communications, Networking and Information Security (WCNIS), pp. 644–649. IEEE (2010)
17. Shamir, A.: Identity-based cryptosystems and signature schemes. In: Blakley, G.R., Chaum, D. (eds.) CRYPTO 1984. LNCS, vol. 196, pp. 47–53. Springer, Heidelberg (1985). https://doi.org/10.1007/3-540-39568-7_5
18. Aravind, A.R., Chakravarthi, R., Natraj, N.A.:Optimal mobility based data gathering scheme for life time enhancement in wireless sensor networks. In 2020 4th International Conference on Computer, Communication and Signal Processing (ICCCSP), pp. 1–5. IEEE (2020)
19. Gopinath, S., Gurumoorthy, K.B., Lakshmi Narayanan, S., Kasiselvanathan, M.: Cluster based optimal energy efficient routing protocol for wireless sensor networks. Revista Geintec-Gestao Inovacao E Tecnologias **11**(2), 1921–1932 (2021)
20. Kraemer, F.A., Braten, N., Tamkittikhun, P.D.: Fog computing in healthcare - a review and discussion. IEEE Access **99**, 9206–9222 (2017)
21. Monteiro, A., Dubey, H., Mahler, L., Yang, Q., Mankodiya, K.: FIT: a fog computing device for speech tele-treatments. In: Proceedings of the IEEE International Conference on Smart Computing (SMARTCOMP), pp. 1–3 (2016)
22. Verma, P., Sood, S.: Fog assisted IoT enabled patient health monitoring in smart homes. IEEE Internet Things J. **5**(3), 1789–1796 (2018)
23. Hamdan, Y.B.: Construction of statistical SVM based recognition model for handwritten character recognition. J. Inf. Technol. **3**(02), 92–107 (2021)
24. Anand, C.: Comparison of stock price prediction models using pre-trained neural networks. J. Ubiquit. Comput. Commun. Technol. (UCCT) **3**(02), 122–134 (2021)
25. Basir, R., et al.: Fog computing enabling industrial internet of things: state-of-the-art and research challenges. Sensors **19**, 4807 (2019)
26. Al Hamid, H., Rahman, S., Hossain, M., Almogren, A., Alamri, A.: A security model for preserving the privacy of medical big data in a healthcare cloud using a fog computing facility with pairing-based cryptography. IEEE Access **5**, 22313–22328 (2017)
27. Natraj, N.A., Bhavani, S.: A certain ınvestigation on secure localization routing protocol for WSN. J. Theoret. Appl. Inf. Technol. **95**(22) (2017)

Automatic Recognition of Plant Leaf Diseases Using Deep Learning (Multilayer CNN) and Image Processing

Abdur Nur Tusher[1](✉), Md. Tariqul Islam[2], Mst. Sakira Rezowana Sammy[1], Shornaly Akter Hasna[1], and Narayan Ranjan Chakraborty[1]

[1] Department of Computer Science and Engineering (CSE), Daffodil International University (DIU), Dhaka 1207, Bangladesh
{abdur15-11632,sakira15-11448,shornaly15-11732}@diu.edu.bd, narayan@daffodilvarsity.edu.bd

[2] Department of Electronics and Communication Engineering (ECE), Khulna University of Engineering and Technology (KUET), Khulna 9203, Bangladesh

Abstract. Bangladesh is highly dependent on Agricultural production on it's economic strength, however different types of diseases have made an enormous damage on the growth of agricultural crops. Bacterial Blight and leaf brown spot (common rust) are the most common diseases affecting the guava, mango, rice, corn and peach plant. Thus, it is become very essential to detect leaf diseases early to protect from damaging the entire crop. The farmers have no enough knowledge about the leaf disease and they used manual process in order to identify disorder. So, the detection accuracy is not good enough and time consuming. An automated and accurate identification system has become essential to overcome this problem. In this paper, a novel technique to diagnose and classify guava, mango, rice, corn and peach diseases has been proposed. One of the effective and modern method for finding the disorder and providing appropriate treatment is deep learning. We have mainly focused on CNN algorithm for training the dataset and found 95.26% accuracy rate. Identification of the diseases would help Bangladesh to grow its economy as it will increase the production rate of guava, mango, rice, corn and peach plants.

Keywords: CNN · Computer vision · Plant leaf disease · Machine learning · Image processing · Deep learning · Bacterial blight · Brown spot · Common rust

1 Introduction

Bangladeshi economical sector stablished base on agricultural product development and a huge proportion of public specially village people direclty or indirectlyt depends on the agricultural work. The backbone of the Bangladesh economy is agriculture which largely depends on the crops grown here in different seasons throughout the year. Guava, Mango, Rice, Corn and Peach is said to be very important crops of Bangladesh. Rice is the main food in Bangladesh, thats why, it is the main agricultural crop in our country.

J. I.-Z. Chen et al. (Eds.): ICIPCN 2022, LNNS 514, pp. 130–142, 2022.
https://doi.org/10.1007/978-3-031-12413-6_11

The cultivation season of rice can be divided into three section (Aush, Aman and Boro) where the duration of Aush is between December and January, boro is between March and April and aush is between July and August. Traditional guava in Bangladesh grows during September and stay for two months and then it is over but the Thai orchards grow guava for almost entire year. In terms of mango, June–July which is hot seasons consider as the peak time for mango production. On the other hand, the corn is mainly cultivated in the season of monsoon and in this time the weather temperature is high rainfall also in increasing pattern. The season is started in March and April and lasted in May and June.

There exist some negative factors which effects largely on farming production. The biggest issues that damages and reduces the quantity and level of quality are leaf diseases which is significant constraints in agricultural production in Bangladesh. There is heavy need of fruits in our country in order to fulfill the requirement of vast population in our country. Hence, farmers are very interested to use pesticides to improve the production of crops that damages the cultivation ecosystems. Sometimes presticides is needed to protect their crops from dangerous insects, but without proper knowledge of our farmers, this use make huge destruction of the effective plants. In order to solve this problem, we can detect the diseases from the plant by using image processing technique. The main target of this research work is to find the guava, mango, rice, corn and peach leaf diseases and suitable cure. We have used deep-learning (CNN) approach in order to train and test the model and we used five types of plant's leaf for supporting this approaches (guava, mango, peach, corn and rice). The disease is for Guava-guava_bacterial_blight, anthracnose, fruit, rot, algal leaf and fruit spot; for Mango-mango_algal_leaf_spot and Powdery mildew Anthracnose; for Rice- rice_brown_spot and bacterial blight; for Peach- peach_bacterial_spot and Shot hole disease; and for Corn- corn_common_rust, corn_cercospora_leaf_spot gray_leaf_spot, Anthroacnose leaf blight, Eyespot and Goss's Wilt.

The bacterial leaf blight disease of guava is caused by Erwinia psidii and because of this disease, the guava tree's twigs, branches and leaves are damaged plentifully. Algal leaf spot is one of the serious diseases which occur during high temperature and rainfall and because of this diseases the mother plant's growth hindrance largely. When the nutritional level of plant is low and drainage facilities is very poor, these types of disease mainly affect the plant. This disease can spread from one to other tree easily through water and wind. As a result, this causes a huge loss of production of mangos. The common and very serious disease of rice is brown leaf spot and this disease mainly occur in leaves, seeds and panicles. This disease become more dangerous when the land is dry and when the fertilizer of land is low and cultivate continuously. The symptom of this disease, little, circular, yellow_brown or brown spots can be found on the seedlings and the leaf sheaths are destroyed. After that the round shape turn into oval spots, purple come from dark_brown, light brown into grey centres and the leaf is surrounded by dark brown margin. When the land is dry, this disease become more dangerous than wetland and causes a huge loss of rice production.

One of the common diseases in corn is Foliar fungal disease (Eyespot, Goss's Wilt). During this disease, the very young leaves faced small Lesions and an oval irregular, water-soaked shape with yellow and riddish brown border color [14, 15]. The maize also faced regularly the powdery, cinnamon brown colored, circular to elongated small pustules on the leaf of the both sides. Day by day the pustules covered the plant with necrotic tissues and the pustules color is orange and initially denoted this on the top surfaces of the leaf. During the heavy infection of leaves with these diseases causes huge damage of production. The scientific name of Shot hole disease of peach is known as 'Coryneumblight' which is a dangerous fungal disease. This disease creates holes which is BB-sized in peach leaves, also creates peach fruit rough areas and branch concentric lesions. The shot hole disease is caused by "Wilsonomyces carpophilus". All the plants ground parts and fruits are affected during the insect attacks, even the stems also affect. However, this disease caused enormous loss in leaves and our systems main motive is to detect and find out some appropriate steps which is very easy and effective for farmer to apply in order to protect their plants.

The system that we made will surely support our farmers to find the plant disorder very efficiently as well as very fastly and by detecting the disease in initial stage the farmers can get rid of those disease. Because our farmers are not educated enough that's why they cannot detect disease using scientific technologies, thus, the farmers who live in our country is unable to use CNN. Instead of using modern technology, they use handmade measuring tools to detect the disease and for making decision use their eye view and guess which is not always correct. But different developed countries around the world, for detecting plant diseases and harvesting their crops with the help of recent scientific technology like Artificial Intelligence, CNN, Image Processing, Deep Learning etc.

2 Literature Review

This part of this paper deals with the previous research work related with plant leaf disorder findings and the authors views to protect these diseases.

Tariqul Islam M., Tusher A.N. (2022) [1] have published a conference paper using CNN model for finding plant (grape, potato and strawberry leaf) diseases. From this paper, it is clear that farmers are faced a huge loss of their crop production due to different kinds of crop disease. Farmers have no advance technology, as a result they cannot identify plant disease accurately. For solve this problem, the authors presented CNN algorithm. They have used around 5500 images and the accuracy level 93.63%. Here they used 5400 images for train the model and 477 images for test the system. Sukanya S. Gaikwada et al. in [2] have presented a conference paper where they used CNN algorithm for identifying guava leaf disease (fungi_affected diseases). In this paper, he presented that the farmer who cultivated guava commercially, faced huge loss due to guava leaf diseases. As a result, they discourage to produce guava commercially. In order so solve this problem, the authors introduced CNN algorithms here. They used around 4000 images as source data and get accuracy 66.3%. In order to corelative resolution authors used AlexNet and SqueezdNet as pre trained approache. Kien Trang et al. in [3] have presented a conference paper where they used plant village dataset for detecting the plant diseases and they focused on image processing mainly (contrast enhancement, transfer learning), neural network as well. Here, they tried to identify plant leaf diseases very short time and accurately so that the farmer can reduce their huge production losses. Authors mainly used neural network and image processing technique. In this presentation, they showed their model accuracy is 88.46% which is higher than the pre trained model accuracy. They also focused on mango dataset for finding mango diseases. S. Ramesh et al. in [4] proposed a modern algorithm (machine learning) in their conference paper. In this paper, authors tried to represent the production loss of rice and the farmer faced huge economical loss in their cultivation due to different types of diseases. Farmers have no advance technology, so they cannot detect diseases accurately. For solving this problem, they tried to introduce machine learning based approach. For completing their work, they used around 300 images as dataset and their obtained accuracy 90%. Umar Ayub in 2018 presented an international conference, where he introduced the Pakistani farmers facing crop diseases and he used Data Mining model [5]. Their main focused is to present Pakistani farmers production losses due to different diseases and the diseases is introduced in leaf for insect attract. In this paper they used three approach such as supporting vector machine, decision tree and neural network.

The main motive of this research is to identify the disease accurately and giving expected solution of the diseases for Guava, Mango, Rice, Corn and Peach plant. We have used defect leaf pictures for our proposed model and provides a very easy detection process of different diseases which is very time saving accurate. Our model (CNN) has the ability to give the outcome which is related with real time approach.

3 Proposed Methodology

For analyzing a picture pixel wise, image processing is the most powerful technique. While the diseases finding with eye only not only very difficult but sometime impossible form affected leaves, by using this mehtod, people can easily find the disorder accurately and effectively. This technology is very simple for those who have knowledge about Guava, Mango, Rice, Corn and Peach diseases. The system takes healthy and unhealthy images of leaf as it's input and after completing it's work gives expected outcome automatically and successfully.

3.1 Flowchart of the Methodology

Image Achievement. For completing research work accurately and effectively, input data or dataset is the fundamental requirement. Since we used image as input data, the resigeing and enhancement of data is an important job for us at the work starting point. The level of correctness of the outcome of research largely depend on the volume of collected data set and if the collected dataset is very big, the result will be very close of the expected outcome. We have collected around 9500 data at our research starting stage and took Guava, Mango, Rice, Corn and Peach field as the most common orinine of the images. Then the collected data was resiged and took as.gif,.bmp,.jpg etc. format in order to getting appropriate result (Fig. 1).

Fig. 1. Flowchart

Fundamental Image Processing. We have subdivided source initially collected images into some folders and 9547 images consider as the appropriate with this model. The entire system divided into two part one is training and the another one is testing where training part took 80% of the total data and testing took remaining data. For this system, we selected guava_bacterial_blight (86 images), guava_healthy (277 images), mango_algal_leaf_spot (260 images), mango_healthy (170 images), rice_brown_spot (523), rice_healthy (1488 images), corn_cercospora_leaf_spot gray_leaf_spot (408 images), corn_common_rust (949 images), corn_healthy (926 images), peach _bacterial_spot (1838 images), peach_healthy (288 images) for completing the operation of training. During the operation of testing, we have also considered guava _bacterial_blight (46 images), guava_healthy (100 images), mango_algal_leaf_spot (197 images), mango_healthy (100 images), rice_brown_spot (400), rice_healthy (400 images), corn_cercospora_leaf_spot gray_leaf_spot (101 images), corn_common_rust (235 images), corn_healthy (230 images), peach_bacterial_spot (455 images), peach_healthy (70 images). We considered 256 * 256 pixels value for reshaping each data. After that we also enhance the image quality and denoising the noise image in order to obtain expected result using the methods of image processing. The dataset considering for our research work shows as example in Fig. 2.

System Architecture. The system can be designed both Using single level or multilevel CNN model. In order to finding good outcome, we considered multilevel approach and the entire system description given here. In the first layer, '1' is consider as ReLu activation function, input shape is (96, 96, 3), filter size is '96', kernel size is 8 × 8, padding is 'SAME' and the strides is (1 × 1).

$$ReLU(X) = MAX(0, X) \tag{1}$$

Fig. 2. Combined dataset

In second layer, the value of ReLu activation funcion is '1', input shape is (96, 96, 3), filter shape is '96', kernel size is 5 * 5, padding is 'SAME' and strides is (1 * 1). The ReLu activation function which we used in the second layer is,

$$\sigma(z) = \frac{e^{z_i}}{\sum_{j=1}^{K} e^{z_j}} \; for\; i = 1, \ldots, k \tag{2}$$

The value of our model learning rate is 0.001 and this value consider as the optimization of ADAM (Fig. 3).

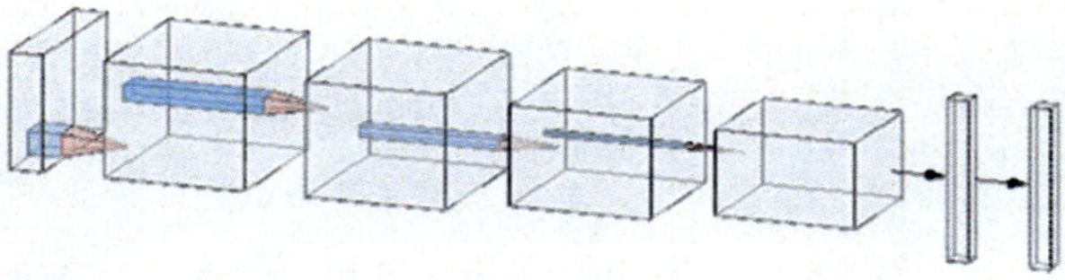

Fig. 3. Proposed convolutional neural network

Optimizer and Learning Rate. It is clear that when we select the optimization approach, computer vision and deep learning output has been changed considerably. The paper which is presented by Adam describe the effectiveness and power of optimization. According this paper, using sub function sub sample data is evaluated and numerous objective funcion is associated in order to produce sub samble. In terms of gradient steps, the algorithm showed it's power of effectiveness and increment [10, 11]. Our model from ADAM optimozer is used 0.001 as the learning rate.

$$V_t = (1 - \beta_2)\Sigma_{i=1}^{t} \beta_2^{t-1} \cdot g_i^2 \tag{3}$$

At present, the Neural network and Cross_Entropy algorithm have been done good jobs in terms of predicton and classification where the Cross_Entroy approach gives better result than classification. Because of the cross-entropy error and the change of weitht, the training is not showed every where. Our method provided a loss function which is given below,

$$L_i = \Sigma_j t_{i \cdot j} \log(p_{i,j}) \tag{4}$$

Image Enhancement. Through the enhancement approach, one image can be divided into different partition and the requirement of this enhancement describes below [12, 13].

- To find the image presentation model which is more simplified and modern.
- To alter the image domain such as shape, phase and magnitude in order to produce huge data.
- To rotate the image with maximum width and height range is 40 and 60 respectively. The range of height and zooming range is 1/5 and the width shifting range is 1/155. The more appropriate data related to this range is given in Fig. 4.
- To control the range of sheer angle considering the direction of counterclockwise and this aprovides the sheard of the image.
- In order to rescale the data, each image is multiplied with appropriate numeric value and the image is in RGB catagory where the coefficient range of each image is 0–255. However, the range of this value is very high with our proposed model. Hence, we target 0 amd 1 as a target range of value and the value is scaled with 1/255 for obtaining that value.

Fig. 4. Image enhancement

3.2 Training the Model

Various types of dataset is used for the training operation and the batch size is approximately 32. The reduction of learning rate is the main tools of the validation accuracy in the time of processing. The working process become manual when 35 epochs of supervision among validation accuracy and reduction of learning rate is completed.

Layer Visualization. This is the process of present the symbol visually with the softly changes images and the visualization is included in Fig. 5.

Fig. 5. Layer visualizatin

4 Result and Discussion

The main task of this section is to present the result of different section of the model like training, testing, validation, error detection and so on. As a result, this part of this paper has super importance than the other section and the description of the result of our proposed algorithm is included here carefully.

4.1 Statistical Analysis

From the model, we are able to get training and validation accuracy 66.76% and 29.88% respectively. The accuracy of the model is improved with the completion of run one by one and after 7th run completion the training and validation accuracy become 91.21% and 82.94% respectively. During this time, the learning rate of the model provides subsiding manner and the value is at 0.0004. When the model completed it's 30 successful run, the validation and training accuracy become 93.06% and 94.81% respectively and the learning rate is 1.0000e−06. During the completion of 50th (final) run, the final accuracy of training and validation are 95.26% and 98.01% respectively.

4.2 Accuracy Graph

In the accuracy graph, we focused mainly four factor training and validation loss, training and validation accuracy. Since we used huge volume of data and all the data are not pure some are noisy, some error we found form our output. So, we used accuracy graph in order to present the loss and accuracy properly, easily and effectively. The entire accuracy graph has two part (upper and lower section). From the upper part, it is clear that this section represent loss function and lower section present accuracy function. The graph present two function training and validation where blue line shows training function and pink color shows validation function. The graph also showed that initially validation loss is very high and accuracy is very low. But with time, validation loss is decreased and accuracy is increased. Same pattern is also maintained by the training line graph. After 10 epochs, we see that all the line become stable and the loss become low and accuracy become high and maintain this pattern till last epochs (Fig. 6).

Fig. 6. Accuracy and loss graph for training and validation

4.3 Confusion Matrix

The error table or error matrix table is sometimes called confusion matrix and this is the best way to present the model performance. This table is created with help of true and false image data and used seperately for all the image data. Table 1 represents the true and false images for each diseases.

The diagonal positional value of the confusion matrix is clearly higher than the any other positional value and the shape of the diagoanl position is (11 $\times$ 11). In terms of color, it is clear that the diagonal position color is more dipper (Blue) than the other positions. This dipper color clearly indicate that the diagonal position maximize the value and provide the more accurate result. Figure 7 represents the confusion matrix and Table 2 presents the classification report.

Table 1. Accurate and error data.

Disease	True	False	Total
Corn_cercospora_leaf_spot gray_leaf_spot	82	19	101
Corn_common_rust	203	5	208
Corn_healthy	230	0	230
Guava_bacterial_blight	27	19	64
Guava_healthy	96	4	100
Mango_algal_leaf_spot	184	13	197
Mango_healthy	61	39	100
Peach_bacterial_spot	432	23	455
Peach_healthy	70	0	70
Rice_healthy	229	171	400
Rice_brown_spot	341	59	400

Fig. 7. Confusion matrix

4.4 Comparison of Result with Other Models

The fundamental and important task for us was to analyze the relevant findings understand and get useful knowledge which has the direct or indirect connection with our proposed model. For that we have compared some their research model to our model to get more clear views of the proposed model. The comparision result clearly showed us that the proposed model provided the result which is more accurate than the other model result and the total views presented in Table 3.

Table 2. Classification report

Disease	Precision	Recall	F1-score	Support
Corn_cercospora_leaf_spot gray_leaf_spot	0.96	0.81	0.88	101
Corn_common_rust	0.98	0.98	0.98	235
Corn_healthy	0.94	1.00	0.97	230
Guava_bacterial_blight	1.00	0.59	0.74	46
Guava_healthy	0.85	0.96	0.90	100
Mango_algal_leaf_spot	0.74	0.93	0.83	197
Mango_healthy	0.79	0.38	0.51	100
Peach_bacterial_spot	0.99	0.95	0.97	455
Peach_healthy	0.74	1.00	0.85	70
Rice _healthy	0.80	0.57	0.67	400
Rice_brown_spot	0.66	0.85	0.75	400
Accuracy			0.84	2334
Macro avg	0.86	0.82	0.82	2334
Weighted avg	0.85	0.84	0.83	2334

Table 3. Accuracy comparison among different dodels

Work	Accuracy (%)
Sharada et al. [6]	85.53
Prem et al. [7]	89.93
Kanabur et al. [9]	79.50
Jyoti and Tanuja [8]	93.00
Tariqul et al. [1]	93.63
Proposed model	95.26

5 Conclusion

The unwanted but unavoidable damage has been faced by our agricultural industry which diminished our fruits production. The farmer cannot detect plant leaf diseases accurately and timely, as a result they faced huge production loss and economical loss. In order to solve this problem, identify plant diseases accurately and effectively, we introduced our algorithm which has the capability to identify plant leaf diseases automatically. Most of our country's farmers are illetrate and they have no advance technology. That's why they used analog and ancient technique to identify leaf diseases which is not only uneffective but also time consuming. By using our technology, farmer can detect their crop disease and increases their production level. We used very powerful model (multiple CNN) for detecting the plant diseases and the result provided our system appriciateable and the accuracy level is 95.26%. In future, we will be created an android application so that farmer can easily used our model. In this apps farmer only need to snap the infected leaf Picture and the apps automatically identify the disease and provide appropriate cure.

References

1. Tariqul Islam, M., Tusher, A.N.: Automatic detection of grape, potato and strawberry leaf diseases using CNN and ımage processing. In: Nanda, P., Verma, V.K., Srivastava, S., Gupta, R.K., Mazumdar, A.P. (eds.) Data Engineering for Smart Systems. Lecture Notes in Networks and Systems, vol. 238. Springer, Singapore (2022). https://doi.org/10.1007/978-981-16-2641-8_20J
2. Gaikwad, S.S.: Identification of fungi infected leaf diseases using deep learning techniques. Turk. J. Comput. Math. Educ. **12**(6), 5618–5625 (2021)
3. Trang, K., TonThat, L., Gia Minh Thao, N. Tran Ta Thi, N.: Mango diseases ıdentification by a deep residual network with contrast enhancement and transfer learning. In: 2019 IEEE Conference on Sustainable Utilization and Development in Engineering and Technologies (CSUDET), pp. 138–142 (2019). https://doi.org/10.1109/CSUDET47057.2019.9214620
4. Ramesh, S., Vydeki, D.: Rice blast disease detection and classification using machine learning algorithm. In: 2018 2nd International Conference on Micro-Electronics and Telecommunication Engineering (ICMETE), pp. 255–259 (2018). https://doi.org/10.1109/ICMETE.2018.00063
5. Ayub, U., Moqurrab, S.A.: Predicting crop diseases using data mining approaches: classification. In: 2018 1st International Conference on Power, Energy and SmartGrid (Icpesg). https://doi.org/10.1109/Icpesg.2018.8384523
6. Kambale1, G., Bilgi, N.: A Survey Paper On Crop Disease Identification And Classification Using Pattern Recognition And Digital Image Processing Techniques. May (2017)
7. Prem Rishi Kranth, G., Hema Lalitha, M., Basava, L., Mathur, A.: Plant disease prediction using machine learning algorithms. Int. J. Comput. Appl. **182**(25) (2018)
8. Jyoti, J.B., Tanuja, S.Z.: Cotton plant leaf diseases identification using support vector machine. Int. J. Recent Sci. Res. **8**(12), 22395–22398 (2017)
9. Kanabur, V., Harakannanavar, S.S., Purnikmath, V.I., Hullole, P., Torse, D.: Detection of leaf disease using hybrid feature extraction techniques and CNN classifier. In: International Conference on Computational Vision and Bio Inspired Computing, pp. 1213–1220. Springer, Cham (2019)
10. Akila, M., Deepan, P.: Detection and classificationof plant leaf diseases by using deep learning algorithm. In: International Journal Of Engineering Research & Technology (Ijert) Issn: 2278-0181 Published By, www.Ijert.Org Iconnect—2k18 Conference Proceedings (2018)
11. Deb, S., Islam, S.M.R., RobaiatMou, J., Islam, M.T.: Design and implementation of low cost ECG monitoring system for the patient using smart device. In: 2017 International Conference on Electrical, Computer and Communication Engineering (ECCE), pp. 774–778). IEEE, February (2017)
12. Islam, M.T., Islam, S.M.R.: A new ımage quality ındex and it's application on MRI image (2021)
13. Sungheetha, A., Rajesh Sharma, R.: Classification of remote sensing image scenes using double feature extraction hybrid deep learning approach. J. Inf. Technol. **3**(02), 133–149 (2021)

14. Karuppusamy, P.: Building detection using two-layered novel convolutional neural networks. J. Soft Comput. Paradigm **3**(01), 29–37 (2021)
15. Islam, S.M.R., Islam, M.T., Huang, X.: A new approach of image quality index. In: 2017 4th International Conference on Advances in Electrical Engineering (ICAEE), pp. 223–228 (2017). https://doi.org/10.1109/ICAEE.2017.8255357

Comparative Analysis of Feature and Intensity Based Image Registration Algorithms in Variable Agricultural Scenarios

Shubham Rana[1](✉), Salvatore Gerbino[1], Pragya Mehrishi[2], and Mariano Crimaldi[3]

[1] Department of Engineering, University of Campania 'Luigi Vanvitelli', Via Roma, 29, 81031 Aversa, Campania, Italy
{shubham.rana,salvatore.gerbino}@unicampania.it

[2] Department of Physical Geography and Geoecology, Charles University, Albertov, 6, 12843 Praha 2, Czechia
mehrishp@natur.cuni.cz

[3] Department of Agricultural Sciences, University of Naples Federico II, Via Università, 100, 80055 Portici, Naples, Italy
mariano.crimaldi@unina.it

Abstract. Image registration has widespread application in fields like medical imaging, satellite imagery and agriculture precision as it is essential for feature detection and extraction. The extent of this paper is focussed on analysis of intensity and feature-based registration algorithms over Blue and RedEdge multispectral images of wheat and cauliflower field under different altitudinal conditions i.e., drone imaging at 3 m for cauliflower and handheld imaging at 1 m for wheat crops. The overall comparison among feature and intensity-based algorithms is based on registration quality and time taken for feature matching. Intra-class comparison of feature-based registration is parameterized on type of transformation, number of features being detected, number of features matched, quality and feature matching time. Intra-class comparison of intensity-based registration algorithms is based on type of transformation, nature of alignment, quality and feature matching time. This study has considered SURF, MSER, KAZE, ORB for feature-based registration and Phase Correlation, Monomodal intensity and Multimodal intensity for intensity-based registration. Quantitatively, feature-based techniques were found superior to intensity-based techniques in terms of quality and computational time, where ORB and MSER scored highest. Among intensity-based methods, Monomodal intensity performed best in terms of registration quality. However, Phase Correlation marginally scored less in quality but fared well in terms of computational time.

Keywords: Multi spectral (MS) · Feature-based · Intensity-based · Speeded Up Robust Features (SURF) · Maximally Stable Extremal Regions (MSER) · KAZE · Oriented Fast and Rotated Brief (ORB) · Monomodal · Multimodal · Phase correlation · Mutual information (MI) · Sum of squared differences (SSD) · Normalized cross correlation · Control points (CP) · Scale-invariant (SI) · Feature transform (SIFT) · Modified difference local binary (MDLB)

J. I.-Z. Chen et al. (Eds.): ICIPCN 2022, LNNS 514, pp. 143–160, 2022.
https://doi.org/10.1007/978-3-031-12413-6_12

1 Introduction

Image registration is the process of alignment of images belonging to the same scene in an overlayed fashion to geometrically align them for purpose of feature detection and feature extraction in applications like medical imaging, remote sensing or even astrophotography. They can either be images captured over the same scene in different wavelength or temporal images of the same scene. The registration is done using two components: a fixed reference image and a moving image. The goal is to geometrically align the moving image over the fixed image. When the images are captured using different sensors (different wavelengths or spectra) over the same scene, the nature of registration applicable is multi-modal.

Multiple lens-based multispectral (MS) imaging cameras use different sensors and spectral filter combinations to achieve desired image resolution and feature combinations. The sensor used for this study and its specifications are illustrated in Fig. 1 and Table 1.

Fig. 1. MicaSense RedEdge-M MS camera

Table 1. Technical specifications of MicaSense RedEdge-M

Weight	170 g
Dimensions	3.7 inches × 2.5 inches × 1.8 inches
External power	4.2 V minimum – 15.8 V maximum
Bands composition	Red, Green, Blue, Near IR and Red Edge
Spatial resolution	8.2 cm/pixel at 120 m altitude
Frame rate	1 image/second
Wavelength (nm)	Red (668 nm), Green (560 nm), Blue (475 nm), Near-IR (840 nm), Red Edge (717 nm)

In the context of agriculture precision, registration is required for extraction of certain crop characteristics which are based on several spectral band ratios indices. In order to render accurate results with respect to a specific patch of crop, weed or soil under study, the analysis has a prerequisite of pixel-to-pixel overlapping through a stack of multi-spectral images. This is only possible through image registration.

This study is aimed at evaluation and comparison of four algorithms of Feature-based image registration Speeded Up Robust Features (SURF), Maximally Stable Extremal Regions (MSER). KAZE and Oriented Fast and Rotated Brief (ORB) and three methods of Intensity-based image registration (Phase Correlation, Monomodal intensity and Multimodal intensity) of the application Registration Estimator under Image Processing and Computer Vision toolbox of MATLAB R2021b. The underlying goal is to guide the user about optimal parameters necessary for registering ground as well as drone-based multispectral images for an agricultural scene. Feature-based image registration methods search for corresponding points, curves and surface model. Whereas, Intensity-based image registration methods search for image grayscale values without subjugation to relative sparsely extracted information [1]. The modus operandi of intensity-based methods is based on searching in a defined space of transformations. The images are non-georeferenced and multispectral blue and RedEdge images acquired over cauliflower and wheat crop fields at 3 m with the drone and 1 m handheld altitude above ground level respectively.

There is a need to understand what kind of specific registration is suitable for registration of images observed in different wavelengths and at different altitudes, particularly in the field of agriculture. Since crops have high spectral intra-class and inter-class heterogeneity, it opens a room for exploring what hyperparameters affect the quality and time taken for registration across feature-based and intensity-based methods.

2 Literature Review

The current study has employed 7 techniques of image registration, as available in the Image Processing and Computer Vision Toolbox based application Registration Estimator of Matlab 2021b. The conceptual details about each method is described in the following sub-section. Feature-based image registration is composed of five stages: detection of features (identification of points of interest), matching of features (point-to-point mapping of reference image and subject image), rejection of outliers (mismatched points are rejected) and resampling through a geometric transformation model before an image is reconstructed (Fig. 2) [2]. The Intensity-based registration is composed of five stages: Interpolation, determination of similarity measures, optimization and geometric transformation for the reconstruction of the image (Fig. 3).

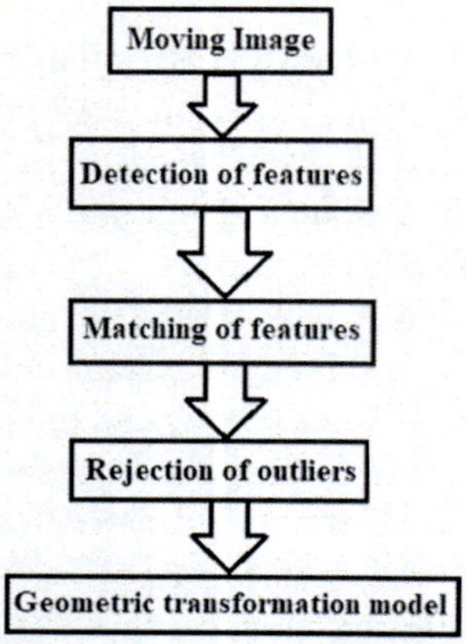

Fig. 2. Feature-based image registration

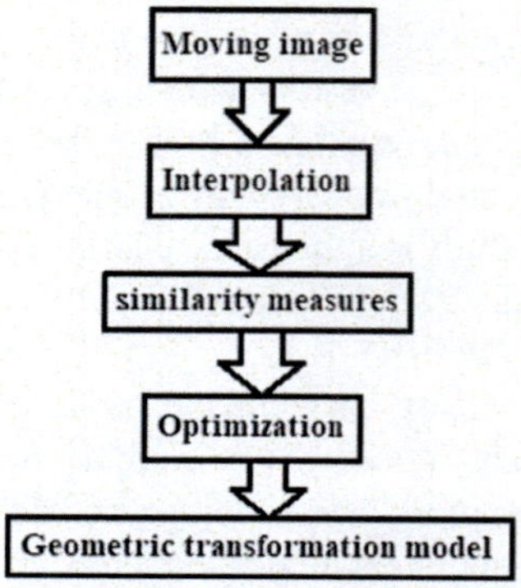

Fig. 3. Intensity-based image registration

2.1 Overview of Feature-Based Image Registration Algorithms

Feature-based image registration methods detect distinct image features, making use of high pass filters like sharp edges and regions with homogenous intensity levels. The moving image is subjected to single global transformation in order to provide the best alignment with corresponding features in the fixed image.

The Feature-based image registration starts with detection of mutual features which are found overlapping in subject image (moving) and reference image (fixed). Subsequently, the features are matched and corresponding costs among the features of the image pair are determined. Next, the unmatched features are rejected as outliers. Lastly, a geometric transformation such as projective, similarity, affine or similarity is applied to obtain the registered image.

The Intensity-based image registration requires a moving image to be interpolated either using bilinear interpolation or cubic convolution. Now, the similarity measures are determined during which the optimization criteria for intensity parameters are sought. In this step, the tolerance of gradient magnitude, minimum-maximum range step length, maximum epochs, relaxation factor and pyramid levels are determined. In the last stage, a transformation model based on either geometrical alignment of centres or centre of mass is applied to obtain the registered image.

2.1.1 Speeded Up Robust Feature Detector

The speedup and rotation-invariant point feature (SURF) algorithm [3] is based on scale and rotation-invariant detectors and descriptors. The features are recorded from coarse to fine textural characteristics through hierarchical multilayer pyramid analysis approach which makes integration of scale and space robust in the images [4]. This attribute makes it superior to SIFT from the perspective of feature extraction [5]. The approximation approach on surrounding pixels envisages edge detection and ameliorates real-time performance. Speedup Robust Features (SURF) was introduced in 2008 and is based on Gaussian scale-space image analysis [6]. Through use of integral images, its inherent detector supplements feature detection speed via projection of determinant through Hessian Matrix. Its intrinsic 64-bin descriptors are derived from box-based convolutional filters [7]. They describe and characterize features using Haar wavelets which are dispersed in proximity [8]. However, rotation and scale uniformity are visible characteristics, but a moderate variance is observed with affine transformation [9]. In order to accustom larger field-of-view changes, the descriptors bear a possibility of extension up to 128 bins. As compared to SIFT, SURF has been found computationally inexpensive. Equation (1) gives a mathematical insight about the Hessian Matrix, H at point $z = (x, y)$ at a scale factor σ:

$$H(z, \sigma) = \begin{bmatrix} L_{xx}(z, \sigma) & L_{xy}(z, \sigma) \\ L_{xy}(z, \sigma) & L_{yy}(z, \sigma) \end{bmatrix} \tag{1}$$

where $L_{xx}(z, \sigma)$ represents convolution of second order derivative in the image at point x, and, respectively for $L_{xy}(z, \sigma)$ and $L_{yy}(z, \sigma)$.

2.1.2 Maximally Stable Extremal Regions

The Maximally Stable Extremal Regions (MSER) algorithm is based on rotational invariant feature detector, i.e. based on complex moments. This results in invariance when image intensities are subjected to affine transformation [10]. Therefore, it is appropriate for image retrieval as it is independent of regions-of-interest per image [2, 11]. MSER applies the covariance matrix for the preservation of adjacent features [12]. Using thresholding at certain intervals, it only selects extreme regions in an image area which are potentially and virtually stable [13]. The stability function for a region under operation is calculated by the normalized area of the connected component in proportionality to the area under change [14]. Due to the absence of smoothing phenomenon, large structures as well as fine details are easily detected [15]. Its scale normalization property is based on recorded extreme regions in an image [16]. There are plenty of applications based on MSER like object recognition, object tracking, mapping of colour and object detection in volumetric images. It is also inclusive of sub-sampling capability and blurring functions [17, 18]. The detector is limited to poor performance in case geometry of levelling line fluctuates due to blur transformation [19]. All extreme regions in an image can be enumerated as O(n log log n), where n is the total number of pixels in the image [20].

2.1.3 Oriented FAST Rotated BRIEF

[21] devised Oriented FAST and Rotated BRIEF (ORB) in 2011. Its algorithm is a conglomeration of improved FAST (Features from Accelerated Segment Test) based feature detection [22] and BRIEF (Binary Robust Independent Elementary Features) [23] description methods with normalized direction. In comparison to SIFT and SURF, this technique upgrades the efficiency of computation and offers upscaled real-time performance [24]. FAST is used for corner detection in each underlying layer of the scale pyramid. Using points detected from the corners, top-quality points are filtered out using Harris Corner score where a corner point is only indicated provided that the neighbour substantially differentiates a pixel [25]. Due to BRIEF descriptor's instability with rotation, a modified descriptor has been employed to balance trade-off [26]. The brightness difference between pixels is determined using detected corner points of

FAST [27]. Directional information is absent mainly due to large difference in quantity and unpredictability. Mathematically, Harris corner score is calculated as illustrated in Eqs. (2) and (3)[2]:

$$H = \det(W) - f(trace(W))^2 \tag{2}$$

$$W = \sum i(x, y)\begin{bmatrix} F_x^2 & F_xF_y \\ F_xF_y & F_y^2 \end{bmatrix} \tag{3}$$

Here, H represents Harris corner score, f ranges between 0.04 and 0.06, W represents a 2 × 2 matrix, $i(x, y)$ is the window image function and F represents variation of horizontal and vertical feature points.

2.1.4 KAZE

KAZE features based algorithm was developed by [28] in 2012. An improvised successor, Accelerated-KAZE (AKAZE) algorithm [29], also based on non-diffusive filtering was presented in 2013. Its framework consisted of non-linear scale spaces built using Fast Explicit Diffusion (FED). This property makes image blurs versatile to feature points thereby preserving boundary regions. KAZE detectors are based on the Hessian matrix determinant whereas AKAZE uses Modified Local Difference Binary (MDLB). Scharr filtering makes AKAZE superior to KAZE, both in terms of rotational invariance and scale changes. Feature points are chosen among maximum values of the detector responses [2].

2.2 Overview of Intensity-Based Image Registration Algorithms

Intensity-based registration techniques correlate image intensity in the spatial or frequency domain. The moving image undergoes a single global transformation to maximize the correlation of its intensity with the intensity of the fixed image.

It is an iterative workflow and is based on specific pair of images, a metric, an optimizer, and a transformation model. The metric defines image similarity measures for the evaluation of registration accuracy. The quality of registration depends on the similarity metric which returns a scalar value that describes the level of correlation among the images. The optimizer is delegated with the task of minimizing or maximizing the similarity measure. The transformation model defines the nature of 2-D transformation that determines the alignment of misaligned (moving) and reference (fixed).

After the type of transformation is determined, the user is given the option to specify an internal transformation matrix. Altogether, specific image transformation is applied to the misaligned image using bilinear interpolation. Successively, the metric compares the transformed misaligned image with the reference image and a scalar metric value is computed.

2.2.1 Phase Correlation

Phase correlation registers images in the frequency domain [30]. It is robust to image brightness and noise as compared to other intensity-based image registration methods. Phase congruency feature has been found adaptive to different illumination conditions and variations in contrast levels [31]. In a nutshell, the structural features and contour of the image elements is attributed to phase information.

Phase congruency is a feature detection method based on the local phase of a reference image. This affirms the presence of features such as corners and edges where the Fourier components are maximal in phase. This model is similar to how the human visual mechanism perceives image features [30]. Its behaviour of invariance to changes in contrast and illumination is attributed to the independence of signals amplitude [31]. For a given signal $t(x)$, the Fourier equivalent ist$(x) = \sum_n A_n \cos(\phi_n(x))$. Here, A_n is amplitude of nth Fourier component and ϕ_n is its local phase at position x. Mathematically, the Phase correlation becomes:

$$PC_1(x) = max_{\varphi(x)\in[0,2\pi]}\left\{\frac{\Sigma_n A_n(x)\cos\left(\varnothing n(x) - \varnothing^-(x)\right)}{\Sigma_n A_n(x)}\right\} \tag{4}$$

2.2.2 Monomodal Intensity

Monomodal intensity-based image registration is used to register images within similar brightness and contrast ranges that have been captured using a single sensor. They are mainly used in applications based on successive imaging. For example, registration of Magnetic Resonance Imaging scans. It is based on mutual information (MI) which determines the statistical relationship between two image variables. It was proposed by Viola and Wells in 1997 [32]. The mutual information is calculated as illustrated in Eqs. (5) and (6):

$$MI(X, Y) = E(X) + E(Y) - E(X, Y) \tag{5}$$

$$NMI(X, Y) = \frac{E(X) + E(Y)}{E(X, Y)} \tag{6}$$

where X and Y are images and MI represents the mutual information. NMI is Normalized mutual information, $E(X)$ and $E(Y)$ are individual entropies and $E(X, Y)$ are joint entropies of X and Y. Although normalized mutual information is similar in performance to mutual information, it is indicative for inspection of the influence due to change of overlapping parts during monomodal registration [33]. Since it is invariant to overlap, it has been found more stable and robust in monomodal intensity-based registration of various dimensions and across different modalities.

Monomodal image registration is often found limited to temporal images, where the images have same pixel sizes in respective dimensions. The interpolation effects have been found austere for monomodal registration. These perturbations are caused by local maxima created due to variable entropy with rapidly changing overlap of images. Considering a test case for image registration, where the reference and misaligned image are same, resulted in poorest interpolations recorded by the function for translation along every dimension [33]. So, the ideal case is monomodal image registration without interpolation.

2.2.3 Multimodal Intensity

Multimodal intensity-based image registration registers images using different combination of brightness and contrast range, which are captured using two different types of sensors. To quote an example, they can be different camera models or state of the art medical imaging components like Computed Tomography and Magnetic Resonance Imaging. The images may be sourced from a single sensor using different exposure settings. Another scenario could be MRI images acquired in a single session in a burst fire setting mode. This is the basis for multiple modalities. This method considers area or template matching of a predefined size for the detection of control points (CP) between two images. Once the size of template window is determined, the corresponding window searches for the matching area over the misaligned image using certain similarity measures. Alignment is only determined based on the geometric centre or the centre of mass control point after the area is matched for two images.

Some of the mostly used similarity measures for multimodal registration include normalized cross correlation (NCC), sum of squared differences (SSD) and the mutual information (MI) [31]. SSD works on direct computation of intensity differences between two images to determine control points. Despite being computationally efficient, SSD has shown sensitivity to radiometric changes. NCC, on the other hand, is widely applied in image registration for satellite remote sensing as it is invariant to uniform intensity variations [34]. However, NCC has been found prone to non-uniform radiometric perturbations [34]. MI computes the joint histogram of each template window, thereby making the similarity measure computationally expensive and sensitive to window size. Due to the limitation of these similarity measures towards radiometric distortion, multimodal intensity-based method has not been successful in remote sensing image registration.

3 Experiments and Results

3.1 Experimental Setup

MATLAB-2022a based **Registration Estimator** has been used for performing the experiments presented in this article. Specifications of the work machines are: AMD Ryzen 9 5900HX @ 4.6 GHz, 16 MB Cache and 16.00 GB RAM. All remaining parameters are used as registration estimator's default for testing of image datasets (Table 2). Observations are shown in Figs. 6, 7 and 8.

Table 2. Experiment settings and dataset details

No. of test images	4
Images bands under study	Blue (Fig. 4) and RedEdge (Fig. 5)
Crops under study	Cauliflower and wheat
Date of cauliflower imagery	25 November 2020
Altitude for cauliflower imagery	3 m
Altitude for wheat imagery	1 m
Date of wheat imagery	13 January 2022
Image dimensions	1080 × 960
Horizontal and vertical resolution	96 dpi
Focal length	6 mm
Image overlap	80%
Work machine specs	AMD Ryzen 9 5900HX @ 4.6 GHz, 16 MB Cache, 16 GB RAM
Application	Registration Estimator of Image Processing and Computer Vision Toolbox, Matlab 2021b
Study area location	Portici, Campania Region, Italy

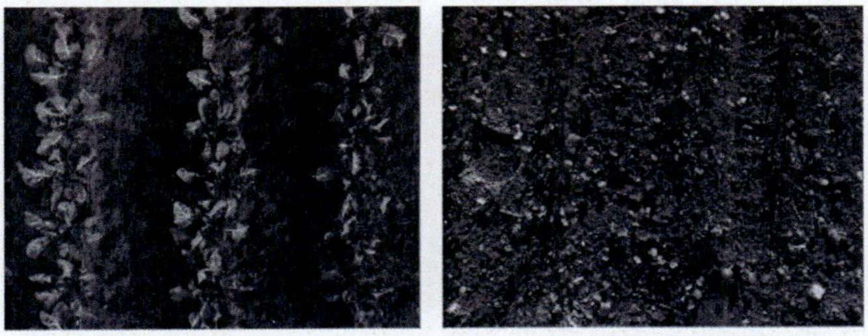

Fig. 4. Blue image band of cauliflower (left) and wheat (Right)

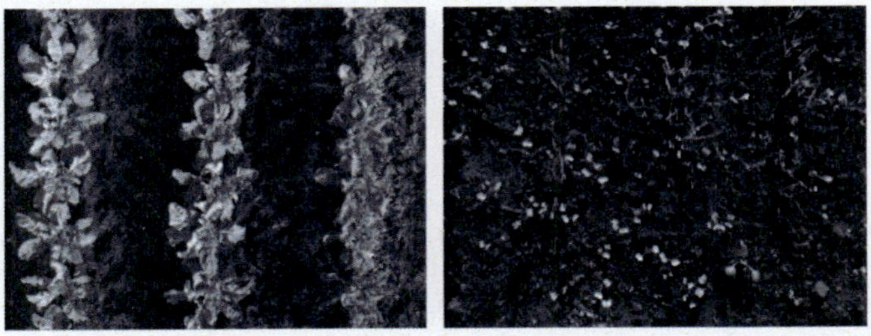

Fig. 5. Red-edge image band of cauliflower (left) and wheat (Right)

Fig. 6. (a) Comparison of feature matching time across feature-based registration algorithms over cauliflower crop (3 m) and wheat crop (1 m). (b) Detailed comparison of feature matching time across feature-based registration algorithms using different transformation models over cauliflower crop (3 m) and wheat crop (1 m). (c) Comparison of registration quality across feature-based registration algorithms over cauliflower crop (3 m) and wheat crop (1 m). (d) Detailed comparison of registration quality across feature-based registration algorithms using different transformation models over cauliflower crop (3 m) and wheat crop (1 m)

Fig. 6. continued

4 Results and Discussion

A. Feature matching time of crop image objects captured at different altitudes using Feature-based registration algorithms

All four Feature-based image registration algorithms, namely, ORB, MSER, SURF and KAZE were found to be computationally efficient for feature matching time in the crop imagery of cauliflower, which was acquired at 3 m altitude using drone. Parametrically, the Rigid transformation model with Edge diffusion and SharpEdge diffusion of KAZE fared best in terms of computational time for registration of cauliflower images at 3 m.

Three Feature-based registration algorithms, namely, MSER, SURF and KAZE were found to be computationally efficient for feature matching time in the crop imagery of wheat, which was acquired at 1 m in a handheld manner. Parametrically, the Projective transformation model of SURF and the Similarity transformation model of MSER were recorded as the best in terms of computational time for registration of wheat images at 1 m.

B. Quality of registration for crop image objects captured at different altitudes using Feature-based registration algorithms

The best quality of registration in case of cauliflower images was achieved using ORB, particularly with the Projective transformation model. The second-best quality of registration was exhibited by MSER, particularly with the Affine transformation model.

In case of wheat, the best quality of registration was achieved with two algorithms, namely, ORB and MSER. Parametrically, the Projective transformation model of ORB and the Affine transformation model of MSER scored best in terms of computational time.

Fig. 7. (a) Comparison of template matching time across intensity-based registration algorithms over cauliflower (3 m) and wheat (1 m). (b) Detailed comparison of template matching time across intensity-based registration algorithms using different transformation models over cauliflower (3 m) and wheat (1 m). (c) Comparison of registration quality across intensity-based registration algorithms over cauliflower (3 m) and wheat (1 m). (d) Detailed comparison of registration quality across intensity-based registration algorithms using different transformation model

Fig. 7. continued

Fig. 8. (a) Comparative analysis of feature and intensity-based image registration methods on registration time over cauliflower crop (3 m) and wheat crop (1 m). (b) Comparative analysis of feature and intensity-based image registration methods on registration quality over cauliflower crop (3 m) and wheat crop (1 m)

C. Template matching time of crop image objects captured at different altitudes using Intensity-based registration algorithms
The best timings in terms of the area / template matching, particularly for the images of cauliflower, captured at 3 m altitude, were recorded for Phase Correlation. Parametrically, the Translation-based transformation model scored best followed by the Similarity transformation model in terms of computational time.

The best timings in terms of the area / template matching, particularly for the images of wheat, captured at 1 m altitude, were recorded for Phase Correlation. Parametrically, the Translation-based transformation model scored best in terms of computational time.

D. Quality of registration for crop image objects captured at different altitudes using Intensity-based registration algorithms
The best quality of registration in case of cauliflower images was achieved using Monomodal Intensity-based image registration, particularly with the Centre of mass setting for alignment of image centres and Rigid transformation model. The second-best parameter was the Translation-based transformation model with Centre of mass setting for alignment of image centres,

The best quality of registration in case of wheat images was achieved using Monomodal Intensity-based image registration, particularly with the Geometric setting for alignment of image centres and Affine transformation model. The second-best parameter was the Affine transformation model with Centre of mass setting for alignment of image centres.

E. Comparative analysis of feature and intensity based image registration methods
Overall, the best image registration quality and computational efficiency for cauliflower imagery (3 m) was observed for ORB Feature-based registration with Similarity transformation model.

The best image registration quality and computational efficiency for wheat imagery (1 m) was observed for MSER Feature-based registration with Affine transformation model.

5 Conclusion and Future Scope

This study represented a detailed comparison of Feature and Intensity-based image registration methods as provided by the Registration Estimator tool of Image Processing and Computer Vision toolbox in Matlab 2021b. The subjects chosen were multispectral image bands from different imaging conditions of wheat and cauliflower crops. Although, all four Feature-based image registration methods outperformed Intensity-based methods in terms of image registration quality and total time of computation, a selection of specific transformation models and coupled diffusion parameters were identified for achieving desirable registration with highest registration quality and computational budget. It was also observed that Monomodal Intensity-based transformation model with Geometric and Centre of mass alignment of image centres can be a potential qualitative method if its time complexity is reduced. The scope of this study is to provide a graphical interpretation for parametrized comparison of Feature and Intensity-based image registration algorithms at inter-class and intra-class level.

6 Supplementary Material

Table S1. Feature based image registration algorithms

	Cauliflower (3 m)				Wheat (1 m)			
Parameter	Quality (PSNR)	Feature matching time (ms)	Number of features matched	Number of features detected	Quality (PSNR)	Feature detection time (ms)	Number of features matched	Number of features detected
ORB								
Rigid	**0.4619**	29.52	197	36415, 23659	**0.43039**	314.51	1157	25658, 61673
Similarity	0.72201	**7.2697**	197	36415, 23659	0.72921	**256.46**	1157	25658, 61673
Projective	0.71489	**34.186**	197	36415, 23659	0.**76694**	257.76	1157	25658, 61673
Affine	**0.72394**	8.3369	197	36415, 23659	0.75468	**384.31**	1157	25658, 61673
SURF								
Rigid	**0.44475**	**58.018**	201	4249, 2485	**0.36976**	**173.28**	201	2847, 5972
Similarity	0.7026	9.9602	201	4249, 2485	0.71035	14.24	201	2847, 5972
Projective	0.71288	26.065	201	4249, 2485	0.74452	**10.854**	201	2847, 5972
Affine	**0.71754**	**9.8376**	201	4249, 2485	**0.76135**	28.454	201	2847, 5972
MSER								
Rigid	**0.45344**	**31.941**	59	5070, 3026	**0.32607**	**53.964**	99	6666, 10940
Similarity	0.71356	**8.351**	59	5070, 3026	0.72582	**0.008565**	99	6666, 10940
Projective	**0.71779**	27.227	59	5070, 3026	0.75748	0.012164	99	6666, 10940
Affine	0.71462	9.4386	59	5070, 3026	**0.76563**	0.021529	99	6666, 10940
KAZE								
Projective, Region	0.70109	16.464	68	1336, 460	0.70213	**0.081074**	88	1043, 1236
Projective, SharpEdge	0.71494	12.488	68	1367, 487	**0.73492**	0.046675	56	1015, 1337
Projective, Edge	0.66823	12.811	56	1273, 433	0.72544	0.051516	56	1007, 1258
Rigid, Region	0.69393	11.418	68	1336, 460	**0.38355**	0.057383	88	1043, 1236
Rigid, SharpEdge	**0.71196**	**8.131**	68	1367, 487	0.38386	0.044364	56	1015, 1337
Rigid, Edge	0.69481	9.8659	56	1273, 433	0.40686	0.044314	56	1007, 1258
Similarity, Region	0.45803	30.177	68	1336, 460	0.70255	0.060518	88	1043, 1236
Similarity, SharpEdge	0.45636	29.735	68	1367, 487	0.6856	**0.021031**	56	1015, 1337
Similarity, Edge	**0.45587**	36.762	56	1273, 433	0.6917	0.021508	56	1007, 1258
Affine, Region	0.61404	47.146	68	1336, 460	0.7069	0.028588	88	1043, 1236
Affine, SharpEdge	0.69633	33.079	68	1367, 487	0.71943	0.023426	56	1015, 1337
Affine, Edge	0.661	**51.92**	56	1273, 433	0.714	0.021313	56	1007, 1258

Table S2. Intensity based image registration algorithms

	Cauliflower (3 m)		Wheat (1 m)	
Parameter	Quality (PSNR)	Feature matching time (ms)	Quality	Feature matching time (ms)
PHASE CORRELATION				
Translation	**0.45557**	**167.89**	**0.65286**	**1440.6**
Similarity	**0.69697**	**901.51**	**0.65335**	**22787.3**
MONOMODAL INTENSITY				
Geometry, Similarity	**0.40731**	**17866.6**	0.65335	22698.4
Geometry, Affine	0.4594	15040.5	**0.75526**	25979
Geometry, Rigid	0.43203	**2794.2**	**0.2079**	22437.7
Geometry, Translation	0.67747	15464.1	0.24884	5178.1
Centre of mass, Similarity	0.43203	2904.9	0.47492	22857
Centre of mass, Affine	0.45979	14686	0.75517	**25987.3**
Centre of mass, Rigid	**0.72391**	17314.6	0.44872	21591.7
Centre of mass, Translation	0.71506	15239.9	0.39707	**4896.1**
MULTIMODAL INTENSITY				
Geometry, Similarity	0.49199	12408.8	**0.61846**	16647.9
Geometry, Affine	0.41677	12337.9	**0.26958**	16533
Geometry, Rigid	0.45764	12409.5	0.40302	16738.2
Geometry, Translation	0.46275	12396.5	0.41996	**16235.4**
Centre of mass, Similarity	**0.49417**	17113.6	0.48238	16673.2
Centre of mass, Affine	0.32194	17288.3	0.39371	16732.2
Centre of mass, Rigid	**0.089066**	**10577.9**	0.39819	**16766.1**
Centre of mass, Translation	0.45981	**18021.4**	0.4165	16422

Acknowledgement. The completion of this analysis would not have been accomplished without the support of our team at Department of Agriculture Sciences, University of Napoli Federico II. We are grateful for all the data resources that they have been sharing with us for research purpose. It's a heartfelt thanks from us for all the opportunities and resources that have been provided by them.

References

1. Boda, S.: Feature-Based Image Registration (2009)
2. Alam, M.S., Morshidi, M.A., Gunawan, T.S., Olanrewaju, R.F.: A comparative analysis of feature extraction algorithms for augmented reality applications. In: 2021 IEEE 7th International Conference of Smart Instrumentation, Measurement and Application, ICSIMA 2021, pp. 59–63 (2021). https://doi.org/10.1109/ICSIMA50015.2021.9526295
3. Sedaghat, A., Mohammadi, N.: ISPRS journal of photogrammetry and remote sensing uniform competency-based local feature extraction for remote sensing images. ISPRS J. Photogramm. Remote Sens. **135**, 142–157 (2018). https://doi.org/10.1016/j.isprsjprs.2017.11.019
4. Sharma, V., Mir, R.N.: A comprehensive and systematic look up into deep learning based object detection techniques: a review. Comput. Sci. Rev. **38**, 100301 (2020). https://doi.org/10.1016/j.cosrev.2020.100301
5. Banerjee, A., Mistry, D.: Comparison of feature detection and matching approaches: SIFT and SURF. GRD Journals-Global Res. Dev. J. Eng. **2**, 7–13 (2017)
6. Bay, H., Ess, A.: Speeded-up robust features (SURF). Comput. Vis. Image Underst. **110**, 346–359 (2008). https://doi.org/10.1016/j.cviu.2007.09.014
7. Bay, H., Ess, A., Tuytelaars, T., Gool, L.V.: Speeded-up robust features (SURF). Comput. Vis. Image Underst. **14** (2007). https://doi.org/10.1016/j.cviu.2007.09.014
8. Ma, J., Jiang, X., Fan, A., Jiang, J., Yan, J.: Image matching from handcrafted to deep features: a survey. Int. J. Comput. Vision **129**(1), 23–79 (2020). https://doi.org/10.1007/s11263-020-01359-2
9. Qin, C., Hu, Y., Yao, H., Duan, X., Gao, L.: Perceptual image hashing based on weber local binary pattern and color angle representation. IEEE Access. **7**, 45460–45471 (2019). https://doi.org/10.1109/ACCESS.2019.2908029
10. Matas, J., Chum, O., Urban, M., Pajdla, T.: Robust wide-baseline stereo from maximally stable extremal regions. Image Vis. Comput. **22**, 761–767 (2004). https://doi.org/10.1016/j.imavis.2004.02.006
11. Nist, D., Stew, H.: Scalable recognition with a vocabulary tree. In: IEEE computer society conference on computer vision and pattern recognition, p. 8 (2006). https://doi.org/10.1109/CVPR.2006.264
12. Korkmaz, Sevcan Aytac; Esmeray, F.: Classification with random forest based on local tangent space alignment and neighborhood preserving embedding for MSER features: MSER_DFT_LTSA-NPE_RF. Int. J. Mod. Res. Eng. Technol. **3**, 7 (2018)
13. Dutta, K., Das, N., Kundu, M., Nasipuri, M.: Text localization in natural scene images using extreme learning machine. In: International conference on advanced computational and communication paradigms, ICACCP 2019, pp. 8–13. IEEE (2019). https://doi.org/10.1109/ICACCP.2019.8882986
14. Sun, Y., Li, H., Sun, L.: A novel wide-baseline stereo matching algorithm combining MSER and DAISY. In: International conference on computer science and application engineering, p. 5 (2018). https://doi.org/10.1145/3207677.3277960
15. Huang, Po-Hsun; Chen, Y., Fuh, C.: String finding based on application connected to server (2017). https://www.csie.ntu.edu.tw/~fuh/personal/StringFindingBasedonApplicationConnectedtoServer.pdf
16. Murugesan, M., Thilagamani, S.: Efficient anomaly detection in surveillance videos based on multi layer perception recurrent neural network. Microprocess. Microsyst. **79**, 103303 (2020). https://doi.org/10.1016/j.micpro.2020.103303
17. Ahmed, S.B., Naz, S.; Razzak, M.I., Yusof, R.B.: A novel dataset for English–Arabic scene text recognition (EASTR)-42K and its evaluation using invariant feature extraction on detected extremal regions. IEEE Access. **7**, 20 (2019). https://doi.org/10.1109/ACCESS.2019.2895876

18. Ahmed, S.B, Razzak, M.I., Yusof, R.: Cursive Script Text Recognition in Natural Scene Images. Springer (2020). https://doi.org/10.1007/978-981-15-1297-1
19. Gupta, N., Rohil, M.K.: Image feature detection using an improved implementation of maximally stable extremal regions for augmented reality applications. Int. J. Image Data Fusion. **9**, 43–62 (2018). https://doi.org/10.1080/19479832.2017.1391337
20. Matas, J., Chum, O., Urban, M., Pajdla, T.: Robust wide baseline stereo from maximally stable extremal regions. In: British machine vision conference, pp. 384–393 (2002)
21. Rublee, E., Garage, W., Park, M.: ORB : an efficient alternative to SIFT or SURF. In: International conference on computer vision, pp. 2564–2571. IEEE (2011)
22. Rosten, E., Drummond, T.: Machine learning for high-speed corner detection. In: Leonardis, A., Bischof, H., Pinz, A. (eds.) ECCV 2006. LNCS, vol. 3951, pp. 430–443. Springer, Heidelberg (2006). https://doi.org/10.1007/11744023_34
23. Calonder, M., Lepetit, V., Strecha, C., Fua, P.: BRIEF : binary robust independent elementary features. In: European conference on computer vision, pp. 778–792 (2010)
24. Qiao, X., Ren, P., Dustdar, S., Chen, J.: A New Era for Web AR with Mobile Edge Computing. IEEE (2018). https://doi.org/10.1109/MIC.2018.043051464
25. Jing, J., Gao, T., Zhang, W., Gao, Y., Sun, C.: Image feature information extraction for interest point detection: a comprehensive review. Comput. Vis. Pattern Recogn. 1–34 (2021)
26. Bal, B., Erdem, T., Kul, S., Sayar, A.: Image-based locating and guiding for unmanned aerial vehicles using scale invariant feature transform, speeded-up robust features, and oriented fast and rotated brief algorithms. Concur. Comput. Pract. Exp. **24**, 11 (2021). https://doi.org/10.1002/cpe.6766
27. Zhang, Z., Wang, L., Zheng, W., Yin, L., Hu, R., Yang, B.: Endoscope image mosaic based on pyramid ORB. Biomed. Signal Process. Control. **71**, 103261 (2022). https://doi.org/10.1016/j.bspc.2021.103261
28. Fitzgibbon, A., Lazebnik, S., Perona, P., Sato, Y., Schmid, C. (eds.): ECCV 2012. LNCS, vol. 7577. Springer, Heidelberg (2012). https://doi.org/10.1007/978-3-642-33783-3
29. Alcantarilla, P.F., Bartoli, A.: Fast explicit diffusion for accelerated features in nonlinear scale spaces. In: British machine vision conference, p. 1 (2013)
30. Oppenheim, A. V., Lim, J.S.: Importance of phase in signals. In: Proceedings of the IEEE (1981). https://doi.org/10.1109/PROC.1981.12022
31. Ye, Y., Shan, J., Member, S., Bruzzone, L., Shen, L.: Robust Registration of multimodal remote sensing images based on structural similarity. IEEE Trans. Geosci. Remote Sens. **55**, 2941–2958 (2017). https://doi.org/10.1109/TGRS.2017.2656380
32. Viola, P., Iii, W.M.W.: Alignment by maximization of mutual information. Int. J. Comput. Visionm. **24**, 137–154 (1997). https://doi.org/10.1023/A:1007958904918
33. Gao, Z., Gu, B., Lin, J.: Monomodal image registration using mutual information based methods. Image Vis. Comput. **26**, 164–173 (2008). https://doi.org/10.1016/j.imavis.2006.08.002
34. Hel-or, Y., Hel-or, H., David, E.: Fast template matching in non-linear tone-mapped images. IEEE Trans. Pattern Anal. Mach. Intell. 1355–1362 (2011). https://doi.org/10.1109/ICCV.2011.6126389

Non-invasive Diagnosis of Diabetes Using Chaotic Features and Genetic Learning

Shiva Shankar Reddy[1(✉)], Nilambar Sethi[2], R. Rajender[3], and V. Sivarama Raju Vetukuri[4]

[1] Research Scholar, Department of CSE, Biju Patnaik University of Technology, Rourkela, Odisha, India
shiva.shankar591@gmail.com
[2] Department of CSE, GIET University, Gunupur, Odisha, India
nilambar@giet.edu
[3] Department of CSE, LIET Vizianagaram, Vizianagaram, Andhra Pradesh, India
rajender.renuguntta@gmail.com
[4] Department of CSE, SRKR Engineering College, Bhimavaram, Andhra Pradesh, India
vvsramaraju@srkrec.ac.in

Abstract. Diagnosis of Diabetes Mellitus (DM) involves an invasive procedure. A pinch of blood sample is extracted by piercing a small needle into the body. This blood sample is fed to an electronic apparatus for detecting the blood glucose range. It is painful and repeated to keep track of the human body's temporal blood glucose level. Researchers are investigating alternative procedures to detect DM without injecting a needle into the human body. This work proposes a novel scheme for diagnosing DM, which is non-invasive and thus not painful. The proposed work takes the digital image of the human retina as input for the purpose. The input images are subjected to appropriate pre-processing beforehand, extracting meaningful features. During feature extraction, the focus is on identifying the chaotic geometric features formed due to the several non-uniform alignments of thin blood vessels inside the image. The chaotic geometry reveals intra-variability among the two classes under consideration (DM and Healthy). Feature vectors are generated to contain this intra-variability. Further classification is performed using a genetic learning method that involves a backpropagation neural network with modified learning weight updation through a Genetic Algorithm. Satisfactory results are obtained by implementing the scheme on a suitable dataset. The overall accuracy rate stands at 81.5%, ideal for emerging solution-oriented research work.

Keywords: Diabetes mellitus · Data mining · Non-invasive diagnosis · Image processing · Genetic algorithm

1 Introduction

Diabetes is a disease that arises when the human body's blood glucose/glucose level is abnormally high. Diabetes develops when the pancreas, a gland in the human body, is unable to produce enough insulin (Type 1 diabetes) or when the insulin produced is unable to be used by the body's cells (Type 2 diabetes). When humans consume glucose

J. I.-Z. Chen et al. (Eds.): ICIPCN 2022, LNNS 514, pp. 161–170, 2022.
https://doi.org/10.1007/978-3-031-12413-6_13

then releases the absorption process. Insulin is a blood molecule that travels from the bloodstream to cells, instructing them to consume glucose and convert it to energy. When the pancreas fails to produce enough insulin, the cells cannot retain which is released into the bloodstream. Blood glucose/glucose expands at an unacceptable level in the blood [1]. Excessive hunger, severe thirst, and constant urination indicate high glucose levels in the human body. The standard glucose levels in the human body are 70 to 99 mg per deciliter. Diabetes is diagnosed when the blood glucose level exceeds 126 mg/dl. If a person's blood glucose concentration is between 100 and 125 mg/dl, they have prediabetes [2]. Complications such as cardiovascular disease, renal failure, stroke, and nerve damage can develop if the human body's glucose level increases too high [3, 4]. Diabetes does not have a long-term cure [5]. The macro-vascular confusion harms the massive veins of the heart, mind, and legs. The micro-vascular disorder causes problems with the small veins, causing problems with the kidneys, eyes, feet, and nerves [6].

Effective diabetes management is possible if the disease is detected early enough. Maintaining a healthy wellness routine and changing your eating habits can help you to avoid Diabetes [7]. If a patient has prediabetes, losing weight and engaging in physical activity can lower the risk of type 2 diabetes. The CDC National Diabetes Prevention Program is a lifestyle change programme that can help prediabetes individuals modify their lifestyles and avoid becoming type 2 diabetes [8]. The healthcare business collects massive data, including clinic records, patient clinical records, and clinical evaluation outcomes. The infection's anticipation is broken down by a specialist's experience and facts for early infection conclusion, but this might be incorrect and defenceless. As a result, the manual options can serve as a warning. The hidden example of information can go unnoticed, swaying in separate directions; as a result, patients are denied effective therapy.

For the early detection of diabetes, a more precise automated distinguishing proof is essential [9–11]. Data mining and artificial intelligence (AI) have been developing, trustworthy, and supporting apparatuses in the healthcare field in recent years. The AI method makes a difference by pre-processing and selecting the relevant highlights from the medical care data using the information mining techniques [12]. Information mining and AI computations can assist in identifying the hidden example of data using the most up-to-date method; thus, a solid exactness choice is possible. Information mining is a technique that employs several strategies, including AI, insights, and an information base framework, to extract an example from a massive amount of data [13]. ML, according to Nvidia, uses a variety of techniques to extract information from parsed data and create predictions [14].

2 Related Work

Alam et al. [15] concentrated their work on enriched Diabetes management by its timely forecast as it is one of the significant ailments. Like many works in the area, they used mining techniques, including ML, for Diabetes early detection. They have used the national diabetes institute dataset from the UCI repository and selected appropriate features utilising data analysis tools. Then they considered data mining algorithms and were applied to the dataset. The performance of each algorithm was measured with relevant metrics. Finally, all the considered algorithms were compared and found that artificial neural networks (ANN) have performed better than their counterparts. Perveen et al. [16] focused their work on predicting risk factors for diabetes as it affects the health and wealth of a large portion of the people in the world. Unlike many other works in the area, they have utilised ensemble techniques like Adaboost and individual mining techniques like J48 to obtain diabetes risk factors for different people of different age groups, male and female categories. "CPCSSN database" is considered as the dataset for their work. Different algorithms were used and compared based on the area under the curve performance parameter and finally concluded that Ada-boost performed better and suggested that Ada-boost with varying learners of the base could also be applied to other domains.

Sisodia et al. [17] highlighted the importance of diabetes prediction as it is one of the most dangerous diseases if not treated in time by predicting it early. They have chosen ML methods like Decision Tree were applied to the considered Pima Indians Diabetes Database (PIDD) to perform this task of diabetes prediction. The applied algorithms were compared using appropriate performance metrics like accuracy. Accuracy is computed based on correctly predicted records, and Naive Bayes was found to be 76. Neha and Shruti [18] undertook the problem of optimal prediction of diabetes, especially in the Indian domain, as diabetes is prominent in Indians, and this disease has many side effects. They have considered the critical factors affecting the onset of diabetes, like a person's lifestyle. Appropriate ML models are trained and tested on the subsets of data collected by way of questionnaires both in offline and online mode. They have concluded that the Random Forests algorithm outperformed other counterparts. Lukas [19] highlighted their work investigating the importance of health care and ML techniques' usage for predicting health problems. Different classification techniques were used to identify the essential topics of significance in research publications. Additional topics of importance like agriculture, marketing and health were considered for comparison. They have identified health care identification using ML as a more important topic than the other topics considered.

N. Sethi et al. [20] focused on predicting early readmission in diabetic patients. The main goal is to utilise a more powerful ML system to help doctors and patients forecast hospital readmission. The Pima Indian Diabetes Dataset is utilised as input for comparison and to determine the best performing algorithm using various ML procedures. The ML approaches studied in this paper are LR, DT, RF, Adaptive Boosting (AB), and Gradient Boosting (GB). These methods were tested on a diabetic dataset from the UCI ML repository, including data from 130 hospitals between 1999 and 2008. R Rajender et al. [21] presented a detailed assessment of various Diabetes predictions, probability of occurrence, co-existence of related diseases, and several other aspects involving applications of DM and ML, particularly for diabetes and associated disorders. In addition, a comparative analysis of selected works was carried out, with relevant recommendations given as a result. N. Sethi et al. [22] researched Diabetes to forecast the dangers and side effects. This study attempts to solve the problem of predicting a diabetic patient's early readmission to the hospital. The Deep belief network, a DL technique, will be used to solve this problem in this study. The new approach and existing methods are implemented in R. To determine its effectiveness, and the proposed DL method is compared to the considered methods utilising assessment criteria such as Precision, Accuracy Specificity, NPV, and F1-score.

Many traditional approaches fail to detect Hard Executes (HE) in diabetic retinopathy images, which are used to assess the severity of the illness. The suggested research incorporates DL via CNN to extract the characteristics to address this issue. The proposed CNN framework enabled the early diagnosis of diabetes by detecting HE in an eye's blood vessel. The proposed framework can also determine a diabetic status [23]. Diabetes is a risk factor for several other chronic diseases. Diabetic nephropathy is a chronic disorder that affects patients with diabetes and their kidneys. To predict diabetic nephropathy, researchers compared various existing ML classification methods with a deep learning methodology called deep belief network. Accordıng to the Results Analysıs, The Cart Decision tree has acquired a better value for AUPR alone. However, DBN outperformed in terms of AUROC, Gini coefficient, and Jaccard index, scoring 0.8203, 0.6406, and 0.7777, respectively [24].

R. Rajender et al. [25] developed a high accuracy model for predicting diabetic retinopathy. Predictive models are trained using machine learning techniques such as DT, RF, Adaptive boosting, and Bagging. This paper proposes a method called "SVM with Gaussian Kernel for Retinopathy Prediction." Accuracy, Youden's J index, concordance, Somers' D statistic, and balanced accuracy are five assessment metrics used to evaluate the suggested algorithm against the essential methods. The proposed approach obtained better outcomes for all evaluated metrics based on the findings obtained. As a result, SVM with a Gaussian kernel is proposed to be utilised for diabetic retinopathy prediction. R. Rajender et al. [26] tried to categorise the different types of Diabetes (Type I and Type II). It also seeks to evaluate the current level of risk connected with the patient. Four distinct algorithms were used to categorise the data as diabetic or non-diabetic: DT, NB, SVM and Adaboost-M1. After then, the voting expert uses a comparative approach to choose the best scheme out of all of them.

Raghavendra S and Santosh Kumar have experimented on PIMA using random forest. The authors have obtained better accuracy by dropping only one feature [27]. The dropping ratio can be improved further. Rajni and Amandeep have implemented the RB-Bayes technique and got 72.9% accuracy over PIMA [28]. Rosita Sofiana and Sutikno have suggested an optimised backpropagation algorithm and observed that the training process improves 12.4 times faster than standard backpropagation [29].

It is learned from the literature that geometric features are less explored towards the prediction of the said ailment. The use of optimisation mechanisms and geometric features can be well examined for enhanced performance of any model. This paper focuses on such a research initiative and thus reflects the novelty.

3 Proposed Work

The proposed work focuses on the chaotic geometric features extracted from the human retina's digital image. These complex geometric features are retrieved through computing the variation of complex fractal geometric formations in the picture. A DM class sample of the image contains swollen, ruptured, and bulky nature of tiny blood vessels that usually got formed due to thickened blood flow in the narrow passages of the small blood vessels. The healthy retinal images barely possess these characteristics. Thus, this pattern variability can be an informative feature for discriminating between DM and healthy classes of samples.

Detail feature extraction mechanism and classification strategy follow:

3.1 Extraction of Chaotic Geometry Features

Suitable pre-processing is first applied to the input image to get the segmented ROI. The sample ROI thus obtained is shown in Fig. 1 for the sample retinal image of a diabetic patient. Algorithm 1 represents the procedure for extracting the features. The steps are discussed below in detail. The algorithm's input is a digital image of the retina (say I_{in}). This algorithm follows the zoning approach.

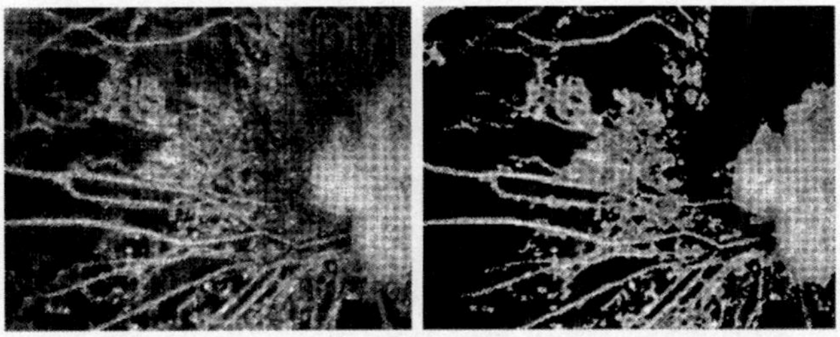

Fig. 1. Pre-processed and segmented retinal image of diabetic patient.

Initially, the image is converted to equivalent gray level representation. The aggregated gray values (θ) are calculated among all the pixels in I_{in}. It is to be noted that the image size is standardised to the dimension 64 × 64. Now, the gray level image (I_{in}) is split into four folds uniformly. Total of 4 × 4 = 16 regions are thus obtained namely, $I_{sub}(sub \in \{1, 2, ..., 16\})$. Each region's aggregated gray values content is calculated individually ($Gray_{sub}$). The probabilities of these regional aggregated gray contents are computed by dividing each value by that of the original image aggregated gray content value. Mathematically:

$$\theta_i = \frac{Gray_{sub}}{R_I} \quad (1)$$

where

$$\sum_{sub=1}^{16} Gray_{sub} = unity \quad (2)$$

Next, the statistical parameters (mean and variance) can be defined as:

$$\mu_{sub} = \frac{Gray_{sub}}{|Pixel_{sub}|} \quad (3)$$

and,

$$\sigma_{sub} = \sum_{j=1}^{|Pixel_{sub}|} |p_j - \mu_{sub}||Pixel_{sub}| \tag{4}$$

where $|pixel|$ denotes the total number of pixels. Further steps are executed with the parametric values obtained through the above-mentioned equations. In this context, now calculate the followings:

$$Gray_{sub}|Pixel_{sub}| = \theta_i \times \propto (I_{in}) \tag{5}$$

where α is the function that maps the original input image aggregation to the regional level image zone in the ratio of 4:1, this is because the original input image is split into four folds. Now, concatenate all the generated $Gray_{sub}|Pixel_{sub}|$ values to get the final feature representation for the first iteration. The execution is re-instantiated from the first step, and the process continues until final feature vectors are generated.

3.2 Genetic Learning

This module takes the features generated in the previous section as the input. It utilises the concept of backpropagation neural networks (BPNN) and genetic algorithms to obtain an optimised learning mechanism. The weight updation through the error learning module is replaced by a slightly modified fitness function enabled error learning module. For the weight updation that occurs during each epoch in the BPNN process, each updation is done subject to the optimal weight value generated through the fitness calculation.

Algorithm 1. Genetic feature ($I_{in}, Gray_{sub}$)

1: Input retinal image I_{in},
2: Convert I_{in} to corresponding gray scale image,
3: SPLIT I_{in} on four-fold basis to get sixteen non-overlapping zones as $Gray_{sub}$; $sub = \{1, 2, 3, \ldots, 16\}$,
4: Compute $\theta_i = \frac{Gray_{sub}}{R_I}$,
5: Compute $\sum_{sub=1}^{16} Gray_{sub} = unity$,
6: Compute statistical parameters $\mu_{sub} = \frac{Gray_{sub}}{|Pixel_{sub}|}$ and $\sigma_{sub} = \sum_{j=1}^{|Pixel_{sub}|} |p_j - \mu_{sub}||Pixel_{sub}|$,
7: Perform compressed mapping as: $Gray_{sub}|Pixel_{sub}| = \theta_i \times \propto (I_{in})$,
8: Generate variability vector value as $\delta(D) = \frac{N^2(D-N(D)^2)}{N(D)^2}$ where N represents a number of pixels whose corresponding intensity is more than the value obtained in the previous step

For this purpose, the fitness function utilised here is presented in the equation given below:

$$fitness = \sum_{i=1}^{S} |T - C| \tag{6}$$

T and S are the target output and actual computed output at the present epoch, respectively. For the computation, the population taken initially comprises the weight values for all the network layers, the target output and the computed actual outputs.

4 Experimental Evaluation

A simulation of the proposed scheme is carried out to validate its significant importance. For this purpose, the dataset is manually collected from the Vijayawada Diabetes Hospital. A total of two hundred fundus image samples are collected. Among these, a hundred samples are of healthy people, and the rest belong to diabetic patients. After applying appropriate pre-processing, the feature extraction is implemented through the algorithm, as cited in the previous section. Genetic learning is implemented for training the model. The trained model utilised a hundred samples during training. The rest of the hundred samples validate the scheme's efficiency. A significant rate of accuracy is observed for the proposed scheme. Overall, the comparison of the scheme is not feasible due to the lack of literature reports in the same context. However, a comparison is performed among four distinct classifiers and plotted in Fig. 3. The chosen classifiers are BPNN, ID3, support vector machine (SVM), and fitness learning. Outperforming in the fitness learning module is evident from the Fig. 3. It is also to be noted that the proposed method is invariant to the scale and resolution parameters of the input samples.

(600 dpi) (200 dpi)

Fig. 2. The segmented output of the processed sample at two different resolutions.

The experimentation is also carried out on the various resolution of the images (two of the samples shown in Fig. 2), and no change is observed in terms of the outcome prediction. However, slight changes are observed in the area of the geometric ROI under consideration. However, it is also observed that the accuracy rate starts falling below the resolution of 150 dpi. Selected approaches from the current work are compared with the proposed models based on the prediction accuracy and shown in Table 1.

Fig. 3. Performance comparison among four distinct classifiers, including genetic learning.

Table 1. Results of existing diabetes predictive models.

Serial no.	Methodology	Reference no.	Results (accuracy %)
1	BPNN	[30]	81
2	Decision tree, SVM and Naïve Bayes	[17]	73.82, 65.1 and 76.30
3	DNN, SVM	[31]	77.8, 77.6
4	PCA + ANN	[15]	75.7
5	PCA + kmeans + LR	[32]	79.94
6	PCA & minimum redundancy maximum relevance	[33]	77.21
7	RMLP – resampling version of MLP	[34]	79.30
8	Gaussian fuzzy decision tree	[35]	75
9	Particle Swarm Optimization with ANN	[36]	80
10	Cultural Algo + ANN	[37]	79
11	Genetic learning	(Current work)	81.5

5 Conclusion

A robust scheme is presented for diagnosing DM, which is non-invasive. Image processing and pattern recognition through chaotic geometry features are the major concerns used for the purpose. The digital fundus images of the human eye retina are considered the input. Through suitable feature extraction of chaotic geometry from these images, the said target is achieved for recognising healthy or diabetic patients. The said work is non-invasive and can be a fine alternative to the traditional DM identification modules. Further, the proposed scheme proved robust at a preliminary rate of the overall accuracy of 81.5%. Future work will focus on enhancing the recognition accuracy through further mathematical refinement of the scheme.

References

1. https://www.webmd.com/diabetes/diabetes-causes
2. https://www.mayoclinic.org/diseases-conditions/prediabetes/diagnosistreatment/drc-20355284
3. https://www.niddk.nih.gov/healthinformation/diabetes/overview/symptomscauses
4. https://www.diabetes.co.uk/diabetescare/blood-sugar-level-ranges.html
5. https://www.healthgrades.com/right-care/diabetes/is-there-a-cure-fordiabetes
6. https://www.betterhealth.vic.gov.au/health/conditionsandtreatments/diabetes-long-term-effects
7. Kumari, M., George, R.: A descriptive study on prevalence and risk factors of diabetes mellitus among adults residing in selected villages of district sirmour. PhD dissertation (2020)
8. https://www.cdc.gov/diabetes/basics/prediabetes.html
9. Ligang, Z., Lai, K.K., Yu, L.: Credit scoring using support vector machines with direct search for parameters selection. Soft Comput. **13**, 149–155 (2009)
10. Chaki, J., Ganesh, S.T., Cidham, S., Theertan, S.: Machine learning and artificial intelligence based diabetes mellitus detection and self-management: a systematic review. J. King Saud Univ. Comput. Inf. Sci. **7** (2020)
11. Contreras, I., Veh´ı, J.: Artificial intelligence for diabetes management and de- cision support: literature review. J. Med. Int. Res. **20**, 5 (2018)
12. Goutham, S., Ravi, V., Kp, S.: Diabetes detection using deep learning algorithms. ICT Expr. **4**, 11 (2018)
13. Andonie, R., Kovalerchuk, B.: Neural networks for data mining: constraints and open problems. Evol. Intell. **1**, 449–458 (2004)
14. https://blogs.nvidia.com/blog/2016/07/29/whats-difference-artificialintelligence-machine-learning-deep-learning-ai/
15. Alam, T.M., et al.: A model for early prediction of diabetes. Inf. Med. Unlock. (2019)
16. Perveen, S., Shahbaz, M., Guergachi, A., Keshavjee, K.: Performance analysis of data mining classification techniques to predict diabetes. Proc. Comput. Sci. **82**, 115–121 (2016). https://www.sciencedirect.com/science/article/pii/S1877050916300308
17. Sisodia, D., Sisodia, D.S.: Prediction of diabetes using classification algorithms. Proc. Comput. Sci. **132**, 1578–1585 (2018). https://www.sciencedirect.com/science/article/pii/S1877050918308548
18. Tigga, N.P., Garg, S.: Prediction of type 2 diabetes using machine learning classification methods. Proc. Comput. Sci. **167**, 706–716 (2020) https://www.sciencedirect.com/science/article/pii/S1877050920308024
19. Priyambodo, L.: Trend of supervised learningmodels based articles. Int. J. Comput. Appl. Inf. Technol. **11**, 282–286 (2019)
20. Reddy, S.S., Sethi, N., Rajender, R.: A Comprehensive analysis of machine learning techniques for incessant prediction of Diabetes mellitus. Int. J. Grid Distrib. Comput. **13**(1), 1–22 (2020)
21. Reddy, S.S., Sethi, N., Rajender, R.: A review of data mining schemes for prediction of diabetes mellitus and correlated ailments. In: 2019 5th International Conference on Computing, Communication, Control and Automation (ICCUBEA), pp. 1–5. IEEE, 19 September 2019
22. Reddy, S.S., Sethi, N., Rajender, R.: Evaluation of deep belief network to predict hospital readmission of diabetic patients. In: 2020 second ınternational conference on Inventive Research in Computing Applications (ICIRCA), pp. 5–9. IEEE, 15 July 2020
23. Sungheetha, A., Sharma, R.: Design an early detection and classification for diabetic retinopathy by deep feature extraction based convolution neural network. J. Trends Comput. Sci. Smart Technol. (TCSST) **3**(2), 81–94 (2021)

24. Reddy, S., Sethi, N., Rajender, R.: Diabetes correlated renal fault prediction through deep learning. EAI Endor. Trans. Pervas. Health Technol. **6**(24), e4 (2020)
25. Reddy, S.S., Sethi, N., Rajender, R.: Discovering optimal algorithm to predict diabetic retinopathy using novel assessment methods. EAI Endor. Trans. Scal. Inf. Syst. **8**(29), e1 (2021)
26. Reddy, S.S., Rajender, R., Sethi, N.: A data mining scheme for detection and classification of diabetes mellitus using voting expert strategy. Int. J. Knowledge-Based Intell. Eng. Syst. **23**(2), 103–108 (2019)
27. Raghavendra, S., Santosh, K.J.: Performance evaluation of random forest with feature selection methods in prediction of diabetes. Int. J. Electr. Comput. Eng. **10**(1), 353 (2020)
28. Rajni, R., Amandeep, A.: RB-Bayes algorithm for the prediction of diabetic in Pima Indian dataset. Int. J. Electr. Comput. Eng. **9**(6), 4866 (2019)
29. Sofiana, R.: Optimization of backpropagation for early detection of Diabetes mellitus. Int. J. Electr. Comput. Eng. (2088-8708) **8**(5) (2018)
30. Joshi S, Borse M.: Detection and prediction of diabetes mellitus using backpropagation neural network. In: 2016 International conference on micro-electronics and telecommunication engineering (ICMETE), pp. 110–113. IEEE, 22 September 2016
31. Wei, S., Zhao, X., Miao, C.: A comprehensive exploration to the machine learning techniques for diabetes identification. In: 2018 IEEE 4th World Forum on Internet of Things (WF-IoT), pp. 291–295. IEEE, 5 February 2018
32. Zhu, C., Idemudia, C.U., Feng, W.: Improved logistic regression model for diabetes prediction by integrating PCA and K-means techniques. Inf. Med. Unlock. **17**, 100179 (2019)
33. Zou, Q., Qu, K., Luo, Y., Yin, D., Ju, Y., Tang, H.: Predicting diabetes mellitus with machine learning techniques. Front. Genet. 515 (2018)
34. Rao, N.M., Kannan, K., Gao, X.Z., Roy, D.S.: Novel classifiers for intelligent disease diagnosis with multi-objective parameter evolution. Comput. Electr. Eng. **1**(67), 483–496 (2018)
35. Varma, K.V., Rao, A.A., Lakshmi, T.S., Rao, P.N.: A computational intelligence approach for a better diagnosis of diabetic patients. Comput. Electr. Eng. **40**(5), 1758–1765 (2014)
36. Patil, R., Tamane, S.C.: PSO-ANN-based computer-aided diagnosis and classification of diabetes. In: Smart trends in computing and communications 2020, pp. 11–20. Springer, Singapore
37. Patil, R., Tamane, S., Patil, K.: An experimental approach toward type 2 diabetes diagnosis using cultural algorithm. In: ICT Systems and Sustainability 2021, pp. 405–415. Springer, Singapore

Analytic for Cricket Match Winner Prediction Through Major Events Quantification

V. Sivaramaraju Vetukuri[1](✉), Nilambar Sethi[2], R. Rajender[3], and Shiva Shankar Reddy[4]

[1] Department of CSE, Biju Patnaik University of Technology, Rourkela, India
sivaramaraju.vetukuri@gmail.com
[2] Department of CSE, GIET University, Gunupur, Odisha, India
nilambar@giet.edu
[3] Department of CSE, LIET, Vizianagaram, Andhra Pradesh, India
rajender.renuguntta@gmail.com
[4] Department of CSE, SRKR Engineering College, Bhimavaram, Andhra Pradesh, India
shiva@srkrec.ac.in

Abstract. Cricket is rated as one of the most famous game across the globe. It has all the possibility to grab the attentions of audience and candidates and in the other hand attentions from researchers as well. Numerous monetary organizations are focusing on this interesting game for profit making and are involving man power and resources towards the task of cricket data analytic. In this paper, an attempt has been made to quantify the active events that count towards the final result of a cricket match. Mainly the batsman scoring efficiency (BSP) and the effective bowling skill (EBS) are taken into consideration for the purpose. These two indicators are used to compute the real time efficiency (RTE) of a particular cricket team. A comparison among the RTE measures of two teams playing each other can be utilized to predict the winner of an ongoing match. User defined functions and equations are framed through this proposed work. Simulation of the proposed framework is carried out on sufficient number of samples. The statistical cricket data samples are chosen from ODI (one day international) matches only. The simulation outcomes reveal a satisfactory justification in for of the validity of the proposed work.

Keywords: Cricket · Analytic · Quantification · Modelling · Machine learning

1 Introduction

The sport of cricket is loved among billions of people. Independent of geographic region, it is played all over the place. It has three unique configurations namely one-day international (ODI), Twenty-twenty (T-20), and test match. This interesting game has drawn in numerous monetary organizations and organizations too to contribute with a target to help enormous benefit through the onlookers from both disconnected and online classifications. Here, the web-based onlookers allude to the crowd enjoying a live telecast available in the scene (arena) and disconnected observers allude to the crowd partaking

J. I.-Z. Chen et al. (Eds.): ICIPCN 2022, LNNS 514, pp. 171–181, 2022.
https://doi.org/10.1007/978-3-031-12413-6_14

in a cricket match from the solace of their couch via TVs and web facilities. Accordingly, specialists are effectively taking into consideration their emphasis on various efficiency measures and tools. Particularly, the majority of the examination works are targeting conceiving expectation plans and devices with most extreme precision. Notwithstanding, at this point, no such profoundly precise apparatus exists. Sub- sequently, effectively huge number of analysts are taking part in the exploration for anticipating a few perspectives connected with a cricket match.

2 Literature Review

Machine learning is extensively being used in diverse applications like text classification, weather forecasting, video analytics, medical analytics, and many more. An exemplary use can be observed in [1] and [2]. In the former work the authors present a novel approach for text labeling with the use of simple capsule network and multiple modern classifiers. The later work interestingly utilizes three stages to achieve sentiment analysis. The first stage involves the use of deep learning. It is followed by data merger and final assembler of overall outcomes to obtain the target analysis output. A more profound examination on the two unique methodologies have been outlined separately in this part. The primary methodology examines about standard strategies revealed up until this point which uses a few AI plans for their calculations to anticipate perspectives like victor expectation, run time score forecasts, group determination, wellness assessment and so on. The subsequent methodology examines about client characterized heuristics and various techniques for a similar reason. In [3], special factual examination has been made to survey the rival candidate's shortcoming and subsequently contriving measurable techniques to expand own groups winning system. K-implies grouping being the unaided classifier assumes the significant part. There are four key elements taken for something very similar. Every one of the candidates are gathered into 3 different classifications; in particular batsmen, bowlers, and defenders. For each part, they have fabricated a likelihood thickness model for deciding the extraordinary candidates, unfortunate candidate, and game dominating candidates. At last, from the rival group they analyzes own group fitting candidates and asses the future techniques. K-means clustering has likewise been utilized in [4, 5], and [6]. In a similar bearing, In [7], center has been put around candidates efficiency. candidates execution and group execution are together considered as the critical variable for the expectation of a cricket math. They have introduced an ordering procedure which is estimated with the continuous coordinate that needs specific difficulties to get done with a success. Current circumstance of a game is taken in a typical scale in corresponding to individual candidate's work list accomplished up to this point. Likewise, the candidate of the match and competition is additionally being anticipated involving this plan for the alluded dataset. Coordinate reproduction with winning forecast, candidate of the match expectation all together considered for ten years. Generally pace of precision has been accounted for to be 86%. In [8], the authors introduce a digging procedure for following the games exercises of candidates. Utilizing something very similar, they are promptly dissecting the hinder wellbeing lack of athletes and giving required goals through additional exercises and nourishment data. Likewise, affiliation rule mining plan has been effectively utilized for information examination of Indian crew information.

For the sake of SnapShot, an incorporated stage has been established in [9], which pictures sports facts and provides expertise in putting shots in other games as well. They have utilized a heatmap methodology for the equivalent.

In [10], investigates the execution of candidates in a group and uses association rule mining to achieve positioning. This is done for national candidates and mentors who are assigned to them. According to the developers, the scheme is still undergoing adjustments. In [11] proposes a method for mining affiliation criteria based on PCA. This is for cricket matches alone. They've developed a structure for setting out relationships in a common pattern of cricket insights using examples. This system is intended to help in making and further developing instructing techniques. In [12] proposed automated and predictive DM algorithms for predicting various characteristics of T-20 matches. This analysis takes into account the current match circumstance and forecasts the innings-by-inning score as well as the overall score. In [13], two potential scenarios for predicting the winning team based on numerous parameters were evaluated. These scenarios are based on the game's two standard formats, ODI cricket and T-20. Two alternative methods are taken in the strategies described here. One method predicts the winner of a one-day match, while another predicts the winner of a Twenty20 match.

In [14] hybrid method that selects efficient players by combining the concepts of genetic algorithms (GA) and recurrent neural networks (RNN). A suitable preprocessing is applied to each player's history statistics, and an initial feature matrix (FM) is constructed for each participant. This FM is supplied into the mathematical function proposed for use in GA. For the minimising of loss-factor, the GA employs a unique fitness function. As a result, a more refined feature matrix is produced. This modified FM is then exposed to RNN to calculate an individual player's final score. In [15] the researchers has written review on ODI, T-20 and Test matches by comparing the match analysis. In [16], a similar affiliation rule digging has been executed for vital getting ready for groups during ICC-2015. A few unequivocal boundary parameters have been dissected. In cricket, the significance of affiliation rule probing for the anticipation of distinct opinions has also been mentioned [17,18] presents a one-of-a-kind approach for determining the cricket crew commander, which is based on PCA. Winner forecasting is another application of PCA. In [19] proposes using information research to programmatically select candidates for a group with the lowest cost assessment. For this reason, traditional NN approaches are used. By and large, the rate of precision builds up to a simple 60%. It also selects eleven applicants for the group rather than the full team's candidates.

In [20], an information mining plan has been introduced for determination of programming colleagues in view of their multi-overlap properties. In this work, an experimental information inquiry on eight key-credits linked to a portion is completed. For the equivalent, they've created a heuristic. This approach could also be applied to group selection for games and sports. A group determination system for IPL (Indian Premier League) matches was proposed in [21]. This is only implied in the context of cricket's T-20 format. This research could aid in the financial benefit analysis for IPL group owners. As a result, K- means clustering was applied to the pool of applicants who were interested in the recent cricket world cup matches. A technique for candidate determination is presented in [22]. For the reason, neural networks approach has been used. They took the statistical game insights for 15 years starting 1985. Moderate preparation and testing has been done on four unique groups of information.

In [16] looked at the powerplay attributes during a cricket match, regardless of the match structure. In the investigation, the difference between the score if there is a powerplay during the game and the score if there isn't a powerplay during the game was taken into account. The nature of the match without powerplay, different powerplay designs, advantages of the powerplay for the batting group, advantages of the powerplay for the bowling crew, and different power- play designs are some of the topics discussed in this work. Regardless of whether the match is an ODI or a T-20, powerplay plans differ. Taking into account no powerplay is likewise a hypothetical condition that may or may not be relevant for any model of research. A cricket outcome indicator was introduced in [23]. This strategy is used to forecast the ODI's outcome. Some of the criteria studied for the work include the nature of the match (day or day/night match), the file of the innings (first/second innings), and the health of the groups.

In this paper they have used NB, SVM, and RF classifiers. By combining the results of these three classifiers, a COP-specific apparatus has been developed (cricket result indicator). Regardless, measuring the precise elements under consideration is a tedious task. Furthermore, this research did not predict the outcome of a T-20 arrangement match. An estimating model for the runtime expectation of the result of an ODI cricket match was proposed in [24]. As the primary tool, logistic regression has been used. Due to the use of a cross- validation process, the work achieves the expected result with a minimal number of highlights. They have obliterated feature points of lesser relevance.

In [25] makes a recommendation for determining the significance and utility of business betting for cricket matches. They claim that if netting is done according to the fall of wickets during the play, a benefit of 20% can be obtained. For this purpose, the Monte-Carlo simulation was used.

For the test cricket match format, a forecasting device was developed in [26]. A test cricket match is a five-day cricket competition between two teams, with each day consisting of about 90 overs. For the expectation of the final result, they used a probabilistic methodology. This study looked at twelve alternative preconditioning borders. Logistic regression is also used as a foundation method. In [27], online social data was used to predict top level candidates and groups in cricket. A match's future pattern is developed based on the information flowing via web media. In [28], two separate themes were combined to create a model for forecasting the outcome of an ODI match. Because the first and second innings of a match (50 overs each) have different boundaries, the results are computed at runtime. For this reason, straight relapse and gullible Bayes have been utilised. A 68% precision rate has been calculated, which increases to 91% in bit by bit increments. Genetic algorithm based calculation has been widely used in [29–32].

3 Proposed Work

In this work, a justified mathematical modeling is proposed to find the RTE value for cricket teams which can help a system predict the match winner. The various parameters utilized are further defined and details for the same are presented in a sequel below.

3.1 Modeling Instantaneous Strength

The strength of a particular cricket team at present instance can be given as:

$$strength = \frac{\sum_{\propto=1}^{r} P(\propto, \beta)\theta_{\alpha}}{\sum_{\propto=1}^{r} P(\propto, \beta)\theta} \tag{1}$$

where, r is the number ODI matches played by the team under consideration exactly in the last one year. This value might vary from team to team. The summation is taken over all these past matches for one year.θ being a polar variable represents the teams winning status (1 for win and 0 for loss). P is the function that is used for priority factoring so that recent competencies of the team gets more weightage and subsequent matches get less priority. The function $P(\propto, \beta)$ can be expressed as:

$$P(\propto, \beta) = (1 - \beta)^{\alpha-1} \tag{2}$$

The value of α ranges in [0, 1].

While framing this parameter, the number of matches played by a particular team (r) is kept varying and time line is kept fixed. However, as per the literature, many of the works have kept the number of matched played as a fixed (5, 10, etc.). The legitimacy of the proposal in this work for taking the number as variable is justified as the fitness of candidates and strength of a team should not be evaluated on fixed number of past matches. This is because a team might have played last five matched in last three years and another team might have played the same number of matched in last three months. Thus, the fitness effect will be different in both the cases. Hence, we prefer the approach of keeping time line fixed but number of all matches played in that entire timeline as a variable count.

For a particular cricket team (say c^{th} team), the instantaneous strength (IS) can be formulated as:

$$IS_C = \frac{\sum_{\propto=1}^{r} (1 - \beta_c)^{\alpha}\theta_{\alpha}}{\sum_{\propto=1}^{r} (1 - \beta_c)^{\alpha}} \tag{3}$$

where, $0 < \beta_c < 1$ and,

$$\theta_{\alpha} = \begin{cases} 1 & \textit{If team is winner for } c^{th} \textit{ match} \\ 0, & \textit{otherwise} \end{cases} \tag{4}$$

For the modeling of the strength parameter, averaging of the prioritized match result values in last matches played in a year by the team is utilized. The modeled outcome for specific input ranges in [0, 1]. A team can get a strength value of 1 only if it has won all the matches that it played in last one year of duration. Similarly, if a team has not won a single match in last one year, then the corresponding output would be 0. However, as we have modeled the prospective outcome through a decreasing rate of priority values in the reverse timeline, hence the strength computation output would also depends on the order in which a team has won/lost matches. Say, team A and team B has played same number of matches (each five) in the last one year duration. The winning order for A is win, win, lost, win, lost and that of B is lost, lost, win, win, win, then B will be having higher strength as it has won recent matches (recent match wins fetch more weight).

3.2 Modeling Real Time Efficency (RTE)

The EBS and BSP parameters corresponding to a specific team (c^{th} team) are the two major features for the scope. Weighted integration of these two inputs can tell about the RTE. Mathematically,

$$RTE = EBS_{\omega} \times \eta_{EBS} + BSP_{\omega} \times \eta_{BSP} \tag{5}$$

where, η_{EBS} and η_{BSP} represent the normalized values for EBS and BSP respectively. Normalizing enables the values for the parameters to fall in the range [0, 1].

The normalized expressions for cth match for these two parameters can be given as:

$$\eta_{BSP} = \frac{BSP_c - \min(BSP)}{\max(EBS) - \min(BSP)} \tag{6}$$

and,

$$\eta_{EBS} = \frac{\max(EBS) - EBS_c}{\max(EBS) - \min(EBS)} \tag{7}$$

The weight factor for normalization for both the parameters under consideration can be expressed as:

$$BSP_{\omega} = \frac{N_c}{\sqrt{\eta_{BSP}}} \tag{8}$$

and,

$$EBS_{\omega} = \frac{N_c}{\sqrt{\eta_{EBS}}} \tag{9}$$

It is to be noted that both these weight values sum to unity. The new standardization constant N_c can be calculated as:

$$N_c = \left(\frac{1}{\eta_{BSP}} + \frac{1}{\eta_{EBS}}\right)^{-1} \tag{10}$$

The above equation ascertains the uniformity among the contribution from both of the parameters BSP and EBS.

3.3 Parameter Modeling (BSP)

Consider Smn is the score hit by mth candidate during nth match; Amn is the number of balls delivered to mth candidate during nth match. Then, the scoring efficiency can be computed by:

$$BSP_{mn} = S_{mn}\frac{candidate\ Strike\ Rate}{Total\ Strike\ Rate} \tag{11}$$

where, the numerator in the above equation refers to the strike rate of individual batsman (i.e. the score against total number of balls faced times hundred). The denominator refers to the total strike rate from the match under consideration.

3.4 Parameter Modeling (EBS)

This modeling is for computing resultant for the bowling efficiency in total for a team. Consider a bowler has delivered B number of balls (six valid balls constitute an over), has taken T number of wickets, and has lost S value of scores to opposition batsman in a specific match. Lets define bowling mean as S/B, bowling economy as S/ (B/6.0), and attack rate as B/T. Thus, the EBS for a specific bowler can be modeled as:

$$EBS_m = \frac{3 * S_m}{T_m + \frac{B_m}{6.0} + T_m\left(\frac{S_m}{B_m}\right)}\pi r^2 \tag{12}$$

4 Experimental Analysis

Suitable experimental analysis of the proposed work is carried out through exact simulation by referring statistical dataset of four various teams. The samples are considered on the matched played at the ICC cricket world cup organized in the year 2019 (List as shown in Fig. 1).

Fig. 1. International teams for the ICC World cup 2019 (ODI); Source: ICC, ESPN- cricinfo.

Ten different nations participated in the grand event. For the purpose, the samples are chosen for the teams of four different nations namely, India, Australia, New Zealand, and England (Host). The RTE modeling for all the teams are done at each of the matches and corresponding outcome predictions are also done. Especially, the semi-finals were played among these teams. Now, let us consider simulating the match played between England and Australia as 1st semi-final in on 9th July, 2019 at Manchester stadium.

Table 1. Modeling the strength values for all the qualifying teams for ICC world cup 2019.

Country	BSP	EBS	Strength
India	25.45	21.33	0.825
Australia	24.56	20.44	0.726
England	26.24	22.12	0.955
New Zealand	25.42	21.3	0.862
Pakistan	22.36	18.24	0.736
Srilanka	22.15	18.03	0.658
South Africa	22.05	17.93	0.544
Bangladesh	21.28	17.16	0.516
West Indies	19.65	15.53	0.485
Afganistan	19.48	15.36	0.438

As an instance, while simulating the match played between England and Australia, the description of the findings as follows. At this stage both of these teams are facing each other in the semi-final stage. So far, strength values for England and Australia are calculated to be 0.812 and 0.965 respectively with correlated calculation as presented in Eq. (1). Although England has won 6 matches and Australia has won 7 matches so far, but due to the reducing weight scheme as discussed earlier, the strength value for England comes out to be more. In Table 1, the strength calculations output until the final day of the tournament for all the teams is presented. Now, based on the further calculation as proposed, the RTE values for both of the teams reveals that England should be the winner of the match under consideration.

The RTE values for both comes out to be 0.955 and 0.726 respectively. It is observed that the prediction through our model becomes true as indeed England makes it way to the final match. The modeled predictions for each of the qualifier matches are presented for all the teams in Table 2. Along with this table, the predicted outcome and the true outcome of the various matches are also presented for the first round of qualification.

Table 2. Predicted outcome versus true outcome for all the qualifier matches in ICC world cup 2019.

Country	Result	India	Australia	England	New Zealand	Pakistan	Srilanka	South Africa	Bangladesh	West Indies	Afghanistan
India	Predicted		India	England	New Zealand	India	India	India	India	India	India
	True		India	England		India	India	India	India	India	India
Australia	Predicted	India		England	New Zealand	Australia	Australia	Australia	Australia	Australia	Australia
	True	India		Australia	Australia	Australia	Australia	South Africa	Australia	Australia	Australia
England	Predicted	India	Australia		England	England	England	England	England	England	England
	True	England	Australia		England	Pakistan	Srilanka	England	England	England	England
New Zealand	Predicted	New Zealand	Australia	England		New Zealand	New Zealand	New Zealand	New Zealand	New Zealand	New Zealand
	True		Australia	England		Pakistan	New Zealand	New Zealand	New Zealand	New Zealand	New Zealand
Pakistan	Predicted	India	Australia	England	New Zealand			Pakistan	Pakistan	Pakistan	Pakistan
	True	India	Australia	Pakistan	Pakistan			Pakistan	Pakistan	West Indies	Pakistan
Srilanka	Predicted	India	Australia	England	New Zealand	Srilanka		South Africa	Srilanka	Srilanka	Srilanka
	True	India	Australia	Srilanka	Srilanka			South Africa		Srilanka	Srilanka
South Africa	Predicted	India	Australia	England	New Zealand	South Africa	South Africa		South Africa	South Africa	South Africa
	True	India	South Africa	England	New Zealand	Pakistan	South Africa		South Africa		South Africa
Bangladesh	Predicted	India	Australia	England	New Zealand	Pakistan	Srilanka	South Africa		West Indies	Bangladesh
	True	India	Australia	England	New Zealand	Pakistan		South Africa		West Indies	Bangladesh
West Indies	Predicted	India	Australia	England	New Zealand	Pakistan	Srilanka	South Africa	West Indies		West Indies
	True	India	Australia	England	New Zealand	Pakistan	Srilanka		Bangladesh		West Indies
Afganistan	Predicted	India	Australia	England	New Zealand	Pakistan	Srilanka	South Africa	Bangladesh	West Indies	
	True	India	Australia	England	New Zealand	Pakistan	Srilanka	South Africa	Bangladesh	West Indies	

5 Conclusion

Suitable mathematical modeling for analytical prediction of ODI format of cricket matches is presented. The model thus presented takes into consideration the batsman and the bowler efficiency as an event wise input. Quantification of these events are well modeled through mathematical approaches. Finally, a real time efficiency (RTE) for a particular team under consideration is computed. When an attempt is to be made to predict the match outcome, the RTE values among the two teams playing that particular match is compared and simply the winner is predicted. It is shown through simulation of the famous ICC world cup 2019, the proposed modeling shows satisfactory rate of accuracy at 85%. However, on a pilot simulation on the T-20 match format, promisingly average rate of prediction (76%). Future work aims at developing refined and efficient models for the T-20 format of matches.

References

1. Manoharan, J.S.: Capsule network algorithm for performance optimization of text classification. J. Soft Comput. Paradigm (JSCP) **3**(01), 1–9 (2021)
2. Pandian, A.P.: Performance evaluation and comparison using deep learning techniques in sentiment analysis. J. Soft Comput. Paradigm (JSCP) **3**(02), 123–134 (2021)
3. Parameswaran, K.: Vector quantization, density estimation and outlier detection on cricket dataset. In: 2013 International Conference on Computer Communication and Informatics, pp. 1–5. IEEE (2013)
4. Satao, P., Tripathi, A., Vankar, J., Vaje, B., Varekar, V.: Cricket score prediction system (csps) using clustering algorithm. Int. J. Curr. Eng. Sci. Res. **3**(04), 43–46 (2016)
5. Awan, M.J., et al.: Cricket match analytics using the big data approach. Electronics **10**(19), 2350 (2021)
6. Singh, M.P., Ahmad, M.: Performance prediction of players in sports league matches. Int. J. Sci. Res. (IJSR) **4**(04), 1–7 (2015)
7. Jhanwar, M.G., Pudi, V.: Quantitative assessment of player performance and winner prediction in ODI cricket. Int. Inst. Inf. Technol. Hyderabad-500032, Indıa. (2017)
8. Fister, I., Fister, D., Fong, S.: Data mining in sporting activities created by sports trackers. In: 2013 İnternational Symposium on Computational and Business İntelligence, pp. 88–91. IEEE (2013)
9. Pileggi, H., Stolper, C.D., Boyle, J.M., Stasko, J.T.: Snapshot: visualization to propel ice hockey analytics. IEEE Trans. Visual Comput. Graph. **18**(12), 2819–2828 (2012)
10. Thakare, I.S., Suyal, S.R., Pandav, K.Y.: Performance evaluation for sports team selection using data mining techniques. AADYA-Natl. J. Manag. Technol. **1**(5), 102–108 (2015)
11. UmaMaheswari, P., Rajaram, M.: A novel approach for mining association rules on sports data using principal component analysis: for cricket match perspective. In: 2009 IEEE International Advance Computing Conference, pp. 1074–1080. IEEE (2009)
12. Sivaramaraju Vetukuri, V., Rajender, R., Sethi, N.: A multi-aspect analysis and prediction scheme for cricket matches in standard T-20 format. Int. J. Knowl.-based Intell. Eng. Syst. **23**(3), 149–154 (2019)
13. Sivaramaraju, V., Sethi, N., Rajender, R.: Heuristics for winner prediction in international cricket matches. Stat. Optim. Inf. Comput. **8**(2), 602–609 (2020)
14. Vetukuri, V.S., Sethi, N., Rajender, R.: Generic model for automated player selection for cricket teams using recurrent neural networks. Evol. Intel. **14**(2), 971–978 (2020). https://doi.org/10.1007/s12065-020-00488-4
15. Raju, V.S., Sethi, N., Rajender, R.: A review of data analytic schemes for prediction of vivid aspects in ınternational cricket matches. In: 2019 5th International Conference On Computing, Communication, Control And Automation (ICCUBEA), 19 September 2019, pp. 1–4. IEEE (2019)
16. Bhattacherjee, S., Sahoo, J., Goswami, A.: Association rule mining approach in strategy planning for team İndia in icc world cup 2015. In: 2015 Second International Conference on Advances in Computing and Communication Engineering, 1 May 2015, pp. 616–621. IEEE (2015)
17. Abraham, A., Dutta, P., Mandal, J.K., Bhattacharya, A., Dutta, S.: Emerging technologies in data mining and information security. In: Proceedings of IEMIS-2018 (2018)
18. Shah, P., Patel, M.N.: Ranking the cricket captains using principal component analysis. Int. J. Physiol. Nutr. Phys. Educ. **3**(2), 477–483 (2018)
19. Al-Shboul, R., Syed, T., Memon, J., Khan, F.: Automated player selection for sports team using competitive neural networks. Int. J. Adv. Comput. Sci. Appl. **8**(8), 457–460 (2017)

20. Bialkowski, A., Lucey, P., Carr, P., Yue, Y., Sridharan, S., Matthews, I.: Large-scale analysis of soccer matches using spatiotemporal tracking data. In: 2014 IEEE İnternational Conference on Data Mining,14 December 2014, pp. 725–730. IEEE (2014)
21. Sankaran, S.: Comparing pay versus performance of IPL bowlers: an application of cluster analysis. Int. J. Perform. Anal. Sport **14**(1), 174–187 (2014)
22. Iyer, S.R., Sharda, R.: Prediction of athletes performance using neural networks: an application in cricket team selection. Expert Syst. Appl. **36**(3), 5510–5522 (2009)
23. Silva, R.M., Manage, A.B., Swartz, T.B.: A study of the powerplay in one-day cricket. Eur. J. Oper. Res. **244**(3), 931–938 (2015)
24. Pathak, N., Wadhwa, H.: Applications of modern classification techniques to predict the outcome of ODI cricket. Procedia Comput. Sci. **1**(87), 55–60 (2016)
25. Asif, M., McHale, I.G.: In-play forecasting of win probability in one-day international cricket: a dynamic logistic regression model. Int. J. Forecast. **32**(1), 34–43 (2016)
26. Norton, H., Gray, S., Faff, R.: Yes, one-day international cricket in-play trading strategies can be profitable! J. Bank. Finance **1**(61), S164–S176 (2015)
27. Akhtar, S., Scarf, P.: Forecasting test cricket match outcomes in play. Int. J. Forecast. **28**(3), 632–643 (2012)
28. Ahmad, H., Daud, A., Wang, L., Hong, H., Dawood, H., Yang, Y.: Prediction of rising stars in the game of cricket. IEEE Access. **14**(5), 4104–4124 (2017)
29. Bharathan, S., Sundarraj, R.P., Abhijeet, S., Ramakrishnan, S.: A self-adapting intelligent optimized analytical model for team selection using player performance utility in cricket. In: 9th MIT Sloan Sports Analytics Conference, MIT, Boston, pp. 1–11 (2015)
30. Ahmed, F., Deb, K., Jindal, A.: Evolutionary multi-objective optimization and decision making approaches to cricket team selection. Swarm Evol. Memetic Comput. SEMCCO (2011)
31. Kumarasiri, S.I.: Optimal one day international cricket team selection by genetic algorithm. J. Sports Anal. **36**(4) (2017)
32. Jayanth, S.B., Anthony, A., Abhilasha, G., Shaik, N., Srinivasa, G.: A team recommendation system and outcome prediction for the game of cricket. J. Sports Anal. **4**(4), 263–273 (2018)

An Investigation of COVID-19 Diagnosis and Severity Detection Using Convolutional Neural Networks

V. Dhanya and Senthilkumar Mathi(✉)

Department of Computer Science and Engineering, Amrita School of Computing, Coimbatore, Amrita Vishwa Vidyapeetham, Coimbatore, India
cb.en.p2cse20012@cb.students.amrita.edu, m_senthil@cb.amrita.edu

Abstract. Severe acute respiratory syndrome coronavirus 2 (SARS-CoV-2) causes coronavirus disease 2019 (COVID-19). It is a contagious disease that has infected more than millions of people around the globe. COVID-19 can be diagnosed based on the amount of infection in the lungs. Apart from the gold standard reverse transcription-polymerase chain reaction test, X-rays and CT scans can be used to diagnose and detect COVID-19 severity. Due to technological advancements, deep learning models play a crucial role in COVID-19 analysis because of their efficiency and accuracy. Hence, the present work investigates various research works to detect COVID-19 using a convolutional neural network. It compares architectures such as Inception, MobileNet, DenseNet, and new architectures like CovidNet and CovidSDNet were developed. Some of the investigations in COVID-19 detection were conducted by performing data augmentation and ensembling techniques with and without transfer learning. Most research works used accuracy, precision, recall, and F1-score as the performance metrics for evaluation. The numerical comparison analysis shows that the earlier works achieved an accuracy of about 89 to 98 percent. However, in most investigated research, multiclass classification is performed to classify the given CT scan or X-ray image into COVID-19, normal or pneumonia. In the current situation, there is a possibility that some people with COVID-19 might have other respiratory diseases as well. Hence, the investigations suggest that multi-label classification with convolutional neural networks can be suitable to determine the combination of respiratory problems present along with COVID-19.

Keywords: Reverse transcription-polymerase chain reaction · Ground-glass opacities · COVID-19 · Computerized tomography scans · X-ray · Convolutional neural networks

J. I.-Z. Chen et al. (Eds.): ICIPCN 2022, LNNS 514, pp. 182–196, 2022.
https://doi.org/10.1007/978-3-031-12413-6_15

1 Introduction

A group of viruses known as the coronavirus can cause diseases in animals and humans. According to the reports, it is believed that the first COVID-19 transmission to humans occurred in Wuhan, China [1]. Since then, the virus has been chiefly spread through person-person contact [2]. COVID-19 is transmitted directly by infected people, surfaces in the immediate environment, or by objects used on the infected persons [3]. COVID-19 is characterized by fever, shortness of breath, cough, tiredness, loss of taste or smell, and speech or mobility in severe conditions [4]. Identifying asymptomatic patients and early detection of disease helps to reduce the spread of the disease [5, 6]. RT-PCR (reverse transcription-polymerase chain reaction) test is conducted to diagnose COVID-19, but the RT-PCR test suffers from false-positive or false-negative results [7]. Coronavirus mainly affects the lungs of the victims [8]. Hence, the other alternative methods for COVID-19 detection are lung X-rays and lung CT scans [9–11]. In the present COVID situation, it is advisable to have more than one approach for COVID detection. CT scans and X-Rays provide much faster results than the gold standard RT-PCR tests [12].

Deep learning models generate more accurate and efficient results [13, 14]. They are faster, cheaper, and safer as they reduce the need for human interaction. A convolutional neural network holds the merit of being an excellent feature extractor for image classification. CT scans play a significant role in the early detection of COVID-19. The lung involvements and ground-glass opacities (GGO) can be detected and analyzed well using CT scans even before the patients have developed clinical symptoms and are diagnosed with RT-PCR test. The sensitivity of CT-Scan is much better than the sensitivity of X-Ray. CT and chest X-Ray findings of COVID-19 include ground-glass opacities, crazy paving patterns, intralobular septal thickening, interlobular septal thickening, consolidations, bilateral and multifocal distributions [15].

There are not many CT-Scan datasets present for mild cases of COVID-19. Data insufficiency in mild and moderate cases is one of the challenges people face in this work. But this could be overcome by data augmentation [16, 17]. Data augmentation, like rotation and slicing of images, can be considered the solution. Figure 1 depicts the COVID CT scan, Non-COVID CT scan, Non-Covid X-Ray image, and COVID X-Ray image. Section 2 deals with the overview of the convolutional neural network and the existing segmentation models. Sect. 3 discusses the survey on the research work related to COVID-19 diagnosis and severity detection. A discussion on the surveyed research work is provided in Sect. 4. Section 5 contains the conclusion of this research work.

Fig. 1. COVID-19 CT scan (a), Non-COVID-19 CT scan (b), (c) Non-COVID-19 X-Ray, (d) COVID X-Ray

2 Convolutional Neural Networks – An Overview

In recent days, artificial neural networks have played a critical role in improving machines' skills in making decisions like humans. The convolutional neural network is a kind of artificial neural network specially intended to process the pixel input in the case of image processing and image recognition. It acts as a suitable feature extractor [18, 19].

The components of convolutional neural network models are discussed here. The fundamental block of the convolutional neural network is the convolutional layer [20]. Each filter builds an activation map by convolving with the image. Filters move over the image based on the striding value. The feature map is extracted using either 2D or 3D filters. Filters play a great role in sharpening and detecting the edges, shapes, and texture. Pooling layers are used to reduce the dimension. Min, max, and average are the different types of pooling. Max pooling is the one that is most commonly used. The strong values are retained in max-pooling while the low values are discarded. It helps in preventing overfitting. The activation function is a non-linear transformation function fed into the convolutional neural network to assist the model in studying complicated patterns within the data. The sigmoid activation function ranges from 0 to 1 [21]. The sigmoid activation function is used mostly in multi-label classification. ReLU (Rectified Linear Unit activation) function is the most commonly used activation function [22]. ReLU activation reduces the vanishing gradient problem [23]. The softmax activation function has coined a place for itself in multiclass classification. Optimizers try to reduce the overall loss and enhance the accuracy. Adam optimizer holds the merits of being one of the best adaptive optimizers.

SegNet and U-Net are the most commonly used segmentation models for lung and lesion segmentation. SegNet is preferred for binary classifications, while U-Net has opted for multiclass classifications. Filtering and masking boost the performance of the models. Lung segmentation [24] and lesion segmentation help improve the accuracy of the classification [25–27].

A convolutional neural network called U-Net was specially designed for segmenting biological images [28]. U-Net is comprised of an encoder and decoder. It consists of three sections, namely bottleneck, contraction, and expansion. The encoder acts as the contraction phase, while the decoder acts as the expansion phase. U-Net has an equal number of contraction and expansion blocks.

SegNet is a model for semantic segmentation. It comprises an encoder, a decoder, and a classification layer that performs pixel-wise [29]. The encoder is commonly a VGG or ResNet. The decoder's job is to semantically project the encoder's learned discriminative features (lower resolution) onto the pixel space (higher resolution) to provide a dense classification.

3 Literature Survey

This section deals with a brief survey on the research works related to the detection and severity identification of COVID-19 using X-Rays and CT Scans. These models use convolutional neural networks as the critical deep learning feature for predicting COVID-19.

R. M. Rawat et al. [30] compared the performance of four different CNN architectures, namely InceptionV3, MobileNet, Xception, and DenseNet121, in predicting COVID-19 using Chest X-Ray images. The dataset opted for is a balanced dataset consisting of X-Ray images of COVID-19, pneumonia, and normal. These four models were initially trained on the ImageNet dataset, consisting of 1000 different classes and 14 million images. In this experiment, a mini-batch gradient descent technique was used. The batch size was fixed as 200, and 10 training epochs with seven steps per epoch were used. The learning rate was 0.5 in the drop-out layer to prevent overfitting. The categorical cross-function was used along with the Adam optimizer. The experiments suggested that the precision, recall, and F1-score are better for MobileNet. It has correctly classified the maximum number of images, while Xception has the least number of correct classifications.

E. F. Ohata et al. [31] experimented with the posterior-anterior and anterior-posterior chest X-ray images to detect COVID-19. The architectures VGG, Inception, ResNet, NASNet, Xception, MobileNet, and DenseNet, and different configurations were initially trained on the ImageNet dataset. Based on the performance of these architectures, the top five architectures were opted for, and the fully connected layers were removed. Two datasets, A and B, with 194 COVID-19 images and 194 healthy images, were used. Pre-processing steps include resizing the images accordingly to the architecture, the training images were rotated magnified, and the width and height were changed. Bayes, Multilayer Perceptron, SVM (linear and RBF kernels), KNN, and Random Forest were used for classification. Here, Random forest is based on the decision tree, multilayer perceptron is constructed on neural networks, Bayes is based upon probability, and SVM is optimal hyperplane based. The classification was performed by repeating model training and testing with ten-fold cross-validation. Random search classifiers were configured for 20 iterations of the search. The experiment was repeated 144 times. The work achieved 98.462% accuracy and 98.461% F1-score in dataset A using MobileNet with SVM. Similarly, an accuracy of 95.64% and an F1-score of 95.633% were obtained in Dataset B using DenseNet 201 with MLP.

D. Haritha et al. [32] proposed a new approach to detect COVID-19. Pre-processing includes resizing the images to 224×224 and transfers learning using GoogleNet. They proposed that a 1×1 convolution reduces the number of parameters such as weights and biases; thus, the depth of the network increases. A 1×1 convolutional layer, when introduced before 3×3 and 5×5 layers, reduces the computational cost and channels. The output from the filters was concatenated with 3×3 max pooling to extract the features. It is a binary classification problem, so binary cross-entropy and softmax activation function is used. Global average pooling was performed to minimize overfitting. The final layer is the fully connected dense layer with two neurons. The experimental result obtained is 99% training accuracy and 98% validation accuracy. The precision and recall obtained for the training set are 1.00, 0.97, while the precision, recall and F1-score obtained for the validation set are 0.9891, 0.9815 and 0.9852, respectively.

J. Rodríguez et al. [33] developed a novel approach to finding the severity of COVID-19 in patients using CT- Scans. In this experiment, they compared the performance of two architectures, long-term convolutional nets and Inflated 3D ConvNet. LTC recovers local volumetric features, and the multiscale structure increases the channel representation.

It focuses more on relevant patterns while the spatial resolution is reduced. It has six successive 3D-CNN layers followed by a non-linearization layer and fully connected layers. I3D learns the low-level volumetric features. It has three convolutional layers, four max-pooling layers, nine inception modules and an average pooling layer before the classification layer. Severity is found using the Grad-CAM method. Severity was based on a scale of 0 to 25, with five categories with 4% for each point. It includes three main steps: parenchyma segmentation, severity classification, and visual attention map computation.

In Parenchyma segmentation, automatic segmentation was performed using 2D U-Net trained with 50 annotated 224 × 224 resized images, corresponding to 2167 CT slices. Adam optimizer with a learning rate of 0.001 and batch size of 8 over 100 epochs was used along with the sparse categorical cross-entropy function. Data augmentation was performed on the training dataset. Axial and sagittal flips, sagittal shifts, and rotations were performed. Each CT scan was subsampled and fixed with 20 slices. Binary cross-entropy and Adam optimization were done with a learning rate of 0.0001 and 0.5 drop-out layer. The experiment was repeated with transfer learning on ImageNet and kinetics dataset. Softmax activation and Adam optimizer with a 0.00001 learning rate are used. Ten training epochs, each with a batch size of 4, have been experimented. It was found that the pre-trained models show a great F1 score than the ones started from scratch. Inflated 3D ConvNet performed better than long-term convolutional nets.

R. Bhadra and S. Kar [34] introduced a lightweight deep, multi-layered CNN model with five layers and 3 × 3 convolutional kernels and a unit stride sliding window and single padding. The model includes a 2 × 2 max pooling layer and ReLU activation function. Three-dimensional tensors were converted to a one-dimensional feature vector with 4096 neurons. The dense layer has 128 neurons 0.5 drop-out rate, and the output layer has three neurons and softmax activation function and categorical cross-entropy. The initial learning rate has opted as 0.0001, which was decreased by ten after every ten epochs. The batch size was fixed as 10 for 30 epochs. The experimental results of this model achieved 99.5% of training accuracy, 97.6% of validation accuracy, and 99.1% of testing accuracy. This proposed model outperformed VGG16, VGG19, MobileNet, InceptionV3, and Resnet50.

M. Foysal and A. B. M. Aowlad Hossain [35] introduced a new methodology to detect COVID-19. They have developed three CNN models. The output from all the models is ensembled to provide the final classification. Model 1 and 3 have three convolutional blocks, and maximum pooling layers follow each block. While Model 2 has four convolutional blocks followed by Max Pool layers. The drop-out layer reduces the overfitting. Dense layers and fully connected layers are used in all the models. Batch normalization is performed to accelerate the process. ReLU is used to overcome the vanishing gradient problem. The experiment can achieve an accuracy of 95.4% in Model1, 93.6% in Model2, and 94.4% in Model3.

A. A. w. A. Musleh and A. Y. Maghari [36] proposed a novel approach to detect COVID-19 from X-Ray images. The architecture is based on the CheXNet algorithm. CheXNet is a 121-layer condensed model trained with 1,00,000 chest images with 14 categories. A Convolutional 2D layer with 64 output filters in a 3 × 3 filter matrix with a ReLU activation function was used. 2D max-pooling was performed with a Hidden layer of 2 × 2 matrix. Flattening is performed to convert it into a single array. A fully connected 128 output layer and sigmoid activation function are used in the dense layer. Adam optimizer and Binary cross-entropy are used. The experiment was repeated twice. The first experiment was conducted with 276 images, and an accuracy of 84% and a loss of 45% were obtained. Five hundred images were used in the second experiment, and the accuracy and loss obtained were 89.7% and 24.8%, respectively.

G. Aparna et al. [37] introduced CNN architecture to predict COVID-19. The model consists of a 2D convolutional layer with 128 filters for kernel sizes 3, 5, and 7. Max pooling, Convolutional layer with 64 and 32 filters, kernel size 3 × 3, and ReLU activation function are used. Flattening is used to convert into a single array. The connected dense layer has 128 and 64 neurons, and 0.5 is the drop-out rate. Two batches are used with a batch size of 2 for 20 epochs in the training phase. The performance of the suggested model is compared with the pre-trained model, and the loss rate was found to be very less in this suggested approach. The loss rate of VGG 16 is 0.25, the loss rate of MobileNet is 1.53, and the loss rate of this developed model is 0.04. In other terms, the work attained an accuracy of 98.3% through this proposed approach, while the accuracy obtained in VGG-16 is 93.3%, and 91.6% was obtained in MobileNet.

S. Tang et al. [38] presented a combination of ensemble method and deep learning to detect COVID-19. They have used multiple deep learning models with a single deep neural network. The models were developed with CovidNet architecture as the base. The 1 × 1 convolutional layer will project features from higher to lower dimensions, and it is then followed by an expansion dimension which enlarges and focuses the features. A 3 × 3 depth-wise convolutional layer is used. Another 1 × 1 convolutional layer follows this to reduce the dimension, and then dimension expansion is performed. The research opted for a cosine annealing learning rate and generated multiple model snapshots. The work consists of two phases: snapshot model training and model ensembling. The averaging method has opted for the ensemble technique. The accuracy obtained from this method is 95%, while the accuracy from single CovidNet architecture is 93.3%.

J. D. Arias-Londono et al. [39] rescaled images to 224 pixels. It is normalized using Z-Score. Mean and standard deviation are calculated for RGB. It is then later converted to greyscale. The main important part of this approach is the Segmentation cropping using U-Net-Segmentation followed by masking. Grad-CAM was used to opt for the areas of significant interest. The work was fixed with the number of epochs as 24, batch size as 32, and learning rate as 2^−5. They experimented with the proposed method, and the results were analyzed at each step. The accuracy achieved through this experiment is 91.5%, while the recall obtained was 87.4%.

S. Tabik et al. [40] proposed a COVID detection and severity classification. They created a smart dataset through noise cancellation, data augmentation, and transformation. Also, they performed segmentation-based cropping, and 5% of the pixel is added after cropping to ensure that the important information is not lost. SGD optimizer and GAN is used for inherent class transformation. The discrimination of positive and negative classes was performed using GRAD-CAM. They removed the last layer in the architecture and added 512 neurons layer with ReLU activation function, and another two to four neuron layer was added with softmax activation. They achieved 97.72% ±

0.95% in severe cases, 86.90% ± 3.20% in moderate cases, and 61.80% ± 5.49% in mild cases of COVID-19.

Kang, Myeongkyun et al. [41] developed a new approach to classify the given CT scan as COVID-19 or Bacterial Pneumonia. The main three components of this methodology are Lung and lesion segmentation, deep feature extraction and k-means clustering models. The segmentation process was done using Mask-cascade-RCNN-ResNeSt-200. To avoid misclassification, patches less than 13mm are not considered. The classification was done using ResNet50 model. Clustering is performed through the K-means cluster, which clusters the lesions into 20 groups. A histogram construction follows this, and it is analyzed using a two-sample t-test as COVID-19 and Bacterial Pneumonia, and the other is severe and non-severe. Classification accuracy of lesion level majority voting and Support Vector Machine was compared. In this research work, an accuracy of 91.2% was obtained for COVID-19 and Bacterial Pneumonia classification, an accuracy of 95% for severity classification.

K. Garlapati et al. [42] performed COVID-19 detection using X-ray images. They performed data augmentation, followed by image resizing. This work focuses on image filtering and edge detection. In this work, the performance of Gaussian, Sobel X, Sobel Y, Laplacian and HOG were compared. Sobel X and Sobel Y have opted for the classification. Edge detection plays a great role in reducing the processing time as it reduces the amount of data by preserving the region of interest. The edge detection performed smoothing and image sharpening, and a support vector machine was used for the final classification. The accuracy obtained in this method is 99.17%.

Fig. 2. Architectures used in COVID-19 prediction

Figure 2 represents the existing and novel hybrid architectures used in the previous research to predict COVID-19. A summary of the above-listed works is tabulated in Table 1

Table 1. Summary of the related research works

Work	Modality	Data processing	Architecture	Metrics
R. M. Rawat et al. [30]	X-Ray images	Transfer learning on ImageNet	1. InceptionV3 2. MobileNet 3. Xception 4. DenseNet121	MobileNet – F1-Score: COVID-19: 0.9913 Normal:0.9987 Pneumonia: 0.9230
E. F. Ohata et al. [31]	X-Ray images	1. Image resizing, 2. Data augmentation, 3. Transfer learning on ImageNet	MobileNet + SVM (Linear) DenseNet201 + MLP	Acc: 98.62%, F1-score: 98.461%, Acc: 95.641%, F1-score: 95.633%
D. Haritha et al. [32]	X-Ray images	1. Image resizing 2. Transfer learning on GoogleNet	GoogleNet	Accuracy: 99% training and 98% validation Precision: 0.9891 Recall: 0.9815 F1-score: 0.9852
Rodríguez et al. [33]	CT - scans	1. Normalization 2. Parenchyma segmentation using 2D U-Net 3. 4-fold cross-validation 4. GradCAM for severity	Inflated 3D ConvNet and Long-Term convolutional nets	Inflated 3D ConvNet: Precision:100% Recall: 96% F1-score:98%
R. Bhadra et al. [34]	X-Ray images	1. Rescaling images 2. 10-fold cross-validation	Deep multi-layered CNN	Validation Accuracy: 97.6% Precision: 1.00 Recall: 1.00 F1-score:1.00
M. Foysal et al. [35]	CT –scans	Ensemble methodology is used	3 Deep CNN models of different layers	F1-score: 0.956 Sensitivity: 0.970 Accuracy: 0.96
A. A. w. A. Musleh et al. [36]	X-Ray images	NA	CheXNet	Accuracy: 89.7%
G. Aparna et al. [37]	X-Ray images	1. Image resizing 2. Greyscale conversion	CNN Model	Accuracy: 98.3% F1-Score: 0.983 Recall: 0.967 Precision:1.00

(*continued*)

Table 1. (*continued*)

Work	Modality	Data processing	Architecture	Metrics
S. Tang et al. [38]	X-Ray images	1. Ensemble methodology 2. Pre-training the models	EDL-COVID based on CovidNet	Accuracy: 95%, Sensitivity: 96%
J. D. Arias-Londoño et al. [39]	X-Ray images	1. Image rescaling 2. Z score normalization 3. Mean and SD is calculated for RGB 4. Converted to greyscale 5. Image Cropping 6. Lung Segmentation 7. Grad-CAM	Based on CovidNet	Accuracy: 91.5% Recall: 87.4%
S. Tabik et al. [40]	X-Ray images	1. Noise cancellation 2. Data augmentation 3. Data transformation 4. Transfer learning on ImageNet weights 5. Segmentation based cropping	CovidSDNet	Severe: 97.72% ± 0.95%, Moderate: 86.90% ± 3.20%, Mild: 61.80% ± 5.49%
Kang, Myeongkyun et al. [41]	Chest CT images	1. Lung and Lesion Segmentation 2. Deep Feature Extraction 3. K-means clustering	Mask-cascade-RCNN-ResNeSt-200, ResNet50 model	Classification Accuracy:91.2% Severity Classification: 95%
K. Garlapati et al. [42]	X-Ray images	1. Image resizing and data augmentation 2. Image filtering 3. Edge Detection	Support Vector Machine Model	Accuracy: 99.17% Precision: 99.24% Recall: 98.13% F1-Score: 98.68%

4 Numerical Comparison Analysis and Discussions

From the earlier research works, it is inferred that the ensemble techniques increase the accuracy. Models with ensembling methods have more accuracy than individual models. It is preferred that the datasets are not biased. Pre-training a model plays a prominent role in improving accuracy. The models are mostly pre-trained on the ImageNet dataset. Novel CNN models and hybrid models improve accuracy. The most predominantly used architectures in these works are CovidNet, CovidSDNet, CheXNet, InceptionV3, DenseNet, MobileNet, GoogleNet, etc. Ensembling techniques, combining RT-PCR test results with CT scans, filtering and masking the lung lobes in CT scans help to reduce the false-positive rates.

X-Ray images are used in most of the works. The pre-processing techniques opted in the above research works are data augmentation, transformation, and normalization. Lung segmentation and lesion segmentation have increased the models' performance to a greater extent. Metrics used for comparison are accuracy, precision, recall, and f1-score. Accuracy is defined as how close the measured values are to each other. Precision is defined as what percentage of those entire predicted positive is positive. The recall is also known as true positive rate and also as sensitivity. It is defined as what percentage

Fig. 3. Dataset analysis

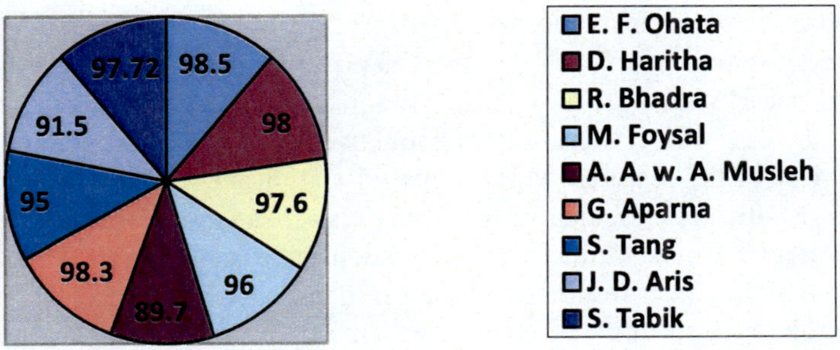

Fig. 4. Accuracy representation for COVID-19 prediction

Fig. 5. Precision, recall, and F1-score analysis for COVID-19 prediction

of the overall positive results is forecasted to be positive. F1-score plays a prominent role in the imbalanced dataset because it is the harmonic mean of recall and precision. The accuracy obtained in this research work ranges from 89 to 98 percent, along with significant precision, recall and F1-score. Figure 3 indicates the number of images present in the datasets used to predict COVID-19 in the research works. Figure 4 describes the accuracy obtained in the research works on COVID-19 detection. Figure 5 indicates the precision, recall, and f1-score achieved in COVID-19 prediction.

5 Conclusion

In the present ongoing pandemic situation, a much faster, cheaper, and more efficient method of COVID-19 detection is required. Early detection helps in reducing the spread of the disease. Deep learning models play a vital role in addressing all the entities above. CT scans and X-rays can be used as an alternative for COVID-19 detection and severity classification. This research work provides a survey on the research works in COVID-19 prediction and severity classification. The research works were mostly related to classifying the given CT scan or X-ray into COVID-19, normal, or pneumonia using convolutional neural networks. X-rays are used in most works when compared with CT scans. In the case of CT scans, it is essential to select the correct slice for training the model. Most of the research works achieved significant accuracy in performing classification. Multiclass classification or binary classification is performed in all the investigated research works. Multiclass classification alone is insufficient if the COVID-19 positive patient has other respiratory diseases. In this case, multi-label classification plays a great role. Patients might be infected with multiple different lung diseases at the same time. Multiclass classification would indicate the presence of only one disease. But

Multi-label classification indicates the presence of other diseases based on the infection in the lungs. It helps determine if the person is affected by any critical combination of respiratory diseases.

References

1. Yadav, P., Menon, N., Ravi, V., Vishvanathan, S.: Lung-GANs: unsupervised representation learning for lung disease classification using chest CT and X-Ray Images. IEEE Trans. Eng. Manag. (2021)
2. Rajeshkumar, A., Mathi, S.: Smart solution for reducing COVID-19 risk using internet of things. Indonesian J. Electr. Eng. Comput. Sci. 474–480 (2022)
3. Sumari, P., Syed, S.J., Sheng, L.H.: Light deep learning model architecture for chest x-ray based covid-19 detection. In: 2021 International Conference on Software Engineering & Computer Systems and 4th International Conference on Computational Science and Information Management (ICSECS-ICOCSIM), pp. 301–305 (2021). https://doi.org/10.1109/ICSECS52883.2021.00061
4. Wang, R.Y., Guo, T.Q., Li, L.G., Jiao, J.Y., Wang, L.Y.: Predictions of COVID-19 ınfection severity based on co-associations between the SNPs of co-morbid diseases and Covıd-19 through machine learning of genetic data. In: 2020 IEEE 8th International Conference on Computer Science and Network Technology (ICCSNT), pp. 92–96 (2020). https://doi.org/10.1109/ICCSNT50940.2020.9304990
5. Kesavan, S.M., Al Naimi, I., Al Attar, F., Rajinikanth, V., Kadry, S.: Res-UNet supported segmentation and evaluation of COVID19 lesion in lung CT. In: 2021 International Conference on System, Computation, Automation and Networking (ICSCAN), pp. 1–4 (2021). https://doi.org/10.1109/ICSCAN53069.2021.9526434
6. Wang, X., et al.: A weakly-supervised framework for COVID-19 classification and lesion localization from chest CT. IEEE Trans. Med. Imaging **39**(8), 2615–2625 (2020). https://doi.org/10.1109/TMI.2020.2995965
7. Radha, D.: Analysis of COVID-19 and pneumonia detection in chest x-ray ımages using deep learning. In: 2021 International Conference on Communication, Control and Information Sciences (ICCISc), pp. 1–6 (2021). https://doi.org/10.1109/ICCISc52257.2021.9484888
8. Rahman, T., et al.: Development and validation of an early scoring system for prediction of disease severity in COVID-19 using complete blood count parameters. IEEE Access **9**, 120422–120441 (2021). https://doi.org/10.1109/ACCESS.2021.3105321
9. Dino, H.I., Zeebaree, S.R.M., Hasan, D.A., Abdulrazzaq, M.B., Haji, L.M., Shukur, H.M.: COVID-19 diagnosis systems based on deep convolutional neural networks techniques: a review. In: 2020 International Conference on Advanced Science and Engineering (ICOASE), pp. 184–1892020). https://doi.org/10.1109/ICOASE51841.2020.9436542
10. Mann, P., Jain, S., Mittal, S., Bhat, A.: Generation of COVID-19 chest CT scan ımages using generative adversarial networks. In: 2021 International Conference on Intelligent Technologies (CONIT), pp. 1–5 (2021). https://doi.org/10.1109/CONIT51480.2021.9498272
11. Li, J., Zhang, D., Liu, Q., Bu, R., Wei, Q.: COVID-GATNet: a deep learning framework for screening of COVID-19 from chest x-ray ımages. In: 2020 IEEE 6th International Conference on Computer and Communications (ICCC), pp. 1897–1902 (2020). https://doi.org/10.1109/ICCC51575.2020.9345005
12. Singh, S., Sapra, P., Garg, A., Vishwakarma, D.K.: CNN based Covid-aid: Covid 19 detection using chest X-ray. In: 2021 5th International Conference on Computing Methodologies and Communication (ICCMC), pp. 1791–1797 (2021). https://doi.org/10.1109/ICCMC51019.2021.9418407

13. Vaishnavi, K.P., Ramadas, M.A., Chanalya, N., Manoj, A., Nair, J.J.: Deep learning approaches for detection of Covid-19 using chest x-ray ımages. In: 2021 Fourth International Conference on Electrical, Computer and Communication Technologies (ICECCT), pp. 1–8 (2021). https://doi.org/10.1109/ICECCT52121.2021.9616623
14. Chaudhary, S., Sadbhawna, S., Jakhetiya, V., Subudhi, B.N., Baid, U., Guntuku, S.C.: Detecting Covid-19 and community acquired pneumonia using chest CT scan ımages with deep learning. In: ICASSP 2021 - 2021 IEEE International Conference on Acoustics, Speech and Signal Processing (ICASSP), pp. 8583–8587 (2021). https://doi.org/10.1109/ICASSP39728.2021.9414007
15. Kishore, S.L.S., Sidhartha, A.V., Reddy, P.S., Rahul, C.M., Vijaya, D.: Detection and diagnosis of Covid-19 from chest x-ray ımages. In: 2021 7th International Conference on Advanced Computing and Communication Systems (ICACCS), pp. 459–465 (2021). https://doi.org/10.1109/ICACCS51430.2021.9441862
16. Thomas, A., Suresh, M.: Assessment of COVID-19 prevention and protection measures in hospitals. Clean. Eng. Technol. 100440 (2022)
17. Diallo, P.A.A.K., Ju, Y.: Accurate detection of COVID-19 using k-efficientnet deep learning ımage classifier and K-COVID chest x-ray ımages dataset. In: 2020 IEEE 6th International Conference on Computer and Communications (ICCC), pp. 1527–1531 (2020). https://doi.org/10.1109/ICCC51575.2020.9344949
18. Fan, X., Feng, X., Dong, Y., Hou, H.: COVID-19 CT ımage recognition algorithm based on transformer and CNN. Displays (2022), 102150 (2022)
19. Chansong, D., Supratid, S.: Impacts of kernel size on different resized ımages in object recognition based on convolutional neural network. In: 2021 9th International Electrical Engineering Congress (iEECON), pp. 448–451 (2021). https://doi.org/10.1109/iEECON51072.2021.9440284
20. Narin, A.: Detection of Covid-19 patients with convolutional neural network based features on multiclass x-ray chest images. In: 2020 Medical Technologies Congress (TIPTEKNO), pp. 1–4 (2020). https://doi.org/10.1109/TIPTEKNO50054.2020.9299289
21. Lau, M.M., Hann Lim, K.: Review of adaptive activation function in deep neural network. In: 2018 IEEE-EMBS Conference on Biomedical Engineering and Sciences (IECBES), pp. 686–690 (2018). https://doi.org/10.1109/IECBES.2018.8626714
22. Arora, D., Garg, M., Gupta, M.: Diving deep in deep convolutional neural network. In: 2020 2nd International Conference on Advances in Computing, Communication Control and Networking (ICACCCN), pp. 749–751 (2020). https://doi.org/10.1109/ICACCCN51052.2020.9362907
23. Mercioni, M.A., Holban, S.: The most used activation functions: classic versus current. In: 2020 International Conference on Development and Application Systems (DAS), pp. 141–145 (2020). https://doi.org/10.1109/DAS49615.2020.9108942
24. Bougourzi, F., Contino, R., Distante, C., Taleb-Ahmed, A.: CNR-IEMN: a deep learning based approach to recognize Covid-19 from CT-Scan. In: ICASSP 2021 - 2021 IEEE International Conference on Acoustics, Speech and Signal Processing (ICASSP), pp. 8568–8572 (2021). https://doi.org/10.1109/ICASSP39728.2021.9414185
25. Zhang, Q., Liao, L., Yuan, H., Zhu, J., Zhang, X.J.: Exploiting shared knowledge from non-COVID lesions for annotation-efficient COVID-19 CT lung ınfection segmentation. IEEE J. Biomed. Health Inform. **25**(11), 4152–4162 (2021). https://doi.org/10.1109/JBHI.2021.3106341

26. Hasanzadeh, N., Paima, S.S., Bashirgonbadi, A., Naghibi, M., Soltanian-Zadeh, H.: Segmentation of COVID-19 ınfections on CT: comparison of four UNet-based networks. In: 2020 27th National and 5th International Iranian Conference on Biomedical Engineering (ICBME), pp. 222–225 (2020).https://doi.org/10.1109/ICBME51989.2020.9319412
27. Yang, Z., Zhao, L., Wu, S., Chen, C.Y.-C.: Lung lesion localization of COVID-19 from chest CT image: a novel weakly supervised learning method. IEEE J. Biomed. Health Inform. **25**(6), 1864–1872 (2021). https://doi.org/10.1109/JBHI.2021.3067465
28. Wu, Y.-H., et al.: JCS: an explainable COVID-19 diagnosis system by joint classification and segmentation. IEEE Trans. Image Process. **30**, 3113–3126 (2021). https://doi.org/10.1109/TIP.2021.3058783
29. Sharma, M., Sau, P.C.: Blood vessel segmentation using segnet. In: 2019 4th International Conference on Information Systems and Computer Networks (ISCON), pp. 620–624 (2019). https://doi.org/10.1109/ISCON47742.2019.9036307
30. Rawat, R.M., Garg, S., Jain, N., Gupta, G.: COVID-19 detection using convolutional neural network architectures based upon chest x-rays ımages. In: 2021 5th International Conference on Intelligent Computing and Control Systems (ICICCS), pp. 1070–1074 (2021). https://doi.org/10.1109/ICICCS51141.2021.9432134
31. Ohata, E.F., et al.: Automatic detection of COVID-19 infection using chest X-ray images through transfer learning. IEEE/CAA J. Autom. Sin. **8**(1), 239–248 (2021). https://doi.org/10.1109/JAS.2020.1003393
32. Haritha, D., Swaroop, N., Mounika, M.: Prediction of COVID-19 cases using CNN with X-rays. In: 2020 5th International Conference on Computing, Communication and Security (ICCCS), pp. 1–6 (2020). https://doi.org/10.1109/ICCCS49678.2020.9276753
33. Rodríguez, J., et al.: A Covid-19 patient severity stratification using a 3D convolutional strategy on CT-scans. In: 2021 IEEE 18th International Symposium on Biomedical Imaging (ISBI), pp. 1665–1668 (2021). https://doi.org/10.1109/ISBI48211.2021.9434154
34. Bhadra, R., Kar, S.: Covid detection from cxr scans using deep multi-layered cnn. IEEE Bombay Sect. Signature Conf. (IBSSC) **2020**, 214–218 (2020). https://doi.org/10.1109/IBSSC51096.2020.9332210
35. Foysal, M., Aowlad Hossain, A.B.M.: COVID-19 detection from chest CT ımages using ensemble deep convolutional neural network. In: 2021 2nd International Conference for Emerging Technology (INCET), pp. 1–6 (2021). https://doi.org/10.1109/INCET51464.2021.9456387
36. Musleh, A.A., Maghari, A.Y.: COVID-19 detection in x-ray ımages using CNN algorithm. In: 2020 International Conference on Promising Electronic Technologies (ICPET), pp. 5–9 (2020). https://doi.org/10.1109/ICPET51420.2020.00010
37. Aparna, G., Gowri, S., Bharathi, R., V.J.S, J.J., A. P.: COVID-19 prediction using x-ray ımages. In: 2021 5th International Conference on Trends in Electronics and Informatics (ICOEI), pp. 903–908 (2021). https://doi.org/10.1109/ICOEI51242.2021.9452740
38. Tang, S., et al.: EDL-COVID: ensemble deep learning for covid-19 case detection from chest x-ray images. IEEE Trans. Industr. Inf. **17**(9), 6539–6549 (2021). https://doi.org/10.1109/TII.2021.3057683
39. Arias-Londoño, J.D., Gómez-García, J.A., Moro-Velázquez, L., Godino-Llorente, J.I.: Artificial intelligence applied to chest X-Ray images for the automatic detection of COVID-19. a thoughtful evaluation approach. IEEE Access **8**, 226811–226827 (2020). https://doi.org/10.1109/ACCESS.2020.3044858
40. Tabik, S., et al.: COVIDGR dataset and COVID-SDNet methodology for predicting COVID-19 based on chest X-ray images. IEEE J. Biomed. Health Inform. **24**(12), 3595–3605 (2020). https://doi.org/10.1109/JBHI.2020.3037127

41. Kang, M., et al.: Quantitative assessment of chest CT patterns in COVID-19 and bacterial pneumonia patients: a deep learning perspective. J. Korean Med. Sci. **36**(5), e46 (2021). https://doi.org/10.3346/jkms.2021.36.e46
42. Garlapati, K., Kota, N., Mondreti, Y.S., Gutha, P., Nair, A.K.: Detection of COVID-19 using x-ray ımage classification. In: 2021 5th International Conference on Trends in Electronics and Informatics (ICOEI), pp. 745–750 (2021). https://doi.org/10.1109/ICOEI51242.2021.9452745

An Efficient Key Frame Extraction from Surveillance Videos for Real-World Anomaly Detection

P. Mangai[1](✉), M. Kalaiselvi Geetha[2], and G. Kumaravelan[3]

[1] Department of Computer Science and Engineering, Annamalai University, Annamalainagar, Tamilnadu, India
pannirselvammangai@gmail.com

[2] Department of Computer Science and Engineering, Annamalai University, Annamalainagar, Tamilnadu, India

[3] Department of Computer Science, Pondicherry University, Karaikal, Puducherry, India

Abstract. Technology development has entrenched video analysis in numerous applications. Video analysis is important to obtain the data contents rather than its characteristics. Instead of examining the entire data content analyzing the essential portions as key frame reduces the computation time and memory. Extracting the important features from the dataset is termed as key frame extraction and this research work presents an efficient methodology to extract important frames from the video instances using HSV histogram and k-means clustering algorithm. The raw video data is preprocessed and then data points are clustered using k-means clustering. The proposed k-means clustering based key frame extraction technique is experimented with using UCF Crime dataset and the key frames are extracted from seven categories of videos. Detailed analysis of actual features and extracted features are comparatively analyzed and the observations are presented in the experimental section.

Keywords: Anomaly detection · Key frame extraction · Histogram · k-Means clustering · Video surveillance

1 Introduction

The tremendous growth in hardware and software technology improves the performance of video-based applications. Video applications are not only limited to surveillance also it has been adopted in daily life such as hotels, banks, schools, etc., for real-time monitoring [1]. However, video surveillance systems need human intervention to operate. Intelligent techniques are required to obtain necessary features and semantic information. The application of data mining in video applications extracts useful and important information in a dataset that contains videos. This aids in the automated understanding of video solutions and increases the effectiveness of video processing [2]. However, processing such huge volume of video data upsurges computation rate in addition with memory while indexing the files. So, image processing concepts are introduced in video

J. I.-Z. Chen et al. (Eds.): ICIPCN 2022, LNNS 514, pp. 197–212, 2022.
https://doi.org/10.1007/978-3-031-12413-6_16

analysis to extract the key frames and foreground extraction process to reduce the computational burden [3]. The key frame extraction process is mainly categorized into four types.

- Shot boundary-based extraction – The essential frame is obtained from the shot in the fixed position frame [4].
- Video contents-based extraction – The essential frame is obtained from the rate of changes in the video [5]
- Motion analysis-based extraction – The essential frame is obtained from the global motion of the video [6]
- Clustering based extraction [7] – The essential frame is obtained from the clustered features.

Each method listed above have their own merits and demerits. The technique based on shot boundary is simple, however it could not able to express the contents properly in the obtained frame. The technique based on video content extracts numerous key frames which are not appropriate for analysis. The technique based on motion analysis selects the minimal key frame which leads to information loss. The technique based on clustering represents the contents in the video better than all approaches.

This research work aims to obtain the essential key frames from the real-world video for anomaly detection. The frame speed and object motion in the video reflects the moving targets and extracting the key frame from the motion is essential to analyze and categorize the anomaly. This research work utilizes k-means clustering technique along with the HSV histogram and groups diverse motion segments in the data. The main contribution of the research work is summarized as follows.

- Proposed an algorithm for obtaining essential frames from real-time video using HSV histogram and k-Means clustering.
- Standard UCF Crime dataset is analyzed under 7 categories and essential frames are obtained.

The remaining portions of the literature is structured as: detailed literature analysis is provided in Sect. 2, Sect. 3 presents the proposed key frame extraction technique and its simulation analysis, observations are presented in Sect. 4, the observations are summarized as conclusion in the last section.

2 Related Works

A vast survey has been made to analyze the existing video key frame mining methods and the observations are summarized shortly. An unsupervised key frame extraction model reported in [8] used sparse representations and low rank strategy to obtain the essential frames in the video contents. The distinct characteristics of a video frame are defined as sparsity and the low rank constraints used in this research work improves the system robustness against noise factors. To preserve the consistency of sparse coders a Laplacian matrix is introduced in the research work for similar local features. However sparse models are less aware due to their spatio-temporal structures. To overcome this gram matrix determinant is reported in [9] to extract key frames from videos. The research model provides an efficient difference of convex programming algorithm to obtain global solution for the cost function.

Key frame extraction for static video summarization is reported in [10] that uses greedy, numerical and genetic algorithm. Out of three algorithms, the greedy algorithm provides a better balance between speed and overall performance. The challenges in real-time video data processing and issues are reported through automated key frame extraction model in literature [11]. Initially, the issues in video frame feature selection, segmentation and frame extraction are analyzed and developed an automatic model for extracting the essential frames. The features such as texture, color, gradients are considered in the complex videos in the extraction process. Though the performance is better for day light videos, it lags in extracting key frames on dark videos.

Key frame extraction for optimal static video summarization model reported in [12] extracts the key frames based on adaptive mean shift algorithm. The most representative key frames are identified and summarized to produce a video stream that reduces the human opinion in video analysis. Hidden Markov states with autoregressive process for time series data modelling and key frame extraction is reported in [13] obtains essential feature frames considering the predefined states in the shots. The conditional probability is used to obtain the necessary frames for further analysis. However, defining the states based on probability will reduces the accuracy in the key frame extraction process.

An image saliency extraction model for key frame extraction is reported in [14] that extracts the key frames from large scale data. The presented research model obtains essential frames considering the deep information assisted saliency extraction technique to process motion pictures. An automatic synthetic aperture radar based scattering video key frame selection model is reported in [15]. The extraction model utilizes sub aperture energy gradient along with object tracker and statistical model to differentiate the key frames. High robustness and reliability are the major merits of the research work and it makes the application suitable for complex video analysis.

A background frame extraction approach reported in [16] utilizes joint Gaussian conditional abstraction approach to obtain video frames weight composition from robust background images. The inter and intra frame relationships are obtained for all the frames based on the temporal and spatial coherence. Since the contextual objects in the frame are static and motionless this research work presents a motionless extractor model to obtain frame changes considering the threshold parameter. Pixelwise difference is obtained to extract the immobile regions through anticipated weights that will produce apparent, consistent and robust background images as key frames.

Spatio-temporal consistency-based salinity detection for object extraction in video is presented in [17]. The detection model separates the foreground and background automatically by observing the irrelevant moving objects. The reliability between the inter frame and the intra frame gradient is obtained as spatiotemporal gradient field. Then the adjacent frame similarity is obtained based on temporal consistency and finally, all the information is fused to obtain salient objects. The extraction of key frames for video summaries aids in rapid comprehension of the content. The key frame extraction model reported in [18] maps all the video data into summary space and by analyzing the space structure a unique description is obtained for each frame. Using Hash algorithm, similarity between the frames is obtained and finally, key frames are organized based on temporal order to construct the video summary.

The above literature analysis discussed various abstraction methods. Though existing methodologies provides better performances, the extraction process requires a complex mathematical model or probability solutions to extract the key frames which increase the system complexity. Since it is a basic process for a classification model adding complex application in the initial stage for key frame extraction will reflect in overall system complexity. Due to this an effective key frame extraction is necessary for real-world video processing applications.

3 Proposed Work

The proposed efficient key frame extraction model for anomaly detection is presented in this section. The proposed approach employs k-means clustering to extract the key frames. HSV histogram is incorporated in the initial steps for feature improvement and minimizes the computation difficulty. The procedure starts with preprocessing the input video data. It is essential to resize the video data frame. Since most of the video is recorded with 1280 × 720 pixel quality, processing such detailed video consumes more time and memory also it will increase the computation burden to the system. So, the input videos are preprocessed and converted into 320 × 240 pixels to minimize the processing time. Once the preprocessing steps are completed, the color features in the frames are analyzed.

The important features in an image is presented as color features and it is considered as essential feature while extracting the essential frames. In most of the image processing applications the color space is represented based on HSV features as it is similar to color perception of human. In the proposed work, the color features are used to accumulate the HSV histogram. HSV represents the elements Hue for H, Saturation in an image for S and Brightness in the image for V. In the proposed work analysis, initially, the hue elements are split into eight different parts, similarly saturation and brightness are split into three parts. Totally 72 subspaces (3 × 3 × 8) are obtained as HSV space. To derive the color components, different weights are assigned to HSV components which are based on human visual sensitivity. From this, a feature vector is represented in one-dimensional space as follows

$$l = 9H + 3S + V \tag{1}$$

The accumulative histogram is obtained by counting the entire pixel and sub-space pixels. Based on Euclidean distance the difference between the video images are calculated and it is given as

$$d\left(h_i, h_j\right) = \sqrt{\sum_{k=0}^{71} \left(h_{ik} - h_{jk}\right)^2} \tag{2}$$

where the distance between i^{th} and j^{th} frame is represented as $d\left(h_i, h_j\right)$ and color accumulation histogram vector for i^{th} and j^{th} is represented as h_i, and h_j. The similarity is measured using the obtained feature vectors and it is expressed as a ratio between object similar degree and overall objects. Mathematically it is expressed as follows.

$$f(h_i) = \sum_{d_{ij<vis}} \left(1 - \frac{d(h_i, h_j)}{\omega}\right) \tag{3}$$

where the sensing range is defined as $d_{ij<vis}$ and the group similarity is represented as ω. The one-dimensional feature vector and similarity group data are further processed using a typical k-means clustering technique that performs clustering using distance between the points. The grouping or clustering of dataset points to represent the data into a defined cluster is the process performed in k-means. Figure 1 depicts the simple illustration of the k-Means clustering process.

The average of all the points represents the data when it is mapped into a co-ordinate system is calculated or measured for grouping in the k-means clustering.

Fig. 1. Simple illustration of k-Means clustering.

The desired clusters can be obtained using statistical models and the obtained clusters are further used to analyze the data. Though various techniques are available for key frame extraction, the reason for selecting k-means for the proposed approach is its simple process and high convergence speed. K-means is efficient and scalable even for large data sets. The complexity of k-means is $O(nkt)$ in which the samples are represented as n, number of classes is represented as k and t denotes the iteration. Generally, k and t range is always lesser than n.

The cluster number and initial centroids must be determined before the clustering process. Usually, a random number is selected for centroid and number of clusters. However, these factors have great connection with results, so, the proposed technique centroid and cluster numbers are obtained as follows.

Step. 1: Initially, the average motion is obtained to measure the segment condition, and then it is quantified by multiplying with 100 for all segments.
Step. 2: The initial centroids are obtained by computing average V and assume that as distance d.
Step. 3: Now compute the maximum distance d_i from the initial centroid for each image frame.
Step. 4: The condition for analysis is $d_i > d$. if it satisfies, then the present image frame is selected for further process and its centroid is considered as the new centroid.
Step. 5: Increase the k value by 1 and repeat until $d_i < d$.
Step. 6: The final d is the initial centroid and k represents the cluster number.
Step. 7: Obtain the criterion function for independent clustering as follows using mean square deviation

$$E = \sum_{i=1}^{j} \sum_{p=N_i} |p - k_i| \tag{4}$$

where the mean of class N_i is represented as k_i and the object in the space is given as p.
Step. 8: The object similarity is obtained through Euclidean distance and greater the similarity if the objects are too close to each other. For an n dimensional vector, the Euclidean distance is given as

$$d(a, b) = \sqrt{(a_1 - b_1)^2 + (a_2 - b_2)^2 + \ldots (a_n - b_n)^2} \tag{5}$$

Figure 2 depicts the process flow of proposed model. Initially, the process starts with preprocessing the input video which reduces the video dimension. Then by applying HSV histogram the color features are obtained from the input frames. The number of clusters and centroids are calculated to obtain essential key frames. Based on the minimum similarity distance the image frames are grouped accordingly and the centroids are repositioned. The process is repeated until no image frames are left to group. Then in each cluster, the distance is computed from the centroid for each image frame. If the distance is larger, then present frame is selected as a key frame. This process is repeated until the desired condition met. The distance between objects is measured based on Euclidean distance and it is grouped based on the minimum distance.

Fig. 2. Process flow of proposed extraction model

4 Results and Discussion

The proposed key frame extraction technique performance is verified through experimentation performed in python installed on Linux platform version 18.04 Ubuntu. The processing unit comprises an intel i5 processor with 8GB RAM. For standard benchmark evaluation, Proposed model experimentation considers UCF Crime dataset [19]. The dataset includes real-world surveillance videos of 1900 data which covers 13 anomalies such as Arrest, Assault, Road Accident, Abuse, Arson, Stealing, Burglary, Robbery, Shoplifting, Explosion, Vandalism, Shooting, and Fighting. Due to computational complexity, 7 anomaly categories are considered in the proposed approach such as Shooting, Arson, Fighting, Abuse, Robbery, Stealing and Vandalism are considered for analysis. Figure 3 depicts the details of UCF Crime dataset and each category frames.

Fig. 3. Frames in each category from UCF Crime dataset

Seven categories of data are used in the proposed model experimentation. Figure 4 depicts the sample key frame images obtained for the abuse category. This category has 50 sample videos and Fig. 4 depicts the key frames for sample video 2.

Fig. 4. Sample key frames extracted from UCF crime dataset for aAbuse category

The abuse category has a total of 193497 frames for 50 videos and using proposed method total 26681 key frames are extracted from 50 videos. The sample images given in Fig. 4 are extracted from video 2 which has 865 original frames in which 375 key feature frames are extracted using the proposed key frame extraction technique. Figure 5 depicts the key frame extracted count for all the 50 videos.

Fig. 5. Key frames extracted for abuse category

Figure 6 depicts the sample key frame images obtained for the robbery category. This category has 150 sample videos and the Fig. 6 depicts the key frames for sample video 5. The robbery category has a total of 422594 frames for 150 videos and using proposed method total 67616 key frames are extracted from 150 videos.

Fig. 6. Sample key frames extracted from UCF crime dataset for robbery category

The sample images given in Fig. 6 are extracted from video 5 which has 2759 original frames in which 390 key feature frames are extracted using the proposed key frame extraction technique. Figure 7 depicts the key frame extracted count for all the 150 videos.

Fig. 7. Key frames extracted for robbery category

Figure 8 depicts the sample key frame images obtained for the arson category. This category has 53 sample videos and the Fig. 8 depicts the key frames for sample video 3. The arson category has a total of 171861 frames for 53 videos and using proposed method total 22119 key frames are extracted from 53 videos.

Fig. 8. Sample key frames extracted from UCF crime dataset for Arson category

The sample images given in Fig. 8 are extracted from video 3 which has 4013 original frames in which 450 key feature frames are extracted using the proposed key frame extraction technique. Figure 9 depicts the key frame extracted count for all the 53 videos.

Fig. 9. Key frames extracted for Arson category

Figure 10 depicts the sample key frame images obtained for the stealing category. This category has 114 sample videos and the Fig. 10 depicts the key frames for sample video 7. The stealing category has a total of 545757 frames for 114 videos and using the proposed method total 48642 key frames are extracted from 114 videos.

Fig. 10. Sample key frames extracted from UCF crime dataset for stealing category

The sample images given in Fig. 10 are extracted from video 7 which has 3090 original frames in which 444 key feature frames are extracted using the proposed key frame extraction technique. Figure 11 depicts the key frame extracted count for all 114 videos.

Fig. 11. Key frames extracted for stealing category

Figure 12 depicts the sample key frame images obtained for the vandalism category. This category has 50 sample videos and the Fig. 12 depicts the key frames for sample video 2. The vandalism category has a total of 147142 frames for 50 videos and using the proposed method total 12485 key frames are extracted from 50 videos.

Fig. 12. Sample key frames from UCF Crime dataset for vandalism category

The sample images given in Fig. 12 are extracted from video 2 which has 1841 original frames in which 157 key feature frames are extracted using the proposed key frame extraction technique. Figure 13 depicts the key frame extracted count for all the 50 videos.

Fig. 13. Key frame extracted for vandalism category

Figure 14 depicts the sample key frame images obtained for the shooting category. This category has 54 sample videos and the Fig. 14 depicts the key frames for sample video 3. The shooting category has a total of 154692 frames for 54 videos and using the proposed method total 14338 key frames are extracted from 54 videos.

Fig. 14. Sample key frames extracted from UCF crime dataset for the shooting category

The sample images given in Fig. 14 are extracted from video 3 which has 2277 original frames in which 261 key feature frames are extracted using the proposed key frame extraction technique. Figure 15 depicts the key frame extracted count for all the 54 videos.

Figure 16 depicts the sample key frame images obtained for the fighting category. This category has 54 sample videos and Fig. 16 depicts the key frames for sample video 3. The fighting category has a total of 264547 frames for 51 videos and using proposed method total 44405 key frames are extracted from 51 videos.

Fig. 15. Key frames extracted for shooting category

Fig. 16. Sample key frames extracted from UCF Crime dataset for fighting category

The sample images given in Fig. 16 are extracted from video 3 which has 3102 original frames in which 489 key feature frames are extracted using the proposed key frame extraction technique. Figure 17 depicts the key frame extracted count for all the 51 videos.

Fig. 17. Key frames extracted for fighting category

Fig. 18. Comparison of actual frames and extracted frames in the proposed approach

Figure 18 depicts an overall comparison of actual frames and extracted key frames using the proposed method for all the seven categories used in the experimentation. From the Fig. 18 it is clear that maximum frames are neglected due to less information and only the essential frames are detected in the presented extraction technique. Processing these key frames will reduce the computation time and memory of the system, so that the proposed extraction technique will be suitable for any real time surveillance applications.

5 Conclusion

An efficient key frame extraction from surveillance videos for real-world anomaly detection is presented in this research work. Extracting the key frames from the large volume of video data reduces the computation complexity of the system and memory requirements. However, implementing a complex algorithm for key frame extraction will improve the accuracy but it might affect the system performance. So, in this research work, a simple but efficient key frame extraction is presented using k-means clustering. HSV histogram is used along with k-means clustering for efficient extraction of key frames. The proposed technique experimented with UCF Crime dataset and seven video categories are processed to extract the essential frames. Further, this research work can be extended as an application for anomaly detection and classification based on the extracted key frames.

References

1. Niu, G., Chen, Q.: Learning a video frame-based face detection system for security fields. J. Vis. Commun. Image Represent. **55**, 457–463 (2018)

2. Huang, C., Wang, H.: A novel key-frames selection framework for comprehensive video summarization. IEEE Trans. Circuits Syst. Video Technol. **30**(2), 577–589 (2020)
3. Ding, L., Tian, Y., Fan, H., Chen, C., Huang, T.: Joint coding of local and global deep features in videos for visual search. IEEE Trans. Image Process. **29**, 3734–3749 (2020)
4. Naveen Kumar, G.S., Reddy, V.S.K.: Detection of shot boundaries and extraction of key frames for video retrieval. Int. J. Knowl.-Based Intell. Engi. Syst. **24**, 11–17 (2020)
5. Chen, M., Han, X., Kamruzzaman, M.M.: Quality-guided key frames selection from video stream based on object detection. J. Vis. Commun. Image Represent. **65**, 1–15 (2019)
6. Wang, Z., Zhu, Y.: Video key frame monitoring algorithm and virtual reality display based on motion vector. IEEE Access **8**, 159027–159038 (2020)
7. Zhao, H., Wang, W.J., Wang, T., Chang, Z.B., Zeng, X.Y.: Key-frame extraction based on HSV histogram and adaptive clustering. Math. Prob. Eng. 1–11 (2019)
8. Gu, X., Lu, L., Yang, Z.: Sentiment key frame extraction in user-generated micro-videos via low-rank and sparse representation. Neurocomputing **410**, 441–453 (2020)
9. Tan, B., Li, Y., Ding, S., Paik, I., Kanemura, A.: DC programming for solving a sparse modeling problem of video key frame extraction. Digit. Signal Process. **83**, 214–222 (2020)
10. Mademlis, I., Tefas, A., Pitas, I.: A salient dictionary learning framework for activity video summarization via key-frame extraction. Inf. Sci. **432**, 319–331 (2018)
11. Chen, L., Wang, Y.: Automatic key frame extraction in continuous videos from construction monitoring by using color, texture, and gradient features. Autom. Constr. **81**, 355–368 (2017)
12. Hannane, R., Elboushaki, A., Afdel, K.: MSKVS: adaptive mean shift-based keyframe extraction for video summarization and a new objective verification approach. J. Vis. Commun. Image Represent **55**, 179–200 (2018)
13. Loukas, C., Varytimidis, C., Rapantzikos, K., Kanakis, M.A.: Keyframe extraction from laparoscopic videos based on visual saliency detection. Comput. Meth. Prog. Biomed. **165**, 13–23 (2018)
14. Zhong, Q., Zhang, Y., Zhang, J., Shi, K., Yu, Y., Liu, C.: Key frame extraction algorithm of motion video based on priori. IEEE Access **8**, 174424–174436 (2020)
15. Zhang, Y., Zhu, D., Bi, H., Zhang, G., Leung, H.: Scattering key-frame extraction for comprehensive VideoSAR summarization: a spatiotemporal background subtraction perspective. IEEE Trans. Instrum. Meas. **69**(7), 4768–4784 (2020)
16. Wang, H.C., Lai, Y.C., Cheng, W.H., Cheng, C.Y., Hua, K.L.: Background extraction based on joint gaussian conditional random fields. IEEE Trans. Circuits Syst. Video Technol. **28**(11), 3127–3140 (2018)
17. Guo, Y., Li, Z., Liu, Y., Yan, G., Yu, M.: Video object extraction based on spatiotemporal consistency saliency detection. IEEE Access **6**, 35171–35181 (2018)
18. Li, X., Zhao, B., Lu, X.: Key frame extraction in the summary space. IEEE Trans. Cybern. **48**(6), 1923–1934 (2018)
19. Sultani, W., Chen, C., Shah, M.: Real-world Anomaly Detection in Surveillance Videos Cornell University Library, vol. 1, pp. 1–10 (2018). http://arxiv.org/abs/1801.04264

The XGBoost Model for Network Intrusion Detection Boosted by Enhanced Sine Cosine Algorithm

Nadheera AlHosni[1], Luka Jovanovic[2], Milos Antonijevic[2], Milos Bukumira[2], Miodrag Zivkovic[2], Ivana Strumberger[2], Joseph P. Mani[1], and Nebojsa Bacanin[2](✉)

[1] Modern College of Business and Science, Muscat, Sultanate of Oman
{20201838,drjosephmani}@mcbs.edu.om

[2] Singidunum University, Danijelova 32, 11000 Belgrade, Serbia
{luka.jovanovic.191,milos.bukumira.20}@singimail.rs,
{mantonijevic,mzivkovic,istrumberger,nbacanin}@singidunum.ac.rs

Abstract. Network intrusion detection systems are created with the purpose of detecting and identifying threats and vulnerabilities of a target network. One of the most cardinal challenge with such systems is that they frequently generate a considerable high rate of false positives and false negatives and that has a profound influence on its efficiency. However, this issue can be mitigated by applying machine learning models and algorithms. Therefore, in the research shown is in this paper, drawbacks of network intrusion detection systems are addressed by applying well-known XGBoost classifier tuned with improved sine cosine metaheuristics. It is known that machine learning models such is XGBoost should be optimized for each practical problem (dataset) and in this study was shown that XGBoost hyper-parameters' tuning can be efficiently conducted by applying improved sine cosine algorithm. Proposed hybrid framework is validated against benchmarking NSL-KDD dataset and compared to XGBoost without tuning, as well as with few other metaheuristics approaches used for XGBoost tuning, including original sine cosine algorithm. Obtained research findings prove that proposed method is able to improve classificthe position of the current solution that is shown ation accuracy and precision compared to other methods including in the analysis.

Keywords: Network security · Intrusion detection · NSL-KDD data-set · Sine-cosine algorithm · XGBoost

1 Introduction

Securing networks and data is considered one of the greatest challenges in the modern technology world, as there are several risk levels that can affect the workflow and growth of any organization [39]. Network security may be defined as a process or number of steps taken in order to keep resources confidential, accessible, as well maintain their integrity. An organizations network safety is never fully guaranteed, as the security

J. I.-Z. Chen et al. (Eds.): ICIPCN 2022, LNNS 514, pp. 213–228, 2022.
https://doi.org/10.1007/978-3-031-12413-6_17

level might change at any time. With this in mind it is apparent that a system needs to grow and adapt to constantly changing demands. To maintain security in a system there are four processes which should be considered as risk prevention. These are assessment, protection, detection, and response [15]. Accordingly, security technologies are always striding to improve available strategies by applying and improving different algorithms, methods and approaches.

An effective approach to improving security that this paper will be focusing on is the deployment of network intrusion detection system (NIDS). These refer to a group of network security classification systems that aim to discover and minimize attack vectors, as well as attempt detect and preempt newly developed attacks [1]. An intrusion can be considered a series of steps that allows for unauthorized access to specific details or resources relating to a network or computer system in an organization, with the aim to destroy the integrity of the system, compromising the confidentiality or availability of resources and services. Detection, however, refers to an application or method that allows an organization to detect illegal activities and discover security breaches [42].

When considering any form of intrusion detection system (IDS) it is important to note that they are comprised of the two prime categories, anomaly identification and misuse detection [2]. The main goal of anomaly detection is to explore and compare activities with known stored signatures of exploits. Accordingly misuse detection leverages prior knowledge of previous malicious actions in order to spot traces of an attack [24].

The gradient boosting (XGBoost) machine learning model presents a decision-tree-based ensemble learning approach that works to provide a gradient boosting library [22]. It may be implemented in any training data system, in order improve performance using a mixture mathematical approaches. Currently the XGBoost model is considered as one of the best performing machine learning methods [40]. Nonetheless, despite admirable capabilities, XGBoost, as most of machine learning models, presents many performance governing parameters that require adequate tuning to ensure optimal operation. As manual adjustment rapidly becomes unfeasible, methods for automation of this process have been developed a prominent research area called hyper-parameter tuning. Moreover, since each machine learning model has many parameters that should be adjusted for each practical dataset, this challenge is NP hard. Various algorithms have been applied to hyper-parameter tuning, with differing degrees of success.

A notable group of algorithms that show great promise when applied to optimization problems are nature inspired algorithms that model behaviors of groups of entities observed in nature. These algorithms often, referred to as swarm intelligence algorithms, make use of metaheuristic methods in order to address computational problems considered difficult. Although a optimal outcome is never guaranteed, swarm algorithms may often handle NP hard problems providing acceptable results, withing reasonable computational periods with realistic resource usage. While many swarm intelligence metaheuristics are indeed inspired by behaviors observed in nature, examples exist that make use of abstract mathematical concepts that guide optimization. One such notable exception is the sine cosine algorithm (SCA) [31], that makes use of trigonometric functions to look for optimal solutions. While the base algorithms shows admirable results when compared to other contemporaries, further improvements are feasible.

The incentive behind the approach suggested in this work is to extend the original since cosine algorithm and improve on the excellent performance offered by the base algorithm, as well as combining it with XGBoost method in order to improve overall classification ability [21]. With this in mind, the no free lunch (NFL) theorem states that it is impossible to have a universal optimization algorithm capable of addressing all optimization problems [43], accordingly further improvements of existing optimization algorithms are always possible. On this basis, the proposed research focuses on addressing an important and difficult challenge in intrusion detection by leveraging the XGBoost methodology. In order to attain desired performance the enhanced SCA is used to optimize XGBoost hyper-parameter's values. In order to test and demonstrate the proposed models effectiveness it is applied to a well-known NSL-KDD data-set, which is often used for benchmarking purposes and validation of intrusion detection methods.

Based on the stated above, the essential contributions of this research can be considered three fold:

- the development of a novel sine cosine algorithm designed specifically to address the shortcomings of the original one;
- application of the proposed methods to hyper-parameters' optimization for XGBoost to increase accuracy when tackling intrusion detection and
- based on the attained results, proposed method obtains better performance metrics than other contemporary algorithms.

The reminder of this work is structured according to the following. Section 2 presents an in depth dive in to works that deal with similar topics, followed by a discussion of the XGBoost methodology. Section 3 gives a description of methods proposed for tackling problems addressed in the this research. Details of the conducted experimentation, the attained results as well as a comparative analysts with other contemporary swarm intelligence algorithms is given in Sect. 4. Finally a closing word on the conducted research as well as plans for future work is presented in Sect. 5.

2 Preliminaries and Related Work

This section presents a look at the background of the XGboost method as well as a swarm intelligence. This is followed by a review of literature relating to their application and research works.

2.1 XGBoost Model

The XGBoost model features algorithms that support linear classification problems resolution, additionally it performs and manages Taylor expansions in order to ensure accurate results for cost functions by presenting the second derivative [28]. The following section represent the XGBoost principles. The first principle in the XGBoost model is applying the additive training method. This method helps in optimizing the objective function. The optimization process works based on the previous step, as the latter step relies on it. This model can be represented as t-th objective function according to Eq. (1)

$$obj^{(t)} = \sum_{i=1}^{n} l(y_i, \hat{y}_i^{t-1} + f_t(x_i)) + \Omega(f_t) + constant \tag{1}$$

where ℓ represent the term of loss in the t-th round, the constant means the term of constant, and the Ω refer to regularization term. As it is represented in Eq. (2):

$$\Omega(f_t) = \gamma \cdot T_t + \lambda \frac{1}{2} \sum_{j=1}^{T} w_j{}^2 \tag{2}$$

with the customization parameters being γ and λ.

The simplicity's of the tree structure depends on a large amount on customization parameter values. Additionally, the over-fitting problems can be solved by applying the second-order Taylor expansion on Eq. (1), this process represented in Eq. (3):

$$obj^{(t)} = \sum_{i=1}^{n} \left[l(y_i, \hat{y}_i^{t-1}) + \dot{g}_i f_t(x_i) + \frac{1}{2} h_i f_t{}^2(x_i) \right] + \Omega(f_t) + constant \tag{3}$$

where g represent the first derivative, h represent the second derivative. These can be calculated according to Eqs. (4) and (5) respectfully:

$$g_i = \partial_{\hat{y}_i^{t-1}} l(y_i, \hat{y}_i^{t-1}) \tag{4}$$

$$h_i = \partial^2{}_{\hat{y}_i^{t-1}} l(y_i, \hat{y}_i^{t-1}) \tag{5}$$

by inserting Eqs. (2), (4) and (5) into Eq. (3) and computing the derivative.

Finally, solution weight and loss function score results can be acquired from Eqs. (6) and (7) respectfully:

$$w_j^* = -\frac{\sum g_i}{\sum h_i + \lambda} \tag{6}$$

$$obj^* = -\frac{1}{2} \sum_{j=1}^{T} \frac{(\sum g_i)^2}{\sum h_i + \lambda} + \gamma \cdot T \tag{7}$$

The XGBoost methodology presents a promising approach when dealing with complex issues providing excellent results when applied to practical problems. As such researchers have put it to use to tackle real obstacles [27,34]. However as a trade off for this admirable performance, it presents a large number of parameters requiring acquitted adjustment in order to meet desired performance criteria. This process is known as hyper-parameter tuning and it has becomes a vital step in optimization. As such, many methods have been applied in an attempt to address it. A promising contender when dealing with hyper-parameter tuning are metaheuristic algorithms, as research shows they provide good results with reasonable resource usages. A notably well performing subgroup of algorithms applied to optimization being swarm intelligence algorithms. Accordingly many research groups have applied swarm intelligence algorithms to the problem of hyper-parameter optimization withing the XGboost framework [45].

2.2 Swarm Intelligence

By utilizing large numbers of simpler rule abiding agents, swarm intelligence algorithms allow for an emergence of problem solving abilities on a grander scale. The mechanics by which this is achieved may vary to some degree depending on the algorithm, usually centered on a stochastic metaheuristic approach. Due to the intrinsic randomness, a single execution of a swarm intelligent algorithm does not guarantee an optimal solution, however each successive iteration increases the chances of an optima being located. Capable of addressing NP hard problems within reasonable time frames and realistic resource utilization, swarm intelligence algorithms show great results when dealing with optimization problems.

Algorithms modeling many different behaviors exist. Some notable nature inspired examples are the artificial bee colony (ABC) algorithm that models the swarm pollen gathering mechanism observed in bees [29], the emperor penguin optimizer (EPO) that mimics huddling behaviors that allow emperor penguins to survive the harsh arctic conditions [25], the firefly algorithms (FA) inspired by mating rituals of various a light-emitting incest species [44]. Some algorithms are inspired by group hunting strategies observed in nature such as of gray wolves optimizer (GWO) [33], or the Whale optimization algorithm (WOA) [32]. However algorithms inspired by more abstract mathematical concepts exist notable among which is the sine-cosine algorithm (SCA), that makes use of trigonometric functions to aid searches [31]. Each algorithms presents a differed set of strengths and weaknesses, accordingly a problem appropriate algorithm needs to be selected in order to attain the best results. These algorithms have been applied to a diverse range of problems in many fields including medicine [14, 16, 19, 35, 47, 50], wireless technology [4, 11, 46, 48, 51], task scheduling [6, 13, 20, 49], artificial neural networks optimizing process [3, 7–10, 12, 18, 23, 26, 30, 41], as well as many others [5, 17, 41].

A notable optimization problem on interest being hyper-parameter tuning withing the GXBoost framework [37, 38] has proven fertile grounds for applications of swarm intelligence algorithm.

3 Proposed Method

3.1 The Original Sine-Cosine Algorithm (SCA)

Population-based optimization techniques use a group of random solution to solve the optimization problems. It is frequently used to evaluate the optimization function in order to improve a new set of rules. However, there is no guarantee a solution will be discovered in a single run. No matter what the challenges are or which algorithm is used, the issue persist for all stochastic population optimizations. The optimization process is separated into two stages: the exploration versus exploitation. In the optimization algorithm a random solution is merged into a set of high rate randomized solutions that works to locate promising regions in the search space. During the exploitation stage some changes might occur in the random solutions. Additionally, there are less random variations compared to the exploration stage [31]. The two position updating phases are shown in Eqs. (8) and (9):

$$X_i^{t+1} = X_i^t + \mathfrak{r}_1 \times sin(\mathfrak{r}_2) \times |\mathfrak{r}_3 P_i^t - X_i^t| \tag{8}$$

$$X_i^{t+1} = X_i^t + \mathfrak{r}_1 \times cos(\mathfrak{r}_2) \times |\mathfrak{r}_3 P_i^t - X_{i_i|}^t \tag{9}$$

where X_i^t represent the position of the current solution that is shown in $\mathfrak{i} - \mathfrak{th}$ iteration, $\mathfrak{r}_1, \mathfrak{r}_2, \mathfrak{r}_3$ are the random numbers, and P_i represent the position of the destination point in the $\mathfrak{i} - \mathfrak{th}$ dimension and the || denote absolute values. The two equations can be combined in to Eq. (10):

$$X_i^{t+1} = \begin{cases} X_i^{t+1} = X_i^t + \mathfrak{r}_1 \times sin(\mathfrak{r}_2) \times |\mathfrak{r}_3 P_i^t - X_i^t| & \mathfrak{r}_4 < 0.5 \\ X_i^{t+1} = X_i^t + \mathfrak{r}_1 \times cos(\mathfrak{r}_2) \times |\mathfrak{r}_3 P_i^t - X_{i_i}^t| & \mathfrak{r}_4 \geq 0.5 \end{cases} \tag{10}$$

with $\mathfrak{r}_4$ representing a random number in [0,1].

The four main parameters in Eq. 10 are $\mathfrak{r}_1$, $\mathfrak{r}_2$, $\mathfrak{r}_3$ and $\mathfrak{r}_4$. Where the $\mathfrak{r}_1$ parameter represent the movement direction or the next position regions, as this movement might occur in or outside of the solution space and destination. The second parameter $\mathfrak{r}_2$ defines the distance towards or away from the destination that the movement should occur in. Parameter $\mathfrak{r}_3$ represents the random weights of the destination which is randomly emphasized when ($\mathfrak{r}_3 > 1$) or not emphasized when ($\mathfrak{r}_3 < 1$). Accordingly, the attained values have an impact on the resulting destination. Finally parameter $\mathfrak{r}_4$ changes whether the sine or cosine functions is used in Eq. (10) [31].

Pseudocode for the SCA that brings these concepts together can be seen in Algorithm 1.

Algorithm 1. Pseudocode of SCA algorithm

```
initialize a set of search agents (X)
while t < maximumnumberofiterations do
    evaluate each of the search agents by the objective function
    update the best solution obtained so far (P = X*)
    update r1, r2, r3, r4 update the positions of search agents
end while
return The best solution obtained so far as the global optimum
```

3.2 The Enhanced Sine-Cosine Algorithm (ESCA)

Although SCA algorithm has established itself as a powerful optimizer, as every other metaheuristics approach, it has some known deficiencies. Thorough simulations with benchmark functions have shown that basic SCA has a powerful exploration mechanism. However, in later rounds of the algorithm's execution, the original implementation of the SCA algorithm lacks exploitation power. To counter this weakness of the basic SCA implementation, the research presented in this paper proposes the following approach. In last 33% of the iterations, when it is necessary to exploit promising areas of the search domain, the worst solution in each iteration is replaced by the new, randomly generated solution. This new solution is generated in a guided-random manner, where the parameters of the new solution are generated within the boundaries of the best solution, and the median value of all solutions.

3.3 Proposed eSCA-XGBoost Model

In the suggested eSCA-XGBoost model, the set of XGBoost parameters has been optimized by using the eSCA metaheuristics. The six parameters determined to be optimized are given in Table 1. These parameters are chosen on the grounds of the previous publications on the topic, for example [28], as they were determined empirically to have the most influence on the XGBoost model behavior.

Table 1. XGBoost model hyper-parameters

Parameters	Default values	Range	Explanation
eta	0,3	[0,1]	Learning rate, reduces weight each step
max_depth	6	[0,∞]	Maximum tree depth, larger values result in more specific and local samples
min_child_weight	1	[0,∞]	Minimum leaf weight larger values help prevent local optimization
gamma	0	[0,∞]	Governs loss function
subsample	1	(0,1]	Controls random sampling Proportion to prevent overfitting
colsample_bytree	1	(0,1]	Controls feature sampling proportion

Consequently, eSCA method is encoded as a vector of six components, every component representing one of the XGBoost hyper-parameters from Table 1. Four components are continuous by nature (eta, gamma, sub-sample, colsample_bytree), remaining two are integer (max_depth and min_child_weight), leading to the conclusion that this task is a typical mixed variables NP-hard challenge. During the exploration, due to the exploration equations of the eSCA algorithm, integer variables are being translated to continuous, and finally translated back to integer values with help of simple sigmoid transfer function.

The fitness value for each solution is being computed by building the XGBoost model based on the solution's parameters, and validating the performances on the training dataset. For the global best solution (the individual that obtained the best fitness on the training dataset), the constructed XGBoost model is being validated on the testing dataset, and these results have been reported in the section with simulation results.

4 Experiments and Comparative Analysis

Training and testing of the proposed model is conducted by applying the proposed method to the publicly available NSL-KDD dataset that can be acquired from the following URL https://www.unb.ca/cic/datasets/nsl.html. This dataset is formulated to suit intrusion detection system testing and evaluation and a detailed description of the features [36], while a summary of the general dataset details is provided in Table 2.

For assessing the proposed models performance several metrics need to be introduced. False negatives (FN), false positives (FP) are combined with true negatives (TN) as well as true positives (TP) to adequately gauge NIDS accuracy. This is done according to Eq. (11).

$$ACC = (TP + TN)/(TP + FP + TN + FN) \tag{11}$$

Furthermore, due to a uneven number of samples for each class in both training and testing sets, a disbalance is inherent in the NSL-KDD dataset. Accordingly precision Eq. (12) and recall Eq. (13) metrics are utilized for the evaluation, along with F-score Eq. (14).

$$Precision = \frac{TP}{TP + FP} \tag{12}$$

$$Recall = \frac{TP}{TP + FN} \tag{13}$$

$$F_1 = 2 \cdot \frac{Precison * Recall}{Precision + Recall} = \frac{TP}{TP + \frac{1}{2(FP+FN)}} \tag{14}$$

Table 2. NSL-KDD dataset summary

Property	Description
Number of records	126,620
Number of features	41
Number of classes	2 (normal uses and attacks)
Groups of attacks	4 (Probe, DoS, U2R and R2L)
Types of attacks	38 in total (21 in training set)
Number of sets	2 (a training and a testing set)

The events represented in the dataset consist of normal network use, denial of service (DoS) attacks, probe attacks, user to root (U2R) attacks, as well as remote to local (R2L) user attacks. Additionally, the dataset has predefined sets for both training and testing, the structure of which is represented in Table 3.

Due to varying data types within the dataset, a continuous range is constricted by standardizing data-points using Eq. (15)

Table 3. NSL-KDD dataset structure

Event type	Training set		Testing set	
Normal use	67,343	53.46%	9,711	43.08%
DoS	45,927	36.46%	7,456	33.07%
Probe	11,656	9.25%	2,421	10.74%
U2R	52	0.04%	200	0.89%
R2L	995	0.79%	2,756	12.22%
Total	125,973		22,544	

$$d'_{ij} = \frac{d_{ij} - \frac{1}{M}\sum_{i=1}^{M} d_{ij}}{\frac{1}{M}\sum_{i=1}^{M}\left|d_{ij} - \frac{1}{M}\sum_{i=1}^{M} d_{ij}\right|} \tag{15}$$

where M is the total number of records, d represents a single data-point during the i-th feature in the j-th record, additionally d' stands for the corresponding points standard value.

Following the standardization process is the normalization step done according to Eq. (16)

$$d''_{ij} = \frac{d'_{ij} - d_{min}}{d_{max} - d_{min}} \tag{16}$$

with d'' being the normalized value of the appropriate d' data-point, d_{max} and d_{min} representing maximum and minimum values for the given j-th feature.

The testing process follows methods laid out in previous works [28] dealing with NIDS optimization, using different algorithms, in order to provide valid grounds for a comparative analysis. For testing purposes a populations size of 100 agents has been applied, in 800 iterations and 8000 fitness function evaluations (FFE). This is done so as to provide valid grounds for a comparative analysis with previous works dealing with the same problem [28].

The introduced model is evaluated based on precision, recall, f-score and P-R curve. The ROC curve is replace with the P-R curve in order to capture the binary event estimation impact. These events occur due to the limited number of U2R attack samples compared to other types of sample. In order to improve performance evaluation P-R curve based values for precision (AP), mean average precision (mAP) and macro-averaging calculations are used.

The performance results of the suggested model are shown in comparison to the original XGBoost approach, as well as other contemporary algorithms used for optimization. These include the original SCA-XGBoost, PSO-XGBoost hybrids, as well as an additional well known metaheuristic algorithms, the artificial bee colony (ABC) [29], tested when applied to XGBoost hyper-parameter optimization. For the sake of a fair and valid comparison, the experiment mirrors the approach and conditions proposed in previous works that originally suggest the PSO-XGboost model [28]. Moreover, for the purpose of this study, experiments with PSO-XGBoost were recreated and results shown in [28] are validated. The results achieved by the proposed methods are shown in Table 4, alongside results for the basic XGBoost method, the PSO-XGBoost optimizer and the original SCA-XGBoost optimizer, with the optimal attained results marked in bold text. Furthermore, a visual comparison between AP values attained by the original XGBoost and the proposed eSCA-XGBoost methods can be seen in Fig. 1. It is possible to conclude from the presented results that the proposed eSCA-XGBoost obtains superior results, even for minority classes (such as U2R and R2L), consequently outperforming all other approaches included in the comparative analysis.

Table 4. The dataset testing set optimal parameters confusion matrix.

		Normal	Probe	Dos	U2R	R2L	Average/Total
XGBoost	Precision	0.63	0.75	**0.96**	0.75	0.67	0.76
	Recall	0.97	0.71	0.67	0.03	0.00	0.72
	F-Score	0.76	0.73	0.79	0.06	0.00	0.67
	Support	9711	2421	7458	200	2754	22544
PSO-XGBoost	Precision	0.66	**0.81**	0.94	**1.00**	**0.95**	0.81
	Recall	0.96	0.52	0.84	0.01	0.05	0.74
	F-Score	0.76	0.64	0.87	0.01	0.09	0.70
	Support	9771	2421	7458	200	2754	22544
ABC-XGBoost	Precision	0.66	0.79	0.94	0.95	0.89	0.80
	Recall	0.97	0.64	0.86	0.08	0.55	0.76
	F-Score	0.76	0.69	0.87	0.10	0.57	0.75
	Support	9771	2421	7458	200	2754	22544
SCA-XGBoost	Precision	0.67	0.79	0.93	0.92	0.85	0.79
	Recall	0.97	0.63	0.87	0.15	0.62	0.76
	F-Score	0.77	0.68	0.88	0.19	0.64	0.72
	Support	9771	2421	7458	200	2754	22544
eSCA-XGBoost	Precision	**1.0**	0.79	0.91	0.89	0.86	**0.89**
	Recall	**1.0**	**0.92**	**0.91**	**0.21**	**0.79**	**0.76**
	F-Score	**1.0**	**0.85**	**0.91**	**0.34**	**0.82**	**0.78**
	Support	9711	2421	7458	200	2754	22544

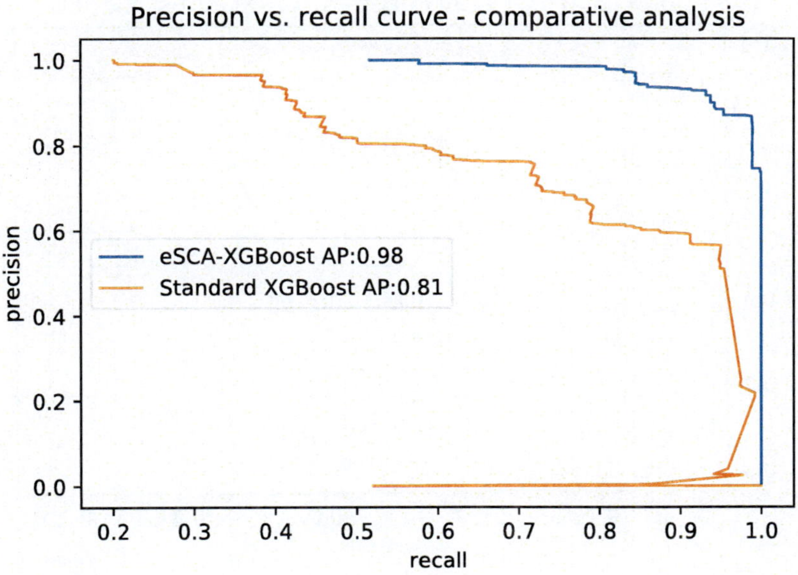

Fig. 1. PR curve comparative analysis

Values shown in Table 5 cover AP results from the P-R curves of the compared algorithms for all available event and class types. A visual representation of the PR curves, for each class of intrusion that the eSCA-XGBoost model has been applied to, can be seen in Fig. 2. Finally, the optimal values for XGBoost hyper-parameter determined using the proposed methods are shown in Table 6.

Table 5. Comparison of AP values for each class.

	XGBoost	PSO-XGBoost	ABC-XGBoost	SCA-XGBoost	eSCA-XGBoost
Normal	0.89	0.89	0.89	0.90	**1.0**
Probe	0.75	0.79	0.77	0.78	**0.81**
Dos	0.88	**0.94**	0.90	0.93	0.91
U2R	0.11	0.15	0.16	0.24	**0.33**
R2L	0.42	0.48	0.52	0.55	**0.84**

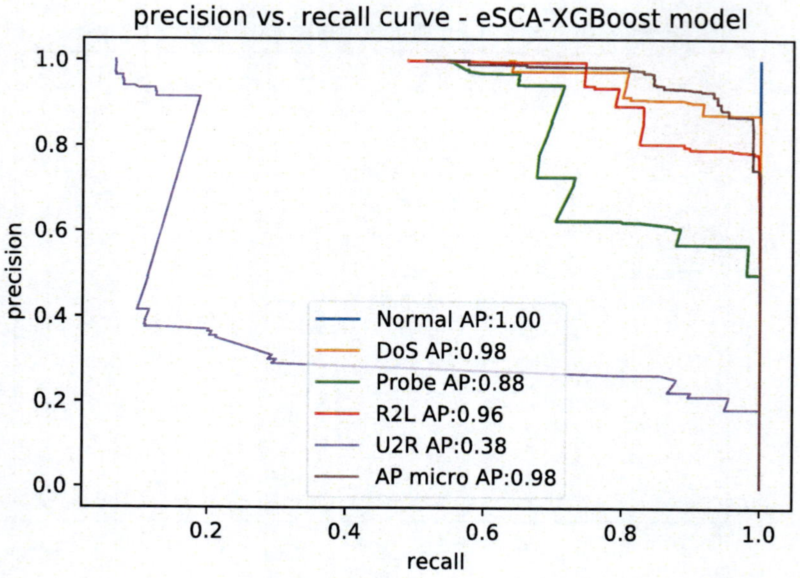

Fig. 2. PR curve model eSCA-XGBoost model

Table 6. XGBoost parameter values after optimisation by eSCA

Parameter	Determined value
booster	gbtree
max_depth	2
learning_rate	0.9488
sample_type	uniform
normalize_type	tree
rate_drop	0.5
n_estimators	300
min_child_weight	1.7453
subsample	0.6
colsample_bytree	0.8771
random_state	23
seed	23
num_class	5
objective	multi:softprob

5 Conclusion

Enhancing the security field is a common interest in today's technology. This experiment research introduce a method that helps in improving the network intrusion detection method using the XGBoost model with the help of the sine-cosine algorithm. Based on the known drawbacks of the basic SCA algorithm, a enhanced eSCA algorithm was devised, and utilized to optimize the hyper-parameters of the XGBoost model. The XGBoost was selected among other machine learning models (including decision tree and AdaBoost) as it obtained the best performances during initial testing on the given dataset. The limitations of the proposed method are that extensive testing should be performed with different machine learning models hybridized with a variety of swarm intelligence metaheuristics, that is significantly out of the scope of the research presented in this paper. To conclude, the research experiment shows that the XGBoost model effect in a great way and has the higher mAP and macro value, and this adjective is missing in other models. What is more, the U2R and R2L final results represented an excellent AP values for each class in the minority classes. As there are some vulnerabilities or issues that may need to be addressed, and will be considered in future research which aimed in developing the network security.

References

1. Ahmed, M., Pal, R., Hossain, M.M., Bikas, M.A.N., Hasan, M.K.: NIDS: a network based approach to intrusion detection and prevention. In: 2009 International Association of Computer Science and Information Technology-Spring Conference, pp. 141–144. IEEE (2009)
2. Anderson, J.A.: An Introduction to Neural Networks. MIT Press, Cambridge (1995)
3. Bacanin, N., Alhazmi, K., Zivkovic, M., Venkatachalam, K., Bezdan, T., Nebhen, J.: Training multi-layer perceptron with enhanced brain storm optimization metaheuristics. Comput. Mater. Continua **70**(2), 4199–4215 (2022). https://doi.org/10.32604/cmc.2022.020449, http://www.techscience.com/cmc/v70n2/44706
4. Bacanin, N., Arnaut, U., Zivkovic, M., Bezdan, T., Rashid, T.A.: Energy efficient clustering in wireless sensor networks by opposition-based initialization Bat algorithm. In: Smys, S., Bestak, R., Palanisamy, R., Kotuliak, I. (eds.) Computer Networks and Inventive Communication Technologies. LNDECT, vol. 75, pp. 1–16. Springer, Singapore (2022). https://doi.org/10.1007/978-981-16-3728-5_1
5. Bacanin, N., Bezdan, T., Tuba, E., Strumberger, I., Tuba, M.: Monarch butterfly optimization based convolutional neural network design. Mathematics **8**(6), 936 (2020)
6. Bacanin, N., Bezdan, T., Tuba, E., Strumberger, I., Tuba, M., Zivkovic, M.: Task scheduling in cloud computing environment by grey wolf optimizer. In: 2019 27th Telecommunications Forum (TELFOR), pp. 1–4. IEEE (2019)
7. Bacanin, N., et al.: Artificial neural networks hidden unit and weight connection optimization by quasi-refection-based learning artificial bee colony algorithm. IEEE Access **9**, 169135–169155 (2021)
8. Bacanin, N., Bezdan, T., Zivkovic, M., Chhabra, A.: Weight optimization in artificial neural network training by improved monarch butterfly algorithm. In: Shakya, S., Bestak, R., Palanisamy, R., Kamel, K.A. (eds.) Mobile Computing and Sustainable Informatics. LNDECT, vol. 68, pp. 397–409. Springer, Singapore (2022). https://doi.org/10.1007/978-981-16-1866-6_29

9. Bacanin, N., Petrovic, A., Zivkovic, M., Bezdan, T., Antonijevic, M.: Feature selection in machine learning by hybrid sine cosine metaheuristics. In: Singh, M., Tyagi, V., Gupta, P.K., Flusser, J., Ören, T., Sonawane, V.R. (eds.) ICACDS 2021. CCIS, vol. 1440, pp. 604–616. Springer, Cham (2021). https://doi.org/10.1007/978-3-030-81462-5_53
10. Bacanin, N., Stoean, R., Zivkovic, M., Petrovic, A., Rashid, T.A., Bezdan, T.: Performance of a novel chaotic firefly algorithm with enhanced exploration for tackling global optimization problems: application for dropout regularization. Mathematics **9**(21), 1–33 (2021). https://doi.org/10.3390/math9212705, https://www.mdpi.com/2227-7390/9/21/2705
11. Bacanin, N., Tuba, E., Zivkovic, M., Strumberger, I., Tuba, M.: Whale optimization algorithm with exploratory move for wireless sensor networks localization. In: Abraham, A., Shandilya, S.K., Garcia-Hernandez, L., Varela, M.L. (eds.) HIS 2019. AISC, vol. 1179, pp. 328–338. Springer, Cham (2021). https://doi.org/10.1007/978-3-030-49336-3_33
12. Bacanin, N., Zivkovic, M., Bezdan, T., Cvetnic, D., Gajic, L.: Dimensionality reduction using hybrid brainstorm optimization algorithm. In: Saraswat, M., Roy, S., Chowdhury, C., Gandomi, A.H. (eds.) Proceedings of International Conference on Data Science and Applications. LNNS, vol. 287, pp. 679–692. Springer, Singapore (2022). https://doi.org/10.1007/978-981-16-5348-3_54
13. Bacanin, N., Zivkovic, M., Bezdan, T., Venkatachalam, K., Abouhawwash, M.: Modified firefly algorithm for workflow scheduling in cloud-edge environment. Neural Comput. Appl. **34**, 9043–9068 (2022)
14. Basha, J., et al.: Chaotic Harris Hawks optimization with quasi-reflection-based learning: an application to enhance CNN design. Sensors **21**(19), 6654 (2021)
15. Bejtlich, R.: The Tao of Network Security Monitoring: Beyond Intrusion Detection. Pearson Education, Boston (2004)
16. Bezdan, T., Milosevic, S., Venkatachalam, K., Zivkovic, M., Bacanin, N., Strumberger, I.: Optimizing convolutional neural network by hybridized elephant herding optimization algorithm for magnetic resonance image classification of glioma brain tumor grade. In: 2021 Zooming Innovation in Consumer Technologies Conference (ZINC), pp. 171–176. IEEE (2021)
17. Bezdan, T., Petrovic, A., Zivkovic, M., Strumberger, I., Devi, V.K., Bacanin, N.: Current best opposition-based learning SALP swarm algorithm for global numerical optimization. In: 2021 Zooming Innovation in Consumer Technologies Conference (ZINC), pp. 5–10. IEEE (2021)
18. Bezdan, T., et al.: Hybrid fruit-fly optimization algorithm with k-means for text document clustering. Mathematics **9**(16), 1929 (2021)
19. Bezdan, T., Zivkovic, M., Tuba, E., Strumberger, I., Bacanin, N., Tuba, M.: Glioma brain tumor grade classification from MRI using convolutional neural networks designed by modified FA. In: Kahraman, C., Cevik Onar, S., Oztaysi, B., Sari, I.U., Cebi, S., Tolga, A.C. (eds.) INFUS 2020. AISC, vol. 1197, pp. 955–963. Springer, Cham (2021). https://doi.org/10.1007/978-3-030-51156-2_111
20. Bezdan, T., Zivkovic, M., Tuba, E., Strumberger, I., Bacanin, N., Tuba, M.: Multi-objective task scheduling in cloud computing environment by hybridized Bat algorithm. In: Kahraman, C., Cevik Onar, S., Oztaysi, B., Sari, I.U., Cebi, S., Tolga, A.C. (eds.) INFUS 2020. AISC, vol. 1197, pp. 718–725. Springer, Cham (2021). https://doi.org/10.1007/978-3-030-51156-2_83
21. Chen, T., Guestrin, C.: XGBoost: a scalable tree boosting system. In: Proceedings of the 22nd ACM SIGKDD International Conference on Knowledge Discovery and Data Mining, pp. 785–794 (2016)
22. Chen, Z., Jiang, F., Cheng, Y., Gu, X., Liu, W., Peng, J.: XGBoost classifier for DDoS attack detection and analysis in SDN-based cloud. In: 2018 IEEE international Conference on Big Data And Smart Computing (BigComp), pp. 251–256. IEEE (2018)

23. Cuk, A., et al.: Feedforward multi-layer perceptron training by hybridized method between genetic algorithm and Artificial Bee Colony. In: Data Science and Data Analytics: Opportunities and Challenges, p. 279 (2021)
24. Depren, O., Topallar, M., Anarim, E., Ciliz, M.K.: An intelligent intrusion detection system (IDS) for anomaly and misuse detection in computer networks. Expert Syst. Appl. **29**(4), 713–722 (2005)
25. Dhiman, G., Kumar, V.: Emperor penguin optimizer: a bio-inspired algorithm for engineering problems. Knowl.-Based Syst. **159**, 20–50 (2018)
26. Gajic, L., Cvetnic, D., Zivkovic, M., Bezdan, T., Bacanin, N., Milosevic, S.: Multi-layer perceptron training using hybridized Bat algorithm. In: Smys, S., Tavares, J.M.R.S., Bestak, R., Shi, F. (eds.) Computational Vision and Bio-Inspired Computing. AISC, vol. 1318, pp. 689–705. Springer, Singapore (2021). https://doi.org/10.1007/978-981-33-6862-0_54
27. Jabeur, S.B., Mefteh-Wali, S., Viviani, J.-L.: Forecasting gold price with the XGBoost algorithm and SHAP interaction values. Annals of Operations Research , 1–21 (2021). https://doi.org/10.1007/s10479-021-04187-w
28. Jiang, H., He, Z., Ye, G., Zhang, H.: Network intrusion detection based on PSO-XGBoost model. IEEE Access **8**, 58392–58401 (2020)
29. Karaboga, D.: Artificial bee colony algorithm. Scholarpedia **5**(3), 6915 (2010)
30. Milosevic, S., Bezdan, T., Zivkovic, M., Bacanin, N., Strumberger, I., Tuba, M.: Feed-forward neural network training by hybrid bat algorithm. In: MDIS 2020. CCIS, vol. 1341, pp. 52–66. Springer, Cham (2021). https://doi.org/10.1007/978-3-030-68527-0_4
31. Mirjalili, S.: SCA: a sine cosine algorithm for solving optimization problems. Knowl.-Based Syst. **96**, 120–133 (2016)
32. Mirjalili, S., Lewis, A.: The whale optimization algorithm. Adv. Eng. Softw. **95**, 51–67 (2016)
33. Mirjalili, S., Mirjalili, S.M., Lewis, A.: Grey wolf optimizer. Adv. Eng. Softw. **69**, 46–61 (2014). https://doi.org/10.1016/j.advengsoft.2013.12.007, https://www.sciencedirect.com/science/article/pii/S0965997813001853
34. Ogunleye, A., Wang, Q.G.: XGBoost model for chronic kidney disease diagnosis. IEEE/ACM Trans. Comput. Biol. Bioinform. **17**(6), 2131–2140 (2019)
35. Prakash, S., Kumar, M.V., Ram, R.S., Zivkovic, M., Bacanin, N., Antonijevic, M.: Hybrid GLFIL enhancement and encoder animal migration classification for breast cancer detection. Comput. Syst. Sci. Eng. **41**(2), 735–749 (2022)
36. Protić, D.D.: Review of KDD CUP'99, NSL-KDD and Kyoto 2006+ datasets. Vojnotehnički glasnik **66**(3), 580–596 (2018)
37. Qin, C., Zhang, Y., Bao, F., Zhang, C., Liu, P., Liu, P.: XGBoost optimized by adaptive particle swarm optimization for credit scoring. Math. Probl. Eng. **2021**, 6655510 (2021)
38. Ryu, S.E., Shin, D.H., Chung, K.: Prediction model of dementia risk based on XGBoost using derived variable extraction and hyper parameter optimization. IEEE Access **8**, 177708–177720 (2020)
39. Sheng, M., Liu, H., Yang, X., Wang, W., Huang, J., Wang, B.: Network security situation prediction in software defined networking data plane. In: 2020 IEEE International Conference on Advances in Electrical Engineering and Computer Applications (AEECA), pp. 475–479. IEEE (2020)
40. da Silva, L.M., Silveira, M.R., Cansian, A.M., Kobayashi, H.K.: Multiclass classification of malicious domains using passive DNS with XGBoost: (work in progress). In: 2020 IEEE 19th International Symposium on Network Computing and Applications (NCA), pp. 1–3. IEEE (2020)
41. Strumberger, I., Tuba, E., Bacanin, N., Zivkovic, M., Beko, M., Tuba, M.: Designing convolutional neural network architecture by the firefly algorithm. In: 2019 International Young Engineers Forum (YEF-ECE), pp. 59–65. IEEE (2019)

42. Wang, Y., et al.: An exhaustive research on the application of intrusion detection technology in computer network security in sensor networks. J. Sens. **2021**, 5558860 (2021)
43. Wolpert, D.H., Macready, W.G.: No free lunch theorems for optimization. IEEE Trans. Evol. Comput. **1**(1), 67–82 (1997)
44. Yang, X.S., Slowik, A.: Firefly algorithm. In: Swarm Intelligence Algorithms, pp. 163–174. CRC Press (2020)
45. Zhang, X., Nguyen, H., Bui, X.N., Tran, Q.H., Nguyen, D.A., Bui, D.T., Moayedi, H.: Novel soft computing model for predicting blast-induced ground vibration in open-pit mines based on Particle Swarm Optimization and XGBoost. Nat. Resour. Res. **29**(2), 711–721 (2020)
46. Zivkovic, M., Bacanin, N., Tuba, E., Strumberger, I., Bezdan, T., Tuba, M.: Wireless sensor networks life time optimization based on the improved firefly algorithm. In: 2020 International Wireless Communications and Mobile Computing (IWCMC), pp. 1176–1181. IEEE (2020)
47. Zivkovic, M., et al.: Covid-19 cases prediction by using hybrid machine learning and beetle antennae search approach. Sustain. Cities Soc. **66**, 102669 (2021)
48. Zivkovic, M., Bacanin, N., Zivkovic, T., Strumberger, I., Tuba, E., Tuba, M.: Enhanced grey wolf algorithm for energy efficient wireless sensor networks. In: 2020 Zooming Innovation in Consumer Technologies Conference (ZINC), pp. 87–92. IEEE (2020)
49. Zivkovic, M., Bezdan, T., Strumberger, I., Bacanin, N., Venkatachalam, K.: Improved Harris Hawks optimization algorithm for workflow scheduling challenge in cloud–edge environment. In: Pandian, A.P., Fernando, X., Islam, S.M.S. (eds.) Computer Networks, Big Data and IoT. LNDECT, vol. 66, pp. 87–102. Springer, Singapore (2021). https://doi.org/10.1007/978-981-16-0965-7_9
50. Zivkovic, M., et al.: Hybrid genetic algorithm and machine learning method for COVID-19 cases prediction. In: Shakya, S., Balas, V.E., Haoxiang, W., Baig, Z. (eds.) Proceedings of International Conference on Sustainable Expert Systems. LNNS, vol. 176, pp. 169–184. Springer, Singapore (2021). https://doi.org/10.1007/978-981-33-4355-9_14
51. Zivkovic, M., Zivkovic, T., Venkatachalam, K., Bacanin, N.: Enhanced dragonfly algorithm adapted for wireless sensor network lifetime optimization. In: Jeena Jacob, I., Kolandapalayam Shanmugam, S., Piramuthu, S., Falkowski-Gilski, P. (eds.) Data Intelligence and Cognitive Informatics. AIS, pp. 803–817. Springer, Singapore (2021). https://doi.org/10.1007/978-981-15-8530-2_63

Developing a Tool to Classify Lethal Weapons by Analyzing Images

Md. Muktadir Mukto[1], Md. Maiyaz Al Mahmud[1], Ikramul Haque[1], Omar Tawhid Imam[2], Ahmed Wasif Reza[1], and Mohammad Shamsul Arefin[3,4](✉)

[1] Department of Computer Science and Engineering, East West University, Dhaka 1212, Bangladesh
wasif@ewubd.edu
[2] Department of EEE, Bangladesh University of Engineering and Technology, Dhaka, Bangladesh
[3] Department of CSE, Daffodil International University, Dhaka 1341, Bangladesh
[4] Department of CSE, Chittagong University of Engineering and Technology, Chattogram 4349, Bangladesh
sarefin@cuet.ac.bd

Abstract. This paper shows how modified VGG16 and a customized model can be used for a weapon classifier system. The basic concept is to detect weapon type by its shape, length, angle, rotation degree, etc. The layered structure of VGG16 has been primarily used along with some extra layers for step-by-step classification. The images have been preprocessed with image augmentation using parameters such as shear, size, and rotation. The VGG16 model has been used for its prediction accuracy with less time complexity. The VGG19 model has better accuracy compared to the VGG16 but the time complexity is higher. The better accuracy of VGG19 is insignificant and negligible. VGG16 has 1×1 convolutional layers. VGG16 has small-sized convolution filters allowing the model to have a large number of weight layers. Instead of using a large number of hyper-parameters, the model uses a 3×3 convolution layer and a 2×2 max pool layer making the layer arrangement more consistent throughout the architecture. The trained mode acquired a test accuracy of 96.5% when tested with new input images.

Keywords: Weapon classifier system · VGG16 · VGG19 · Augmentation · Convolution layer · Classification

1 Introduction

Weapons and firearms have been used for public security and to protect a country's sovereignty. Military gears and equipment are being exported or imported almost all the time. For a country's military to function properly good firearms are required. Based on the situation the requirement of different types of firearms also differs. Non-lethal weapons should be used for public suppression and lethal weapons have to be used to fight invaders or to protect territorial borders. During an emergency, classifying and choosing what weapon to use in which situation can be difficult. That is why a weapon classifier

J. I.-Z. Chen et al. (Eds.): ICIPCN 2022, LNNS 514, pp. 229–242, 2022.
https://doi.org/10.1007/978-3-031-12413-6_18

system is needed. An automated weapon classifier can detect and classify weapon types by image processing. From the image, the detailed shape of a weapon can be detected and classified based on it. For example, a Machine gun will have a large drum or box magazine and is a much larger size than an Assault rifle. From this detailed observation, the difference between the weapons can be recognized. During a confrontation based on personnel role, a suitable firearm can be suggested by the weapon classifier system.

In this paper, a method has been used to create a weapon classifier system that can be implemented in military pay-load. An integrated system can classify the weapon as soon as a weapon has been added. In military confrontation when new pay-loads arrive the weapon may not be arranged according to their class. As a result, much time may be wasted when arranging and classifying the weapons manually. A weapon classifier system can classify and arrange the weapons based on their types so that soldiers do not have to go through every weapon to find the appropriate weapon for the task.

The dataset that has been used in this paper contains images of five types of weapons. A handgun is a small ranged weapon. The weapon grip can be held in one hand and thus this can be considered a one-handed weapon. The magazines can be loaded in the grip itself. Handguns can be easily classified for their smaller size. Machine guns have a somewhat similar design to a handgun but are larger, have longer barrels, and magazines are separated from the grip. The magazine can be box-shaped or drum-shaped. Some machine guns can have a scope attached to their rail. RPGs or Rocket Propelled Grenades are very distinct-looking weapons. The weapon has a grip but no magazine. Instead, only one grenade/rocket can be loaded in an RPG at a time. All short melee weapons can be classified as knives with a single grip and a sharp blade on top. The grenades are smaller throwable explosives that contain highly explosive materials and chemicals. This weapon most of the time has a round or pineapple-shaped body with a safety pin attached to its head. For better performance, the system must be able to identify each of these weapons based on their shape.

Our method can be used to implement a weapon classifier system that can be implemented in military pay-load. An integrated system can classify the weapon as soon as a weapon has been added. To develop this system VGG-16 has been used. VGG16 is a convolutional neural network system. ImageNet dataset contains more than 14 million images that belong to 1000 classes. The test accuracy of VGG-16 is 97.5% in ImageNet. By making a replacement of large kernel-sized filters it improves AlexNet with multiple 3×3 kernel-sized filters. VGG16 is a very useful NN system for image classification. A weapon dataset has been used containing several folders with weapon images in every folder. Every folder refers to different types of weapons. Images extracted from these folders can train the system. Afterward, for the new image, the system will be tested and by the accuracy result, the system can be evaluated.

2 Related Work

In [1] Kaniz Fatema, Razu Ahmed, and Mohammad Shamsul Arefin present a system that can be used to recognize books by their titles on book spines. The images of books are segmented using Canny Edge Segmentation. OCR engine is used to identify the book titles. After extracting the book titles the system can detect the books from a database. This paper helps us to familiarize ourselves with image processing project development.

Roberto Olmos, Siham Tabik, and Francisco Herrera design a pistol detecting system in [2].With a faster R-NN, the system achieves the highest accuracy. A video stream is used to extract image data. Even on low-quality Youtube videos, the quicker RNN model outperformed the previous model. Faster RCNN may sometime fail to detect weapons when there are multiple weapons in one frame.

J. Dever; N. da Vitoria Lobo; M. Shah offer a method for analyzing silhouettes and recognizing a traditional armed robbery holdup stance in [3]. The primary idea is to combine anomalous behavior detection with weapon identification. The victim of an armed robbery can be identified by his raised arm, while the criminal can be identified by the weapon in his hand.

In [4] Himanshu Buckchash and Balasubramanian Raman describe their project to create object identification algorithms in security cameras that can detect specific objects such as dangerous blades. The suggested approach has three stages. Foreground segmentation, FAST-based prominent feature detection, and Multi-Resolution Analysis (MRA). In this method, differently shaped knives can affect the performance of the model generating some false positives.

CNN (Convolutional Neural Network) was used to detect knives, blood, and weapons from an image by Mohammad Nakib, Rozin Tanvir Khan, Md. Sakibul Hasan, and Jia Uddin [5]. To arrive at a detection result, this model employs the Rectified Linear Unit (ReLU), Convolutional Layer, Fully Connected Layer, and CNN dropout function. For the tested dataset, the suggested model obtains a 90.2% accuracy.

Aaron Damashek and John Doherty [6] present an attempt to use Chamfer Matching to identify firearms in x-ray pictures. This method extends on the fundamental Chamfer method by integrating results from sub-polygon templates utilizing voting and machine learning to address occlusion issues.

Micha Grega, Andrzej Matiolański, Piotr Guzik, and Mikoaj Leszczuk present algorithms in [7] that can detect a handgun or knife in an image and inform the human operator. The proposed technology can be utilized to detect crimes in real-time using CCTV cameras.

In [8], A.Glowacz, M. Kmiec, and A. Dziech described a unique use of Active Appearance Models to detect knives in images. By the use of a computer vision technique which is generally used to locate an object in an experimental image, the existence of a knife in an image can be detected by characteristics of sharp edges. Knife edges can be used to identify knives or other sharp items in a photograph. This can be applied to security software.

The detection of a weapon from a CCTV frame is described in [9] Rohith Vajhala, Rohit Maddeneni, and Preethi Raj by employing the Histogram of Oriented Gradients (HOG) as a vector and artificial neural networks using the back-propagation method for classification. An armed weapon is more dangerous than an unequipped weapon. This concept has been used in this research.

Surajit Saikia, Fidalgo, Enrique Alegre, and Laura Fernandez-Robles describe a Faster R-CNN (Region-based Convolutional Neural Network) based real-time system in [10] that recognizes objects in an indoor environment. Katrina dataset was used to validate the proposed system. The average accuracy was 74.33%, 0.12 s were required to completely recognize the objects in the image.

Abdalrahman Al-Qubaa and Gui Yun Tian address weapon detection using a time and frequency feature extraction approach based on a system that combines walk-through metal detectors and a Giant Magneto-Resistive sensor array in [11]. For potential automated classification, the study additionally employs and compares two classification algorithms. The strategies described have the potential to significantly improve automatic weapon identification and categorization.

In [12] an algorithm is developed that can detect a person with an exposed firearm and alert the CCTV operator to potentially dangerous situations. The limitations and difficulties of such an image analysis application, as well as a discussion of the proposed algorithm's construction and numerical conclusions in terms of sensitivity and specificity, were all addressed.

Umadevi V Navalgund and Priyadharshini K suggest a CCTV crime detection approach based on VGG19 in the paper [13]. As compared to another method that has been used, GoogleNet InceptionV3, the proposed method outperforms it.

Volkan Kaya, Servet Tuncer, and Ahmet Baran suggest using Deep Learning and CNN for detecting and classifying different weapon types in [14]. It has been demonstrated that the custom model gives better accuracy than VGG16 and ResNet 50 with 98% accuracy.

In [15], authors Mitchell Singleton, Benjamin Tailor, and Jacob Tailor describe a system that can detect and classify weapons from real-time video surveillance via security cameras using Google TensorFlow API. In this case, the authors suggest using the MobileNet NN algorithm along with TensorFlow and OpenCV. Unlike typical CNN, the MobileNet separates the conventional layer into 3*3 depth-wise and 1*1 convolution. As a result, this model has less computational time than typical CNN. The model detects weapons accurately by detecting some key points. In various cases where many key points are detected in the same place, the model may perform badly.

3 System Architecture and Design

The proposed framework is depicted in Fig. 1. It shows the prediction process after taking input images including our custom VGG16 layer framework.

3.1 Dataset Description

The dataset that has been used is the weapon dataset (Fig. 2). The dataset has been obtained from Kaggle. The dataset contained less number of images than required to properly train and test the model. As such new images were downloaded from the internet and were added to the database. The images that were downloaded were of good quality. Much of the images were blurry and contained extra background. To train the model weapon images with different angles were selected. Additionally, faulty images were individually inspected and modified to enhance quality. Images with the worst quality were discarded. This dataset contains five folders. In each folder, there are 250 images of a specific type of weapon. The overall number of images is 1250. Each folder represents a weapon type containing only the weapons of that type.

Fig. 1. The system architecture of the proposed framework.

Fig. 2. Sample images of different weapon types.

The weapons can be categorized as:

- Knives
- RPG
- Grenades
- Machine guns
- Handguns and pistols

3.2 Data Preprocessing

In this proposed method, the weapon images have been pre-processed before being used for training the model. At first manual pre-processing was done where we went through images one after another to identify any faults in the images. Faulty images such as images with unusual shapes, backgrounds, and writings were separated from the main dataset. Additional images have been added to fill the void.

In general coding, the data has been preprocessed. The augmentation technique has been used to bring uniqueness to the dataset. The original images could result in biases in weapon direction, size, and shape. The augmentation process has been done to do some position tweaking on existing images and introduce some more unique images to the model for better training. The following parameters have been used:

- shear_range = 0.2 (‘*Shear*’ means that the image will be distorted along an axis, mostly to create or rectify the perception angles.)
- zoom_range = 0.2 (The amount of zoom in or zoom out)
- horizontal_flip = True (Flipping the image in horizontal)

Algorithm 1 : Extract and Pre-processed image

Input: Weapon image folder path
Require: Pre-processed image

```
Begin
Provide image folder path
Pre-processed image
For each image of train and test do
  Rescale the image
  Set Shear range and Zoom range
  Do Horizontal flip
  Set target size to (224, 224)
  And set batch size and class mode
end for
print class indices
end
```

3.3 VGG16 Model Implementation

VGG16 convolutional neural network has been used for the overall prediction process. The image sizes are set to (224*224). To build the image hierarchy imagenet has been used which gives weight to data like wordnet. The weights will be moved to untrainable layers. Using softmax normalization layer and class number for making the layers dense. Then the model can be trained with optimized parameters.

Algorithm 2: VGG16 Model Implementation

Input: Extracted and Pre-processed image
Require: Vgg16 Model

1. **Begin**
2. Set Image Size (224, 224)
3. Initializing the pre-trained VGG16 model with weights 'imagenet' and input
4. **for** each vgg16 layer **do**
5. set layers trainable false
6. **end for**
7. Flatten the model
8. Dense the model with activation 'softmax' and class number
9. Compile the model use
   loss='categorical_crossentropy',
   optimizer='adam',
   metrics=['accuracy']
10. use ModelCheckpoint to save the model
11. use EarlyStopping to stop the training process as soon as monitored metric stops improving
12. fit the model with epoch number
13. **end**

3.4 Customized Model Implementation

A customized model has been used for further implementation and testing. The model is set as a sequential model. Convolution of different channels and maxpool layers with different layer sizes have also been added. A dense softmax layer has been set with 5 units. With every epoch, the model has been trained with optimized parameters. The convolution layer in the VGG-16 model applies a filter in every layer for feature extraction. Multiple layered convolutions can help the system determine features that would be lost otherwise during flattening. Feature maps generated from the convolution layer are calculated in the max-pooling layer. The maximum values of the maps are calculated. The maximum presence of a feature is calculated for the next process. Afterward Flatten is used to flatten multi-dimension to a single dimension and pass the data. Parameters for customized model (Table 1):

Table 1. Parameters for customized model

Layer (type)	Output shape	Param
conv2d (Conv2D)	(None, 224, 224, 64)	1792
conv2d_1 (Conv2D)	(None, 224, 224, 64)	36928
max_pooling2d (MaxPooling2D)	(None, 112, 112, 64)	0
conv2d_2 (Conv2D)	(None, 112, 112, 128)	73856
conv2d_3 (Conv2D)	(None, 112, 112, 128)	147584
flatten_1 (Flatten)	(None, 1605632)	0
dense_1 (Dense)	(None, 5)	8028165

Algorithm 3: Customized Model Implementation

Input: Extracted and Pre-processed image
Require: Customized Model

1. **Begin**
2. Set Image Size (224, 224)
3. Initializing the model by specifying that the model is sequential
4. Add 2 x convolution layer of 64 channel of 3x3 kernel and same padding
5. Add 1 x maxpool layer of 2x2 pool size and stride 2x2
6. Add 2 x convolution layer of 128 channel of 3x3 kernel and same padding
7. Flatten the model
8. 1 x Dense Softmax layer of 5 units
9. Compile the model use
   loss='categorical_crossentropy',
   optimizer='adam',
   metrics=['accuracy']
10. use ModelCheckpoint to save the model
11. use EarlyStopping to stop the training process as soon as monitored metric stops improving
12. fit the model with epoch number
13. **end**

4 Implementation and Experimental Result

4.1 Experimental Setup

The proposed system has been implemented on a machine having Windows 10, Core i7 1.08–1.99 GHz with 8 GB RAM. Python 3.6(version) is used for developing it. Google Colab has been used for running the codes.

4.2 Implementation

Random weapon images can be given as input for classification. The following figures show the classification result after applying the proposed method to the image. Figure 3 shows custom VGG16 with different layer functionalities.

Fig. 3. Customized VGG16 layers.

Through processing the image in layers and hidden layers Customized VGG16 can predict the type of the weapon from the learned model.

4.3 Performance Evaluation

The performance of the model depends on how accurately the system can detect the input weapon type or class. For this experimentation, several weapon images have been used as input. The result is given as follows (Table 2)

Prediction array = ['Grenade', 'Knife', 'Machine_Guns', 'Pistol_Hand_Guns', 'RPG']

The model works perfectly for known data. It successfully classified the images of known types of weapons. The first five inputs show how efficiently the model can predict the image types by the learned model.

Table 2. Different types of weapons

Sample Image	Type	Number of test Samples	Correctly detected Samples	VGG16 Accuracy
	Grenade	20	20	100%
	Multiple Grenade	10	10	100%
	RPG	20	20	100%
	Multiple RPG	10	10	100%
	Knife	20	20	100%
	Multiple Knife	10	10	100%

(*continued*)

Table 2. (*continued*)

	Machine guns	20	19	95%
	Multiple Machine guns	10	7	70%
	Pistol Hand Gun	20	20	100%
	Multiple Pistol Hand Gun	10	8	80%

However, there are some limitations. Given an image of an unknown weapon as input, the system fails to categorize it. The system tries to match the unknown image with the types provided in the data set and classify based on them only. Thus there can be some false positives. So for better performance, the model must be trained with more data and different data for every class. The VGG16 model performed better than the customized model. To improve the customized model more image data should be provided for the train. To check the model performance 100 input images were used with 20 images in each category of 5 categories. It was observed that for the customized model 96% validation accuracy was achieved while training the model and model accuracy was 96.9% but when new images were given as input to predict image category the accuracy was 57%. But it was also observed that the VGG16 model achieved an accuracy of 100% and validation accuracy of 100%. To check the model performance of VGG16 the same input images were used and it yielded almost 100% prediction accuracy.

4.4 Comparison with Other Existing Frameworks

The proposed method uses VGG16 to classify weapons. In [14] a similar work has been done with a proposed method based on a deep learning model. Several other models have been used in the model such as VGG16, ResNet 50, and ResNet 101. The Accuracy and loss graphs are given below.

Fig. 4. VGG16.

Fig. 5. ResNet 50.

Fig. 6. ResNet 101.

Fig. 7. Deep learning

The proposed VGG16 and Customized Model graphs showing validation accuracy, accuracy, loss, and validation loss are given below:

Fig. 8. The proposed VGG16 and customized model.

Accuracy is a way to assess the performance of a model. It refers to the number of correct predictions of a model. The loss indicates how poorly a model performs by calculating the deviation of prediction from the actual result. Accuracy and loss are inversely proportional to each other. In Fig. 4 we can see the accuracy of the proposed VGG 16 model reaches almost 0.9 meaning the model performed well in classification. The loss of the Test set can reach 1.5. In ResNet 50 and ResNet 101, the accuracy and loss do not stay stable and vary a lot in every epoch as shown in Fig. 5 and Fig. 6. For ResNet 50 the accuracy can go as low as 0.85 while for 101 it can be as low as 0.75. These figures explain the performance of the ResNet model cannot be determined. The deep learning method performance depicted in Fig. 7 is somewhat similar to that of VGG-16. In Fig. 8 the performance of both default VGG-16 and custom VGG-16 models can be observed. While the VGG-16 model with default parameter performs as observed previously, the test validation accuracy of the customized model varies a lot. This indicates that the customized model can dodge false positives while performing somewhat similarly to the default model.

5 Conclusion

In this paper, VGG-16 and a custom model have been used to build a weapon classifier system. After some pre-processing, the model was trained by the weapon dataset for the weapon classifying process. For new input images, the system has shown very good performance with good accuracy. But it has some limitations such as it can properly ignore an unknown weapon. Thus to avoid false positives it is recommended to use relevant weapon images for training and testing. Also, a large number of different data can improve performance.

References

1. Fatema, K., Ahmed, M.R., Arefin, M.S.: Developing a system for automatic detection of books. In: Chen, J.-Z., Tavares, J.M.R.S., Iliyasu, A.M., Du, K.-L. (eds.) ICIPCN 2021. LNNS, vol. 300, pp. 309–321. Springer, Cham (2022). https://doi.org/10.1007/978-3-030-84760-9_27
2. Olmos, S.T., Herrera, F.: Soft Computing and Intelligent Information Systems research group, on automatic handgun detection alarm in videos using deep learning: Roberto (2022)
3. Dever, J., da Vitoria Lobo, N., Shah, M.: Automatic visual recognition of armed robbery. In: Proceedings of the 2002 International Conference on Pattern Recognition, vol. 1, pp. 451–455 (2002). https://doi.org/10.1109/ICPR.2002.1044755
4. Buckchash, H., Raman, B.: A robust object detector: Application to detection of visual knives. In: Proceedings of the IEEE International Conference on Multimedia & Expo Workshops (ICMEW) 2017, pp. 633–638 (2017). https://doi.org/10.1109/ICMEW.2017.8026214
5. BRAC university on Crime scene prediction by detecting threatening objects using convolutional neural network. Submitted 13 Apr 2017
6. Damashek, A., Doherty, J.: Detecting Guns Using Parametric Edge Matching Aaron. Stanford University, Stanford, CA, USA
7. Grega, M., Matiolański, A., Guzik, P., Leszczuk, M.: Automated detection of firearms and knives in a CCTV image. Sensors **16**, 47 (2015). https://doi.org/10.3390/s16010047
8. Glowacz, A., Kmieć, M., Dziech, A.: Visual detection of knives in security applications using active appearance models. Multimedia Tools Appl. **74**(12), 4253–4267 (2013). https://doi.org/10.1007/s11042-013-1537-2
9. Vajhala, R., Maddineni, R., Yeruva, P.R. Weapon Detection In Surveillance Camera Images (Dissertation) (2016). http://urn.kb.se
10. Saikia, S., Fidalgo, E., Alegre, E., Fernández-Robles, L.: object detection for crime scene evidence analysis using deep learning. In: Battiato, S., Gallo, G., Schettini, R., Stanco, F. (eds.) ICIAP 2017. LNCS, vol. 10485, pp. 14–24. Springer, Cham (2017). https://doi.org/10.1007/978-3-319-68548-9_2
11. Al-qubaa, A., Tian, G.Y.: Weapon detection and classification based on time-frequency analysis of electromagnetic transient images (2012). © copyright by authors, published under agreement with iaria - www.Iaria.Org
12. Grega, M., Łach, S., Sieradzki, R: Automated recognition of firearms in surveillance video. IEEE International Multi-Disciplinary Conference on Cognitive Methods in Situation Awareness and Decision Support (CogSIMA), pp. 45–50 (2013).https://doi.org/10.1109/CogSIMA.2013.6523822
13. Navalgund, U.V., Priyadharshini, K.: Crime intention detection system using deep learning. In: Proceedings of the 2018 International Conference on Circuits and Systems in Digital Enterprise Technology (ICCSDET), pp. 1–6 (2018). https://doi.org/10.1109/ICCSDET.2018.8821168
14. Kaya, V., Tuncer, S., Baran, A.: Detection and classification of different weapon types using deep learning. Appl. Sci. **11**, 7535 (2021). https://doi.org/10.3390/app11167535
15. Singleton, M., Taylor, B., Taylor, J., Liu, Q.: Gun identification using tensorflow. In: Meng, L., Zhang, Y. (eds.) MLICOM 2018. LNICSSITE, vol. 251, pp. 3–12. Springer, Cham (2018). https://doi.org/10.1007/978-3-030-00557-3_1

Selfie2Business - An Application to Identify Objects and Recommend Relevant Service Providers

Joseph Dominic Cherukara(✉), Anirudh S. Ayya, Abhishek Pai, Rohan Varghese Biju, and Nitin V. Pujari

Department of C.S.E., PES University, Bangalore, Karnataka, India
jd.cherukara@gmail.com

Abstract. Due to the rise in the number of electronic devices, 2.5 million terabytes of data is generated every day. This ranges from device logs and browsing history, to social media posts in the form of text, images, and videos, websites, and news articles. Images alone contain a wealth of data. This data is usually used by companies to glean valuable insights into market trends, usage patterns, and so on. In this paper, we propose giving users a way to utilize the information in images/selfies to search for products and services. This is achieved by using an object detection neural network to detect objects in an image and suggesting relevant products based on the results.

Keywords: Selfie2Business · Yolov3 · Yolov4-tiny · Object-detection · Convolution

1 Introduction

Nowadays, E-commerce websites have brought about a revolution in the way we shop. This rampant rise can be attributed to the fact that most E-commerce websites tend to provide a versatile shopping experience to customers as the customers can browse through a variety of products. In some cases, the large number of products available on these sites can give the customer a very hard time in choosing their desired product. This is because the recommender systems on most E-commerce sites make use of traditional methods such as collaborative filtering, content-based systems and keyword matching to provide suggestions to the customer. There has also been a tremendous growth in the amount of image data that has been generated in the past few years. These images are used in several applications to generate meaningful insights and patterns. Many E-commerce sites also make use of high-resolution images to provide a visual description of the intended product. This has helped various sites and online businesses to boost their revenue, bringing in significant customer interest. In this work, we explore the use of an image recognition and classification model's results as inputs for a recommender system. The proposed application aims at

J. I.-Z. Chen et al. (Eds.): ICIPCN 2022, LNNS 514, pp. 243–259, 2022.
https://doi.org/10.1007/978-3-031-12413-6_19

providing another dimension to E-commerce sites by making use of images as an input to generate product recommendations. The application can provide more clarity on the look and feel of the product that a customer is looking to purchase. The application would also provide an interactive and easy to use interface. An added advantage to this application is that it can reduce the number of irrelevant recommendations considerably. Image based search systems cannot completely replace the existing systems. However, visual search systems can be augmented with existing systems. This can help many online businesses reach new customers and expand their markets.

2 Literature Review

The advent of Convolution Neural Networks has brought about a massive shift for neural networks particularly when working with image data. Since then, CNNs have been used as the basis for various image-related applications, ranging from simple digit identification to complex medical imaging classification.

Joseph Redmon and Ali Farhadi [6] propose a Real-Time Object Detection model named Yolov3. Yolov3 uses a neural network with successive 3×3 and 1×1 convolutional layers for feature extraction. The feature extraction network has a total of 53 convolutional layers. For classification of the bounding boxes, Yolov3 uses independent logistic classifiers, allowing the model to perform well even when there are overlapping labels. Binary cross-entropy loss was used for training. Yolov3 predicts bounding boxes at 3 different scales. The authors use additional convolutional layers. The last of these additional layers produces the bounding box and class predictions. K-means clustering is used to determine the bounding box priors across the 3 different scales.

The Yolov3 model along with another similar model - Yolov4 tiny forms the basis of the proposed application.

The application proposed in this paper would be a mobile application that would have to leverage the use of Deep Convolutional Neural Networks for Object Detection. To accomplish this, there can be two approaches - Train a model that can run in the constrained environment of a mobile device or train a model to run on a server dedicated for object-detection.

Robert J. Wang, Xiang Li, Charles X. Ling [7] propose an efficient architecture named PeleeNet, which uses conventional convolution and can run on mobile devices with a GPU. The proposed architecture is named PeleeNet and consists of a stem block and four stages of feature extractors. An average pooling layer with a stride of 2 is the last layer in all stages (except in the last stage). The model was trained to detect dogs in an image and able to classify dogs into 120 different categories based on their breeds. 1,50,466 images were used for training and 6000 images were used for validation. The images were taken from the ILSVRC 2012 dataset. The model achieved a mAP@0.75 of 22.9 when tested against the MS COCO dataset, which is better than YoloV2 (19.2) but it is surpassed by YoloV3 (34.3).

Yuxuan Cai et al. [2] proposes another efficient architecture that can be run on mobile devices. The proposed architecture aims to achieve a state-of-the-art object detection model that does not sacrifice accuracy for speed in the constrained environment of a mobile device. To achieve high accuracy and parallelism, the paper proposes a novel pruning scheme names block-punched pruning. The weights for a layer are divided into equal-sized blocks. In each block, a group of weights is pruned at the same location of all filters and all channels. The Reweighted Group Lasso pruning method is also proposed. The proposed architecture also adopts a GPU-CPU Collaborative scheme where both the GPU and CPU are utilized for calculations to prevent underutilization of available resources. The proposed model (mAP of 49) outperforms YoloV3-tiny (mAP of 29) and YoloV4-tiny (mAP of 40.2) but is surpassed by YoloV4 (mAP of 57.3).

Both these models tackle the first approach of training a model to run on a mobile device. However, both models optimally only when there is a GPU on the device. However, this is not the case for most ordinary mobile devices. Hence, the second approach of training a model to run on a dedicated object-detection server was taken.

3 Dataset

Two Models were built as part of the proposed system. The first was a Convolutional Neural Network to classify a given image as a selfie or not. The Selfie-Image-Detection-Dataset [1] was used to train the CNN was obtained from Kaggle. The second model was a neural network used to detect various objects in the image. 15 objects were chosen that frequently appeared in selfies and has potential to generate revenue. 1000 samples of each object were chosen from the Open Images [4,5] dataset to form the training set for this model. In order to produce search suggestions, the 1 Billion Word Language Model Benchmark [3] dataset was used.

3.1 Data Pre-processing

The selfie detection dataset did not need any pre-processing and was used as such.

The dataset used to train the object detection networks was a subset of Open Images v6 dataset and hence needed some amount of processing. The Fiftyone python package was used to create a subset of the dataset. The subset created included negative samples of the various classes as well. These samples needed to be removed. To do this, all the empty annotation files were found. For each of these files, the corresponding image file was also found. Once a pair of files was found, both the annotation file and the image were deleted.

4 Methodology

A mobile app was developed for the user to be able to interact with the application. The application was built using React Native and can function on mobile devices running on Android OS (7.0 and above). To ensure modularity in the application, the various functionalities of the application was separated into different screens of the app. This ensures higher cohesion as each screen is designed to handle a single responsibility. The following are the various components of the app

1. Registration and Login: A new user is allowed to sign up and an existing user can log in to access the application
2. Image Upload: The user can select an image to be processed.
3. Object Selection: If any of the objects detected in the image are part of our list of classes, the user would be prompted to select a shape for a bounding box to single out the chosen object.
4. Service providers: Once the user has chosen an object, a list of service providers (specific to the chosen object) is shown, from which the user can choose a service provider from where he/she would buy the product.
5. Search queries: Based on the object that has been recognized, a list of relevant queries is generated. Once the user has selected their service provider, clicking on any of the queries would redirect them to the service provider's results page.

The user registers through the user interface. A major issue with applications that involve users having to log in, is password management. Improper management of user's passwords can have unfortunate consequences. Passwords should not be sent as plaintext over the internet, as this can be attacked by inspecting the packets being sent by the client. To overcome this, the password entered by the user (to log in to the application) is hashed before it is sent to the server for authentication. This is accomplished using the Node package called Bcrypt. Bcrypt implements the Blowfish cipher to hash passwords. The Blowfish cipher has a 64-bit block size and uses a key from 32 to 448 bits in length. The Bcrypt implementation of the blowfish cipher uses a 128-bit salt and 192-bit hash. While signing up, the new user's password is hashed (using Bcrypt) and sent to the database to be stored. While logging in, the password entered by the unauthenticated user is hashed and sent to the database. The stored hashed password is compared to the hashed password sent by the user. This is done to ensure that a malicious user does not get access to the hashed password. Each time the user tries to log in, the hashed values are compared in the database.

For the purposes of the application, a fast, cloud based database was ideal. MongoDB is a cloud hosted, free, NoSQL database. The document model of MongoDB allows for a lot of flexibility in the type of data stored. Data in MongoDB is stored in the form of JSON, making it easy to use with a React Native without having to work with heavy Object Relational Mappers. Using a database on the cloud provides reliability and scalability based on the requirements of the application. All these reasons make MongoDB ideal for the needs of the application. Post this, the user is presented with the image upload page where the user uploads the selfie image either by opening the smartphone camera on spot to take a selfie or by choosing from the gallery. Once the image is uploaded, the image is sent as an HTTP post request to a server hosted on an AWS EC2 instance. The server consists of the selfie detector and the object detector. The image passes through the selfie detector, which detects if the image is a selfie or not. The image then passes through the object detector, which detects the objects in the image. We have chosen 15 classes (objects) as part of the object detector. The classes are as follows

Bicycle Helmet	Headphones
Cat	Human Face
Chair	Laptop
Clock	Luggage and Bags
Dog	Necklace
Earrings	Shirt
Glasses	Table
Hat	

The object detector returns a JSON string containing the co-ordinates of the top left corner of the bounding boxes, along with the width and height and count of each object that was detected. It also contains the results of the selfie detection. In case of multiple detections for the same object, the data of the highest confidence value of the object class is returned in the JSON string. In addition to this, it also contains all the search suggestions of all the objects detected in the image. All this data is returned to the app front end through the node proxy server. The original image sent to the server is deleted after the objects are detected since they are of no use anymore. The app then plots the bounding boxes over the image based on the values in the JSON string. The user then applies the constraints of the number of objects that must be visible on the screen. The user also selects the shape of the bounding enclosure for the object. After this, the user clicks on the object, for which a list of related search suggestions is shown. The user is free to choose the service provider (E-commerce site) from which they wish to purchase the product. Upon selecting a particular product from the list of options, the user is redirected to the service provider. For the objects Dog and Cat the search suggestions include various pet products such as food, shampoo, toys, beds etc. For Human Face, various skin-care products are included in the results.

The following image shows the system architecture of the application. (see Fig. 1)

Fig. 1. Selfie2Business application system architecture

Our application consists of three major parts:

- Selfie Detector
- Object Detector
- Search Query Listing

4.1 Selfie Detector

The Selfie detector is used to detect whether a given image is a Selfie or not. To do this, a simple Convolutional Neural Network (CNN) was used. The model takes as input an image of size 224×224 pixels. Following are the layers of our model

- Convolution 2D layer ($222 \times 222 \times 16$)
- Max Pooling ($111 \times 111 \times 16$)
- Convolution 2D layer ($109 \times 109 \times 32$)
- Max Pooling ($54 \times 54 \times 32$)
- Convolution 2D layer ($52 \times 52 \times 64$)
- Max Pooling ($26 \times 26 \times 64$)
- Convolution 2D layer ($24 \times 24 \times 64$)

- Max Pooling ($12 \times 12 \times 64$)
- Convolution 2D layer ($10 \times 10 \times 64$)
- Max Pooling ($5 \times 5 \times 64$)
- Dense Layer with ReLU as activation function ($1 \times 1 \times 512$)
- Dense Layer with Sigmoid as activation function ($1 \times 1 \times 1$)

The following figure shows a diagrammatic representation of the structure of the Network. (see Fig. 2)

Fig. 2. Selfie detector convolutional neural network structure

The Selfie-Image-Detection-Dataset [1] was used to train this CNN. The dataset has 78,619 images in total. These are split into Train, Validation, and Test sets (80%-10%-10%) The CNN was implemented and trained using Python and TensorFlow. The Adam optimizer was used in training with a learning rate $= 0.001$, beta1 $= 0.9$, beta2 $= 0.999$ and epsilon $= 1\text{e}{-}07$. Since the output is a binary variable and the activation function of the output layer is a Sigmoid function, Binary Cross entropy was used as the loss function. The model was trained for a total of 100 epochs. Checkpoints were stored every time the model showed improvement in the validation accuracy. Once the training was completed, the epoch with the highest validation accuracy (Epoch - 98: validation accuracy - 98.105%) was chosen as the best.

The trained CNN is hosted on an AWS EC2 instance along with the object detection models. This instance acts as the detection server. The image that the server receives from the client can be of any size. To allow the image to be passed through the CNN, the image is scaled to 224×224. Once this is done, it is fed through the selfie detection CNN. The CNN outputs a floating-point value between 0 and 1. If the output value is $\leq$0.5, the image is classified as not a selfie and if the output is >0.5, the image is classified as a selfie. The output of the selfie detection CNN is combined with the output of object detection in the form of a JSON string and then sent back to the client.

4.2 Object Detector

Once an image is classified whether it is a selfie or not by the selfie detector, the objects present in it should be detected. For this, we have implemented the object detection model using the YOLO framework. We have used two kinds of YOLO models:

- YOLOv3
- YOLOv4 tiny

The technologies used were Python3, OpenCV, YOLOv3 and YOLOv4 tiny. The Open Images v6 [4,5] dataset was used to train all the models for object detection. Using these models, we came up with four different approaches to go about object detection:

- A single YOLOv3 for all 15 classes
- A YOLOv3 for each object arranged in a parallel manner
- A YOLOv4 tiny for all 15 classes
- A YOLOv4 tiny for each object arranged in a parallel manner

Approach 1. In this approach (see Fig. 3), one YOLOv3 model is used to detect objects in all 15 classes. The yolov3 model is a variant of Darknet, which is a 53-layer network trained on Imagenet. For the task of detection, 53 more layers are stacked onto it, producing a 106 layer fully convolutional underlying architecture for yolov3. The detection is done by applying 1×1 detection kernels on feature maps of three different sizes at three different places in the network. The shape of detection kernel is $1 \times 1 \times (B \times (5 + C))$. Here B is the number of bounding boxes a cell on the feature map can predict, 5 is for the 4 bounding box attributes and one object confidence and C is the number of classes. Yolov3 uses binary cross-entropy for calculating the classification loss for each label, while object confidence and class predictions are predicted through logistic regression. The model was trained using a total of 13124 images (with roughly 1000 annotations for each class). The model was trained for 30000 epochs.

Fig. 3. Object detector architecture overview: Approach 1

Approach 2. In this approach (see Fig. 4), rather than using a single yolov3 model to detect all 15 object classes, 15 yolov3 models are used, each trained to detect objects from one of the 15 classes. Each of the 15 models was trained using a set of images of a particular object class. Each training set had around 1000 annotations of the object to be detected. Each model was trained on the dataset for 6000 epochs and the best performing weight file for each object was chosen.

Fig. 4. Object detector architecture overview: Approach 2

Approach 3. This approach (see Fig. 5) is similar to Approach 1 in that a single neural network is used to detect all 15 classes of objects. It differs from Approach 1 as instead of a yolov3 network, a yolov4-tiny network is used. Yolov4-tiny (29 convolutional layers) is a significantly smaller network when compared to yolov3 (106 convolutional layers). This allows for much faster detection and lower training times. The model was trained on the same dataset as used for Approach 1. The training was run for 30000 epochs.

Fig. 5. Object detector architecture overview: Approach 3

Approach 4. This approach (see Fig. 6) is similar to Approach 2 in that it uses 15 neural networks each trained to detect 1 of the 15 object classes. It differs from Approach 2 in that it uses yolov4-tiny networks instead of yolov3. Each of the 15 models was trained using the same dataset as used for the corresponding network in Approach 2. Each model was trained for 6000 epochs.

Fig. 6. Object detector architecture overview: Approach 4

4.3 Search Query Listing

A corpus named statmt.org was used as a dataset to extract adjective-noun pairs. This is done using a Python package known as spaCy with the help of a module known as noun-chunks. Out of these noun chunks, only the ones which contain the names of the objects that we had planned to detect through our Object Detector are filtered out and used as search queries.

5 Results

5.1 Selfie Detection

The training accuracy of the model was found to be 98.92%. The highest validation accuracy was obtained in the 98th epoch and was found to be 98.015% and the test accuracy was found to be 89.35%. Out of a total of 7862 samples, the true positives were 3499. The true negatives were 3526. The false positives were found to be 405 and the false negatives were found to be 432. The precision, recall and F1-scores were found to be 89.62, 89.01 and 89.31 respectively. Below is the confusion matrix depicting the above results. (see Fig. 7)

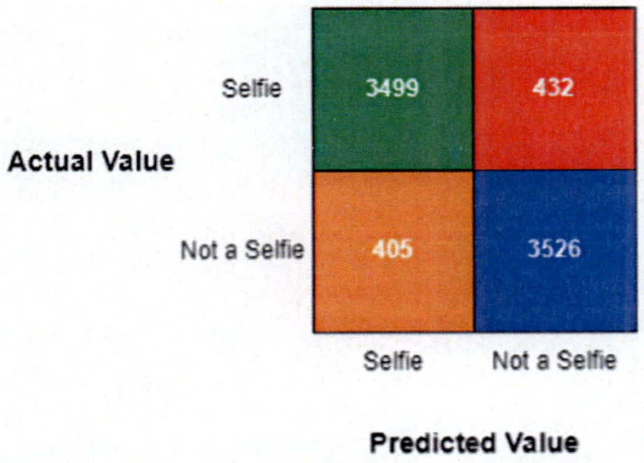

Fig. 7. Selfie detector convolutional neural network confusion matrix

5.2 Object Detection

Approach 1. We have one YOLOv3 model for all 15 classes. The average precision for each class was found for 12000, 14000 and 19000 epochs. Table 1 displays the values.

We observe that most of the best average precision values of the classes are present in 14000 epochs. Due to this we consider the 14000 epochs weight file from the YOLOv3 model for the final evaluation.

Table 1. Results of object detector Approach 1

Class	12000 epoch	14000 epoch	19000 epoch	Best epoch
Bags	70.76	77.51	79.54	19000
Bicycle helmet	58.22	68.19	65.26	14000
Cat	73.36	76.14	72.6	14000
Chair	22.48	24.29	22.2	14000
Clock	91.17	90.8	95.59	19000
Dog	75.21	70.3	74.08	12000
Earrings	85.05	88.72	85.87	14000
Glasses	47.1	46.38	45.81	12000
Hat	63.89	67.24	58.99	14000
Headphones	89.85	90.92	90.77	14000
Human face	25.83	32.33	28.08	14000
Laptop	69.57	72.86	70.44	14000
Necklace	80.44	73.79	74.03	12000
Shirts	69.07	68.46	70.56	19000
Table	34.15	29.61	30.86	12000

Approach 2. We have a network of 15 YOLOv3 models in parallel. Each YOLOv3 is responsible for detecting a particular object. We take the best possible epoch for each object. That is, the epoch having the highest average precision value for each object. Table 2 shows the results of the average precision values for each object in each epoch

Table 2. Results of object detector Approach 2

Class	4000 epoch	5000 epoch	6000 epoch	Best epoch
Bicycle helmet	61.03	63.78	67.21	6000
Cat	53.43	58.36	51.29	5000
Chair	18.13	15.94	16.74	4000
Clock	95.68	90.02	95.89	6000
Dog	30.95	42.29	33.72	5000
Earrings	79.84	85.52	90.34	6000
Glasses	34.76	34.1	36.32	6000
Hat	57.54	56.02	57.79	6000
Headphones	82.89	85.55	83.86	5000
Human face	10.54	9.9	9.7	4000
Laptop	65.37	52.85	64.25	4000
Luggage and bags	75.8	70.75	84.3	6000
Necklace	53.36	57.44	59.34	6000
Shirt	59.65	57.34	59.61	4000
Table	17.38	18.71	15.83	5000

For the final evaluation, we consider the best performing epoch for each object from the YOLOv3 single class model.

Approach 3. We have one YOLOv4 Tiny model for all 15 classes. The average precision for each class was found for 10000, 20000, 30000 epochs. Table 3 shows the same.

Table 3. Results of object detector Approach 3

Class	10000 epoch	20000 epoch	30000 epoch	Best epoch
Bags	46.93	58.41	63.33	30000
Bicycle helmet	49.18	58.08	55.91	20000
Cat	64.3	64.71	57.66	20000
Chair	21.02	22.25	20.04	20000
Clock	76.57	74.51	83.66	30000
Dog	59.92	60.89	50.61	20000
Earrings	62.04	69.32	68.82	20000
Glasses	29.48	36.77	29.98	20000
Hat	52.3	60.87	61.23	30000
Headphones	58.18	68.45	77.85	30000
Human face	12.84	18.78	21.23	30000
Laptop	67.58	68.45	71.01	30000
Necklace	55.45	56.72	65.37	30000
Shirts	56.67	55.13	47.99	10000
Table	31.51	31.66	28.58	20000

We find that 20000 and 30000 epochs have the same number of classes which are best detected. Since there is a tie, we are free to choose between 20000 and 30000. The 20000-epoch weight file is taken into consideration for the final model evaluation.

Approach 4. Each YOLOv4 Tiny is responsible for detecting a particular class. The models are arranged in a parallel fashion. We take the best possible epoch weights for each object, that is the epoch having the highest average precision values for each object. Table 4 depicts the results, shows the average precision values for each object in each epoch.

For the final evaluation, we consider the best performing epoch for each object.

Table 4. Results of object detector Approach 4

Class	4000 epoch	5000 epoch	6000 epoch	Best epoch
Bags	71.54	74.99	71.77	5000
Bicycle helmet	59.69	68.08	64.38	5000
Cat	45.12	49.37	51.02	6000
Chair	18.97	21.46	19.03	5000
Clock	94.47	92.54	93.81	4000
Dog	32.86	34.32	32.29	5000
Earrings	70.6	78.61	83.31	6000
Glasses	31.66	32.22	29.77	5000
Hat	52.26	53.6	49.08	5000
Headphones	73.71	82.64	83.39	6000
Human face	9.79	9.62	8.74	4000
Laptop	68.6	60.64	67.72	4000
Necklace	51.77	60.44	60.3	5000
Shirts	51.5	52.04	52.17	6000
Table	15.08	13.31	13.73	4000

Out of a total of 3790 samples, the true positives were 1687. The false positives were found to be 801 and the false negatives were found to be 1302. The precision, recall and F1-scores were found to be 0.68, 0.56 and 0.62 respectively. True negatives don't exist for this case. Below is the confusion matrix depicting the above results. (see Fig. 8)

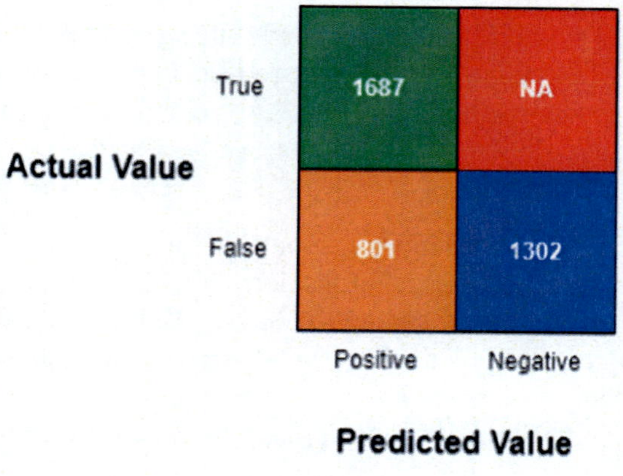

Fig. 8. Best performing object detector confusion matrix

6 Discussion

The results from the four approaches of object detection were tabulated and compared. The best performing epoch weight file were taken from each of the models and the average precision for each class was considered to decide the same. Table 5 depicts the final results.

Table 5. Final object detector results

Class	Approach 1	Approach 2	Approach 3	Approach 4	Best Approach
Bicycle Helmet	68.19	67.21	58.08	68.08	Approach 1
Cat	76.14	58.36	64.71	51.02	Approach 1
Chair	24.29	18.13	22.25	21.46	Approach 1
Clock	90.8	95.89	74.51	94.47	Approach 2
Dog	70.3	42.29	60.89	34.32	Approach 1
Earrings	88.72	90.34	69.32	83.31	Approach 2
Glasses	46.38	36.32	36.77	32.22	Approach 1
Hat	67.24	57.79	60.87	53.6	Approach 1
Headphones	90.92	85.55	68.45	83.39	Approach 1
Human face	32.33	10.54	18.78	9.79	Approach 1
Laptop	72.86	65.37	68.45	68.6	Approach 1
Luggage and bags	77.51	84.3	58.41	74.99	Approach 2
Necklace	73.79	59.34	56.72	60.44	Approach 1
Shirt	68.46	59.65	55.13	52.17	Approach 1
Table	29.61	18.71	31.66	15.08	Approach 3

The models in all 4 approaches seem to perform poorly on some objects, namely - chair, human face, glasses and table. This is most likely explained by the fact that there is a large amount of variance in the overall appearance of these objects. For example, tables come in all sorts of shapes with varying number of legs and so on. This variance in appearance can make it extremely hard for a neural network to generalize especially when given limited amount of training data. However, it is observed that of the 4 approaches, the Approach 1 and Approach 3 (One neural network detecting all 15 object classes) show the best results for these classes.

Approach 1, the YOLOv3 model that takes all 15 classes into consideration in one single model, outperforms the other models and hence is the best model. Majority (11) classes have the highest average precision with this model. The mean average precision value for this was found to be 65.17% and the best epoch was found to be 14000. Hence, we choose this as our final model for object detection.

The outputs of the final object detection neural network along with the original images are shown in Table 6.

Table 6. Object detection outputs

7 Conclusions and Future Work

The proposed application was successfully developed. In addition to this, various design approaches for building an Object Detector model were experimented with, and the best approach was chosen. Adjective-noun pairs were also generated using the noun-chunks module in spaCy to aid the generation of search phrases.

As a next step, the accuracy of both the Selfie Detector and the Object Detector can be improved. The existing Object Detector models can also be extended to ensure that more number of objects can be identified. Another improvement that could be made is to compare detected objects against a pre-existing dataset of objects in the same category, and provide further classification of the object.

References

1. Bhatt, J.: Selfie-image-detection-dataset (2020). https://www.kaggle.com/jigrubhatt/selfieimagedetectiondataset
2. Cai, Y., et al.: YOLObile: real-time object detection on mobile devices via compression-compilation co-design (2020). https://arxiv.org/abs/2009.05697
3. Chelba, C., Mikolov, T., Schuster, M., Ge, Q., Brants, T., Koehn, P.: One billion word benchmark for measuring progress in statistical language modeling. CoRR abs/1312.3005 (2013). http://arxiv.org/abs/1312.3005
4. Krasin, I., et al.: OpenImages: a public dataset for large-scale multi-label and multi-class image classification (2017). Dataset available from https://storage.googleapis.com/openimages/web/index.html
5. Kuznetsova, A., et al.: The Open Images Dataset V4: unified image classification, object detection, and visual relationship detection at scale. IJCV **128**, 1956–1981 (2020)
6. Redmon, J., Farhadi, A.: YOLOv3: an incremental improvement. arXiv (2018)
7. Wang, R.J., Li, X., Ling, C.X.: Pelee: a real-time object detection system on mobile devices (2019). https://arxiv.org/abs/1804.06882

A Systematic and Novel Ensemble Construction Method for Handling Data Stream Challenges

Rucha Chetan Samant(✉) and Suhas H. Patil

Department of Computer Engineering, College of Engineering, Bharati Vidyapeeth Deemed to be University, Pune, India
ruchasamant25@gmail.com, shpatil@bvucoep.edu.in

Abstract. Data stream mining is a difficult job because it must deal with its speed, variety, and ever-changing nature. As the industry's current demand is for data processing at high speed and accuracy, it has become an important task for developers and researchers. One solution to this problem is to use an ensemble approach. Along with speed, concept change is a data stream issue that drift detectors can handle. Similarly, there could be imbalances in the data. Taking these three issues into account, this paper proposes a novel approach to resolving the problem by combining a data stream balancer with an ensemble and a drift detector. The proposed solution is tested using a cutting-edge data balancer, a boosting-based ensemble, and various drift detectors. The experiments are carried out on cutting-edge real-world imbalanced datasets, and the results are discussed in terms of performance metrics.

Keywords: Boosting method · Data stream · Imbalance data · Drift detector · Ensemble · Classification · Adaptive learning

1 Introduction

A data stream is a type of data generated by a variety of web applications. Because it is generated through the internet media, its nature is changeable, fast, and unending. User actions, event records, recorded photos, and numerous signals could all be included in this data. It is usually represented as a continuous stream S = {E1, E2,....En}, where E1 is an Event with two major parameters. E1 = {(A1, A2,...An), C} where A1 through An are Event properties and C is the Event's Category or Label.

Data streams have unique qualities that distinguish them from traditional databases. First, there's the nature of it: Data streams are generated by a variety of activities that occur on the internet, such as a user's click on a webpage or some buying activity, live news streaming, photos shot by an IOT device, and so on. As a result, the data generated might not be in a specific format. It might be organised, semi-organized, or even unorganised [1].

The data stream generating speed is the second attribute. Its speed is quite great because its source of generating is the internet. Every millisecond, a large amount of

J. I.-Z. Chen et al. (Eds.): ICIPCN 2022, LNNS 514, pp. 260–273, 2022.
https://doi.org/10.1007/978-3-031-12413-6_20

data is generated. It may consist of repetitive or similar data, but every piece of it is saved for later use.

The quantity of the data stream is the third exciting but difficult feature. Because of its high rate of creation, it generates a large amount of data. Because the information generated cannot be kept in its entirety, it must be processed online or in batch mode.

1.1 Methods for Producing Data Streams

Social Media: Users on numerous social media platforms contribute data by uploading twists, photographs, news, or opinions. This information is presented in a different way.

Online News Channels: Live news updates provide a variety of information as well as temporal data.

Internet of Things (IoT) Gadgets: Smart city initiatives, smart home applications, and healthcare services are just a few examples of IoT devices being used for monitoring and control. This data is extremely sensitive and requires processing in real time or near real time.

1.2 Major Issues Related to Data Streams Handling

Storage: There is no maximum limit to the size of a data stream because it is generated in a quick manner with a big volume. For decision-making, strategy planning, and other purposes, many business verticals and organisations required this data stream information. However, storing all of the data becomes too expensive and inconvenient. As a result, batch processing or online processing, as well as analysis capabilities, are required.

Processing Speed: Because data streams are created at a high rate, they must be processed at a rapid pace to extract information. A single processor or analyzer is incapable of processing data at such a high rate [2]. It is more appropriate to tackle it using internal parallel processing methods in this case.

Imbalanced Data: Due to data generation is continuous, data preprocessing is a critical task. Since the nature of the data is unknown in advance, the standard data balancing algorithm cannot be used.

Drift: Regardless of the situation, data stream creation sources continue to produce it. As a result, it contains all of the information about that particular event or activity. It has varying attribute values or its resulting patterns are always changing. Analyzing evolving data is difficult since the labels or final class may vary over time. This shift in data is known as data drift in the mining process, and it requires a highly robust and adaptable mechanism to detect it.. Before studying drift handling techniques, familiarise yourself with the different forms of drifts in data streams and how they are formed.

1.3 Types of Data Drifts

Abrupt Drift: A drift is when one thought is fully supplanted by another. It might happen unexpectedly or abruptly. The first type of drift is sudden drift, which occurs when a distribution's label or category changes abruptly. Because abruptly changing distributions change much faster than gradual changes, the beneath dependencies also change abruptly, so if we consider the value of a dataset attribute at a specific time stamp, it drastically changes for the next timestamp with no relation between them. As a result, there is a sudden drift. For example,a person's shopping preferences can shift depending on market trends.

Gradual Drift: Another type of drift is gradual drift, in which changes occur gradually yet steadily. Initially, the two concepts are present, but as time passes, another class emerges and gradually gains prominence. It can be thought of as a crop-cultivating system. Wheat cultivating farmers begin yielding sugarcane harvests at the same time, and as sugarcane becomes a more profitable crop, they abandon wheat cultivation and begin full-scale sugarcane production.

Progressive Drift: It is called progressive drift when a total change of the below concept occurs incrementally, giving rise to a second concept through a change in the previous one. This form of drift system gradually becomes habitual for the new class. Drift can be incremental or decremental, which distinguishes it from progressive change. The change in whether can be example of this type of drift. Temperature varies gently as the season transitions from winter to summer.

Repeating/Periodical Drift: Another sort of drift that is significantly different in nature from the others is recurring or periodic drift. Similar concepts or class appear after a certain length of time, but the frequency and duration of such changes are unknown. For example, In the case of the stock market, it frequently remains high for prolonged periods of time before suddenly falling. Its gain is high after some period. When it will go down or up and for how long is unknown in this case.

2 Methods to Handle Data Stream Issues

The general concept of data stream mining is represented in Fig. 1. Data is provided into the stream mining technique via web apps. Initially, the stream is pre-processed or sent to the processing system in block or parallel fashion. The stream mining technique must be adaptable in order to cope up with changes in the input data patterns. A parallel processing strategy is also used for quick processing. Drift detection methods are sometimes used to locate drifts in it [3, 4]. Bagging and boosting-like techniques are more suited to dealing with high-speed data [5]. Following the application of these strategies, useful knowledge is gained.

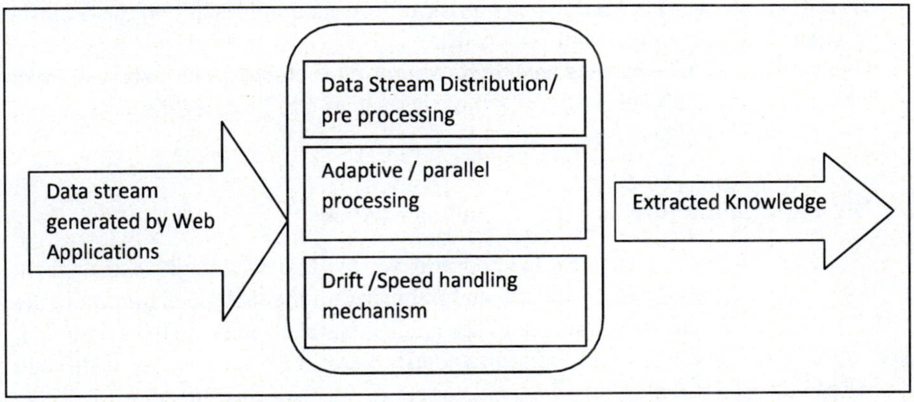

Fig. 1. Generalised process of data stream mining

2.1 Preprosessing of Data Stream

In data stream mining, streams are always considered as attributes and labels, such as S = {$A_{1 \ldots n}$, C}, where $A_{1 \ldots n}$ are characteristics and C is the distribution's class or label. Drift is classified as conceptual drift or real drift depending on how the distribution or class label is changed. The vividness of the data stream is one of the primary qualities that have led to the development of a new adaptive learning solution. In many online applications, the data stream environment suffers from the problem of unequal class distribution, also known as skewed data.

When there are only two types of classes, unequal distribution impedes the overall operation of the mining technique. In the case of multiple classes, the class with the minority may be ignored. The useability of minority classes is suppressed, as is overall mining accuracy. Imbalanced data is a term used to describe this type of distribution. Many researchers have conducted extensive research to solve this problem [6–9]. The binary class imbalancing problem is handled more accurately in the Navo Minority Over-sampling Technique (NMOTe) [10] data balancing technique, but it takes longer than Synthetic Minority Over-sampling Technique (SMOTE) [11]. The mining process is hampered by the imbalanced distribution of data streams, as minority classes are ignored by majority classes. Equation 1 can be used to calculate the class distribution ratio BR(α), which can be used to identify data class balance.

$$BR(\alpha) = \frac{Majority\ classes\ examples}{Minority\ classes\ examples} \quad (1)$$

At the data level, there are three solutions.

I. Oversampling algorithm: This method generates more samples of minority classes in order to justify class distribution. The Synthetic Minority Over-sampling Technique (SMOTE) is a well-known algorithm that is used in the literature [10–12].
II. Under sampling algorithm:This method would seem to be identical to the previous one in that it removes examples of majority classes and attempts to balance the total

class distribution. However, there is a risk of information loss due to the removal of some examples of majority classes [8].

III. Hybrid Approach:To balance data, hybrid algorithms employ both over and under sampling techniques sequentially. All of these steps can be categorised as part of the pre-processing task [9].

2.2 Dealing with Data Drifts: Drift Handling Methods

Several studies have been conducted to uncover new ways to deal with data stream drift, and many methods have been developed and published in the drift detection literature [3, 13, 14]. A rapid response or a wait-and-see method based on working behaviour drift detection strategy might be used. Designing a drift detector for data stream is difficult task as type of drift is not known. It's challenging to design a drift detector for a data stream since the nature of drift is unknown. In the literature, the following tactics are mostly used.

Online data monitoring is carried out in the rapid response technique, and action is made depending on performance criteria. The performance measure plays a crucial role in rapid response models. Acceptance ratio, model flexibility, and drift type uncertainty are all important criteria to consider while constructing a rapid model.

In the second scenario, the data stream is monitored for a fixed or customizable amount of time, and its performance is then assessed to detect changes. The time duration for collecting samples and the comparison parameters in two samples are both hazy in the wait and see model [5]. Because drift might be repeating, abrupt, or gradual, the distribution of incoming data streams cannot be kept constant at a set interval.

2.3 Ensemble Designed

Ensemble design is divided into three broad categories: bagging, boosting, and hybrid methods. Typically, boosting is used to address overfitting issues, whereas bagging is used to address underfitting issues. Bagging ensembles function in parallel mode, implying bias variance balancing. Boosting ensembles, on the other hand, are linked together to reduce bias error.

The Online Bagging and Boosting method proposed by Oza and Russell [5] has gained popularity and accuracy in dealing with data stream problems. Leverage bagging [15] improved the performance of online bagging algorithm by randomising the data instances and the online bagging output prediction. Randomization occurs in two steps: increasing resampling and using output detection codes. The Poisson distribution is used in classifiers to count the number of events that occur within a given time interval. By increasing the resampling value and applying random weights to all instances rather than just incorrectly classified instances, the leverage bagging method computes the Poisson distribution value.

Instead of working on an online stream, the idea is to divide it into chunks and work on each one separately. The chunk-based method for processing a data stream is to divide data stream into blocks, which include instances that are sequential in time [16]. These blocks are sometimes non-overlapping and sometimes overlapping. These techniques are appropriate for both sudden and incremental drifts.

The authors, W. N. Street et al. developed the first blocked-based ensemble algorithm, popularly known as the Streaming Ensemble Algorithm (SEA) [17]. The SEA method uses decision trees built with Quinlan's C4.5 algorithm to design all of its base classifiers. This approach was novel at first due to the use of different blocks of data for separate classifiers, one pass data processing, and the use of the decision trees algorithm, but it has limitations in prediction accuracy and memory size that must be overcome.

The authors, H. Wang et al. [18] introduced the Accuracy Weighted Ensemble (AWE) algorithm, which learns ensemble base models from sequential chunks of the data stream. The incoming data streams are divided into equal parts using this method. The classifier's weight is predicted using the classification error on the current training set. There are many ensemble design exist which based on chunk based or block based concepts.

To solve the problem of data stream mining, many ensemble designing approaches have been developed and implemented. We conducted a detailed investigation of distinct features and parameters utilised for ensemble building by several authors in a prior paper [19]. Because ensemble always works by combining all results or calculating cumulative results, it always achieves higher accuracy than other single systems.

3 Proposed Method

Based on the prediction of the distribution D of the joint probability P (A, Li) between the instance A and Category Li; the ensemble classifiers E attempt to predict the category of the arriving instance. The distribution of joint probability is observed at different time stamps, and if this changes over two intervals, concept drift occurs. Based on this theory following steps are used to developed ensemble model:

Step 1. Many different types of data streams are investigated, but only imbalanced data streams are considered in the proposed method for experiments. The incoming data stream is pre-processed with the data balancing algorithm SMOTE [11] before being fed into the boosting-based ensemble classifier for prediction. As mention in Eq. 1, the BR (α) index for each dataset is calculated and imbalanced datasets are pre-processed using SMOTE algorithm.

Step 2. The general premise of ensemble is that it is always better to distribute work among numerous classifiers and examine aggregate results rather than working with single classifiers. The boosting based ensemble classification approaches have been shown to be beneficial and fast when applied to data streams. Thus, the ensemble classification technique is used in this study to handle data streams. The Boosting-Like Online Learning Algorithm (BOLE) [20] has been found to be more appropriate than all other boosting ensembles. We conducted a thorough study and compared state-of-the-art boosting, bagging and heterogeneous algorithms, and BOLE is selected for ensemble construction based on this comparative analysis [21]. For voting in BOLE, ten base hoeffding tree classifiers are utilised. In most ensembles, a poor classifier with an error rate of less than 50% is not allowed to participate in the final vote. However, in the case of BOLE, this method is altered, allowing additional classifiers to participate. Negative voting to classifiers should also be avoided. As a result, the proposed method applied the BOLE algorithm to an unbalanced data stream to see if any improvements in outcomes could be made.

Each incoming data instance has equal weights at the start of the boosting method. After each repetition, the instance weight is changed based on classification accuracy. Every instance has correctly classified D_{ic} or incorrectly classified D_{iw} weight with it, according to the classifiers' results. Based on these data, an error is calculated as shown in Eq. 2.

$$em = \frac{\text{Diw}}{\text{Dic} + \text{Diw}} \tag{2}$$

The weight of classifiers is calculated based on the error rate, and for subsequent sample instances, all classifiers are sorted according to their weights. Every classifier's weights are calculated as indicated in Eqs. 3 and 4.

$$Weight(m) = \log\left(\frac{1}{\beta m}\right) \tag{3}$$

$$\text{Where as } \beta m = \frac{em}{1 - em} \tag{4}$$

Step 3. The Drift Detection Method (DDM) [3] has proven to be useful in detecting different types of drifts, and it is also used in the proposed model. The distance between consecutive classification errors is used in this method. During the learning process, the prediction correctness improves automatically as the error decreases. This proposed method used the DDM drift detection method with three parameters changed as follows. First, it considers the number of minimum instances to be 30, and then it sets the warning level to 2 and the drift detector's out control level to 3. These changes are made to ensure the system's long-term stability. Similarly to see effect of windowing technique ADWIN [13] drift detector is also used.

Figure 2 depicts the operation of the proposed ensemble classifier, which begins with stream pre-processing and is followed by a boosting-based ensemble classifier. Classifier predictions are aggregated while drift detector results are used to identify drift and feed backward for improvement.

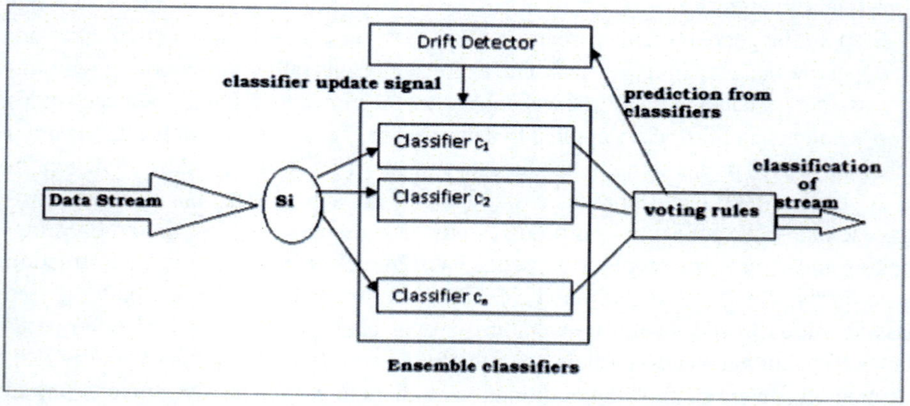

Fig. 2. Proposed architecture of boosting ensemble based classifier

4 Result Analysis

4.1 Experimental Setup

The Massive Online Analysis (MOA) framework [22] has inbuilt DDM algorithm whereas BOLE is also available but added by modifying MOA code. The Intel core i5 processor with Windows 10 and 8GB RAM is used for implementation. Whereas the DataStream balancing is done through WEKA [23] framework using inbuilt Synthetic Minority Oversampling Technique (SMOTE) [11] algorithm.

4.2 Results Discussion

As previously stated, data stream balancing is required in imbalanced data sets; however, because the size of the data stream is unknown, the standard data balancing algorithm cannot be used. Similarly, an algorithm's performance is measured by its ability to accurately predict results. The same performance measures, however, cannot be applied in the case of an imbalanced dataset. So, at first stage we tried to balance the data stream using stream balancer as follows.

Data stream balancing is done as illustrated in Fig. 3, the Adult dataset is binary class and total imbalanced, so the SMOTE [11] algorithm is applied to it. As a result, the previously completely unbalanced dataset is now more balanced and useful for classification. Five different imbalanced datasets are used in the experiment. The ADULT, KDDCUP and Cover type datasets can be found at https://archive.ics.uci.edu/ml/datasets. Whereas Phoneme and Satimage datasets are downloaded from https://datahub.io/machine-learning/phoneme and https://datahub.io/machine-learning/satimage#data sites respectively. ADULT, KDDCUP and Phoneme are binary class data sets that are highly imbalanced. Cover type and Satimage, on the other hand, has multiple classes and is highly imbalanced. The balancing algorithm has little effect on cover type dataset,

Fig. 3. Adult data set before balancing.

but it is considered for testing the applicability of balancing for datasets with multiple classes.

Figure 3 depicts the distribution of class instances in Adult data sets. It is a binary class with 11687 instances of salary greater than $50,000 and 37155 instances of salary less than $50,000. The SMOTE algorithm is used to balance this dataset as follows:

1. First, a sample from a small class is picked. A sample from class 1 will be chosen in the following example.
2. Create synthetic examples that are similar to the sample using the K-nearest neighbour technique.
3. The number of instances of class 1 grew to 23374 after using this approach, as seen in Fig. 4.

Fig. 4. Adult data set after balancing.

Ensemble receives the balanced datasets in stream format for further processing. BOLE [20] is used as the basic ensemble in this experiment. It has a built-in drift detecting mechanism called DDM. To test the effect of drift detector parameters, a change is made in DDM and used with BOLE. Thus first version with the DDM drift detection method with three parameters changed is used as follows. First, it considers the number of minimum instances to be 30, and then it sets the warning level to 2 and the drift detector's out control level to 3. These changes are made to ensure the system's long-term stability. Thus to make changes to the original BOLE, we changed DDM parameters as above and renamed it BOLE2.

The second drift detector employed is Adaptive Windowing (ADWIN) [13], which operates via a windowing mechanism. It only has one parameter, alpha, and it is set to 0.02. Hence, we checked BOLE with a different drift detector, ADWIN, and dubbed it BOLE3. Table 1 displays the results of the based method as well as the changed drift detector method. The accuracy of ensemble model is calculated as per formula given in Eq. 5.

$$Accuracy = \frac{true\ number\ of\ correct\ prediction}{total\ number\ of\ prediction\ done\ for\ dataset} \tag{5}$$

Table 1. Accuracy recorded by original BOLE method, BOLE with changed DDM parameters and BOLE with ADWIN drift detector for balanced datasets (BD) as well as imbalanced datasets(ID).

Ensemble methods/ dataset	BOLE + ID	BOLE + BD	BOLE2 + ID	BOLE2 + BD	BOLE3 + ID	BOLE3 + BD
ADULT	81.82	82.3	81.4	81.8	82.21	**82.65**
KDDCUP	94.14	94.86	95.6	**96.08**	95.48	95.97
Phoneme	75.55	**76.77**	75.47	76.59	75.47	76.59
Satimage	79.55	**80.56**	79.11	79.66	79.11	78.87
Cover type	87.68	**87.68**	85.14	85.14	62.75	62.75

The accuracy of prediction in the ADULT dataset improved from 81.82% to 82.3% in the original BOLE, climbed by 0.4% from imbalanced dataset 81.4 to 81.8 in BOLE2, and recorded the maximum accuracy in all three approaches by 82.65% in BOLE3. Similarly, the ADWIN drift detector performs better than DDM in ADULT dataset classification, as evidenced by the results.

KDDCUP is a binary class unbalanced dataset; after balancing, BOLE improved its classification accuracy from 94.14 to 94.86%, while BOLE2 improved to 96.08% and BOLE3 to 95.97%. Changing the drift detector parameters has benefits in the case of the KDDCUP dataset.

Phoneme dataset comprises two classes, one with 3818 instances and the other with 1586 occurrences; after balancing, the second class climbed to 3172 instances and the dataset became balanced. The accuracy of the original BOLE has improved to 76.77% in the balanced Phoneme dataset, and it is 76.59% for BOLE2 and BOLE3 methods.

Satimage and cover type are used as examples to investigate multi-class imbalanced datasets. Satimage is divided into six classes, with the first having the most occurrences (1531) and the last having the fewest (625). After performing the balancing procedure trice, it became balanced with 1531 to 1250 instances. In both datasets, balanced data performed well for the BOLE approach, with accuracy of 80.56 and 87.68% for Satimage and cover type datasets, respectively.

It is obvious from these results that when the dataset is balanced, classification accuracy improves. The balancing techniques helped to improve predictive accuracy for datasets that were very imbalanced.

However, in the case of multi-class datasets, original BOLE method has improved its performance by working on balanced data. It should be noted that all of the datasets are real-world datasets, so drift detection is a major issue.

This study uses time as another parameter to do a comparison between the original BOLE and the altered parameters BOLE2 and BOLE3. On balanced and imbalanced datasets, Table 2 displays the processing time (seconds) recorded by all three approaches.

When comparing balanced vs. imbalanced data, balanced data takes longer to analyse since it has more instances to classify. When we look at the utility of drift detection methods, BOLE2 has a faster time for ADULT, KDDCUP, and Cover type datasets, with 2.71 s, 5.51 s, and 8.44 s respectively. Phoneme and Satimage have BOLE3 outcomes in the shortest time (0.06 s and 0.69 s, respectively).

Table 2. Time recorded by original BOLE method, BOLE with changed DDM parameters (BOLE2) and BOLE with ADWIN drift detector (BOLE3) for balanced datasets (BD) as well as imbalanced datasets(ID).

	BOLE + ID	BOLE2 + ID	BOLE3 + ID	BOLE + BD	BOLE2 + BD	BOLE3 + BD
ADULT	2.84	**2.71**	9.81	3.33	**2.57**	14.27
KDDCUP	5.81	**5.51**	15.68	8.01	**7.78**	23.26
Phoneme	0.26	0.32	**0.06**	0.27	0.22	**0.21**
Satimage	1.79	0.72	**0.69**	2.03	**0.81**	4.15
Cover type	20.44	**8.44**	23.8	19.92	**8.42**	30

Figure 5 depicts a graphical representation of the time recorded by these three methods on imbalanced datasets. In terms of time, the graphs show that modified drift detectors perform better in binary classification. BOLE2 and BOLE3 took less time than the original BOLE process.

Figure 6 depicts a comparison of the original BOLE ensemble approach with the BOLE2 and BOLE3 methods for balanced datasets. When it came to time analysis, cover type datasets took longer to analyse by all three algorithms, whereas phoneme ran quite quickly. The drift in the cover type dataset may be affecting classifier performance and increasing time.

Fig. 5. Time analysis of original BOLE and BOLE with modified parameter DDM and ADWIN for 5 different imbalanced datasets.

Fig. 6. Time analysis of original BOLE and BOLE with modified parameter DDM and ADWIN for 5 different balanced datasets.

5 Conclusion and Future Research Work

This paper presented a combined approach to dealing with the problem of a data stream with an imbalanced class datasets. To deal with the high speed of data generation, a boosting-based ensemble is used here. Similarly, the role of drift detectors in improving results accuracy is also investigated.

The purpose of this study is to determine the impact of data pre-processing and different drift detectors on the development of an ensemble method for data stream classification. The research is carried out in two ways: first, to assess the influence of imbalanced datasets on classification accuracy, and second, to assess the impact of drift detector method modifications on ensemble construction.

From the accuracy and time analysis, it is obvious that data balancing always aids in enhancing ensemble performance. It was discovered that processing balanced data took a little longer. If processing time is a concern, balanced data is always a superior choice for analysis.

The employment of different drift detectors has been successful in the ADULT and KDDCUP datasets. These two datasets were most likely used in different ensemble tests. Phoneme and Satimage are commonly employed in balancing algorithms, according to our knowledge. We can deduce from these findings that more datasets utilised in ensemble technique testing should be investigated further.

This study revealed the limitations of balancing in the case of multiclass data, necessitating further research into improving the balancing technique. The ensemble can also be developed using new boosting and resampling techniques. Similarly, different types of drift data sets will be tested, each with a different drift detector and boosting strategy. In the future, ensembles with different data distribution strategies and effective drift detector construction can be developed.

References

1. Domingos, P., Hulten, G.: Mining high-speed data streams. In: Proceedings of the Sixth ACM SIGKDD International Conference on Knowledge Discovery and Data Mining, pp. 71–80 (2000). https://doi.org/10.1145/347090.347107
2. Zhang, S., Zhou, A.C., He, J., He, B.: BriskStream: Scaling data stream processing on shared-memory multicore architectures. In: Proceedings of the ACM SIGMOD International Conference on Management of Data, pp. 705–722 (2019). https://doi.org/10.1145/3299869.3300067
3. Gama, J., Medas, P., Castillo, G., Rodrigues, P.: Learning with drift detection. In: Bazzan, A.L.C., Labidi, S. (eds.) SBIA 2004. LNCS (LNAI), vol. 3171, pp. 286–295. Springer, Heidelberg (2004). https://doi.org/10.1007/978-3-540-28645-5_29
4. Brzezinski, D., Stefanowski, J.: Reacting to different types of concept drift **25**(1), 81–94 (2014)
5. Oza, N.C., Russel, S.J.: Online bagging and boosting. In: Proceedings of the 8th International Workshop on Artificial Intelligence and Statistics, pp. 105–112 (2001)
6. Chhotu, D.M., Mankar, J.R.: A Survey of Data Balancing Technique for Multi-Class Imbalanced Problem **2020**(2), 2–4 (2020)
7. Alfhaid, M.A., Abdullah, M.: Classification of imbalanced data stream: Techniques and challenges. Trans. Mach. Learn. Artif. Intell. **9**(2), 36–52 (2021). https://doi.org/10.14738/tmlai.92.9964
8. Susan, S., Kumar, A.: The balancing trick: Optimized sampling of imbalanced datasets—A brief survey of the recent State of the Art. Eng. Reports **3**(4), e12298 (2021). https://doi.org/10.1002/eng2.12298
9. Chakraborty, T.: Imbalanced ensemble classifier for learning from imbalanced business school dataset. Int. J. Math. Eng. Manag. Sci. **4**(4), 861–869 (2019). https://doi.org/10.33889/IJMEMS.2019.4.4-068

10. Chakrabarty, N., Biswas, S.: Navo minority over-sampling technique (NMOTe): A consistent performance booster on imbalanced datasets. J. Electron. Inform. **2**(2), 96–136 (2020). https://doi.org/10.36548/jei.2020.2.004
11. Chawla, W.P.K.N.V., Bowyer, K.W., Hall, L.O.: SMOTE: Synthetic minority over-sampling technique. J. Artif. Intell. Res. **16**(1), 321–357 (2002)
12. Satpathy, S.: Overcoming Class Imbalance using SMOTE Techniques. Data Science Blogathon. https://www.analyticsvidhya.com/blog/2020/10/overcoming-class-imbalance-using-smote-techniques/
13. Bifet, A., Gavaldà, R.: Learning from time-changing data with adaptive windowing. In: Proceedings of the 7th SIAM International Conference on Data Mining, pp. 443–448 (2007). https://doi.org/10.1137/1.9781611972771.42
14. Bifet, A., et al.: Early drift detection method. In: Proceedings of the 4th ECML PKDD Internatioal Workshop on Knowledge Discovery from Data Streams, vol. 6, pp. 77–86 (2006)
15. Bifet, A., Holmes, G., Pfahringer, B.: Leveraging bagging for evolving data streams. In: Balcázar, J.L., Bonchi, F., Gionis, A., Sebag, M. (eds.) ECML PKDD 2010. LNCS (LNAI), vol. 6321, pp. 135–150. Springer, Heidelberg (2010). https://doi.org/10.1007/978-3-642-15880-3_15
16. Brzezinski, D., Stefanowski, J.: Combining block-based and online methods in learning ensembles from concept drifting data streams. Inf. Sci. (Ny) **265**, 50–67 (2014). https://doi.org/10.1016/j.ins.2013.12.011
17. Street, W.N., Kim, Y.S.: A streaming ensemble algorithm (SEA) for large-scale classification. In: Proceedings of the Seventh ACM SIGKDD International Conference on Knowledge Discovery Data Mining, vol. 4, pp. 377–382 (2001). https://doi.org/10.1145/502512.502568
18. Wang, H., Fan, W., Yu, P.S., Han, J.: Mining concept-drifting data streams using ensemble classifiers. In: Proceedings of the Ninth ACM SIGKDD International Conference on Knowledge Discovery and Data Mining (KDD '03). Association for Computing Machinery, New York, NY, USA, pp. 226–235 (2003). https://doi.org/10.1145/956750.956778
19. Samant, R.C., Thakore, D.D.M.: A rigorous review on an ensemble based data stream drift classification methods. Int. J. Comput. Sci. Eng. **7**(5), 380–385 (2019)
20. de Barros, R.S.M., de Carvalho Santos, S.G.T., Junior, P.M.G.: A boosting-like online learning ensemble. In: Proceedings of the International Jt. Conference on Neural Networks, pp. 1871–1878 (2016). https://doi.org/10.1109/IJCNN.2016.7727427
21. Samant, R.C., Patil, S.H.: Adequacy of effectual ensemble classification approach to detect drift in data streams. In: Proceedings of the 2022 International Conference for Advancement in Technology (ICONAT), pp. 1–6 (2022). https://doi.org/10.1109/ICONAT53423.2022.9725854
22. Bifet, A., Holmes, G., Kirkby, R., Pfahringer, B.: MOA: Massive online analysis. J. Mach. Learn. Res. **11**, 1601–1604 (2010)
23. Witten, I.H., Frank, E.: Data Mining: Practical Machine Learning Tools and Techniques, 2nd edn. (2005)

Survey on Various Performance Metrices for Lightweight Encryption Algorithms

Radhika Rani Chintala[1(✉)] and Somu Venkateswarlu[2]

[1] Koneru Lakshmaiah Education Foundation, Guntur, Andhra Pradesh, India
radhikarani_cse@kluniversity.in
[2] Woxsen University, Hyderabad, India

Abstract. Lightweight cryptography is the most emerging area in resource constraint applications. In lightweight cryptography, encryption algorithm must be lightweight w.r.t. to resources and should also offer good security against possible attacks. Cryptography algorithms are categorized depending on mode of their implementation or on the architecture. Most of the lightweight cryptography algorithms falls under block ciphers and follow symmetric architecture and can be either hardware or software implemented. Area and energy are essential factors taken into consideration for hardware implemented designs. Whereas code size and throughput are essential factors for software implemented designs. Block ciphers implemented in software are less expensive and offers high flexibility compared to hardware implementation. Hardware implemented block ciphers are relatively simpler to implement compared to software implementation and they mainly focus on optimization of hardware resources. In this article, we represent various hardware and software metrics available for evaluating the performance of lightweight encryption algorithms.

Keywords: Lightweight cryptography · Resource constraint · Performance metrics · Hardware metrics · Software metrics

1 Introduction

Lightweight cryptography is an encryption algorithm or a protocol designed for resource constraint environments that comprises of sensors, RFID tags, healthcare devices, hands-free smart cards, etc. ISO/IEC 29192 is a lightweight cryptographic standardization project, and its properties were clearly mentioned depending on the target platforms. For hardware implementations, the important lightweight properties to consider are chip size and energy consumption. For software implementation, the important lightweight properties to consider are code size and RAM size. Sufficient security is provided by lightweight cryptographic algorithms.

There are various emerging domains where the high constrain devices are interconnected and work together to carry out few tasks. Sensor networks, Health-care monitoring devices, Automotive systems, IoT devices, Smart-grid, Distributed control and Cyber physical systems are examples of these domains where the security and privacy

J. I.-Z. Chen et al. (Eds.): ICIPCN 2022, LNNS 514, pp. 274–283, 2022.
https://doi.org/10.1007/978-3-031-12413-6_21

are highly crucial in all of these domains [1]. The performance of conventional cryptographic algorithms approved by NIST- National Institute of Standards and Technology [2], might not be satisfactory when they are developed to fit into resource constrained environments. Due to these reasons, a project on lightweight cryptography was started by NIST in order to study more about the issues and develop a standard policy for lightweight cryptography algorithms.

Lightweight cryptography focuses on extremely wide range of resource-limited devices like RFID tags and sensor devices. They can be implemented both on software and hardware using different communication technologies. Due to the constraints on processor speed, code size, energy consumption, throughput, etc., conventional cryptographic algorithms are harder to implement on resource-limited environments. Trade-offs of lightweight cryptography include speed, cost of implementation, energy consumption, performance and security. The objective of lightweight cryptography is to use low processing resources, low power supply, less energy consumption and less memory and provide a better security solution that can operate on resource-limited devices. Lightweight cryptographic algorithm is expected to be simple and works faster than that of conventional cryptographic algorithms.

In order to overcome most of the difficulties of conventional cryptography, lightweight cryptographic algorithms are proposed. This includes constraints associated with limited memory, physical size, energy drain and less processing requirements. During the past few years, several of the lightweight cryptographic algorithms are designed and are developed. These algorithms are mainly used for the applications involving resource-constraint devices [3].

Lightweight cryptography is a block cipher based encoding method [4] that provides data confidentiality in highspeed lightweight environments like sensor devices. A lightweight encryption algorithm is considered as a best cipher when it provides appropriate security and balances tradeoff between the performance and design cost. To measure the efficiency of lightweight encryption algorithms, certain metrics must be used. Different performance metrics are available centered on type of the implementation.

2 Literature Survey

Immense research was carried on analyzing the performance of different lightweight encryption algorithms. Performance metrics that are discussed by various authors included the parameters related to area, code size, and clock cycles needed for encryption as well as for decryption operation. Few researchers just considered the security parameter while measuring encryption algorithm's performance.

Authors of [5] proposed a performance metric for evaluating the efficiency of a lightweight encryption method and applied it on Hight algorithm. By considering metrics like power, energy and LEs(logical elements), optimal design is determined. Of all these, the performance metric to be considered depends on the applications. Authors have considered energy and area as the essential factors for hardware implemented designs and thus proposed a performance metric that is dependent on design area and energy factors. Block ciphers with minimum energy and area consumption are treated as highly efficient.

Authors of [6] discussed about the security-performance tradeoff of lightweight block ciphers for resource constrained applications in industrial WSNs. They have proposed a software performance metric that merely depends on block size, code size, RAM size, and clock cycles necessary for encryption as well as for decryption operation. Throughput refers to the no. of bits encrypted per second and is measured in kbps. Finally, they have proposed a combined metric. Block ciphers with low combined metric are treated as highly efficient algorithms.

Sohel Rana et al. [7] have presented a survey on various lightweight encryption algorithms, and analyzed their performance on different metrics like key size, cycle count, code size, and RAM size. Authors of [8] discussed regarding some benchmark projects such as ECRYPT II, eBACS and BLOC that are used to calculate the performance of cryptographic algorithms and also the lightweight block ciphers depending on a variety of hardware and software platforms. ENCRYPT II is European network excellence project of cryptology[9].

Eisenbarth et al. [10] have implemented and assessed the performance of twelve lightweight block ciphers w.r.t RAM size, code size, and count of cycles for encryption and decryption. For this they used 8-bit AVR microcontroller. eBACS(ENCRYPT Benchmark of Cryptographic system) is a standard benchmarking for cryptography and evaluated performance of several cryptographic primitives on servers and personal computers,considering the speed metric. Another research project called BLOC analyzed block ciphers implemented for resource limited environments. At the time of project, Cazorla et al. [11] evaluated five conventional and twelve lightweight block ciphers meant for wireless sensor nodes. For this purpose, authors used 16-bit Texas MSP microcontroller and expressed that the results of simulation are not appropriate when equated to real implementation.

Deepti Sehrawat et al. [12] discussed about the parameters and tools needed to measure the performance of security algorithms that are implemented in software. RAM usage, code size, encryption cycle count, decryption cycle count are essentail factors considered for evaluating a lightweight block cipher. Based on these factors they have presented the performance metrics such as energy measured in μJ (micro joules) and combined metric. Smaller metric value indicates better implementation of cipher algorithm.

M. Matsui et al. [13] categorized lightweight implementations into three variants depending on the values attained from the combined metric discussed in [14]. First category falls under Ultra Lightweight implementation, which requires ROM capacity of 4 KB and RAM capacity of 256 bytes. Second category falls under Low-cost implementation, which requires ROM capacity of 4 KB and RAM capacity of 8 KB. Third category falls under Lightweight implementation, which requires ROM capacity of 32 KB and RAM capacity of 8 KB. Mohd. B. J. et al. [15] discussed about various metrics to measure the performance of lightweight encryption algorithms centered on software or hardware implementation. They have discussed about two software metrics namely Throughput (measured in bits/sec) and Synthetic metric. A number of hardware metrics were also discussed. Hardware metrics that were discussed are Design area, Area-per-bit.

Authors of [16] has proposed a metric called Performance efficiency. Researchers felt that the efficiency metric doesn't suit for lightweight block ciphers, and proposed a metric called FOM (Figure-of-Merit). Authors in [17] have proposed the metrics such as Area, Power and Energy, which are the essential hardware performance metrics of lightweight block ciphers that are targeted specially for low energy-resource-constraint devices. Energy per bit is one more hardware metric, where the energy is normalized w.r.t no.of bits in a block. Authors of [18] proposed is a hardware metric that calculates energy cost, area cost and time to encrypt a single bit.

Hatzivasilis G. et al. [19] presented few evaluation metrics related to performance factors, cost and security. In several cases, key length indicates the security level and is quantified in bits. Another metric discussed is throughput which is measured in kbps at a certain frequency. Mostly, hardware implementations of lightweight encryption algorithms use a frequency of 100 kHz and software implementations use a frequency of 4 MHz.

One More general metric presented in [19] is latency. Latency is defined as the no. of clock cycles needed to process a single block. Next comes the Power metric which is measured in µW (micro watts) for hardware implementation. Another generic metric is Energy per bit which indicates the energy expended for processing a single bit. This metric is same for both types of implementations and measured in µJ. Memory usage is another metric which is measured as the requirement of RAM and ROM for an algorithm. Tradeoff between the design size and performance is measured using the Efficiency metric. Authors of [19] proposed a metric Hardware Efficiency for hardware implementation, and Software Efficiency metric for software implementation. The higher efficiency value implies an effective implementation of encryption algorithm.

3 Performance Metrices for Lightweight Encryption Algorithms

A variety of metrics have been proposed by the researchers to measure the performance of lightweight encryption algorithms. Performance metrics are proposed centered implementation type of encryption algorithm.

3.1 Hardware Performance Metrics

The hardware metrics available are defined below:

Throughput. The throughput of an encryption algorithm is measured using the parameters block size, clock cycles and frequency [20]. It is calculated using the following equation.

$$Throughput = \frac{Block\ size}{Clock\ Cycles} * Frequency \tag{1}$$

Larger the throughput value, better is the performance of an encryption algorithm.

Area. Design area is measured differently for different implementations. In ASIC implementation, design area is measured in terms μm2 for physical design tools, whereas pre layout design the area is measured in terms of GE(Gate Equivalent) [15]. Usually, GE estimates the complexity of hardware design, and 1 GE is equal to 2-input NAND gate. In FPGA implementation, design area is measured based on the utilization of resources that depend on the vendor. LE is the measure for design area in Altera Quartus II design, where an LE contains lookup table(LUT) and register. CLB(Configurable logic block) is the measure for design area in Xilinx design, where a CLB contains various logic cells which in turn contains a lookup table and D flipflop. However, there are tables that converts the design area from one vendor representation to other vendor representation. Smaller the design area, efficient is the encryption algorithm.

Area-per-Bit. This is another hardware metric [15], where the design area is normalized to one bit. It defines the cost of area that is required to encrypt a single bit. This allows fair comparison of block ciphers w.r.t to area. Area per bit is defined as:

$$Areaperbit = \frac{Design\,Area}{Block\ size} \tag{2}$$

Smaller the area-per-bit metric value, efficient is the lightweight encryption algorithm.

Throughput-to-Area. Throughput-to-Area of an encryption algorithm is measured using the throughput and area metrics [20]. It is calculated using the following equation.

$$Throughput\ to\,Area = \frac{Throughput}{Area} \tag{3}$$

Larger the throughput-to-area value, better is the performance of an encryption algorithm.

Energy. Energy is one most important metric considered for lightweight encryption algorithms used in resource-constraint applications [18]. It is measured in joules and uses the parameters block size, frequency, power and clock cycles. It is defined as:

$$Energy = \frac{Power * Cycles\,per\ block}{Frequency} \tag{4}$$

Lesser the energy consumption, better is the efficiency of encryption algorithm.

Energy-per-Bit. Energy per bit is a hardware metric, where the energy is normalized w.r.t no.of bits in one block [18]. This is a best metric to measure the actual cost of energy and is measured utilizing the parameters energy and block size. It is calculated using the following equation.

$$Energy\,per\ bit = \frac{Energy\ consumed\ per\ block}{Block\ size} \tag{5}$$

Lesser the energy-per-bit value, better is the efficiency of encryption algorithm.

Hardware Efficiency. Hardware efficiency of an encryption algorithm is measured using the parameters throughput and GE complexity [19]. It is calculated using the following equation.

$$Hardware\ Efficiency = \frac{Throughput}{Complexity\ in\ GE} \tag{6}$$

Larger the hardware efficiency value, better is the performance of an encryption algorithm.

Performance Efficiency. Performance efficiency is a hardware metric [16] and is described as:

$$Performance\ Efficiency = \frac{Block\ size}{Time\ to\ encrypt\ one\ block * Area} \tag{7}$$

or

$$Performance\ Efficiency = \frac{Block\ size * Frequency}{Clock\ cycles\ per\ block * Area} \tag{8}$$

Larger the performance efficiency value, better is the performance of an encryption algorithm.

Figure of Merit. Researchers of [16] felt that the efficiency metric doesn't suit for lightweight block ciphers, and proposed a metric called FOM (Figure-of-Merit). FOM is defined as:

$$FOM = \frac{Throughput}{GE^2} \tag{9}$$

Higher FOM value indicates efficient lightweight encryption algorithm.

Hybrid Metric. This metric is a combination of multiple metrics [20]. It calculates energy cost, time and area cost to encrypt a single bit and can be defined as:

$$Hybrid\ Metric = \frac{Throughput}{Area * Energy\ per\ bit} \tag{10}$$

Higher the hybrid metric values, more efficient is the lightweight encryption algorithm.

MSEC. The metric MSEC is the Metric for Security Versus Energy Consumption [21] dependent on the factors such as energy consumption and security of an algorithm. MSEC value can be computed by dividing the secured years left with normalized energy, where secured years left is computed as: year till which the cipher provides adequate protection minus the current year.

$$MSEC = \frac{No.\ of\ secured\ years\ left}{Normalized\ Energy} \tag{11}$$

Higher the MSEC value, better is the secured and energy efficient lightweight algorithm. Negative MSEC value indicates that the algorithm is no more safe to use.

3.2 Software Performance Metrices

The software performance metrics available mainly based on the code size of an encryption algorithm. Few such metrics are described below.

Throughput. One software metric is Throughput [15] that is measured in bits/sec and is defined as:

$$Throughput = \frac{Block\ size}{Block\ encryption\ time} \tag{12}$$

As throughput is the function of design frequency, throughput can also be defined as:

$$Throughput = \frac{Block\ size * Frequency}{Cycle\ count} \tag{13}$$

Larger the throughput value, better is the performance of an encryption algorithm.

Combined Metric. The performance of an encryption algorithm is measured using combined metric [12] that is mainly based on the parameters such as code size in bytes and cycle count of an algorithm. It is measured using the following equation.

$$Combined\ Metric = Code\ size * Cycle\ count \tag{14}$$

Smaller the metric value, better is the encryption algorithm.

Synthetic Metric. Multiple non-correlate metrics are combined to form a synthetic metric [6] that measures various performance aspects. One such synthetic metric is measured based on the parameters such as code size, cycle count and block size of an encryption algorithm and is defined as below.

$$Synthetic\ Metric = \frac{Code\ size * Cycle\ count}{Block\ size} \tag{15}$$

Smaller the synthetic metric value, better is the encryption algorithm.

Software Efficiency. Software efficiency of an encryption algorithm is measured using parameters such as throughput and code size of an algorithm [19]. It is calculated using the following equation.

$$Software\ Efficiency = \frac{Throughput}{Code\ size} \tag{16}$$

Larger the software efficiency value, better is the performance of an encryption algorithm.

Table 1 gives the the description of paramters and metrices used in the performance equations.

Table 1. Parameters and metrices used in performance equations.

Parameter/Metric	Description
Throughput	Defines the encryption speed. It indicates the no. of bits encrypted in a given time and usually measured in kbps
Block size	No. of bits in a block
Clock cycle	Indicates the CPU or processor speed
Frequency	Used to measure processing speed. Usually measured in Hz
Area	Indicates the amount of design area required to encrypt a block of data. Usually measured in GE
Area-per-bit	Indicates the amount of design area required to encrypt a bit of data. Usually measured in GE/bit
Throughput-to-Area	Indicates the no. of bits encrypted per area. Usually measured in kbps/GE
Energy	Indicates the amount of energy consumed for encrypting a block of data. Usually measured in joules/byte
Energy-per-bit	Indicates the amount of energy consumed for encrypting a bit of data. Usually measured in joules/bit
Hardware esfficiency	Similar to throughput-to-area. Indicates the no. of bits encrypted per area
MSEC	Indicates the performance of encryption algorithm w.r.t security and energy consumption
Code size	Indicates the memory requirement of an encryption algorithm. Usually measured in bytes

4 Conclusion

In this paper, a variety of performance metrics are discussed that are meant to measure the effectiveness of lightweight block ciphers. The metrics were categorized based on the implementation type. Four metrics are discussed for software implemented lightweight block ciphers and ten metrics are discussed for hardware implemented block ciphers. All the metrics described measures performance of lightweight encryption algorithm based only on the implementation factors. For an encryption algorithm, one important factor to consider is the security parameter. So, there must be a metric to measure the performance of a lightweight encryption algorithm based on the implementation parameters as well as security parameter. One such hardware metric that measures the tradeoff amongst security as well as energy consumption discussed is MSEC. Testing the performance based on any one metric doesn't give the actual effectiveness of an algorithm. Hence, it is suggestable to measure the effectiveness of lightweight encryption algorithm by considering multiple metrics.

References

1. Chintala, R.R., Narasinga Rao, M.R., Somu, V.: Review on security issues in human sensor networks for healthcare applications. Int. J. Eng. Technol. **7**, 269–274 (2018)
2. Buchanan, W.J., Li, S., Asif, R.: Lightweight cryptography methods. J. Cyber Secur. Technol. **1**, 187–201 (2017)
3. Chintala, R.R., Jagan, L.S., Harika, C.L.: Lightweight encryption algorithms for wireless body area networks. Int. J. Eng. Technol. **7**, 64–66 (2018)
4. Arora, N., Gigras, Y.: Block and stream cipher based cryptographic algorithms: A survey. Int. J. Inf. Comput. Technol. **4**, 189–196 (2014)
5. Mohd, B.J., Thaier, H., Zaid, A.K., Khalil, M.A.Y.: Modeling and optimization of the lightweight HIGHT block cipher design with FPGA implementation. Secur. Commun. Netw. **9**, 2200–2216 (2016)
6. Pei, C., Xiao, Y., Liang, W., Han, X.: Trade-off of security and performance of lightweight block ciphers in industrial wireless sensor networks. EURASIP J. Wirel. Commun. Netw. **2018**(1), 1–18 (2018). https://doi.org/10.1186/s13638-018-1121-6
7. Sohel Rana, M., Wadud, A.H., Azgar, A., Kashem, M.A.: A survey paper of lightweight block ciphers based on their different design architectures and performance metrics. Int. J. Comput. Eng. Inf. Technol. **11**, 119–129 (2019)
8. Shin, S., Kim, M., Kwon, T.: Experimental performance analysis of lightweight block ciphers and message authentication codes for wireless sensor networks. Int. J. Distrib. Sens. Netw. **13**, 1–13 (2017)
9. Bernstein, D.J., Lange, T.: eBACS: ECRYPT benchmarking of cryptographic systems (2016)
10. Eisenbarth, T., et al.: Compact implementation and performance evaluation of block ciphers in ATtiny devices. In: Mitrokotsa, A., Vaudenay, S. (eds.) AFRICACRYPT 2012. LNCS, vol. 7374, pp. 172–187. Springer, Heidelberg (2012). https://doi.org/10.1007/978-3-642-31410-0_11
11. Cazorla, M., Marquet, K., Minier, M.: Survey and benchmark of lightweight block ciphers for wireless sensor networks. IDEA **64** (2013)
12. Sehrawat, D., Gill, N.: A review on performance evaluation criteria and tools for lightweight block ciphers. Int. J. Adv. Trends Comput. Sci. Eng. **8**, 630–639 (2019)
13. Matsui, M., Murakami, Y.: Minimalism of software implementation. In: Moriai, S. (ed.) FSE 2013. LNCS, vol. 8424, pp. 393–409. Springer, Heidelberg (2014). https://doi.org/10.1007/978-3-662-43933-3_20
14. Priyanka, A.A., Pal, S.K.: A survey of cryptanalytic attacks on lightweight block ciphers. Int. J. Comput. Sci. Inf. Technol. Secur. **2**, 472–481 (2012)
15. Mohd, B.J., Hayajneh, T., Vasilakos, A.V.: A survey on lightweight block ciphers for low-resource devices: Comparative study and open issues. J. Netw. Comput. Appl. **58**, 73–93 (2015)
16. Rolfes, C., Poschmann, A., Leander, G., Paar, C.: Ultra-lightweight implementations for smart devices – security for 1000 gate equivalents. In: Grimaud, G., Standaert, F.-X. (eds.) CARDIS 2008. LNCS, vol. 5189, pp. 89–103. Springer, Heidelberg (2008). https://doi.org/10.1007/978-3-540-85893-5_7
17. Kerckhof, S., Durvaux, F., Hocquet, C., Bol, D., Standaert, F.-X.: Towards green cryptography: a comparison of lightweight ciphers from the energy viewpoint. In: Prouff, E., Schaumont, P. (eds.) CHES 2012. LNCS, vol. 7428, pp. 390–407. Springer, Heidelberg (2012). https://doi.org/10.1007/978-3-642-33027-8_23
18. Chintala, R.R., Narasinga Rao, M.R., Somu, V.: Performance metrics and energy evaluation of a lightweight block cipher in human sensor networks. Int. J. Adv. Trends Comput. Sci. Eng. **8**, 1487–1490 (2019)

19. Hatzivasilis, G., Fysarakis, K., Papaefstathiou, I., Manifavas, C.: A review of lightweight block ciphers. J. Cryptogr. Eng. **8**(2), 141–184 (2017). https://doi.org/10.1007/s13389-017-0160-y
20. Diehl, W., Farahmand, F., Yalla, P., Kaps, J.P., Gaj, K.: Comparison of hardware and software implementations of selected lightweight block ciphers. In: Proceedings of the 27th International Conference on Field Programmable Logic and Applications (FPL), pp. 1–4 (2017)
21. Chintala, R.R., Narasinga Rao, M.R., Somu, V.: Performance analysis of EELWE algorithm using MSEC method. Eur. J. Mol. Clin. Med. **7**, 3225–3233 (2021)

A Hybrid Approach to Facial Recognition for Online Shopping Using PCA and Haar Cascade

S. V. Shri Bharathi(✉), Tulabandu Aadithya Kiran, Nallamilli Dileep Kanth, Bolla Raghu Ram Reddy, and Angelina Geetha

Hindustan Institute of Technology and Science, Chennai, India
shribharathi01@gmail.com

Abstract. This paper proposes a model, designed to add an extra layer of security using facial recognition to make the payment method more secure. In this facial recognition, the algorithm used is the combination of both Principal Components Analysis (PCA) and Haar Cascade algorithms, so that the face recognition will be more accurate and faster. The Haar cascade algorithm is used in the detection of faces and PCA is used to analyse the face and match with the existing data stored in pixel format, so that it can be matched at the time checkout and proceed to the payment page.

Keywords: Face-recognition · Markov gaussian mixture model · Zafeiriou

1 Introduction

Face-recognition algorithms in the literature may be classified into two major categories: photo-based and video-based. Video-based approaches may combine the look and dynamic aspects of the face, while image-based methods often employ facial appearance functions. Face recognition costs are cheaper even when utilizing high-performing deep-learning-based approaches under more realistic and hard settings, such as facial expression, head function and illumination modifications, occlusion, poor resolution, and noise. The importance of seamless biometric functions has grown as a solution to these issues. A person's physical or behavioral characteristics, such as age, gender, ethnicity, hair shade, eye colour, stride, and facial movements, may be used to identify them using soft biometric traits. Using facial movements, soft biometric capabilities have been utilised to estimate age, gender, and face recognition. Humans utilize facial motion to identify faces, despite the fact that it is more slowly learnt. Lighting and appearance modifications, including spectacles, beards, and cosmetics, have little effect on facial dynamics. Furthermore, facial dynamics of emotional emotions have been demonstrated to be age-independent and powerful across time.

J. I.-Z. Chen et al. (Eds.): ICIPCN 2022, LNNS 514, pp. 284–295, 2022.
https://doi.org/10.1007/978-3-031-12413-6_22

During one of the first experiments to utilize facial dynamics to measure popularity of faces, researchers examined the frequency of a few specific facial motion devices while subjects were watching a movie or being interviewed. Action unit largely based dynamic functions may be employed for facial recognition. Adaptive hidden Markov models with temporal dynamics of the face were used in another research to increase the popularity of a person. They then developed an Eigen face-like person reputation gadget that used behaviour recordings from rigid head movements and physiological information. When combining behaviour and physical variables, the rate of face reputation rises. This was shown using a Bayesian classifier and a Gaussian Mixture Model (GMM). Zafeiriou and Pantic found that the interchange of biometric information taking place on the face during a spontaneous grin contains information. Between the neutral frame of the grin and the peak frame, the facial movement was shown as a thick movement discipline. Promising results have been obtained in experimental investigations using a library of 563 recordings of individuals laughing, collected from 22 different sources. In summary, the video-based face reputation strategies in the literature, which make the most facial dynamics use four exceptional methods:

i) Face movement units as a unit of measurement.
ii) Methods that make advantage of optical float or a face-specific dense motion.
iii) Facial landmarks-based methods.
iv) Approaches including spatial-temporal algorithms, which describe facial movements implicitly, but only a few of them have an awareness of emotional facial expression dynamics and report results on datasets with a limited range of themes.

2 Literature Survey

Many applications in access control, law enforcement, safety, surveillance, internet communication, and laptop pleasure have drawn attention to face recognition. The contemporary face popularity structures have made significant progress, however they only function well in controlled conditions and deteriorate significantly when presented with real-world situations. Lighting, stance, and occlusion/expression are all left to the artist's discretion in real-world scenarios. In other words, there are a ton of challenging circumstances and chances ahead. Recently, a small number of academics have begun to examine the reputation of a person's face in an unrestricted setting. As a reader's guide rather than an in-depth experimental evaluation, this study serves as a supplement to the cited publications.

Hence, the purpose of this study is to address the significant challenges involved in the modification of current facial recognition algorithms to construct a successful system that can be used in the real world. After that, it goes through the work that has been done so far, highlighting the most effective algorithms and providing an overview of their achievements and shortcomings in solving the challenge. For facial recognition, it also offers a few probable futures. That's why face-recognition research may get off to a great start using this approach, since helpful tactics can be identified, and errors can be avoided. Mohammed & Aly Saleh Hassaballah [1] 2020.

Recent research by A. Dantcheva, P. Elia, and A. Ross [2] investigated the possibility of collecting secondary information from basic biometric developments, including face, fingerprints, hand shape, and iris. Non-public characteristics like gender, age, ethnicity, hair color, height, weight, and many more were included in this supplemental record. In monitoring and indexing biometric databases, these features were referred to as "soft biometric." A number one biometric gadget (e.g., merging face with gender data) may be improved by including these features into a fusion framework. The latter was particularly beneficial in bridging the semantic gap between human and system descriptions of biometric data. An overview of mild biometric was provided, as well as a discussion of several methods for extracting them from visual and video data that have been put out so far. A taxonomy for organizing and categorizing smooth biometric properties was also presented, as are a list of the strengths and limitations of each feature in the context of an operating biometric machine. Finally, it discussed open research questions on this topic. Biometrics scholars and practitioners were the target audience for the study.

Video monitoring, human-computer interaction, anonymous custom-designed advertising, and picture retrieval are just a few of the many uses for automated gender assessment. Algorithms often use face features to identify gender. Using dynamic features gleaned from smiles, a novel method for gender estimation was proposed, and demonstrated that (a) facial dynamics incorporate clues for gender dimorphism, and (b) that even as for adult individuals' appearance functions are more accurate than dynamic capabilities, for subjects under 18 years old facial dynamics can outperform appearance capabilities by Dantcheva, Antitza. (2018) [3]. The look-based gender estimation performance was significantly improved by fusing the suggested dynamics-based technique with state-of-the-art look- based totally methods. It is clear from the results that smile-dynamics provided relevant and complimentary information on appearance gender.

Bauer (2019) referred to the patient LF, who was unable to recognize familiar faces after a stroke. Despite the inability to vocally recognize familiar faces, a psychophysiological study revealed that the capacity to process facial identification in the unconscious had not been lost. The use of behavioral duties to confirm covert facial recognition has been proven in subsequent investigations. Overt popularity was missing from studies on the afflicted person PH, which found no impact of conventional face familiarit on matching [4], interfering, priming, or learning new tasks (De Haan, Young and Newcomb, 2019.

The "covert popularity" phenomena have been conceptualized using a variety of approaches. Until present, no patient was exposed to all of the techniques. The employment of behavioral obligations that were previously associated with PH was examined in this study of LF, who exhibits psychophysiological indications of covert recognition and the application of such behavioral duties. The results provided strong behavioral evidence that a person's reputation may be maintained even if they aren't aware of it. These results suggest that comparable processes should be tapped into by each methodology and have important consequences for theoretical fads in covert face identification. Psychophysiological and behavioral evidence of face popularity should be combined in a conceptual paradigm, as presented.

T. Gevers and Hamdi Dibeklioglu (2019) [5], suggested that nonverbal communication would be incomplete without the use of a smile. The ability to distinguish between spontaneous and staged emotions in the visual processing of social signals is also crucial. This is why, the dynamics of eyelid, cheek, and lip nook motions were utilized to distinguish between spontaneous and staged pleasure grins in this research. More than one database were analyzed to see how various fusion ranges affect those maneuvers' discriminative power. Consequences became better. The largest database of spontaneous and prepared smiles to date, as well as fresh empirical and conceptual insights on smile dynamics were also provided. A total of 1240 samples were collected, covering 400 distinct themes. In addition, it has the distinct advantage of having a wide age range ranging from eight years old to seventy-six. There were age-related changes in smile dynamics based on large-scale tests.

3 Methodology

In the existing systems of online shopping webpages, when user clicks proceed from cart then it will be redirected to the payment gateway page, and from there the users can pay for their items. The main problem in payment page is if the user has money in website wallet, then password will not be requested and hence money can be easily stolen once their login credentials get leaked. Hence the facial recognition is proposed for additional security. The main reason of using the facial recognition is because most of the devices have webcam or front camera and so it doesn't require an external device like fingerprint, rfid scanner etc. In fingerprint scanner, the user can cast another user's fingerprint to authenticate and there may be a possibility that fingerprint sensor does not work. Moreover, the current facial recognition systems have a limitation in that they can only recognise roughly six photos of the same person at a time. As a result, it takes longer to recognise and requires more storage space. This proposed Facial Recognition model uses both PCA and Haar Cascade algorithms that makes the facial recognition model accurate and efficient.

PCA algorithm is used in reducing the variables in face recognition, and Eigenfaces which are a linear combination of weighted eigenvectors, are used to represent each image in the training set. These eigenvectors are found in the covariance matrix of a training image set. To make this model more efficient, Haar cascade is used in addition to PCA algorithm so that the face can be detected faster. As this method can be used in most of the devices it can easily implemented in real world. Recognition of face, a physical property of the individual, is the only implemented stage in the current system. The old norms of authentication have not been employed by this method. Traditional approaches have a major drawback of the necessity to remember lengthy passwords.

Authentication techniques that depend only on a person's physical attributes heighten the issue of identity theft, which is already a problem with the conventional systems. It is possible to overcome the limits of the current technology by incorporating physical aspects with the user's "smile" behavior. After the users register, their faces are utilized to train the machine learning model in the pickle file format. Computer libraries may be used for this purpose. Emotions of the user are captured using FER, a face-recognition library trained to recognize trained faces. The payment will be successful if the user and emotion of the detected face can be determined. Proposed methodology's steps are summarized as follows:

- Algorithms for face identification in videos begin by spotting and tracking a person's face in the video. As the position of the head, occlusion, and light vary, the face area should be detected.
- Second, dynamic characteristics from face videos are extracted and utilized to recognize individuals. First, 68 landmark points are identified on the face, and then 27 facial distances are computed using the landmark points, which are predicted to move during the smile movement.

There are several benefits to the suggested strategy.

- Authentication system based on a hybrid biometric recognition technology.
- A highly secured method for payments.
- Method that is quick and dependable.

4 Implementation

The modules of the proposed system are as follows:

4.1 Admin

Administrators have access to the project at a high level. Admin duties include updating the website to show what products are now available. Adding a new order state, such as "pending" or "successful". Developer-supplied login credentials may be used by administrators. Essentially, he's in charge of the whole process. After logging in, the administrator may see a list of all the users and their associated data. He has the ability to create new product categories. Additionally, he has the option to post the products on his website in order to broaden the audience that would be able to see them. After a user places an order, the administrator may review the order and either approve or reject it, depending on his preference.

4.2 User

The user is the primary purchaser of products and is ultimately responsible for their payment, which is a critical part of the project. The user must first sign up before doing any of the aforementioned actions. To add products and things to the basket, "Add to Cart" button is clicked once the registration process is completed. The user may sign up by inputting his personal information, and the app will also snap a picture of him. His username and password are all that is required for him to log in when the registration is complete. He may look at the products on the website, and if he/she likes any of them, he/she can add them to his/her shopping basket and then proceed to make an order.

4.3 Cart

The cart shows the goods that the user has added.

4.4 Input Design

It is the interface between the user and the information system. There are a variety of ways to enter data into a computer, including reading it from a written or printed document or having individuals input it directly into the system. This includes defining specifications and processes for data preparation. The goal of input design is to maintain the process as easy as possible while also reducing the quantity of input necessary, reducing mistakes, and preventing delays. The input is set up in such a manner that it offers convenience and secure while also protecting user privacy. The following factors were taken into account by Input:

Objectives
When a user-oriented description of the input is translated into a computer-based system, the process is known as Input Design. In order to eliminate data entry mistakes and to point management in the right path for obtaining accurate information from the computerised system, this design is critical. To manage massive amounts of data, user-friendly displays must be created. The purpose of input design is to make data entering as simple and error-free as possible. All data manipulations are possible due to the layout of the data entering panel. In addition, you may use it to go over your recorded data. Validation will be performed as soon as the data is input. Using displays, data may be input. Messages are sent at the right time so that the user doesn't get lost in the shuffle. The goal of input design is to provide a user-friendly input arrangement.

4.5 Output Design

A high-quality output is one that satisfies the needs of the end user and effectively communicates the information. Outputs are used in any system to convey the outcomes of processing to the user and to other systems. In output design, it is decided how the information is to be displaced for immediate use and also the hard copy output. All most crucial facts can be obtained from here. The link between the system and the user is improved via efficient and intelligent output design. When designing computer output, it is important to proceed in an orderly and well-thought-out way, and to ensure that each output piece is built so that users find the system easy and effective to use. A computer output analysis should identify exactly what is required to satisfy the needs of a project (Fig. 1).

Fig. 1. Overview of admin and user modules

This algorithm has two phases which involve PCA and Haar Cascade algorithms. The PCA algorithm uses eigen vectors of covariance matrix that corresponds to the largest Eigenvalues.

The phases involved in registration process are:

Face Recognition

Opencv is one of the libraries given by Python for image and video processing. Opencv is used to collect video and recognise faces because it has a large library that aids in offering numerous image and video operators. OpenCV allows to make a video object that can be used to capture an image from a video. It can also be used to grab video from a webcam and then perform operations on it (Fig. 2 and 3).

Fig. 2. Face capture

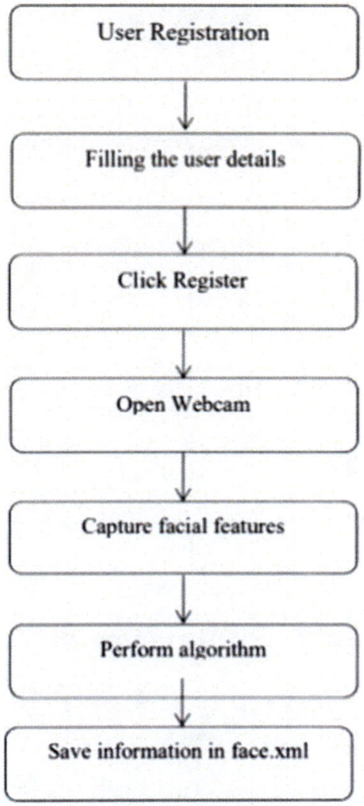

Fig. 3. Registration phase

In registration phase, the user needs to register with his/her detail like First name, last name, email id and password. Once the password match with confirmation password, then webcam opens and captures the face with the help of opencv and proceed to face registration phase.

In face registration, the model takes 32 parameters like forehead, chin, eyes, nose, lips, length of forehead and length and width of face. With these parameters, the face data will be stored in the file with extraction.xml file. Each user registration has different files with their first name respectively.

After this, the user will be directed to the shopping page where there are variety of products to view and shop according to their desire.

The facial recognition model is shown below (Fig. 4):

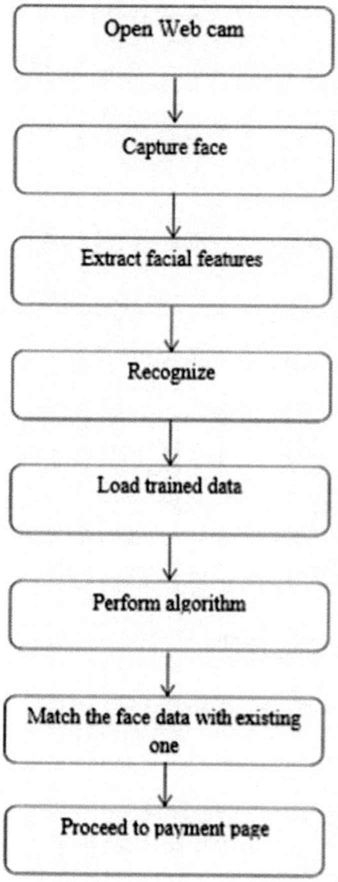

Fig. 4. Steps in reorganization

Once the user clicks proceed from cart, then webcam opens and finds for face. When it finds the face, the dataset is loaded and the captured face and the face taken in registration phase are matched. Once the face data is matched, then the payment page is opened, and once payment is successful then it automatically shows the message that the order is placed. If the face captured is not matched with the existing data, the order will be canceled.

5 Experimental Results

The facial recognition has been tested with a group of 60 and it is observed that there are no false acceptances i.e., no order is placed with unrecognized faces by algorithm error. Each person is allotted with different accounts and tested with 10 orders each and the success rate is 98 percent. With this test, it is proved that this algorithm is efficient and can be used in real world scenario so that it adds a layer of security to the customer. The following are the experimental results obtained by the project.

Fig. 5. User registration

When the user is in the registration process, user need to go through all the details mentioned and then the face data is been taken using the inbuilt camera and stores the image in the database provided (Fig. 5).

Fig. 6. Products page

This is the product page where the products which are added by the admin appear so that the user can add the desired items to cart (Fig. 6).

Once the face data is matched at checkout page, the order will be placed successfully and will be reflected at the admin end (Fig. 7).

After the order is placed, the user can check the order details whenever required. Moreover, the order status can also be checked (Fig. 8).

Fig. 7. Order placed

Fig. 8. Order details

6 Conclusion

In this proposed system, a new way of ideology has been identified with PCA and Haar cascade for facial recognition with high accuracy and efficiency to add another layer of security using facial recognition at the time of checkout in online shopping. As most of the devices nowadays has inbuilt web cam, it is easier to implement this model in real world scenario. In future, more authentication methods could be employed so that it can be used in old devices to increase the count of users.

References

1. Barr, J.R., Bowyer, K.W., Flynn, P.J., Biswas, S.: Face recognition from video: A review. IEEE Int. J. Pattern Recogn. Artif. Intell. **26**(5), 1–53 (2020)

2. Hassaballah, M., Aly, S.: Face recognition: Challenges, achievements and future directions. IET Comput. Vis. **9**(4), 614–626 (2018)
3. Grm, K., Struc, V., Artiges, A., Caron, M., Ekenel, H.K.: Strengths and weaknesses of deep learning models for face recognition against image degradations. IET Biometrics **7**(1), 81–89 (2019)
4. Dantcheva, A., Elia, P., Ross, A.: What else does your biometric data reveal? A survey on soft biometrics. IEEE Trans. Inf. Forensics Secur. **11**(3), 441–467 (2019)
5. Dibeklioglu, H., Alnajar, F., Ali Salah, A., Gevers, T.: Combining facial dynamics with appearance for age estimation. IEEE Trans. Image Process. **24**(6), 1928–1943 (2018)
6. Dantcheva, A., Bremond, F.: Gender estimation based on smile dynamics. IEEE Trans. Inf. Forensics Secur. **12**(3), 719–729 (2020)
7. Hadid, A., Pietikinen, M.: An experimental investigation about the integration of facial dynamics in video-based face recognition. Electron. Lett. Comput. Vis. Image Anal. **5**(1), 1–13 (2017)
8. Hadid, A., Dugelay, J.L., Pietikinen, M.: On the use of dynamic features in face biometrics: Recent advances and challenges. SIViP **5**, 495–506 (2019)
9. Cohn, J.F., Schmidt, K., Gross, R., Ekman, P.: Individual differences in facial expression: Stability over time, relation to self-reported emotion, and ability to inform person identification. In: Proceedings of the IEEE International Conference on Multimodal Interfaces, pp. 491–496 (2020)
10. O'Toole, A.J., Roark, D.A., Abdi, H.: Recognizing moving faces: A psychological and neural synthesis. Trends Cogn. Sci. **6**, 261–266 (2021)
11. Knight, B., Johnston, A.: The role of movement in face recognition. Vis. Cogn. **4**, 265–274 (2017)
12. Schmidt, K.L., Cohn, J.F.: Dynamics of facial expression: Normative characteristics and individual differences. In: Proceedings of the IEEE International Conference on Multimedia and Expo, pp. 728–731 (2019)
13. Liu, X., Cheng, T.: Video-based face recognition using adaptive hidden markov models. In: Proceedings of the IEEE Conference on Computer Vision and Pattern Recognition (2020)
14. Mattaand, F., Dugelay, J.: A behavioural approach to person recognition. In: Proceedings of the IEEE International Conference on Multimedia and Expo, pp. 1461–1464 (2006)
15. Matta, F., Dugelay, J.: Video face recognition: A physiological and behavioural multimodal approach. In: Proceedings of the IEEE International Conference on Image Processing, pp. 497–500 (2017)

Analysis of IoT Cloud Security Computerization Technology Based on Artificial Intelligence

P. A. Padmaavathy[1](✉), S. Suganya Bharathi[2], K. Arun Kumar[3], Ch. V. Sivaram Prasad[4], and G. Ramachandran[5]

[1] Department of Management, Karpagam Academy of Higher Education, Coimbatore, India
plccampus@gmail.com
[2] Departments of Master of Business Administration, SA Engineering College, Chennai, India
[3] Department of ECE, CVR College Of Engineering, Hyderabad, Telangana, India
[4] Department of Mathematics, Basic Science Humanities, Aditya Engineering College, Surampalem, Kakinada, Andhra Pradesh, India
[5] Department of Electronics and Communication Engineering, Vinayaka Mission's Kirupananda Variyar Engineering College, Vinayaka Mission's Research Foundation (Deemed to be University), Salem, Tamilnadu, India

Abstract. The concept of artificial intelligence enhanced sensor technologies has been aided by the creation of industrial robots, which serve as carriers of artificial intelligence. Based on this study background, the paper introduces the mobile robot climber's network model, hardware system, and application software. Simultaneously, the paper focuses on the mechanism compound technique and obstacle avoidance optimization technique for the climber robot. Simultaneously, the paper focuses on the mechanism compound method and obstacle avoidance control algorithm for the climber robot.

Internet - of - things computing focuses on "home" and brings high importance industries together to promote smart security services such as household appliances, interactive media, home health care, and access controls in maintaining a safe and healthy, secure, resources, sustainable, and convenient home living environment.

Keywords: Artificial İntelligence · Computerization · Cloud computing · Networking robots · Security · IoT

1 Introduction

At the moment, the application of artificial intelligence in the development of industrial automation control systems in India has been greatly improved, but there are still some issues with the technical level and quality of artificial intelligence. This necessitates the hiring of qualified personnel as well as the advancement and optimization of related technologies. Simultaneously, the staff must correctly recognise the characteristics of artificial intelligence in order to ensure that the benefits of artificial intelligence are utilised to the greatest extent possible, ultimately improving the performance of the

J. I.-Z. Chen et al. (Eds.): ICIPCN 2022, LNNS 514, pp. 296–304, 2022.
https://doi.org/10.1007/978-3-031-12413-6_23

automated control system. [1, 2] Only in this manner can we truly encourage the development of industrial automation control systems. Overall, the application of artificial intelligence to the application and development of industrial automation control systems has strong feasibility, while also being beneficial to the development of any industry and promoting the progress of India's industry.

Intelligent mobile robots are a type of robot system that can perceive the environment and its own state using sensors and perform target-oriented autonomous movement in an obstacle-filled environment, completing specific tasks. Planetary exploration has gradually become a research hotspot as the aerospace industry has grown in recent years. According to the existence of paths, mobile robot path planning can be divided into global path planning and local path planning. Global path planning is a well-known planning method, also known as model-based planning. Local path planning, on the other hand, is a planning method in which the environment is unknown or partially unknown, requiring sensors to determine the size, shape, and location of obstacles.

It is also known as sensor-based path planning. In order to address the issue that the traditional global path planning algorithm only considers the shortest path and ignores the steering cost, this paper refers to the wave propagation algorithm and proposes a shortest path propagation algorithm based on field scanning. By field scanning, the algorithm first generates a step conversion matrix, and then searches for a path in the step conversion matrix [2–5]. When searching for a path, the path direction is prioritised to be consistent with the previous search direction, ensuring that the shortest path is searched while the least occurrence Tum to behaviour is searched.

2 Mobile Robot Path Planning Method

2.1 Global Strategy for Path Planning

Based on the representation of the environmental model, global path planning can be divided into graph-based methods (such as Voronoi graph method, Q-M graph method, etc.) and grid-based methods (such as Dijkstra algorithm, A algorithm, wave propagation algorithm, etc.). The grid method is the most widely studied and applied path planning method because it is simple to implement computer modelling, storage, processing, updating, and analysis.

The task of global path planning is to find a feasible or optimal path from the starting point to the target point that meets certain performance requirements based on the environmental model, with the so-called optimal standard referring to the shortest path, the shortest time-consuming path, or the least cost [6].

The wave propagation algorithm is a path planning method that simulates how waves propagate on the water's surface. The wave propagation algorithm assumes that the wave spreads outward from the target point. When the wave reaches the starting point, the search path is the reverse sequence of the shortest wave propagation path from the target point to the starting point. The wave I propagation algorithm works similarly to other global path planning algorithms. Although the shortest path based on grid movement can be found, it is not always the best path. In general, global path planning treats the robot as a mass point.

As a result, when searching the path, only the walking distance cost of the robot is considered, and the steering cost of the robot at the path's turning point is ignored. As a result, existing path planning algorithms essentially use the shortest path as the optimal path standard (because path length represents the power consumption or time of the robot's movement). However, there may be more than one shortest path, and each path's smoothness varies. Because the path in the grid environment is made up of straight segments, the degree of tortuousness can reflect the smoothness of the path.

If the robot's moving path consistency constraint is considered in the global path planning, the global path can be made smoother to reduce unnecessary turning on the path, which is convenient for the actual robot's tracking execution. In order to address the issue that the traditional global path planning algorithm only considers the shortest path and ignores the steering cost, this paper refers to the wave propagation algorithm and proposes a shortest path propagation algorithm based on field scanning.

By field scanning, the algorithm first generates a step conversion matrix and then searches for a path in the step conversion matrix. When searching for a path, the path direction is prioritised so that it is consistent with the previous search direction, ensuring that the shortest path is searched while the fewest occurrences are searched. Now consider behaviour.

2.2 Artificial Intelligence Robot's Environments

The robot global path planning algorithm proposed in this paper performs path planning directly on the map of the relevant environment, avoiding the traditional path planning algorithm's requirement to establish a network connection model as well as the shortcomings of poor real-time performance caused by search and increasing search speed. The robot's t mechanism is more adaptable to changing environments. It can walk on a variety of ground conditions and has a high ability to escape from traps. At the same time, the robot's autonomous control and emergency response capabilities have been enhanced.

3 IoT-Based Systems Security Model

The growing use of Internet of Things (IoT) technology in the industrial sector has posed new issues for the data security of such systems. When IoT devices are used to develop SCADA systems, standard protocols and public networks are used to communicate data. Industrial control systems have considerable security threats, and commercialized off-the-shelf gadgets constitute a new platform for them.There are some effective models for analysing the security of information systems, so they do not considering the architecture of the Internet of Things. The layered attributed metagraph model is suggested and discussed for the security of IoT-based systems.

A security model for IoT-based systems has been presented, which, unlike previous models, includes the architectural elements of web-based SCADA systems as well as their hardware and software restrictions.The resulting model is a nested attribute metagraph that may be used for additional analysis and optimizations, such as visualisation of data flows in the platform, system states across time, and so on.The development of models for different components and tools for their visualisation and analysis depending on matrix operations on metagraphs, as well as a model for characterising security vulnerabilities (CVSS) for each node and the entire system, are all possibilities for future research.

4 A Machine Learning-Based Survey on IoT Security

The Internet of Things (IoT) is a network that allows machines such as sensors and appliances to communicate with one another without the need for human intervention. These wireless sensor networks are a collection of interconnected devices that are exposed to a variety of security threats.As a result, IoT security becomes critical. Machine learning inspires a slew of IoT security initiatives.Here, we look at a variety of intimidation techniques that have a greater chance of assaulting IoT, as well as machine learning techniques that can be used to combat them.

A variety of machine learning algorithms that are utilised for IoT security are discussed here.SVM is less sophisticated than Neural Network, despite the fact that they are both methodologies that result in increased accuracy.There are still several security challenges in the IoT.Because IoT devices have limited resources, machine learning works as a bridge between good security and low computing complexity.

5 IoT-Cloud Security Issues

The limited resources of IoT devices pose a hindrance to the rapid expansion and development of IoT technology in all areas of life.However, by combining IoT and cloud computing, the expansion of IoT technologies can be accelerated.As a result, a new computer area known as IoT Cloud has evolved.That is, data acquired by IoT technologies is recorded and stored in cloud infrastructure, allowing IoT technologies to be free of resource constraints.As a result, several new security and privacy concerns have emerged.The security challenges relating to IoT cloud are discussed in this study.

6 Challenges in the IoT Cloud

Despite the fact that the IoT cloud benefits both users and providers, it still has significant difficulties that endanger its use.The variety of IoT technologies, clouds, system software, and network protocols from various manufacturers creates a more difficult environment, which could lead to a lack of portability and portability within IoT cloud [7–10]. Furthermore, cloud elasticity and scalability are necessary in the IoT cloud.If, for example, IoT cloud infrastructure resources are insufficient to satisfy the rising demand for IoT technologies, service interruption or unavailability may occur [11, 12]. The security risks in the IoT virtual environment are more variable than the security issues in traditional cloud computing. It is not possible to run anti-virus on IoT devices, for eg, due to the limitation resources of IoT technology.

The security problems of the IoT cloud are discussed in this study.The fundamentals of the IoT cloud are discussed, followed by a discussion of the security issues that consumers may face while utilising smart devices can be connected to the cloud.In addition, solutions from the literature are researched and given[12–17]. Open security research challenges that require immediate attention from the scientific community are presented, along with some potential solutions that could work well with the IoT cloud paradigm.Finally, we hope that our study will be a useful contribution to allowing the secure interconnection of IoT and cloud computing.

7 IoT, AI, and Software Intelligence in the Smart Home

In this session, the speaker will discuss "smart home with soft-intelligence, IoT, and AI," which includes a variety of situations that differ from traditional uses that need time-consuming manual procedures.In order to develop a smart house, I will suggest four technical areas, including home networking, context awareness, recognition, and user interface, which I refer to as "enabling technologies and effects." The speaker will describe a set of practical design, development, and services for implementing some interesting applications using various networked devices such as sensors in home computing contexts in order to make it easier to grasp.Smart home services are thought to deliver a better user experience while also inspiring new and future apps and services in ubiquitous home wireless environments.

8 IoT, AI, and Software Intelligence in the Smart Home

The Internet of Things has advanced significantly in recent years, and it now has a wide spectrum of uses in the smart home, Internet of Automobiles, and Industrial Internet of Things. The perception layer, transport layer, and application layer of most developing smart devices have very simple structures, and security vulnerabilities may exist at these layers. Currently, the majority of security research frameworks for IoT devices necessitate the use of extra high-performance devices. Simultaneously, developing "mining" malware and other attack methods that directly pillage the device's processing resources have garnered less attention. This study develops and maintains a smart home security analysis system in response to the aforementioned issues.Our tests suggest that the system can detect and fight against contactless assaults that smart homes may face while having a minor impact on home network connectivity. The system can better reconcile the disparity between home kit network security requirements with device performance constraints.

9 Network Protection

In essence, the term "security" refers to being in a non-threatened environment. Individuals can go about their daily lives without fear of disrupting their regular condition. Extending network security means that the network system on the Internet platform can operate properly in a secure environment, and users do not have to worry about data leakage or computer software and hardware damage.

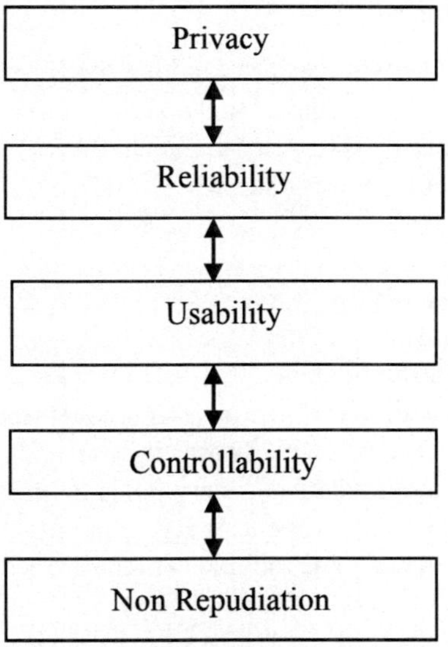

Fig. 1. Block diagram network security

The Fig. 1 definition of network security is broad, but in everyday life and at work, people usually refer to network security as the security of a computer network, or, in other words, network security as the smooth operation of a computer in a secure communication environment. A computer network connects numerous independent computers to allow data to be exchanged and transmitted between them. One of the primary functions of computers in the modern day is to facilitate the exchange of data resources and to disseminate important data information to people. As a result, network security is critical, as it directly influences whether network resources and data can be sent safely, as well as whether people's personal privacy is safeguarded.

10 Network Security Stored Difficulty in Understanding Investigation System Software Installation Core Tech

10.1 Software Development

The network security hidden danger investigation system, which is based on Internet of Things technology, is designed to assist users and network supervisors in investigating network security hidden dangers. Its embedded system may provide network security hidden danger investigation instructions to users, and then assist network security supervisors, network security hidden danger investigation professionals, and users in investigating each network security hidden danger individually.

At the same time, the system can track and respond to network security occurrences, with the ultimate goal of correctly and promptly recognising network security threats and increasing Internet users' network security awareness.

10.2 Hardware Development

Four modules make up the network security hidden danger investigation system based on Internet of Things technology. These four modules have various purposes, but their goal is to uncover hidden network security threats and lower the likelihood of network security incidents. The module incorporates a RAM storage unit as well as a high-speed processor.

1. Module of storage capacity flash storage unit that can create a network security risk information database and design a network security risk investigation standard based on current network security incidents.
2. Computer module with a high level of trust The module is a built-in trusted password module chip that runs on the ARM hardware platform and can communicate with the main CPU. As a result, the module's primary job is to improve the system's security and reliability through enhancing security and reliability.
3. A module for communication. The module, which is based on the Internet platform, may connect to the Internet via WiFi, 3G, 4G, or Bluetooth features and send back the acquired real-time data to the network security supervision department and the user end, allowing relevant departments and staff to plan ahead.
4. Power source The system's battery module is the foundation for all software and hardware operations. It uses an intrinsically safe battery that can offer high security and stable working current for each module, ensuring that the network security concealed danger investigation mission is not disrupted.

The article can monitor the network security environment in real time and feed back network security data information to users and network security regulators, depending on the precise circumstances of current network security occurrences. The author believes that the network security hidden trouble investigation system developed in this paper, which is based on Internet of Things technology, can improve cooperation among individuals, governments, and network security regulatory departments, and is the most effective way to prevent network security incidents at their source.

11 Conclusion

A router-based home surveillance analytics system is designed and implemented in this article. Our tests show that the smart home security analysis system can detect and defend against assaults efficiently, and that it can employ plug-ins to cover future weaknesses. The next phase will be to incorporate machine learning approaches into the system to improve the accuracy of abnormal traffic detection.

References

1. Atlam, H.F., Alenezi, A., Alharthi, A., Walters, R.J., Wills, G.B.: Integration of cloud computing with internet of things: Challenges and open issues. In: Proceedings of the 2017 IEEE International Conference on Internet of Things (iThings) and IEEE Green Computing and Communications (GreenCom) and IEEE Cyber, Physical and Social Computing (CPSCom) and IEEE Smart Data (SmartData), pp. 670–675. IEEE (s2017)

2. Li, S., Choo, K.-K.R., Sun, Q., Buchanan, W.J., Cao, J.: Iot forensics: Amazon echo as a use case. IEEE Internet Things J. **6**(4), 6487–6497 (2019)
3. Oriwoh, E., Jazani, D., Epiphaniou, G., Sant, P.: Internet of things forensics: Challenges and approaches. In: Proceedings of the 9th IEEE International Conference on Collaborative Computing: Networking, Applications and Worksharing, pp. 608–615. IEEE (2013)
4. Hagebring, F., Farooqui, A., Fabian, M., Lennartson, B.: On optimization of automation systems: Integrating modular learning and optimization. IEEE Trans. Autom. Sci. Eng. https://doi.org/10.1109/TASE.2022.3144230
5. Nader, S.I., Das, A., Al Mamun, A., Deb Nath, P., Chowdhury, G.M.: Cost-efficient smart home automation and security system based on IoT and GSM. In: Proceedings of the 2022 International Conference for Advancement in Technology (ICONAT), pp. 1–5 (2022). https://doi.org/10.1109/ICONAT53423.2022.9726112
6. Yu, P., Long, Y., Yan, H., Chen, H., Geng, X.: Design of security protection based on ındustrial ınternet of things technology. In: Proceedings of the 2022 14th International Conference on Measuring Technology and Mechatronics Automation (ICMTMA), pp. 515–518 (2022). https://doi.org/10.1109/ICMTMA54903.2022.00109
7. Tran, T.K., Yahoui, H., Heng, S., Cheang, V.: Automation for the future workforce in Southeast Asian countries - Factori 4.0 project. In: Proceedings of the 2022 Joint International Conference on Digital Arts, Media and Technology with ECTI Northern Section Conference on Electrical, Electronics, Computer and Telecommunications Engineering (ECTI DAMT & NCON), pp. 380–383 (2022). https://doi.org/10.1109/ECTIDAMTNCON53731.2022.9720336
8. Kumbhar, V., Chavan, M.: Multidisciplinary project-based learning in ındustrial automation. In: Proceedings of the 2022 Joint International Conference on Digital Arts, Media and Technology with ECTI Northern Section Conference on Electrical, Electronics, Computer and Telecommunications Engineering (ECTI DAMT & NCON), pp. 412–416 (2022). https://doi.org/10.1109/ECTIDAMTNCON53731.2022.9720333
9. Lin, L., Yang, H., Wang, Y., Shi, C., Lu, Y.: Design and ımplementation of electric power UI automation construction system. In: Proceedings of the 2022 IEEE 6th Information Technology and Mechatronics Engineering Conference (ITOEC), pp. 1276–1284 (2022). https://doi.org/10.1109/ITOEC53115.2022.9734502
10. Cheng, F.T.: Evolution of automation and development strategy of ıntelligent manufacturing with zero defects. In: Industry 4.1: Intelligent Manufacturing with Zero Defects, pp.1–23. IEEE (2022). https://doi.org/10.1002/9781119739920.ch1
11. Suo, S., Huang, K., Kuang, X., Cao, Y., Chen, L., Tao, W.: Communication security design of distribution automation system with multiple protection. IEEE Int. Conf. Consum. Electron. Comput. Eng. (ICCECE) **2021**, 750–754 (2021). https://doi.org/10.1109/ICCECE51280.2021.9342482
12. Kotenko, I., Parashchuk, I.: Evaluation of information security of industrial automation systems using fuzzy algorithms and predicates. Int. Russ. Autom. Conf. (RusAutoCon) **2021**, 261–266 (2021). https://doi.org/10.1109/RusAutoCon52004.2021.9537332
13. Gill, A.K., Zavarsky, P., Swar, B.: Automation of security and privacy controls for efficient ınformation security management. In: Proceedings of the 2021 2nd International Conference on Secure Cyber Computing and Communications (ICSCCC), pp. 371–375 (2021). https://doi.org/10.1109/ICSCCC51823.2021.9478126
14. Junmei, W., Chengkang, Y.: Automation testing of software security based on burpsuite. In: Proceedings of the 2021 International Conference of Social Computing and Digital Economy (ICSCDE), pp. 71–74 (2021). https://doi.org/10.1109/ICSCDE54196.2021.00025
15. Lesi, V., Jakovljevic, Z., Pajic, M.: Security analysis for distributed IoT-based ındustrial automation. IEEE Trans. Autom. Sci. Eng. https://doi.org/10.1109/TASE.2021.3106335

16. Setzler, T., Mountrouidou, X.: IoT metrics and automation for security evaluation. In: Proceedings of the 2021 IEEE 18th Annual Consumer Communications & Networking Conference (CCNC), pp. 1–4 (2021). https://doi.org/10.1109/CCNC49032.2021.9369533
17. Alsuhaym, F., Al-Hadhrami, T., Saeed, F., Awuson-David, K.: Toward home automation: An IoT based home automation system control and security. Int. Congr. Adv. Technol. Eng. s(ICOTEN) **2021**, 1–11 (2021). https://doi.org/10.1109/ICOTEN52080.2021.9493464

Artificial Intelligence Based Real Time Packet Analysing to Detect DOS Attacks

Sai Harsh Makineedi(✉), Soumya Chowdhury, and Vaidhehi Manivannan

SRM Institute of Science and Technology, Chennai, India
msaiharsh@gmail.com

Abstract. A Denial-of-Service attack is a common network attack. Hence, the research into the early detection of DOS attacks is very crucial. However, there has yet to be a detection approach that is both accurate and quick to detect the attack. In light of this, this research presents a neural network-based DOS detection approach. The dataset of collected packets, feature extraction, and classification comprises the three parts of this article. The dataset consists of both malicious and non-malicious raw captured packets; in the classification stage, packets are categorized as malicious or non-malicious; in the feature extraction stage, different attributes of packets are extracted using Natural Language Processing; in the implementation stage, these features are used as input to the machine learning model. The experimental findings demonstrate that the proposed DOS attack detection model has a high level of accuracy and can identify common DOS assaults in a reasonable amount of time.

Keywords: Machine learning · Neural networks · Denial of service · Wireshark · Packet capturing · Artificial intelligence

1 Introduction

A Denial of Service (DoS) exploit is a network exploit that makes a machine or service unavailable to its primary audience. This is accomplished in a Denial-of-Service exploit by flooding the target with traffic, making it impossible to serve the legitimate users, or by causing a machine crash, preventing the legitimate user from accessing the resource they intended. There are different types of DOS attacks, for example ICMP flood takes advantage of the misconfiguration of devices in the network by delivering bogus packets which ping each machine on the target network. As a result, the network is enabled to handle traffic in the network. This kind of assault is often named as "Smurf attack" or the "Ping of Death" another example is the "SYN flood" which sends a connection request but never completes the handshake. Detecting a Denial of Service (DoS) exploit is difficult because there is a need to accurately distinguish normal traffic from malicious traffic. At present, the research on defense against DOS attack is based on standard datasets which are highly pre-processed and does not use raw packets directly captured from packet sniffing tools. This is a problem because doing all this processing in real-time is very difficult which is why despite having good models for detecting DOS attacks

J. I.-Z. Chen et al. (Eds.): ICIPCN 2022, LNNS 514, pp. 305–320, 2022.
https://doi.org/10.1007/978-3-031-12413-6_24

we don't see practical products using them. Hence in this paper, we will first create a DOS attack dataset which consists of raw packets captured directly from a sniffing tool and then we will use the raw data captured to train our model to detect DOS attacks.

This paper will first explain how we captured and created the DOS attack dataset using various tools in order stimulate various kinds of DOS attacks. Then we will be working on how are we going to pre-process the raw packet information using NLP techniques for very quickly getting it ready for Machine Learning (ML) models and neural networks. Finally the paper will highlight how both machine learning and neural networks perform for different types of detecting Denial of Service attacks.

2 Literature Survey

The findings of Gulshan et al. (2010) [21] provide useful comparison between different artificial intelligence based intrusion detection system. From this paper we concluded that the multi classifier based technique like ensemble methods works the best for instruction detection.

The Jiangtao et al. (2019) [22] provides us a two stage model for Distributed Denial of Service attack detection. We also followed this structure and divided our model into the two stages of feature extraction and AI model.

The article provided by V.Kanimozhi and T.PremJacob (2019) [23] presents a Artificial Neural Network to detect botnet attacks. This Artificial Neural Network achieved a very high accuracy. Based on the results of this paper we decided to use neural network for our research.

This paper presented by Shenfield et al. (2018) [24] provides an Artificial Neural Network based technique to identify malicious network traffic which uses various sources of network traffic data like dynamic link library files, log files etc. We also decided to use logs created during attack and during normal usage traffic based on this paper.

Sarraf et al. (2020) [25] presented work on intrusion detection and classification using k-nearest neighbourhood, artificial neural network, random forest and XGBoost models. The detection of intrusion is done using binary classification, once an intrusion is detected different models are implemented to classify the attacks. UNSW-NB15 dataset is used to train and test the models. The work provides an comprehensive knowledge about the efficiency of different models for intrusion detection and classification.

Abas Aboras and Mohammed Kamal Hadi (2021) [26] presented a survey on network attack detection, initially it provides an elaboration on different network attacks (e.g. TCP SYN denial of service attack, ICMP flood attack, UDP flood attack, Port scan attack). Then it explains the network attack detection process and the current two types of network attack detection (abuse detection and anomaly detection). Finally, the paper provides an detailed account on network security strategies and designs for intrusion detection system.

Y Wu et al. (2020) [27] presented work on attack detection using various deep learning methods. The work also presents an account for the fundamental problems faced by network security and attack detection. A detailed account were presented in regards of various deep learning methods which were narrowed down to the ones used for attack detection (e.g. autoencoders, generative adversarial network, recurrent neural network, and convolutional neural network). Finally, a summary for current benchmark datasets is provided.

Rizgar et al. (2018) [28] presented work on the impact of HTTP and SYN flood attacks on Apache2 and IIS 10.0. The performance are measured with the key metrics of average response time, average CPU usage and standard deviation as a responsiveness, efficiency and stability of the web servers. At the beginning, the paper compares HTTP flood and SYN flood and explains their speciality and what differentiates them. The work elaborates on different tools used for the attack (e.g. Hping3 and HOIC) and tools used to calculate the performance of the servers under attack (e.g. System Activity Report and Get-counter).

3 Implementation

3.1 Experimental Setup

At set-up of four machines as shown in Fig. 1 are used to generate normal and malicious traffic, Machine 1 is used as a target machine, Machine 2 and Machine 3 is used to generate normal traffic and Machine 4 is used to generate malicious traffic. All the four machines can communicate with each other using Switch0.

Fig. 1. Machine setup

Wireshark [17] is the most widely used network protocol analyzer. It supports most computer platforms, including Windows, OS X, Linux, and UNIX, and offers a large and sophisticated feature set. "Wireshark" captures the data coming or going through the "NICs" on its device by using an "underlying packet capture library called npcap".

On Machine 1 Wireshark is configured to capture packets in "Promiscuous mode" which ensures that every "data packet" that is "transmitted" is received and read by a "Network Adapter". Once packet capture is triggered on the eth1 interface in the Wireshark application, the eth1 interface of the "Network Interface Card" of Machine 1 is configured into "promiscuous mode" by Wireshark, and using the underlying npcap

Fig. 2. Captured packets using Wireshark

library it captures the details of the packets which are sent or received on eth1 interface of the NIC (Fig. 2).

Wireshark has a lot more versatility than other IDS/IPS devices. The captured packets may be stored as a.pcap file and analysed later. Another intriguing feature is the option to convert the capture file to more intelligible formats like plain text, CSV, and others. All packet logs recorded on Machine 1 are converted to CSV format, including both regular and malicious traffic (Fig. 3).

Fig. 3. Exporting captured packets to CSV

Hping3 [18], a command line and open source tool, is used as an attacker spawner. It is capable of generating a variety of attacks, including "TCP" and "UDP". "TCP SYN flood", "TCP FIN flood", "TCP RST flood", "TCP PSH+ACK flood", and "UDP flood" are all created in a certain order from machine 4 to machine 1, and the packets are captured on machine 1 (Table 1).

Table 1. Hping3 tool arguments used

Argument	Description
-i	Indicate the interface to use
--flood	Sends packets as fast as possible without taking care to show incoming replies
-1	Uses ICMP mode
--rand-source	Sends packets with random source IP
-p	Indicates destination ports
-2	Uses UDP mode
-F	Set FIN tcp flag
-S	Set SYN tcp flag
-R	Set RST tcp flag
-P	Set PSH tcp flag
-A	Set ACK tcp flag

Low Orbit Ion Cannon (LOIC) [19] is a frequently used open source tool for network stress testing, DoS and DDoS assaults.

It has the ability to cause TCP SYN floods, UDP floods, and HTTP floods. It focuses its assault on a single IP address, namely the target computer's (machine 1) IP address (Fig. 4).

Fig. 4. LOIC executing attack

Slowrois [20] is a GET-based DoS technique that takes up server resource by allowing a single server to destroy another web server with limited bandwidth and associated services and ports as a side effect. To prevent sockets from being closed, HTTP requests, including delayed HTTP headers, are transmitted to the victim's web server at regular intervals.

Hping3, LOIC, and Slowrois are configured to generate malicious packets of different types of attacks, all these software are executed on Machine 4 providing the IP address of the target machine i.e. Machine 1 to the attack simulators. Machine 1 captures the packets and creates the logs with the help of the "Wireshark Packet Sniffing Tool".

3.2 Data Pre-processing

So now we have many csv files having http attack, http normal, tcp attack, tcp normal, udp attack, udp normal, icmp attck, icmp normal which have been directly captured by wireshark.

We load these files protocol by protocol in the python [15] environment using pandas [16] then first we remove all the packets belonging to any other protocol than the one we are procession currently or any outgoing traffic.

Now for each protocol we have two different dataframes, one containing the packets during the attack and another containing the packets of the same protocol during normal traffic.

After that we make the number of rows in both the attack and normal data same to make sure that our data is not biased.

Finally we add the attack attribute to the dataframe to signify whether the packet is malicious on not. To do this we simply add a column named attack first in the attack dataframe and fill the entire column with ones then add a column named attack in normal dataframe and fill it with zeros. We will repeat this for all the four protocols.

Then we combined all these different attack and normal dataframe into one dataframe using pandas then shuffled it and exported it to csv. So our dataset is ready.

3.3 Dataset

Once the pre-processing part is completed we have a CSV file containing all the packet information both malicious and non-malicious, with the label (Fig. 5).

The Time field provides the timestamp of the packect when it was received by the system, the Source and Destination fields shows the Source IP address and Destination IP address of the packets, Protocol field defines the underlying protocol of the packet, Length field defines the length of the packet, Info field elaborates of the specific information captured by wireshark from the packet. The Info field will have different values for different protocol packets. Finally, the Attack field acts as the label to define whether the packet is malicious or not.

To detect DOS attack we are using the Info field which varies from protocol to protocol.

For ICMP packets the Info field provides the details of request id, sequence number and time to live.

Fig. 5. Dataset obtained after preprocessing phase

For HTTP packets the Info fields elaborates on the HTTP version, the request type and the request details.

For TCP packets the Info fields provides information on source and destination port number, TCP flag, sequence number, window size, maximum segment size, Sack-Permitted, Time Stamp value, Timestamp Echo Reply and window scaling.

For UDP the Info provided us with information on Source and destination port number and length of the payload.

3.4 Feature Extraction

Now for creating our model the first challenge we faced it how to use raw packet data for creating our model, as can be seen in the sample of our dataset provided all the information which we need to create our model like sequence number, source port, destination port, request id, time to live, window size, flag etc. are all in one attribute of our dataset called info.

So we decided to process this attribute of our dataset called info and create separate attributes in our dataset for all the values of importance to us mentioned above.

We performed this processing in our python [1] environment and then feed the processed data to our machine learning [2] models but we quickly realised a very big road block it's the fact that there are thousands of packets coming to our system every second and if we have to pass every packet through this pre-processing then the internet speed will go down unacceptably low.

Now to after quite some trials we concluded that we cannot do this breaking down of the info attribute into different attributes within acceptable time frame.

Then we thought of processing the entire info attribute as a single unit but we still need to highlight the values of importance like source port, destination port, sequence number etc. So we decided to use Natural Language Processing (NLP) [3] techniques in order to get the important details from the info attribute quickly.

We used NLP approach "Term Frequency—Inverse Document Frequency" (TFIDF) [4]. TFIDF is a commonly utilised and well-known Natural language processing technique.

This method gives each word a score in order to assess its importance in the text. This approach is widely used in the context of "information retrieval" and "text mining".

The term frequency is used to quantify the number of times a term appears in a document. The length of the document and the broadness of the phrase are important factors. The longer the paper, the higher the word frequencies, but we can't say that it's more important than a shorter one. As a consequence, we normalise the frequency by dividing it by the total words in a document.

The normalised TF value's final value is always in the range of 0 to 1, inclusive 1 because the term frequency is unique to each document and word, it may be expressed as follows for a term (t) in document (d):

$$\mathrm{tf}(\mathrm{t}, \mathrm{d}) = \text{number of t in d/number of words in d} \tag{1}$$

The "document frequency" is the number of times a word appears in a specific document. We consider a phrase to be an occurrence if it appears in the text at least once; we don't need to know how many times it appears. We must divide the occurrence frequency by all of the occurrences, just as we did with word frequency.

Therefore, the document frequency for a term (t) in N documents is formulated as follows:

$$\mathrm{df}(\mathrm{t}) = \text{Occurrence of t in N documents/N} \tag{2}$$

The reciprocal of the frequency of the document that represents the relevance of the phrase t measures is "Inverse Document Frequency".

As the name suggests, it can be calculated by inverting the document frequency as follows:

$$\mathrm{idf}(\mathrm{t}) = \mathrm{N}/\mathrm{df} \tag{3}$$

However, the preceding definition raises certain issues. First, when the corpus size is huge, the IDF value skyrockets. As a result, IDF logging is utilised to mitigate the damage.

Second words can be missing in a document, so df will be 0, but we cannot divide by 0, so 1 will be added to the denominator.

Therefore the inverse document frequency for a term (t) in N documents is formulated as follows:

$$\mathrm{idf}(\mathrm{t}) = \log(\mathrm{N}/(\mathrm{df} + 1)) \tag{4}$$

Putting everything together gives the final formula we use to Calculating TFIDF for our data set is:

$$\mathrm{tfidf}(\mathrm{t}, \mathrm{d}) = \mathrm{tf}(\mathrm{t}, \mathrm{d}) * \log(\mathrm{N}/(\mathrm{df} + 1)) \tag{5}$$

Using this method we can easily highlight the important parts of info attribute by applying just a mathematical formula which is very fast to execute.

After passing our dataset through TFIDF we get a sparse matrix [5] which can be directly provided as an input to our machine learning model but we also have the attribute length of packet in our dataset so we will be using sklearn [6] to horizontally stack the packet length attribute to our sparse matrix.

Since our data is ready we decided to split it into train and test data. We decided to keep 25% of the data for testing which was 14972 different packets and the training data consisted of 44914 different packets.

3.5 Model Creation, Training and Testing

Now we are ready to implement machine learning models so first we decided to try logistic regression [7]. We decided so because logistic regression is a best place to start for a binary classification problem [8].

The "sigmoid function" is used in" logistic regression" to assign predicted values to fractions. Each real value is mapped to another value between 0 and 1. We utilize Sigmoid in machine learning to correlate predictions with probabilities (Fig. 6).

Fig. 6. Sigmoid function

Logistic regression creates a decision boundary in order to give a specific class instead the probability which we get from the above graph.

We used the logistic regression provided in sklearn library [9] for our project.

After implementing logistic regression and analysing the results we decided to go for creating our own neural network [10] to improve performance accuracy.

A neural network consists of a set of techniques that aims to identify the underlying correlations in a data set by simulating how the real brain functions. Artificial neural networks is another name for it.

"Artificial neurons" are a collection of linked units or nodes in an ANN that approximately replicate the biological neuron. Each link, resembling synapses in a human brain, may give a message to other neurons.

A threshold may be set in neurons such that a message is transmitted if the extra message surpasses it. Neurons generally clump together in different layers. On their inputs, various layers can conduct distinct transformations. Messages are transferred from the 1st (input) layer to the ending (output) layer, perhaps after numerous times passing the layers.

We used keras [11] library to create our neural network. For doing this we first need to set the random seed for numpy and tensorflow in order to set the initial weights of our model.

Then we created an input layer as per our input data shape which is 44914×35102 using keras input layer. After that we added five dense layers with 12, 24, 48, 24 and 12 neurons each having an activation function relu [12]. The final output layer has one neuron and an activation function of sigmoid because we want a binary output and sigmoid always gives an output close to zero or one. This model is depicted in Fig. 7.

Finally, we use the loss function binary cross entropy [13] and the optimizer adam [14] to compile all of the layers together.

In order to fit our data into a neural network we cannot use sparse matrix so we converted it into a sparse tensor and converted y_train into numpy.

We had tried varying the number of neurons and the number of layers. The above mentioned neural network works the best increasing the number of layers will not cause any significant change in the accuracy.

Our neural network achieved a great accuracy on the training data.

Fig. 7. Architecture of neural network

Now to test our model we again converted the x_test from sparse matrix into sparse tensor then we also convert y_test to a numpy array.

Then we can use the neural network to predict the output for the entire test set and store it in y_pred then compare it to y_test to calculate the accuracy, precision and recall using sklearn.

These two lists can also be used to calculate the true positive and the false positive rate, which is then used to draw the ROC curve with the help of matplotlib library.

We also used these two lists to create the confusion matrix using sklearn and plot it using seaborn library.

4 Results and Discussion

First Logistic Regression Classifier outlined in the implementation section above was applied to the dataset containing both malicious and normal traffic.

The Table 2 below shows the accuracy, pression and recall for the logistic regression model:

Table 2. Logistic regression matrix

Accuracy	78.04%
Precision	71.30%
Recall	94.17%

As the table shows the accuracy produced by the logistic regression model is 78.04% which is calculated as follows

$$\text{Accuracy} = \text{Number of packets correctly predicted/Number of total packets in the test dataset} \quad (6)$$

Similarly precision is 71.30% which was calculated as follows:

$$\text{Precision} = \text{Number of malicious packets labelled as attack/Total number of packets labelled as attack} \quad (7)$$

Similarly recall is 94.17% which was calculated asfollows:

$$\text{Recall} = \text{Number of malicious packets labelled as attack/Total number of malicious packets in the test set} \quad (8)$$

The Fig. 8. below visualizes the confusion matrix of the logistic regression model.

Fig. 8. Confusion matrix of logistic regression

The "logistic regression classification model" generates a Receiver Operator Characteristic (ROC) curve, as shown in Fig. 9. ROC curve is frequently used to investigate classifier sensitivity vs. specificity trade-off at various thresholds. A classification model's overall discrimination may be measured using the area under the ROC curve (an area under the ROC curve with a greater value indicates that the classifier is better at differentiating between the 2 categories).

Fig. 9. ROC curve of logistic regression

The logistic regression classifier model has an area under the curve of 0.78, as seen in the picture.

Now the neural network model which was created as mentioned in the implementation section was applied to the dataset containing both malicious and normal traffic.

In Fig. 10 we would like to show the training graph of the neural network which shows us the loss and accuracy during the training phase.

Fig. 10. Training curve of neural network

The Table 3 below shows the accuracy, precision and recall for the artificial neural network model:

Table 3. Neural network matrix

Accuracy	98.14%
Precision	99.06%
Recall	97.23%

As shown by the the table, accuracy, precision and recall for neural network model is 98.14%, 99.06% and 97.23% respectively. All the values were calculated using the same formulas mentioned above.

The Fig. 11 below shows the confusion matrix for the neural network model.

Fig. 11. Confusion matrix of neural network

Finally we would also like to present the ROC curve for the artificial neural network model (Fig. 12).

Fig. 12. ROC curve of neural network

As we can see the area under the curve is 0.98 because the true positive rate is 0.972 and the false positive rate is 0.009. Hence our model is highly accurate.

5 Conclusion

The Artificial neural network along with the Natural Language Processing based pre-processing presented in this paper really enhances the detection accuracy as well as the detection speed of artificial intelligence based DOS attack detection systems. Processing the raw network packets so that they can be understood by the neural network for detection of attack is a difficult and time consuming task. This paper proposes treating packets as text and using NLP to process the text. By doing this the processing overhead is heavily reduced. Since the model proposed in this paper deals with text it can be used in combination with any packet capture system. Further this approach of using Natural Language Processing for quick processing of raw network packet data can be applied to different kinds of network attacks in order to create practical artificial intelligence based intrusion detection systems.

References

1. van Rossum, G.: Python reference manual. In: Department of Computer Science [CS] R 9525 (1995)
2. El Naqa, I., Murphy, M.J.: What is machine learning? In: El Naqa, I., Li, R., Murphy, M. (eds.) Machine Learning in Radiation Oncology, pp. 3–11. Springer, Cham (2015). https://doi.org/10.1007/978-3-319-18305-3_1
3. Chowdhary, K.R.: Natural language processing. In: Chowdhary, K.R. (ed.) Fundamentals of Artificial Intelligence, pp. 603–649. Springer, New Delhi (2020). https://doi.org/10.1007/978-81-322-3972-7_19
4. Havrlant, L., Kreinovich, V.: A simple probabilistic explanation of term frequency-inverse document frequency (tf-idf) heuristic (and variations motivated by this explanation). Int. J. Gener. Syst. **46**(1), 27–36 (2017)
5. Bunch, J.R., Rose, D.J. (eds.): Sparse Matrix Computations. Academic Press (2014)
6. Komer, B., Bergstra, J., Eliasmith, C.: Hyperopt-sklearn. In: Hutter, F., Kotthoff, L., Vanschoren, J. (eds.) Automated Machine Learning. The Springer Series on Challenges in Machine Learning, pp. 97–111. Springer, Cham (2019). https://doi.org/10.1007/978-3-030-05318-5_5
7. Kleinbaum, D.G., Dietz, K., Gail, M., Klein, M., Klein, M.: Logistic Regression. Springer, New York (2002)
8. Kumari, R., Srivastava, S.Kr.: Machine learning: a review on binary classification. Int. J. Comput. Appl. **160**(7), 11–15 (2017)
9. Bisong, E.: Logistic regression. In: Building Machine Learning and Deep Learning Models on Google Cloud Platform, pp. 243–250. Apress, Berkeley (2019)
10. Wang, S.-C.: Artificial neural network. In: Wang, S.-C. (ed.) Interdisciplinary Computing in Java Programming. The Springer International Series in Engineering and Computer Science, vol. 743, pp. 81–100. Springer, Boston (2003). https://doi.org/10.1007/978-1-4615-0377-4_5
11. Manaswi, N.K.: Understanding and working with Keras. In: Deep Learning with Applications Using Python, pp. 31–43. Apress, Berkeley (2018)

12. Agarap, A.F.: Deep learning using rectified linear units (relu). arXiv preprint arXiv:1803.08375 (2018)
13. Ho, Y., Wookey, S.: The real-world- weight cross-entropy loss function: modeling the costs of mislabeling. IEEE Access **8**, 4806–4813 (2019)
14. Zhang, Z.: Improved adam optimizer for deep neural networks. In: 2018 IEEE/ACM 26th International Symposium on Quality of Service (IWQoS), pp. 1–2. IEEE (2018)
15. Van Rossum, G., Drake Jr., F.L.: Python tutorial, vol. 620. Centrum voor Wiskunde en Informatica, Amsterdam (1995)
16. McKinney, W.: Pandas: a foundational Python library for data analysis and statistics. Python High Perform. Sci. Comput. **14**(9), 1–9 (2011)
17. Banerjee, U., Vashishtha, A., Saxena, M.: Evaluation of the capabilities of Wireshark as a tool for intrusion detection. Int. J. Comput. Appl. **6**(7), 1–5 (2010)
18. Manso, P., Moura, J., Serrão, C.: SDN-based intrusion detection system for early detection and mitigation of DDoS attacks. Information **10**(3), 106 (2019)
19. Sauter, M.: "LOIC will tear us apart" the impact of tool design and media portrayals in the success of activist DDOS attacks. Am. Behav. Sci. **57**(7), 983–1007 (2013)
20. Arafat, M.Y., Alam, M.M., Alam, M.F.: A practical approach and mitigation techniques on application layer DDoS attack in web server. Int. J. Comput. Appl. **131**(1), 13–20 (2015)
21. Kumar, G., Kumar, K., Sachdeva, M.: The use of artificial intelligence based techniques for intrusion detection: a review. Artif. Intell. Rev. **34**(4), 369–387 (2010)
22. Pei, J., Chen, Y., Ji, W.: A DDoS attack detection method based on machine learning. J. Phys. Conf. Ser. **1237**(3), 032040 (2019)
23. Kanimozhi, V., Prem Jacob, T.: Artificial intelligence based network intrusion detection with hyper-parameter optimization tuning on the realistic cyber dataset CSE-CIC- IDS2018 using cloud computing. In: 2019 International Conference on Communication and Signal Processing (ICCSP), pp. 0033–0036. IEEE (2019)
24. Shenfield, A., Day, D., Ayesh, A.: Intelligent intrusion detection systems using artificial neural networks. Ict Express **4**(2), 95–99 (2018)
25. Sarraf, J., Vaibhaw, Chakraborty, S., Pattnaik, P.K.: Detection of network intrusion and classification of cyberattack using machine learning algorithms: a multistage classifier approach. In: Pattnaik, P.K., Sain, M., Al-Absi, A.A., Kumar, P. (eds.) SMARTCYBER 2020. LNNS, vol. 149, pp. 285–295. Springer, Singapore (2021). https://doi.org/10.1007/978-981-15-7990-5_28
26. Aboras, A., Hadi, M.K.: A survey of network attack detection research. Int. J. Eng. Res. Technol. (IJERT) **10**(08) (2021)
27. Wu, Y., Wei, D., Feng, J.: Network attacks detection methods based on deep learning techniques: a survey. Secur. Commun. Netw. (2020)
28. Zebari, R.R., Subhi, R.M.Z., Jacksi, K.: Impact analysis of HTTP and SYN flood DDoS attacks on apache 2 and IIS 10.0 Web servers. In: 2018 International Conference on Advanced Science and Engineering (ICOASE), pp. 156–161. IEEE (2018)

Decision Trees and Gender Stereotypes in University Academic Desertion

Sylvia Andrade-Zurita(✉), Sonia Armas-Arias, Rocío Núñez-López, and Josué Arévalo-Peralta

Facultad de Ciencias Humanas y de la Educación, Universidad Técnica de Ambato, Ambato, Ecuador
{sylviajandradez,sp.armas,carmenrnunezl,jarevalo3676}@uta.edu.ec

Abstract. Recently, Decision trees are referred as an essential tool for interpreting and understanding the trend in the academic desertion of male and female students. This research advancement has been utilised by the proposed study to provide a deeper understanding on the gender stereotypes of the University of Ecuador. This research work is proposed by using a quantitative technique with a descriptive, interpretative scope of the data mining findings acquired in the years 2014 and 2015 in the enrollment and gender-wise student graduation in the years 2021 and 2022. As a result of the high rate of non-graduate students, the degree of male students has a stronger tendency in most faculties, despite the fact that the number of women enrolled is 56.6%. It is verified that the utilized supervised technique based on decision trees allows identifying the Faculties of the Technical University of Ambato. It is quite positive for the education of women because 2.1% of all women obtained their third level degree in the denominations of engineering and bachelor's degrees. It is considered that Faculties such as Civil and Mechanical Engineering, as well as Systems, Electronic and Industrial Engineering should increase the number of women in their enrollment and ensure their qualifications through affirmative action policies.

Keywords: Data mining · Decision trees · Artificial intelligence · Gender stereotypes · Higher education · Academic desertion

1 Introduction

Data analytics is a technique that is used in different real-time applications and organizations to make better decisions and anticipate the future through the generation of knowledge [1], the storage of data via artificial intelligence, chatbots, mobile technology, and data intelligence. This facilitate the specific actions of users and people that require machine learning and predictive analysis.

In everyday life, another sector that has benefited from Open Banking is the banking sector via digital banking, which consumes less time and also [2] the functionality of mobile applications for banking applications during the COVID 19 pandemic [3] justified their existence by simplifying the different transactions and making users feel closer to the institution by delivering efficient customer service.

J. I.-Z. Chen et al. (Eds.): ICIPCN 2022, LNNS 514, pp. 321–331, 2022.
https://doi.org/10.1007/978-3-031-12413-6_25

According to [4], desertion considers a set of variables that determine whether the students will graduate or abandon their studies.

Artificial intelligence [AI] in healthcare is currently the most sophisticated technique for predicting early treatment. Also, the data collected in medical records are becoming a useful resource to build preventive care protocols.

Artificial Intelligence [AI] is based on the neural network model, which allows the understanding of data through cognitive computing, deep learning algorithms and patterns [5], further education and research is also required for finding the patterns in large data sets, and thereby optimize and predict the necessary results to improve the management and administration in case of Higher Education Institutions (HEI). This is very useful in detecting the trends and making decisions to prevent or solve specific problems.

There are related works in which the use of expert systems and data mining techniques allow the prediction models to be established with which we help the person in charge in taking them. [6] show the results of an investigation whose purpose is to evaluate the technical efficiency of HEIs in Colombia between the years 2011-and 2013, through the application of data envelope analysis and data mining techniques.

[7] propose a model to detect possible dropouts in Higher Education in a public university, where two proposals are stated for the quantification of dropouts: The first is established as the proportion of students who graduate in a determined time corresponding to the duration of the race, and the second is simply the number of students who drop out of school. To reduce desertion, these investigations are proposed to improve the early detection mechanisms of potential deserters. To develop their research, they used different methods, such as logistic regression, k nearest neighbor, decision trees that include random forests, Bayesian networks, neural networks, etc. [8].

Similarly, the empirical study on predictors of desertion and permanence in the Medicine career: Evidence from the Ecuadorian Higher Education System [9] based has established its results on the motive and personal reasons for desertion. However, they constitute a reference that supports the 26% drop out of university students declared by [10].

This is the starting point for the investigation because the comparative analysis of the phenomenon from the gender stereotypes variable (man, woman) through a predictive model has been scarcely addressed, preventing the contribution of a legendary perception of inequality that persists in the environment. Statistics and diagnostic tools do not fix reality, therefore university has recognized normally with no solution alternative.

2 State of the Art

2.1 Gender Stereotype

A gender stereotype is generally due to a mental representation that has been socially constructed by each individual, used to classify people according to the roles they have to fulfill. For [11], stereotypes constitute an unconscious distinction based on characteristics specific to each segmented group as follows: men, women, each with their family, professional, and social roles. While research carried out by [12] updated by [13] on gender stereotyping in the legal field, indicates that it is always harmful because it threatens the freedom to make personal and professional decisions, including life projects, a situation that acts to the detriment of the quality of life of individuals.

Despite any diagnosis and analysis, the practice of gender stereotypes and their application is daily because it is part of history, organization, understanding of relationships, social communication, it is a circumstantial part of the social structure that establishes a form of expected behavior for men and another for women, these generalizations according to [14] attribute characteristics and behaviors to the individual by the simple action of pigeonholing him within the group, regardless of his reasoning, because it is a matter of culture that goes through the history of colonization with implicit manifestations, especially those related to women classified as docile, complacent, supportive and currently the intelligent one is added [15].

This statement should remind us that stereotypes do not change due to the effect of time but rather due to understanding and practice in relationships, which forces us to carry out predictive analyzes that allow us to have a different view of the different cultures, histories and situations than is customary.

Until the year 2020, when the COVID 19 pandemic began, world organizations in the various latitudes of the world were adequately organized with various entities working towards the same objective "the search for balanced environments where women and men enjoy the same rights and obligations", now in the year 2022 the survivors find themselves in post-pandemic contexts where there are political, social, economic, personal, geographical scenarios that have exacerbated the gender-stereotyped reality, attributing to women the same roles but much more grounded for its fulfilment.

One of the fields where great achievements were made before the pandemic is in women's rights to freedom of study. About this aspect in Ecuador, there is a constitution that guarantees rights that serves as the basis for the promulgation of laws in favor of equal opportunities where it is established that "Education is a right of people throughout their lives. his life and an inescapable and inexcusable duty of the State.

It constitutes a priority area of public policy and state investment, a guarantee of equality and social inclusion, and an essential condition for a good living [16], as a derivation to the promulgation of this right, in the Law of Higher Education reads a whole set of articles that favor the application of the mandate in practice.

Under this state structure, Ecuadorian citizens have the necessary regulations for their free admission to higher education centers where an equal selection process is applied for men and women, according to the law, admission to the university on equal terms. conditions, this situation is corroborated by the statistics that each university has in its enrollment records.

But what happens when the statistics of the graduates are also reviewed? and at first glance a difference is detected, clearly noticing that in the course of the studies there is university academic desertion.

2.2 Regarding Desertion

Regarding desertion or abandonment and the high cost that corresponds to the state for each individual who deserts, [17] points out that basically, it is a personal decision to interrupt their academic activities, stressing that it is inherent to several economic factors, social, domestic violence, gender, among others [18, 19] coincide in noting in their research that students who choose to drop out have irrational beliefs and that it is these beliefs that act to keep stereotypes alive.

Knowledge of the subject allows establishing that desertion is an educational problem related to family, cultural, labor, and social spheres, to understand its breadth and depth, analysis is carried out based on projections that magnify the problem and shed light to propose solutions.

The study carried out by [20], builds a predictive model of student dropout that concludes that dropout has the most influential causes of level and grades, despite having sex and gender data, they discard the variable. Likewise, the results obtained by [21], in their study, determine that the sex/gender factor demonstrates a high level of prediction, in contrast to the work of (Sifuentes 2018), who concludes that sex/gender is a factor of low prediction.

A model is predictive because it is developed using a large number of machine learning algorithms. Authors such as [22] and [7], classify the techniques according to the process approach, which is: supervised, unsupervised, semi-supervised and learning reinforcement. [19] defines the supervised learning as a concise model of the distribution of the class labels concerning the predictor features The classifier that is used to assign the class labels to test instances where the features of the predictor values are known, but the value of the class label is unknown On the other hand, unsupervised learning consists of determining models that focus on finding hidden patterns in the data, without having a training set.

Semi-supervised learning, according to Chapelle, Scholkopf, and Zien, eds. (2009), lies in creating a model from a training system with missing information, where the result with incomplete data is learned and eliminated. Finally, Sutton and Barto (1998) describe reinforcement learning as a mathematical or statistical analysis model for learning based on external feedback given by a body of thought or the environment. This means that Data mining is the set of techniques and technologies that allow exploring large databases, automatically or semi-automatically, to find repetitive patterns, trends or rules that explain the behaviour of data in a given context.

Among the classification methods and supervised learning techniques are artificial neural networks, decision trees, Bayesian algorithms, logistic regression, support vector machines, nearest neighbor search, or assembly techniques of these algorithms.

2.3 Predictive Models

[23] propose a predictive model using data mining techniques through a web interface that facilitates the identification of students vulnerable to school dropout at the Technological University of Izúcar de Matamoros (UTM), in Mexico.

[24] designed a classification model to detect early dropouts in the Faculty of Engineering of La Salle University, by applying the CRISP-D (Cross Industry Standard Process for Data Mining) methodology.

Likewise, the literature from 1982 to 2017 analyzes the applications of machine learning and data mining to address the problem with methods such as decision trees, neural networks, support vector machines (SVM), k nearest neighbor (KNN), logistic regression (LR), among others, with which prediction rules are generated based on a set of management indicators that can be used in the design of educational policies to determine the reasons for some inefficiencies of HEIs. We cite a related work in which the use of expert systems and data mining techniques allow prediction models to be established, with which the person in charge is helped in making them [25].

The authors [26] show the results of an investigation whose purpose is to evaluate the technical efficiency of HEIs in Colombia between the years 2011–2013, through the application of data envelope analysis and data mining techniques., the two works seek to solve a specific problem and coincide in helping the responsible authorities in decision-making.

3 Methodology

It is a research with a quantitative approach, descriptive scope, and historical interpretation, whose general objective is to predict academic desertion due to the gender stereotypes factor.

For this project, variables were chosen that identified each student of the ten Faculties of the Technical University of Ambato, studies previously prepared by researchers from other universities were taken as a reference, as is the case of the Complutense University of Madrid, where he carried out a study to determine academic success/failure, using multiple linear regression and logistic regression techniques [23].

In this part of the work, the variables that are considered to affect the dropout rate of students will be described. In addition, the conceptual model has presented that allowed understanding of the interaction of the variables to later develop the Data Analysis Model. It is important to take into account a rigorous separation of the different types of desertion for its study. The authors [26] explain that student dropout can be understood from two points of view: temporal and spatial (Table 1).

Table 1. Types of academic dropout

No	Temporary concept	Spatial concept
1	Early dropout: when a student leaves a degree before being accepted	Change career within the same institution
2	Early dropout: when the degree is dropped during the first four semesters	Change in educational institutions
3	Late dropout: understood as dropout after the fifth semester	Dropping out of the higher education system, where there is the possibility of re-entry in the future, either to the same or another campus in the country

For the analysis, the concept is taken that desertion is a decision to abandon a career, which can be temporary or definitive. With the data collected, it is difficult to track whether the students who have enrolled have temporarily dropped out, therefore, definitive dropout is conceived. The process of collecting and processing information graphed allows understanding of the applied process Fig. 1.

Fig. 1. Phase 1 data collection

In the development of the instrument, international surveys on gender stereotypes, academic desertion and aspects focused on the research project were taken into account, the instrument was validated by Cronbach's Alpha statistic with a high level of reliability and finally, it was applied to the study population that were the students of the faculties of the Technical University of Ambato (Fig. 2).

Fig. 2. Phase 2 process for knowledge development

In Phase 2 of Processing for the development of knowledge, the tabulation of the surveys began to obtain the corresponding data, later the coding was carried out in the statistical software, and the data were cleaned and loaded for processing. respective, then the relevant attributes were identified to finally apply the decision tree focused on gender stereotypes and academic desertion in university students.

For this purpose, the sex/gender that identified each student of the ten Faculties of the Technical University of Ambato was chosen as the only variable.

Table 2. Total number of students enrolled disaggregated by sex/gender

Period	Men	%	Women	%	LGTBI	%	Total
April/16–September/16	5406	43,39	7054	56,61			12460
October/15–March/16	5555	43,65	7170	56,35			12725

Table 2 shows the number of students enrolled by the gender variable concerning the semesters October 2014, March 2015 and April–September 2015, starting from these data of enrolled students, we proceed with the identification of students. graduates April 2021–Sept 2021 and Oct 2021–Mar 2022.

Table 3. Total number of graduated students

Period	Female	%	Male	%	Total
April/16–September/16	574	8,13%	307	5,68%	881
October/15–March/16	851	11,87%	539	9,70	1390

Table 3 shows the total number of graduate students taking into account the same sex/gender variable in the period April–September 2021 and October 2021–Mar 2022.

4 Results

The decision trees that were generated using the investigated data will identify the trends according to the gender variable (male-female), as well as the degree acquired by the students and the faculties to which they belong and being able to make predictive decisions.

The decision tree in Fig. 3 shows the sex/gender variable as the main node 0, where the number of students enrolled is identified, with 56.6% being women and 43.4% being men, none of the respondents mentions being part of an LGBTI group, node 1 shows the number of graduate and non-graduate students where the following data is obtained; graduate students 65.2% are women and 34.8% are men, no one mentions belonging to a GLBTI group; Regarding non-graduates, 56.0% are women and 44% are men; therefore it is concluded that despite having enrolled a greater number of women, in the same way, there is a greater number of women who do not graduate at the same time they drop out of the careers they enter (Fig. 4).

Of a total of 12,460 students who enrolled, in Node 1 called 1.9 where 21.2% belong to the Faculty of Health Sciences, 15.6% to the Faculty of Human Sciences and Education, 13.6% to the Faculty of Accounting and Auditing, 10% to the Faculty of Administrative Sciences; 8.6% to the Faculty of Jurisprudence and Social Sciences; 8.3% to the Faculties of Systems, Electronic and Industrial Engineering and Civil and Mechanical Engineering, 5.3% to the Faculty of Food Engineering, 4.7% to the Faculty

Fig. 3. Number of graduates by gender variable

Fig. 4. Graduated students by faculty

of Design, Architecture and Arts; 4.3% to the Faculty of Agricultural Sciences, as a result it is evident that there is a high tendency in the Faculty of Health Sciences; thereforeNode 2 (2.9) shows that this faculty has a high demand for non-graduates with 22.3%, followed by the Faculty of Accounting and Auditing with 14%; in Node 1, 3 and 4 (1.6, 2.3, 2.2) it is evident that the degree of the Faculties such as Human Sciences and Education has a high demand for Graduates with 22%, followed by the Faculty of Civil and Mechanical Engineering where there is a trend of 10.5% but their degree is based more on men than on women.

Nodes 5 and 6 (1.5 and 1.3) identify men and women with degrees identify men and women graduates from the faculty to which they belong, from which the following data is obtained: 307 men, representing 2.5% of the total population, are graduates, with a trend in Science faculties Human and Education with 29%; in Mechanical Civil Engineering 15%; Health Sciences 11.4%; Accounting and Auditing 10.7%; Regarding women, 266, which represents 2.1% of the total, obtained their undergraduate degree, being the Faculties of Accounting and Auditing where there are more women graduates with 44%; Human Sciences and Education 13.9%; and Administrative Sciences with 9.4%; It is worth mentioning that even so, the percentage of women with degrees is not higher than that of men.

Concerning the Non-degree Nodes 7 and 8 (2.7 and 2.1) of the students who entered the University as enrolled, there is a high percentage as evidenced in the decision tree; where the following data is obtained: 40.9 which corresponds to 5099 male students did not graduate, being the faculties of Health Sciences, Administrative Sciences, and Food Engineering that have a higher percentage of non-graduates; Regarding women, 48%, which are 5,975 students, are not qualified, being the Faculties of Health Sciences with 29.3%; Accounting and Auditing with 18.6%; Administrative Sciences 11.3%.

This research addressed current issues focused on the development of decision trees in gender stereotypes and university academic desertion, these issues are currently very worrying in various aspects of technological, educational and, above all, administrative research. The results allow solving specific issues for the institution but are common for higher education, it is worth mentioning that the majority of Higher Education Institutions should make decisions with the support of technological tools focused on research and the generation of knowledge through mining. of data and the construction of decision trees where the hierarchy of the most relevant characteristics allows recognizing the level of desertion in the degree of the students.

Similarly, in the review of the literature for the construction of the state of the art, it was shown that one of the most used techniques in the predictive analysis is decision trees since these generate more relevant characteristics to determine the level of academic desertion.

In the study carried out, it is about creating and applying strategies in this case to academic desertion based on the gender variable, it became more common in the times of the COVID -19 Pandemic, through decision trees, the authorities will be able to work on the factors that provide the problem of non-graduation of students effectively and efficiently since the model predicts the characteristics where a greater tendency to dropout is obtained.

5 Conclusion

It is concluded that there is a high rate of non-graduate students within the Faculties of the Technical University of Ambato, the same ones who drop out of the careers in which they began their enrollment; In the same way, the degree of men has a greater tendency in most of the faculties, although the number of women enrolled is higher by 56.6% compared to that of men.

The supervised technique based on decision trees allows identifying the Faculties of the Technical University of Ambato where the student desertion of men and women is high because they have a high rate of non-graduates. It is quite positive for the education of women because 2.1% of all women obtained their third level degree in the denominations of engineering and bachelor's degrees.

Finally, decision-making is reached where the authorities of each of the Faculties of the Technical University of Ambato, based on the results obtained in the decision trees, must consider that Faculties such as Civil and Mechanical Engineering, as well as Systems Engineering, Electronics and Industrial should increase the number of women in their enrollment and ensure their qualifications through affirmative action policies.

At the same time, it is intended to make correct decisions so that this rate drops by more than 50%, through this type of technique the solutions will be taken into account for a contingency plan that promotes alternatives focused on the graduation of students.

6 Future Research Work

As future work, it is desired to make predictions in the correct selection of the students' career, since in Ecuador there is a limitation when entering the University by exams that measure requirements other than that of a career or aptitudes that students have innately.

In the same way, it is desired to make decisions based on the degree of the students, since there are many students who do not manage to complete this process.

References

1. Anjarwati, R., Setiawan, S., Laksono, K.: Experiential meaning as meaning making choice in article writing: a case study of female and male writers. Heliyon **7**, e06909 (2021)
2. UNESCO: Uso de las TIC en la educación (2020). https://es.unesco.org/themes/tic-educacion/accion
3. COBIS: COBIS, 31 December 2019. https://blog.cobiscorp.com/aplicaciones-moviles
4. Giovagnoli, P.I.: Determinantes de la deserción y graduación universitaria (2001)
5. Korporate Technologies Group: El impacto de la Analítica de Datos en la Experiencia de Cliente, 06 May 2022. https://grupokorporate.com/el-impacto-de-la-analitica-de-datos-en-la-experiencia-de-cliente/
6. Zhang, Y., Zhang, Y., Oussena, S., Clark, T., Kim, H.: Using data mining to improve student retention in higher education: a case study. In: International Conference on Enterprise Information Systems (2010)
7. Meza, M., Gutierrez, G., Becerrea, M.: Deserción escolar en alumnos de la Universidad Michoacán de San Nicolás de Hidalgo. La experiencia de la Facultad de Contaduría y Ciencias Administrativas de Michoacán, México. Educación, gestion del conocimiento y creacion de valor (2022)
8. Orea, S.V., Vargas, A.S., Alonso, M.G.: Minería de datos: predicción de la deserción escolar mediante el algoritmo de árboles de decisión y el algoritmo de los k vecinos más cercanos. Recursos Digitales Para La Educacion y La Cultura, pp. 33–39 (2010)
9. Torres, S.: Determinantes de la deserción y permanencia en la carrera de Medicina: Evidencias del Sistema de Educación Superior ecuatoriano. Revista andina de Educación **5**(1), 000516 (2022)

10. Secretaria de Educación Superior Ciencia, «Senescyt», La deserción universitaria en el país alcanza el 26%, 07 May 2018. https://www.expreso.ec/guayaquil/desercion-universitaria-pais-alcanza-26-1456.html
11. Andrade, M.: El estudio de los estereotipos de género desde la Psicología, Ambato (2022)
12. Zampas, F.: Poder Judicial en el abordaje de los estereotipos, Ecuador (2013)
13. Geneva, O.: Poder Judicial en el abordje de estereotipos de nocivos de género en casos relativos a la salud y los derechos sexuales y reproductivos: una reseña de jurisprudencia (2018)
14. Cook, R.C.: Estereotipos de Género, Perspectivas Legales Transnacionales, Portafalia (2010)
15. Castillo, R.M.: Análisis de estereotipos de Género actuales. Anales de Psicología (2014)
16. Asamblea Nacional del Ecuador, Constitución 2008, Quito (2008)
17. Montoya, N.J.: Posibles Causas de deserción escolar en los jovenes y niños del Colegio Departamental General Santander Sede San Benito de Sibaté, Bogotá (2019)
18. Abarca, S.: Deserción estudiantil en la educación superior: caso de la Universidad de Costa Rica. Actualidades Investigactivas en Eduación, pp. 1–22 (2005)
19. Hernández, M.A.: El problema de la deserción escolar en la producción ciéntifica Educativa. Revista Internacional de Ciencias Sociales y humanidades, SOCIOTAM, pp. 89–112 (2017)
20. Cuji, B.G.: Modelo predictivo de deserción estudiantil basado en árboles de decisión, Espacios, pp. 1–9 (2017)
21. Osorio, A.B.-C.: Deserción y graduación estudiantil universitaria: una aplicación de los modelos de supervivencia. Universia, pp. 31–57 (2012)
22. Apostolou, B., Dorminey, J., Hassell, J.: Accounting education literature review. J. Accounting Educ., 1000725, 2021 (2020)
23. Panhalkar, A., Doye, D.: Optimization of decision trees using modified African buffalo algorithm. J. King Saud Univ. Comput. Inf. Sci. (2021, in Press). Corrected Proof
24. Felizzola, H.A., Arias, Y.A.J., Pedroza, F.V., Pastrana, A.: Modelo de predicción para la deserción temprana en la Facultad de Ingeniería de la Universidad de la Salle. Encuentro Internacional de Educación en Ingeniería (2018)
25. Chapelle, O., Scholkopf, B., Zien, E.A.: Semi-supervised learning. IEEE Trans. Neural Netw. **20**(3), 542–552 (2009)
26. Assuad, C., Tvenge, N., Martinsen, K.: System dynamics modelling and learning factories for manufacturing systems education. Procedia CIRP **88**, 15–18 (2020)
27. Bahadur Pal, K., et al.: Education system of Nepal: impacts and future perspectives of COVID-19 pandemic. Heliyon **7**(9), e08014 (2021)
28. Kozlova, D., Pikhart, M.: The use of ICT in higher education from the perspective of the university students. Procedia Comput. Sci. **192**, 2309–2317 (2021)
29. Páez-Quinde, C., Infante-Paredes, R., Chimbo-Cáceres, M., Barragán-Mejía, E.: Educaplay: una herramienta de gamificación para el rendimiento académico en la educación virtual durante la pandemia covid-19. Catedra **5**(1), 32–46 (2022)

Exploring Public Attitude Towards Children by Leveraging Emoji to Track Out Sentiment Using Distil-BERT a Fine-Tuned Model

Uchchhwas Saha[1(✉)], Md. Shihab Mahmud[1], Mumenunnessa Keya[1], Effat Ara Easmin Lucky[1], Sharun Akter Khushbu[1], Sheak Rashed Haider Noori[1], and Muntaser Mansur Syed[2]

[1] Department of Computer Science and Engineering, Daffodil International University, Dhaka, Bangladesh
{uchchhwas15-10842,shihab15-10961,mumenunnessa15-10100, effat15-10793,sharun.cse}@diu.edu.bd, drnoori@daffodilvarsity.edu.bd

[2] Department of Computer Engineering and Science, College of Science and Engineering, Florida Institute of Technology, Florida, USA
msyed2011@my.fit.edu

Abstract. Sentiment analysis is a computational method that extracts emotional keywords from different texts through initial emotion analysis (e.g., Happy, Sad, Positive, Negative & Neutral). A recent study by a human rights organization found that 30% of children in Bangladesh are being abused on online in the COVID-19 epidemic by various obscene comments. The main goal of our research is to collect textual data from social media and classify the way children are harassed by various abusive comments online through the use of emoji in a text-mining method and to expose to society the risks that children face online. Another goal of this study is to set a precedent through a detailed study of child abuse and neglect in the big data age. To make the work effective, 3373 child abusive comments are collected manually from online (e.g. Facebook, Newspapers and various Blogs). At present, there is still a very limited number of Bengali child sentiment analysis studies. Fine-tuned general purpose language representation models, such as the BERT family model (BERT, Distil-BERT), and glove word embedding based CNN and Fast-Text models have been used to successfully complete the study. We show that Distil-BERT defeated BERT, Fast-Text, and CNN by 96.09% (relative) accuracy, while Bert, Fast-Text and CNN have 93.66%, 95.73%, and 95.05%, respectively. But observations show that the accuracy of the Distil-BERT does not differ much from the rest of the models. From our analysis, it can be said that the pre-trained models performed outstanding and in addition, child sentiment analysis can serve as a potential motivator for the government to formulate child protection policies and build child welfare systems.

Keywords: Child sentiment analysis · Emoji · BERT · Distil-BERT · Fast-Text · Convolutional Neural Network · Natural language processing

J. I.-Z. Chen et al. (Eds.): ICIPCN 2022, LNNS 514, pp. 332–346, 2022.
https://doi.org/10.1007/978-3-031-12413-6_26

1 Introduction

Recently Sentiment Analysis (SA) is perhaps the most hot topics in the research field. It is a kind of NLP to follow the public's opinion towards a particular law, strategy or commercialism, etc. It is mainly proposed to evaluate various opinions, reviews, texts, etc. in different languages [1]. SA is mainly classified into two parts: Positive and Negative but sometimes it's extended to be less positive, less negative, more positive, more negative, neutral etc., and it depends on datasets. The analysis of emotions or opinion is a part of machine learning. Various supervised and unsupervised algorithms have been used for training and testing [2]. Analysis of sentiments gives the exact results of a judgement or opinion. Imagine a customer or person conducting a review of SA that how this person actually says. Most SA articles are based on the English language but recently some articles have been made in other languages such as Arabic, Persian, Bengali, Thai etc. [3]. Child sentiment, it's basically a child's emotion such as happy, sad or neutral. Child sentiment is quite different from adult sentiment because for the same sentence, a child's reaction will be different from an adult person's reaction. Child Sentiment Analysis using in the research field which is a powerful tool for understanding the emotions of children. Children's emotions can be analyzed in a variety of languages, including English, Thai, Arabic etc. [4]. Social Network Services (SNS) has become a part of our day-to-day life. In social media, children are neglected and bullied by others' comments, messages etc. Kids have recently faced many other problems such as bullying, cyber bullying, threat by others but cannot talk about the problems. When children want to talk about their problems, the maximum time they can't express it properly or people can't understand the problem. This analysis works very well to detect what is true and what is false [5]. Child sentiment is analyzed in different ways. Sometimes using positive, negative and neutral, sometimes using rating systems (one to five starts, one to ten scale etc.). These methods work very well for child sentiment analysis or sentiment analysis. Recently some papers used emojis for labeling. Emoji uses like a pattern have expanded rapidly. The language of this document is English and it incorporates emojis. People's emotions are high because they used emoji in their datasets. Emoji were first available on mobile devices in 2010 [6]. This analysis tested five different methods to analyze twitter posts: NB, ME, SVM, SNN and the LSTM based RNN. Every emoticon (emoji) could not be typed using a keyboard and is recorded as a Unicode character. In social media, emojis and emoticons are commonly used to show emotion, feelings and thoughts [7]. In order to enable internet devices, Emojis appeared at the end of the twentieth century in japan. SA is the field in which working about viewpoints, feelings. reviews, attitudes and emotions that individuals are analyzed from a text. In exploring big or short sentences, comments, blogs etc., emoji provide important information [8].

In this research, we try to analyze child sentiment with different sentences which are heard or said by a child and it's mainly analyzed by emojis. We see there are many articles about this topic in different languages but there is no article in the Bengali language. That's why we motivate and work in Bengali language child sentiment analysis with emojis. We used child sentiment and explained child emotions in Bengali sentences. For our Bengali text child sentiment analysis, we experimented with four separate deep learning techniques: BERT, Distil-BERT, Fast-Text, Convolutional Neural Network (CNN) (Table 1).

Table 1. The data table represents the category of child sentiment in emojis.

Emoji	Child Sentiment	Sentences
	Positive	নিষ্পাপ শিশু কিশোরদের মন খুবই সরল কোমল ও পবিত্র
	Negative	অটিস্টিক শিশুরা কখনই তার সমবয়সী শিশুদের সাথে মিশবে না
	Neutral	জলবায়ুর প্রভাবে বাংলাদেশের শিশুদের ভবিষ্যত অনিশ্চিত

2 Literature Review

We started this study by reading literature reviews related to it. Specifically, our experiments are based on three literary strands: Sentiment Analysis (SA), Child Sentiment Analysis, Emojis in Child sentiment analysis with both Bengali and English sentences or texts.

2.1 Sentiment Analysis

In recent eras, sentiment analysis is becoming a hot topic in natural language processing, Duyu Tang et al. [9]. Including a summary of effective deep learning methods, its goal is to find, extract, and organize sentiments from user-generated content in social media, blogs, and product reviews. Geetika Gautam et al. [10], introduce sentiment analysis as a system which is able to quantify the subjectivity and polarity of a sentence or passage to positive, negative and neutral. This paper, viewed twitter data analysis by some machine learning algorithms and semantic analysis. Dataset used online and the primary objective is to analyze many reviews using this dataset and semantic analysis (WordNet) gave the best result. Aliaksei Severyn et al. [11], used polarities at the same message and phase level to describe the deep learning model for twitter sentiment analysis. A new model is the primary contributor to the convolution neural parameter weights network, which is important to train an accurate version while preventing the necessity to inject further functions. Erik Boiy et al. [12], proposed a machine learning approach based on sentiment analysis focusing on a person's emotion. Dataset made by three different languages and used three different machine learning methods such as SVM, MNB and ME. The best result came from English text which was 83%. Tapasy Rabeya et al. [13], Introduced an emotion sensing model for Bengali sentence level identification. This article experiments about Bengali people's emotion from Bengalis text. With 77.16% accuracy and two basic feelings of happiness and sadness considered, a suggested method produces an outcome. Ali Hasan et al. [14], performed a comparison analysis with political views by using supervised ML algorithms such Naive Bayes (NB) and SVM and with three classes dataset made from twitter API and got best accuracy. K. M. Azharul et al. [15], presented a sentimental analyzer which accepts Bengali opinions about Bengali topics. Dataset was collected from Bengali text and constructing sentiment classification from Bengali text and took five classes. Tanvirul Islam et al. [16], performed an evolutionary strategy to comparative study of identifying violent and aggressive text in Bangla. Firstly, they collected data from Facebook and YouTube then created a balanced dataset. Some supervised Machine learning methods were used in this article such as SVM, MNB, MLP, K-NN, SGD and the best outcome came from the SVM classifier and it was 88%. Kesinee Boonchuay et al. [17], used the text acquisition classifier to improve the category of sentiment for Thai teaching assessment with a dataset of their own. Their work used two methods of classification text embedding. The results represent the best system results in the Fast Text categorization; the second approach used text vectors to improve results in relation to TF-IDF for the closest to K-nearest neighbors and Naïve Bayes. Raphael Tang et al. [18], analyzed distillation through an usual BiLSTM method of the BERT understanding. The distilled model performs similar outcomes with ELMo with much smaller datasets and less running time. The results have been shown BiLSTM are much more expressive for natural language tasks as thought previously. Conducting tests on the benchmark for the evaluation of general language understanding and collecting six natural languages comprehension tasks in three categories. Xin Li et al. [19] explained all about the BERT model which is the most popular pre-training language model in recent times and investigated the possibility of combining the BERT embedding element with multiple neural methods. The experiments show the BERT based methods are better compared. Manish Munikar et al. [20], worked with the most famous public datasets for fine-gained feeling classification work. Their model exists without advanced architectural style in the other popular models of this task. They also showed the efficiency of transfer learning in natural language processing, and how it can be applied to computer programming.

2.2 Child Sentiment Analysis

Nhan Cach dang et al. [21], research on social networking viewpoints, including such Facebook or twitter, has become a significant method to learn about the views and implementations of users. Overview of DL methods in SA and have been shown in recent times to be a great alternative to NLP's barriers. Marta R. Costa-Jussa et al. [22], tried to use three different models SVM, RF and BERT and discussed about experimental analysis on short phase categorization for Spanish harmful language. Posts were selected from some sources like twitter, blogs, forums, medical forums, internet sites etc. Pranav malik et al. [23], worked on two particular datasets, one of which is twitter's toxic youth discussions. A combined model (FastText and BERT) has been used to categorize toxic discussion between many young people.

2.3 Emoji and Child Sentiment Analysis in Bengali

In sentimental analytical analysis, the use of Emoji characters seemed to have a greater effect than the negative views on general emotions of positive opinions. Child sentiment analysis of Bangla language with emoji is a very new research work field. There are many papers and articles on the analysis of childhood feelings with emoji in English, German, Arabic, Persian etc., but hopefully it's the first work to be done with emoji in Bengali sentences.

2.4 Emoji and Child Sentiment Analysis in English

Tuan Tran et al. [24], tried to analyze an emoji-based sentiment analysis and collected data from the High Times Magazine Facebook page and found that "LIKE" and "LOVE" are the most repeatedly used reactions. And it appeared that the relationship between the number of comments and reactions is comparable to overall emoji types, except "SAD". Travis LeCompte et al. [25], Used machine learning algorithm to analyses Twitter data for emotion classification. In this article, MNB and SVM classification methods for labeling twitter data and both methods worked better. Mohammed O. Shihal et al. [26], The review looks at the utilization of Emoji characters on social networks and the effects on text mining and SA. In twitter dataset utilization of Emoji characters in SA highly affects the overall opinions of the good sentiments more than negative sentiment, it finds. Sanjaya Wijeratne et al. [27], Emoji experience negation has been used to calculate the semantic similarity of emojis and used two one kind of datasets, one is Twitter and the other is Google News and also used emoji embedding fashions to measure emoji similarity. Mayu Kimura et al. [28], tried to develop a technique for mechanically setting up an emoji feeling lexicon primarily dependent on the co-event connections among both opinion stages and emojis. Data collected from twitter API. Sadam Al-Azani et al. [29], collected data from Arabic twitter and it was once publicly available. Four strategies for textual aspects extractions are explored and the Minimal Optimization-based SVM (SMO-SVM) classifier is utilized for assessment with the PUK kernel function. Toshiki Tomhira et al. [30], Developing another plan that gains from sentences the utilization of emoji as names, gathering Japanese tweets from Twitter as the corpus and it's primarily founded on two neural networks – CNN and RNN. Yuxiao Chen et al. [31], proposed another model for twitter opinion assessment with more interest in emojis. Data collected from the internet and made a unique dataset and it's aided by slithering tweets by means the REST Programming interface. Performed two components to exploit bi-sense emoji inserting for the emotion exploration task. Jibin Zheng et al. [32], The BERT+BiGRU method is intended for the FSS, which accomplishes the impact of insightful interaction among students and instructors. In this paper, Gathered an outsized number of online education platform web asset surveys to pre-train the word2vec model. Kottilingam Kottursamy [33], This paper represents that emotion recognition by various deep learning models. Developed a new model called eXnet (Expression Net) for improved accuracy. A. Pasumpon Pandian [34], Performance measure of Various deep learning algorithms in different purposes in SA with feature extraction. T Ganesan [35], Using different types of social media data analysis and it's understanding the learning experiences, learning improvement etc. to students. Milan Tripathi [36], this paper mainly says that using various traditional classifiers (NB, SVM, LSTM) to detect sentiment analysis of Nepali Covid19 tweets. Akey Sungheetha [37], the model used is a Transfer Capsule Network, that has the option to transmit information acquired at the data level to the relation to the issue in order to categorize based on the emotion identified in the text. Mainak Ghosh [38], Without any previous training, a technique is presented that categorizes tweets depending on their own negative posts. An aspect-based method is used to reveal an unsupervised negative sentiment analysis.

3 Methodology

As we talked about before, SA is a part of NLP study. There are a lot of ML and DL approaches that we can see in SA. Some of are supervised and other is unsupervised. In this part we briefly described our proposed four models. Three is pre-trained word embedding and another is glove word embeddings. BERT, DistilBERT and FastText is already pretrained models that we have used.

The workflow of our work is given below- (Fig. 1)

Fig. 1. Proposed workflow for the model.

3.1 Data Collection

Data collection could even be a scientific process of gathering observations data. The very first step of sentiment analysis is gaining a large frame of high quality Bengali textual data for training and testing. We got about 3373 Bengali sentences from different platforms and social media like Facebook, newspaper, blogs etc. This dataset consists of total 3373 Bengali sentences and there is 1121 positive ( ), 1120 negative ( ), and 1132 neutral ( )sentences. The dataset is unique because it is its own built dataset for child sentiment analysis. In this dataset we can use three classes (Fig. 2).

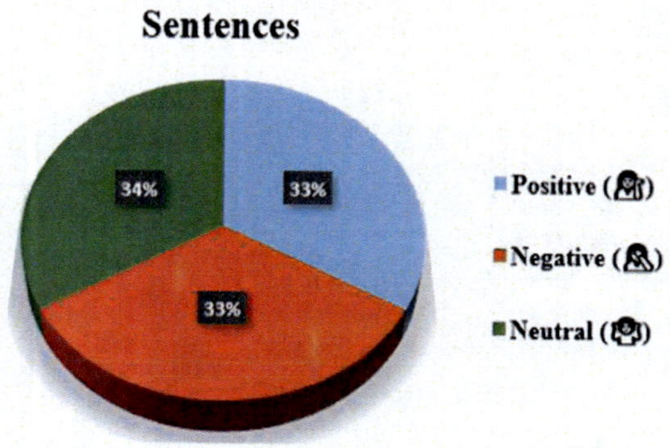

Fig. 2. Pie chart: the percentage of positive, negative & neutral data.

3.2 BERT Model

BERT is a NLP open source Machine Learning framework which was developed by google in 2018. BERT has capabilities to fine tune data to the specific language context and it brings much better model performance over legacy methods [39]. We use the ktrain python library, a light-weight wrapper for the DL library TensorFlow Keras, to construct and run BERT. The method has 12 layers, 768 measurement and 12 heads. We used a batch measurement of 7, maximum size of 500, studying rate of 0.00002 and epoch of 100 in this model (Fig. 3).

Fig. 3. The architecture of pre-training BERT.

3.3 DistilBERT Model

The DistilBERT in an architecture based on BERT small, quick, cost-effective and light transformer. We used a hugging face pre-trained transformer model 'DistilBert-base-multilingual-cased' as DistilBERT with ktrain. The pre-trained model is trained in 104 different languages including Bangla. It has 6 layers, 768 dimensions, 12 heads and 134M parameters. We used an algorithm of maximum length 500 and learning rate of 0.00002. Then, we finally trained our model with a batch size of 7 and 100 epochs (Fig. 4).

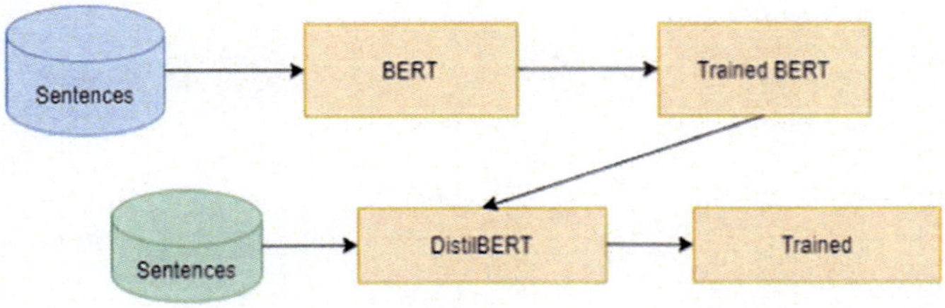

Fig. 4. The architecture of DistilBERT for child sentiment classification.

3.4 FastText Model

Fast Text is a lighter weight, accessible program that allows learners to know text representations and classifications. FastText is a natural language processing (NLP) framework for text categorization and word embeddings and it has two layers, one is embedding layer and another is linear layer [40]. The hardware works on basic, generic terms. Methods may subsequently be reduced to fit even on portable devices. FastText may want to stand for a library constructed by using the Facebook research team because of environment friendly instruction of word representations yet sentences classification. FastText is well-known for its speed and accuracy in training. This FastText model has a maximum rate of 1000 and studying rate of 0.0002 to train our dataset. Then we trained it with a batch size of 32 and 300 epochs with the assistance of the ktrain library.

3.5 Glove+CNN Model

Convolutional neural networks (CNN) are layered synthetic neural networks capable of detecting complex data features, such as extracting features from image data and text data. In previous years, CNN have shown revolutionary implications in some NLP tasks, one unique task is the classification of sentences, for example, the classification of short sentences into a set of predefined categories. In our model we utilized a pre-prepared glove model as word embedding and made embeddings matrix. Using Conv1D in the subsequent layer makes a convolution bit that is convolved with the layer input one spatial measurement to create a tensor of outcomes and the dense layer is the last layer for convolution networks which is fully connected. Our proposed model, we use 'Adam' as an optimizer, 'binary_crossentropy' as a loss function and an activation called 'SoftMax' in the dense layer. Lastly, trained our model with a batch size of 512 and epochs is 45.

3.6 Parameter Tuning for Models

This table shows which parameters we can use in our model to achieve the best result. In parameter settings we use GridSearchCV to identify the best configuration of our models. This graph we used the most batch in Glove+CNN model which is 512 and the lower batch is 6, used in BERT (Fig. 5).

Fig. 5. Parameter tuning for the models- BERT, DistilBERT, FastText & Glove+CNN.

4 Result Discussion

The Deep Learning algorithms are training on word embedding techniques & pretrained glove embedding [41] training techniques and it's results shown in below. The accuracy of Bidirectional Encoder Representation from Transformers (BERT) is 93.66%. Accuracy of DistilBERT is 96.09%. FastText gained 95.73% and the last one was Convolutional Neural Network (CNN) which is gaining 95.05%.

Fig. 6. The graph shows the accuracy of BERT, DistilBERT, FastText & Glove+CNN using word embedding and pretrained glove embedding.

From Fig. 6, Among all the algorithms of Deep Learning, DistilBERT gives the best performance than other algorithms. FastText is very close to CNN but the other model is showing lower accuracy.

In this Table 2, In the given table, we can see the precision, recall and f1_score value for BERT, DistilBERT, FastText, and CNN are different for three defined emoji. These results are different from each other in different cases. Moreover, these values are also different for defined emojis in each model [42]. It is a classification report of Confusion Matrix. We need to calculate it to determine the false positive and false negative rate. Precision relates with false positives, recall relates with false negative and f1-score relates both of these, it's called weighted average of precision and recall.

Table 2. Precision, Recall, F1_score for the standard matrix values based on the parameters of confusion matrix.

	Emoji	Precision	Recall	F1_score	Accuracy
BERT		0.91	0.97	0.94	
		0.95	0.96	0.96	93.66%
		0.96	0.88	0.92	
		0.93	0.98	0.95	
DistilBERT		0.98	0.96	0.97	**96.09%**
		0.98	0.95	0.96	
		0.96	0.95	0.95	
FastText		0.94	0.97	0.96	95.73%
		0.98	0.95	0.96	
		0.95	0.91	0.95	
Glove+CNN		0.96	0.94	0.95	95.05%
		0.95	0.96	0.96	

In Fig. 7 ((a), (b)), show the macro and weighted average of the confusion matrix. All the classes equally contribute to the final average, it's called macro average. For Weighted average, every class's performance according to the average is weighted by using its size. Below we show the Macro and Weighted Average of the confusion matrix for our four separated model.

Fig. 7. (a) Represent macro average and (b) Represent the weighted average for the Standard matrix values based on the parameters of confusion matrix.

In the below Table 3 show that FPR is determined by the number of actual negatives predicted incorrectly by the models. A False Negative Rate (FNR) is when the classifier trains the negative class incorrectly. The proportion of expected negatives that are true negatives is defined as the NPV. FDR is mainly performed in our model that the overall amount of invalid rejections includes both false positive (FP) and true positive (TP). In our proposed model, the amplitude of the variation between a prediction and real value of an analysis is referred to as absolute error and MAE is the average of absolute error. Mean Squared Error (MSE) is a square difference for true outcomes and predicted outcomes. Root Mean Squared Error (RMSE) is calculated for our model to find the squared difference between the actual result and the prediction results. In our models, we faced some problems about multiclass or imbalance class. Cohen Kappa Score (CKS) is an excellent measure to solve this problem [43].

Table 3. Receive characteristics and the performance measure matrix for BERT, DistilBERT, Fasttext & Glove+CNN.

Model	FPR	FNR	NPV	FDR	MAE	MSE	RMSE	CKS
BERT	0.01	0.01	0.98	0.01	0.08	0.03	0.18	0.9
DistilBERT	0.01	0.02	0.97	0.01	0.07	0.05	0.16	0.89
Fast-Text	0.03	0.01	0.98	0.03	0.21	0.06	0.25	0.93
Glove+CNN	0.02	0.03	0.96	0.02	0.16	0.04	0.22	0.92

Figure 8 visualizes the Sensitivity which mainly used in our models to calculate the proportion of true positive results and the predicted positive results which we got from our models. Specificity is the same as sensitivity but in specificity we use negative results.

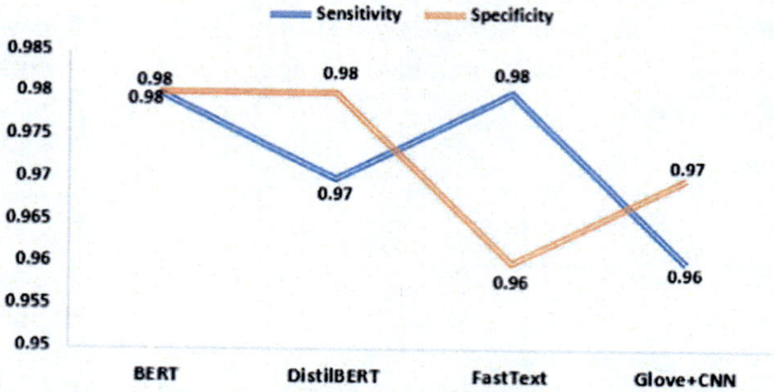

Fig. 8. Sensitivity & specificity analysis exhibited by the graphical representation for BERT, DistilBERT, FastText & Glove+CNN.

Fig. 9. (a) Represent the Pre-trained BERT Model, (b) Represent the Pre-trained DistilBert Model, (c) Represent the Pre-trained FastText Model, (d) Represent the Glove Word Embedding Pre-trained CNN Model, the Four Figure of the Confusion Matrix Presents Performance of the Model for the Prediction of the Child-Sentiment of Emoji Based.

Figure 9 ((a), (b), (c), (d)) are shows that the confusion matrix of our performed models. Evaluating the performance of the classification model is called a confusion matrix or N $\times$ N matrix, where N is the quantity of target classes we have utilized. This gives us a comprehensive view for model execution and what kind of mistakes there are. This matrix compares predicted value and target value by deep learning models. In our model, we use a 3 $\times$ 3 confusion matrixes for showing our model performances.

In Fig. 10 explains that the training accuracy, loss and the testing accuracy, loss for our proposed four models (Table 4).

Fig. 10. Demonstrate the BERT, DistilBERT, FastText & Glove+CNN model history based on per epoch rate.

Table 4. The predicted testing result of the models.

Original Sentence	শিশু শব্দটার সাথে মিশে আছে ভালোবাসা আদর এবং মমতা	**Positive**	
BERT	শিশু শব্দটার সাথে মিশে আছে ভালোবাসা আদর এবং মমতা	**Positive**	
DistilBERT	শিশু শব্দটার সাথে মিশে আছে ভালোবাসা আদর এবং মমতা	**Positive**	
FastText	শিশু শব্দটার সাথে মিশে আছে ভালোবাসা আদর এবং মমতা	**Positive**	
Glove+CNN	শিশু শব্দটার সাথে মিশে আছে ভালোবাসা আদর এবং মমতা	**Positive**	

5 Conclusion

The motivation behind this paper is to work on the presentation of sentiment classification of Child Sentiment with emojis using word embedding. First, we collect raw data from a variety of sources and clean it, because machines can accurately interpret preprocessed data. Three pretrained models and one glove word embedding are used. DistilBERT algorithms provide the best overall performance, according to the results. In the meantime, the second approach used pre-trained word embedding, which provides better performance over BERT and CNN. Nevertheless, the output of pre-trained BERT is poor to that of DistilBERT, CNN and FastText.

6 Future Works and Limitations

Performs with neural network-based models such as ANN, DNN, LSTM, RNN, BiLSTM and others might also be conducted in the future. Main limitations on this paper is that we use only three classes because classification or utilizes only 3373 data. In the future we will attempt to make use of more classes yet greater data because of our upcoming work.

References

1. Saad, S., Saberi, B.: Sentiment analysis or opinion mining: a review. Int. J. Adv. Sci. Eng. Inf. Technol. **7**(5), 1660 (2017)
2. Varghese, R.: A survey on sentiment analysis and opinion mining. Int. J. Res. Eng. Technol. (2013)
3. Vinodhini, G., Chandrasekaran, R.M.: Sentiment analysis and opinion mining: a survey. Int. J. Adv. Res. Comput. Sci. Softw. Eng. **2,** 282–292 (2012)
4. Pasupa, K., Netisopakul, P., Lertsuksakda, R.: Sentiment analysis of Thai Children's stories. Artif. Life Robot. **21**(3), 357–364 (2016)
5. Li, Z., Kawamoto, J., Feng, Y., Sakurai, K.: Cyberbullying detection using parent-child relationship between comments. In: Proceedings of the 18th International Conference on Information Integration and Web-based Applications and Services (2016)
6. Hankamer, D., Liedtka, D.: Twitter sentiment analysis with emojis (2019)
7. Yo, B., Rayz, J.: Understanding emojis for sentiment analysis. In: The International FLAIRS Conference Proceedings, vol. 34 (2021)

8. Novak, P. K., Smailovic, J., Sluban, B., Mozetic, I.: Sentiment of emojis. PLoS ONE **10**(12), e0144296 (2015)
9. Tang, D., Qin, B., Liu, T.: Deep learning for sentiment analysis: Successful approaches and future challenges. Wiley Int. Rev. Data Min. Knowl. Discov. **5**(6), 292–303 (2015)
10. Gautam, G., Yadav, D.: Sentiment analysis of Twitter data using machine learning approaches and semantic analysis. In: Seventh International Conference on Contemporary Computing (IC3), India (2014)
11. Severyn, A., Moshiti, A.: Twitter sentiment analysis with deep convolutional neural networks. In: The 38th International ACM SIGIR Conference (2015)
12. Boiy, E., Moens, M.: A machine learning approach to sentiment analysis in multilingual web texts. Inf. Retr. J. **12**(5), 526–558 (2009)
13. Rabeya, T., Ferdous, S., Suhita, H., Chakraborty, N.R.: A survey on emotion detection: a lexicon based backtracking approach for detecting emotion from Bengali text. In: 10th International Conference of Computer and Information Technology (ICCIT) (2017)
14. Hasan, A., Moin, S., Karim, A., Band, S.S.: Machine learning-based sentimental analysis for Twitter accounts. Math. Comput. Appl. **23**(1), 11 (2018)
15. Azharul Hasan, K.M., Islam, Md.S., Mashrur-E-Elahi, G.M, Izhar, M.N.: Sentiment recognition from Bangla text. In: Technical Challenges and Design Issues in Bangla Language Processing (2013)
16. Islam, T., Ahmed, N., Latif, S.: An evolutionary approach to comparative analysis of detecting Bangla abusive text. Bull. Electr. Eng. Inform. **10**(4), 2163–2169 (2021). International Conference on Innovation in Engineering and Technology (ICIET)
17. Boonchuay, K.: Sentiment classification using text embedding for Thai teaching evaluation. Appl. Mech. Mater. **886**, 221–226 (2019)
18. Tang, R., Lu, Y., Liu, L., Mou, L., Vechtomova, O., Lin, J.: Distilling task-specific knowledge from BERT into simple neural networks. In: Computation and Language (cs.CL); Machine Learning (cs.LG) (2019)
19. Li, X., Bing, L., Zhang, W., Lam, W.: Exploring BERT for end-to-end aspect-based sentiment analysis. In: Proceedings of the 5th Workshop on Noisy User-generated Text (W-NUT 2019) (2019)
20. Munikar, M., Shakya, S., Shrestha, A.: Fine-grained sentiment classification using BERT. In: 2019 Artificial Intelligence for Transforming Business and Society (AITB), Kathmandu, Nepal, vol. 1 (2019)
21. Cach Dang, N., Moreno-Gracia, M.N., De la Prieta, F.: Sentiment analysis based on deep learning: a comparative study. Electronics **9**(3), 483 (2020)
22. Costa-jussa, M.R., Gonzalez, E., Moreno, A., Cumalat, E.: Abusive language in Spanish children and young teenager's conversations: data preparation and short text classification with contextual word embeddings. In: Proceedings of the 12th Language Resources and Evaluation Conference 2020 (2020)
23. Malik, P., Aggrawal, A., Vishwakarma, D.K.: Toxic speech detection using traditional machine learning models and BERT and fastText embedding with deep neural networks. In: 2021 5th International Conference on Computing Methodologies and Communication (ICCMC) (2021)
24. Tran, T., Nguyen, D., Nguyen, A., Golen, E.: Sentiment analysis of marijuana content via Facebook emoji-based reactions. In: 2018 IEEE International Conference on Communications (ICC), pp. 793–798 (2018)
25. LeCompte, T., Chen, J.: Sentiment analysis of tweets including emoji data. In: 2017 International Conference on Computational Science and Computational Intelligence (CSCI), USA (2017)
26. Shiha, M.O., Ayvaz, S.: The effects of emoji in sentiment analysis. Int. J. Comput. Electr. Eng. **9**, 360–369 (2017)

27. Wijeratne, S., Balasuriya, L., Sheth, A., Doran, D.: A semantics-based measure of emoji similarity. In: 2017 IEEE/WIC/ACM International Conference on Web Intelligence (WI), Germany (2017)
28. Kimura, M., Katsurai, M.: Automatic construction of an emoji sentiment lexicon. In: Proceedings of the 2017 IEEE/ACM International Conference on Advances in Social Networks Analysis and Mining (2017)
29. Al-Azani, S., El-Alfy, E.-S.M.: Combining emojis with Arabic textual features for sentiment classification. In: 2018 9th International Conference on Information and Communication Systems (ICICS) (2018)
30. Tomhira, T., Otsuka, A., Yamashita, A., Satoh, T.: What does your tweet emotion mean? Neural emoji prediction for sentiment analysis. In: Proceedings of the 20th International Conference on Information Integration and Web-Based Applications & Services, pp. 289–269 (2018)
31. Chen, Y., Yuan, J., You, Q., Luo, J.: Twitter sentiment analysis via bi-sense emoji embedding and attention-based LSTM. In: Proceedings of the 26th ACM International Conference on Multimedia (2018)
32. Zheng, J., Wang, J., Ren, Y., Yang, Z.: Chinese sentiment analysis of online education and internet buzzwords based on BERT. J. Phys Conf. Ser. **1631**, 012034 (2020)
33. Kottursamy, K.: A review on finding efficient approach to detect customer emotion analysis using deep learning analysis. J. Trends Comput. Sci. Smart Technol. **3**(2), 95–113 (2021)
34. Pandian, A.P.: Performance evaluation and comparison using deep learning techniques in sentiment analysis. J. Soft Comput. Paradigm (JSCP) **3**(02), 123–134 (2021)
35. Ganesan, T., Anuradha, S., Harika, A., Nikitha, N., Nalajala, S.: Analyzing social media data for better understanding students' learning experiences. In: Hemanth, J., Bestak, R., Chen, J.I.-Z. (eds.) Intelligent Data Communication Technologies and Internet of Things. LNDECT, vol. 57, pp. 523–533. Springer, Singapore (2021). https://doi.org/10.1007/978-981-15-9509-7_43
36. Tripathi, M.: Sentiment analysis of Nepali COVID19 tweets using NB, SVM AND LSTM. J. Artif. Intell. **3**(03), 151–168 (2021)
37. Sungheetha, A., Sharma, R.: Transcapsule model for sentiment classification. J. Artif. Intell. **2**(03), 163–169 (2020)
38. Ghosh, M., Gupta, K., Susan, S.: Aspect-based unsupervised negative sentiment analysis. In: Hemanth, J., Bestak, R., Chen, J.I.-Z. (eds.) Intelligent Data Communication Technologies and Internet of Things. Lecture Notes on Data Engineering and Communications Technologies, vol. 57, pp. 335–344. Springer, Singapore (2021). https://doi.org/10.1007/978-981-15-9509-7_29
39. Boukabous, M., Azizi, M.: A comparative study of deep learning-based language representation learning models. Indones. J. Electr. Eng. Comput. Sci. **22**(2), 1032 (2021)
40. Tomihira, T., Otsuka, A., Yamashita, A., Satoh, T.: Multilingual emoji prediction using BERT for sentiment analysis. Int. J. Web Inf. Syst. **16**(3), 265–280 (2020)
41. Emon, E.A, Rahman, S., Banarjee, J., Das, A.K., Mittra, T.: A deep learning approach to detect abusive Bengali text. In: 2019 7th International Conference on Smart Computing & Communications (ICSCC), Malaysia, pp. 1–5 (2019)
42. Lucky, E.A.E., Sany, M.M.H., Keya, M., Khushbu, S.A. Noori, S.R.H.: An attention on sentiment analysis of child abusive public comments towards Bangla text and ML. In: 2021 12th International Conference on Computing Communication and Networking Technologies (ICCCNT), pp. 1–6 (2021). https://doi.org/10.1109/ICCCNT51525.2021.9580154
43. Mahmud, M.S., Jaman Bonny, A., Saha, U., Jahan, M., Tuna, Z.F., Al Marouf, A.: Sentiment analysis from user-generated reviews of ride-sharing mobile applications. In: 2022 6th International Conference on Computing Methodologies and Communication (ICCMC), pp. 738–744 (2022). https://doi.org/10.1109/ICCMC53470.2022.9753947

Real Time Classification of Fruits and Vegetables Deployed on Low Power Embedded Devices Using Tiny ML

Vivek Gutti and R. Karthi(✉)

Department of Computer Science and Engineering, Amrita School of Engineering, Amrita Vishwa Vidyapeetham, Coimbatore, India
cb.en.p2aid20021@cb.students.amrita.edu, r_karthi@cb.amrita.edu

Abstract. The appearance of fruits and vegetables has a significant impact on their market value and consumer preference. Manual identification and sorting are time-consuming and costly. Many machine learning and deep learning models have been proposed in the past for fruit categorization which were deployed on GPUs. In this work, deep learning models for fruit classification are built on GPU and tested to be deployed on the ESP-32 microcontroller. This paper examines machine learning approaches based on CNN, Mobile net models for low cost and low power embedded devices to detect fruits and vegetables. Fruits 360 data set is used in this study where models are built to classify fruits into multi classes. The MobileNetV1 model fared well in distinguishing fruits and vegetables for 17 classes and achieved 94% testing accuracy on a google colab (GPU) environment. The model was deployed onto the ESP32 Cam environment and tested for its performance. The accuracy in prediction classes by the model on ESP32 was 77% with an inferencing speed of 51 ms and memory usage of 66.1 kb. Experimental results show that a low-cost ESP32 microcontroller can be deployed for agricultural product classification.

Keywords: Arduino · ESP32 Cam microcontroller · Convolution neural networks · Transfer learning · Tiny machine learning

1 Introduction

Because of recent hardware developments, complex machine learning algorithms can now be executed on low-power embedded devices such as Arduino Nano BLE 33 Sense, ESP 32 Cam, Raspberry Pi Pico, and others. Embedded machine learning is used in many industrial applications, including agriculture, healthcare, and ocean life conservation. Tiny ML is one such framework, which allows complex neural network models to be executed with only a few milliwatts of power. The Tiny ML framework is used to create machine learning applications for forecasting, identification and classification, recognition, and control on edge devices [1]. Traditionally machine learning models run on servers with a lot of processing and storage power [2]. The demand to deploy these systems at the edge has grown with the rise of IoT and edge computing [3]. Because of their high computational and memory footprint, deep neural networks are difficult

J. I.-Z. Chen et al. (Eds.): ICIPCN 2022, LNNS 514, pp. 347–359, 2022.
https://doi.org/10.1007/978-3-031-12413-6_27

to execute at the edge. Image classification is difficult at the edge and deployment is in high demand because many apps rely on it. In predictive maintenance of machinery, monitoring of machine health, security system [4] IOT-based solutions are proposed. They are also used In the agricultural sphere, edge ml helps farmers and vendors sort and classify fruits and vegetables at a low cost, and monitors the health of plants [5].

Low-power MicroControllers [MC] hardware like the Arduino Nano BLE 33 Sense, ESP32 Cam, Raspberry Pi Pico, and other embedded controllers can be utilized to build machine learning applications as edge devices. Arduino Nano BLE 33 Sense is a 64 MHz 32-bit ARM Cortex-M4F microcontroller with 1 MB of programmed memory and 256 KB RAM. This microcontroller has built-in Bluetooth and adequate processing capability to run TinyML models. Brightness, proximity, color, gesture, motion, vibration, temperature, humidity, pressure sensors, and a microphone are all included in the Arduino Nano 33 BLE Sense. The ESP32 Cam is a low-power microcontroller with integrated Wi-Fi and a dual-mode Bluetooth module [6]. It has an ESP32-S module, 520 KB of SRAM, and a clock speed of 160 MHz.

In this research, Tiny ML approach is used for object identification using camera-based image processing techniques to classify fruits and vegetables using a low power ESP-32 MC deployed with a transfer learning model [7]. The Mobile Net model, which runs on an ESP32 MC is designed to automatically classify agricultural products. Manually recognizing fruits and vegetables would take significant time and effort, so ML-based solutions using low-cost devices are tested for their performance. ESP 32 MC is deployed with a transfer learning model to accurately classify numerous fruits and vegetables at a low cost. The prototype built can be used by vegetable, fruit retailers, and farmers to identify types of fruits and vegetables during billing and access food quality in packaging, export, and marketing applications.

2 Literature Survey

Currently, there are many fruits and vegetables classification techniques but all the past works are deployed on cloud, local computers, or microcontrollers like raspberry pi with at least 8 GB RAM. We proposed a classification system to identify fruits and vegetables which is deployed on a low-power microcontroller which is 520 Kb SRAM. Rajasekar et al. proposed a fruit identification system using an RGB sensor on Arduino Nano BLE 33 Sense where they classified banana, apple and orange using the color sensor. They used tflite based logistic regression model which gave an accuracy of 93.11%. The board has a proximity sensor and color sensors that read the RGB values of the images that have been collected and used for the experiment to classify whether it is banana, apple, or orange. Shariff et al. proposed an application for categorizing fruits and a quality maintenance system based on Raspberry pi and Convolutional Neural Network (CNN) models. Authors choose banana, apple, papaya, dates, grapes, and lemon which are classified based on the quality of the individual fruit whether it is diseased or whether it is good. The performance was tested using Inception v3 and got an accuracy of 91%. Tripathi et al. reported a survey on computer vision solutions for horticulture products to identify and classify fruits and vegetables, disease identification in fruits, and quality assessment of products. The survey concluded that the SVM classifier has given the highest accuracy for most of the experiments. Hossain et al. proposed a framework that is built on top of two separate deep learning architectures. The first is a six-layer convolutional neural network light model, and the second is a finetuned pre-trained deep learning model with visual geometry. The model was trained on GPU and is not deployed on MC. Jose et al. proposed a lightweight CNN for classifying fruits to speed up the checkout process in businesses. Dataset was created by taking images of fruits by placing them outside on steel tables and inside a plastic cover. Different models were tried by fine-tuning the CNN architecture to improve the accuracy by adding extra features such as histogram and image centroids. The model was deployed and tested on GPU. Nikitha et al. proposed deep learning models to detect diseases in fruits and vegetables and grade them using the inception v3 model for three products. From the above survey, we infer that for the real-time classification of fruits and vegetables, few models have been tested on real microcontroller platforms.

3 Design of Proposed System

3.1 ESP 32 Cam

The ESP32-CAM is a microcontroller with a built-in camera that can be used for monitoring, recognition, video, and image capture supported by Arduino programming. It comes with a small secure digital memory card and 2 MP OV2640 camera, multiple GPIOs for attaching peripherals, and the ability to save images acquired by the camera. The reason for opting for ESP32 Cam is, that it is a widely popular microcontroller and many applications currently require affordable boards that are integrated with the camera. ESP32 Chip is enhanced such that it can even perform image processing. Its specifications include a RAM of 520 Kb and a built-in flash memory of 4 MB that supports the interface of UART and PWM. It has an inbuilt Wifi module and Bluetooth module with BLE for transferring and connecting devices.

3.1.1 Interfacing of Microcontroller ESP32 Cam with FTDI Module

FTDI Module

To upload the code to the microcontroller an external USB adapter is required and an FTDI Module is used. It is commonly used to communicate with PC and microcontroller development boards that don't have USB ports. It can run on 3.3 or 5 V DC and contains Tx/Rx and other breakout points. After connecting the ESP32 Cam with FTDI Module the FTDI module is connected to the PC. The pin diagram and interfacing are shown in Fig. 1.

Fig. 1. (a): Pin diagram and (b): interfacing ESP32-CAM - FTDI module

3.1.2 Converting the Model to TFlite

After building the model in GPU the model cannot be uploaded directly to the microcontroller because of its memory constraints. Firstly the developed model needs to be converted into a TFlite [14] model and then it should be encoded into a byte array format in a C++ header file with an extension of .h. The header file should be added with a few other libraries to initialize several variables in the Arduino code. The model, interpreter which is responsible for predictions should be initialized, and tensors which hold the model's input and output should be allocated. The interpreter should invoke the tensor to make inferences.

After connecting the ESP32 cam to the PC, the model.h file which will be generated after converting the trained model to tflite is added to the Arduino web editor. The WIFI SSID and password are added to the Arduino web editor code and compiled. The wiring connection from GND and IO0 are set as shown in Fig. 1(a) and the code is deployed on ESP 32 MCU. After upload, the wiring connection from GND and IO0 is removed and the RST button of the ESP32 camera has to be reset. When the serial monitor is opened IP address of the device will be displayed. The IP address can be used in any of the web browsers to navigate and access the page and use it for the classification of fruits and vegetables.

3.2 Building a Classifier Model

3.2.1 Data Collection

The fruit-360 dataset [15] is one of the most conventional datasets, with 90,460 images of 131 different fruits and vegetables. Objects were placed in the shaft of a low-speed motor with 3 rpm, a 21-s video documentary was recorded and the orientation of all angles of the fruits and vegetables are captured The images were captured using a Logitech C920 camera. A white paper was used as a background for the images. The background was even changed by using various lighting conditions, The dataset was separated into train and test sets for further processing and all the images are of size 100 × 100 pixels (Fig. 2).

For this experiment, 17 fruits and vegetables are selected from this dataset. They are apple, avocado, banana, blueberry, cauliflower, corn, eggplant, guava, grapes, ginger, lemon, onion, orange, papaya, pomegranate, tomato, and watermelon. The reason behind choosing these fruits and vegetables is due to their availability in the local area. The model

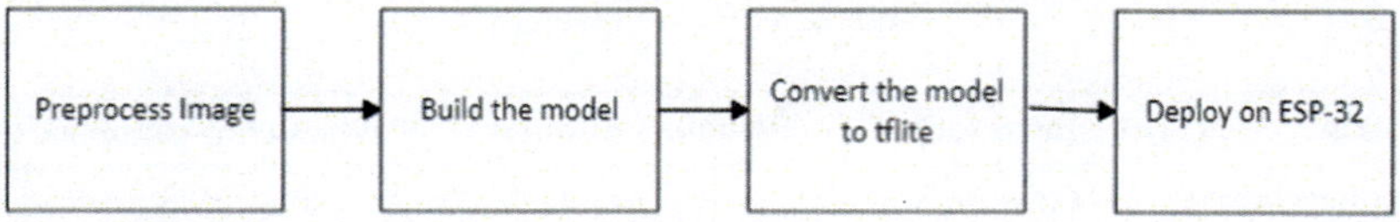

Fig. 2. Block diagram of the proposed work

needs to be deployed into a microcontroller and as the memory is constrained the number of classes is limited to 17. Experiments were conducted with 7 classes, where the selected items are Apple, Banana, Blueberry, Papaya, Guava, Orange, and Watermelon. Table 1 shows the type of fruits/vegetables and the count of images available in the dataset.

Table 1. 17 Classes and the count of images

Fruit/vegetable	Apple	Avocado	Banana	Blueberry	Cauliflower	Corn	Egg plant	Ginger	Grapes
Count	656	570	656	616	936	600	624	398	656
Fruit/vegetable	Lemon	Onion	Orange	Papaya	Pomegranate	Tomato	Water melon	Guava	
Count	656	600	639	658	658	897	632	656	

3.2.2 Data Pre-processing

Images from the repository are 100 pixels wide and they were resized to 96 × 96 pixels. Data augmentation is done on each of the classes to increase the size of the dataset. Data augmentation is done by varying the following parameters. Increase the zoom range to 0.2 and the sheer range to 0.3, flip the images horizontally, and retain the brightness range between 0.5 and 1.5. After pre-processing, a 96 × 96 pixel RGB image is used as input for the model.

3.2.3 Models Development

In this paper, CNN and transfer learning-based Mobile Net Models were used to build the classifier model.

a. CNN Architecture

The CNN architecture used for fruit classification is outlined below. The input to the model is an RGB image with a size of 96 × 96 pixels. Three convolution layers are used with the Relu activation function and are followed by a max-pooling layer. Next, a drop-out layer followed by a flattening layer is used to convert the output to a 1D array. The data is passed to a dense layer and finally, the softmax activation function is used to classify the fruit. Table 2 shows the details of CNN architecture.

The number of convolutional layer parameters can be calculated as ((kernel height * kernel width * depth) + 1) * number of filters. The first layer is a convolutional layer, with 32 filters with a kernel size of (3, 3) and a stride of 1. The filters will

Table 2. CNN architecture

Layer name	Input shape	Number of filters	Number of parameters
Convolution layer1	$96 \times 96 \times 3$	32	896
Maxpooling layer1	$48 \times 48 \times 3$	32	0
Convolution layer2	$48 \times 48 \times 3$	32	9248
Maxpooling layer2	$24 \times 24 \times 3$	32	0
Convolution layer3	$24 \times 24 \times 3$	64	18496
Maxpooling layer3	$12 \times 12 \times 3$	64	0
Dropout layer	$12 \times 12 \times 3$	64	0
Flatten layer	9216×1	0	0
Dense layer1	64×1	0	589888
Dense layer2	17×1	0	1105

create new feature maps by using the stride in all the directions. Convolutional layer 1 has $(3 * 3 * 3 + 1) * 32 = 896$ parameters. A max-pooling layer with (2, 2) pooling size is given after the convolution layer 1. The maximum value from each 2×2 block is chosen. A convolution Layer 2 with $(3 * 3 * 32 + 1) * 48 = 9248$ parameters is given to the network followed by another max pooling layer with (2, 2). Convolutional layer 3 is followed with 64 filters, kernel size is (3, 3) and stride of 1. Total number of parameters in this convolution layer 3 is $(3 * 3 * 32 + 1) * 64 = 18496$ parameters. Convolution Layer3 is followed by a max-pooling layer with (2, 2) pooling size and a dropout layer is added so that the network flows only on the few connections. Convolutional layer 3 is flattened into a single vector and passed as input to a dense layer with 9216 units. This acts as a layer with Rectified Linear Unit (ReLU) as an activation function. Total number of parameters in this layer is $(9216 * 64 + 1) * 1 = 589888$ parameters. This is followed by a final output layer with a softmax activation function to convert CNN output to the probabilities of each class. The final output layer is a SoftMax layer with 17 outputs representing the probabilities for each of the classes.

After training the CNN model with default parameters it is important to tune the hyperparameters to get a generalized model without overfitting and underfitting the data. As there are limited memory constraints in ESP32 cam it is important to get good accuracy on the device. Tuning is done and the final parameters selected are specified in Table 3, where the classifier produced an accuracy of 99% with 30 Epochs.

Table 3. Hyperparameters of CNN model

Loss function	Optimizer	Learning rate	Epochs	Batch size
Categorical Cross-entropy	Adam	0.0005	30	32

b. Transfer Learning

The technique of training and predicting a new dataset using a pre-trained model learned on a previous dataset is known as transfer learning [16]. VGG16, VGG19, Google Net, Mobile Net, Residual Network, and others are examples of large architectural models that are trained on large datasets. Mobile net models [17] are selected for implementation due to the size constraints as these models need to be deployed in ESP-32 CAM.

3.2.4 TF Lite and TF Lite Interpreter

The tensor flow model is optimized to reduce the size by using the TF Lite converter. A smaller model will use less memory on the MCU platform. Quantization is one such optimization technique that optimizes the weights and biases of a neural network model. The weights and biases are stored as 32-bit floating-point integers. To fit it into the ESP controller, these numbers are quantized to 8-bit values to reduce their precision. Although this reduces the size of the model, it may also impair its accuracy. On the MCU, a customized Tensor Flow Lite model is executed using space-saving methods. To create the lite mode, the Tensor Flow Lite library is installed in the Arduino IDE, and the Tensor Flow Lite components needed must be included in the project's sketch file. Before the interpreter can be set up, two stages must be fulfilled. Edge Impulse Software [18] is also used to train and convert the models on low power microcontrollers which leverages the Tensor flow Framework and was used to convert the model to tflite.

3.2.5 Image Capture in Real-Time and Inference

Once the fruit is placed in front of ESP-32 cam, click on the capture image and click on the run inference button. The inference output is shown as probability values for the image to belong to a specified class. The image is assigned to the class with maximum probability.

4 Experimental Setup

To perform experimentation, three classifier models CNN, Mobile Net V1, Mobile Net V2 0.5 alpha are built. The classifier models are built for two platforms: GPU and MCU. First, the classifier model is built on a Google Co-Lab environment using GPU. Next, the built model is converted to a TF-Lite (TensorFlow-Lite) model for it to be deployed on ESP32 MCU. Analysis has been done using 17 fruits and vegetables with a total of 11108 images. Another experiment was done with 7 fruits with 4513 images to understand the performance with a lesser number of classes, where the data was split as 80-20 Train-Test split for individual classes.

4.1 Analysis of the Classifier with 17 Classes

To evaluate the performance of the model accuracy, precision, recall, and F1-score are used. The 17 categories of fruits and vegetables classified with performance metrics are listed in Table 4. The comparison of models on GPU and MCU is discussed below.

Table 4. Performance metrics of models on GPU

Classifier	Training accuracy	Validation accuracy	Average precision	Average recall	Average F1-score
CNN	87.69	84.76	93	92	92
Mobile Net V1 (α = 0.1)	99.21	94.65	84	82	77
Mobile Net V2 (α = 0.05)	94.88	99.25	86	87	87
Mobile Net V2 (α = 0.25)	98.05	95.34	83	81	82
Mobile Net V2 (α = 0.35)	98.10	99.05	99	99	99

By looking at the precision, recall and F1 score of the models, apple, banana, oranges, pomegranate, and watermelon were classified correctly by CNN, MobileNetV1 and Mobile Net V2 models. Cherry, tomato, eggplants and grapes, and Onions are having less F1 score with less than 70%. Cherry is misclassified as apple and black grapes are being misclassified as Eggplants.

Figure 3(a), (b) and (c) show precision, recall and F1 score of CNN, Mobile Net V1 (0.1) and Mobile net (0.05) model on when tested on colab environment.

Fig. 3. **(a):** CNN **(b):** Mobile Net V1 **(c):** Mobile Net V2 Model for 17 classes

After building the mobile net models, the model is tuned by varying the alpha to get good accuracy for classifying the fruits and vegetables. On the GPU platform compared to all models, Mobile Net V2 0.35 alpha model was able to classify products correctly with a training accuracy of 98.10% and testing accuracy of 99.05%.

To deploy the above-built classifier on MCU, the model is converted to a TF-Lite model. The tflite model is deployed on the MCU. The models are compared according to the most important metrics of TinyML requirements which are accuracy, inferencing time, peak RAM usage, and flash usage. These metrics help to understand which model is more suitable to be used on the deployment. The performance of the TFLite models is shown in Table 5.

After converting the models to TF lite, the model is deployed onto the ESP32 cam. CNN Model gave good accuracy compared to Mobile Net models but the RAM usage and Flash usage exceed the limit of the board which is 4 MB. Compared to all the model's

Table 5. Performance metrics of models after converting to TF Lite for MCU

Model	Accuracy	Inferencing time (ms)	Peak RAM usage (KB)	Flash usage (MB)
MobileNetV2 (α = 0.25)	91.1	321	130.7	311.9
MobileNet V2 (α = 0.35)	88	859	346.8	576.65
MobileNet V1 (α = 0.1)	**77**	**51**	**66.1**	**193**
CNN model	93	2739	369.3	616.1
MobileNet V2 (α = 0.05)	84	350	283.7	163

inferencing time of Mobile Net V1 with alpha as 0.1, has an inferencing time of 51 ms using only 66.1 Kb of RAM and giving an accuracy of 77%.

4.2 Analysis of Classifier with 7 Classes

Another experiment was conducted by reducing the number of classes from 17 to 7 to check the model performance. Fruits used for this experiment are Apple, Banana, Blueberry, Orange, Papaya, Guava and Watermelon. Table 6 shows the type of fruits/vegetables and the count of images considered for analysis.

Table 6. 7 class and count of images

Fruit/vegetable	Apple	Banana	Blueberry	Orange	Papaya	Watermelon	Guava
Count	656	656	616	639	658	632	656

The performance of the classifier when modeled on GPU platform is shown below. Table 7 shows the metrics obtained by the models when deployed in GPU platform.

Compared to the 17 class the 7-class model gives higher precision, recall and F1 score for all the classes.

Figure 4(a), (b) and (c) shows the precision, recall, and F1 score of CNN, mobilenet V1 and Mobile Net v2 models on Google colab. After building the mobile net models, the model is tuned by varying the alpha to get good accuracy for classifying the fruits and vegetables. On the GPU environment, Mobile Net V2 0.05 alpha model was able to classify correctly with a training accuracy of 99.32% and testing accuracy of 99.55%. To deploy the above-built classifier to MCU, the model is converted to a TF-Lite model. The tflite model is deployed on the MCU. The models are compared using accuracy, inferencing time, peak RAM usage, and flash usage. The performance of the TFLite models is shown in Table 8.

Table 7. Model performance on GPU

Classifier	Training accuracy	Validation accuracy	Average precision	Average recall	Average F1-score
CNN	99.55	98.76	96	97	96
Mobile Net V1 (α = 0.1)	99.21	98.25	97	96	97
Mobile Net V2 (α = 0.05)	96.32	99.55	96	97	97
Mobile Net V2 (α = 0.25)	98.25	98.34	98	98	98
Mobile Net V2 (α = 0.35)	99.10	99.35	99	99	99

Fig. 4. **(a):** CNN **(b):** Mobile Net V1 **(c):** Mobile Net V2 Model for 7 classes

Table 8. Model performance after converting to TF Lite for MCU

Model	Accuracy	Inferencing time (ms)	Peak RAM usage (KB)	Flash usage (MB)
MobileNetV2 (α = 0.25)	100%	250	283.7	163
MobileNetV2 (α = 0.35)	96.4%	859	346.8	576.5
MobileNet V2 (α = 0.1)	100%	291	293.2	212.9
CNN model	99.9%	202	106.3	225.9
MobileNetV1 (α = 0.1)	**98.1%**	**51**	**66.1**	**193**

All the models gave good accuracy, but CNN and Mobile Net V2 (alpha 0.25 and alpha 0.35) models did not fit into ESP 32 cam as the RAM usage and flash usage exceed the limit of the board which is 4 MB. Compared to all the models the inferencing time Mobile Net V1 with alpha as 0.1 is 51 ms using only 66.1 Kb of RAM and giving an accuracy of 98.1%. The trained model detects different types of fruits and vegetables in low-powered and low memory Microcontroller environments.

4.3 Real Time Inferencing from ESP32

Below are the outputs of the experiment of the 17 class. MobileNet V1 0.1 was deployed on the ESP32 cam and tested with sample images. A real banana has been placed near the ESP32 Cam and the Mobile Net V1 0.1 model has classified it correctly with the probability of 91%. Figure 5 shows the parameter setting and the output probability value inferred by the model for each of the classes. To test the performance of the 7-class model, the watermelon image is considered and the model has classified it correctly with a probability of 94%. Figure 6 shows the parameter setting and the output probability value inferred by the model for each of the classes.

Fig. 5. Banana classification using17 class model

Fig. 6. Watermelon classification using 7 class model

5 Conclusion

Tensor flow Lite appears to be an excellent open-source platform for implementing machine learning algorithms on the Embedded board. ESP32 MCU is used as an inferencing device to classify fruits and vegetables. The Mobile Net V1 model is built and deployed onto the ESP32 Cam microcontroller board which is a low-power and low cost microcontroller. Seventeen different fruits and vegetables were used to train the model. The Mobile Net V1 model with alpha 0.1 trained on the CPU had a test accuracy of 94%, and when deployed on the ESP32 cam board, it had an accuracy of 77%. with an inference speed of 51 ms. Another experiment has been done on 7 classes to see the performance of model. The mobile net v1 model with alpha 0.1 gave a test accuracy of 99% when trained on GPU and after deploying it on the microcontroller it gave an accuracy of 98%. As the size of the class is high, a larger model would be required to identify the classes. As the utilized MCU has a memory constraint issue for processing large data in the real world this can be enhanced and tested using strong, powerful, and advanced micro-controllers.

References

1. David, R., et al.: TensorFlow lite micro: embedded machine learning for TinyML systems. Proc. Mach. Learn. Syst. **3**, 800–811 (2021)

2. Thanga Manickam, M., Karthik Rao, M., Barath, K., Shree Vijay, S., Karthi, R.: Convolutional neural network for land cover classification and mapping using landsat images. In: Saini, H.S., Rishi Sayal, A., Govardhan, R.B. (eds.) Innovations in Computer Science and Engineering: Proceedings of the Ninth ICICSE, 2021, pp. 221–232. Springer, Singapore (2022). https://doi.org/10.1007/978-981-16-8987-1_24
3. Dai, W., Nishi, H., Vyatkin, V., Huang, V., Shi, Y., Guan, X.: Industrial edge computing: enabling embedded intelligence. IEEE Ind. Electron. Mag. **13**(4), 48–56 (2019)
4. Ankitdeshpandey, Karthi, R.: Development of intrusion detection system using deep learning for classifying attacks in power systems. In: Pant, M., Kumar Sharma, T., Arya, R., Sahana, B., Zolfagharinia, H. (eds.) Soft Computing: Theories and Applications. Advances in Intelligent Systems and Computing, vol. 1154, pp. 755–766. Springer, Singapore (2020). https://doi.org/10.1007/978-981-15-4032-5_68
5. Gajjar, R., Gajjar, N., Thakor, V.J., Patel, N.P., Ruparelia, S.: Real-time detection and identification of plant leaf diseases using convolutional neural networks on an embedded platform. Vis. Comput., 1–16 (2021). https://doi.org/10.1007/s00371-021-02164-9
6. Dokic, K.: Microcontrollers on the edge – is ESP32 with camera ready for machine learning? In: El Moataz, A., Mammass, D., Mansouri, A., Nouboud, F. (eds.) Image and Signal Processing: 9th International Conference, ICISP 2020, Marrakesh, Morocco, June 4–6, 2020, Proceedings, pp. 213–220. Springer, Cham (2020). https://doi.org/10.1007/978-3-030-51935-3_23
7. Chung, D.T.P., Van Tai, D.: A fruits recognition system based on a modern deep learning technique. J. Phys. Conf. Ser. **1327**(1), 012050 (2019)
8. Rajasekar, L., Ganesh Babu, C., Sharmila, D.: Identification of fruits and vegetables using embedded sensor. IOP Conf. Ser. Mater. Sci. Eng. **1084**(1), 012095 (2021)
9. Shariff, S.U., et al.: Fruit categorization and disease detection using ML. Int. J. Sci. Technol. Res. **9**(11), 219–227 (2020)
10. Tripathi, M.K., Maktedar, D.D.: A role of computer vision in fruits and vegetables among various horticulture products of agriculture fields: a survey. Inf. Process. Agric. **7**(2), 183–203 (2020)
11. Hossain, M.S., Al-Hammadi, M., Muhammad, G.: Automatic fruit classification using deep learning for industrial applications. IEEE Trans. Industr. Inf. **15**(2), 1027–1034 (2018)
12. Rojas-Aranda, J.L., Nunez-Varela, J.I., Cuevas-Tello, J.C., Rangel-Ramirez, G.: Fruit classification for retail stores using deep learning. In: Mora, K.M.F., Marín, J.A., Cerda, J., Carrasco-Ochoa, J.A., Martínez-Trinidad, J.F., Olvera-López, J.A. (eds.) Pattern Recognition: 12th Mexican Conference, MCPR 2020, Morelia, Mexico, June 24–27, 2020, Proceedings, pp. 3–13. Springer, Cham (2020). https://doi.org/10.1007/978-3-030-49076-8_1
13. Nikhitha, M., Sri, S.R., Maheswari, B.U.: Fruit recognition and grade of disease detection using Inception V3 model. In: 2019 3rd International Conference on Electronics, Communication and Aerospace Technology (ICECA), pp. 1040–1043. IEEE (June 2019)
14. Warden, P., Situnayake, D.: TinyML: Machine Learning with TensorFlow Lite on Arduino and Ultra-Low-Power Microcontrollers. O'Reilly Media (2019)
15. https://www.kaggle.com/datasets/moltean/fruits
16. Aswathi, T., Swapna, T.R., Padmavathi, S.: Transfer learning approach for grading of diabetic retinopathy. J. Phys. Conf. Ser. **1767**(1), 012033 (2021)
17. Sinha, D., El-Sharkawy, M.: Thin MobileNet: an enhanced MobileNet architecture. In: 2019 IEEE 10th Annual Ubiquitous Computing, Electronics & Mobile Communication Conference (UEMCON), pp. 0280–0285. IEEE (October 2019)
18. https://www.edgeimpulse.com

A Deep Neural Networks-Based Food Recognition Approach for Hypertension Triggering Food

Papon Sarker(✉), Shaikh Hasibul Islam, Khadiza Akter, Lamia Rukhsara, and Rashidul Hasan Hridoy

Department of Computer Science and Engineering, Daffodil International University, Dhaka, Bangladesh
{papon15-11580,shaikh15-11444,khadiza15-11588,lamia.cse, rashidul15-8596}@diu.edu.bd

Abstract. In today's world, high blood pressure has become a major issue for many people. The prevalence of this hypertension is increasing as a result of the consumption of certain risky foods. The purpose of our research is to classify the images of high-risk foods for hypertension patients to improve their quality of life by removing these hazardous foods from their daily eating. A dataset of 40995 food images from 15 distinct classes is used to generalize deep learning models using the transfer learning technique to fine-tune the used pre-trained models. In this case, MobileNetV2 achieved an accuracy of 95.84% across 2094 test images. Whereas Xception and VGG19 achieved accuracy rates of 81.31% and 89.97%, respectively. In comparison to other algorithms, MobileNetV2 has produced better results in less time and less misclassification. The results of the experiments illustrate that the proposed structure is capable of correctly classifying risky food images.

Keywords: Hypertension · Food classification · Deep neural networks · Transfer learning · Xception · VGG19 · MobileNetV2

1 Introduction

Blood pressure that is higher than usual is referred to as high blood pressure or hypertension. It is a critical health issue that raises the chance of health disorders like kidney, heart, and brain. According to WHO, high blood pressure is the world's leading cause of death, affecting one out of every five women and one out of every four men, totaling more than a billion people. Hypertension is most common in struggling nations, with two-thirds of instances occurring during the last few decades. The difference between primary (essential) and secondary hypertension is arbitrary. A long-term increase in blood pressure induced by several hereditary and environmental variables is known as

J. I.-Z. Chen et al. (Eds.): ICIPCN 2022, LNNS 514, pp. 360–373, 2022.
https://doi.org/10.1007/978-3-031-12413-6_28

essential hypertension. Regardless of the type of blood pressure monitoring or the diagnostic criteria utilized, its prevalence grows with age. High blood pressure produced by another medical disease is referred to as secondary hypertension. Dietary variables have a significant influence on the prevention of both direct and indirect hypertension. Avoiding high-blood-pressure items in one's regular diet reduce the risk of hypertension, which is important for reducing the severity of high blood pressure and enhancing a sufferer's quality of life. In this work, 15 hypertension triggering food items were chosen to create a dataset for the classification, including pizza, sandwiches, crackers, alcohol, chocolate bar, cake, red meat, salt, bacon, pickled, cheese, sausage, yogurt, ketchup, tacos and burritos. Fat, salt, and high calories food are responsible for hypertension. All those foods are higher in fat, calories, and salt. Sodium is found in pizza, sandwiches, and canned soup, a credible source of cardiovascular disease and stroke disease which can raise blood pressure also. Many people unknowingly consume too much sodium. Red meat and eggs contain arachidonic acid, and polyunsaturated lipid. The body's metabolization of red meat may also generate substances that raise blood pressure even further. Processed meats such as sausage and bacon, on the other hand, are also linked to high blood pressure. Excessive alcohol consumption can also raise blood pressure. High blood pressure patients might considerably benefit from our hypertension food detection approach when eating at restaurants, hotels, airplanes, hospitals, retirement homes, and crippled care centers. People will be kept safe from the fear of this if they can use our rapid recognition approach to quickly spot hypertension foods. When visiting new places, this food identification technique will help people to reduce the worry that comes with eating hypertensive meals. Nonetheless, in several countries, dietary advice for hypertension prevention is ambiguous. This method may be able to assist them in lowering their blood pressure and improving their overall excellence of life. Computer vision approaches have lately shown incredible effectiveness in detecting skin disorders, leaf diseases, and a range of other conditions [17, 18]. In this study, convolutional neural networks (CNNs) are used to present a fast and accurate object classification technique for classifying hypertension-triggering foods to assist patients with high blood pressure. A hypertension triggering food (HTF) dataset has been created, and it serves as a crucial foundation for the development of CNN models like MobileNetV2, VGG19, and Xception. By supplying enough images for training, the image augmentation (IA) process was employed to restrict the overfitting of our CNN models. 34857 training images, 4094 validation, and 2044 testing images make up the used food dataset. All CNN models in this work were created using the transfer learning (TL) approach. During the training, Xception took a longer time than the others; on the contrary, VGG19 had the shortest training time for each epoch but took longer to recognize unseen new images. In this research, we have found that MobileNetV2 is the most accurate of the three models, with a training accuracy of 96.94%. MobileNetV2 exceeds all other models in terms of validation and testing accuracy as well, with 95.09% for validation and 95.84% for testing accuracy.

The formation of our document is outlined below. In the Sect. 2, we have outlined the associated works whereas, in the Sect. 3, we have explained our data set and CNN model. In the same way in Sect. 4, we have presented the result with an explanation and our research comes to an end in Sect. 5.

2 Literature Review

Because of excellent performance, using of deep learning methods is growing rapidly. Nowadays researchers are using deep learning in various types of research works. To classify junk-meal images, Shroff et al. originated DiaWear, a food recognition program [1]. It was divided into four sections. A vector of size, color, shape, texture, and context-based information is calculated for each food item and fed into a feedforward artificial neural network (ANN). 80.00%, 90.00%, 95.00%, and 90.00% of accuracy are obtained by them for fries, apple pies, hamburgers, and chicken nuggets, respectively. To categorize food images, Anthimopoulos et al. employed the bag-of-features (BoF) model [2]. A sophisticated technique for identification and optimization is used for 5,000 images of food of 11 categories. They stated that their system had been optimized to compute dense local characteristics. 78.00% accuracy is achieved using hierarchy k-means clustering and linear support vector machine (SVM). A smartphone-based deep learning image identification mechanism is proposed to supply diabetic patients with alternative diet recipes [3]. After doing 20000 iterations, they attained 70.00% accuracy. They stated that their precision was only second to none due to the number of iterations and wanted to work with more datasets and image classes in the future. Bossard et al. combined a substantial percentage of cuisine categories with a huge image [4]. 101000 images and 101 cuisine categories were used in their study. The random forest (RF) strategy was utilized to cluster the superpixels of the training dataset to train the component models. The average accuracy was 50.76%. To learn spatial correlations between constituents, Yang et al. proposed using pairwise features [5]. Semantic texton forests (STF) were used to get pixel-level soft labels. They deal with 61 different food item categories as well as seven main food type categories. A system for automatic meal categorization using BoF was proposed by M Kanchana et al. for a total of 4,868 images of 11 categories [6]. The BoF is also used to categorize food images and calculate carbohydrate content [7]. To extract the key point, the visual vector from an image is extracted, the researchers employed the dense STF method, k-means clustering, and SVM, and reached 88.00% accuracy. Z Shen et al. used InceptionV3, InceptionV4, and V4-101 models to classify food items. InceptionV4 gives them the highest accuracy of 91.70%. They used CNN architecture and 50,000 images of the dataset [8]. Siyuan Lu et al. have used nine fruits for food classification. 91.44% accuracy is attained by them using six-layer CNN architecture, which is better than voting-based SVM, wavelet entropy (WE), and genetic algorithm (GA), achieved 86.56%, 89.78%, and 82.33% accuracy, respectively [9]. Guoxiang Zeng got 95.60% accuracy by using the VGG model. To classify fruits and vegetables, he applied the CNN model. He used 12,173 images from a dataset that included 26 different types of fruits and vegetables and took 80.00% and 20.00% for both training and validation, respectively [10]. To diagnose leaf diseases, Jagadeesh Basavaiah used RF and Decision Tree (DT) algorithms. They work with four different kinds of leaf diseases. The RF gives the maximum accuracy of 94.00%, whereas the DT has a 90.00% accuracy. For training and testing, they used Hu Moments, color histograms, local binary, and Haralick pattern features [11]. Probabilistic neural network (PNN), fuzzy, artificial neural networks (ANN), SVM, back propagation neural network (BPNN), and *k*-nearest neighbors (*k*NN) classifiers were applied for the classification of disease [12]. They worked with segmentation, enhancement, feature extraction, background removal, and classification. Hokuto Kagaya et al. employed an SVM classifier and a CNN to categorize ten food products which obtained 89.70% and 93.8% efficiency, respectively [13]. Punnarumol Temdee et al. employed InceptionV3 to obtain 75.20% accuracy in identifying 40 food types and examined different training steps outcomes, finding that after both 8,000 and 10,000 training phases, efficiency goes 73.40% [14]. For disease categorization investigation, an adaptive neuro-fuzzy inference system (ANFIS) was used [15]. A grey-level co-occurrence matrix (GLCM) was used to calculate the feature. They claimed to have a 90.70% accuracy rate for tomato disease detection and a 98.00% accuracy rate for eggplant disease detection. Besides these, CNNs are now widely used in other several studies such as traffic forecasting and abnormal incident detection [19, 20]. Till now, no study is conducted for recognizing hypertension triggering foods.

3 Materials and Methods

3.1 Dataset and Image Augmentation

Working with an in-house dataset was our sole goal, so we built the HTF dataset of a total of 40995 images of 15 different foods, where 31367 images were built using image augmentation (IA) techniques from 9628 gathered images. IA is a method of creating fresh training data from existing data. IA was employed to combat overfitting during the training cycle of our CNNs. IA processes including rotation transformations (90 and 270 degrees), horizontal flip, and top-bottom flip were employed to create the HTF dataset. Figure 2 shows a procedure for creating images using IA as an example. We have reshaped our images according to our model demanded shape as 299 $\times$ 299, and 224 $\times$ 224 dimensions. The HTF food dataset is partitioned into three sets such as training (85%), validation (10%), and test (5%). Each class of the HTF dataset is represented in Fig. 1.

Fig. 1. Dataset sample: 1) Alcohol, 2) Bacon, 3) Cake, 4) Cheese, 5) Chocolate-Bar, 6) Crackers, 7) Ketchup, 8) Pickles, 9) Pizza, 10) Red Meat, 11) Salt, 12) Sandwich, 13) Sausage, 14) Tacos and Burritos, and 15) Yogurt

Fig. 2. Red meat sample image augmentation: 1) Original image, 2) Horizontal flip, 3) 90-degree rotation, 4) Flip top-bottom, 5) 270-degree rotation

3.2 CNN Based Models

Pre-trained CNN architectures including VGG19, MobileNetV2, and Xception were utilized to discriminate highly linguistic meals by using the transfer learning approach, and various experimental tests were conducted to evaluate the models' performance. By delivering better results and enhancing performance, transfer learning saves time. In transfer learning, the acquired knowledge during teaching a CNN model can be used to teach another CNN model. During training, a few layers of a CNN model are taught to describe the issue features. Transfer learning, on the other hand, allows the last few layers of a CNN's learned architecture to be removed and replaced with a new pile for the objective job. The VGG19, which won the ILSVRC-2014, has 19 layers [16]. It has roughly 138 million (M) parameters and can train with 224 × 224 pixels images. The design is composed of five convolutional layer blocks with three fully linked layers. After each convolution, a rectified linear unit (ReLU) activation has been performed, and in rare instances, a max-pooling process is utilized to minimize the spatial component. For the research, the VGG19 architecture's last completely linked layer is coupled to the layer SoftMax with fifteen neurons. MobileNetV2 is an object segmentation and detection system that makes optimal use of depth-wise separable convolution layers. It's a 53-layer deep CNN architecture with 3.4 M parameters and 224 × 224-pixel inputs. MobileNetV2's design includes a 32-filters fully convolutional layer with 19 bottleneck layers. ReLU is chosen as the non-linearity since it is adaptable with low-accuracy computation [17]. During training, we utilize kernel size 3 × 3, which really is typical in contemporary networks, as well as dropout and batch normalization. Francois Chollet's Xception architecture is a variation on the Inception concept [17]. A rectangular stack of depth-wise separated convolution layers is joined together by residual connections in this architecture. The depth-wise separated convolution attempts to reduce computation time and memory consumption. It's a 71-layer deep CNN architecture that accepts input photos with a resolution of 299 × 299 pixels. With the exception of the first and last modules, Xception has 36 convolutional layers distributed into 14 modules, all of which have linear residual connections. To divide the learning of channel and space-wise properties, Xception employs separable convolution.

3.3 Experiments

To train and fit pre-trained CNN models and test model's performance, we employed training and validation, and test sets of HTF. All layers of Xception, VGG19, and MobileNetV2 were configured as trainable throughout the training phase, and the final entirely connected layers were substituted with 15 outputs to recognize photos of 15 distinct hypertension-inducing meals. For activation, we employed softmax and as a loss function, categorical cross-entropy was utilized in the last layer of our used three models to assume the damage of CNN models in order that the weights could be modified to diminish the loss in the future assessments. During the training phase, Adam and SGD were utilized to reduce overfitting. A range of performance criteria, including sensitivity, specificity, accuracy, and precision, was employed to inspect the outcomes of models. The confusion matrix's True Positive (TP), True Negative (TN), False Positive (FP), and False Negative (FN) values were utilized to create the metrics shown below. The fraction of accurately detected positives concerning all true positives is referred to as sensitivity (Sen). The fraction of true negatives accurately categorized is known as specificity (Spe). The percentage of images properly categorized out of all images is referred to as accuracy (Acc). The fraction of correctly identified positives to all positive recognitions is known as precision (Pre). These metrics and their expanded computations are presented between Eq. 1–8 for multi-class classification using macro-averaging [18].

For class hi,

$$Sen(hi) = \frac{TP(hi)}{TP(hi) + FN(hi)} \tag{1}$$

$$Spe(hi) = \frac{TN(hi)}{TN(hi) + FP(hi)} \tag{2}$$

$$Acc(hi) = \frac{TP(hi) + TN(hi)}{TP(hi) + TN(hi) + FP(hi) + FN(hi)} \tag{3}$$

$$Pre(hi) = \frac{TP(hi)}{TP(hi) + FP(hi)} \tag{4}$$

$$AverageSen = \frac{1}{classes}\sum_{hi=1}^{classes} Rec(hi) \tag{5}$$

$$AverageSpe = \frac{1}{classes}\sum_{hi=1}^{classes} Spe(hi) \tag{6}$$

$$AverageAcc = \frac{1}{classes}\sum_{hi=1}^{classes} Acc(hi) \tag{7}$$

$$AveragePre = \frac{1}{classes}\sum_{hi=1}^{classes} Pre(hi) \tag{8}$$

4 Result and Discussion

The core ambition of our work is to establish a reliable food recognition method for hypertension-inducing foods using cutting-edge CNN models. Several experiments using the food dataset have been undertaken to assess the performance of Xception, VGG19, and MobileNetV2. The training, validation, and test accuracies were obtained on the HTF dataset using pre-trained CNN models with the TL technique. These accuracies of the three models we obtained are shown in Table 1.

Table 1. Training, Validation, and Test accuracies of our CNN models.

Model name	Training accuracy	Validation accuracy	Testing accuracy
Xception	83.91%	81.97%	81.31%
VGG19	90.61%	88.64%	89.97%
MobileNetV2	96.94%	95.09%	95.84%

MobileNetV2 outperformed other CNN models in this study, with training and test accuracy of 96.94% and 95.84%, respectively. The training and test accuracy of Xception is the lowest. MobileNetV2 has categorized 85 images incorrectly, whereas Xception and VGG19 have identified 382 and 205 images incorrectly, respectively. Class-wise classification performance has been studied to more clearly highlight the performance of the three employed models that were mentioned in this work. TP, TN, FP, and FN values were generated for fifteen classes of the HTF dataset using the confusion matrix. We have shown class-wise TP, TN, FP, FN values, and performance metrics for the Xception model in Table 2 below.

Table 2. TP, FP, FN, TN and performance table for Xception model

Class	TP	FP	FN	TN	Sen	Spe	Acc	Pre
Alcohol	297	29	37	1681	88.92%	98.30%	96.77%	91.10%
Bacon	183	26	21	1814	**89.71%**	98.59%	97.70%	87.56%
Cake	112	33	25	1874	81.75%	98.27%	97.16%	77.24%
Cheese	47	34	24	1939	66.20%	98.28%	97.16%	58.02%
Chocolate bar	76	26	22	1920	77.55%	98.66%	97.65%	74.51%
Cracker	110	31	24	1879	82.09%	98.38%	97.31%	78.01%
Ketchup	58	23	21	1942	73.42%	98.83%	97.85%	71.60%
Pickles	75	26	34	1909	68.81%	98.66%	97.06%	74.26%
Pizza	72	13	20	1939	78.26%	**99.33%**	**98.39%**	84.71%
Red meat	38	33	32	1941	54.29%	98.33%	96.82%	53.52%
Salt	158	15	25	1846	86.34%	99.19%	98.04%	**91.33%**
Sandwich	182	19	25	1818	87.92%	98.97%	97.85%	90.55%
Sausage	66	17	21	1940	75.86%	99.13%	98.14%	79.52%
Tacos and burritos	82	36	23	1903	78.10%	98.14%	97.11%	69.49%
Yogurt	106	21	28	1889	79.10%	98.90%	97.60%	83.46%

The VGG19 model's sensitivity, specificity, accuracy, and precision values were calculated for each class of food dataset using TP, TN, FP, and FN values obtained from the confusion matrix, those shown in Table 3. On a class-by-class basis, we emphasize each class's classification performance. In the sandwich class, VGG19 has a maximum sensitivity of 95.38%. VGG19 has the highest specificity of 99.57% in the salt class. VGG19 has the highest accuracy for several classes. The VGG19 model has the highest accuracy in the cake, sandwich, and yogurt classes. All three classes have the same accuracy of 99.02%. With 93.56%, the alcohol class is the most accurate by providing precision accuracy for the VGG19 model.

The most accurate model of our study is MobileNetV2. This model provides the best accuracy within less time and the misclassification number of this model is too low considering the other two models. In the below Fig. 3, we presented a normalized confusion matrix for the MobileNetV2 model.

Table 3. TP, FP, FN, TN and performance table for the VGG19 model

Class	TP	FP	FN	TN	Sen	Spe	Acc	Pre
Alcohol	305	21	22	1696	93.27%	98.78%	97.90%	**93.56%**
Bacon	191	18	16	1819	92.27%	99.02%	98.34%	91.39%
Cake	133	12	8	1891	94.33%	99.37%	**99.02%**	91.72%
Cheese	67	14	8	1955	89.33%	99.29%	98.92%	82.72%
Chocolate bar	91	11	28	1914	76.47%	99.43%	98.09%	89.22%
Cracker	127	14	11	1892	92.03%	99.27%	98.78%	90.07%
Ketchup	69	12	15	1948	82.14%	99.39%	98.68%	85.19%
Pickles	85	16	12	1931	87.63%	99.18%	98.63%	84.16%
Pizza	74	11	14	1945	84.09%	99.44%	98.78%	87.06%
Red meat	58	13	13	1960	81.69%	99.34%	98.73%	81.69%
Salt	165	8	14	1857	92.18%	**99.57%**	98.92%	95.38%
Sandwich	186	15	5	1838	**95.38%**	99.19%	**99.02%**	92.54%
Sausage	69	14	13	1948	84.15%	99.29%	98.68%	83.13%
Tacos and burritos	103	15	17	1909	85.83%	99.22%	98.43%	87.29%
Yogurt	116	11	9	1908	92.80%	99.43%	**99.02%**	91.34%

Fig. 3. Confusion matrix of MobileNetV2: C1) alcohol C2) bacon C3) cake C4) cheese C5) chocolate bar C6) cracker C7) ketchup C8) pickles C9) pizza C10) red meat C11) salt C12) sandwich C13) sausage C14) tacos and burritos C15) yogurt

In Table 4, we emphasize categorization performance on a class-by-class basis alone with TP, TN, FP, and FN values. MobileNetV2 has the highest sensitivity of 98.02% in the pizza class. In the pickles and alcohol classes, however, MobileNetV2 has the greatest specificity and accuracy, with 99.90% and 98.16%, respectively. With 99.61%, the pizza and sandwich classes are the most accurate for the MobileNetV2 model.

Table 4. TP, FP, FN, TN, and performance table for MobileNetV2 model

Class	TP	FP	FN	TN	Sen	Spe	Acc	Pre
Alcohol	320	6	7	1711	97.86%	99.65%	99.36%	**98.16%**
Bacon	202	7	5	1830	97.58%	99.62%	99.41%	96.65%
Cake	140	5	5	1894	96.55%	99.74%	99.51%	96.55%
Cheese	75	6	7	1956	91.46%	99.69%	99.36%	92.59%
Chocolate bar	94	8	7	1935	92.59%	99.59%	99.27%	92.16%
Cracker	135	6	3	1900	97.83%	99.69%	99.56%	95.74%
Ketchup	73	8	6	1957	92.41%	99.59%	99.32%	90.12%
Pickles	99	2	7	1936	93.40%	**99.90%**	99.56%	98.02%
Pizza	79	6	2	1957	**98.02%**	99.69%	**99.61%**	92.94%
Red meat	63	8	6	1967	91.30%	99.59%	99.32%	88.73%
Salt	166	7	5	1866	97.08%	99.63%	99.41%	95.95%
Sandwich	197	4	4	1839	98.01%	99.78%	**99.61%**	98.01%
Sausage	78	5	7	1954	91.76%	99.74%	99.41%	93.98%
Tacos and burritos	115	3	6	1920	95.04%	99.84%	99.56%	97.46%
Yogurt	123	4	8	1909	93.89%	99.79%	99.41%	96.85%

To demonstrate how effectively deep learning models work with datasets, Table 5 displays erroneous classifications for every class of HTF datasets. For the pickles class, MobileNetV2 put up a very impressive performance. The Xception model, on the other hand, has performed the highest misclassifications in the tacos and burritos class.

In the study, from overall performance MobileNetV2 gives us better accuracy for our dataset. In Fig. 4, we have presented training and validation accuracy and loss for our used three deep learning models. Training loss is lowest for MobileNetV2 where the highest loss gain by the Xception model. This performance graph is obtained for 50 epochs.

Table 5. Misclassification number of our model per class.

Class Name	Xception	VGG19	MobileNetV2
Alcohol	29	21	6
Bacon	26	18	7
Cake	33	12	5
Cheese	34	14	6
Chocolate bar	26	11	8
Cracker	31	14	6
Ketchup	23	12	8
Pickles	26	16	**2**
Pizza	13	11	6
Red meat	33	13	8
Salt	15	8	7
Sandwich	19	15	4
Sausage	17	14	5
Tacos and burritos	36	15	3
Yogurt	21	11	4

Fig. 4. Graph of training and validation accuracy and loss: 1) MobileNetV2 accuracy, 2) MobileNetV2 loss

The result of this study was compared to those of numerous studies introduced for different types of foods classification, as shown in Table 6. In the case of identifying hypertension-triggering foods using deep learning techniques to help hypertension sufferers, our one is the first attempt.

Table 6. Comparison table for deep learning algorithms for food identification

Paper	Method	Class number	Accuracy
Shroff et al. [1]	ANN	4	95.00%
Anthimopoulos et al. [2]	BoF	11	78.00%
Kaiz Merchant et al. [3]	CNN	101	70.00%
M Kanchana et al. [6]	BoF	11	[a]*NM*
P Velvizhy et al. [7]	BoF	11	88.00%
Shen et al. [8]	InceptionV4	101	91.73%
Siyuan et al. [9]	CNN	9	91.44%
Guoxiang Zeng et al. [10]	VGG	26	95.60%
Hokuto Kagaya et al. [13]	CNN	10	93.80%
P Temdee et al. [14]	InceptionV3	40	75.20%
Our Study	**MobileNetV2**	15	**95.84%**

[a] *NM*: Not Mentioned

5 Conclusion

Hypertension is a big problem in today's society. People are curious about measuring blood pressure and eating a balanced diet to avoid hypertension. This research introduces a new approach for detecting hypertension-related foods. Using MobileNetV2, Xception, and VGG19, an enhanced computer vision-based classification system is provided for effectively identifying hypertension-related foods. Among the three models, MobileNetV2 gives the highest training, test, and validation accuracy, 96.94%, 95.84%, and 95.09%, respectively. IA techniques were used to create 31367 images from 9628 source images in a collection of fifteen distinct hypertension-triggering foods. The TL approach was applied in this research with MobileNetV2, Xception, and VGG19. MobileNetV2 has gained the best accuracy of 99.61% for the pizza and sandwich class. With 99.02% accuracy, VGG19 is the most accurate in the cake, sandwich, and yogurt classes. With a score of 98.39%, Xception has the best accuracy in the pizza class. When compared to the other two models, the Xception model has a higher misclassification number of 382. MobileNetV2, which has a high recognition rate in this research, has been shown to work well in various forms of food recognition in experiments. Our research will be beneficial to hypertensive patients. Patients would be able to protect themselves from financial damage by avoiding hypertension-related foods and maintaining good health. Our research will encourage individuals to be aware of and prevent hypertension. We intend to include our models in a mobile application. At the moment, our CNN model can recognize 15 different hypertension-related foods. We have a plan to develop a more effective food identification method, capable of recognizing more hypertension-related foods in less time, thereby assisting hypertension sufferers. We will create a lengthier dataset for future research. We'll concentrate on increasing the efficiency of our CNN model in the future.

References

1. Shroff, G., Smailagic, A., Siewiorek, D.P.: Wearable context-aware food recognition for calorie monitoring. In: 12th IEEE International Symposium on Wearable Computers, pp. 119–120. IEEE (2008)
2. Anthimopoulos, M.M., Gianola, L., Scarnato, L., Diem, P., Mougiakakou, S.G.: A food recognition system for diabetic patients based on an optimized bag-of-features model. IEEE J. Biomed. Health Inform. **18**(4), 1261–1271 (2014)
3. Merchant, K., Pande, Y.: ConvFood: a CNN-based food recognition mobile application for obese and diabetic patients. In: Shetty, N.R., Patnaik, L.M., Nagaraj, H.C., Hamsavath, P.N., Nalini, N. (eds.) Emerging Research in Computing, Information, Communication and Applications: ERCICA 2018, Volume 1, pp. 493–502. Springer, Singapore (2019). https://doi.org/10.1007/978-981-13-5953-8_41
4. Bossard, L., Guillaumin, M., Van Gool, L.: Food-101 – mining discriminative components with Random Forests. In: Fleet, D., Pajdla, T., Schiele, B., Tuytelaars, T. (eds.) Computer Vision – ECCV 2014: 13th European Conference, Zurich, Switzerland, September 6–12, 2014, Proceedings, Part VI, pp. 446–461. Springer, Cham (2014). https://doi.org/10.1007/978-3-319-10599-4_29
5. Yang, S., Chen, M., Pomerleau, D., Sukthankar, R.: Food recognition using statistics of pairwise local features. In: 2010 IEEE Computer Society Conference on Computer Vision and Pattern Recognition, pp. 2249–2256. IEEE (2010)
6. Kanchana, M., Bharath, M., Syed Jaffar, K.: Automatic food recognition system for diabetic patients. Int. J. Innov. Res. Sci. Technol. **1**, 47–51(2015)
7. Velvizhy, P., Kannan, A.: Automatic food recognition system for diabetic patients. In 2014 6th International Conference on Advanced Computing (ICoAC), pp. 329–334. IEEE (2014)
8. Shen, Z., Shehzad, A., Chen, S., Sun, H., Liu, J.: Machine learning based approach on food recognition and nutrition estimation. Procedia Comput. Sci. **174**, 448–453 (2020)
9. Lu, S., Lu, Z., Aok, S., Graham, L.: Fruit classification based on six layer convolutional neural network. In IEEE 23rd International Conference on Digital Signal Processing (DSP), pp. 1–5. IEEE (2018)
10. Zeng, G.: Fruit and vegetables classification system using image saliency and convolutional neural network. In IEEE 3rd Information Technology and Mechatronics Engineering Conference (ITOEC), pp. 613–617. IEEE (2017)
11. Basavaiah, J., Anthony, A.A.: Tomato leaf disease classification using multiple feature extraction techniques. Wirel. Pers. Commun. **115**(1), 633–651 (2020)
12. Goswami, M., Maheshwari, S., Poonia, A., Songara, D.: Taxonomy of leaf disease detection and classification. In: Sa, P.K., Bakshi, S., Hatzilygeroudis, I.K., Sahoo, M.N. (eds.) Recent Findings in Intelligent Computing Techniques. AISC, vol. 708, pp. 557–563. Springer, Singapore (2018). https://doi.org/10.1007/978-981-10-8636-6_59
13. Kagaya, H., Aizawa, K., Ogawa, M.: Food detection and recognition using convolutional neural network. In: Proceedings of the 22nd ACM International Conference on Multimedia, MM 2014, pp. 1085–1088. Association for Computing Machinery, New York, NY, USA (2014). https://doi.org/10.1145/2647868.2654970
14. Temdee, P., Uttama, S.: Food recognition on smartphone using transfer learning of convolution neural network. In: Global Wireless Summit (GWS), Cape Town, South Africa, pp. 132–135 (2017). https://doi.org/10.1109/GWS.2017.8300490
15. Sabrol, H., Kumar, S.: Plant leaf disease detection using adaptive neuro-fuzzy classification. In: Arai, K., Kapoor, S. (eds.) CVC 2019. AISC, vol. 943, pp. 434–443. Springer, Cham (2020). https://doi.org/10.1007/978-3-030-17795-9_32

16. Simonyan, K., Zisserman, A.: Very deep convolutional networks for large-scale image recognition. arXiv preprint arXiv:1409.1556 (2014)
17. Hridoy, R.H., Akter, F., Afroz, M.: An efficient computer vision approach for rapid recognition of poisonous plants by classifying leaf ımages using transfer learning. In: 12th International Conference on Computing Communication and Networking Technologies (ICCCNT), pp. 01–07. IEEE (2021)
18. Hridoy, R.H., Rakshit, A.: BGCNN: a computer vision approach to recognize of yellow mosaic disease for black gram. In: Smys, S., Bestak, R., Palanisamy, R., Kotuliak, I. (eds.) Computer Networks and Inventive Communication Technologies. LNDECT, vol. 75, pp. 189–202. Springer, Singapore (2022). https://doi.org/10.1007/978-981-16-3728-5_14
19. Kumar, T.S.: Video based traffic forecasting using convolution neural network model and transfer learning techniques. J. Innov. Image Process. (JIIP) **2**(03), 128–134 (2020)
20. Sharma, R., Sungheetha, A.: An efficient dimension reduction based fusion of cnn and svm model for detection of abnormal incident in video surveillance. J. Soft Comput. Paradigm (JSCP) **3**(02), 55–69 (2021)

Novel 1D and 2D Convolutional Neural Networks for Facial and Speech Emotion Recognition

Pavan Nageswar Reddy Bodavarapu[1](✉), B. Gowtham Kumar Reddy[2], and P. V. V. S. Srinivas[2]

[1] CSE Department, Koneru Lakshmaiah Education Foundation, Viyayawada, India
pavannageswarreddy@gmail.com
[2] Koneru Lakshmaiah Education Foundation, Viyayawada, India

Abstract. As humans, we express most naturally via speech. Speech emotion recognition systems are defined as a set of techniques for processing and classifying speech signals in order to detect the emotions that are inherent in them. For emotion recognition in audio files, a novel 1D convolutional neural network (Audio_EmotionModel) is designed, which contains 5 1D convolutional layers, 5 max pooling layers and 3 dropout layers. The Audio_EmotionModel is used for RAVDESS dataset for emotion recognition in audio datasets. For Emotion recognition in images (obtained by splitting videos into images), a novel 2D Convolutional neural network is designed (Image_EmotionModel), which contains 4 2D convolutional layers, 2 max pooling layers, 3 dropout layers and 2 batch normalization layers. The Image_EmotionModel is used for RAVDESS dataset for emotion recognition in video datasets (videos are converted into frames). The results clearly indicate, the two proposed models perform better than various state-of-art models. Human Recognition [29] achieved only 40.9% while recognizing the emotions, clearly the proposed Audio_EmotionModel outperformed the human recognition by nearly more than 50% .

Keywords: Emotion recognition · Speech recognition · Convolutional neural network · Deep learning

1 Introduction

As humans, we express most naturally via speech. Speech emotion recognition systems are defined as a set of techniques for processing and classifying speech signals in order to detect the emotions that are inherent in them. Prosodic and spectral characteristics are the most often used features in SER systems because they enable a larger spectrum of emotion and produce superior results. Adding characteristics from other modalities, such as those that rely on visual or linguistic cues, can improve the findings even further [1]. They describe a language-individualistic emotion detection system for identifying human emotional states in voice signals in this work. For the sake of creating and evaluating the system's viability, a corpus of emotional speech from different subject speaking

J. I.-Z. Chen et al. (Eds.): ICIPCN 2022, LNNS 514, pp. 374–384, 2022.
https://doi.org/10.1007/978-3-031-12413-6_29

multiple languages has been gathered [2]. This study proposes an ensemble visual-audio emotion detection multi-tasking framework focused on combining learning with various features. To address the issue that current characteristics are unable to distinguish between distinct emotions, they extract two types of features, i.e., SVM classifiers and CNN are intended for manual features because to the variety of features [3].

A weighted linear mixture was utilised to merge both monomodal feeling assessments at the decision level. The average prediction error of the combined result was 17.6% lower than the individual error of auditory and visual emotion identification, respectively. The accuracy between emotion estimations and manual reference rose by 12.3% and 9.0% [4]. The emotional gap between feelings and audio-visual characteristics makes emotion identification difficult. Networks are a type of network. The suggested method's performance on three public audio-visual emotional databases, including the acted RML database, the acted eNTERFACE05 database, and the spontaneous BAUM-1s database, is encouraging [5]. This study offers an emotion identification system based on emotional Big Data and a deep learning methodology. A voice signal is first analysed in the frequency domain to produce a Mel-spectrogram, which may then be regarded as a picture in the proposed model. Multiple input emotional databases, one being Big Data, are used to assess the proposed solution [6].

The relevance of SER in the area of Human-Computer Interaction is reflected in its development as a focal focus of research into speech processing. This study presents a novel method to speech recognition based on a cloud model in addition to the traditional technique for detecting emotion in speech. When the RAVDESS database was used, and a 95.3 percent detection rate was attained [7]. This is to be expected, given the various benefits of utilising an adaptive and compact 1D CNN rather than a traditional deep equivalent. Aside from 1D to 2D data translation, 2D deep CNNs often require huge datasets [8].

2 Related Work

Chunjun Zheng et al. [12] In the realm of speech processing, voice emotion detection is a difficult and intensively researched issue. The feature set and model design of effective speech have a direct impact on the accuracy of speech emotion identification, thus research into features and models is crucial. Andreea Birhala et al. [13] The authors present a multimodal fusion approach for emotion identification that combines auditory and from a temporal window with visual modalities varying temporal offsets for each modality. Rory Beard et al. [14] In affective computing and human-computer interaction, emotion recognition is critical. In general, human perception of a subject's emotion is dependent on verbal and visual data acquired in the initial seconds of engagement with the subject. Natural human conversation is complex and multi-modal by nature. Esam Ghaleb et al. [15] The authors present a unique multimodal temporal deep network architecture that embeds video clips into a metric space utilising their audio-visual content, reducing their gap and exploring their complimentary and additional information.

Esam Ghaleb et al. [16] People use a variety of channels to communicate their emotions, including visual and audible ones. As a result, multimodal learning can greatly improve automatic emotion identification. However, in audio-video emotion detection, their interdependencies and relationships are not fully explored. Nick Rossenbach et al. [17] Recent developments in text-to-speech technology have resulted in the creation of multi-speaker end-to-end TTS systems that are adaptable. Author augment current attention-based automated speech recognition system with synthetic audio produced by a TTS system educated only on the ASR corpora. Houwei Cao et al. [18] author provide an audio-visual data collection that is ideal for research into multi-modal emotion expression and perception. For audio-only, visual-only, and voice files, the human recognition of intended emotion is 40.9 percent, 58.2 percent, and 63.6 percent, respectively. Visual-only perception has the greatest average intensity levels of emotion.

Amiya Kumar et al. [19] Speech is a natural way of experiencing emotions that gives detailed information about a person's various cognitive states. For speaker independent situations, the technique of employing a mixture of characteristics obtained an average accuracy rate of 82.26 percent, according to the total findings of the trials. Mohit Srivastava et al. [20] With the passage of time, human contact has spread to a variety of other disciplines, including engineering, cognition, and medicine. Speech analysis has also been a major topic of discussion. This method of contact with machines is being used by those concerned to bridge the gap between the physical and digital worlds. Zhao et al. [20] Speech emotion detection is a relatively young topic of study with a wide range of potential applications both human computer and human interaction systems. Human speech has a wide range of temporal and spectral properties that can be retrieved.

3 Proposed Work

3.1 RAVDESS Dataset

Ravdess is an audio-visual dataset, which contains both speech and song files. The total number of files in this dataset are 7356. This database is created by 24 artists (12 male and 12 females), who vocalized two statements in various emotions. The total number of files in Audio-only are 2452 and the total number of files in Video-only are 4904. The videos are converted into frames using a python script. Next, the corresponding images are converted to grayscale and then resized into 224×224 pixels. The total number of images after converting videos into frames are 49871. The number of images in Fear, Angry, Neutral, Sad, Disgust, Happy are 8182, 7874, 8694, 8583, 8154, 8384 respectively (Figs. 1 and 2).

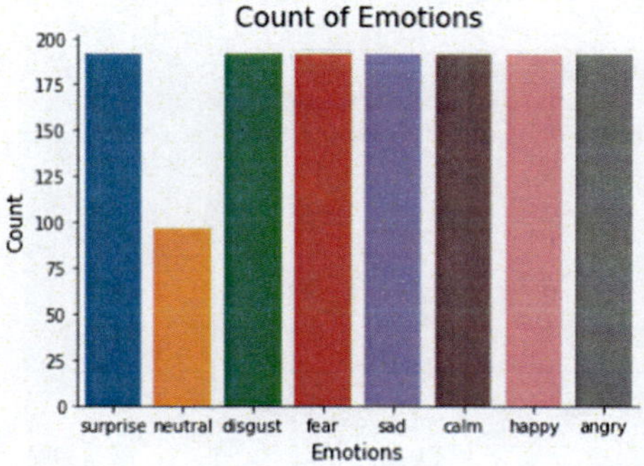

Fig. 1. Number of emotions in each category of Ravdess Dataset (Audio-only)

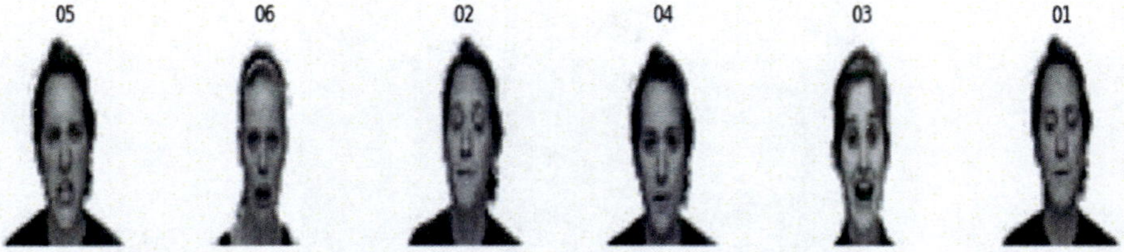

Fig. 2. Sample images in ravdess dataset after converting videos into frames.

3.2 Feature Extraction for Audio Dataset

Extracting prominent features from the audio dataset is important for accurate recognition of emotions. There are various features extracting techniques. The 5 feature extraction techniques used in this work are Zero Crossing Rate, Chroma_stft, Mel Frequency Cepstral Coefficients (MFCC'S), RMS (root mean square) value, MelSpectogram to train the model. Zero crossing rate is defined as the rate of sign-changes of the signal during the duration of a particular frame. MFCC'S is defined as the forming a cepstral representation where the frequency bands are not linear but distributed.

3.3 Model Architecture

For emotion recognition in audio files, a novel 1D convolutional neural network (Audio_EmotionModel) is designed, which contains 5 1D convolutional layers, 5 max pooling layers and 3 dropout layers. Total number of parameters in this 1D convolutional neural network are 152,614. The number of trainable parameters are 152,614 and number of Non-trainable parameters are 0. There are 2 dense layers, first dense layer output shape is (none, 256) and the second dense layer output is (none, 6). The number of trainable parameters at each 5 convolutional layers are 1024, 98432, 24640, 12352 and 6176 respectively. Here the dropout layers are used to avoid any overfitting of the proposed model on the audio datasets. The proposed model is used for RAVDESS dataset for emotion recognition in audio datasets (Fig. 3).

```
Model: "sequential_1"
_________________________________________________________________
Layer (type)                 Output Shape              Param #
=================================================================
conv1d_1 (Conv1D)            (None, 160, 256)          1024
_________________________________________________________________
max_pooling1d_1 (MaxPooling1 (None, 53, 256)           0
_________________________________________________________________
conv1d_2 (Conv1D)            (None, 51, 128)           98432
_________________________________________________________________
max_pooling1d_2 (MaxPooling1 (None, 17, 128)           0
_________________________________________________________________
dropout_1 (Dropout)          (None, 17, 128)           0
_________________________________________________________________
conv1d_3 (Conv1D)            (None, 15, 64)            24640
_________________________________________________________________
max_pooling1d_3 (MaxPooling1 (None, 5, 64)             0
_________________________________________________________________
dropout_2 (Dropout)          (None, 5, 64)             0
_________________________________________________________________
conv1d_4 (Conv1D)            (None, 3, 64)             12352
_________________________________________________________________
max_pooling1d_4 (MaxPooling1 (None, 3, 64)             0
_________________________________________________________________
conv1d_5 (Conv1D)            (None, 1, 32)             6176
_________________________________________________________________
max_pooling1d_5 (MaxPooling1 (None, 1, 32)             0
_________________________________________________________________
flatten_1 (Flatten)          (None, 32)                0
_________________________________________________________________
dense_1 (Dense)              (None, 256)               8448
_________________________________________________________________
dropout_3 (Dropout)          (None, 256)               0
_________________________________________________________________
dense_2 (Dense)              (None, 6)                 1542
=================================================================
Total params: 152,614
Trainable params: 152,614
Non-trainable params: 0
_________________________________________________________________
```

Fig. 3. Model Architecture for emotion recognition in audio files.

For Emotion recognition in images (obtained by splitting videos into images), a novel 2D Convolutional neural network is designed (Image_EmotionModel), which contains 4 2D convolutional layers, 2 max pooling layers, 3 dropout layers and 2 batch normalization layers. Total number of parameters in this 2D convolutional neural network are 32 M. The Number of trainable parameters are 32.11M and number of Non-trainable parameters are 640. There are 2 dense layers, first dense layer output shape is (none, 1024) and the second dense layer output is (none, 6). The number of trainable parameters at each 4 convolutional layers are 320, 18496, 73856 and 295168 respectively. Here the dropout layers are used to avoid any overfitting of the proposed model on the video dataset (videos are split into frames). The proposed model is used for RAVDESS dataset for emotion recognition in video datasets (videos are converted into frames) (Fig. 4).

Fig. 4. Model Architecture for emotion recognition in images.

4 Experimental Results

Table 1. Comparison of different techniques on audio-only ravdess dataset

S. No	Reference	Classifier	Accuracy obtained
1	Zheng [9]	DNN's	64.52%
2	*Ghaleb* [12]	SVM's	60.1%
3	*Samantaray* [16]	Bagged ensemble of SVM's	75.69%
4	Proposed model (Audio_EmotionModel)	CNN's	95.5%

Ravdess is an audio-visual dataset, which contains both speech and song files. This database is created by 24 artists (12 male and 12 females), the total number of files in Audio-only are 2452. There are 8 emotions in Audio-only dataset namely (1) Surprise (2) Neutral (3) Disgust (4) Fear (5) Sad (6) Calm (7) Happy (8) Angry. The dataset is divided in the ratio of 80:20 for training and testing the model. Since convolutional neural networks require large amount of annotated data for training and avoiding overfitting problem, 80% of dataset is used for training and the remaining 20% is divided in two equal half's for validation and testing the model. We proposed a novel one dimensional convolutional neuural network model for emotion in audio files.

The Audio_EmotionModel achieved an 95.5% accuracy on RAVDESS audio-only dataset. Novel Deep neural network model [9] obtained 64.52% accuracy on audio-only ravdess dataset. SVM [12] achieved an accuracy of 60.1% on this dataset. Where as, Bagged ensemble of SVM's [16] achieved 75.69% accuracy on audio-only ravdess dataset. Clearly, the above results indicate that the Audio_EmotionModel outperformed the remaining methods for emotion recognition in audio-only ravdess dataset. The below confusion matrix represents the number of labels correctly predicted by Audio_EmotionModel for each class of emotion. Except Neutral emotion other classes of emotions are classified almost accurately. Out of all the classes of emotions, the proposed model misclaasified some samples of neutral emotion into sad and calm. Table 1 represents the performance metrics of Audio_EmotionModel on audio-only RAVDESS Dataset (Figs. 5 and 6).

Fig. 5. Confusion Matrix for Audio-only

Fig. 6. Accuracy Loss of Audio_EmotionModel

Table 2. Performance metrics of Audio_EmotionModel on audio-only RAVDESS dataset.

Emotion/Metrics	Precision	Recall	F1-Score
Angry	0.98	0.97	0.98
Calm	0.97	0.97	0.97
Disgust	0.98	0.92	0.95
Fear	0.95	0.98	0.96
Happy	0.94	0.98	0.96
Neutral	0.90	0.95	0.92
Sad	0.97	0.90	0.93
Surprise	0.95	0.96	0.95

Table 3. Accuracy and loss comparison of various models on video-only RAVDESS dataset.

S. No	Model name	Train accuracy	Test accuracy	Train loss	Test loss
1	ResNet152	0.73	0.75	0.84	0.94
2	MobileNet V2	0.84	0.85	0.52	0.49
3	DenseNet121	0.94	0.93	0.20	0.22
4	Image_EmotionModel	0.97	0.97	0.16	0.26

Ravdess is an audio-visual dataset, which contains both speech and song files. This database is created by 24 artists (12 male and 12 females), the total number of files in Video-only are 4904. We developed a python script, which can convert the videos into frames. Each frame is considered as one image. Later, these corresponding images are converted into gray-scale format from RGB format. Next, these the gray-scale images are re-sized to 224×224 pixels and grouped into respective emotion directories. The images are grouped into 6 classes of emotions i.e., Fear, Angry, Disgust, Happy, Sad and Neutral. The size of this dataset is 49871 images. he number of images in Fear, Angry, Neutral, Sad, Disgust, Happy are 8182, 7874, 8694, 8583, 8154, 8384 respectively. We proposed a novel two dimensional neural network model(Image_EmotionModel) for emotion recognition in these images. The Image_EmotionModel consists of 4 2D convolutional layers, 2 max pooling layers, 3 dropout layers and 2 batch normalization layers. The dataset is divided in the ratio of 80:20 for training and testing purpose respectively. The Image_EmotionModel achieved 0.97 accuracy on these Video-only RAVDESS Dataset. The train loss and test loss of this model are 0.16 0.26 respectively. We also compared the proposed model with existing state-of-art models on this dataset. The ResNet152 obtained 0.75 accuracy with train and test loss being 0.84 0.94 respectively. Where as, MobileNet V2 obtained 0.84 0.84 accuracy on train and test set respectively. Among the existing state-of-art models, DenseNet121 achieved 0.94 0.93 accuracies on train and test set respectively. The results in Table 5 clearly indicate that the Image_EmotionModel had outperformed the ResNet152 and MobileNet V2 in terms of accuracy and loss on train and test set. The proposed model achieved slightly higher accuracy than DenseNet121. The proposed model is simple in architecture with less layers, when compared to DenseNet121, and performed better than the DenseNet121 (Tables 2 and 3) and (Figs 7, 8, 9 and 10).

Fig. 7. Accuracy and loss of ResNet152

Fig. 8. Accuracy and loss of MobileNetV2

Fig. 9. Accuracy loss of DenseNet12

Fig. 10. Accuracy loss of image_EmotionModel

5 Conclusion

For emotion recognition in audio files, a novel 1D convolutional neural network (Audio_EmotionModel) is designed, which contains 5 1D convolutional layers, 5 max pooling layers and 3 dropout layers. The Audio_EmotionModel is used for RAVDESS dataset for emotion recognition in audio datasets. For Emotion recognition in images (obtained by splitting videos into images), a novel 2D Convolutional neural network is designed (Image_EmotionModel), which contains 4 2D convolutional layers, 2 max pooling layers, 3 dropout layers and 2 batch normalization layers. The Image_EmotionModel is used for RAVDESS dataset for emotion recognition in video datasets (videos are converted into frames). Clearly, the above results indicate that the Audio_EmotionModel outperformed the remaining methods for emotion recognition in audio-only ravdess dataset. The Audio_EmotionModel achieved an 95.5% accuracy on RAVDESS audio-only dataset. Human Recognition [9] achieved only 40.9% while recognizing the emotions, clearly the proposed Audio_EmotionModel outperformed the human recognition by nearly more than 50%.

References

1. Akçay, M.B., Oğuz, K.: Speech emotion recognition: emotional models, databases, features, preprocessing methods, supporting modalities, and classifiers. Speech Commun. **116**, 56–76 (2020)
2. Bhatti, M.W., Wang, Y., Guan, L.: A neural network approach for human emotion recognition in speech. In: 2004 IEEE International Symposium on Circuits and Systems (IEEE Cat. No. 04CH37512), vol. 2, pp. II–181. IEEE (2004)
3. Hao, M., Cao, W.-H., Liu, Z.-T., Min, W., Xiao, P.: Visual-audio emotion recognition based on multi-task and ensemble learning with multiple features. Neurocomputing **391**, 42–51 (2020)
4. Kanluan, I., Grimm, M., Kroschel, K.: Audio-visual emotion recognition using an emotion space concept. In: 2008 16th European Signal Processing Conference, pp. 1–5. IEEE (2008)
5. Zhang, S., Zhang, S., Huang, T., Gao, W., Tian, Q.: Learning affective features with a hybrid deep model for audio–visual emotion recognition. IEEE Trans. Circuits Syst. Video Technol. **28**(10), 3030–3043 (2017)
6. Hossain, M.S., Muhammad, G.: Emotion recognition using deep learning approach from audio–visual emotional big data. Inf. Fusion **49**, 69–78 (2019)
7. Alshamsi, H., Kepuska, V., Alshamsi, H., Meng, H.: Automated speech emotion recognition on smart phones. In: 2018 9th IEEE Annual Ubiquitous Computing, Electronics & Mobile Communication Conference (UEMCON), pp. 44–50. IEEE (2018)
8. Kiranyaz, S., Ince, T., Abdeljaber, O., Avci, O., Gabbouj, M.: 1-d convolutional neural networks for signal processing applications. In: ICASSP 2019–2019 IEEE International Conference on Acoustics, Speech and Signal Processing (ICASSP), pp. 8360–8364. IEEE (2019)
9. Zheng, C., Wang, C., Jia, N.: An ensemble model for multi-level speech emotion recognition. Appl. Sci. **10**(1), 205 (2020)
10. Birhala, A., Ristea, C.N., Radoi, A., Dutu, L.C.: Temporal aggregation of audio-visual modalities for emotion recognition. In: 2020 43rd International Conference on Telecommunications and Signal Processing (TSP), pp. 305–308. IEEE (2020)
11. Beard, R., et al.: Multi-modal sequence fusion via recursive attention for emotion recognition. In: Proceedings of the 22nd Conference on Computational Natural Language Learning, pp. 251–259 (2018)
12. Ghaleb, E., Popa, M., Asteriadis, S.: Multimodal and temporal perception of audio-visual cues for emotion recognition. In: 2019 8th International Conference on Affective Computing and Intelligent Interaction (ACII), pp. 552–558. IEEE (2019)
13. Ghaleb, E., Popa, M., Asteriadis, S.: Metric learning-based multimodal audio-visual emotion recognition. IEEE Multimedia **27**(1), 37–48 (2019)

14. Rossenbach, N., Zeyer, A., Schlüter, R., Ney, H.: Generating synthetic audio data for attention-based speech recognition systems. In: ICASSP 2020–2020 IEEE International Conference on Acoustics, Speech and Signal Processing (ICASSP), pp. 7069–7073. IEEE (2020)
15. Cao, H., Cooper, D.G., Keutmann, M.K., Gur, R.C., Nenkova, A., Verma, R.: Crema-d: Crowd-sourced emotional multimodal actors dataset. IEEE Trans. Affect. Comput. **5**(4), 377–390 (2014)
16. Samantaray, A.K., Mahapatra, K., Kabi, B., Routray, A.: A novel approach of speech emotion recognition with prosody, quality and derived features using SVM classifier for a class of North-Eastern Languages. In: 2015 IEEE 2nd International Conference on Recent Trends in Information Systems (ReTIS), pp. 372–377. IEEE (2015)
17. Srivastava, M., Agarwal, A.: Classification of emotions from speech using implicit features. In: 2014 9th International Conference on Industrial and Information Systems (ICIIS), pp. 1–6. IEEE (2014)
18. Dhaouadi, S., Abdelkrim, H., Saoud, S.B.: Speech Emotion Recognition: Models Implementation & Evaluation. In: 2019 International Conference on Advanced Systems and Emergent Technologies (IC_ASET), pp. 256–261. IEEE (2019)
19. Shaqra, F.A., Duwairi, R., Al-Ayyoub, M.: Recognizing emotion from speech based on age and gender using hierarchical models. Proc. Comput. Sci. **151**, 37–44 (2019)
20. Zhao, J., Mao, X., Chen, L.: Speech emotion recognition using deep 1D & 2D CNN LSTM networks. Biomed. Signal Process. Control **47**, 312–323 (2019)

Performance Evaluation of Morphological Features in Ear Recognition

Abhisek Hazra[1,2](✉), Sourya Saha[1], Sankhayan Chowdhury[1], Nabarun Bhattacharyya[2], and Nabendu Chaki[1]

[1] University of Calcutta, Kolkata, West Bengal, India
[2] Centre for Development of Advanced Computing, Salt Lake Electronics Complex, Kolkata, West Bengal, India
abhisek.hazra@cdac.in

Abstract. With the advances in processor design, password, and PIN-based authentication methods are proven to be inadequate as a means of security. The raw computing power makes it possible to use nearly brute force approaches to innovative techniques for cracking passwords. The conventional authentication methods are relatively easier to crack, as compared to biometric-based authentication. Biometric traits cannot be easily mimicked. Ear biometrics, comparatively a newer entrant in the block, seems to be promising because it does not require any active participation from the user. In ear-recognition literature, mathematical morphological features have been explored very little, even though they carry significant discriminatory information among the classes. Further, they exhibit relatively better performance upon partially occluded data as contours and shape information neglect such occlusions. This work aims to measure the impact and importance of such features on publicly available bench-marked ear data sets like AMI, University of Sheffield, and CP ear databases.

Keywords: Morphology · Ear recognition · Shape features · Ear biometric · Classification · Feature selection

1 Introduction

Biometric recognition systems provide the advantage of preventing security attacks, concerning the identification or verification of individuals based on the analysis of unique biological traits in those individuals [1]. Humans tend to recognize other individuals based on how they look, how they sound while they speak, their movement patterns, and many more unique traits [2]. Biometric traits are unique to individuals and it is difficult to fool a computer because a computer system performs much better in recognizing the patterns, compared to manual authentication. Such an arrangement has led to a declined rate of fraudulence that gets committed in domains where identity matching is a serious security concern. So, the need for biometric authentication is a necessity now, owing to its success rates in the reduction of fraudulent claims to access a restricted entry.

J. I.-Z. Chen et al. (Eds.): ICIPCN 2022, LNNS 514, pp. 385–396, 2022.
https://doi.org/10.1007/978-3-031-12413-6_30

Biometric recognition systems such as fingerprint recognition systems and facial recognition systems have served as a successful means for the authentication of individuals. But these systems tend to become more unreliable when the subjects fail to cooperate for providing their biometrics to the system. In the case of a facial recognition system, the person to be authenticated might be wearing a mask or eyeglasses, with different hair colours or styles, with a grown beard and/or with a moustache or might not be facing the camera and of course, high use of cosmetics. For fingerprint recognition systems, the subject might fail to provide proper fingerprints due to improper positioning of fingers, dirt on fingers, or intentional attempts. Apart from these, face or fingerprint spoofing cases are also demanding attention from researchers in the same fields. Ear biometric systems do not require the participation of the person to be authenticated and the subject can be stealthily photographed. However, the most significant problem with ears is the occlusion made by the earrings, hairs, head covers, changes in (head) pose, and motion blurring which are still open research problems in this area, especially for the unconstrained dataset. There are plenty of scopes for improvements in this area. For even better results, this can be combined with a facial recognition system. Moreover, the ear is somewhat a robust, stable structure and physical changes in the ear shape occur before 8 years of age and after that 70 years because of ear elongation and ageing factors [3] which is advantageous for recognition.

Several morphological features for the recognition of the ear from 2D images have been defined in an earlier research work [4]. In the present work, the effort has been made to complete the pending tasks of [4] by extracting and utilizing those features, described earlier in [4] for several benchmarked data sets (AMI, CP, and the University of Sheffield) which resemble real-world data closely. The validation of such features on benchmarked data is a crucial factor that can be considered while choosing the morphological features for real-time applications. There have been several works in the literature where different types of features have been utilized. In these works, several feature descriptors have been used which can provide good recognition results for a certain data set. One point which demands our attention here is that very few approaches have been tested on multiple datasets to prove their global trustworthiness. Also, shape-based features haven't been given enough attention though they may be easily hand-crafted with little time, space complexity and easily interpretable. Also, the selection of a reduced dimension of feature set with the same or better effectiveness in terms of accuracy measure demands a lot of attention though it was rarely addressed. Hence, the goal of this work is to determine how well the morphological feature scheme performs on multiple benchmarked ear data sets. To achieve this, a range of feature selection strategies followed by some classification algorithms were used. The best classification strategy was decided after an analysis of the performances of the schemes. All these steps were applied to multiple data sets to determine the acceptability of the shape-based features on different benchmarked ear data sets.

The rest of the paper has been structured in different sections. The proposed methodology has been described in Sect. 2, while the Sect. 3 contains the results of the experimentation process. An analysis and discussion of these results are presented in Sect. 4 of the paper. Finally, in Sect. 5, the concluding remarks are noted along with the scope of possible future extensions of this work.

2 Adopted Methodology

In this work, we propose to build and validate an ear recognition model on different benchmarked ear datasets based on morphological features mentioned in [4]. Throughout the research phase, the developmental and experimental tasks were carried out in the MS-Windows platform using Python 3.0, with the help of some libraries like openCV-python, scikit-learn etc. and Weka ML workbench. Noise removal of the ear images started with cropping each ear image into a 50×180 strip like the IIT Delhi ear-dataset. All images were brought to grayscale if they were in some other colour space. A series of adjustments and enhancements were made to the intensity and contrast of the images followed by smoothing and binarisation of sample images. The binary templates were used to derive the morphological features. It is evident from the workflow of the system (see Fig. 1), that three benchmarked data sets (AMI, University of Sheffield, CP) were used for testing the model and wider acceptability of our scheme. However, the AMI ear data set [5] brought forth a few different pre-processing steps whereas the University of Sheffield ear data set [6] and the CP Ear data set [7] came with the same set of pre-processing strategies since it was observed that the pre-processing steps adopted for CP and University of Sheffield data set were not producing good enough results for AMI which contains 4 subsets. Data preparation and noise removal steps are described below.

2.1 AMI Ear Dataset

Only right ear images for each subject from all the AMI subsets were used. As, per subject, there is a single left ear image in the data set, so we omitted it due to a lack of sample data for training the machine learning model. Cropping was done on each image into strips of 50×180 pixels with the ear profile visible. Then the following steps were adopted for noise removal - Gamma correction, grayscale conversion, histogram equalization, median filtering using 5×5 mask, otsu's thresholding, morphological opening with 3×3 structuring element, binary mask creation for noise removal from top left and bottom left corners of the binarised image (see Fig. 2).

Fig. 1. Workflow of the proposed system.

2.2 University of Sheffield and CP Ear Dataset

Cropping was performed on each image into strips of 50×180 pixels. After that, the pre-processing steps which were applied are gray-scaling, intensity adjustment, histogram equalization, median filtering (5×5 kernel), and binarisation using Otsu's thresholding. The next step was to extract the shape-oriented features from the processed images. Morphology of an object refers to its form or structure or shape. Shape-based features have the important characteristic that they are invariant to different kinds of transformation. Further, efficient morphological features are also invariant to affine transformations. Such features exhibit statistical independence and are robust. They are robust in the sense that their values do not change when the same feature is extracted from the same object several times. Since morphological features depend on the shape of the object, the presence of noise does not affect the morphology of the object, hence these features show robustness against noise. Even if parts of the object are occluded in some way, the shape features from the rest of the part of the image may be

Fig. 2. [Left →Right] Gamma correction, Gray scaling, Histogram equalisation, Median filtering, binary thresholding, morphological opening (3×3 kernel), binary mask creation, and final image from a subset of AMI ear data set.

used efficiently to classify the object correctly [4]. These were the reasons for choosing morphological features so that similar elements from a data set might be defined in the same manner no matter what kind of transformation it is in. In the paper [4], a set of forty-four morphological features were computed, which were utilized in this work as well as for analyzing the global acceptability of the morphological features tested on benchmarked data sets like AMI, CP, and University of Sheffield.

A strong emphasis was also given to selecting the most discriminatory features by applying efficient feature selection algorithms [8] to select a subset from the original feature set that proved to be sufficient as well as necessary to improve the prediction accuracy [9] of the ear recognition system. We used three feature selection strategies (see Fig. 1) which are depicted below.

- **OneR based Attribute Evaluation**
  Invented by Holte [10–12], it evaluates the features based on error rate [12]. For each feature in the training data, one rule is built and the rule with the lowest error is selected [10], operating on categorical data [11]. If the features are numerical, it considers them continuous and employs a technique to split the entire range of values into multiple disjoint gaps [10].
- **CFS Subset Evaluation**
  Correlation-based Feature Selection (CFS) evaluates subsets of features by computing the correlation between subsets of features and the class. That subset is regarded as the best in which the features have a high correlation with the class and low correlation among themselves using a heuristic function based on correlation [11]. It is a filtering strategy that produces ranking or merit values for subsets of features based on the heuristic function and the selected subset of features is the one with the best merit [10]. Just like the above method, CFS operates to find the correlation among features which

are categorical and the numeric features that are discretised [11]. The merit measure for a feature subset is computed as follows [10,11].

$$Merit_s = \frac{kr_c}{\sqrt{(k + (k+1)r_f)}} \quad (1)$$

whereas $Merit_s$ is the merit measure of subset S of features with k features in the subset, kr_c is the mean of correlation between features and class and r_f is the mean of correlation between features of the subset. The numerator denotes the degree of prediction of a subset of features and the denominator denotes how redundant the features of the subset are among themselves [10].

- **Relief based Attribute Selection**
  This technique performs feature evaluation based on how well a feature value can differentiate items that belong to different groups but have commonalities with each other. For each of the features, the algorithm randomly selects a sample and its k-nearest neighbours from the same as well as each of the different classes. A score is then generated for each of the features which are computed as the total of weighted differences in the same class and different classes. A differential expression for a feature will exhibit higher differences with items of different classes and will have a higher score. Lower differences with items of different classes will produce lower scores [10].

The various classification algorithms that have been used in this work have been stated(see Fig. 1). The classifiers along with the groups they belong to are as follows

- Bagging with Random Forest (meta-Classifiers)
- Naïve Bayes (Bayesian Classifiers)
- Multilayer Perceptron or MLP (function-based Classifiers)
- J48 (tree-based Classifiers)
- Random Forest (tree-based Classifiers)

The performance of biometric systems oversees the assessment of accuracy and other quantifiable characteristics of the system. Feature templates from each contributor from the ear data set were used to train the classifiers mentioned atop. Different genres of classifiers were imitated in this task to emphasize the most favourable one in terms of recognition veracity. Each of the classifiers was individually tested to obtain the individual merit of the classifiers on each of the data set.

3 Experimental Results

In this section, the results of classification using different classification strategies coupled with the different feature selection strategies [13,14] were discussed. For each data set, we looked at the percentages for correct and incorrect recognition for each of the five classifiers using three different feature selection strategies.

Table 1. Recognition results for the CP ear database.

Database	Feature selection	Classification algorithm	Correct recognition	Incorrect recognition
CP	OneR (Ranker search)	Bagging (Random forest)	77.451	22.549
		Naive Bayes	73.5294	26.4706
		J48	53.9216	46.0784
		Random forest	82.3529	17.6471
		Multilayer perceptron	76.4706	23.5294
	CFS(Greedy search)	Bagging (Random forest)	76.4706	23.5294
		Naive Bayes	70.5882	29.4118
		J48	61.7647	38.2353
		Random forest	81.3725	18.6275
		Multilayer perceptron	76.4706	23.5294
	Relief (Ranker search)	Bagging (Random forest)	77.451	22.549
		Naive Bayes	71.5686	28.4314
		J48	60.7843	39.2157
		Random forest	84.3137	15.6863
		Multilayer perceptron	77.451	22.549

The red texts (Table 1) indicated the value of recognition accuracy for the classifier which performs best for each feature selection strategy. Then, we noted the performance of the best performing classifier for each feature selection strategy in terms of some ML measures (TPR, FPR, TNR, FNR, Precision, Recall, and F1-score). It was observed (Table 1) that random forest was the meticulous one for CP Ear Dataset.

From Fig. 3, it is indisputable that the TPR of the Relief-based attribute evaluation method is the highest. Thus, Relief based Attribute Evaluation method performs the best on the CP Ear dataset with a TPR of 0.843 with the random forest classifier. The experimental results obtained from the University of Sheffield Ear Dataset also conformed to the facts obtained from the CP dataset. Thus, Relief based Attribute Evaluation method performs the best (Fig. 4) on the University of Sheffield Ear Dataset with a TPR of 0.864. Repetitively, Relief based Attribute Evaluation method performs the best on AMI subsets 1, 2, 3 with a TPR of 0.72, 0.627, and 0.68 respectively (see Figs. 5, 6, and 7) whereas it is quite evident that the TPR of the CFS Subset attribute evaluation method is the highest in case of AMI subset 4 (see Fig. 8).

Fig. 3. TPR, TNR, FPR, FNR, Precision, Recall and F1-Score for CP ear data set.

Fig. 4. TPR, TNR, FPR, FNR, Precision, Recall and F1-Score for University of Sheffields ear data set.

4 Discussions and Guidelines

From the results of the experimentation performed on the image databases, we hereby arrived at the following inferences for the morphological features mentioned below

- The maximum recognition accuracy of 84.3137% on the CP Data set, 86.3636% on the University of Sheffield database whereas 72%, 62.6667%, 68% on the AMI subset-1, subset-2, subset-3 may be achieved using Relief Based feature selection in conjunction with Random Forest Classifier.

Fig. 5. TPR, TNR, FPR, FNR, Precision, Recall and F1-Score for AMI subset-1 ear data set.

Fig. 6. TPR, TNR, FPR, FNR, Precision, Recall and F1-Score for AMI subset-2 ear data set.

- Subset-4 of the AMI Ear Dataset has a maximum recognition accuracy of 72% using Bagging with Random Forest Classifier and CFS Subset Attribute Evaluation Strategy.

From the above results, a clear trend was visible that most of the datasets achieve the best results with Random Forest or Bagging with Random Forest along with the Relief Based Attribute Evaluation. The significant drop in the accuracy of the subsets of the AMI Ear Data set can be attributed to the pre-processing strategy used. A single binary mask has been used to remove noise from all the images of each of the subsets of the AMI Ear Data set. As the 'No

Fig. 7. TPR, TNR, FPR, FNR, Precision, Recall and F1-Score for AMI subset-3 ear data set.

Fig. 8. TPR, TNR, FPR, FNR, Precision, Recall and F1-Score for AMI subset-4 ear data set.

Free Lunch Theorem' from Wolpert and Macready [15] suggests that there cannot be a single global strategy that will work finest with all problem definitions. Secondly, as the AMI Ear Data set contains only a single left ear image compared to 6 right ear images for each subject, the model struggled to correctly classify the single left ear image due to the dearth of left ear images for the subject. For this reason, we used the six right ear images per subject for all the subsets of the AMI Ear Data set.

Random Forest Classifier seemed to outperform other classifiers because this classifier builds several decision trees for a data set by selecting subsets of the training data and making predictions on them. The final prediction is based on the majority votes of all the predictions. As a subset of training data is selected to build the decision trees, the trees are uncorrelated concerning each other. Even though some trees produce wrong predictions, the final prediction will be based on the majority of the group of predictions. Thus, if the majority of the other trees predict rightly, the result is a correct prediction. The success of the Random Forest over other methods is its low level of correlation between different decision trees.

5 Conclusions

It is noticeable from the above discussions that the morphological features perform well on the majority of the data set. There has been a strong need to develop noise removal approaches specific to any of the benchmark data sets. This is to ensure that the performance of classification for all the models can be improved equally. Besides, the state of the art scenario demands that a deep learning approach for the classification of ear data sets needs to be explored. The objective is to optimize the performance of these features on the data sets. This had been to help design a global ear recognition scheme that would work in real-time for different ear data sets. This work also involved manual cropping of ear images. Thus, an automatic segmentation scheme is also required to devise an efficient real-time ear recognition system. The focus had been on exploring better distinguishable features for simple computation, similar to the morphological features discussed in [4]. This has improved the recognition accuracy to a greater extent for all the benchmark data sets. In future, ear recognition in unconstrained environments needs to be addressed with greater emphasis as no real-time application will facilitate image acquisition in constrained environments.

References

1. Faundez-Zanuy, M.: Biometric security technology. Aerosp. Electron. Syst. Magaz. IEEE **21**, 15–26 (2006). https://doi.org/10.1109/MAES.2006.1662038
2. Pato, J.N., Millett, L.I., National Research Council. Biometric Recognition: Challenges and Opportunities. The National Academies Press, Washington, D.C. (2010). https://doi.org/10.17226/12720
3. Iannarelli, A.: Ear Identification. Forensic Identification Series. Fremont, Paramont Publishing Company, California (1989)
4. Hazra, A., Choudhury, S., Bhattacharyya, N., Chaki, N.: An intelligent scheme for human ear recognition based on shape and amplitude features. In: Chaki, R., Chaki, N., Cortesi, A., Saeed, K. (eds.) Advanced Computing and Systems for Security: Volume 13. LNNS, vol. 241, pp. 61–75. Springer, Singapore (2022). https://doi.org/10.1007/978-981-16-4287-6_5
5. Gonzalez, E., Alvarez, L., Mazorra, L.: AMI Ear Database (2022). https://ctim.ulpgc.es/research_works/ami_ear_database/#cita

6. Daniel, B.G., Nigel, M.A.: Characterizing virtual eigen signatures for general purpose face recognition, face recognition: from theory to applications. In: Wechsler, H., Phillips, P.J., Bruce, V., Fogelman-Soulie, F., Huang, T.S. (eds.) NATO ASI Series F, Computer and Systems Sciences, vol. 163, pp. 446–456 (1998)
7. Carreira-Perpinan, M.A.: Compression neural networks for feature extraction: application to human recognition from ear images, (in Spanish). M.Sc. thesis, Faculty of Informatics, Technical University of Madrid, Spain (1995)
8. Hobbs, L., Hillson, S., Lawande, S., Smith, P.: Oracle 10g Data Warehousing, Elsevier, 1st Edn (2004). ISBN 978-1-55558-322-4. https://doi.org/10.1016/B978-1-55558-322-4.X5000-0
9. Dash, M., Liu, H.: Feature selection for classification. Intell. Data Anal. 1(1–4), 131–156 (1997). https://doi.org/10.1016/S1088-467X(97)00008-5
10. Yildirim, P.: Filter based feature selection methods for prediction of risks in Hepatitis disease. Int. J. Mach. Learn. Comput. **5**(4), 258–263 (2015). https://doi.org/10.7763/IJMLC.2015.V5.517
11. Janabi, D., Kadhim, B.: A Comparative Study for Attribute Selection Methods, S. Al. Data Reduction Techniques (2018)
12. Holte, R.C.: Very simple classification rules perform well on most commonly used datasets. Mach. Learn. **11**, 63–90 (1993). https://doi.org/10.1023/A:1022631118932
13. Kotsiantis, S.B., Zaharakis, I.D., Pintelas, P.E.: Machine learning: a review of classification and combining techniques. Artif. Intell. Rev. **26**, 159–190 (2006). https://doi.org/10.1007/s10462-007-9052-3
14. Alpaydin, E.: Introduction to Machine Learning, Massachusetts. The MIT Press, Cambridge (2020)
15. Wolpert, D.H., Macready, W.G.: No free lunch theorems for optimization. IEEE Trans. Evolution. Comput. **1**(1), 67–82 (1997). https://doi.org/10.1109/4235.585893

GQNN: Greedy Quanvolutional Neural Network Model

Aansh Savla, Ali Abbas Kanadia(✉), Deep Mehta, and Kriti Srivastava

Dwarkadas J. Sanghvi College of Engineering, Mumbai, India
aakanadia@gmail.com, kriti.srivastava@djsce.ac.in

Abstract. Quantum computation, particularly in the field of machine learning, is a rapidly growing technology. Major advantage of Quantum computing is its speed to perform calculations. This paper proposes a novel model architecture for feature extraction. It extracts the features from a colored spectrogram as an extension to the already existing Quanvolutional Neural Network which works on only grayscale images or 2-dimensional representation of spectrograms. The proposed model architecture works on all the three layers of an image (RGB) and uses random quantum circuits to extract features from them and distribute them into several output images and out of them one is selected which contains the most important and pertinent features from the original image helping the training of the CNN model used ahead. COVID-19 use case is used for performance evaluation. Normally the testing methods used to detect virus are expensive, examples include RT-PCR test, CT scan images. These methods require a medical professional to conduct the test while being in the proximity of the patient. Also, the testing kits once used cannot be used again. One of the most evident changes in a Covid 19 patient is the change in his/her coughing and breathing pattern. This work analyzed the spectrograms of the audio samples of coughing and breathing patterns of Covid 19 patients using the proposed model architecture and provided subsequent results. Finally to generalize our model's applicability, the model is also run-on Alzheimer disease dataset and corresponding results are provided.

Keywords: Quantum Computing · Qubits · Bloch sphere · Spectrograms · Quanvolutional Neural Network

1 Introduction

Coronavirus outbreak in 2019 was declared as a pandemic by WHO. There is a need for diagnosing the Covid-19 patient which first involves testing the patient. For detecting Covid-19, the tests include Computed Tomography (CT) images, real time fluorescence polymerase chain reaction (RT-PCR) test. However, CT scan images are costly. These tests include expert professionals for both the tests. Also tests like RT-PCR takes 4 to 8 h which is time consuming. Hence there is a need for an efficient, less time consuming, cheaper method to detect whether a patient is Covid-19 positive or not. One of the factors which is easily available in a Covid-19 patient is breathing and coughing pattern. The audio-recording of a patient coughing can be used to detect the Covid-19.

J. I.-Z. Chen et al. (Eds.): ICIPCN 2022, LNNS 514, pp. 397–410, 2022.
https://doi.org/10.1007/978-3-031-12413-6_31

Alzheimer's is a neurodegenerative disease which is a common occurrence in people with old age. Because it is an irreversible condition, early discovery is critical in slowing the disease's progression. The existing techniques used for AD detection require medical images. Procurement of medical images are difficult and costly. One of the effects of AD is speech disorder, so their audio samples and its spectrograms can be used to predict the onset of the disease with the help of artificial intelligence techniques.

The spectrograms give a detailed representation of an audio signal in frequency or time domain and give better results as proven by Boddapati [7]. Spectrograms are 2 dimensional graphs with color as the third dimension. Time runs from left to right. Vertical axis is used to represent frequencies which are interpreted as pitch or tone. The amplitude or energy or loudness is represented by the third-dimension color.

Deep learning has many applications in the domain of image processing which can be used in the medical field. However, the images involved in the medical domain are much more complicated and we need efficient feature extraction to save time and space. Also getting data in the medical field is a tedious task. It needs a lot of permissions and care must be taken while getting the data. The availability of such data is less. Classical deep learning methods are not efficient when learnt on less training dataset. Quantum machine learning, with its properties such as entanglement and superposition, can help to improve the performance of machine learning algorithms by evaluating all possible scenarios concurrently. Quantum machine learning is a novel computing technology that is expected to outperform traditional machine learning. A quantum machine learning model performs well in case of a hybrid model where data is classical and encoded using a quantum circuit like Henderson [15] which proposed hybrid Quanvolutional Neural Network. The quanvolutional layer is a new sort of transformational layer added to the regular CNN design that extends the possibilities of CNN. It comprises of N quanvolutional layers that convert input data to produce feature maps. The quanvolutional layers extract features by using random quantum circuits to transform spatial subsections of data. As this algorithm is Hybrid in nature, it has no QRAM requirements which makes it easy for simulating it on classical computer. Also, since the algorithm works on local subsection, so the number of qubits required are less. This research focuses on the hybrid quantum-classical algorithm 'The Quanvolutional Neural Network'. It also aims to investigate a new type of model by extending the capabilities of QNN algorithm and to modify it for extracting the features of a colored image by considering even the 3rd dimension of the image. Feature extraction of spectrograms is performed using new architecture. Experiments were conducted using Pennylane. Results show that the modified algorithm is advantageous in terms of classification than its classical version.

Table 1. Comparison of Spectrograms for different diseases

Disease	Normal Person Spectrogram	Patient Spectrogram
Cough		
Abnormal Heart Rhythm		
Parkinson Disease		
Autism Spectrum Disorder		
Alzheimer Disease		

Table 1 contains the comparison of Spectrograms for different diseases [18, 19, 20], [21, 22]. It is evident that spectrogram of the patient having disease have quite noticeable trends. Hence Spectrograms are quite usable in detection of various types of diseases.

2 Literature Review

Various techniques have been developed for the testing of Covid-19. Some of them include RT-PCR test, CT scan images etc. but coughing and breathing patterns prove to be efficient for this test. Various audio feature extraction techniques were learnt during the study. Hanieh Chatrzarrin [2] has shown how wet and dry coughs can be differentiated based on frequency at specific times of coughs. Many papers have been published on different techniques for feature extraction from audio signals [3–5]. MFCC proves to be the most efficient for audio signals.

Madhurananda Pahar [6] has done a comparative analysis using MFCC as the base feature extraction technique and using different classifiers such as LR, SVM, MLP and CNN. After having done an in-depth analysis on audio signals, it was found that spectrograms of audio signal prove to be more information rich as compared to its corresponding audio signal.

Venkatesh Boddapati [7] created a deep convolutional neural network that could accurately classify spectral images after being trained. In their experiments, they performed classification experiment on three different image representations of sound-Spectrograms, Mel-Frequency Cepstral Coefficients (MFCC), and Cross Recurrence Plot (CRP). The maximum categorization accuracy was found when sound spectrograms were used. Furthermore, Lazhar Khriji [8] used a deep Long Term Short Memory (LSTM) model for deep feature extraction of audio signals of breathing and coughing of patients. They achieved an accuracy of approx. 80% which is better as compared to many other existing techniques. Spectrograms have also been used to recognize speech emotion using deep convolutional neural networks [9]. To understand the basics of Quantum Computing, many base papers were referred during the course of study. It is necessary to understand the feasibility of using Quantum Computing in the Machine Learning domain [10, 11]. As we encounter many Gates in classical computing, there are various Quantum Gates. Also, the Rotation Gates in Quantum Computing are introduced due to the concept of geometrical representation of qubits [12]. The audio data was trained and tested on Quantum Neural Network given by Avinash Chalumuri et al. [13]. This network performs better in simple dataset like Iris dataset, while using large volume of audio data makes it harder in training this network. The accuracy almost goes constant after some iterations and there is no training after that point.

Detecting Covid-19 using images gives better results as presented by Erdi ACAR [14] which used CT images for detecting Covid-19 positivity. Maxwell Henderson et. al. proposed Quanvolutional Neural Network [15] which stated the quanvolutional layer is an expansion of classical CNNs with an added transformational layer. Any number of quanvolutional layer can be stacked and the order of these layers depends on the end user itself. A quanvolutional filter creates a feature map by altering the input tensor's spatially local portion. It uses a quantum circuit, which can be organized or random, instead of element-by-element matrix multiplication. It employs a random circuit for simplicity. The results of the experiments demonstrated that applying a quanvolutional transformation to classical data can improve the accuracy of a network that can be utilized for classification.

In the paper 'A Tutorial on Quantum Convolutional Neural Networks' [16], a structured circuit is used for encoding process of the images. MNIST dataset is downscaled and used for the experimental purpose. The paper compares 3 models viz fully connected, CNN, QCNN. It is found that the QCNN outperforms the fully connected layer and has comparable performance with CNN. Hence QCNN can be used in more complex data which needs to be researched. Chao-Han Huck Yang [17] in their work has used the original QNN model for preprocessing of spectrograms. They haven't preprocessed the spectrogram image, but the 2-dimensional array received before the plotting of the spectrogram. The authors have proposed their own classical architecture of attention recurrent neural networks for the purpose of training the preprocessed images. This paper also experiments with different quanvolutional filter sizes and concludes that filter of size 3×3 gives the worst accuracy and filter of size 2×2 gives the best preprocessing result. Their model shows results for spoken word recognition with stable performance.

The literature review clearly demonstrates that most research use greyscale images of spectrograms in the analysis process which are 2-dimensional in nature. The paper hypotheses that in the analysis of spectrograms, the color parameter contains valuable information and cannot be excluded from the analysis process. There are no existing methods of preprocessing which can directly be applied to spectrograms for the purpose of feature extraction. Quanvolutional Neural Network does feature extraction but does not include the color parameter. Greyscale images are basically 2 dimensional arrays where every value in the array represents a pixel value of the image. But colored images are 3 dimensional arrays and the size of the 3rd dimension is 3 for the three RGB (Red, Green and Blue layers) layers. It can be said that colored images have a set of 3 2-dimensional arrays stacked where a value in the same position in all the three arrays represent the RGB value of the pixel in that position in the image. A quantum preprocessing method which can be applied to colored 3-dimensional images can be used to solve this problem. The Quanvolutional Neural Network uses all preprocessed 4 channel images for further training. However, a research is to be done that whether all 4 channels are necessary for the training of CNN.

3 Methodology

This paper deals with preprocessing the spectrogram images of the coughing audio signals of the patients. First the audio signals are converted to their corresponding spectrograms, which are then preprocessed using the proposed GQNN network. Finally, a classical CNN is used for prediction and results.

Dataset used for this study is from the Pfizer Digital Medicine Challenge. Dataset consists of audio files of breathing and coughing patterns of patients. Audio files from ESC-50 and Audio Set were used to build this dataset [1]. Details of the Alzheimer's Dataset is explained in the Sect. 4.3.

3.1 Proposed Design Overview

The proposed architecture shown in Fig. 1 is simply an extension of the already established Quanvolutional Neural Network [16]. The model proposes some changes in QNN so that images having dimensions greater than 2 i.e. colored images can also be incorporated in this model. Like the QNN, the model also serves as a pre-processing model for images using random quantum circuits, the results from which can be used as an input to Classical CNNs which use it for training and giving the results. The quanvolutional filters convolve the entire image. The pixels present in the filter are used for preprocessing, and when it is completed the filter shifts to the right by a particular stride amount (stride value in the proposed model is 2) and uses those pixels in the filter for preprocessing. When the filter reaches the end of the image, it goes all the way back to the left and shift a stride amount below and continues the same process until the filter reaches the bottom right part of the image. So the filter traverses the image in left to right and top to bottom manner. Unlike the original architecture, this model takes into consideration all the three layers of which an image is built from (The Red, Green and Blue layers of the RGB representation of an image). Each of the quanvolutional filter gives a feature map but the process is unlike that of the Classical Convolutional Networks. The pixels present in the filter are used as an input to a random quantum circuit. The outputs of these circuits for all the layers of the same spatial subsection of the images is taken into consideration together and fitted in the feature map. The ordering of the quantum output to the feature map is done in such a way that the order of the RGB doesn't break. This is achieved by placing the first output from the red layer in the first channel of the feature map.

In the next channel, the first output from the green layer is placed in the same spatial location and after that the first output from the blue layer is placed in the next channel. This process goes on for all the outputs from the random quantum circuit considering a quanvolutional layer spanning the same spatial location for all three layers. In this way the feature map holds multiple 2-dimensional tensors, but when clubbed together in groups of 3, give multiple colored images which have features extracted from the original image. All the output-colored images are examined and compared with the original image for multiple examples and a single image channel is selected which contains the most important and pertinent features from the original image. This entire process helps in the actual training of the model which is done through classical CNNs.

Thus, the model extracts features from a colored image and distributes it into 4 channels producing 4 colored images. This reduces the size of the input image which helps in reducing storage cost and also the training time of CNN.

3.2 Classical CNN Models

The structure of the Classical CNN model used as a part of the hybrid GQNN:

Covid Dataset

For this dataset, a convolutional neural network with the following structure is used: CONV2 (16 × 3 × 3) – MAXPOOL (2 × 2) – CONV2 (32 × 3 × 3) – MAXPOOL (2 × 2) – CONV2 (64 × 3 × 3) – MAXPOOL (2 × 2) – CONV2 (64 × 3 × 3) – MAXPOOL (2 × 2) – Flatten – Dense (256) – Sigmoid (1)

Alzheimer's Dataset

For this dataset, the structure of AlexNet model is used.

Fig. 1. Greedy quanvolutional pre-processing layer

4 Experimental Results and Analysis

First, the audio recordings are transformed into spectrogram pictures. The spectrogram images are then given as input to the proposed model for preprocessing. The output consists of images which have features extracted from the main image. The preprocessed images are then fed to a classical CNN model which is used for the training process. This work is assessed on metrics like accuracy, precision, recall and f1-score.

4.1 Experimental Setup

Pennylane is a framework for implementing Quantum Machine Learning codes in python. This framework is used for implementing the Quanvolutional layer in our work. It provides inbuilt function for designing a random quantum circuit. Instead of executing the quanvolution layer on practical quantum computer we use default simulator provided by the Pennylane framework. Hence the time for pre-processing depends more on the hardware specification of the system. We used 2 systems to run the preprocessing of the spectrogram images. The first system takes 60 min and the second system takes 10 min to preprocess a single spectrogram image.

1. **Windows Laptop:** 16 GB RAM, No dedicated GPU, i5 processor
2. **iMac PC:** 16 GB RAM, GPU, i9 processor.

4.2 Results Analysis

After pre-processing of a single spectrogram image as shown in Fig. 2 through the GQNN model, 4 images are obtained which are shown in Fig. 3, 4, 5 and 6.

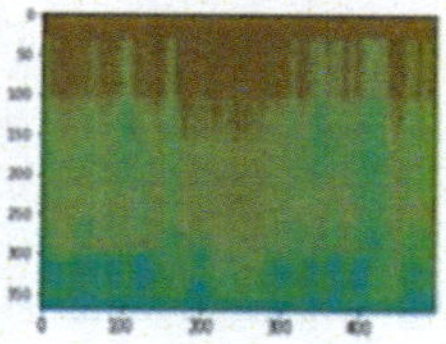

Fig. 2. Original Covid spectrogram

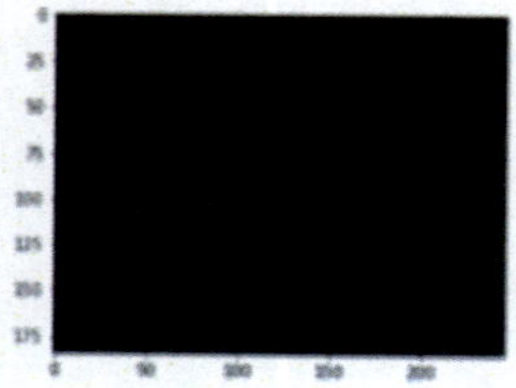

Fig. 3. Channel 1: Covid spectrogram

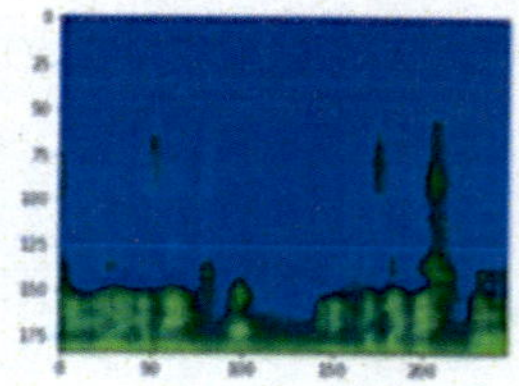

Fig. 4. Channel 2: Covid spectrogram

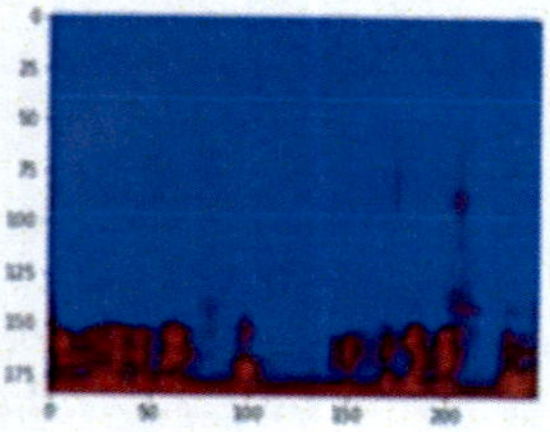

Fig. 5. Channel 3: Covid spectrogram

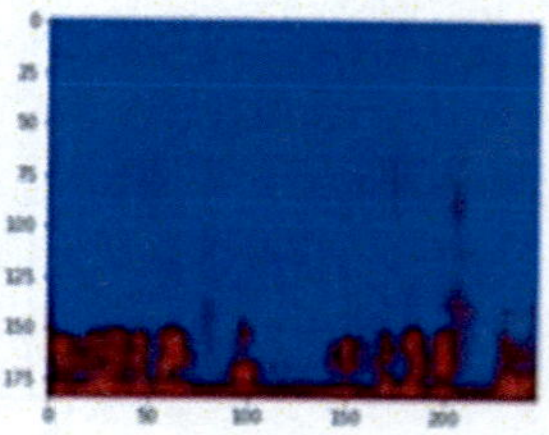

Fig. 6. Channel 4: Covid spectrogram

Instead of using all the four images for further training of CNN, only Channel 2 image is used since it visually resembles more to the original image. This helps in reducing the space required to store an image and reduces the training time of the Convolution Network.

50 images were used for the analysis. The dataset split for the experiment was 35:15, i.e., 35 train images and 15 test images. Table 2 consists of the results of different parameters achieved during the work. Comparison was conducted between 2 approaches. In the

first approach, spectrogram images were pre-processed using Greedy Qunavolutional pre-processing layer and then the CNN architecture was applied on it to obtain results, the entire process being called Greedy Quanvolutional Neural Network (GQNN). Second approach was directly using the original spectrogram image as an input to the CNN architecture. Figure 7 and 8 show the Confusion Matrix for GQNN and CNN Model respectively obtained after the experiment. Confusion Matrix represents the number of True Positives, True Negatives, False Positives and False Negatives obtained after running predictions on the testing dataset.

Table 2. Results on Covid dataset

Parameter	Quantum model	Classical model
Accuracy	73%	43%
Precision	0.78	0.43
Recall	0.78	0.18
F1-score	0.78	0.26

Fig. 7. Confusion matrix (GQNN model on Covid dataset)

Fig. 8. Confusion matrix (CNN model on Covid dataset)

4.3 Generalization of Model

Initially, when the model was applied on Covid-19 dataset, only a single channel was used which provided better results than the classical CNN. For generalization of the model, it is tested on Alzheimer's disease dataset [23]. For testing the model on Alzheimer's disease, VBSD dataset is used. It comes from wearable IoT devices. The age of the patients is around 65–92 years. Sampling frequency of each audio dataset is 44.1 kHz and duration is 1 s. Alzheimer's disease causes speech with pauses and unclearness because of the disease's effect on the memory which causes the patient memory barriers and difficulty in finding words. The spectrogram mostly contains features such as pauses, speech rate etc. The dataset directly contains the spectrograms of these audio samples. Inside the training dataset there are 250 spectrograms of normal patient and 254 spectrograms of AD patients. However, for experimental purpose only 100 samples containing 50 spectrograms of normal patients and 50 spectrograms of AD patients are used. After pre-processing of a single Alzheimer spectrogram image as shown in Fig. 9 through the QCNN model, 4 images are obtained which are shown in Fig. 10, 11, 12 and 13.

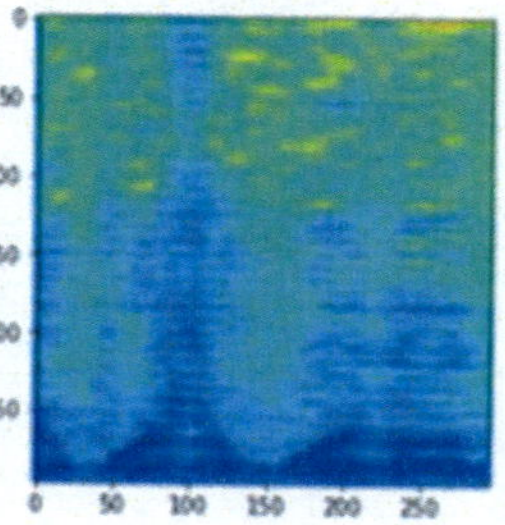

Fig. 9. Original Spectrogram for Alzheimer's disease

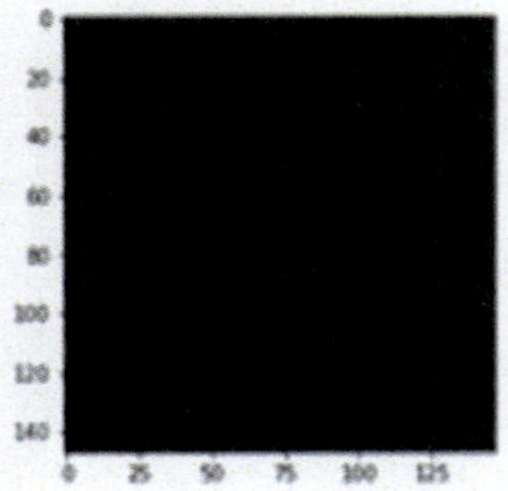

Fig. 10. Channel 1: Alzheimer spectrogram

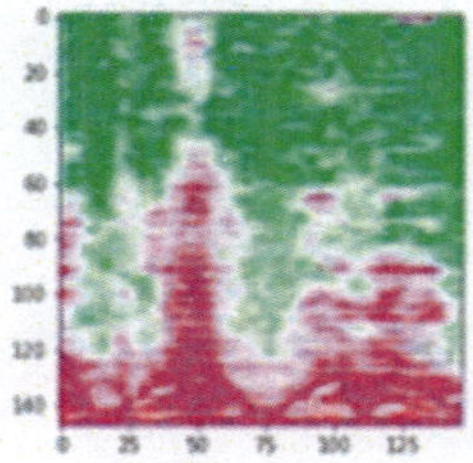

Fig. 11. Channel 2: Alzheimer spectrogram

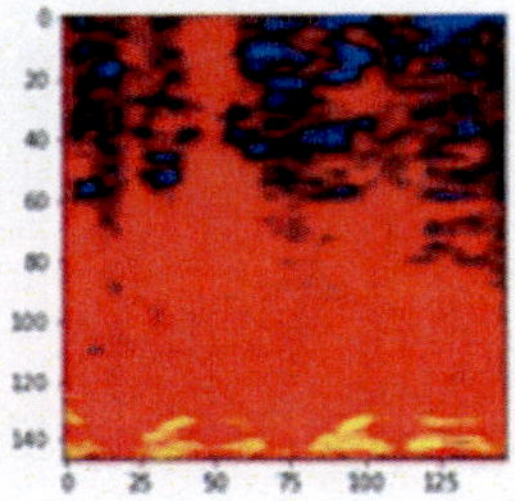

Fig. 12. Channel 3: Alzheimer spectrogram

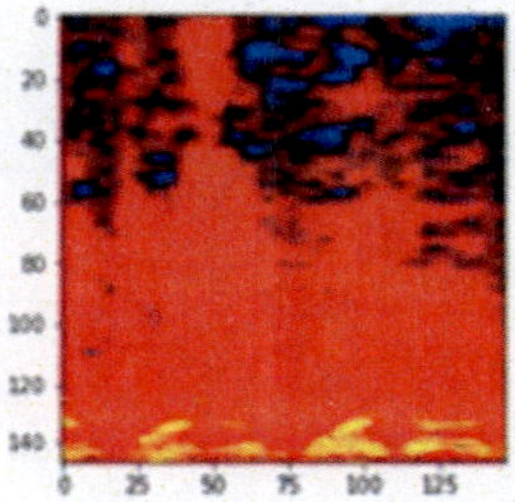

Fig. 13. Channel 4: Alzheimer spectrogram

However, the results obtained on choosing a single channel were not satisfactory. Since the original image splits into 4 channels the features are also distributed among these 4 channels. On observing it is found that 2nd and 3rd channel images are different and hence contain different features of the original image. Since single channel is not satisfactory, 2 channels were chosen for Alzheimer's dataset which provided better results. This provides the convolutional network to learn more features and improves generalizability of the model. Table 3 contains comparative results with different parameters using both the models. Figure 14 and 15 show the Confusion Matrix for GQNN and CNN Model respectively obtained after the experiment.

Table 3. Results on Alzheimer's dataset

Parameter	Quantum model	Classical model
Accuracy	65%	40%
Precision	0.75	0.57
Recall	0.69	0.31
F1-score	0.72	0.4

Fig. 14. Confusion matrix (GQNN model for Alzheimer dataset)

Fig. 15. Confusion matrix (CNN model for Alzheimer dataset)

5 Conclusion

In this research, a new quantum computing-based feature extraction technique for multi-dimensional tensors is proposed. This work assesses the proposed model specifically on colored spectrograms. The novel GQNN model considers all three channels (RGB color channels) and produces a set of 4 split images which has features extracted from the original image. Out of the 4 images obtained from pre-processing layer, one or two images are selected from them in a greedy way based on the resemblance of the pre-processed image with the original one. On taking this output image as the input for the further CNN architecture, a max accuracy of 73% was achieved on a train size of 35 samples, which is a better result as that achieved from its classical counterpart. Also for Alzheimer's disease dataset, the accuracy obtained was 65%, again better than Classical CNN. Hence the proposed architecture outperforms the classical CNN in the test cases considered. The model reduces size of the colored spectrogram which reduces the training time of the network, hence if a quantum computer pre-processes such colored spectrograms and stores it which can be further used for training a network then this can reduce storage cost which is extremely useful in streaming data. Also, one of the major finding was that when a dataset of small size was available, pre-processing the images using the proposed Quantum model helps in obtaining better accuracy as opposed to using the original images in the network. Hence quantum machine learning can be used where there is lack of data availability. The model works on any spectrogram dataset and can be used as a feature extraction technique for any kind of colored spectrogram image where color is an important factor. However, the choice of channels depends on the user application.

Though there are various methods still to be researched which involves using a trainable quantum circuit, increasing the training size, and testing on various train-test splits etc., the suggested architecture is predicted to pre-process more complex data and achieve higher accuracy in the NISQ era's practical quantum computer.

References

1. OSF. https://osf.io/tmkud/wiki/home/
2. Chatrzarrin, H., Arcelus, A., Goubran, R., Knoefel, F.: Feature extraction for the differentiation of dry and wet cough sounds. In: 2011 IEEE International Symposium on Medical Measurements and Applications, pp. 162–166 (2011). https://doi.org/10.1109/MeMeA.2011.5966670
3. Likitha, M.S., Gupta, S.R.R., Hasitha, K., Raju, A.U.: Speech based human emotion recognition using MFCC. In: 2017 International Conference on Wireless Communications, Signal Processing and Networking (WiSPNET), pp. 2257–2260 (2017). https://doi.org/10.1109/WiSP-NET.2017.8300161
4. Ranjan, R., Thakur, A.: Analysis of feature extraction techniques for speech recognition system. Int. J. Innov. Technol. Exp. Eng. (IJITEE) **8**(7C2), 197–200 (2019). ISSN 2278-3075
5. Han, W., Chan, C.-F., Choy, C.-S., Pun, K.-P.: An efficient MFCC extraction method in speech recognition. In: 2006 IEEE International Symposium on Circuits and Systems (ISCAS), p. 4 (2006). https://doi.org/10.1109/ISCAS.2006.1692543
6. Pahar, M., Klopper, M., Reeve, B., Theron, G., Warren, R., Niesler, T.: Automatic cough classification for tuberculosis screening in a real-world environment, 23 March 2021 (v1). https://doi.org/10.48550/arXiv.2103.13300
7. Boddapati, V., Petef, A., Rasmusson, J., Lundberg, L.: Classifying environmental sounds using image recognition networks. Procedia Comput. Sci. **112**, 2048–2056 (2017). ISSN 1877-0509. https://doi.org/10.1016/j.procs.2017.08.250
8. Khriji, L., Ammari, A., Messaoud, S., Bouaafia, S., Maraoui, A., Machhout, M.: COVID-19 recognition based on patient's coughing and breathing patterns analysis: deep learning approach. In: 2021 29th Conference of Open Innovations Association (FRUCT), pp. 185–191 (2021). https://doi.org/10.23919/FRUCT52173.2021.9435454

9. Badshah, A.M., Ahmad, J., Rahim, N., Baik, S.W.: Speech emotion recognition from spectrograms with deep convolutional neural network. In: 2017 International Conference on Platform Technology and Service (PlatCon), pp. 1–5 (2017). https://doi.org/10.1109/PlatCon.2017.7883728
10. Ullah Khan, S.: Quantum K means Algorithm, Dissertation (2019)
11. Kopczyk, D.: Quantum machine learning for data scientists, 25 April 2018. https://doi.org/10.48550/arXiv.1804.10068
12. Nielsen, M.A., Chuang, I.L.: Quantum Computation and Quantum Information, December 2010. ISBN 9781107002173
13. Chalumuri, A., Kune, R., Manoj, B.S.: Training an artificial neural network using qubits as artificial neurons: a quantum computing approach. Procedia Comput. Sci. **171**, 568–575 (2020). ISSN 1877-0509. https://doi.org/10.1016/j.procs.2020.04.061
14. Acar, E., Yilmaz, İ.: COVID-19 detection on IBM quantum computer with classical-quantum transfer learning. https://doi.org/10.1101/2020.11.07.20227306
15. Henderson, M., Shakya, S., Pradhan, S., Cook, T.: Quanvolutional neural networks: powering image recognition with quantum circuits, 9 April 2019. https://doi.org/10.48550/arXiv.1904.04767
16. Oh, S., Choi, J., Kim, J.: A tutorial on quantum convolutional neural networks (QCNN), 20 September 2020. https://doi.org/10.48550/arXiv.2009.09423
17. Yang, C.-H.H., et al.: Decentralising feature extraction with quantum convolutional neural network for automatic speech recognition. In: 2021 IEEE International Conference on Acoustics, Speech and Signal Processing (ICASSP) (2021). https://doi.org/10.48550/arXiv.2010.13309
18. Quan, Z., et al.: Cough recognition based on Mel-spectrogram and convolutional neural network. Front. Robot. AI **8** (2021). ISSN 2296-9144. https://doi.org/10.3389/frobt.2021.580080
19. Wibawa, M.S., Maysanjaya, I.M.D., Novianti, N.K.D.P., Crisnapati, P.N.: Abnormal heart rhythm detection based on spectrogram of heart sound using convolutional neural network, pp. 1–4 (2018). https://doi.org/10.1109/CITSM.2018.8674341
20. Tawhid, M.N.A., Siuly, S., Wang, H., Whittaker, F., Wang, K., Zhang, Y.: A spectrogram image based intelligent technique for automatic detection of autism spectrum disorder from EEG. PLoS ONE **16**(6), e0253094 (2021). https://doi.org/10.1371/journal.pone.0253094
21. Xu, Z.-J., Wang, R.-F., Wang, J., Yu, D.-H.: Parkinson's disease detection based on spectrogram-deep convolutional generative adversarial network sample augmentation. In: IEEE Access **8**, 206888–206900 (2020). https://doi.org/10.1109/ACCESS.2020.3037775
22. Liu, L., Zhao, S., Chen, H., Wang, A.: A new machine learning method for identifying Alzheimer's disease. Simul. Model. Pract. Theory **99**, 102023 (2020). ISSN 1569-190X
23. Liu, L.: VBSD Dataset (2020). Available:Github. https://github.com/LinLLiu/AD

Opinion Mining from Student Feedback Data Using Supervised Learning Algorithms

Malti Bansal(✉), Shreya Verma, Kartik Vig, and Kartikey Kakran

Department of Electronics and Communication Engineering, Delhi Technological University (DTU), Delhi 110042, India
maltibansal@gmail.com

Abstract. This paper examines the mining of ideas using learning algorithms monitored to determine student polarity feedback based on previously defined aspects of teaching, learning, course content, examination pattern, laboratory, library facilities, and extra co-curricular activities. The research involves using an aggregate of Machine Learning (ML) and Natural Language Processing (NLP) strategies with open student reaction information gathered from the University's class module survey results. Furthermore, this paper imparts a look at the overall performance of Naïve Bayes algorithms. The survey data are used for processing facts and to train algorithms for binomial separation. Trained algorithms are also able to predict the polarity of a scholar's opinion, based totally on excluded features which include coaching, mastering, direction content, examination sample, laboratory, library centres, more co-curricular sports, and many others.

Keywords: Naïve Bayes algorithm · Dataset · Precision · Learning algorithms · Polarity

1 Introduction

A teacher is the backbone of any university/educational institution. Although institutions have their own procedures for monitoring the effectiveness of their intelligence, the quality of education, etc. the foremost method is to take feedback from students about the teaching skills of staffs. Students provide their feedback in text format to present their reviews or simply rate it from 0 to 10 based on the quality of teaching they have experienced. The data used in this study is the students' response data to generate frequently commented reviews and the names of their opinions [1–5]. Emotional sections such as document, sentence and feature level are available to summarize the required reviews. In this paper, an ML algorithm has been used to analyse emotions. This assists the service provider / curriculum developer and allows the user to select the appropriate curriculum without spending too much time. The process of extracting important information in text format, is called emotion mining. Analysing feedback from students helps educational institutions improve their teaching quality [6–8].

J. I.-Z. Chen et al. (Eds.): ICIPCN 2022, LNNS 514, pp. 411–418, 2022.
https://doi.org/10.1007/978-3-031-12413-6_32

2 Literature Survey

Sentiment analysis is a natural language processing technique that identifies the emotional tone behind a text format. It contains content formats for sensing text and specific information, such as data mining formatting, machine learning, and AI. As the sentiment analysis system provides the organization to collect information in formal and informal formats from online sources such as email, web-chats, forums, and comments, the process can also be used to alleviate various emotional states such as happiness, sadness, love, etc.

Classification Technique: It is also one of the most popular methods of drilling, targeting and providing accurate predictions of the target in different data environments [9–12]. Decision trees are a non-parasitic controlled learning method used for segmentation and regression, in which the main motivation is to create a model that predicts the number of rooted differences by studying the simple decision rules found in the data components. Being able to manage numerical and phase data, some techniques are generally specialized in analyzing datasets that have a single type of variance, but it is a category that is able to classify multiple categories into databases.

Network Classifiers' heads comprise neurons, arranged in different categories, that turn a vector like input into a specific output, each neuron takes an input and does a specific task on it, and then transmits output to the next.

Layer: It is made using Rapid miner, which is used for the division of ideas. A certain weight is already associated with senses. Background data transmitted from each neuron to the other layers acts as training data. Input set, parallel weights in neurons, as well as trained neurons are key parameters for a neural network classifier basically making predictions [13–16].

KNN is an ML-monitored ML algorithm that relies on specified input data to read a specific function that provides the appropriate output when inserting new unlabeled data. It does not have specific parameters, as the slow learning algorithm is widely used for website purposes, where the data is taken from different classes to obtain new sample points. It works in two parts, first placing samples close to k between the test file database and comparing the same points in each document. The similarity effect next to the reference test document helps to capture the weights of the previously defined categories. The second step is to summarize all the weight of the reference document next to k to find out if it is possible for each group [17–20].

Data pre-processing is used to prepare all the data and make it fit the model. After the creation of the model, it is not always possible to always get clean and formatted data for processing. It is necessary to clean and format the data before processing, so hence pre-processing works. In this model, the dataset from its response is processed in advance such as spacing, and commas, to make the data eligible for further processing [21–23]. All English responses are extracted and stored in another file. And some words that are defined as tags are created that explain many meanings of the word. These terms are then classified and stored in a specific class used for model training. Names are called labels and some classes are featured.

3 Methodology

This section describes the database and steps used in the study and analyses the performance of different dividers. 'Data Collection and Sample' is used as a tool to display the amount of data collected and the amount required. This helps the sample to enable the system to take command of the system rating, process, or problem. To suggest sampling, it is important to determine the sample size and sample frequency, where the data is regularly collected to monitor the process. It is important to confirm and regularly update the appropriate sample size and to specify the type of data to be collected. The data collected should be timely as it is easy to collect incorrect data [24–26]. The main purpose is to ask the expected questions that can be used to analyse the data and what additional data is most desirable. While doing this for additional data, it is important to keep a record of additional information (Fig. 1).

Fig. 1. Model architecture

4 Working

The research is performed on a particular university. In survey, a feedback form is provided to students to complete certain fields related to university. This research form serves as the feedback data for the model. Then this response data is processed in advance and data purification is performed. In pre-processing, English responses are extracted and stored in a separate file. Response data is divided into sentences and these sentences are also separated by tokens. Spaces and punctuation marks are removed, with certain token tags associated with NLP. These markers describe in great detail a particular word. For the purpose of supervised training, this data is divided into positive and negative groups and is used for model training. Some selected terms such as labels and classes are featured. After extracting, the separating element occurs when certain elements and labels are extracted from the database using the N-gram method [27, 28]. To understand the diversity of ideas, different words have been chosen to understand the polarity aspect of the meaning of these words and the same meaning of them used in Rapid Miner. After extracting feature data, it is divided into three categories e.g. Good, Medium and Bad Class. To classify the data, the Naïve Bayes classifier is used. With a multiclass scenario, Naïve Bayes works better in this scenario than with other algorithms. Based

Fig. 2. Flow chart

on the polarity of the ideas, words are categorized according to categories. This process is repeated several times until all the text data are separated (Fig. 2).

Naïve Bayes Classifiers:
It is a collection of different algorithms, which works on the Bayes Theorem principle. It is the parent of all algorithms in which all algorithms share the same goal when each pair of elements represents each other.

Separating Data Using SVM:
The SVM model is a basic representation of points in space, which is simplified by the same examples of different categories separated by a wide gap. When SVM line-based editing can do well in indirect partitions, mapping is clearly obtained.

Steps to Test:
Accuracy: - The estimation of the made predictions which are correct for the total predictions made.
Precision: - It is a measure of true positives for true positives and false positives.
Recall: - It is a measure of true positives for true positives and false negatives.

5 Results and Discussion

Using the model [29, 30], the view of the test data became predicted and in comparison with the actual goal price. The confusion matrix is adjusted based totally at the actual and predicted value of advantageous, bad, and intermediate values. This trouble is taken into consideration a 3-magnificence trouble. The goal values are fantastic (A), poor (B) and intermediate (C). Based on the uncertainty of the accuracy of different metrics i.e. accuracy, and f-measure are assessed and the output is projected in Table 1.

Table 1. Outputs of the project

Feature	Positive	Neutral	Negative
Teacher feedback	7	1	9
Course content	10	3	9
Examination pattern	17	1	4
Laboratory	14	2	6
Library facilities	15	4	3
Extra co-curricular zctivities	19	1	2

Fig. 3. Representation of results in Bar graphs

6 Conclusion and Future Scope

In this study, basic pre-processing strategies and e hybrid Naïve Bayes algorithm are used for opinion mining to determine student feedback based on predefined aspects of teaching. The cleaned database is incorporated into the machine learning strategies available in the Multinomial Naïve Bayes. All databases are divided into training databases and test databases. In each algorithm, a model is developed based on a set of training. In future, different affiliation rule mining and machine learning algorithms may be used to locate the most constructive set of rules to extract capabilities and opinionated terms. Moreover, the work may aim to concentrate on the power of opinion words and categorize them into positive and negative clauses. An integrated framework would be projected in an effort to visualize the outcomes of the sentiment-based evaluation and export them. Future research are encouraged to feature more phrases in the lexicon dictionary to enhance the results of students' comments classification accuracy. The system have to additionally be brought with a spelling and grammar checker for verifying each comment. The Naïve Bayes classifier can be implied to explore other fields or other purposes, wherein instructional information extraction which include placement evaluation is required.

References

1. Kim, Y., Street, W., Menczer, F.: Feature selection for unsupervised learning via evolutionary search. In: Proceedings of 6th ACM SIGKDD International Conference on Knowledge Discovery and Data Mining, Washington, DC, pp. 365–369, August 2000

2. Dasgupta, A., et al.: Feature selection methods for text classification. In: Proceedings of the 13th ACM SIGKDD international conference on Knowledge discovery and data mining. ACM (2007)
3. Bansal, M.P.: Performance comparison of MQTT and CoAP protocols in different simulation environments. In: Ranganathan, G., Chen, J., Rocha, Á. (eds.) Inventive Communication and Computational Technologies. LNNS, vol 145, pp. 549–560. Springer, Singapore (2021). https://doi.org/10.1007/978-981-15-7345-3_47
4. Br, H., Tews, E., Rling, G.: Improving feedback and classroom interaction using mobile phones. In: Proceedings of Mobile Learning, pp. 55–62 (2005)
5. Akkoyunlu, B., Soylu, M.Y.: A study of student's perceptions in a blended learning environment based on different learning styles. Educ. Technol. Soc. **11**(1), 183–193 (2008)
6. Bansal, M., Goyal, A., Choudhary, A.: Industrial Internet of Things (IIoT): A Vivid Perspective. In: Suma, V., Chen, J.-Z., Baig, Z., Wang, H. (eds.) Inventive Systems and Control. LNNS, vol. 204, pp. 939–949. Springer, Singapore (2021). https://doi.org/10.1007/978-981-16-1395-1_68
7. Vijayarani, S., Ilamathi, J., Nithya, S.: Preprocessing techniques for text mining - an overview. Int. J. Comput. Sci. Commun. Net. **5**, 7–16 (2015)
8. Pang, B., Lillian L.: Opinion mining and sentiment analysis Found. Trends Inf. Retriev. **2**, 1–135 (2008)
9. Bansal, M., Sirpal, V., Choudhary, M.K.: Advancing e-Government Using Internet of Things. In: Shakya, S., Bestak, R., Palanisamy, R., Kamel, K.A. (eds.) Mobile Computing and Sustainable Informatics. Lecture Notes on Data Engineering and Communications Technologies, vol. 68, pp. 123–137. Springer, Singapore (2022). https://doi.org/10.1007/978-981-16-1866-6_8
10. Bansal, M., Nanda, M., Husain, M.N.: Security and privacy aspects for Internet of Things (IoT). In: 2021 6th International Conference on Inventive Computation Technologies (ICICT), pp. 199–204 (2021). https://doi.org/10.1109/ICICT50816.2021.9358665
11. Ravi, K., Ravi, V.: A survey on opinion mining and sentiment analysis: tasks, approaches and applications. Published Knowl. Based Syst. **89**, 14–46 (2015)
12. Bienkowski, M., Feng, M.: Enhancing Teaching and Learning Through Educational Data Mining and Learning Analytics. Department of Education, Office of Educational Technology, October 2012
13. Bansal, M., Gupta, S., Mathur, S.: Comparison of ECC and RSA algorithm with DNA encoding for IoT security. In: 2021 6th International Conference on Inventive Computation Technologies (ICICT), pp. 1340–1343 (2021). https://doi.org/10.1109/ICICT50816.2021.9358591
14. Bansal, M., Garg, S.: Internet of Things (IoT) based assistive devices In: 2021 6th International Conference on Inventive Computation Technologies (ICICT), pp. 1006–1009 (2021). https://doi.org/10.1109/ICICT50816.2021.9358662
15. Tribhuvan, P.P., et al.: A peer review of feature based opinion mining and summarization. Int. J. Comput. Sci. Inf. Technol. **5**(1), 247–250 (2014)
16. Tripathi, G., Naganna, S.: Feature selection and classification approach for sentiment analysis. Mach. Learn. Appli. Int. J. (MLAIJ) **2**(2) (2015)
17. Bansal, M., Adarsh, N., Kumar, N., Meena, M.: 24×7 smart IoT based integrated home security system. In: Fourth International Conference on Inventive Systems and Control (ICISC), vol. 2020, pp. 477–481 (2020). https://doi.org/10.1109/ICISC47916.2020.9171051
18. Bansal, M., Oberoi, N., Sameer, M.: IoT in online banking. J. Ubiqu. Comput. Commun. Technol. (UCCT) **2**(4), 219–222 (2020)
19. Bansal, M., Sirpal, V.: Fog computing-based Internet of Things and its applications in healthcare. J. Phys. Conf. Ser. **1916**(012041), 1–9 (2021)

20. Chauhan, G.S., Agrawal, P., Meena, Y.K.: Aspect-based sentiment analysis ofstudents' feedback to improve teaching–learning process. In: Satapathy, S.C., Joshi, A. (eds.) Information and Communication Technology for Intelligent Systems. SIST, pp. 259–266. Springer, Singapore (2019). https://doi.org/10.1007/978-981-13-1747-7_25
21. Drus, Z., Khalid, H.: Sentiment analysis in social media and its application: Systematic literature review. Proc. Comput. Sci. **161**, 707–714 (2019)
22. Bansal, M., Priya: Machine Learning Perspective in VLSI Computer-Aided Design at Different Abstraction Levels. In: Shakya, S., Bestak, R., Palanisamy, R., Kamel, K.A. (eds) Mobile Computing and Sustainable Informatics. LNDECT vol 68, pp. 95–112. Springer, Singapore (2022). https://doi.org/10.1007/978-981-16-1866-6_6
23. Bansal, M., Chopra, T., Biswas, S.: Organ simulation and healthcare services: an application of IoT. In: 2021 6th International Conference on Inventive Computation Technologies (ICICT), pp. 205–208 (2021). https://doi.org/10.1109/ICICT50816.2021.9358677
24. Abirami, A.M., Gayathri, V.: A survey on sentiment analysis methods andapproaches. In Proceedings of the 2016 Eighth International Conference on Advanced Computing, pp. 72–76. IEEE (2017)
25. Bansal, M., Prince, Yadav, R., Ujjwal, P.K.: Palmistry using machine learning and OpenCV. In: 2020 Fourth International Conference on Inventive Systems and Control (ICISC), pp. 536–539 (2020). https://doi.org/10.1109/ICISC47916.2020.9171158
26. Bansal, M., Harsh: Reduced instruction set computer (RISC): a survey. J. Phys. Conf. Ser. **1916**(012040), 1–14 (2021)
27. Tan, A.-H.: Text mining: The state of the art and the challenges. In: Proceedings of the PAKDD 1999 Workshop on Knowledge Disocovery from Advanced Databases, Vol. 8, (1999)
28. Kim, S.M., Rafael: Sentiment Analysis in Student Experiences of Learning. Available at ResearchGate.com
29. Bansal, M., Malik, S., Kumar, M., Meena, N.: Arduino based smart walking cane for visually impaired people. In: Fourth International Conference on Inventive Systems and Control (ICISC), vol. 2020, pp. 462–465 (2020). https://doi.org/10.1109/ICISC47916.2020.9171209
30. Bansal, M., Singh, H.: The Genre of applications requiring the use of IoT in day-to-day life. Int. J. Innov. Adv. Comput. Sci. (IJIACS) **6**(11), 147–152 (2017)

Blind Assistance System Using Machine Learning

Naveen Kumar, Sanjeevani Sharma, Ilin Mariam Abraham, and S. Sathya Priya(✉)

Department of Computer Science and Engineering, Hindustan Institute of Technology and Science, Chennai, India
{18113100,18113101,18113117}@student.hindustanuniv.ac.in, sathyap@hindustanuniv.ac.in

Abstract. Blindness is one of the most frequent and debilitating of the various disabilities. There are million visually impaired people in the globe, according to the World Health Organization (WHO). The proposed system is designed to aid visually impaired persons with real-time obstacle detection, avoidance, indoors and out navigation, and actual position tracking. The gadget proposed is a camera-visual detection hybrid that performs well in low light as part of the recommended technique, this method is utilized to detect and avoid impediments, as well as to aid visually impaired persons in identifying the environment around them. A simple and effective method for people with visual impairments to identify things in their environment and convert them into speech for improved comprehension and navigation. Along with these, the depth estimation, which calculates the safe distance between the object and the person, allowing them to be more self-sufficient and less reliant on others. This were able to achieve this model with the help of TensorFlow and pre-trained models. The approach is suggest is dependable, inexpensive, practical, and practicable.

Keywords: Depth estimation · Object detection · Single shot detection · TensorFlow

1 Introduction

The goal of restoring vision to those who have been blinded from the outside is the focus of a comprehensive evaluation in both design and medication. Exploring their environment without colliding with any impediments is one of the most difficult challenges for vision-impaired people. The vision handicapped have used long sticks and guide hounds to complete this test for a long time. These long sticks and guide dogs, on the other hand, will only provide information about nearby obstacles within a limited range, while excluding weather information. This project was created by merging the ssd algorithm into the field of machine learning using the TensorFlow api packages. Having the ability to provide a completely aiding guide by offering vocal feedbacks is one significant factor in all prior systems where object detection is regarded a key step for the visually challenged in their navigation.

J. I.-Z. Chen et al. (Eds.): ICIPCN 2022, LNNS 514, pp. 419–432, 2022.
https://doi.org/10.1007/978-3-031-12413-6_33

Millions of people suffer from a variety of issues, including visual impairment. Even the most basic tasks are challenging for them. Even in the privacy of their own house or the struggle to travel from one point to another in the business without becoming dependent on others. In truth, according to a new study, blindness can be restored to a degree, but in the long run, the majority of people would require assistance to carry out their daily duties.

Even though there are proven ways to help these people, such as a cane to help them travel or a guide dog, they are insufficient and prone to errors. Despite the fact that they can help them avoid any obstacles in their way, they cannot help them see what is right in front of them.

As a result, our contribution to solving this problem statement is to develop a blind aid system that enables object detection and real-time speech conversion so that they can hear it as and when needed, as well as depth estimate for safe navigation. This is encapsulated in a single system that is utilized to make it easier for the blind to function in society.

The goal of this exercise is to find a solution. This would not only assist these exceptional individuals in avoiding obstacles and managing their daily tasks, but it would also allow them to use their brains to see their surroundings. Giving kids real-time voice feedback is one method to help them become more self-reliant and safer. This device will assist blind individuals by tolerating their voice directions and applying picture handling algorithms to perceive items, as well as providing the client with a hearable result so they can explore the bar. This framework will also recognize a few major events, such as a "Washroom" sign, which will provide information. The visually challenged person can recognize the image as soon as it is discerned. Objects, signs, and boards, among other things, are likely to be detected by the framework. This will assist the vision-impaired person in completing duties and organizing daily errands.

Our project's main goal and motive is to achieve this. The rest of the paper is formatted as follows. Section 2 describes the literature review. Section 3 explains the proposed system. Section 3.1 contains the prerequisites. The suggested system is defined in Sect. 4. Section 4 shows how to design a system. The suggested system and its implementation are detailed in Sect. 5. Sections 6 and 7, respectively, provide the discussion and conclusion.

2 Literature Survey

In the paper proposed by CK Lakde, a gadget called a multi sensor probe is used to conduct the duty of people detection anytime the user is walking in a busy area. Infrared radiation is used by the PIR sensor to detect or check the movement of the person. The target distance and velocity provided by the Sonar module are used to calculate the actual distance. Smart gloves and a walking stick were used to find obstacles [1]. The walking stick serves as a reminder to the blind of the obstacle. If a collision, information will be transmitted to the caregivers via an internet portal.

A visually impaired system was invented by Devashish Pradeep Khairnar. The proposed VC (Vision Challenged) Assistant System has four components that assist persons who are visually impaired: Obstacle recognition, avoidance, inside and outside navigating, and actual position sharing are all features available. The recommended device is a hybrid of a smart glove and smartphone software that performs well in low-light situations. The suggested solution includes the use of a smart glove to find and avoid obstacles [2], as well as to aid visually impaired persons in recognizing their surroundings. The smart phone-based obstacle and object detection system detects a variety of items in the surroundings.

Deepak Gaikwad recommended that the system's functioning start with infrared sensors detecting impediments. Infrared sensors detect obstacles in three directions: front, left, and right. It uses sensor circuitry to detect impediments. This is then supplied as an input to the microcontroller [3]. This controller is connected via Bluetooth to an Android phone, which creates speech depending on microcontroller commands. The influence of multipath transmission on multi-server offloading was predicted and investigated. They tested this approach using both real-world and simulation 4 versions.

Ali Khan and Aftab Khan wrote a study with the goal of developing something exclusively for the blind. This study resulted in the development of an obstacle detection system using ultrasonic sensors. The wearable garment has ultrasonic sensors [4], as well as a vibration device and a buzzer. Sensors check the user's surroundings and let them know of any obstructions via vibration and a buzzer sound. An Arduino-based ultrasonic blind walking stick is described in this paper. Because the blockage is not detected with a standard stick, it is useless for visually impaired persons. An ultrasonic sensor is used to detect the length between objects and the Smart Walking Stick. The user may hear the Buzzer and learn about the obstacle when items or barriers come within range of the ultrasonic sensor. It finds impediments in the user's path and allows him to move freely.

Ashish Mishra and Changzhi Li invented a gadget that detects barriers using RADAR architecture. It is based on the RADAR's transmitter and receiver. Miniaturization and portability are the device's advantages. Shubham Suman, Sushruta Mishra, Kshira Sagar Sahoo, and Anand Nayyar worked on echolocation and image processing using an image detecting sensor. The static or dynamic items put in the region are named using these collected photos. They find barriers and distances using an ultrasonic sensor. It has a GPS module that can assist blind people in navigating [5].

When coming to our project, it has been made simple, reliable, and easy to use system which is cost effective and reduces much hassle. With the help of real time audio voice feedback, the blind pepoles are navigating by the inputs which they get. So rather than having an alarm or a buzzer, it is more important to have the object detected and send as real time inputs with the help of depth estimation. It will also say if they are in safe distance from the object and re-route our paths [6].

3 Proposed System

The design aims to replace existing technologies of eyeless navigation systems based on detectors and buzzers with a simpler yet effective way of creating an eyeless backing system based on machine literacy, where it can describe an object while receiving the real-time voice feedback and depth estimation. The proposed system is more efficient and reliable. The system is set up to record real-time frames and execute all calculations.

The object's class will be turned into dereliction voice notes after speech module testing, and will also be sent to the eyeless victims for assistance. In this an alert mechanism is add to find the object, where an approximate will be calculated [15]. If that Eyeless Person is truly close to the frame or is far down at a safer area, nevertheless, it will induce voice- grounded labor's as well as distance units. The discipline of computer vision is concerned with detecting meaningful things in photos and movies (by creating rectangle boxes around them in our case). It can also offer accurate distances and convert labeled texts into voice answers. Our strategy is trustworthy, cost-effective, feasible, and practical. This allows the blind to be self-sufficient in society and to avoid the societal barriers that still exist. An Integrated Machine Learning System that allows eyeless sufferers to recognize and classify common everyday items in real time, generate verbal feedback, and calculate distance, producing alerts whether they're veritably close or veritably far down from the object was one such endeavor on our side. Handicap Research The same approach can be used to enforce medium. Now that the world is changing dramatically and new discoveries in medical science are occurring, it is necessary to improve the status of the visually bloodied as well. To make them more independent in all aspects of their daily lives, as well as from a business one, a change like this was more vital to bring about and implement (Fig. 1).

Fig. 1. Block diagram of the system

The camera input is shown in this diagram will capture the image and send it to storage as well as it is stored and pre-processed and try to identify the image and after identifying the image, it will generate a voice singel which can be heard by the user. Voice feedback will also be delivered once the image has been captured.

3.1 Limitation

The basic method used was object avoidance and object detection. It also includes outdoor location sharing which is quite a tedious task. In exiting system it only have a buzzer to detect the object and alarms will be sent accordingly to the visually impaired. With reference to the journal paper proposed by Deepak, only infrared sensors are used in this paper which is small, cheap and has low power consumption. Also, Bluetooth is not reliable in this [7] system.

In 'Wearable Navigation Assistance system for the blind and visually impaired' by Ali khan, an algorithm for reacting to close obstacles and motion commands gets generated with haptic belts. There was also a system which uses stereo vision and image processing methods. Distance between the camera and the laser must be constant. This may not work effectively on shiny surface as laser intensity may decrease. In regards to the Teleguidance based remote navigation assistance for visually impaired and blind people- usability and user experience proposed by Babra Chaudary, there are only two vibrating actuators to guide the visually impaired left and right [8].

It had been incorporated object detection along with converting them into real time audio feedback which will help them to navigate without much hassle. Along with that, they can also get depth estimation features which will help them in their safe traversal.

4 Methodology and System Design

Our current application has the following design flow. It basically consists of three modules. They are:

1. Object Detection
2. Converting the detected object into Speech
3. Depth Estimation

4.1 Object Detection

Object detection is the first module in our project, and it serves as its foundation. It essentially entails detecting near and far objects using datasets for which the model has been trained. Because our application's target users are visually challenged persons, detection is crucial [14]. It is detecting the object with the help of the system webcam. There are a set of pre-defined boundary boxes with certain height and width. These boxes are defined to capture the scale and ratios of the object classes. SSD (Single Shot Detector) algorithm uses a single layer of convolutional network to detect the object in single shot multi box detector. It uses multi-layer convolutional neural network to classify the boxes to any of the defined classes [8]. The anchor box with the highest degree of overlap with the object handles predicting the object's class. Once the object is trained, this property is used for predicting the detected object. After the object detection is completed, it compares the detected object with the pre-defined trained COCO dataset. After the object gets detected, it compares with the pre-trained dataset to recognize the name of the object and it will be shown to the user. In this project, it is using the COCO dataset to store all the images which can be character-recognized using OCR (Optical character recognition [9]. So once an object is in front of the camera, it can get the outputs from the system with the help of these pre-trained data sets.

Algrothim is for Object dection :

```
Step 1: declare and define a backbone map of 7X7 grid;
Step 2: declare anchor box for each grid //49 anchor boxes in this case;
        for each anchor box:

                get the shape and size for each anchor box;
                declare object class and location;
Step 3: for each anchor box:
        if the overlapping degree of the anchor box is maximum:
                define object class and location;
        permute and extract
                object class  and location from the anchor box details;
        detect the object from
                anchor box array, location, object class, lighting etc.
        for each element from anchor box array:
                get its shape, size, pixel pattern, aspect ratio and lighting;

Step 4: test_cases:
        if aspect ratio is like m:n where m > n then
                object has larger length
        else
                object has larger height

Step 5: if pixel pattern has soft edges
                object is of complex structure and has curved edges
        if the object is in dark lighting
                object is in bedroom //inclines mostly to the bedside objects

Step 6: return the object class and its location;
```

4.2 Converting the Detected Object into Text

This module consists of turning text to speech, which is critical since it aids blind victims in identifying and analyzing who and what is close by and around them. It assists kids in navigating and understanding what is happening around them.

4.3 Depth Estimation

The processes and calculations used to create a depiction of the spatial construction of a scene are known as profundity evaluation or extraction highlight. To put it another way, it's used to figure out how far two objects are apart. Our model is used to assist visually impaired people [13], and it anticipates giving them ahead notification of any problems that may arise. To accomplish, it will look at the distance between the deterrent and the individual in the given situation. A rectangular box is created around the item after it has been identified.

5 Implementation

Compatibility with Python and library set up obstacles are necessary for efficient implementation of this Model. To tell you the truth, this was one of the most difficult sections of the project for me to complete. Thank you to stack overflow and Python Unofficial Binary Releases for providing pre-built docs, which you may download from here if your system supports it.

5.1 Anchor Box

Every grid cell in SSD can have several anchor/earlier containers. These pre-described anchor containers are each responsible for a length and form within a grid cell. SSD use the matching section during training to ensure that the anchor field is properly aligned with the bounding containers of each floor fact item inside an image [10]. The anchor field with the greatest degree of similarity with an object is responsible for forecasting that item's elegance and proximity. Once a community has been trained, these resources are used to teach the local surroundings as well, in order to anticipate the recognized things and their locations. Smaller Every anchor field has a part proportion and a zoom level specified explicitly. However, that not all of the objects are rectangular. Some are narrower, while others are longer and wider to varying degrees. To account for this, the SSD structure allows pre-defined anchor container component ratios to be used. At each zoom/scale level, the usage of the depth ratio of the anchor connected with each grid cell may be tailored to the outstanding component ratios.

5.2 Zoom Level

The anchor bins do not always have to have the same length as the grid cell. The customer is undoubtedly curious about the location of each smaller or larger object within a lattice cell. The zooms parameter is used to define how a horrendous part of the anchor containers has to be raised or decreased with respect to each network cell.

5.3 Depth Estimation

Depth estimates or extraction function refers to the procedures and algorithms used to produce a representation of a scene's spatial form. Miles is used to calculate the distance between things in less basic terms.

Algorithm is for Depth Estimation:

Step 1: detect whether the object is within the frame;

Step 2: if object fits in the frame then:
detect objects from ->

each grid box;

the detection score;

number of detection attempts;

object class;

Step 3: for each box:

evaluate y coordinate from the top;

evaluate x coordinate from the left;

evaluate y coordinate from the bottom;

evaluate x coordinate from the right;

Step 4: calculate mid_x from:

mid_x = avg(left x coordinate,right x coordinate);

mid_y = (boxes[0][i][0] + boxes[0][i][2])/2;

mid_x = avg(top y coordinate,bottom y coordinate);

Step 5: calculate apx_distance from:

apx_distance = (1 - right x coordinate - left x coordinate) ^ 4;

round the apx_distance to first decimal point; //round until one-tenth part

Step 6: plot a dot at the centre of the object;

Step 7: calculate the score from the grid boxes;

Step 8: if score is > 50% then:

object is detected;

step 8: if apx_distance < 0.5 and mid_x > 0.3 and mid_y < 0.7 then:

the object is closer and hence

subject the image to the recognition module;

5.4 Voice Estimation

Following the detection of an object, it is critical to inform the person on his or her way of the presence of that thing. PYTTSX3 is a crucial part of the voice generating module. Pyttsx3 is a Python conversion module for converting text to speech. The text is detected by scanning and analyzing the picture. As a result, Python-tesseract recognizes and "reads" text encoded in [11] pictures. Furthermore, these lines link to a pyttsx. Engine instance, and an application calls the factory method pyttsx.init(). This in turn helps to generate voice for the detected text in our project.

As an output, audio commands are created. "Warning: The object (class of object) is too close to you," it says if the thing is too close. Otherwise, if the item is at a safe distance, a voice is created that states, "The object is at a safe distance." This is done using libraries like Py torch, pyttsx3, Py tesseract, and engine.io.

Py torch is a machine learning library primarily. Py torch is mostly used in the audio field. Py torch aids with the loading of the voice file in mp3 format. It also controls the audio dimension rate. As a result, it may be used to change sound qualities such as frequency, wavelength, and waveform [12]. Looking at Py torch's capabilities can help confirm the wide range of audio synthesis choices available. So, these are the modules it is using to make the text to speech conversion.

6 Result and Discussion

6.1 Test

This project improves performance while also allowing for quick output delivery. This server-based project allows you to break down your paintings into components and recognise the central component of any system. This method allows for the development of reliable and enjoyable software from the inside out.

- Initially, it take real-time images from the webcam of a blind person's mobile phone, and a connection is established between the phone and the computer, after which the photos are sent to the user.
- The device in the computer will examine it with the aid of its APIs and SSD ALGORITHM, and it will determine the picture's self-assurance accuracy. Positive lessons like books, drinks, and the remote had a 98 percent accuracy rate.

- Following the evaluation of the photographs, it generate an output on a computer-based device, It is then converted to speech using voice modules and transmitted to the blind individual via wireless audio assistance devices. The information is then converted to speech using voice modules and supplied to the blind individual via wireless audio assistance devices.
- TensorFlow API is used for building and training the model and to reduce the distributed runtime. It is using COCO dataset which is pre-trained dataset to compare the detected object. SSD applies more convolutional layers to the backbone feature mapping and has each of these convolution layers as output an object detection results. With the help of the webcam system, it is detecting the object. A set of pre-defined boundary boxes with specific height and width are available. These boxes are used to stand for the object classes' scale and ratios. In a single shot multi box detector, the SSD (Single Shot Detector) technique uses a single layer of neural network to detect the object. It classifies the boxes into any of the defined classes using a multilayer convolutional neural network.

6.2 Result

The devices on which it was tested are listed below, and it produced the following result, which was analyzed along with the help matplotlib libraries.

Fig. 2. The cup

The end range is 0.3 units from the frame, and a range warning is provided since it is too near, with the speech output noting that it belongs to the cup class. It emits a warning linke cup when it gets too close to the frame.

The ultimate distance from the frame is 0.8 units, and no distance-based warning is sent since it is at a safer distance, the class recognition voice can be heard, and the class is distant.

The ultimate distance from the frame is 0.9 units, and there is no distance-based warning because it is at a safe distance; instead, a class identification voice is created, and the object's name is heard as bed.

Because it is at a safer distance, no distance basd alert is created, and the class recognition voice may be heard as expected. In a single frame, it can identify many objects.

Fig. 3. The remote

Fig. 4. The bed

Fig. 5. The chair

Fig. 6. The tv

The ultimate distance from the frame is 0.8 units, and there is no distance-based warning because it is at a safer distance. Instead, a class identification voice is created, and the object's name is heard as tv.

The suggested system successfully recognises 90 items, names them, and indicates their correctness. The version also estimates the distance between the item and the digital digicam and provides audio feedback as the user approaches the item. SSD Mobile net V1 and SSD Inception V2 were used to evaluate the data. The SSD Mobile net V1 version, on the other hand, showed a significant reduction in latency and improved object detection speed.

With the refersence of the above Figs. 2, 3, 4, 5 and 6 the table which is given below is made.

Table 1 refers to the accuracy level of the object detected.

Table 1. Objects and accuracy level

S. no	Objects	Accuracy level
1	Cup	99%
2	Remote	98%
3	Bed	96%
4	Chair	96%
5	Tv	96%
6	Person	90%

7 Conclusion

Using machine learning and pre-trained models, it is designed as Blind Assistance System that will assist in the detection of objects in the environment. In order to complete this project, The TensorFlow API, SSD architecture, and COCO Dataset is used in this. It use object detection and depth computation to turn the discovered item into speech. There are a variety of uses for this proposed system. It makes it easier for blind people to acquire, analyses, and translate information. Reference or scope in the future the study's purpose is to let visually impaired persons navigate freely so they can move fast while being safe. The device gives the blind person distance and object detection, speech awareness of the object.

It is optimistic that by integrating these services, it will be able to improve the application's effectiveness in the future.

Making this application into a hardware project to expand it. Creating a chatbot that allows the user to converse and engage. Installing GPS to know where you are in real time and making it an all-in-one system. Adding Web Support to it.

References

1. Santos, M.E.C., Taketomi, T., Sandor, C., Polvi, J., Yamamoto, G., Kato, H.: A usability scale for handheld augmented reality. In: Proceedings of the 20th ACM Symposium on Virtual Reality Software and Technology, pp. 167–176 (2014)
2. Lakde, C.K., Prasad, P.S.: Navigation system for visually impaired people. In: Proceedings of 2015 International Conference on Computation of Power, Energy, Information and Communication (ICCPEIC), pp. 0093–0098 (2015). https://doi.org/10.1109/ICCPEIC.2015.7259447
3. Gaikwad, D., Baje, C., Kapale, V., Ladage, T.: Blind assist system. Int. J. Adv. Res. Comput. Commun. Eng. **6** (2017). ISSN 2278-1021
4. Bose, P., Malpthak, A., Bansal, U., Harsola, A.: Digital assistant for the blind. In: Proceedings of 2017 2nd International Conference for Convergence in Technology (I2CT), pp. 1250–1253 (2017). https://doi.org/10.1109/I2CT.2017.8226327
5. Deepthi Jain, B., Thakur, S.M., Suresh, K.V.: Visual assistance for blind using image processing. In: Proceedings of 2018 International Conference on Communication and Signal Processing (ICCSP), pp. 0499–0503 (2018). https://doi.org/10.1109/ICCSP.2018.8524251
6. Awad, M., Haddad, J.E., Khneisser, E., Mahmoud, T., Yaacoub, E., Malli, M.: Intelligent eye: a mobile application for assisting blind people. In: Proceedings of 2018 IEEE Middle East and North Africa Communications Conference (MENACOMM), pp. 1–6 (2018). https://doi.org/10.1109/MENACOMM.2018.8371005
7. Felix, S.M., Kumar, S., Veeramuthu, A.: A smart personal AI assistant for visually impaired people. In: Proceedings of 2018 2nd International Conference on Trends in Electronics and Informatics (ICOEI), pp. 1245–1250 (2018). https://doi.org/10.1109/ICOEI.2018.8553750
8. Khan, A., Khan, A., Waleed, M.: Wearable navigation assistance system for the blind and visually impaired. In: Proceedings of 2018 International Conference on Innovation and Intelligence for Informatics, Computing, and Technologies (3ICT), pp. 1–6 (2018). https://doi.org/10.1109/3ICT.2018.8855778
9. Paul, J.L., Sasirekha, S., Mohanavalli, S., Jayashree, C., Moohana Priya, P., Monika, K.: Smart Eye for Visually Impaired-an aid to help the blind people. In: Proceedings of 2019 International Conference on Computational Intelligence in Data Science (ICCIDS), pp. 1–5 (2019). https://doi.org/10.1109/ICCIDS.2019.8862066
10. Divya, S., Raj, S., Praveen Shai, M., Jawahar Akash, A., Nisha, V.: Smart assistance navigational system for visually impaired individuals. In: Proceedings of 2019 IEEE International Conference on Intelligent Techniques in Control, Optimization and Signal Processing (INCOS), pp. 1–5 (2019). https://doi.org/10.1109/INCOS45849.2019.8951333
11. Khairnar, D.P., Karad, R.B., Kapse, A., Kale, G., Jadhav, P.: PARTHA: a Visually Impaired Assistance System. In: Proceedings of 2020 3rd International Conference on Communication System, Computing and IT Applications (CSCITA), pp. 32–37 (2020). https://doi.org/10.1109/CSCITA47329.2020.9137791
12. Al-Muqbali, F., Al-Tourshi, N., K. Al-Kiyumi, N., Hajmohideen, F.: Smart technologies for visually impaired: assisting and conquering infirmity of blind people using AI technologies. In: Proceedings of 2020 12th Annual Undergraduate Research Conference on Applied Computing (URC), pp. 1–4 (2020). https://doi.org/10.1109/URC49805.2020.9099184
13. Mohith, S.S., Vijay, S., Sanjana, V., Krupa, N.: Visual world to an audible experience: visual assistance for the blind and visually impaired. In: Proceedings of 2020 IEEE 17th India Council International Conference (INDICON), pp. 1–6 (2020). https://doi.org/10.1109/INDICON49873.2020.9342481

14. Khan, M.A., Paul, P., Rashid, M., Hossain, M., Ahad, M.A.R.: An AI-based visual aid with integrated reading assistant for the completely blind. IEEE Trans. Hum.-Mac. Syst. *50*(6), 507–517 (2020). https://doi.org/10.1109/THMS.2020.3027534
15. Patil, K., Kharat, A., Chaudhary, P., Bidgar, S., Gavhane, R.: Guidance system for visually impaired people. In: Proceedings of 2021 International Conference on Artificial Intelligence and Smart Systems (ICAIS), pp. 988–993 (2021). https://doi.org/10.1109/ICAIS50930.2021.9395973

Detection of EMCI in Alzheimer's Disease Using Lenet-5 and Faster RCNN Algorithm

A. Mohamed Rayaan, M. S. Rhakesh, and N. Sabiyath Fatima(✉)

Department of Computer Science and Engineering, B.S.A. Crescent Institute of Science and Technology, Chennai, Tamil Nadu, India
sabiyathfathima@crescent.education

Abstract. Alzheimer's Disease (AD) is a brain disease in which the brain shrinks and the brain cells die. The most common cause of memory loss (dementia) is the Alzheimer's disease. It will result in a decline in thinking, behavioral, and basic social skills, which will have an impact on an individual's abilities and daily functions. There are several symptoms associated with this disease, but one of the first is forgetting recent events or conversations. In clinical research, an MRI scan is commonly used to diagnose Alzheimer's disease. However, the findings of this prevalent type of diagnosis take longer to arrive. The purpose of this paper is to detect the Early Mild Cognitive Impairment (EMCI), which is the initial stage of the Alzheimer's disease. The detection of EMCI is done by using two advanced CNN algorithms called Lenet-5 and Faster Rcnn. Lenet-5 is used for well scaled image processing as the work involves MRI images and Faster Rcnn excels in finding and differentiating an object as the work's creation involves more than 6000 images. Segmentation of the brain images are imperative in surgical and treatment planning, so this paper can come as a great help. This paper can be applied in the field of medicine, where physicians use it to determine the Alzheimer's disease accurate and time efficient than normal diagnosis. The outcome of this work is that the accuracy is improved which is about 98%.

Keywords: Alzheimer's disease · Early Mild Cognitive Impairment · Lenet-5 algorithm · Faster Rcnn algorithm

1 Introduction

Each and every movement in the body is specifically handled by the brain. Even the responses and the senses are managed by the brain function. Brain can also strengthen the emotions. Alzheimer's disease is a gradual, irreversible form of brain deterioration. Every four seconds, somewhere in the world people are getting affected by the Alzheimer's disease. This disease particularly attacks the brain causing dementia or memory loss, continuously manipulating an individual's potentiality to think. This disease affects the neurons present in the brain. If a person is affected by this disease their average life span is anticipated as only 4–8 years. Every single people out of 10 are

J. I.-Z. Chen et al. (Eds.): ICIPCN 2022, LNNS 514, pp. 433–447, 2022.
https://doi.org/10.1007/978-3-031-12413-6_34

affected by this disease. This disease mainly targets the people of age above 65. This disease is also been diagnosed who are younger. This disease accounts for 60–80% of dementia cases. Brain MRI scanning and analysis is one of the most popular and preliminary Alzheimer's disease detection techniques, along with psychological assessments. Medical professionals review the MRI scans and evaluate potential factors that could disclose the presence of Alzheimer's disease, such as brain matter deterioration, tumors, and so on. At present, the Alzheimer's Disease detection is done using a patient's medical history, thinking ability test, physical test, and neuroimaging diagnostic test. Although manual evaluations of MRI data are effective in detecting the existence of Alzheimer's disease, this approach tends to slow down the speed with which findings are reached. This research presents detection of Alzheimer's disease automatically.

2 Related Works

S Hiroki fuse et al. [1] proposed a method using AP-type Fourier descriptor which is used as a first and second ventricle excluding the system lucidum was analysed. It made use of SVM. The accuracy was about 87.5% and is greater to the accuracy achieved using volume which is 81.5%. The limitations are that it achieved using volume ratio to intracranial volume which is a time-consuming process.

Priyanka Thakore et al. [2] proposed a method to detect and track Alzheimer's disease, where in detection, EEG database is filtered then noise and artifacts are removed using independent component analysis. Using wavelet transformation four features are extracted and classification is done by SVM. In monitoring system, patients are tracked by using GPS and GSM. The limitations are that EEG is a time-consuming process and accuracy may not be determined.

Aarthi Sharma et al. [3] proposed a method for detecting Alzheimer's disease in its early stages. The datasets of MCI, healthy patients, and memory loss patients were used to produce this work. EEG was made use. The limitations are that EEG is a time-consuming process.

A.N.N.P Gunawardena et al. [4] articulated detection to pre-detection of Alzheimer's disease fromimagnetic resonance image data. The detection of Alzheimer's disease is based on the clinical history and tests such as MMSE and PAL. The limitations are that the diagnosis of Alzheimer's disease is based on the examination of clinical history which leads to inaccuracies.

H.M. Tarek Ullah et al. [5] proposed a method of Alzheimer's disease detection (AD) from 3D brain image MRI data using CNN. This approach is fast, costless and is more reliable. CNN is used for image processing and is used to detect AD from 3d MRI image. The limitations are that the 3D image costs lot to create a 3D type dataset.

Hongming Li et al. [6] proposed a method for pre-determination of Alzheimer's disease (AD). This method made use of a deep learning model based on recurrent RNNs to acquire skill in informative representation and secular fluctuation of longitudinal cognitive measures in single patients. The limitation is that it consists of a large dataset which is time consuming.

Solale Tabarestani et al. [7] proposed a longitudinal prediction model that involved (a (MMSE) and (b) stationing a multiclass multimodal brain imaging categorizing process that includes 3 different known stages: cognitively normal (CN), Mild Cognitive Impairment (MCI), and Alzheimer's Disease (AD). The SVM algorithm was employed. The limitations are that the accuracy, F-score, sensitivity was moderate.

F. J. Martinez-Murica et al. [8] proposed a deep decomposition of MRI to explore neurodegeneration in AD. The aim was to predict neuro-psychological test outcomes based on MRI data. A self-supervised decomposition of MRI data using a deep convolutional auto encoder was performed. The prediction achieved R2 rates up to more 0.3. The limitations are that it did not support coloured images and the accuracy was moderate.

Pengbo Jiang et al. [9] proposed a new syndicate literacy articulation. An efficient iterative method was built and experiments were carried out utilising data from the AD neuroimaging project dataset to evaluate the suggested improved articulation. Baseline MRI is used to predict cognitive scores at several time points. The limitations are it did not support coloured images.

Solale Tabarestani et al. [10] proposed a method that consists data of 4 classes: (EMCI), (LMCI), (LMCI), and (NC). ADNI baseline MRI, FDG-PET, and CSF data from 1,458 patients were used to train profile specific atavism to estimate MMSE scores. The limitations are that it is a time-consuming process.

3 Proposed Methodology

3.1 System Architecture

The system architecture comprises of different phases and flows, as represented in Fig. 1. Initially it comprises of an imaging equipment from where the brain images are extracted which are also known as input images for the creation of the model. Then the images undergo image segmentation which is otherwise known as image preprocessing. The feature extraction is done for reducing the images into 4 classes and then the dataset is split into train dataset and test dataset, the train dataset trains the model. The test dataset helps to evaluate the model. Then the model is built using the two algorithms namely Lenet-5 and Faster Rcnn. After the completion of all these modules prediction is done whether the given MRI image is high-flown by Alzheimer's Disease (AD) or not. The prediction is only done when the user gives in the MRI image as an input, where the user gets the output as any one of the four classes namely moderate demented, mild demented, non demented, very mild demented.

Fig. 1. Architecture diagram

3.2 Module Description

1) Data collection

Data collection is the process of collecting datasets. The datasets are extracted from an online source called Kaggle. The datasets are extracted based on the four required categories, which are mild demented, moderate demented, very mild demented, and nonidemented.

2) Pre-Processing

Preprocessing is an important technique which is used to edit the datasets i.e., Reforming extracted datasets into a clean set of data. In this work the image which are extracted from Kaggle undergoes various steps like the image is enhanced, the images are made sure that each image are of same size and same color. Then the images are converted from rgb to grey scale.

3) Feature Extraction

The process of feature extraction is a proportionally reduction strategy which is used for reduction of the extracted datasets for the purpose of processing. A greatnumber of variables, which stipulates a lot of arithmetic assets to process, is a feature of these enormous data sets. In this paper the brain images which are the input images are breakdown into small manageable groups which makes it easier for processing.

4) Model Creation

The extracted datasets consist of about 6000 images which are divided into four classes namely mild demented, very mild demented, moderate demented, and non demented. These datasets are split into train dataset and split dataset for training the model. Eighty percent as the train dataset and twenty percent as the test dataset i.e. The datasets are split into 8:2 ratio. Then the training of the model commences.

5) Prediction

Python prediction function enables to predict the outcome of the trained model. The prediction is done using the anaconda python prompt, a simulator which involves some commands that changes the environment. This command generates a web address, where the input image is to be given. The model then predicts the outcome based on the input image provided with the help of Lenet-5 and Faster Rcnn algorithm. The hardware setup involves a system of Pentium IV 2.4 GHz of Ram: 8 GB and hard disk of 40 GB.

Fig. 2. Input Process Output diagram

The above depicted Fig. 2 is the Input Process Output diagram of detection of EMCI in Alzheimer's disease using Lenet-5 and Faster Rcnn algorithm. The initial step is the data collection which are the datasets abstracted from the website known as Kaggle, the next step is data preprocessing where the datasets are edited then converted into grey scale and made sure that each and every dataset are of same size and same color. The next step is called feature extraction which is an proportionally reduction strategy which is used for reduction of the extracted datasets for the purpose of processing. Here the datasets are split into small manageable groups, which makes it easier for processing. In feature extraction, the features are converted into array using NumPy. The next step is the model creation where the datasets has to be split into train test dataset. The process step consists of two algorithms namely Lenet-5 and Faster Rcnn. All the modules put together, prediction is done when the user gives in the input of an MRI and gets the output.

4 Implementation

This system, the detection of EMCI in Alzheimer's Disease uses two algorithms namely, Lenet-5 algorithm and Faster Rcnn algorithm. Both these algorithms are made together as a single entity and is implemented with the help of other modules. The detection of EMCI in Alzheimer's disease consists of five modules which are used to create the model namely data collection, pre-processing, feature extraction, model creation and prediction.

The system is implemented as follows,

Step 1: The First and foremost step is collecting the data to train the model. The data are collected from a website known as Kaggle and the data is an image dataset with four types of classification of the Alzheimer Disease. (Mild demented, non-demented, moderate demented, very mild demented).

Step 2: The collected datasets are then saved in a particular file with respect to its type of classification.

Step 3: The data sets are undergone editing process because, the datasets are collected from various csv files. The datasets are made sure that they are of same size in both width and height and the borders are removed if present.

Step 3.1: The collected data is now pre-processed by converting the image into grey scale. Here, the datasets are made sure that they are of same colour which is black and white, as the MRI images are also black and white.

Step 4: The next step is feature extraction where the features are extracted. This feature extraction is mainly done because it is very complex to maintain a huge amount of data, so by performing feature extraction the huge amount of data is split into small manageable groups, which makes it easy for processing.

Step 5: The features are extracted by the process of feature extraction; The feature extraction network has a number of convolutional and pooling layer pairs. The convolutional layer consists of a collection of digital filters that perform convolution on the input data. The pooling layer, which also serves as a dimensionality reduction layer, determines the threshold. A number of parameters must be modified during backpropagation, which reduces the number of connections inside the neural network architecture. They are converted into an array using NumPy in python.

Step 6: The Lenet-5 algorithm which is used for well scaled image processing, which helps to analyse every image and the Faster Rcnn algorithm which is used for finding the location of an object and differentiating an object from other are then introduced.

Step 7: The extracted feature is then added to algorithm with the epoch value.

Step 8: Model creation is done aside where the datasets are split into two: Train and test dataset where 80% of the dataset is assigned as train dataset and 20% of the dataset is assigned as test dataset.

Step 9: After that the model is to be created in the.h5 format and saved for the prediction.

Step 10: Prediction is done when the user access python flask which generates a webapp without any user login. The user gives in the input image in the input page.

Step 11: Now that the input from the user is got through Webpage, with that input image the image will be pre-processed and the input image feature will be extracted and the extract feature is converted into array format and then the array will be given to the model for prediction.

Step 12: With the given input image, the output is predicted into any one of the four classes (Mild demented, non-demented, Moderate demented, very mild demented).

4.1 Lenet-5 Algorithm

Lenet-5 algorithm is an advancement of Convolutional Neural Network (CNN) algorithm. The algorithm serves many purposes such that, it helps in well-scaled image processing. This algorithm can process every square of the MRI images which is given as the input. The difference between LeNET and resNET is that resNET is 50 layers deep and it is an artificial neural network while leNET is a simple neural network.

Lenet-5 algorithm is implemented as follows,

Step 1: The dimensions of Alzheimer's images are 28×28 pixels, but they are zero-padded to 32×32 pixels and normalised before being sent to the network. The input image gets smaller as it travels through the network.

Step 2: Each neuron in the average pooling layers computes the mean of its inputs, multiplies it by a learnable coefficient, and finally applies the activation function by adding a learnable bias term.

Step 3: Only three or four 2nd average pooling layers are connected to the majority of neurons in the 3rd convolutional layer.

Step 4: Each neuron outputs the square of the Euclidean distance between its input vector and its weight vector in the output layer. Each output metric predicts how likely an image is to correspond to a given digit class. The cross-entropy cost function is used in this stage.

The pseudocode for Lenet-5 algorithm is

```
START
function CoNv0P(i, Bzt f.)
Assign function Co -0
while Co < 0 do
if Co = 0 and i = 0 then
assign genuine= 0
READ BIAS() )
READ KERNEL(Co Co + 31, K,)
end if
PPMACCoNv(Ko, IBu f.) PREFETCHKERNEL(C0 + 32, Co ± 63,
Assign function  KO c
end while
end function
function CoNv(T., Ty, Ci)
if T< 0, T < =0, T>=0
while Ty < Y do while T. < X do
while Ci < I do
if T = 0 and Ty = 0 then
Assign function  c F 0
READ TTLE(/Bwfy, T. , Ty ,
end if
CoNv0P(Ci, I Bzt fo) PREFETcHTTLE(IBuf o, T, Ty, Ci ± 1) c
end while
end while
end while
end function
STOP
```

The working of Lenet-5 algorithm used in this work is depicted in the Fig. 3. The algorithm executes by performing well-scaled image processing when the MRI images are provided.

Fig. 3. Working of Lenet-5 algorithm

As depicted in Fig. 3, the input images from the test dataset and train dataset are assigned initially, the output of one filter applied to the previous layer is the feature map. A particular filter is painted one pixel at a time across the entire prior layer. The neuron is activated at each place, and the output is collected in the feature map. The images are featured into (28 $\times$ 28) pixels with the help of feature map. Furthermore, pooling is done with the help of a convolutional layer which has seven layers and the dimensions of the images are converted to 2 $\times$ 2 and 5 $\times$ 5 with the help of a convolution layer. Then comes the subsampling. Subsampling is used to minimize the dependency on accurate location in feature maps generated by CNN convolutional layers and it is fully connected with the help of Lenet-5. The Gaussian connection is a way of connecting the complete connected layer to the output layer that should be used exclusively. The Gaussian connection is the Euclidean radial basis function calculated with a set of artificially determined and pre-set weights. With that the prediction is done.

4.2 Faster RCNN Algorithm

Faster Rcnn algorithm is an advancement of Convolutional Neural Network (CNN) algorithm. The algorithm has a variety of uses, such that it helps in finding the location of the particular MRI images needed and differentiating the MRI images from one and other. This algorithm is mainly used because, this work consists of more than 6000 MRI images used for creation.

Faster Rcnn algorithm is implemented as follows,

Step 1: To acquire the feature map, the input image is first fed through the backbone CNN (Feature size: 60, 40, 512). Apart from test time efficiency, the benefits of weight sharing between the RPN backbone and the Fast R-CNN detector backbone are another important reason to use an RPN as a proposal generator.

Step 2: Trained the Region Proposal Network (RPN) by fine-tuning one of the models and after Conv3 layer and training the newly added layers based on anchor boxes.

Step 3: In this step, the detection network is trained. The RPN network trained in Step-1 provides regions across the feature map for training, which are subsequently passed to the RoI pooling layer and finally to the FC layer. However, the model used for this step is different from the previous one. After conv3, all of the layers are fine-tuned.

Step 4: The ROI pooling layer works by a) taking the region from the backbone feature map that corresponds to a proposal; b) dividing this region into a fixed number of sub-windows and c) performing max-pooling over these sub-windows to provide a fixed size output (N, 7, 7, 512) is the size of the ROI pooling layer output, where N is the number of suggestions from the region proposal algorithm. The features are sent into the sibling classification and regression branches after passing through two fully linked layers.

Step 5: Now, the Fast-RCNN which was initialized in step 2 is used to fine-tune the newly added layers of the RPN network after freezing the convolution layers.

Step 6: The FC layers of the detecting network are fine-tuned using this fine-tuned RPN network.

The pseudocode of Faster R-CNN algorithm is

```
START
procedure NMS(B,c)
function Conv(Bnms)
for bi E B do
discard    False
for bj E B do
if same (bi, bi) > Anms then
if score (c, bi) > score(c, bi) then
discard +- True
if not discard then
Conv(Bnms) Conv(Bnms) U bi
return Function Conv(Bnms)
STOP
```

The algorithm executes by differentiating one MRI image from the other and finding the location of a particular MRI.

Fig. 4. Working of Faster Rcnn algorithm

As depicted in Fig. 4, the input image is passed through the backbone of CNN. Filters or Feature detectors are applied to the input image or the feature map output of the previous layers to create feature maps. The internal representations for specific input for each of the Convolutional layers in theimodel will be visualized using feature maps. Next, a region proposal network's (RPN) output is a collection of boxes proposals that will be sent into a classifier and regressor to check for the presence of items. RPN forecasts whether an anchor will be in the background or foreground. RPNs are designed to propose several candidate regions. From which, a selection will be done to know which candidates fits the needs. ROI pooling is then done which is also known as Region of Interest Pooling which is a common process in object detection applications involving convolutional neural networks. The features are sent into the sibling classification and regression branches after they have passed through two completely connected layers.

5 Result and Analysis

The results for the detection of EMCI in Alzheimer's Disease using Lenet-5 and Faster Rcnn algorithm are depicted separately for identifying the performance based on the accuracy of each Algorithms namely Lenet-5 algorithm and Faster Rcnn algorithm. Model accuracy is defined as the number of classifications a model correctly predicts divided by the total number of predictions. It is a method of evaluating a model's performance. The model accuracy of the whole model where the Lenet-5 algorithm and Faster Rcnn algorithm are combined is also depicted for performance. In the end the model accuracy of Lenet-5 algorithm, Faster Rcnn algorithm and the model where both algorithms are combined and is compared for performance.

In the detection of EMCI in Alzheimer's Disease using Lenet-5 and Faster Rcnn algorithm, the model accuracy of Lenet-5 algorithm is depicted in the below graph Fig. 5.

Fig. 5. Lenet-5 algorithm model accuracy in detection of EMCI in Alzheimer's Disease

From the model accuracy graph it is seen that the model accuracy of the Lenet-5 algorithm which is used in the detection of EMCI in Alzheimer's Disease is 97% which is a good accuracy score and it can be concluded that the more is the epoch value then greater is the accuracy. The model accuracy is calculated by the formula:

$$MODEL\ ACCURACY = ACCURACY * EPOCH\ VALUE$$

In the detection of EMCI in Alzheimer's Disease using Lenet-5 and Faster Rcnn algorithm, the model accuracy of Faster Rcnn algorithm is depicted in the below graph Fig. 6.

Fig. 6. Faster Rcnn algorithm model accuracy in detection of EMCI in Alzheimer's Disease.

From the model accuracy graph it is seen that the model accuracy of Faster Rcnn algorithm which is used in the detection of EMCI in Alzheimer's Disease is 80% which is a good accuracy score and can be concluded that the more is the epoch value to greater is the accuracy.

In the detection of EMCI in Alzheimer's Disease using Lenet-5 and Faster Rcnn algorithm, the model accuracy of both Lenet-5 algorithm and Faster Rcnn algorithm is depicted in the below graph Fig. 7.

Fig. 7. Model accuracy of Lenet-5 and Faster Rcnn algorithm combined

From the model accuracy graph it is seen that the model accuracy of both Lenet-5 and Faster Rcnn algorithm combined which is used in the in the detection of EMCI in Alzheimer's Disease is 98% which is a good accuracy score. So, it can be concluded that, if the model accuracy is greater, the model is regarded as a good model.

In the detection of EMCI in Alzheimer's Disease using Lenet-5 and Faster Rcnn algorithm, the model accuracy of Lenet-5 algorithm, Faster Rcnn algorithm and both the algorithms combined is compared in the below graph Fig. 8.

Fig. 8. Model accuracy comparison

From the model accuracy comparison graph it is seen that the model accuracy of the combined algorithms consisting of both Lenet-5 and Faster Rcnn is greater than the model accuracy of Lenet-5 algorithm and Faster Rcnn algorithm separately. The model accuracy of the combined algorithms consisting of both Lenet-5 and Faster Rcnn is 98%.

6 Conclusion and Future Work

In this detection of EMCI in Alzheimer's Disease using Lenet-5 and Faster Rcnn algorithm, the system efficiently predicts the Alzheimer's Disease into four classes namely, mild demented, moderate demented, non demented, very mild demented. This proposed method also has high accuracy of about 98% which means that this system can be used efficiently to determine the EMCI in Alzheimer's Disease. The proposed system is also time efficient than normal diagnosis. Thus, this system can be used in the field of medicine, by the physicians for accurate and faster diagnosis of Alzheimer's Disease. Detection of EMCI in Alzheimer's Disease using Lenet-5 and Faster Rcnn algorithm can be further be developed into a software which can be used in the laboratories. Further it can be developed into an application which can be used even in homes.

References

1. Bi, X., Zhou, W., Li, L., Xing, Z.: Detecting risk gene and pathogenic brain region in EMCI using a novel GERF algorithm based on brain imaging and genetic data **23**(8) August 2021
2. Kam, T.-E., Zhang, H., Jiao, Z., Shen, D.: Deep learning of static and dynamic brain functional networks for early mci detection. IEEE Trans. Med. Imaging **39**(2), 478–487 (2020)
3. Liu, J., Wang, J., Tang, Z., Hu, B., Wu, F., Pan, Y.: Improving Alzheimer's disease classification by combining multiple measures. IEEE/ACM Trans. Comput. Biol. Bioinf. **15**(5), 1649–1659 (2018)
4. Liu, M., Zhang, D., Adeli, E., Shen, D.: Inherent structure-based multiview learning with multitemplate feature representation for Alzheimer's disease diagnosis. IEEE Trans. Bio-Med. Eng. **63**(7), 1473–1482 (2016)
5. Wang, Z., Zheng, Y., Zhu, D.C., Bozoki, A.C., Li, T.: Classification of Alzheimer's disease, mild cognitive impairment and normal control subjects using resting-state fMRI based network connectivity analysis. IEEE J. Transl. Eng. Health Med. **6**, 1–9 (2018)
6. Li, H., Fan, Y.: Early prediction of Alzheimer's disease dementia based on baseline hippocampal MRI and 1-year follow-up cognitive measures using deep recurrent neural networks. In: 2019 IEEE 16th International Symposium on Biomedical Imaging (ISBI 2019) (2019)
7. Peng, J., Zhu, X., Wang, Y., An, L., Shen, D.: Structured sparsity regularized multiple kernel learning for alzheimer's disease diagnosis. Pattern Recognit. **88**, 370–382 (2019)
8. Martínez-Murcia, F.J., et al.: A deep decomposition of MRI to explore neurodegeneration in Alzheimer's disease. In: 2018 IEEE Nuclear Science Symposium and Medical Imaging Conference Proceedings (NSS/MIC) (2018)
9. Jiang, P., Wang, X., Li, Q., Jin, L., Li, S.: Correlation-aware sparse and low-rank constrained multi-task learning for longitudinal analysis of Alzheimer's disease. IEEE J. Biomed. Health Inform. **23**(4), 1450–1456 (2019)
10. Tabarestani, S., Aghili, M., Shojaie, M., Freytes, C., Adjouadi, M.: Profile-specific regression model for progression prediction ofAlzheimer's disease using longitudinal data. In: 2018 17th IEEE International Conference on Machine Learning and Applications (ICMLA) (2018)
11. Fuse, H., Oishi, K., Maikusa, N., Fukami, T.: Detection of Alzheimer's disease with shape analysis of MRI images Japanese Alzheimer's disease neuroimaging initiative. In: 2018 Joint 10th International Conference on Soft Computing and Intelligent Systems (SCIS) and 19th International Symposium on Advanced Intelligent Systems (ISIS) (2018)
12. Thakare, P., Pawar, V.R.: Alzheimer disease detection and tracking of Alzheimer patient. In: 2016 International Conference on Inventive Computation Technologies (ICICT) (2016)
13. Dhaya, R.: Deep net model for detection of Covid-19 using radiographs based on ROC analysis. J. Innov. Image Process. (JIIP) **2**(03), 135–140 (2020)
14. Chen, J.I.Z., Hengjinda, P.: Early prediction of Coronary Artery Disease (CAD) by machine learning method-a comparative study. J. Artif. Intell. **3**(01), 17–33 (2021)
15. Pant, H., Lohani, M.C., Pant, J., Petshali, P.: GUI-based Alzheimer's disease screening system using deep convolutional neural network. In: Smys, S., Tavares, J.M.R.S., Bestak, R., Shi, F. (eds.) Computational Vision and Bio-Inspired Computing. AISC, vol. 1318, pp. 259–272. Springer, Singapore (2021). https://doi.org/10.1007/978-981-33-6862-0_22

Smart Farming Using Data Science Approach

Amit Kumar Goel, Krishanpal Singh(✉), and Rajat Ranjan

Galgotias University, Greater Noida, India
amit.goel@galgotiasuniversity.edu.in, ksolanki1306@gmail.com

Abstract. Smart Farming is a new trend that stresses the use of internet in agricultural production cycle. New technologies like the Internet of Things and Cloud Computing are projected to accelerate this trend by allowing farmers to use more robots and artificial intelligence. This is encapsulated by the Big Data phenomena, which refers to large amounts of data of varying types that may be acquired, evaluated, and used for judgement. The goal of this review is to get an understanding of the current state of Big Data applications in Smart Farming and to identify the corresponding socio-economic concerns that must be solved. A conceptual model for analysis was constructed using a structured approach which can also be used for further studies on this issue. The analysis reveals that Big Data applications in Smart Farming have an impact on the entire food supply chain, in addition to primary production. Big data is being used in farming operations to give predictive insights, drive real-time operational choices, and rethink business processes in order to create game-changing business models. As a result, several experts believe that Big Data will result in significant adjustments in the roles and power relationships among various players in today's food supply chain operations. The stakeholder landscape reveals an intriguing interplay between large digital corporations, venture capitalists, and often small start-ups and newcomers. Several public entities, on the other hand, disseminate open data under the condition that individuals' privacy be protected.

Keywords: Products farming · Moisture level · Temperature notifications

1 Introduction

Farming activities will become increasingly data-driven and data-enabled as smart machines and sensors appear on farms and agricultural data grows in amount and scope. The phenomena of smart farming is being propelled by rapid breakthroughs in the IOT and Cloud Computing. While Precision Agriculture merely considers in-field variability, Smart Farming goes a step further by basing management duties on data as well as location, with context and situation awareness prompted by real-time occurrences. To conduct out agile actions, real-time aiding reconfiguration features are essential, especially in times of abruptly altered operating temperatures or other circumstances. Robots are outfitted with a variety of sensors that measure information in their surroundings that is used to guide the machines' behavior. Big Data technologies play an important, inverse role in this evolution. From simple feedback methods to deep learning algorithms, there

J. I.-Z. Chen et al. (Eds.): ICIPCN 2022, LNNS 514, pp. 448–456, 2022.
https://doi.org/10.1007/978-3-031-12413-6_35

are lot of options. This is augmented by merging it with other outside Big Data sources like weather or market data, as well as benchmarks from other farms. Because of the quick pace of change in this field, it's difficult to establish a unified definition of Big Data, although it's often defined as data sets that are so massive or complicated that traditional data processing applications are insufficient. Because Big Data and Smart Farming are income in developing countries while also reducing world hunger.

2 Literature Survey

A systematic approach was used to conduct the literature review. This was accomplished in three stages. We began by searching two large bibliographic repositories, Web of Knowledge and Scopus, for all groups of two categories of terms, the first of which deals with Big Data and the second group refers to farming. The two databases were chosen for their wall to wall coverage of available literature as well as advanced bibliometric features like related literature or citation suggestions. 613 peer-reviewed publications were retrieved from these two databases. The relevance of these was determined by identifying texts that addressed the research questions. When reviewing the literature, we utilized the search feature to find paragraphs that had the key phrases, then read the content to determine whether they could be relevant to the study questions. Four researchers conducted the screening, with each rating approximately 150 articles and exchanging their findings with the others using the reference management programmed EndNote X7. As a consequence, 20 were deemed most significant, while 94 were deemed relevant. The remaining papers were deemed irrelevant since they only tangentially addressed Big Data or agriculture, and hence were not read or analyzed further.

3 Empower Farmer

Data scientists now have the tools to efficiently collect and analyse massive volumes of data, and studies are underway to see how that knowledge may aid small-scale farmers in the fight against global food shortages [1]. A group launched a project in September 2018 which will run until 2030 to look at data from about 45 million farmers in disadvantaged areas across 50 countries. The project's creators expect that the data will reveal if farm expenditures in different nations are paid off and will aid in the development of farmer-friendly legislation. On a global scale, this effort supports the United Nations' Sustainable Development Goals, which aim to double agricultural output and income in developing countries while also reducing world hunger.

4 COP Diseases

Agricultural pests may eat into a farmer's revenues quickly [2]. Misuse of pesticides, on the other hand, can have negative consequences for humans, plants, and other living creatures. Fortunately, some businesses hire data scientists to assist them in developing user-friendly platforms that determine when and how many pesticides should be applied. Agro smart, a Brazilian firm, is one of them. Its technology uses IOT sensors and AI to assess the type and quantity of pests on a crop. Farmers will receive an associated report, which they can use to plan your pest management strategies. The purpose is to assist farmers in achieving cost-effective pest control while minimizing environmental damage.

5 Climate Change

Changing climate is a looming threat that has already had an impact on agriculture. Data scientists, on the other hand, are hard at work finding out to adjust for the shift. One idea is providing IOT sensors to Taiwanese rice farmers in order for them to obtain critical data on their harvests. It'll all go into the system that will aid farmers in optimizing their cycle times, even if changing climate makes that difficult [3]. Because of climate change, following the conventional farming schedule is no longer sufficient. However, data analysis has the potential to forever alter the future of agriculture.

6 Innovation

Agriculture has been the dominant mode of production since the birth of human civilization. Rearing crops and animals has always been a labor-intensive undertaking. Farming has changed dramatically in terms of methodology, tools, and machinery. Agriculture has undergone a great deal of research to get to where it is now, that's still in the process of improving. Farmers' reliance on their intuition alone to make farming decisions is one of the elements fueling innovation. There seems to be a risk unless the farmers makes an error, the season may end with no harvest. As a result, the farmer must take steps to reduce the risk of such a situation and ensure that cost-effective options are made. Furthermore, the burgeoning area of nanotechnology and the use of various types of sensors can be used to results gained, high-quantity, and high-accuracy data from farms [4]. On the internet, also there is a vast volume of digital data. All of this information can be put to good use to improve agriculture. It has also been noted that customer behavior has evolved in recent years. They are increasingly interested in consuming good food and, as a result, want to understand where and also how the food is grown, packed, modified, and distributed. The most essential element driving agricultural innovation is to ensure that food is available to all members of human population.

7 Data Based-Solution

Farmers make agricultural decisions while juggling a variety of circumstances. They must plan what they will plant, where they will cultivate it, and when they will cultivate it in order to raise a range of crops. They should next decide on how to employ irrigation, fertilizers, and pesticides. The time of reap, harvest, and send the items to market are then determined. This farming approach is indeed an unpredictable technology, and it is critical to get all of the variables just so for optimal profit [5]. Fortunately, in this day and age, farmers can utilize data to help them make those difficult decisions. Farmers can collect data from a variety of sources and analyses it using data analytics to learn more about their land and crops. Data received from sensors inside the farm, such as soil quality, water content, air permeability, and so on (also known as localized data) can be utilized alone or in combination with data from other sources, such as temperature and rainfall, to derive various types of information [6]. All of this information may be combined to continuously assess the situation and make changes as needed. Furthermore, data from modern technologies such as spectroscopes can be utilized on farms to determine the quality of the soil as well as the availability and quality of the fruits grown on the farm. The basic premise of spectroscopy is to pass various wavelengths of visible light through an item in order to acquire various attributes such as heat, mass, brightness, and composition. As a result, the spectroscopic data can also be utilized to automate some farm procedures, such as the trying to open of irrigation sprinklers based on soil moisture content. Furthermore, the data could be used to forecast production using prediction models. Precision agriculture can benefit from the availability of localized and external data. Precision agriculture is based on the principles of sustainability and judicious use of resources without wastes them [7]. Plants can be examined using precision agriculture, and their mineral, fertilizer, and water requirements can be met on an individual basis. Only those plants with insufficient resources will be provided with what they require. As a result, this strategy can save a lot of resources, lowering the overall cost of production. Precision can also be gained in horticulture by implanting Radio Frequency chips in all of the animals. Farm animals can be identified and tracked using these chips. When one sick animal has been recognized, the farmer can begin treating it. Digital agriculture approaches used by both smallholder and large-scale farmers are expected to alleviate world hunger in the next years.

8 Implementation Issue

One of the most significant challenges in implementing data science is the agriculture industry's aversion to change. Farmers are wary of changing their farming practices since it may be highly costly if things go wrong. Switching to digital agricultural practices also necessitates a major investment, which really only majority of the farmers can afford. When opposed to smallholder farmers, big enterprises may get a return rather quickly. Farmers who are uneducated or who operate on a small scale may not even be ready to enact digital farming, and they may also be unable to distinguish between data which has been presented to them. Another worry with data-driven approaches is that they may primarily help well-educated large-scale farmers. The second difficulty is common to all data-based solutions: data collection, cleaning, storage, and delivery over secure means. Agriculture data is no exception. For localized data, a farmer should spend the cost of building sensors and storing the data on a centralized server. The issue with data sets is that it available in a number of formats and time periods. For example, data from a device that provides the water content and the pH value of soil may be retrieved at any time, but data from human tasks like evaluating fruits for ripeness cannot be obtained as easily. As a result, such data must be converted to a moment in time or aggregated so that they can be compared to conventional data of comparable nature and type. To be used in analytics, the data must be examined for interoperability, reusability, usefulness, application, appropriateness, and efficacy. A standard should be developed and utilized as a reference to compare for a certain area or type of plant. Farms also should share the local data with the other producers so that there is an enormous amount of information to analyses. This information should only be shared with people you trust because it could be exploited, commercialized, or used to create unfair competition.

9 Agriculture Niches

Data scientists understand how to use technologies to uncover trends and regularities that would otherwise go unnoticed. As a result, they can draw findings that advance agricultural research by examining specific issues. Researchers have discovered that trace elements improve the metabolic activities of poultry farms, and that carotenoids improve the quality and nutrition of egg yolks [8]. The conclusions reached through searching through data and studies to get at these conclusions demonstrate how seemingly minor changes in agricultural processes may have a significant impact [9]. When animal feed companies, farmers, and others in the agricultural industry use data scientists' insights, they may be able to improve their processes and achieve better results.

10 Yield Prediction

A poor crop yield can lead to a disastrous season for farmers and all those who rely on the crops. IBM has developed a platform that forecasts maize yields two to three months ahead advance, avoiding unpleasant surprises for farmers. Researchers from the University of Illinois, meanwhile, rely on annual forecasts and satellites data to forecast the end of the season earlier than typical. According to lab tests, this modern tech is even more precise than the real-time data provided by the US Agriculture department [10].

11 Impact

In today's environment, digitalization in agriculture has resulted in a slew of new breakthroughs. Micro, a self - aware and self-true system that takes account every farmer's position, weather, and crop data, is one of these initiatives. It leverages big data, deep learning, and smartphone technologies to provide smallholder farmers with knowledge, expertise, and resources. Large-scale farmers and those who are in industrialized countries are automating their farms with contemporary technologies.[5] They've turned their fields into factories, automating every procedure that can be automated with a lot of computation. Smart farming is a term used to describe this form of farming. Data science is being utilized to control the quality of milk output in cows at the TH Milk plant in Vietnam, where each cow is equipped with an RFID chip. Devices detect irritation in the cow's mammary glands, allowing the milking process to be fully automated [11]. The equipment will stop milking and the animal will be recorded and monitored if irritation is observed. Each goat's legs are fitted with a similar chip that registers its movement in Animal. The goat will be examined for disease if it does not walk for an extended period of time or if its sleeping patterns are erratic. A significant amount of money has been invested in digital agriculture as the globe moves toward it. Farm efficiency is the subject of extensive research and development. Small - medium and large-scale farms will benefit from incorporating modern technology [11]. New farming technologies, such as data science and analytics, will transform the business, allowing it to produce higher-quality foods in larger quantities in a sustainable manner, allowing the worldwide goal of increasing agricultural production before 70% by 2050 to be met.

12 System Module

12.1 Admin module.

- 12.1.1 Admin has to first register him then login.
- 12.1.2 He can access everything in the sites.
- 12.1.3. He can update on website anytime he wants.
- 12.1.4. Admin can see any data he wants.

12.2 User module.

- 12.2.1 Customer can register about himself/herself.
- 12.2.2 He can select any data that he want to see.
- 12.2.3 Can register any complaint that he facing online.

13 Data Investigation of Agriculture

13.1 With the internet of things, data mining was utilized to extract meaningful and significant knowledge from additional information about crops. The process of knowledge discovery was. In this section, we'll look at what's going on. It refers to gaining knowledge from others. Additional information this section is split into two parts. They are marked and data reduction [12].

13.2 Pre-processing: This was a crucial step in the finding of knowledge because the quality of the data determined the quality of the knowledge. Information nowadays tends to be unreliable, stale, and incomplete. This technique may contribute to the improvement and precision of the mining process that follows. Information transformation, cleansing, and integration are also included. This paper employed more data from internet connected things gadgets to judge yields, soil temperature, humidity, and moisture level, as well as information for the village's first illustration. This is depicted in Fig. 1. To aid data design, the internet of things was turned to a distinct design.

13.3 Data reduction: To reduce the amount of data, this part may encode the information to a tiny amount. Because the original information's reliability was preserved, eliminating the data reduction required more energy to provide the same results in the analysis process. Numerical reduction was utilized in this research, with parametric methods for loading low representations of data adding histograms. This approach used equal height histograms and equal height histograms.

Fig. 1. IOT information and products

14 IoT in Agriculture

The Internet of Things (IoT) is about linking "dumb" things to each other and the internet to make them "smart." It allows physical items to be detected and controlled from afar, allowing for closer integration of the real world with computer-based systems. IoT allows devices with sensor to connect and interact with one another over the internet. Pumps, sheds, tractors, weather stations, and computers are just some of the devices that may be electronically monitored and managed in real time.

15 Robotics in Digital Farming

Computer vision and motion sensor work together in autonomous farm equipment to avoid impediments while traversing the field. The robots construct a virtual 3D representation of the terrain, which they can navigate freely thanks to high-resolution cameras. Movement, on either hand, is not automatic because it is determined by the criteria they are designed to avoid or navigate through [7]. The particularly of John Deere tractors have indeed been standardized. Automatic Tractor, the company's auto technologies, is based upon GPS, satellite corrected future technologies, and all units sold in the United States come equipped with self-driving technology and a modem. The next phase will be to use artificial intelligence and machine learning to assist the tractors in detecting new impediments or animals in their route.

16 Implementation

The above Fig. 2 shows the implementation our website and how does our user interface will look.

Fig. 2. Web pages

17 Conclusion

This study and introduction essay to the special issue has offered an introduction of particular identification of social science on digital agriculture, demonstrating that this is a growing subject with substantial consequences for digital agricultural policy and practice. This article really hasn't systemically investigated, compared, or synthesized the data in the many subject clusters of social science on digital farming because it is an exploratory review that summarizes past strands of work. This requires the use of a systematic review approach in future studies. The studies in the this special issue supplement the five key clusters by demonstrating contemporary responses in policy terms, practices, and institutional structures to embed digital farming in various sectors and countries. While we have demonstrated the multiplicity of social scientific views used thus far, as well as their compatibility, we believe there still is room for more interdisciplinary, transdisciplinary studies. Interdisciplinary and transdisciplinary can both contribute to a better understanding of the institutional frameworks and stakeholder dynamics in which virtual innovation are generated and the effects they may have. There is also room for methodological innovation, such as switching from analogue to digital sociology or social information science, as computer literate social scientists develop analogue views of digital societies, say firmly. The variety of new topics offered in this article demonstrates that there are numerous avenues of inquiry that can be pursued across and across social science, natural science, and technological scientific disciplines. As digital agriculture progresses past the concept and hype phases, there will be plenty of opportunities to scientifically address these and other questions. As a result, psychology research in conjunction with natural or technological sciences can help to guide the rise of online agriculture in ways that take into account and respond to social dynamics, thereby attempting to maximize the benefits of these emerging technologies while minimizing the potential negative consequences.

References

1. https://blog.3dcart.com/the-ecommerce-process is the site from where we get the idea
2. Noh, Y., Ro, J.Y.: A study on the service provision direction of the national library for children and young adults in the 5G era. Int. J. Knowl. Content Dev. Technol. **11**(2), 77–105 (2021)
3. Balasubramanian, V., Bashan, A.: Document management and Web technologies: Alice marries the Mad Hatter. Commun. ACM **41**(7), 107–114 (1998)
4. Lang, M.: Web-based systems development: the influence of disciplinary backgrounds on design practices. J. Inf. Organ. Sci. **33**, 65–77 (2009)
5. DB-Engines: Solid IT consulting & software development GmbH. Accessed 3 Apr 2020
6. Taylor, M.M.: Methodologies and website development: a survey of practice. Inf. Softw. Technol. **44**, 381–391 (2002)
7. England, E., Finney, A.: Managing Multimedia. Addison Wesley, Cambridge (1996)
8. Tanenbaum, J.M.: WISs and electronic commerce. Commun. ACM **41**(7), 89–90 (1998)
9. Turnbull, D., Barrington, L., Lanckriet, G.: Five approaches to collecting tags for music. In: ISMIR 2008: Proceedings of the 9th International Conference of Information Retrieval, pp. 225–230 (2008)
10. Rentfrow, P.J., Gosling, S.D.: The Do Re Mi's of Everyday Life: the structure and personality preferences. J. Pers. Soc. Psychol. **84**(6), 1236–1256 (2003)
11. Flanagan, D.: JavaScript: The Definitive Guide, p. 1. O'Reilly, Beijing, Farnham (2011). ISBN 978-1-4493-9385-4. OCLC 686709345
12. Valera's, P.A.: A survey of requirements specification in model-driven development of web applications. ACM Trans. Web (TWEB), 10 (2011)

A Survey on Different Methods of Detecting Rheumatoid Arthritis

D. R. Ujval(✉), G. Vignesh, K. S. Vishwas, S. Gowrishankar, and A. H. Srinivasa

Department of Computer Science and Engineering, Dr. Ambedkar Institute of Technology, Bengaluru 560056, Karnataka, India
ujvaldr@gmail.com, gowrishankarnath@acm.org, srinivasaah.cs@drait.edu.in

Abstract. Rheumatoid Arthritis (RA) is basically a chronic inflammatory autoimmune disorder which causes swelling, pain and stiffness in synovial joints. Hands, feet, elbows, shoulders and ankles are some of those joints which are commonly affected by this disease. RA might occur at all ages, more often it is seen in between the ages of 30 and 50. Women are more likely to get affected by this disease than men and causes difficulty in doing regular activities. People with RA find it very difficult to perform daily activities and in severe cases, people lose their jobs. RA is not totally curable, but proper medication and treatment can control it. Any disease, when detected early, can be controlled and cured. Hence, Detection of RA in early stages would help in controlling it. Detection of RA includes various methods such as CBC, SGPT, Hand radiographs, MRI, CT, etc. We made a survey on different research papers regarding the various ways to detect and diagnose RA. The following survey provides details on the various methods implemented to detect RA and we make a comparison on the various detection methods and their feasibility.

Keywords: Rheumatoid arthritis · Autoimmune · Synovial · Machine learning · Deep learning · Convolutional neural network (CNN) · Support vector machine (SVM)

1 Introduction

Over the years, we have heard a variety of diseases around the world which are so common and take in number of people every year into its control and affecting them. Humans on the other side are always into finding a perfect solution to cure these diseases, though he succeeds in many but with some conditions with uncrackable combinations of the disease he stands behind with the line "to be detected earlier to cure it further".

Here we have taken one such disease called Rheumatoid arthritis, which is autoimmune type and mainly attacks the synovial tissues with joints. This disease requires diagnosis in its early stage and if not, results in affecting other organs of the body such as eyes, lungs, blood, skin and even causes heart attack and stroke. So, detecting this

J. I.-Z. Chen et al. (Eds.): ICIPCN 2022, LNNS 514, pp. 457–466, 2022.
https://doi.org/10.1007/978-3-031-12413-6_36

disease earlier has gained its importance and there are many approaches which have been suggested over the years. This paper throws light on some of those works which are proposed over this problem and also linearly it is trying to convey the way research, technique, technology have evolved in detecting this Rheumatoid Arthritis.

In this survey work, the Sect. 1 states about the current situation the world is facing with many diseases and focusing on one such disease called RA and giving an overview about how we have to consider it further. In Sect. 2 and Sect. 3, we have reviewed some of those papers which are related to the detection of RA by categorizing it based on the data namely Statistical Data and Image Data. Here we will look into some of those approaches proposed before and its efficiency in solving this problem. At last, In Sect. 4, we conclude this survey paper in terms of an overview meaning of what we have gained over referring all those papers and how the approaches have evolved over the time.

2 Detection Using Statistical Analysis

Statistical analysis includes the clinical data and analysis includes various computations and calculations on the data. Detection of RA on these clinical data may include the methods to find the joint space width, prediction using ML algorithms on numerical data.

It is imperative to develop techniques for diagnosis of RA as early as possible. Magnetic Resonance Imaging (MRI) of the extremities can provide precise information about the early signs of inflammatory arthritis, specifically Bone Marrow Edema (BME) and synovitis. "Artificial intelligence in Detecting Early RA" [6] signified how the importance of AI in early detection of Rheumatoid Arthritis has grown its interest in recent times [6]. He has mentioned a number of research over this field and finds his interest over using AI techniques such as atlas-based segmentation and fuzzy clustering, to develop a software for detecting very early RA based on automatic quantification of bone marrow edema and tenosynovitis and MRI imagery. This paper is about telling how the detection of RA has evolved its interest towards deep learning and hence forth specifying various upgradation in previous researches has gained much importance in deep learning approach is concluded here.

An effective way to differentiate Osteoarthritis Arthritis (OA) and rheumatoid arthritis is to use x-ray image processing. The degenerative bone disorders are diagnosed using X-ray detection. X-ray scans alone do not detect the type of arthritis. Image processing can aid in improving the diagnosis [12]. Through image processing of x-rays, it is possible to detect arthritis from different regions. The image processing includes, Image Acquisition: Images are acquired through sampling. Color conversion: grayscale conversion of the image is performed. Thresholding: the image is made binary using the threshold function. Compression: the image is resized. Median Filter: the image is made smooth. In order to collect statistical information, the images are analyzed according to the following properties: centroid, extreme, orientation, major axis length, minor axis length, and eccentricity. Feature Extraction: Joint location, orientation, circular features and space detection is done. Image Recognition: classification amongst RA and normal.

With patients having Rheumatoid Arthritis, the radiologists' task of measuring the joint space width (JSW) involves a lengthy and tedious process. A manual assessment lacks accuracy and is observer-dependent, making it difficult to accurately assess joint degeneration in early diagnosis and follow-up studies. The automatic analysis of JSW data is crucial considering standardization, sensitivity, and reproducibility [10]. They focus on joint margin detection and joint location in this paper. Automated joint detection entails Hand mask extraction, which refers to how the intensity varies in the radiographs. Entropy is used as a texture feature to reflect the randomness of a specific region in the radiographs. Multiscale Gaussian blurring was used to detect five peaks and four valleys along the horizontal profiles of the masked image. Extraction of joint features: Joint locations are determined from the midlines found so far. The directional LoG gives strong positive responses for dark joint space and strong negative responses for bright cortical bone and edges. Geometric relationship of finger joints: To restrict the search area in the blob response image, they used knowledge of finger bone lengths. Determining the joint span: The joint span of a PIP and DIP is determined by measuring the upper edge of the lower bone. Upper margin detection and lower margin detection. In this approach, all 70 finger joints were correctly located, from the located joints margins are measured. Using measured values, the presence of rheumatoid arthritis can be known. In comparison with manual joint segmentation, the presented method provides satisfactory results for joint location and margin detection [10].

In cases of extreme pain, RA requires periodic blood testing, including lipids and complete blood counts [1]. They suggested a study about the variation in blood and lipid components of a patient suffering from RA, where many examinations were carried out for 10 months over the patients of age group 40 to 60 years. The pathological laboratory examined blood and serum samples from the patients for ten months. The list of the clinical tests performed in the laboratory are listed as: Blood glucose test, Calcium blood test, Cardiac enzymes test, Cholesterol lipid test, C-Reactive Protein (CRP) test, Erythrocyte Sedimentation Rate (ESR) test, Complete blood and lipid profile tests, Kidney function test, Liver function test, Thyroid function test. After performing the number of tests, the pathological reports conclude the chances of hyperthyroidism, damage in liver and spleen, frequent weakness, chest and joint pain in RA patients [1].

The chance of curing rheumatoid arthritis increases if it is detected and treated early. Rheumatoid arthritis can be diagnosed based on several factors. There are several tests that are typical for judging rheumatoid arthritis, including Rheumatoid Factor, Anti-CCP, SJC, and ESR [3]. The k-means algorithm was used to predict the disease with four factors. K-means algorithm was applied to 60 anonymous data points to analyze them. Out of the 60 parameters, four clusters were selected since they were performing comparative analysis using the four factors as described above, later to figure out the centroid of their cluster at an initial stage they set it randomly and the setting it to the nearest centroid by adding a number of times. Rheumatoid arthritis can be predicted by two of four factors using the K-Means algorithm (Fig. 1).

By adopting this approach over the factor values, average values and cluster estimations the results were concluded with 84% accuracy with this explanatory model [3].

Fig. 1. Visualization of K-means model [4]

As Rheumatoid Arthritis (RA) progresses and improves, joint stiffness becomes a determinant factor [14]. A new method to determine joint range of movements and stiffness in RA patients was proposed by James Connolly and team [14]. The study is focused on developing a hand Range of Motion (ROM) measurement tool that continuously measures joint stiffness using a control glove and software. Using a 5DT Data glove with 14 sensors placed on the metacarpophalangeal (MCP) and Proximal Interphalangeal (PIP) joints, a system was designed and developed. Measurements of joint stiffness were made by measuring the maximum velocity captured during extension and flexion of the hand. The angular movement data and velocity data was collected during an objective routine. The time taken during the entire objective was recorded in addition to the angular movement of the joints. Tabular and graphical analysis was performed on the data collected. Normal people had good angular movement and time taken was very less to extend and fold the hand. RA people with damaged joints took so much time to perform the objective. The angular movement of the joints was very low and the velocity was too less when compared to normal people. Initial results show differences between normal and stiff joints in movement patterns and stiffness that can be calculated from velocity and angle measurements. The purpose of this research was to explore the potential of using a data glove and an application for measuring finger movement and stiffness.

Thermal imaging provides information about temperature variations in human skin on different parts of the body and can help in analyzing human body dysfunctions alternatively to existing diagnostic methods [15]. RA is primarily characterized by inflammation, so thermography can be used as a diagnostic tool to monitor dysfunctions in any part of the body. Sudhir Rathore and team [15] has developed a portable hardware thermographic system for detection of RA. Initially thermal images of the RA affected areas such as knees, wrists etc., are captured with emissivity set to 0.98. Image segmentation was performed on the captured images to find the Region of Interest (ROI) using the FCM algorithm. In the next step, image analysis was done to find the Mean, Variance, Skewness, Kurtosis. Then the decision is made by a Neural Network (NN) to produce reliable results with minimized error. A feed forward neural network with three layers containing four neurons each in the input layer and one neuron in the output layer is applied. As inputs to the Neural Network, we have Mean, Variance, Skewness, and Kurtosis. The NN was first trained and then validated. Finally, a new image is loaded, Statistical parameters are given to NN and a decision is made by comparing the output value to the threshold value. Thermographic images of RA patients show higher levels of temperature in abnormal areas than in normal areas. In conclusion, artificial neural networks for thermal imaging provide an alternative way to analyze information concerning Rheumatoid Arthritis.

3 Detection Using Image Dataset

Image dataset includes X-ray hand radiographs, Ultrasound images of joints, MRI images, Thermal images and so on. Among them, the hand radiographs stand out in detecting the disease in an effective way and it is cost-effective. The following section briefs about the various detection methods and their advantages and disadvantages. The image dataset can detect RA using Machine Learning algorithms, image processing techniques, Deep learning and many more.

3.1 Machine Learning

Among various Machine Learning approaches, a diagnosis system based on Vision API and Auto ML was proposed [2], where the system's one half is about training the model with image dataset and classify them based on grading. The images used for the training data is around 130 for grade 0–2 and for grade-3 around 65. These grading is according to the seriousness in RA patients. The latter part is about taking real time ultrasound images and diagnose them. The proposed system predicted around 70% accuracy for grade 0 to grade 2 whereas for other grades it predicted above 55% (Fig. 2).

Fig. 2. Implementation results [2]

The system was built as a webpage, to be available for easy means for doctors to train and diagnose regarding the ultrasonic images.

In [4] a recognition system to locate the joints in hands by analyzing X-ray images and also to evaluate those images towards Rheumatoid Arthritis by assigning scores with respect to each joint through machine learning. Here, only the recognition system is taken more concern and implemented using the python with OpenCV, where the identification of each joint is also validated at the end of the recognition procedure. The accuracy in recognizing those joints is found to be 90% through this approach.

Modified Total Sharp (mTS) scores are widely used to assess Rheumatoid Arthritis progression. Regarding it, there's an approach estimating it using Support Vector Machine (SVM) [9]. A histogram of oriented gradients (HOG) is used to represent the rough shape of finger joints in the study. Based on HOG, a support vector machine detects finger joints in an X-ray image, and a support vector regression (SVR) method is used to estimate the modified Total Sharp (mTS) score. They have also performed the finger join detection by clustering the evaluated patches on X-ray image. The patches are sorted in descending order by the SVM output. This method detects finger joints with 81.4% accuracy in 45 RA patients' X-ray images, and with an estimation of erosion score at 50.9% accuracy and JSN score at 64.3% accuracy when analyzing 45 RA patients' X-ray images.

3.2 Deep Learning

There is a deep learning approach to assigning the Rheumatoid arthritis score for all joints, specifically the narrowing and erosion scores for each joint [5], where 42 joint area scores for joint space narrowing and 44 joint area scores for joint erosion are summed up. Narrowing scores range from 0 to 4, and erosion scores range from 0 to 5, with the sum of these scores determining the patient's RA severity. The deep learning model is designed and implemented in MATLAB, and the training is done on the dataset collected from the RA2 Dream challenge. They used 1662 training samples to train the model and 416 test samples to evaluate its performance. The accuracy of assigning scores to joints was found to be 90.8% on average, with an error magnitude of about 4.6%.

In addition, Janick Rohrbach provided another deep convolutional neural network-based method for scoring X-ray images of patients with rheumatoid arthritis that is fully automated, fast, and reproducible. The dataset was collected from Swiss Clinical Quality Management and restricted their dataset to only have left hand radiographs. They pre-processed the dataset to have only the images of joints of 150 × 150. They have used the Ratingen Scoring method of assessing bone erosion and this deep learning model is inspired by VGG16 with six blocks of two convolutional layers and a max pooling layer. The number of filters per convolutional layer increases by 32, 64 and 128 every second block (Fig. 3).

Cohen's Quadratic Kappa was used to evaluating the agreement between different scorers. The model and the two human experts have inter-rater reliability scores of 0.675 and 0.580, respectively, and thus outperform the human scorers. This clearly shows that their model is on par with a human expert. The main advantages of a deep learning approach over human scorers are the speed with which the scoring process can be

Fig. 3. Architecture of the neural network. All convolution blocks are identical, differing only in the number of convolutional filters. The two fully connected blocks are also identical except for the number of neurons in the fully connected layer [7].

completed, i.e., milliseconds rather than minutes, and the consistency with which the results can be replicated. (i.e., same images always get the same scores).

Discussing the various diagnosis methods of RA, Kemal Ureter [8] specifies juxta-articular erosion on hand radiographs is one of the seven criteria for RA classification. One of the criteria for RA classification is imaging-proven synovitis, demonstrating the importance of imaging in RA diagnosis. They consider plain radiography to be the most widely used and first-line imaging method for diagnosing and differentiating RA, as well as monitoring the disease's activity, because it is relatively inexpensive and easily accessible. They proposed a CNN model architecture with six convolution layers: a convolution layer, a batch normalization layer, a ReLU, and five max-pooling layers, followed by one fully connected layer with a SoftMax layer. The dataset comprised 180 radiograph images of both hands with different sizes. 81 patients were normal and 99 were RA affected. Data pre-processing and splitting was done on the collected dataset and images were resized to 160×240. Data augmentation was also done. A number of performance metrics were used to evaluate the efficiency of the developed CNN model, including sensitivity, specificity, precision, F1 score, false negative rate, false discovery rate, false positive rate, negative predictive value, and classification accuracy. The confusion matrix was used to evaluate these metrics. The proposed CNN model correctly classified 33 out of 45 patients. However, as shown above, their model failed to identify 5 normal and 7 RA patients. The network's accuracy was 73.33%. The network's sensitivity was 68.18%, while its specificity was 78.26%. The accuracy was 0.75. As a result, the model can aid specialists in their diagnosis. The model may be useful even for non-specialists during the initial examination [8].

There are two approaches for hand X-ray classification of rheumatoid arthritis using CNN models, one is described in Bhupesh Kumar Singh's paper [11] where they collected around 662 images of patients suffering from RA regarding the MCP (Metacarpophalangeal) joints, and the other incorporates 315 images of the PIP (proximal interphalangeal) joints. The collected images were resized to 100 × 100 and considered 80% data for training and 20% for testing. The processed image dataset was fed into the CNN model, which contains one or more number of convolutional layer and pooling layer (Fig. 4).

Fig. 4. Using a convolutional neural network, a framework for automatically classifying hand X-rays has been developed [11].

After training and testing the accuracy was about 95% and they also compared the result with SVM (Support Vector Machine) which was around 60% accuracy and ANN (Artificial Neural Network) around 80% accuracy.

In [13], as a CNN approach over hand radiographs have gathered about 92 Gy scale radiographs images then they reduced the size of those images from 4280 × 3520 pixels to 256 × 204 pixels and values of the pixels was normalized in a range between 0 and 1. As training a model with small amounts of data was a challenging task, they transform the original radiographs with a random combination of rotation, zooming, stretching and flipping as shown below. The dataset was split into two after data augmentation, one for training and the other for validation. Their dataset was trained with such type of models to take smaller number of parameters and having a smaller number of operations. First, with LeNet architecture with 3 * 3 kernels, ReLU as activation function and SoftMax as output function resulted in 93% accuracy. Secondly, with a minimalistic variant of Network in Network was used to reduce computational cost and achieved the same 93% accuracy. Later, with SqueezeNet architecture having same activation function and output function as for LeNet but with 5 * 5 kernel strides achieved 100% accuracy (Table 1).

Table 1. Implications from the survey

Authors	Techniques	Advantages	Disadvantages
[1] Saurav Bharadwaj and Pallab Sharma	Study about variation in blood and lipid components	Clinical findings	So many tests, and cost is high
[2] Takashi Muronosono et al. [2021]	Machine learning (using Vision API and Auto ML) to diagnose RA	Stable accuracy	Usage of ultrasound images
[3] Jihyung Yoo et al	k-means algorithm	Quick results	Statistical data analysis
[4] Koji Makino et al	x-ray image processing using OpenCV	90% accuracy	Location of joints using ROI and scoring each joint
[5] Son Do Hai Dang and Leigh Allison	MATLAB DL model	Decent accuracy	Less dataset
[6] Berend C. Stoel [2019]	CNN on MRI dataset	Predicting RA with high sensitivity	MRI scanning is not cost-effective
[7] Janick Rohrbach et al. [2019]	CNN on X-ray images	Cost effective and balanced accuracy	Image size was 150 × 150
[9] Kento Morita et al. [2017]	SVM with HOG	Good modified Total Sharp (mTS)	Error rate was high
[14] James Connolly et al	5DT Data glove with 14 sensors	Joint stiffness calculation	Dependent on sensor values
[15] Sudhir Rathore et al	ANN on thermographic images	Affordable and cost efficient	Accuracy of thermal images is poor and inflammation cannot alone predict

4 Conclusion

Every paper discussed above has found its uniqueness in the way of approaching the problem, it might be in the type of data used, the learning model deployed or by considering some factors causing Rheumatoid Arthritis. As an overview, these approaches are proving that there are new possibilities to detect rheumatoid arthritis. All these things contribute to our motive of doing this survey paper, which is basically exploring new solutions and finding the gap of opportunity for the readers to fill it with their creative and innovative approach towards solving this global problem.

From our perspective to discuss about the gap of research we can evidently see that there are disadvantages in all the above methods discussed but yet the hunger in finding the best solution is always the drive for us and this survey paper is undoubtedly a supportive tool.

References

1. Bharadwaj, S., Sarma, P.: Symptomization of rheumatoid arthritis in patients on pathological examination: a case study. In: Proceedings of the 2020 IEEE International Students' Conference on Electrical, Electronics and Computer Science (SCEECS), Bhopal, India, pp. 1–3, 22–23 February 2020. ISBN: 978-1-7281-4862-5
2. Muronosono, T., Nishiyama, T., Kawajiri, S., Imai, T., Arai, K., Kobayashi, T.: Research on rheumatoid arthritis detection support system. In: Proceedings of 2021 IEEE 3rd Global Conference on Life Sciences and Technologies (LifeTech), Nara, pp. 122–123, Japan, 9–11 March 2021. ISBN: 978-1-6654-1875-1
3. Yoo, J., Lim, M.K., Ihm, C., Choi, E.S., Kang, M.S.: A study on prediction of rheumatoid arthritis using machine learning. Int. J. App. Eng. Res. **12**(20), 9858–9862 (2017)
4. Makino, K., Koyama, K., Hioki, Y., Haro, H., Terada, H.: Recognition system of positions of joints of hands in an X-ray photograph to develop an automatic evaluation system for rheumatoid arthritis using machine learning. In: Proceedings of the 2020 13th International Conference on Human System Interaction (HSI), Tokyo, Japan, pp. 216–221, 6–8 June 2020. ISBN: 978-1-7281-7392-4
5. Dang, S.D.H., Allison, L.: Using deep learning to assign rheumatoid arthritis scores. In: Proceedings of the 2020 IEEE 21st International Conference on Information Reuse and Integration for Data Science (IRI), Las Vegas, NV, USA, pp. 399–402, 11–13 August 2020. ISBN: 978-1-7281-1054-7
6. Stoel, B.: Artificial intelligence in detecting early RA. Semin. Arthritis Rheum. **49**, S25–S28 (2019)
7. Rohrbach, J., Reinhard, T., Sick, B., Dürr, O.: Bone erosion scoring for rheumatoid arthritis with deep convolutional neural networks. Comput. Electr. Eng **78**, 472–481 (2019)
8. Üreten, K., Erbay, H., Maraş, H.H.: Detection of rheumatoid arthritis from hand radiographs using a convolutional neural network. Clin. Rheumatol. **39**(4), 969–974 (2020)
9. Morita, K., Tashita, A., Nii, M., Kobashi, S.: Computer-aided diagnosis system for Rheumatoid Arthritis using machine learning. In: Proceedings of the 2017 International Conference on Machine Learning and Cybernetics (ICMLC), Ningbo, China, pp. 357–360, 9–12 July 2017. ISBN: 978-1-5386-0408-3
10. Huo, Y., Vincken, K.L., Viergever, M.A., Lafeber, F.P.: Automatic joint detection in rheumatoid arthritis hand radiographs. In: Proceedings of the 2013 IEEE 10th International Symposium on Biomedical Imaging, San Francisco, CA, USA, pp. 125–128, 7–11 April 2013. ISBN: 978-1-4673-6455-3
11. Mate, G.S., Kureshi, A.K., Singh, B.K.: An efficient CNN for hand X-ray classification of rheumatoid arthritis. J. Healthc. Eng. (2021)
12. Hayat, H., Gilani, S., Jamil, M.: Arthritis identification from multiple regions by X-ray image processing. Int. J. Signal Process. Image Process. Pattern Recogn. **10**, 23–32 (2017)
13. Betancourt-Hernández, M., Viera-López, G., Serrano-Muñoz, A.: Automatic diagnosis of rheumatoid arthritis from hand radiographs using convolutional neural networks. Revista Cubana de Física **35**(1), 39–43 (2018). ISSN 2224-7939
14. Connolly, J., Condell, J., Curran, K., Gardiner, P.: A new method to determine joint range of movement and stiffness in rheumatoid arthritic patients. In: Proceedings of 2012 Annual International Conference of the IEEE Engineering in Medicine and Biology Society, pp. 6386–6389 (2012). ISBN: 978-1-4577-1787-1
15. Rathore, S., Bhalerao, S.V.: Implementation of neuro-fuzzy based portable thermographic system for detection of Rheumatoid Arthritis. In: Proceedings of 2015 Global Conference on Communication Technologies (GCCT), pp. 902–905 (2015). ISBN: 978-1-4799-8553-1

Co-F I N D: LSTM Based Adaptive Recurrent Neural Network for CoVID-19 Fraud Index Detection

Anika Anjum(✉), Mumenunnessa Keya, Abu Kaisar Mohammad Masum, Sharun Akter Khushbu, and Sheak Rashed Haider Noori

Department of Computer Science and Engineering, Daffodil International University, Dhaka, Bangladesh
{anika15-11488,mumenunnessa15-10100,abu.cse, sharun.cse}@diu.edu.bd, drnoori@daffodilvarsity.edu.bd

Abstract. On March 8, 2020, the IEDCR reported three cases of the first corona infection in Bangladesh, and there was a lot of fake news surrounding the virus, which the WHO Director-General called "infodemic". Infodemic, additional information about any problem that is usually unbelievable, spreads quickly and makes that problem difficult to solve and it is even more dangerous than the Corona epidemic. The misinformation provided by the media, false information, religious discrimination, miraculous remedies, and vague instructions of the government have created panic among the people of Bangladesh. Many news portals are intentionally or accidentally publishing fake news about the covid vaccine, the rate of infection and survival, the situation in other countries, the symptoms, and what to do after being infected. The most widely reported controversy is China's involvement in the creation and spread of the coronavirus. This article has been proposed in the context of identifying, sorting most of the fake news and misinformation about coronal infodemics in Bangladesh so that the people can take necessary steps accordingly. LSTM-Recurrent Neural Networks have been applied for classification and detection of fake news because RNN can easily detect complex sentences from textual data and LSTM is called a memory network that can easily perform detection work by remembering the sequence of the sentences. RNN has provided the most accuracy between LSTM and RNN models but LSTM has been able to perform the prediction work more accurately than RNN.

Keywords: Fraud detection · Deep learning · Covid-19 · LSTM · RNN · Fake news · NLP

1 Introduction

Nowadays, the improvement of social media, mobile technologies, and communication systems make it easier for spreading the news all over the world. The availability of the internet has made people more modern and communicative. But impetuous development may affect our daily and social life in many ways. Social media plays an important role

J. I.-Z. Chen et al. (Eds.): ICIPCN 2022, LNNS 514, pp. 467–478, 2022.
https://doi.org/10.1007/978-3-031-12413-6_37

in spreading news for human beings all over the sphere as it has become one of the fundamental wellsprings of data among people all over the world. The primary elements behind that are the minimal expense, the pace of access, simplicity of utilization, as well as accessibility on all computerized gadgets together with work stations, smartphones, Intelligent Portable Over Device (iPods), and more. By publishing or spreading fake news, social media misguides and confuses people in many ways. Nowadays, it has become a worldwide problem and a worldwide issue and warning to the present world. Most of the time the content of fake news is arranged in such a way that people can't catch it. Before the US 2016 presidential general election, fake news spread widely [1].

The information industry is greatly affected by this. In the previous year, there has been a mass migration of clients from the more conventional media like papers, TV, and radio to new arrangements: informal organizations, YouTube, webcasts, online bulletin, news applications, and so on [2]. The major reason information media degrade is a developing capacity, because the Internet possesses a moment as well as free admittance to an extensive assortment of information sources not to mention various administrations that permit splitting the news to a large number of individuals all throughout the world. Therefore, the media have begun to respond to the change.

The vast majority of these media have chosen to begin adapting their substance through promoting installed in their contrivances, videos, and so on. The most persistent technique is distributing contrivances with conspicuous features as well as photographs planned to be staked on public platforms so that clients explore their sites in this way amplifying their income. The large quantity of information individual access is normally unsubstantiated and for the most part, accepted as obvious [3]. This sort of approach can prompt risky circumstances.

Fake news is viewed as the greatest danger to trade, reporting, and majority rule governments from one side of the world to the other, with vast coincidental losses. A US $130 billion misfortune in the monetary trade was the immediate result of a phony news report that US President Barack Obama got hurt in an impact [4]. Fake news is also known as garbage news or pseudo-news, is a sort of dramatist detailing or exposure that includes deliberate disinformation or stunts spread through customary news media or online electronic media [5]. Because of the trouble of physically identifying fake news, basic consideration has been given to distinguishing fake news and perceiving solid information from a few analysts, recently. Despite the fact that there are alternate points of view to contemplate this issue, AI techniques are viewed as promising innovations for further developing fake news identification solutions. In this paper, we have used various kinds of deep learning approaches for detecting fake and authentic news detection. We have used RNN, LSTM, and HAN in this paper for detecting fake and authentic news. Our fundamental commitments are the accompanying:

- Evaluating our model on two types of news detection: fake news and authentic news
- Performing deep learning approaches for news detection

The remainder of this paper is organized as follows: A short literature review of deep learning models for fake and authentic news detection is proposed in segment 2; Segment 3 describe the proposed model in elaborately; Experiments, as well as results, are presented in Sect. 4; Finally, Sect. 5 concludes the paper and offer some directions for our future work. The spread of misinformation hit the ceiling during the covid pandemic. We have collected such fake news headlines to build a model that is capable of predicting if news related to covid is fake or not.

2 Literature Review

The task of distinguishing fake news has gone through an assortment of names, from deception to talk, to spam. Comparably every person may have their own instinctive definition of such related ideas, each paper embraces its own definition of these words which conflicts or on the other hand, covers both with different terms and different papers. For this reason, we indicate that the objective of our examination is recognizing news content that is manufactured, that is fake.

Vlachos and Riedel et al. [6] characterized the undertaking of reality checking, gathered a dataset from two mainstream truth checking sites, and considered KNN classifiers for dealing with certainty checking as an order task. Proclaim is a start to finish neural organization model suggested by Popat et al. [7] for exposing counterfeit news and wrong cases. It utilizes proof as well as counter-confirmations separated from the webbing to help instead of invalidating a case. Outwardly highlighting designing along with manual intercessions, creators accomplished a generally speaking 80% grouping exactness on four various datasets, via preparing a Bi-LSTM representation with consideration and provenance embeddings.

Yang et al. [8] have mentioned the TI-CNN model which is prepared with both the content and picture information at the same time. The intricacy neural organization constructs the representation see the whole contribution instant, as well as it is very well prepared and a lot quicker than Long Short Term Memory and numerous other Recurrent Neural Network representations. Khan et al. [9] assessed Support Vector Machine, Logistic Regression, NB, Decision Tree, Adaboost, K-Nearest Neighbor, Long Short Term Memory, Bidirectional Long Short Term Memory, CNN, Convolutional Long Short Term Memory, Hierarchical Attention Networks, Conv-Hierarchical Attention Networks, as well as character-level C-Long Short Term Memory classifiers on three distinctive adjusted datasets. Authors expressed that despite neural organizations executing preferably on bigger datasets, Naive Bayes can also perform as great as neural organizations on more modest datasets.

Miller and Oswalt et al. [10] work with the FNC-1, which was introduced in a widespread contest which was expected to discover programmed strategies for distinguishing counterfeit bulletin. They assembled an organization engineering utilization which is different Bidirectional LSTMs and a consideration system to foresee the consequence of the contrivance to their matched features. The foremost outcome was accomplished by the Bidirectional Long Short Time Memory and Multilayer Perceptron with 57% precision. Ethar et al. [11] suggested a model which is a Bidirectional Long Short

Time Memory connected model which is appealed on the Fake News Challenged-1 dataset with 85.3% precision execution. By using the FNC-1 dataset Ayat Abedalla et al. [12] developed various models to recognize fake news depending on the connection between article feature and article body. Their models are amassed essentially from CNN, LSTM, and Bi-LSTM. They gained an accuracy of 71.2% for the formal testing dataset.

Federico et al. [13] showed a narrative programmed fake bulletin identification representation dependent on mathematical profound learning. Their observation shows that informal organization structure spread significant highlights permitting profoundly the exact (92.7% ROC AUC) fake news recognition.

Álvaro Ibrain Rodríguez et al. [14] considered the plausibility of appealing deep learning strategies to segregate fake bulletin on the web utilization just in their content. They proposed three distinctive neural organization structures that are LSTM, CNN, BERT which is a cutting-edge language model made by Google which accomplishes cutting-edge results. For the LSTM based model, they gained an exactness of 0.91 on both the test set and the validation. CNN-based model arriving at the exactness of 0.937 on the approval parcel and 0.94 on the test set. BERT representation was assessed over the inspection crease acquiring exactness of 0.98 as well as an F1 metric of 0.97. Natali Ruchansky et al. [15] suggested a model named CSI that is made out of three modules: Capture, Score, and Integrate. Therefore, the first module depends on the reaction and text; it's anything but an RNN to catch the worldly pattern of client action on a given article, the second module learns the source trademark in light of the conduct of clients, and the two are incorporated with the third module to arrange an article as fake or not.

Roy et al. [16] chipped away at the LIAR dataset [13] and utilized an outfit model to arrange proclamations into the six classes. The outfit model comprises two unique models that are Bi-LSTM model and the CNN model. A multi-facet perceptron model has been utilized to join these two models.

3 Methodology

In this section, a brief overview of the deep learning approaches used for detecting fake and authentic news is presented. We discuss various deep learning approaches that we experimented with and present deep learning approaches that gave us the best results. Specifically, LSTM, RNN, and HAN approaches are described in Sect. 3. We also provide details into data preprocessing, various methods in this section (Fig. 1).

Fig. 1. Workflow of fake news detection

3.1 Data Processing

From various sources data has been collected, namely- Newspapers, Click baits, and News articles and around 1500 Bangla data has been for the pursuit where the dataset contains the body of the news articles, the headlines of the news articles. The pre-processing methods applied for the dataset to fit the model perfectly in Fig. 2 and the Pre-processing description briefly described in Table 1.

3.2 Training Parameter

In Table 2, a description of our parameter is shown. We set different parameters for getting better performance.

Fig. 2. Components of data pre-processing

Table 1. A brief description of the pre-processing steps

Preprocessing step	Description
Remove punctuation	It's important to transform fragmented and conflicting raw data into machine-understandable configurations. For cleaning data we removed punctuations. Punctuation in regular language gives the syntactic setting to the sentence. Punctuations like a comma, probably won't add a lot of significant worth of incomprehension. We used the reject expression for removing punctuation. We also removed white space from the text data
Remove stop words	We eliminate stop words from the text data accessible. Stop words that are of less significance may occupy important preparation time, and henceforth eliminating stop words as a piece of data preprocessing is a key initial phase in natural language processing. By eliminating the stopwords, we reduce the processing time and save space in any case taken by futile words. We used the NLTK (Natural Language Toolkit) library to eliminate stop words
Word2vec	We use the gensim package and KeyedVectors for mapping among keys and vectors. We load the word2vec format and store it into the word2vec variable. Then we store the word2vec variable into the model variable. We define our max word length as 200000 and fit the text data in the array of News. Then we create a two-dimensional NumPy array of zeros and build the "embedding_matrix". Words not found in the embedding index will be all zero, hence +1

(*continued*)

Table 1. (*continued*)

Preprocessing step	Description
Tokenizer	We use Tokenizer from Keras library. We passed nb words as a parameter into the Tokenizer function. When will call any transformative method - Tokenizer will use only 200000 most common words and at the same time, it will keep the counter of all words. For keeping the counter of all words we call fit_on_texts and each time it will update the internal counters, and when transformations are called, it will use the top words based on the updated counters. We use text_to_sequence to change every text in texts into an arrangement of whole numbers. It fundamentally takes each word in the substance and replaces it with its related number worth from the word_index word reference Then we found the unique tokens from our data
PAD	We use a pad sequence that allows using sentences to different lengths and using truncation to make all of the sentences to the same length. We tokenize with nb_words, creating a special token that is used for words. We specify how the token is over in text data and call tokenizer fit and sequence text data in tokenizer and word_index that tokenizer. Then print the length of the word_index. We set a path for our pad sequences with that sequence and set max_len as 200

Table 2. The LSTM-RNN training parameter description.

Parameter name	Describe
learning_rate	Decides the progression size at every emphasis while pushing toward at least a loss function
batch_size	The number of instances of a succession we need to prepare our model with for that batch
epochs	The number of occasions that the learning calculation will work through the whole preparing dataset
maxlen	The maximum used sentence in the dataset
random_state	Pseudo-random number generator utilized for random sampling
test_size	Split the test size 0.1
Dropout	Slow down overfitting of data
verbose	The training progress for each epoch set the verbose 2
validation_split	Determine a different approval dataset while accommodating a model that can likewise be assessed

3.3 LSTM-RNN Model Estimation with the Parameters

Different parameters and their settings are described in Tables 2 and 3. We set different values for different parameters to get better output for our model. Test size and random size are equal for both models.

Table 3. The LSTM-RNN training parameter description.

Test size	Random state	LSTM			RNN		
		Max nb words	Batch size	Epochs	Max words	Batch size	Epochs
0.1	42	2000	256	50	800	52	15

3.4 Embeddings

Embeddings is a decently low-dimensional space into which we can decipher high-dimensional vectors. Embeddings simplify it to do machine learning on enormous data sources like inadequate vectors tending to words. We used word2vec for producing word embeddings. These representations are slight, two-layer neural organizations that are ready to reproduce the etymological setting of words. Word2vec takes as its input a colossal corpus of text and produces a vector space, normally of two or three hundred estimations, with each fascinating word with regards to the corpus being allotted a looking at vector in the space. Word vectors are arranged in the vector space with the ultimate objective that words that share typical settings in the corpus are found close to one another in the space. We used the tokenizer function from Keras's library to split each news into a vector of words.

3.5 LSTM Model Analysis

LSTM networks are a sort of redundant neural organization fit for figuring out how to demand dependence on game plan assumption issues. LSTMs are the complex domains of deep learning. It is intended to moderate the disappearing and detonating angle issue separated from the secret state vector. Each LSTM cell keeps a cell state vector and at each time step, the following LSTM can decide to peruse from it right to it or reset the cell utilizing an unequivocal gating instrument. Every unit has three doors of a similar shape. The info door whether the memory cell is refreshed, the neglect entryway controls if the memory cell is reset to 0 and the yielding door controls whether the data of the current cell state is made noticeable.

We used the embedding layer to make word vectors for approaching words and stands between the input and the LSTM layer. The principal layer is an LSTM layer with 64 memory units and it brings successions back. The outcome of the Embedding layer is the contribution to the LSTM layer. We used a linear stack of layers as a sequential model where the LSTM layer gets successions and not simply randomly dispersed information.

The last layer is used as a completely associated layer with the "sigmoid" activation function and RMSprop optimizer to increment our learning rate and our calculation could make bigger strides the horizontal way uniting quicker. We used binary cross-entropy to ascertain a score that sums up the normal contrast between the genuine and anticipated probability circulations for predicting class 1. LSTM permits the preservation of the loads that are back and forward engendered through layers. LSTMs are sensitive scales to the input data. The data has been rescaled to the range of 0 to 1.

3.6 RNN Model Analysis

Recurrent neural networks are mainly used for natural language processing tasks. It is the primary calculation that recollects its contribution, because of an inward memory, which makes it consummately appropriate for machine learning issues that include consecutive information. RNN is able to recall those characters due to its inside memory. It builds output, duplicates that output, and loops it back into the network. In an RNN the data spins through a circle. At the point when it settles on a choice, it considers the current information and furthermore what it has gained from the data sources it got already (Table 4).

Table 4. LSTM-RNN model summary

LSTM			RNN		
Layer (type)	Output shape	Parameter	Layer type	Output shape	Parameter
embedding (Embedding)	None, None, 50	164800	dense 1 (Dense)	None, 256	205056
			dropout (Dropout)	None, 256	0
lstm (LSTM)	None, 64	29440	activation (Activation)	None, 256	0
			dense_2 (Dense)	None, 2	514
dense (Dense)	None, 1	65	dropout_1 (Dropout)	None, 2	0
			activation_1 (Activation)	None, 2	0

The summary of model LSTM and RNN are shown in the model summary table. Different layers, output shapes, and parameters are defined for both models. In LSTM, three layers are used including embedding, lstm, and dense where parameters of the layers are included. We got two different output shapes for six RNN model layers. The maximum parameter for the RNN model is obtained by a dense layer.

4 Results and Discussion

After an exhaustive hyperparameter tuning on our best-performing model, we assessed the model on test data. Among our models, we got the best prediction outcome from the lstm model. But RNN gave better accuracy for validation loss and validation accuracy. RNN performs better than LSTM in the validation loss and validation accuracy for epochs. When we test our model, RNN gave the opposite prediction result where LSTM gave the actual prediction outcome. RNN gave around 98% train accuracy but in test loss, it gave the opposite prediction outcome. In Fig. 3, we have shown the loss, accuracy, validation loss, and validation accuracy for the LSTM model. We worked on 50 epochs where our model gave 70% of validation accuracy, 68% of validation loss, 67% of accuracy, 81% of the loss for our model.

Fig. 3. LSTM model loss & accuracy graph.

In Fig. 4, we have shown a loss, accuracy, validation accuracy, and validation loss for the RNN model. We worked on 15 epochs where we got the highest validation loss 74%, validation accuracy 90%, accuracy 99%, and loss of 76%. To fit our model we set the validation split to 0.15 (Table 5).

Fig. 4. RNN model loss & accuracy graph

Table 5. Prediction result of LSTM-RNN model.

Original News	হারাম জীবন যাপনন কনরানা ভাইরাস গজব হনে নননম আসন ◌া চীনানের উপ
Original Prediction	Fake News
Input News	হারাম জীবন যাপনে করোনা ভাইরাস গজব হয়ে নেমে আসল চীনাদের উপর
Prediction For Our Model	
LSTM	Fake News
RNN	Fake News
Original News	ভ্যাক্সিনে স্যালাইন এবং পানি ভরে বাংলদেশকে দিচ্ছে ভারত
Original Prediction	Fake News
Input News	ভ্যাক্সিনে স্যালাইন এবং পানি ভরে বাংলদেশকে দিচ্ছে ভারত
Prediction For Our Model	
LSTM	Fake News
RNN	Authentic News
Original News	করোনা শুধু নাস্তিক দের হয়
Original Prediction	Authentic News
Input News	করোনা শুধু নাস্তিক দের হয়
Prediction For Our Model	
LSTM	Authentic News
RNN	Fake News

5 Conclusion and Future Work

Due to the pandemic situation, most of the people are staying at home and most of the work is continued from home through the internet. The percentage of people using the internet is increasing. In recent times, they spend most of their time online. The expanding number of fake bulletin online is a risk for social orders as well as administrations while has been as of now examined.

In this work, we have applied different deep learning approaches for the recognition of fake news. We inspected in detail two models, that is LSTM and RNN. This representation was appealed to our dataset and outcomes showed that the LSTM model has the most elevated test exactness with 70% validation accuracy followed by the RNN model of about 90% validation accuracy. RNN model has the highest validation loss and validation accuracy where LSTM model has 68% validation loss. Our proposed LSTM can detect fake news successfully but RNN gave the opposite prediction outcomes. RNN model performs better for validation loss and validation accuracy. LSTM model performs better for test outcomes. In the future, we want to work with sentiment analysis in cyberbullying. Most people face different problems due to cyberbullying which affects their life in any way.

References

1. Bovet, A., Makse, H.: Influence of fake news in Twitter during the 2016 US presidential election. Nat. Commun. **10**(1), 7 (2019)
2. Which, S.: Which, if any, is your main source of news? https://www.statista.com/statistics/198765/main-source-
3. Allcott, H., Gentzkow, M.: Social media and fake news in the 2016 election. Technical report, National Bureau of Economic Research, 2016
4. Ahmed, H., Traore, I., Saad, S.: Detecting opinion spam and fake news using text classification. Secur. Priv.
5. Vlachos, A., Riedel, S.: Fact checking: task definition and dataset construction. In: Workshop on Language Technologies and Computational Social Science (Proceedings of the ACL) (2014)
6. Qawasmeh, E., Tawalbeh, M., Abdullah, M.: Automatic identification of fake news using deep learning
7. Abedalla, A., Sadi, A., Abdullah, M.: A closer look at fake news detection: a deep learning perspective
8. Monti, F., Frascal, F., Eynard, D., Mannion, D., Bronstein, M.: Fake news detection on social media using geometric deep learning. arXiv:1902.06673v1 [cs.SI], 10 February 2019
9. Rodríguez, A., Iglesias, L.: Fake news detection using deep learning. arXiv:1910.03496v2 [cs.CL]
10. Ruchansky, N., Seo, S., Li, Y.: CSI: a hybrid deep model for fake news detection
11. Roy, A., Basak, K., Ekbal, A., Bhattacharyya, P.: A deep ensemble framework for fake news detection and classification. arXiv:1811.04670 (2018)
12. Wang, W.: "Liar, liar pants on fire": a new benchmark dataset for fake news detection. arXiv preprint arXiv:1705.00648 (2017)
13. Miller, K., Oswalt, A.: Fake news headline classification using neural networks with attention
14. Popat, K., Mukherjee, S., Yates, A., Weikum, G.: DeClarE: debunking fake news and false claims using evidence-aware deep learning. arXiv:1809.06416
15. Yang, Y., Zhang, L.: Ti-CNN: convolutional neural networks for fake news detection. arXiv:1806.00749
16. Khan, J., Khondokar, M., Islam, T., Iqbal, A., Afroz, S.: A benchmark study on machine learning methods for fake news detection. arXiv:1905.04749

Human Posture Estimation: In Aspect of the Agriculture Industry

Meharaj-Ul-Mahmmud[1], Md. Ahsan Ahmed[1], Sayed Monshurul Alam[1], Omar Tawhid Imam[2], Ahmed Wasif Reza[1], and Mohammad Shamsul Arefin[3,4](✉)

[1] Department of Computer Science and Engineering, East West University, Dhaka 1212, Bangladesh
{2018-2-60-067,2018-2-60-029,2018-2-60-123}@std.ewubd.edu, wasif@ewubd.edu

[2] Department of EEE, Bangladesh University of Engineering and Technology, Dhaka 1000, Bangladesh

[3] Department of CSE, Daffodil International University, Dhaka 1341, Bangladesh
sarefin@cuet.ac.bd

[4] Department of CSE, Chittagong University of Engineering and Technology, Chattogram 4318, Bangladesh

Abstract. Pose estimation is an artificial intelligence and computer vision approach. Human Posture Estimate is a more advanced version of pose estimation technology that graphically depicts the position and orientation of a human body. It's one of the most appealing fields of research, and it's gaining popularity thanks to its practicality and versatility—utilized in a range of industries, including gaming, healthcare, agriculture, augmented reality, and sports. This research project intends to establish a deep learning-based human posture identification system that can be used to identify diverse agricultural operations, with the intention of introducing the concept of automation into the agriculture field. A proprietary dataset of farmer postures is used to run this system. The picture from the dataset is pre-processed before a deep neural network is used to detect body points in the image, and OpenCV creates a graphical representation of the points. The angle between body components is crucial in determining posture, which is derived from various calculations. Finally, the result is compared to a threshold value before being processed. Our model could accurately measure a farmer's or human's posture in three major categories: sitting, bending, and standing with a test accuracy of about 77%.

Keywords: Human pose estimation · Gaussian blur · Agriculture posture estimation

1 Introduction

Pose estimation is a technique in computer vision and artificial intelligence that involves recognizing, associating, and tracking semantic key points of a human body or an object. "Right shoulders," "left knees," and "left brake lights of automobiles" are examples of

J. I.-Z. Chen et al. (Eds.): ICIPCN 2022, LNNS 514, pp. 479–490, 2022.
https://doi.org/10.1007/978-3-031-12413-6_38

semantic key - points. Human Pose Estimation is a better version of pose estimation technology that identifies the location and orientation of a human body graphically. In essence, it is a set of coordinates that may be linked to describe a person's stance. Each skeletal coordinate is referred to as a component (or a joint, or a key point). A pair is a legitimate link between two components (or a limb).

1.1 Background

It's one of the most fascinating areas of study, and it's gaining a lot of traction due to its utility and versatility—used in a variety of sectors, including gaming, healthcare, augmented reality, and sports. People are only seen as a bounding box in standard object detection (a square). Computers can learn to interpret human body language by doing pose detection and tracking. Traditional position tracking systems, on the other hand, aren't quick enough or resistant to occlusions to be practical. Some of the most significant advancements in computer vision will be driven by high-performance real-time posture detection and tracking. Computers will be able to build a finer-grained and more natural knowledge of human behavior by tracking human stance in real-time, for example. This will have a significant influence on a variety of sectors, including autonomous driving. The bulk of self-driving car accidents now is caused by "robotic" driving, in which the self-driving car makes an expected but unexpected stop and a human driver collides with it. Computers can better comprehend and anticipate pedestrian behavior with real-time human posture identification and tracking, allowing for more natural driving. Cultivation is another possible application. Landowners in agriculture may need to undertake various critical judgments, and computer vision-based human pose estimation can help them take positive action by recognizing the posture of employees in the field. Consider a scenario in which some laborers are working in the field and the owner needs to know whether they are operating correctly or not. An Intelligent system can be implemented to assess whether the employees are working appropriately or not and to do so, the system may need to evaluate their posture, and that is where the human posture estimation system can assist the main system.

1.2 Objectives

Through moving our bodies, we living beings accomplish certain important tasks. In the age of AI technology, computers sometimes have to comprehend the work of human beings and to do so, it needs to identify the posture of that human body, and human posture estimation is indeed a technique that can help machines accomplish that objective. As the universe undergoes a revolution thanks to artificial intelligence, the agricultural industry is following suit, however, there are few resources devoted to human pose detection in agriculture. The concept of AI in agriculture is becoming more popular, therefore this research study intends to create a human posture recognition system from the foundational concept of human position estimate in relation to the completed jobs such as yoga postures and classical dance techniques using deep learning. An advanced branch of machine learning, to aid Bangladesh's digitization efforts in the agriculture industry by identifying farmers' engagement posture in three major categories: sitting, bending, and standing.

2 Related Work

The group of authors in previous related works experimented with human postures. Many articles have proposed deep learning algorithms that may be used to recognize human poses.

In [1], authors developed a method for estimating human posture using Deep Neural Networks (DNNs). Their solution, which is a DNN-based regression to joint coordinates with a cascade of such regressors, has the benefit of collecting context and reasoning about position holistically. Using the PCP measure, they compared the findings to those of other LSP techniques. They identified the average value for all comparison algorithms for the four most demanding limbs – lower and upper arms and legs – as well as the average value across these.

In [2], authors provide a comprehensive overview of existing deep learning-based works for 3D pose estimation, summarizes the benefits and drawbacks of various approaches, and provides an in-depth knowledge of this field. They also investigate the most generally used benchmark datasets, on which they do a detailed comparison and analysis. a complete evaluation of existing deep learning-based research for 3D pose estimation, including a summary of their benefits and drawbacks, as well as an in-depth grasp of the field. They also investigate the most regularly used benchmark datasets, on which a full comparison and analysis research was undertaken. In [3, 4] and [5], different techniques for pose estimations were described.

The authors of [6] focused on the state-of-the-art advancements in deep learning-based 2-D human posture estimation. They compared their results with some other approaches like strict-PCP, PCK, and mAP. They concluded that under strict-PCP and PCK, RNN and GAN techniques attained the maximum accuracy.

In [7], they illustrate the most promising research topics for the future years and highlight the state of the art in human pose estimation by defining new qualitatively better criteria for assessment and analysis of pose estimation systems. They discovered that PS had a little better result, although the overall improvement due to retraining is less (46:1 PCPm for retrained vs. 42:3 PCPm the original).

In [8] authors provide a method for detecting numerous people's 2D poses. To recognize body parts with persons in a picture, the suggested technique employs a nonlinear representation known as Part Affinity Fields (PAFs). When a PAF-only refinement is used instead of a PAF and body part location refinement, runtime speed and accuracy are significantly improved. Using an internal annotated foot dataset, they provide the first combined body and foot major point detector. In comparison to processing them consecutively, the integrated detector not only decreases convergence speed but also preserves the precision of every part independently. OpenPose, the very first open-source real-time system for multi-person 2D posture recognition, comprising a body, foot, hand, and face key- points, was released as a result of these studies.

In [9], the researchers show a systematic design for how convolutional networks can be incorporated into the pose machine framework for learning image features and image-dependent spatial models for the task of pose estimation. For that purpose, researchers are trying to implicitly model long-range dependencies between variables in structured prediction tasks such as articulated pose estimation. They achieve this by designing a sequential architecture composed of convolutional networks that directly operate on belief maps from previous stages, producing increasingly refined estimates for part locations, without the need for explicit graphical model-style inference. This paper solves the problem of disappearing gradients during training by introducing a natural learning goal function that imposes intermediate supervision, restoring back-propagated gradients, and modulating the learning method. Standard benchmarks such as the MPII, LSP, and FLIC datasets were utilized to demonstrate state-of-the-art performance and surpass rival approaches.

In [10], authors adapted two ways to train the network: (1) a multi-task framework that simultaneously trains pose regression and body part detectors, and (2) a pre-training strategy that uses a network trained for body part detection to initialize the pose regressor. Researchers also evaluate the structure of a big data set, the Human3.6M Dataset, and obtain that it outshines traditional methods significantly. Finally, they demonstrate that the deep convolutional network has learned the association between output variables and has disentangled relationships between body parts.

In [11] authors explained the way to distinguish human activity in video surveillance using the notion of human posture estimation. The authors of this research offer a unique technique for 2D Human Pose Estimation based on object identification utilizing RGB-D information in this paper. They used Convolutional Posture Machines (CPM) to create belief maps and assess human pose by sensing 14 critical body locations. CPM is then applied to the rescaled area of the discovered item. They established that the suggested technique recognizes target items reliably in occlusion based on the experimental findings. By supplying a correctly identified region as an input to the CPM, it is also feasible to do 2D human posture estimation.

In [12], authors used a normal two-step pipeline, first identifying the 2D location of the N body joints and then inferring 3D pose from these observations. A modern CNN-based detector is employed in the first stage. Most known techniques conduct 2N-to-3N regression of the Cartesian joint coordinates as the second phase. Researchers of this study show that by modeling both 2D and 3D human poses using $N \times N$ distance matrices and phrasing the issue as a 2D-to-3D distance matrix regression, more correct pose estimations can be produced. They use basic Neural Network designs to train such a regressor, which guarantees positivism and homogeneity of the anticipated matrices by design. The method also has the benefit of generally managing missing data and allows the location of non-observed joints to be hypothesized. Research findings on the Humaneva and Human3.6M datasets show continuous improvements above the current state-of-the-art. A qualitative examination of the LSP dataset's photos in the wild using the regressor learned on Human3.6M shows extremely encouraging generalization results.

The work in [13] offers a deep learning model that adapts from 2D joint labels to estimate 3D human posture and camera from monocular photos. The suggested technique has a standard design, but it also has an extra output layer that designs projected 3D joints onto 2D and imposes 3D length limitations on body parts. Pose restrictions are also imposed using an autonomously neural classifier that obtains a previous distribution across 3D positions. The proposed system is evaluated on multiple benchmark datasets, including HumanEva-I and Human3.6M, and is examined with state-of-the-art algorithms for 3D human pose estimation, with comparable results. Furthermore, the methodology exceeds competing methods such as Baseline CNN and Ionescu in circumstances when 3D ground truth data is absent, and this model has strong generalization capabilities, according to academics.

In [14] authors proposed a hybrid approach that collects scene feature information from remote sensing photos twice for improved recognition. In layman's terms, these features are classified as raw and have only a single defined frame, allowing basic detection from remote sensing photos. Using a double feature extraction method, this research provides a hybrid deep learning approach based on feature abstraction approaches to classify remotely sensed visual sceneries. In terms of accuracy and recognition rate, this study produced a novel hybrid framework method that beats any existing model.

In [15] a CNN architecture has been incorporated that can be utilized to detect the existence of buildings even when the data for training is limited. Furthermore, because of its optimization quality and modular structure, this methodology will be effective in improving performance. They have also developed a new CNN architecture that works in tandem with the feature-based classifier to improve the accuracy of building detection using 3D and 2D data. Two new layers have been added to the suggested methodology to improve the robustness and discrimination ability of the CNN architecture. The innovative layers and their impacts have been evaluated in comparison to other novel approaches for detecting buildings in remote sensing photos.

3 System Architecture and Design

Figure 1 illustrates the developed methodology, which consists of four key steps:

1. Image Pre-Processing
2. Generating Body Parts & Coordinates
3. Graphical Representation & Pairing Body Parts
4. Estimating Posture & Revealing Outcomes

Fig. 1. The system architecture of the human pose detection.

3.1 Dataset Description

The dataset for this work is an image dataset which is Farmers Activity Datatset, that was constructed in a custom manner. The whole image dataset contains a collection of around 40 digital photographs of various human positions that serve as inputs to the suggested approach. Figure 2 shows an example input collection containing images of human postures.

Fig. 2. Sample input data set (Farmers Activity Dataset)

A total of 3600 samples were gathered for analysis to categorize positions. Many body parts, such as the nose, neck, shoulder, hip, and knee, are involved. It will try to link the points based on posture pairings, such as Neck to Right Shoulder Pair, Right Shoulder to Right Elbow Pair, Right Elbow to Right Wrist Pair, and so on, after extracting the points from the image. The system calculates the angle between specified pairs to determine the posture when the pairing process is completed. If the system could not match the calculated result to any category, the pose will be denoted as Unknown.

3.2 Data Preprocessing

In this approach, a photo of a person is used as the input and is pre-processed before the feature extraction. The image has been resized to a size of 400x400 pixels, with a height and width of 368 pixels. The picture is then smoothed and noise is removed using Gaussian blur. The next stage is to extract information about bodily components as well as their locations.

Algorithm1: Pre-processing of an Image

Input: A full-body image of a human

Require: Pre-process the image

1. **begin**
2. **for** each image **do**
3. use a mask to remove the background
4. use GaussianBlur to smoothen the edges
5. resize the images to 368x368 pixels
6. **end for**
7. **end**

3.3 Information Extraction

A single individual is usually viewed in a single photograph. We ran each of the preprocessed photos through the deep neural network to retrieve the information. It's a Pose Classifier based on OpenPose. It provides us with the detected body points as well as the portion coordinates. The output of the DNN returns a dictionary. The names of the body parts are the keys in the dictionary, and the locations are the values. Analyzing the dictionary data will be used to recognize the posture.

Algorithm2: Body Part Extraction from an Image

Input: A pre-processed full-body image
Require: Extract body parts from the image

1. **begin**
2. **for** each pre-processed image **do**
3. Feed image to OpenPose DNN to get image details in a dictionary
4. **for** each point found in the body **do**
5. append the point to a dictionary
6. **end for**
7. **for** each point in the dictionary **do**
8. **if** the two ends of a pair exist
9. draw a line from one end to another
10. draw two ellipses at both ends
11. **else**
12. do nothing
13. **end if**
14. **end for**
11. join all elements of the list
12. append the resultant text to another list
13. print the list
14. **end**

3.4 Posture Categorization

We divided the posture into three categories. Standing, bending, and sitting are all options. It will compute the distance between two points and find an angle of human arms position for each key point that is required for classification.

Algorithm3: Posture Categorization

Input: A full-body image of a human
Require: Identify the pose
1. **begin**
2. **for** each pre-processed image **do**
3. **if** Neck is in points and Right Hip or Left Hip is in Point and Left Knee or Right Knee is in Points
4. assign Neck point coordinate to a variable (p1)
5. assign Right Hip or Left Hip point coordinate to a variable (p2)
6. assign Right Knee or Left Knee point coordinate to a variable (p3)
7. Calculate the angle between the lines (line1 is between p1 and p2, line2 is between p2 and p3)
8. **if** the angle is between 60 and 150 degrees
9. **if** Left or Right Ankle is found
9. set pose "Farmer is engaged in a task, which is generally carried out while in a bending position"
10. **else**
11. set pose "Farmer is engaged in a task, which is generally carried out while in a sitting position"
12. **else if** angle is between 150 and 200 degrees
13. set pose "Farmer is engaged in a task, which is generally carried out while in a standing position"
14. **else**
15. set pose to Unknown
16. **end if**
17. **end if**
18. **if** any Knee is not found
19. set pose "Farmer is engaged in a task, which is generally carried out while in a sitting position"
20. **end if**
21. **end for**
22. **end**

---

4 Implementation and Experimental Result

4.1 Experimental Setup

The suggested system was tested on a Windows 10 PC with an Intel Core i5 2.4 GHz processor and 8 GB of RAM. It was created with Python 3.8 (version). Pose extraction is done with a Deep Neural Network.

4.2 Implementation

The inputs of our system are images of human pose that are sorted by fixed size which is 368 × 368 pixels. The simulation results of our proposed technique derived from numerous images from the dataset are shown in the following figures by a single image within the dataset. Figure 3(a) shows the input image, whereas Fig. 3(b) shows the preprocessed image. Figure 3(c) skeleton image of the human pose and Fig. 3(d) extracted data.

Fig. 3. (a) input image, (b) preprocessed image, (c) skeleton of human pose (d) extracted data.

After extracting the human postures from the input image, our system will match the angles with key points and determine whether the person is performing an activity that requires him to be in a standing, bending, or sitting position based on this model.

4.3 Performance Evaluation

We use the following equation to compute the accuracy of the suggested system to assess its efficiency (Table 1).

$$accuracy = \frac{No.of\ corrected\ correspondence}{No.of\ correspondence} * 100\% \tag{1}$$

Table 1. Performance analysis for the proposed system

Activity	No. of activities input based on posture	Correctly detected poses input	Detected accuracy	Misclassify percentage
Performed in a standing position	21	21	100%	0.00%
Performed in a bending position	12	5	41.66%	58.34%
Performed in a sitting position	7	6	85.71%	14.29%

For the evaluation purpose, we took 21 sample pictures of farmers' activity in standing posture, 12 sample pictures of farmers' activity in bending posture, and finally 7 sample pictures of farmers' activity in sitting posture.

The issue we had was seamlessly identifying body components. Body components could not be distinguished when the foreground and backdrop were the same color. We couldn't acquire sufficient photographs that completely satisfied our requirements due to a limitation of relevant images of farmers in specific stances.

Due to a lack of good photos of farmers working in bending posture, the system's accuracy in recognizing bending position was therefore low. Because identifying bending posture necessitates a correct angle between the upper and lower bodies, a good picture must be provided for accurate angle computation.

5 Conclusion

A classifier is used in this article to recognize several types of human postures from a photograph. This is, as far as we know, a zesty procedure. This technique uses a series of images with the farmer and other human activities to properly predict the posture in three major categories as standing, sitting, and bending. The system's problem in execution is that the skeleton won't materialize if a body part isn't visible or the background color is the same as the foreground color. However, the proposed strategy achieved a reasonable degree of accuracy around 77% in the end. If the model could be trained in a reasonably bigger data set, the accuracy could have been better.

References

1. Toshev, A., Szegedy, C.: DeepPose: human pose estimation via deep neural networks. İn: Proceedings of IEEE Computer Social Conference Computer Vision and Pattern Recognition, pp. 1653–1660 (2014). https://doi.org/10.1109/CVPR.2014.214
2. Wang, J., et al.: Deep 3D human pose estimation: a review. Comput. Vision Image Underst. **210**, 103225 (2021). https://doi.org/10.1016/j.cviu.2021.103225
3. Akhter, I., Black, M.J.: Pose-conditioned joint angle limits for 3D human pose reconstruction. In: Proceedings of IEEE Computer Socity Conference on Computer Vision and Pattern Recognition, vol. 07–12-June, pp. 1446–1455 (2015). https://doi.org/10.1109/CVPR.2015.7298751
4. Guler, R.A., Neverova, N., Kokkinos, I.: DensePose: dense human pose estimation ın thewild. In: Proceedings of IEEE Conference Computer Vision and Pattern Recognition, pp. 7297–7306 (2016). http://arxiv.org/abs/1612.01202
5. Andriluka, M., et al.: PoseTrack: a benchmark for human pose estimation and tracking. In: Proceedings of IEEE Computer Society Conference on Computer Vision and Pattern Recognition, pp. 5167–5176 (2018). https://doi.org/10.1109/CVPR.2018.00542
6. Liu, Y., Xu, Y., Li, S.B.: 2-D human pose estimation from ımages based on deep learning: a review. In: Proceedings of 2018 2nd IEEE Advances in Information Management Communication, Electronics and Automation Control Conference, IMCEC 2018, no. Imcec, pp. 462–465 (2018). https://doi.org/10.1109/IMCEC.2018.8469573
7. Andriluka, M., Pishchulin, L., Gehler, P., Schiele, B.: 2D human pose estimation: New benchmark and state of the art analysis. İn: Proceedings of IEEE Computer Society Conference on Computer Vision and Pattern Recognition, pp. 3686–3693 (2014). https://doi.org/10.1109/CVPR.2014.471
8. Cao, Z., Hidalgo, G., Simon, T., Wei, S.E., Sheikh, Y.: OpenPose: realtime multi-person 2D pose estimation using part affinity fields. IEEE Trans. Pattern Anal. Mach. Intell. **43**(1), 172–186 (2021). https://doi.org/10.1109/TPAMI.2019.2929257
9. Wei, S.-E., Ramakrishna, V., Kanada, T., Sheikh, Y.: Pose machines. In: Proceedings of IEEE Conference on Computer Vision Pattern Recognition (2016)
10. Li, S., Chan, A.B.: 3D human pose estimation from monocular images with deep convolutional neural network. In: Cremers, D., Reid, I., Saito, H., Yang, M.-H. (eds.) ACCV 2014. LNCS, vol. 9004, pp. 332–347. Springer, Cham (2015). https://doi.org/10.1007/978-3-319-16808-1_23
11. Park, S., Ji, M., Chun, J.: 2D human pose estimation based on object detection using RGB-D information. KSII Trans. Internet Inf. Syst. **12**(2), 800–816 (2018). https://doi.org/10.3837/tiis.2018.02.015
12. Moreno-Noguer, F.: 3D human pose estimation from a single image via distance matrix regression. In: Proceedings - 30th IEEE Conference on Computer Vision Pattern Recognition, CVPR 2017, vol. 2017-Janua, pp. 1561–1570 (2017). https://doi.org/10.1109/CVPR.2017.170
13. Brau, E., Jiang, H.: 3D human pose estimation via deep learning from 2D annotations. In: Proceedings - 2016 4th International Conference 3D Vision, 3DV 2016, pp. 582–591 (2016). https://doi.org/10.1109/3DV.2016.84
14. Sungheetha, A., Rajesh Sharma, R.: Classification of remote sensing ımage scenes using double feature extraction hybrid deep learning approach. J. Inf. Technol. Digital World **3**(2), 133–149 (2021). https://doi.org/10.36548/jitdw.2021.2.006
15. Karuppusamy, P.: Building detection using two-layered novel convolutional neural networks. J. Soft Comput. Paradigm **3**(1), 29–37 (2021). https://doi.org/10.36548/jscp.2021.1.004

A Survey on Image Segmentation for Handwriting Recognition

Prarthana Dutta(✉) and Naresh Babu Muppalaneni

Department of Computer Science and Engineering, National Institute of Technology, Silchar, Assam, India
prarthana.dutta01@gmail.com, nareshmuppalaneni@gmail.com

Abstract. The machine learning community has helped solve many challenging computer vision-related tasks over the last decades. Many deep learning techniques have also evolved over the years concerning the segmentation of images for various computer vision-related tasks. Literature has witnessed a surge of numerous segmentation techniques in image processing applications. Segmentation has been lately proving as an efficient tool in affecting the recognition rate in any image processing task to a great extent. This is because only by employing the correctly segmented portions of the image can one further accurately process the image for various recognition tasks. We have thus surveyed the different segmentation methods in this brief literature, focusing on the segmentation methods for handwriting recognition. Researchers in this domain are also brought to notice the various challenges encountered.

Keywords: Deep learning · Machine learning · Image processing · Segmentation

1 Introduction

Images captured for processing has to be cleaned and simplified for further processing. A digital image contains a lot of hidden information needed to be analyzed for processing and extracting valuable information. Rather than considering the entire image at a time, it is preferred to break down an image into its constituent parts or subgroups to make the image analysis simpler. These subgroups are called "segments." These individual segments or constituent elements of the image make the image processing task simpler. Each pixel in the constituent segment comprises identical properties to each other. Segmentation can be thus addressed as assigning labels to pixels. The pixels with the same label constitute a segment.

This information is used to train various machine learning models for addressing various problems. The segmentation quality directly influences the recognition task to a great extent as the errors in the segmentation process are found to propagate towards the recognition directly. Thus, when a digital image is broken

J. I.-Z. Chen et al. (Eds.): ICIPCN 2022, LNNS 514, pp. 491–506, 2022.
https://doi.org/10.1007/978-3-031-12413-6_39

down into its constituent parts and utilized for further processing or analysis, and it is called "image segmentation."

The image processing basic pipeline comprises various constituent steps such as preprocessing, feature extraction, segmentation, and recognition or classification. Each step has its own methods and approaches depending on the application domain. Being an essential domain in computer vision, image segmentation can be considered one of the oldest problems being pondered upon. Among all the techniques in image processing, image segmentation is also becoming an intrinsic phase that is also closely related to classification in many applications.

In this literature, we try to overview the various segmentations carried out using deep learning. Literature witnessed several image segmentation algorithms [43]. Some of the worth mentioning ones are thresholding [51], histogram based bunding, region growing [49], k-means clustering [20], watershed methods [48], active contours [32], graph cuts [15], conditional and Markov random fields [57], and sparsity-based methods [44,70], etc.

1.1 Applications of Image Segmentation

Deep learning applications of recognition can be accomplished only when it passes through a proper image segmentation procedure. The image segmentation has several applications, some of them are listed below:

- Facial Recognition [31]
- Number Plate Identification [7,60,64]
- Satellite image analysis [6,7,12]
- Scene Understanding [37]
- Character Recognition [22,42], etc.

The motivation of this survey is to bring into the limelight the current image segmentation techniques employed in domains of computer vision, especially in the handwriting recognition task. Various challenges and shortcomings encountered in literature are also brought to the notice of the reader. The rest of the paper is organized as: Sect. 2 introduces the Optical Character Recognition system and discusses about the levels of image segmentation for handwriting recognition. Section 3 discusses the current works carried in literature for segmentation of handwriting documents and Sect. 4 focuses on the challenges that one may come across while working on segmentation for the same. Finally Sect. 5 and Sect. 6 concludes the survey by briefing the performance metrics measurement for model efficiency and then showing directions toward future aspects in this field.

2 Background Study

We shall direct our focus towards image segmentation for character recognition as one of the application domains.

2.1 Optical Character Recognition (OCR)

In the digital age, where abundant writing tools are available, many people prefer traditional writing methods with paper and pen.

In today's paperless world, there have been instances where tasks are accomplished in handwriting communication. Handwritten mode is found to be more convenient and cost-effective globally across generations. Many documents, such as invoices, taxes, postcards, etc., have fields to be filled up with a pen, which may further require digitization. Preservation of ancient historical texts, scriptures, manuscripts, etc., must be preserved in the digitized form [10]. But to store, access, or even efficiently share them in the future becomes difficult. Hence, many of these documents remain uninvaded and are left untransformed into digital form. Therefore, there has been an emerging need to preserve these documents in digital format for access, retrieve and share in the future.

Humans incur the knowledge of learning characters and words of languages since childhood. But, for computers to emulate the same to identify the content in a language, they must also go through a specific learning process that is possible only by the advent of Artificial Intelligence (AI), Machine Learning (ML) and Deep Learning (DL).

Artificial Intelligence simulates a machine to work, think, and act accordingly like a human in real-life applications. John McCarthy coined the term Artificial Intelligence by defining it as "the science and engineering of making intelligent machines" in the 1950s. The applications and impact of AI has been witnessed in fields such as natural language processing, recognition, computer vision, etc.

Machine Learning can be simply understood as a technique of automatically studying and understanding patterns in huge data collection and applying this learning and experience to make decisions on unseen data. Machine learning has paved the path to better and improved understandings in the future and hence is an essential topic in many research activities seeking perfection and improvements [18,63].

Deep learning is a sub-field and an active area of research in Machine Learning that relies on the collection of algorithms based on machine learning for attaining the high levels of abstractions in data. With the sheer size of the data available today, deep learning can utilize the vast data and obtain abstractions from the data for providing solutions in almost every field such as business, healthcare, recognition, etc. Many innovative models have been designed by researchers employing deep learning techniques for handwriting recognition (Table 1).

Deep Learning Models for Handwriting Recognition. Recognition of a language, text, word, or character has been emerging as an important and necessary field of research in various application domains such as banking, healthcare, multimedia database, etc. So, to train a system to see and read automatically requires the application of Optical Character Recognition (OCR) systems [45,69]. Hence, OCR is a document image analysis process where machine-printed or handwritten documents are fed into a learning process for converting

Table 1. A few of the deep learning models employed in literature for handwriting recognition

Author and Year	Model	Digit/Character recognition	Dataset/Language
Banumathi *et al.* [11] (2011)	Kohonen Self Organizing Map (SOM)	Character	Tamil
Kim *et al.* [33] (2015)	Deep Neural Network	Characters	(Hangul) SERI95*a* and PE92 dataset
Alom *et al.* [2] (2017)	Deep Belief Network and CNN	Digits	(Bangla) CMATERdb 3.1.1 dataset
Ashiquzzaman *et al.* [5] (2017)	Multi-Layer Perceptron and Convolutional Neural Network	Digit	CMATERDB 3.3.1 (Arabic)dataset
Vaidya *et al.* [73] (2018)	Convolutional Neural Network	Character	NIST dataset
Arora *et al.* [4] (2018)	Convolutional Neural Network and Feed Forward neural Network	Digit	MNIST dataset
Boufenar *et al.* [14] (2018)	Deep CNN (Transfer Learning)	Characters	Arabic
Aneja *et al.* [3] (2019)	CNN	Character	Devanagri
Muppalaneni *et al.* [47] (2020)	CNN	Compound Characters	Telugu dataset
Dutta *et al.* [21] (2020)	DigiNet	Digits	Assamese
Prathima *et al.* [59] (2021)	CNN	Characters	Telugu Vowels dataset
Rahman *et al.* [61] (2021)	RNN-LSTM network	Digits	English (MNIST)
Geetha *et al.* [24] (2021)	CNN-RNN hybrid	Digits and Characters	IAM and RIMS
Geetha *et al.* [24] (2021)	Hybrid CNN+LSTM	Characters	IAM and RIMS
Alkhawaldeh *et al.* [1] (2021)	Deep Hybrid transfer Learning (CNN + LSTM)	Digits	Arabic
Hamdan *et al.* [28] (2021)	Statistical SVM	Character	MNIST, CENPARMI, UCOM, and IAM

into machine-readable digital text format. In a nutshell, OCR describes a system that performs the mechanical or electronic conversion of images. It provides full alphanumeric recognition of printed or handwritten text or characters at electronic speed by simply scanning the document. Amongst printed and handwritten document images, recognition is more challenging for handwritten ones since the handwriting varies from person to person depending on various criteria.

OCR can identify both handwritten as well as printed text. As a part of OCR, handwriting recognition, is being extensively studied [56, 71]. It is easy for humans to recognize handwritten or printed characters. Still, it is a difficult task for machines to identify them due to their complex structure and variation of handwriting styles across persons. We human beings can easily recognize printed or handwritten characters, but making a computer system to recognize these characters becomes challenging due to random variation of noise, varying styles, cursive nature, unvarying fonts, and size.

Some of the difficulties and challenges in Handwriting Character Recognition (HCR) are unidentical skew angles, overlapping or touchings, irregular spacings, punctuation, writing styles may differ from the same person or different persons, variation in shape and style, the appearance of skew/slants, unequal height, irregular text size, missing points during irregular movements, left or right bends, etc.

Among a large number of application domains in the fields of AI, ML and DL, handwriting character recognition is one of the active and fascinating areas utilizing image processing and computer vision [18,21,26,54,63]. It is a highly challenging task to implement that works under all scenarios and provides high recognition accuracy.

2.2 Levels of Image Segmentation for Handwriting Recognition

The image segmentation levels in Handwriting Recognition can be of the following levels as shown in Fig. 1:

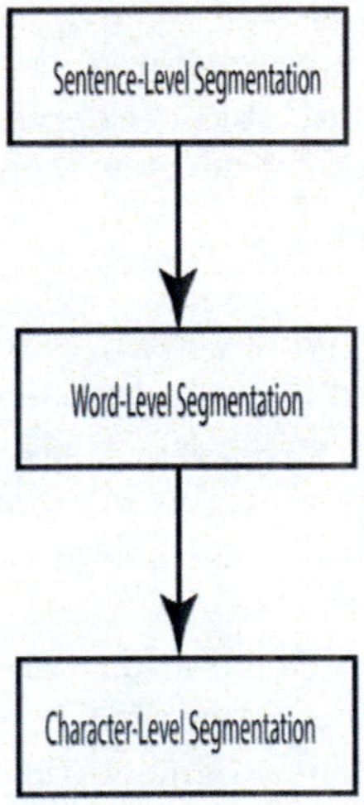

Fig. 1. Levels in image segmentation for handwriting recognition

1. Segmentation of sentences from page (Sentence-level segmentation)
2. Segmentation of words from lines (Word-level segmentation) and
3. Segmentation of characters from words (Character-level segmentation)

Sentence-Level Segmentation: From the image of a page of a written document, when we want to extract each individual line of the page, we perform segmentation at the sentence or line level. Various segmentation methods can be applied at this level, such as horizontal projection. In this approach, the black pixels are counted in each row, and the low horizontal projections are cut out. Errors at this level may occur in segmenting or cutting off portions of the dots and diacritics of the text. Raza *et al.* proposed a text-line segmentation method in [62] for the Urdu language. The proposed method comprised of the following methods:

- Apply cubic interpolation.
- Remove (salt and pepper noise) using a median filter.
- Apply Hogue transform for line detection.
- Apply Otsu's Global thresholding method for binarizing the extracted regions (for extracting the ROI).
- Filtering of the connected components and their centroids.
- Horizontal Projection Profiling (HPP).
- HPP based line segmentation.

After the application of the above line segmentation technique, it was observed that only 2.4% of the handwritten lines were not correctly segmented. Thus out of the 2051 text lines present, 2001 text lines were precisely segmented.

Another work by Papavassiliou *et al.* proposed the use of the Viterbi algorithm by identifying the optimal succession of text and gaps within vertical zones [53]. Here, the document is initially separated along vertical sections, and each zone is separated into "gap-region" and "text-region" accordingly. Further, to recognize the optimal text and gap sessions in each zone by utilizing the Viterbi algorithm.

Word-Level Segmentation: Complications arise when the recognition is performed based on the entire word, since a considerable vocabulary of words is needed to be maintained for the same, which also consumes a lot of time and memory. At the same time, some wrong words might also be present, which do not belong to any vocabulary, and hence cannot be recognized. Therefore there is a need to go for word-level segmentation. Thus, after sentences are extracted from the document, word-level segmentation can be applied to extract independent words from the sentences. Papavassiliou *et al.* proposed a method for word-level segmentation and text-line segmentation in their work [53]. For word-level segmentation, the authors computed the separability distance between two adjacent connected components. Finally, the candidate words are classified as "within-words" or "between-words."

Character-Level Segmentation: After extracting individual words, characters can also be separated out from the words. This level of segmentation is recognized as character-level segmentation [38]. Various character segmentation techniques are employed in literature such as, [8,27,46], etc. The various challenges encountered are discussed in Sect. 4.

3 Literature Survey

Many strategies have evolved in literature for dealing with the segmentation of characters for recognition [16]. Segmentation plays a significant role in determining the accuracy of the model employed for recognition. Many literature studies reveal that the segmentation quality is a requirement for attaining good recognition accuracy. Segmentation-free approaches are employed in recognition tasks of cursive scripts but failed to achieve a desired and satisfactory accuracy. Hence, segmentation-based methods are usually preferred in this context. Segmenting a document text into its constituent parts comprises text line segmentation and word segmentation. Both the segmentations have their drawbacks, which are tried to work upon by researchers and explore the room for further improvements in the segmentation, further improving the recognition accuracy. Some of the difficulties in text-line segmentation include the difference in skew angles, overlapping and touching components, etc. While in word-level segmentation, difficulties may arise, such as skew and slants in the text lines, punctuation marks, ununiform gaps between words or characters, etc.

A new deep learning based approach to segmenting lines and words from documents is initiated by Belabiod and Belaıd [13]. RU-net is employed for line segmentation for separating text lines from the background using a pixel-wise classification mechanism on Arabic documents. For word segmentation, they employed a CNN and Bi-LSTM (bidirectional Long Short Term Memory) for efficient learning with a CTC (Connectionist Temporal Classification) function. The RU-net and CNN+BiLSTM+CTC architectures attained an accuracy of 96.71% for line segmentation and 80.1% for word segmentation. Further post-processing may be required for achieving more precision on the word segmentation using the CNN+BiLSTM+CTC approach.

One of such recent works is that of Malik et al. [39]. They tried to develop an enhanced segmentation-based approach to the Urdu cursive OCR to handle the problems such as unequal text size, line gaps, skewness, etc. The authors explored both handwritten (687-line images) and printed (495 line images) texts. Preprocessing methods such as adaptive thresholding, Otsu's thresholding, Binarization, Canny edge detector, etc., are employed to remove various noises from the original raw images. With the noise-free images, skew lines are detected and corrected using the skew correction algorithm that considers the pixel intensities. It employs the "header and baseline detection technique" for segmenting the text lines. They observed that the line segmentation accuracy is dependent on the level of skew correction. Less skewed lines, better is the segmentation. This segmentation approach attained satisfactory accuracy of 96.7% on handwritten documents and 98.3% on printed documents. Over-segmentation and under-segmentation are addressed as the causes of the error during segmentation. Under-segmentation resulted due to inter-line skewness, while over-segmentation resulted from the presence of dots and diacritics.

A combination of the CNN and RNN has been witnessing to surge in the recognition accuracy in several tasks, including recognizing handwritten characters in a number of languages. One such work on Urdu handwritten cursive text recognition was initiated by Hassan *et al.* [30] by relying on character segmentation. Segmentation of characters is carried out by employing the CNN for feature extraction and feeding these features to a Bi-directional LSTM for classification. Two bi-LSTMs follow seven initial convolutional layers in the network architecture developed by the authors. The Levenshtein distance is used to compute the character recognition accuracy. An average recognition rate of 83.69% is obtained by the model. Though the recognition rate is less than that of models proposed in other works, the dataset used and the model employed are complex ones compared to others. Other works used isolated characters and small-sized datasets, which cannot be employed for real-world applications. Hence, the authors claim that this work can be employed for challenging character recognition shortly.

A technique named SFF (Segmentation Facilitate Feature) is proposed in [34] by Kohli *et al.* for segmenting touched components. In the variants of handwriting by different writers, some word components may get touched, leading to difficulty in word and line segmentation. So to identify such parts in an image, SFF is employed as a segmentation mechanism to segment the touched components, and are then recognized using a CNN. The evaluation is carried out on a Hindi dataset comprising only two touched consonants, where the proposed methodology attained an accuracy of 96.2%.

These are a few of the recent segmentation approaches employed in literature at the sentence, word, and character levels over different languages.

4 Challenges in Image Segmentation for Handwriting Recognition

For partitioning the image of a word into regions consisting of isolated and complete characters are the significance of employing segmentation in character recognition. Contrary to the machine-printed text, the handwritten text is a difficult task that needs support from recognition algorithms as well. Thus the problem of handwritten character segmentation and recognition are closely related [38]. Across the literature, it has been observed that the segmentation algorithms proposed so far can extract text lines, words, and even characters efficiently for recognition. But literature has also witnessed difficult situations where the segmentation fails. The segmentation task comes with several challenges.

- We have experimented with recognizing the Assamese handwritten digits in our work [21]. A sample of the experimented dataset is shown in Fig. 2. Here, we had to deal with different characters with minute differences. Such as the Assamese digits 1, 9 and 5, 6 in Fig. 3.
  If during the segmentation process, the component that contributed to the difference between two or more characters is ignored, then the recognition system might identify the characters as identical.

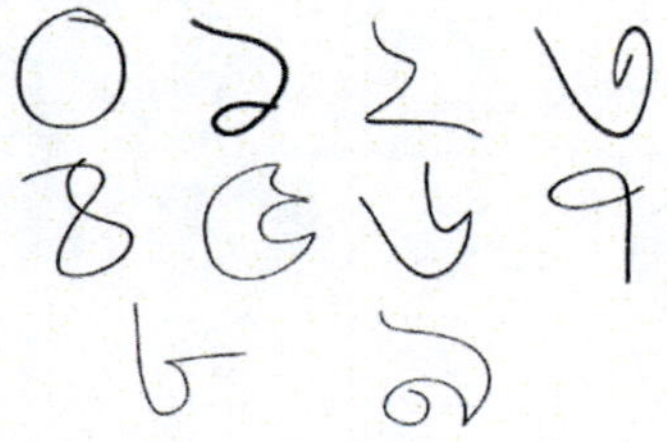

Fig. 2. Sample of the Assamese handwritten digits dataset

Fig. 3. Similarity in the Assamese Digits 5 and 6

- Another challenge encountered in this aspect is that the angular positions of the characters or text may vary while acquiring the handwritten text or characters. For example, the Assamese digit 8 appearing in two different angular positions can be visualized as in the Fig. 4.

Fig. 4. Assamese Digits 8 appearing in two different angular positions

- Some overlapping or touching characters might also be present in the characters to be segmented as shown in Fig. 5 which also might be another challenge to be faced.
- A similar type of challenges could also be encountered in another work in [47] while working on the Telugu compound characters. A sample of their dataset is shown in Fig. 6.

Thus, we observe that some of the reasons for the wrong segmentation or no segmentation or challenges in segmentation could be as follows:

1. Skew angle variation within the text-line (Fig. 4).
2. Skew angle variation between two different text lines.
3. Intra-class variation.

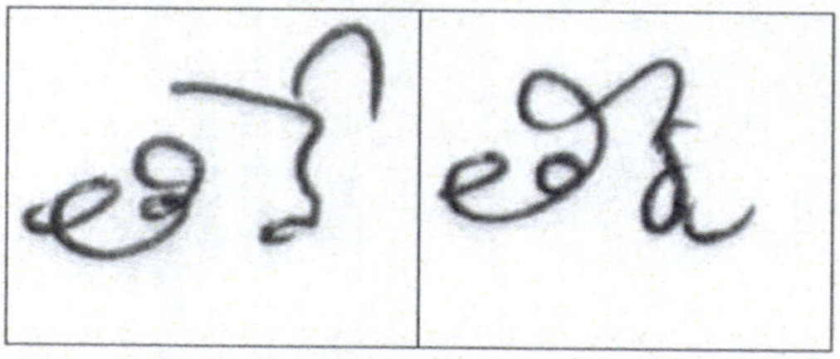

Fig. 5. Two different overlapping and touching Telugu handwritten characters

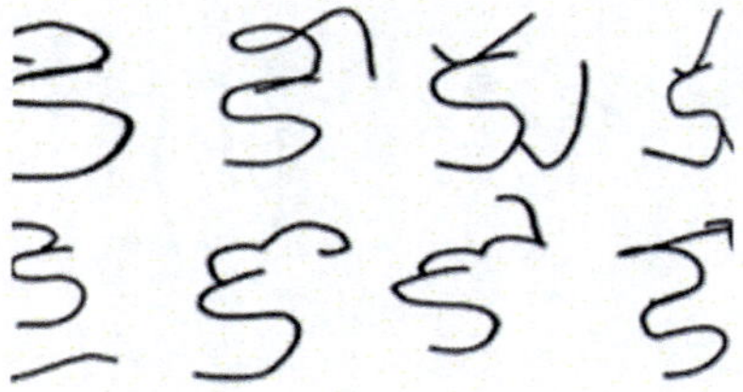

Fig. 6. Sample of the Telugu handwritten gunintam

4. Presence of overlapping or touching characters (Fig. 5).
5. Variation among the characters of the same sentence or word.
6. Variation in the handwriting style.

Diversity in the handwritten images poses serious technical challenges, such as variation in slope and skew and angles, relative size, appearance, etc. Intra-class variation, i.e., characters or digits belonging to the same class may be written in different styles by different writers, which might be recognized wrongly by the classifiers as belonging to other classes. The deep learning models developed so far are focused upon minimizing the intra-class variations by increasing the size of the labeled data, employing augmentations, concentrating on salient regions of the images in the training samples, using variance-based methods, modifying the loss function, etc. [55].

Skew detection in the text document is an essential phase in recognition of handwritten documents. Efficient skew detection is a vital recognition task in image processing. Any tilt present in the text or character to be processed is the skewness present in it. This tilt is the deviation of the baseline of the text from the original horizontal axis. Detecting and normalizing all such type of tilts or deviations in the image or text is an important preprocessing phase in document image analysis. Different text images may contain either large or small skew angles (depending on the extent of deviation). Numerous methods have been proposed in the literature so far for skewness detection, such as, Global and Local Hogue transform, Projection Histogram, Least Square Methos, etc. [19]. Bal and Saha presented an efficient skew detection mechanism in their work [9] based on the orthogonal projection of the $x-$axis. This method computed the pixel values of the first and last segmented line of the text. They tested their mechanism on the IAM English dataset and obtained good recognition accuracy of 96% for all the skew angles present.

Skewness in the text can be present during the data collection phase or while the writer writes the text. It can also be present during the scanning of the document for processing. Some of the other skew detection and normalization approaches employed for handwriting text recognition are shown in Table 2.

Table 2. Skew angle detection and normalization techniques employed for handwriting recognition

Author	Approach
Bal *et al.* [9]	Orthogonal projection of line/words towards the horizontal axis.
Chin *et al.* [19]	Global and Local Houng transform, Projection Histogram, Least Square, etc.
Panwar *et al.* [52]	Orthogonal Projection
O'Gorman *et al.* [50]	Projection profile, Hogue Transform, Horizontal Neighbour, etc.
Gatos *et al.* [23]	Crossed-correlation between Vertical document lines
Salagar *et al.* [66]	Run Length Smoothing Algorithm (RLSM)
Malik *et al.* [39]	Counting Pixel Approach
Pramanik *et al.* [58]	Utilizing salient feature (matra or head line)
Shakuntala *et al.* [67]	Computing Optimum Skew angle by connected components

Images with different orientations also pose a challenge in the segmentation task, thereby reducing the recognition accuracy. This challenge can be handled by employing *Capsule Networks* proposed by Hinton and his team [65]. It has been observed in literature that the Convolutional Neural Networks require a huge amount of training data and sometimes fail to recognize deformed contents in the image. The Capsule network is introduced to take care of these aspects. Presently the capsule network has been extending its application across all domains. Detailed literature could be found in [35,36,41,68] etc.

Many experimental studies show that the Capsule Network architectures outperforms many of the traditional Convolutional Network models in context of handwritten text. Some of them are shown in Table 3.

5 Performance Metric Evaluation for Handwriting Recognition

Immediately after the recognition or classification, a performance evaluation is performed to examine or understand the performance of any proposed model or

Table 3. Capsule Networks employed for handwritten digits and character recognition

Author	Dataset/Language	Network architecture	Recognition accuracy
Yao *et al.* [74]	MNIST	FOD_DCNet	93.53%
Mandal *et al.* [40]	Indic Characters and Digits	AlexNet+CapsNet	96.51%
Haque *et al.* [29]	ISI handwritten database, BanglaLekha Isolated, CMATERdb 3.1.1	ShonkhaNet	99.28%
Sabour *et al.* [65]	MNIST	CapsNet	95%
UÇAR *et al.* [72]	Kannada-MNIST	Capsule-NET	81.63%
Chen *et al.* [17]	MNIST	CapsNet	99.75%
Ghofrani *et al.* [25]	Hoda (Persian/Arabic handwritten digits)	CAPSNET	99.87%

approach. Some of the mostly used performance evaluation metrics for recognition of tasks, including handwriting employed across literature, are listed below:

1. Accuracy: It is a metric for classification models that compute the correct predictions made by the model against all the predictions made.
2. Precision: Ratio of the correctly predicted class to that of all the predicted classes.
3. Recall: Ratio of the correctly predicted class to the true class values.
4. F1 score: This score is used basically for unbalanced data. It is a function of precision and recall. It thus combines the precision and recall into a single metric form by computing the harmonic mean between these two measures.

6 Conclusion and Discussion

We have surveyed the image segmentation methods employed in the task of handwriting recognition. Some of the shortcomings faced in this context are: over-segmentation, under-segmentation, identical shape of the characters, presence of punctuation or dots, etc. Sometimes the small size of the dataset may also lead to wrong results due to insufficient training. Based on this survey, relevant work to solve these issues can be considered in the near future.

References

1. Alkhawaldeh, R.S.: Arabic (Indian) digit handwritten recognition using recurrent transfer deep architecture. Soft. Comput. **25**(4), 3131–3141 (2021)
2. Alom, M.Z., Sidike, P., Taha, T.M., Asari, V.K.: Handwritten Bangla digit recognition using deep learning. arXiv preprint arXiv:1705.02680 (2017)
3. Aneja, N., Aneja, S.: Transfer learning using cnn for handwritten devanagari character recognition. In: 2019 1st International Conference on Advances in Information Technology (ICAIT), pp. 293–296. IEEE (2019)
4. Arora, S., Bhatia, M.S.: Handwriting recognition using deep learning in keras. In: 2018 International Conference on Advances in Computing, Communication Control and Networking (ICACCCN), pp. 142–145. IEEE (2018)

5. Ashiquzzaman, A., Tushar, A.K.: Handwritten Arabic numeral recognition using deep learning neural networks. In: 2017 IEEE International Conference on Imaging, Vision and Pattern Recognition (icIVPR), pp. 1–4. IEEE (2017)
6. Awad, M.: An unsupervised artificial neural network method for satellite image segmentation. Int. Arab J. Inf. Technol. **7**(2), 199–205 (2010)
7. Babu, K.M., Raghunadh, M.: Vehicle number plate detection and recognition using bounding box method. In: 2016 International Conference on Advanced Communication Control and Computing Technologies (ICACCCT), pp. 106–110. IEEE (2016)
8. Bag, S., Krishna, A.: Character segmentation of Hindi unconstrained handwritten words. In: Barneva, R.P., Bhattacharya, B.B., Brimkov, V.E. (eds.) IWCIA 2015. LNCS, vol. 9448, pp. 247–260. Springer, Cham (2015). https://doi.org/10.1007/978-3-319-26145-4_18
9. Bal, A., Saha, R.: An efficient method for skew normalization of handwriting image. In: 6th IEEE International Conference on Communication Systems and Network Technologies, Chandigarh, pp. 222–228 (2016)
10. Balakrishnan, N., Reddy, R., Ganapathiraju, M., Ambati, V.: Digital library of India: a testbed for Indian language research. TCDL Bull. **3**(1) (2006)
11. Banumathi, P., Nasira, G.: Handwritten Tamil character recognition using artificial neural networks. In: 2011 International Conference on Process Automation, Control and Computing, pp. 1–5. IEEE (2011)
12. Barbieri, A.L., De Arruda, G., Rodrigues, F.A., Bruno, O.M., da Fontoura Costa, L.: An entropy-based approach to automatic image segmentation of satellite images. Physica A **390**(3), 512–518 (2011)
13. Belabiod, A., Belaïd, A.: Line and word segmentation of Arabic handwritten documents using neural networks. Ph.D. thesis, LORIA-Université de Lorraine; READ (2018)
14. Boufenar, C., Kerboua, A., Batouche, M.: Investigation on deep learning for off-line handwritten Arabic character recognition. Cogn. Syst. Res. **50**, 180–195 (2018)
15. Boykov, Y., Veksler, O., Zabih, R.: Fast approximate energy minimization via graph cuts. IEEE Trans. Pattern Anal. Mach. Intell. **23**(11), 1222–1239 (2001)
16. Casey, R.G., Lecolinet, E.: A survey of methods and strategies in character segmentation. IEEE Trans. Pattern Anal. Mach. Intell. **18**(7), 690–706 (1996)
17. Chen, F., Chen, N., Mao, H., Hu, H.: Assessing four neural networks on handwritten digit recognition dataset (MNIST). arXiv preprint arXiv:1811.08278 (2018)
18. Chen, K., Seuret, M., Hennebert, J., Ingold, R.: Convolutional neural networks for page segmentation of historical document images. In: 2017 14th IAPR International Conference on Document Analysis and Recognition (ICDAR), vol. 1, pp. 965–970. IEEE (2017)
19. Chin, W., Harvey, A., Jennings, A.: Skew detection in handwritten scripts. In: TENCON 1997 Brisbane-Australia. Proceedings of IEEE TENCON 1997. IEEE Region 10 Annual Conference. Speech and Image Technologies for Computing and Telecommunications (Cat. No. 97CH36162), vol. 1, pp. 319–322. IEEE (1997)
20. Dhanachandra, N., Manglem, K., Chanu, Y.J.: Image segmentation using k-means clustering algorithm and subtractive clustering algorithm. Procedia Comput. Sci. **54**, 764–771 (2015)
21. Dutta, P., Muppalaneni, N.B.: DigiNet: prediction of Assamese handwritten digits using convolutional neural network. Concurr. Comput. Pract. Exp. **33**(24), e6451 (2021)

22. El-Hajj, R., Likforman-Sulem, L., Mokbel, C.: Arabic handwriting recognition using baseline dependant features and hidden Markov modeling. In: Eighth International Conference on Document Analysis and Recognition (ICDAR 2005), pp. 893–897. IEEE (2005)
23. Gatos, B., Papamarkos, N., Chamzas, C.: Skew detection and text line position determination in digitized documents. Pattern Recogn. **30**(9), 1505–1519 (1997)
24. Geetha, R., Thilagam, T., Padmavathy, T.: Effective offline handwritten text recognition model based on a sequence-to-sequence approach with CNN-RNN networks. Neural Comput. Appl. **33**(17), 10923–10934 (2021)
25. Ghofrani, A., Toroghi, R.M.: Capsule-based Persian/Arabic robust handwritten digit recognition using EM routing. In: 2019 4th International Conference on Pattern Recognition and Image Analysis (IPRIA), pp. 168–172. IEEE (2019)
26. Grüning, T., Leifert, G., Strauß, T., Michael, J., Labahn, R.: A two-stage method for text line detection in historical documents. Int. J. Doc. Anal. Recognit. (IJDAR) **22**(3), 285–302 (2019). https://doi.org/10.1007/s10032-019-00332-1
27. Gupta, A., Srivastava, M., Mahanta, C.: Offline handwritten character recognition using neural network. In: 2011 IEEE International Conference on Computer Applications and Industrial Electronics (ICCAIE), pp. 102–107 (2011). https://doi.org/10.1109/ICCAIE.2011.6162113
28. Hamdan, Y.B.: Construction of statistical SVM based recognition model for handwritten character recognition. J. Inf. Technol. **3**(02), 92–107 (2021)
29. Haque, S., Rabby, A.K.M.S.A., Islam, M.S., Hossain, S.A.: ShonkhaNet: a dynamic routing for Bangla handwritten digit recognition using capsule network. In: Santosh, K.C., Hegadi, R.S. (eds.) RTIP2R 2018. CCIS, vol. 1037, pp. 159–170. Springer, Singapore (2019). https://doi.org/10.1007/978-981-13-9187-3_15
30. Hassan, S., Irfan, A., Mirza, A., Siddiqi, I.: Cursive handwritten text recognition using bi-directional LSTMs: a case study on Urdu handwriting. In: 2019 International Conference on Deep Learning and Machine Learning in Emerging Applications (Deep-ML), pp. 67–72. IEEE (2019)
31. Kamencay, P., Zachariasova, M., Hudec, R., Jarina, R., Benco, M., Hlubik, J.: A novel approach to face recognition using image segmentation based on SPCA-KNN method. Radioengineering **22**(1), 92–99 (2013)
32. Kass, M., Witkin, A., Terzopoulos, D.: Snakes: active contour models. Int. J. Comput. Vision **1**(4), 321–331 (1988)
33. Kim, I.J., Xie, X.: Handwritten hangul recognition using deep convolutional neural networks. Int. J. Doc. Anal. Recognit. (IJDAR) **18**(1), 1–13 (2015)
34. Kohli, M., Kumar, S.: Segmentation of handwritten words into characters. Multimedia Tools Appl. **80**(14), 22121–22133 (2021). https://doi.org/10.1007/s11042-021-10638-0
35. Kwabena Patrick, M., Felix Adekoya, A., Abra Mighty, A., Edward, B.Y.: Capsule networks–a survey (2022)
36. Li, J., et al.: A survey on capsule networks: Evolution, application, and future development. In: 2021 International Conference on High Performance Big Data and Intelligent Systems (HPBD&IS), pp. 177–185. IEEE (2021)
37. Li, L.J., Socher, R., Fei-Fei, L.: Towards total scene understanding: classification, annotation and segmentation in an automatic framework. In: 2009 IEEE Conference on Computer Vision and Pattern Recognition, pp. 2036–2043. IEEE (2009)
38. Lu, Y., Shridhar, M.: Character segmentation in handwritten words-an overview. Pattern Recogn. **29**(1), 77–96 (1996)
39. Malik, S., et al.: An efficient skewed line segmentation technique for cursive script OCR. Sci. Program. **2020**, 1–12 (2020)

40. Mandal, B., Dubey, S., Ghosh, S., Sarkhel, R., Das, N.: Handwritten Indic character recognition using capsule networks. In: 2018 IEEE Applied Signal Processing Conference (ASPCON), pp. 304–308. IEEE (2018)
41. Manoharan, J.S.: Capsule network algorithm for performance optimization of text classification. J. Soft Comput. Paradigm (JSCP) **3**(01), 1–9 (2021)
42. Marti, U.V., Bunke, H.: Text line segmentation and word recognition in a system for general writer independent handwriting recognition. In: Proceedings of Sixth International Conference on Document Analysis and Recognition, pp. 159–163. IEEE (2001)
43. Minaee, S., Boykov, Y.Y., Porikli, F., Plaza, A.J., Kehtarnavaz, N., Terzopoulos, D.: Image segmentation using deep learning: a survey. IEEE Trans. Pattern Anal. Mach. Intell. **44**(7), 3523–3542 (2021)
44. Minaee, S., Wang, Y.: An ADMM approach to masked signal decomposition using subspace representation. IEEE Trans. Image Process. **28**(7), 3192–3204 (2019)
45. Mithe, R., Indalkar, S., Divekar, N.: Optical character recognition. Int. J. Recent Technol. Eng. (IJRTE) **2**(1), 72–75 (2013)
46. Mohamed, M., Gader, P.: Handwritten word recognition using segmentation-free hidden Markov modeling and segmentation-based dynamic programming techniques. IEEE Trans. Pattern Anal. Mach. Intell. **18**(5), 548–554 (1996). https://doi.org/10.1109/34.494644
47. Muppalaneni, N.B.: Handwritten Telugu compound character prediction using convolutional neural network. In: 2020 International Conference on Emerging Trends in Information Technology and Engineering (ic-ETITE), pp. 1–4. IEEE (2020)
48. Najman, L., Schmitt, M.: Watershed of a continuous function. Signal Process. **38**(1), 99–112 (1994)
49. Nock, R., Nielsen, F.: Statistical region merging. IEEE Trans. Pattern Anal. Mach. Intell. **26**(11), 1452–1458 (2004)
50. O'Gorman, L.: The document spectrum for page layout analysis. IEEE Trans. Pattern Anal. Mach. Intell. **15**(11), 1162–1173 (1993)
51. Otsu, N.: A threshold selection method from gray-level histograms. IEEE Trans. Syst. Man Cybern. **9**(1), 62–66 (1979)
52. Panwar, S., Nain, N.: A novel approach of skew normalization for handwritten text lines and words. In: 2012 Eighth International Conference on Signal Image Technology and Internet Based Systems, pp. 296–299. IEEE (2012)
53. Papavassiliou, V., Stafylakis, T., Katsouros, V., Carayannis, G.: Handwritten document image segmentation into text lines and words. Pattern Recogn. **43**(1), 369–377 (2010)
54. Pastor-Pellicer, J., Afzal, M.Z., Liwicki, M., Castro-Bleda, M.J.: Complete system for text line extraction using convolutional neural networks and watershed transform. In: 2016 12th IAPR Workshop on Document Analysis Systems (DAS), pp. 30–35. IEEE (2016)
55. Pilarczyk, R., Skarbek, W.: On intra-class variance for deep learning of classifiers. arXiv preprint arXiv:1901.11186 (2019)
56. Plamondon, R., Srihari, S.N.: Online and off-line handwriting recognition: a comprehensive survey. IEEE Trans. Pattern Anal. Mach. Intell. **22**(1), 63–84 (2000)
57. Plath, N., Toussaint, M., Nakajima, S.: Multi-class image segmentation using conditional random fields and global classification. In: Proceedings of the 26th Annual International Conference on Machine Learning, pp. 817–824 (2009)
58. Pramanik, R., Bag, S.: A novel skew correction methodology for handwritten words in multilingual multi-oriented documents. Multimedia Tools Appl. **80**(18), 27323–27342 (2021). https://doi.org/10.1007/s11042-021-10822-2

59. Prathima, C., Muppalaneni, N.B.: Deep learning approach for prediction of handwritten Telugu vowels. In: Reddy, A.N.R., Marla, D., Favorskaya, M.N., Satapathy, S.C. (eds.) Intelligent Manufacturing and Energy Sustainability. SIST, vol. 213, pp. 367–374. Springer, Singapore (2021). https://doi.org/10.1007/978-981-33-4443-3_35
60. Qadri, M.T., Asif, M.: Automatic number plate recognition system for vehicle identification using optical character recognition. In: 2009 International Conference on Education Technology and Computer, pp. 335–338. IEEE (2009)
61. Rahman, A., Roy, P., Pal, U.: Air writing: recognizing multi-digit numeral string traced in air using RNN-LSTM architecture. SN Comput. Sci. **2**(1), 1–13 (2021)
62. Raza, A., Siddiqi, I., Abidi, A., Arif, F.: An unconstrained benchmark Urdu handwritten sentence database with automatic line segmentation. In: 2012 International Conference on Frontiers in Handwriting Recognition, pp. 491–496. IEEE (2012)
63. Renton, G., Chatelain, C., Adam, S., Kermorvant, C., Paquet, T.: Handwritten text line segmentation using fully convolutional network. In: 2017 14th IAPR International Conference on Document Analysis and Recognition (ICDAR), vol. 5, pp. 5–9. IEEE (2017)
64. Roy, A., Ghoshal, D.P.: Number plate recognition for use in different countries using an improved segmentation. In: 2011 2nd National Conference on Emerging Trends and Applications in Computer Science, pp. 1–5. IEEE (2011)
65. Sabour, S., Frosst, N., Hinton, G.E.: Dynamic routing between capsules. In: Advances in Neural Information Processing Systems, vol. 30 (2017)
66. Salagar, R., Patil, P.B.: Application of RLSA for skew detection and correction in Kannada text images. In: 2020 Fourth International Conference on Computing Methodologies and Communication (ICCMC), pp. 785–788. IEEE (2020)
67. Shakunthala, B., Pillai, C.: Enhanced text line segmentation and skew estimation for handwritten Kannada document. J. Theor. Appl. Inf. Technol. **99**(1), 196–206 (2021)
68. Shi, R., Niu, L.: A brief survey on capsule network. In: 2020 IEEE/WIC/ACM International Joint Conference on Web Intelligence and Intelligent Agent Technology (WI-IAT), pp. 682–686. IEEE (2020)
69. Smith, R.: An overview of the tesseract OCR engine. In: Ninth International Conference on Document Analysis and Recognition (ICDAR 2007), vol. 2, pp. 629–633. IEEE (2007)
70. Starck, J.L., Elad, M., Donoho, D.L.: Image decomposition via the combination of sparse representations and a variational approach. IEEE Trans. Image Process. **14**(10), 1570–1582 (2005)
71. Tappert, C.C., Suen, C.Y., Wakahara, T.: The state of the art in online handwriting recognition. IEEE Trans. Pattern Anal. Mach. Intell. **12**(8), 787–808 (1990)
72. Uçar, E., Uçar, M., et al.: Applying capsule network on Kannada-MNIST handwritten digit dataset. Nat. Eng. Sci. **4**(3), 100–106 (2019)
73. Vaidya, R., Trivedi, D., Satra, S., Pimpale, M.: Handwritten character recognition using deep-learning. In: 2018 Second International Conference on Inventive Communication and Computational Technologies (ICICCT), pp. 772–775. IEEE (2018)
74. Yao, H., Tan, Y., Xu, C., Yu, J., Bai, X.: Deep capsule network for recognition and separation of fully overlapping handwritten digits. Comput. Electr. Eng. **91**, 107028 (2021)

Backpropagation in Spiking Neural Network Using Reverse Spiking Mechanism

M. Malathi(✉), K. K. Faiyaz, R. M. Naveen, and C. Nithish

Sri Krishna College of Technology, Coimbatore, India
{m.malathi,18tuit029,18tuit058,18tuit061}@skct.edu.in

Abstract. This paper theorizes the concept of backpropagation in Spiking Neural Networks (SNN) using the reverse spiking mechanism. A Spiking Neural Network utilizes biologically-realistic models of neurons to operate, bridging the gap between neuroscience and machine learning. Leaky Integrate and Fire Neurons form the basis of this Spiking Neural Network, which is a combination of leaky resistors and capacitors. Synaptic currents I(t) serve as input to charge caps to produce potential V(t). Backpropagation in Spiking Neural Network has been a challenge for a long as its source of inspiration is biological neurons that naturally don't backpropagate. In hardware implementation of deep learning algorithms, these are the most demanding metrics. Here we constructed a multilayered SNN following reverse-spike backpropagation and simulated it using Brian Simulator.

Keywords: Spiking neural network · Backpropagation · Leaky integrate and fire neuron model · Reverse spiking

1 Introduction

Over the past five or six decades, several studies in computer science were based on Von Neumann's architecture. Von Neumann's architecture has a limitation, which is that the memory and processor are held separately, which causes latency when a piece of information needs to be processed. Although most modern hardware processes information at a faster rate and can run even complex neural networks, there is still a small amount of latency that exists. But there is no latency in the human brain while processing information. This is because, in the brain, a single neuron acts as a memory and processor, thereby contributing to information processing at a higher rate with less energy consumption. A human brain-inspired neuromorphic processor can do pretty much the same, but we need a new type of neural network to work well on neuromorphic chips since neuromorphic chips work with electric voltages rather than with numbers like existing processors do. So we opt for spiking neural networks, which closely mimic human neurons and respond to spike events. But for any neural network to work well, we need backpropagation, but the human brain doesn't do backpropagation. From a technical standpoint, this problem requires the fusion of methods from both computer science and neuroscience. In recent years, several theories concerning the learning methods of spiking neural networks have been discussed, and they are demanding and fascinating.

J. I.-Z. Chen et al. (Eds.): ICIPCN 2022, LNNS 514, pp. 507–518, 2022.
https://doi.org/10.1007/978-3-031-12413-6_40

Deep learning has made remarkable development in recent years and has become a common technique for completing numerous cognitive tasks such as object detection, speech recognition, and reasoning [17]. Nonetheless, these procedures are frequently time-consuming, complex, and energy-intensive processes. However, our human brain does this quite easily. Despite the fact that the task is difficult, it uses less power and energy. Several studies have shown that spiking neural networks, a neural network that closely mimics the human brain, are capable of performing complex tasks since the advent of neuromorphic computing in recent years.

2 Literature Survey

A variety of studies and research has been carried out on Spiking Neural Networks and Neuromorphic Computing. Spiking NNs are similar in structure to artificial NNs. They have neurons arranged in layers, so each neuron is totally coupled to the neurons within the layer above it. Synapses are the points at which neurons communicate with one another. It is also accompanied by weight to symbolise the connection's strength. Rate-based artificial neural networks are fundamentally different from event-driven SNNs in the way neurons' excitation is represented. Instead of continuous-valued inputs, spiking neural networks analyse data by creating "spike trains" of bits or spikes correlating to logical levels "0" and "1" in the temporal domain. Spikes travel through the network until they approach the output layer. Input's class labelled neuron is the neuron which spikes most in the output layer [23]. Lee, Delbruck, and Pfeiffer 2016 wrote about a backpropagation algorithm that is applicable to spiking neural networks [1]. Because spiking neural networks are non-differentiable, the scholars devised a method for treating spiking neurons' membrane potentials as nonlinear signals. There are discontinuities during spikes that act as noise. As a result, it allows a backpropagation method for errors, similar to an antiquated neural net. Wunderlich and Pehle 2021, introduced a backpropagation algorithm for Spiking Neural Networks in which they utilized the spike times' errors to compute the gradient in an event-driven, temporally and spatially sparse manner [2]. This study describes gradient-based learning algorithms for spiking neural networks and casts light on how they can be implemented. Kheradpisheh and Masquelier 2020, proposed a supervised training rule for a multi-layer SNN based on temporal coding, in which all neurons use one spike per stimulus [3]. In this technique, errors are backpropagated using latencies, but the technique is more traditional. The researchers demonstrate how approximated error gradients can be computed backwards in a feedforward network with multiple layers.

3 The Rudiments of SNN

3.1 Leaky Integrate and Fire (LIF) Neuron

Tens of billions of neurons communicate with one another via synapses in the human brain. Through as many as 1,000 trillion synapses, every neuron may have the ability to communicate with as many as 10,000 other cells [6]. Synapses are points where two neurons come close to each other and release chemicals called neurotransmitters into the inter-synaptic space. Various neurotransmitters attach to receptors on postsynaptic neurons' membranes that are either connected to ion channels or contain ion channels themselves. The ion channels are activated when an ion enters through a membrane and acts as a signaling mechanism. The altered membrane potential then decays and returns to the original value, which is how neurons communicate with one another. Here we use Leaky Integrate and Fire (LIF) neurons to mimic the working of a biological neuron. The LIF theory suggests that neurons are merely parallel combinations of a resistor(leaky) and a capacitor [5]. Here the leaky resistance corresponds to the wavering membrane potential. A synaptic current is fed as an input to the capacitor which produces the potential (V) as an output. When the voltage (V) reaches a threshold voltage (Vth) spike occurs, and the voltage springs back gradually to resting potential just like a biological neuron [7]. The equation for the LIF neuron goes by,

$$dVmem/dt = (RIin - Vmem)/\tau \tag{1}$$

where $\tau = R*C$ is the decay period of the membrane potential and Iin is the input current. For simplicity we keep the equation to be,

$$dVmem/dt = (Iin - Vmem)/\tau \tag{2}$$

Depending on the noise and type of current flowing through the neuron, the above equation can be modified. The below equation is an example of a neuron with noise,

$$dVmem/dt = -Vmem/\tau + \text{sigma} * \text{sqrt}(2/\text{tau}) * \text{xi} \tag{3}$$

where xi as just a Gaussian random variable with mean 0 and standard deviation 1.

3.2 Feedforward

Synaptic connections allow neurons to communicate with each other through spikes [15]. Similarly, the neurons contain some electric potential same as that of biological neurons. As per the equation defined, each neuron's value rises or falls when influenced by the pre or post-synaptic input. When the neuron's value exceeds the threshold, it will send a single impulse indicating a spike to each of the neighboring neurons that the neuron is connected. Soon after the spike the neuron's membrane potential drops below average and resets back to the resting potential. It goes through a refractory period during which it doesn't spike anymore just like the biological neuron [16].

3.3 Rate Coding

The information provided is measured only by the spike rate within the rate encoding interval. Spikes in terms of their temporal characteristics are abstracted. Ratio coding is a result of physiological neurons firing more frequently in response to larger pixels or values or stimuli. These are called "rate neurons" because they convert real-valued input numbers to spikes at each time step [13, 14].

3.4 Latency Coding

Spike time is considered in the timing of latencies, but not the number of spikes. Information is encoded using the time between a certain (internal or external) occurrence and the first spike [13]. This is based on the observation that major sensory events drive upstream neurons to spike earlier. SpikeProp and the Chronotron, for example, have used this encoding in both techniques of machine learning approaches(supervised and unsupervised) [14]. In order to decode stimulus information, neurons within a group of neurons release their first spikes in an orderly fashion, which is known as rank-order coding.

3.5 Temporal Coding

A spike-timing is the simultaneous occurrence of all the electrical spikes in the brain. An entirely temporal code ties timing to a specific (internal or external) event such as the onset of a stimulus or an event or spike in the brain [13, 14].

3.6 Spiking Neural Network Model (SNN)

Spiking neural networks are next-generation neural networks that perform calculations using physiologically accurate neuronal models. Whereas traditional artificial neural networks work with real or integer inputs, SNNs work with spikes instead of real numbers. This means that a single piece of information needs to alternate between the spiked and non-spike dichotomy. The mathematical model of spikes is defined by differential equations that describe a series of biological events. The membrane potential of neurons is the most crucial of them. When a neuron's membrane potential hits a specific threshold, the neuron fires and sends a signal to the neurons around it to reset the cell's potential. Adjacent neurons respond by increasing or decreasing the membrane potential. Synapses allow spikes to be transmitted from one neuron to another. In addition, SNNs often have poor connectivity and use specific network topologies [16].

3.7 Advantages of Spiking Neural Networks

- As opposed to conventional neural networks, SNNs send spike events instead of continuous values, and so they are considered faster than conventional neural networks.
- SNNs take advantage of temporal dynamics by exploiting data as a time-varying sequence of spikes for a predetermined number of timesteps. This method increases information processing productivity.
- It has been proven that SNNs can handle dynamic data and dynamic processes better than traditional neural network methods, so moving object detection, action detection, speech recognition, etc. can be made simpler.
- SNNs use less power, energy, and resources, as it relies on neuromorphic chips, which are known to be better, faster, and more robust at processing information than conventional neural networks.

3.8 Energy Efficient Than Traditional Neural Network

SNNs or stochastic-based SNNs consume considerably less power than traditional DNNs. Multipliers are excluded from the computation, which reduces power consumption. The spike trains' firing rate is largely responsible for stochastic-based SNN's power consumption, with higher firing rates resulting in higher power consumption. Table 1 presents a comparative analysis of the performances of stochastic-based SNN, DNN, SNN, and optimized SNN [21]. Our network architectures are all based on the same training specifications [21]. In these tables, MNIST datasets were used to measure the networks. As a first step, we render a straightforward but rational assumption about energy consumption. In SNN, the amount of operations for each neuron is determined by calculating the number of spikes fed into the neuron or by taking random samples. Using the number of input synapses, the number of multiplications and additions that each neuron is supposed to perform is based on the number of synapses. From [22], we determine the number of Joules required for each operation. We accomplish a relatively similar performance as the original DNN and try to relate it to the state-of-the-art SNN using our proposed scheme, as can be seen in Table 1. Furthermore, stochastic-based SNN consumes less energy than the other networks, with densely integrated DNN, traditional SNN, and data normalized SNN using 38.24, 1.83, and 1.85 percent more energy, respectively, than stochastic-based SNN. Table 1 aims to compare different NN architectures in terms of performance accuracy and energy consumption.

Table 1. Energy-efficient comparisons

Neural network architecture	Accuracy	Consumption of energy
Stochastic SNN	98.66%	17.3
Data-Norm SNN	98.65%	31.7
DNN	98.69%	657.77
SNN	98.47%	31.6

4 Backpropagation Using Reverse Spiking Mechanism

4.1 Backpropagation in ANN

The backpropagation method is one of the essential methods that any machine learning algorithm uses to improve accuracy and model performance. Based on the weights of the network, a backpropagation algorithm calculates the gradient of the loss function by using gradient descent. The loss function, which is the difference between the actual and predicted output values, is determined in conventional ANN topologies by applying gradient descent and back-propagation algorithms across the hidden layers to reduce error. In general, it employs a chain rule of calculus, which is ideal for ANN values that can be differentiated and are continuous in nature. However, spiking neural networks are not differentiable since the SNN values are dichotomous between spiking and non-spiking values, making them incompatible with traditional gradient computation for loss functions[17].

4.2 Backpropagation in SNN

Although backpropagation is widely employed in today's neural networks, organic neurons do not use it. Because Spiking Neural Networks are biologically inspired, the backpropagation notion is novel within this class of networks. In this study, we claim a reverse spiking method that transmits a spike from the output class to the hidden layer neurons for each given input.

The reason for using spikes for backpropagation is that Spiking Neural Networks don't work using weights and biases and also neuromorphic machines work when electric spikes or signals are passed through the artificial neurons. We employ the MNIST digit-classifier dataset, which has 60,000 picture samples with 28 * 28 dimensions each. A 784-neuron Artificial Neural Network (ANN) makes up the input layer [4]. The hidden layer neurons consist of two positions, one is a random initialized value and the other is a temporary memory. For the neurons of that layer, the random initialized values represent their membrane potential. The temporary memory consists of new values that need to be updated into the first position. Each hidden layer neuron has an initial membrane potential value between 0 and 1. Initially, the image array is scaled to a value between 0 and 1 and is passed to the input neural layer. Here we treat the pixel values as membrane potential and synaptic connections are established to the neurons present in the hidden layer. Depending on the equations we define, the neurons may or may not generate spikes. Each sample is provided with a default number of time steps as part of temporal coding, which helps the model improve its accuracy over time while adjusting the hidden layer neuron values. Backpropagation is performed with the mean squared error, which is calculated by subtracting the obtained value from the actual value. The formula to calculate the mean squared error goes by [17],

$$\mathrm{MSE} = 1/n \sum_{i}^{n} (\mathrm{Yi} - \mathrm{Y'i})\text{^}2 \tag{4}$$

where Y_i is the original output and Y_i' is the obtained output. Here we assume 1 or 0 to be Y_i depending on whether the spike is backpropagated from the neuron of the correct index or wrong index. Output neurons consist of accuracy value which is nothing but the number of spikes occurring on each output neuron. This accuracy value is considered Y_i'.

During backpropagation, we calculate the mean of the squared error for every output neuron index and distribute the error evenly to all the hidden layer neurons that are connected to the output layer either directly or indirectly. Neuromorphic chips can keep track of the voltage input from an upstream neuron, which gives us the count for calculating the mean of squared error. Backpropagation in a spiking neural network is achieved by reversing the direction of the current. In this method, the voltage remains constant regardless of whether the current is positive or negative or the resistance increases or decreases.

In reverse spiking, unlike feedforward where a spike occurs only if the membrane potential reaches a certain threshold, the calculated error values despite it being less than the threshold value, are treated as a spike and are backpropagated to the neurons in the hidden layer. In other words, the threshold of the output neuron is made zero while doing backpropagation. This method is called the reverse spiking mechanism. The reason for treating errors as spikes is that, unlike conventional neural networks where backpropagation depends on weights and biases, here the randomly initialized scalar hidden layer values act as a prime factor for backpropagation, so even the slightest error, which can affect the accuracy, is taken into consideration.

The Mean Squared Error value for an output neuron of the correct index holds a positive value and that of wrong indices holds a negative value. All these values are stored in an immediate memory (a.k.a temporary memory) of the neurons in the hidden layer. For each time step, the temporary memory values are replaced with a new value which is the summation of the current and previous value and the average of it is taken at the end of the sequence of time steps, which is then added to the hidden layer value. The temporary memory values are re-initialized to zero after each set of timesteps to make the space available for values from forthcoming backpropagation.The same procedure is followed for each sample. Using this approach, reverse spiking can be achieved from the neurons of the output layer to the neurons of the hidden layer. This process is terminated when reaching the first hidden layer travelling from the output layer. The architecture of SNN is depicted in Fig. 1 shows the input layer, two hidden layers, and output layer. The hidden layer consists of V (Voltage) and TM (Temporary Memory). Initially, this temporary memory is initialized to zero, once the backpropagation is carried out, for each time step, it begins to store the new values until the defined sequence of time steps is completed. This process is carried out for every available sample.

Figure 2 displays the spikes generated during feedforward at every time interval with its membrane potential along with the y-axis and time interval along the x-axis. This image is generated using Brian Simulator, an open-source simulator for spiking neural networks built using python. This figure displays the spikes that occurred on different neurons at the corresponding time interval. Brian Simulator has a built-in functionality named spike monitor which is capable of recording the spikes generated by each neuron at different time intervals.

Figure 3 and 4 visualize the synaptic connections among each layer of neurons. Their corresponding neuron index is also plotted. These graphs show synaptic connections between neurons in the input and hidden layers, as well as neurons in the hidden and output layers. Synaptic connections in Brian Simulator can be formed via a mathematical calculation or by establishing a bespoke probability. Here in Fig. 3, the synaptic connection between the input layer and hidden layer is established with a connection probability of 0.6, which means that each neuron in the input layer has a 60% probability that it will connect to every other neuron in the output layer. Similarly in Fig. 4, the connection probability for synaptic connections is defined as 0.4, which means that every neuron in the hidden layer has a connection probability of 40% with every other neuron in the output layer.

Figure 5 displays the code level implementation of backpropagation using the reverse spiking mechanism. This figure illustrates the synaptic connections from the output layer to the hidden layer in the form of a python dictionary. Here we use a single hidden layer to make the process simple. For a given output neuron index, the error value is calculated and backpropagated to all subsequent neurons that are connected to it. This figure shows that before reverse spiking, the temporary memory of the hidden layer is initialized to zero. But when doing backpropagation, the error values get accumulated to it. Once the entire process is completed for all timesteps, the hidden layer value is changed and the temporary memory value is re-initialized to zero. Here the 'error' variable holds the squared error value and the 'prop_val' holds the mean value that's to be distributed evenly to all subsequent neurons.

Figure 6 displays the obtained accuracy results at every output neuron index for each time step. The accuracy change is recorded at the end of each time step. We could see a change in accuracy while doing backpropagation for every timestep. Since we use backpropagation for 10 timesteps here, the error value is backpropagated 10 times for each data sample.

Fig. 1. Architecture of SNN, comprising of an input layer, two hidden layers, and an output layer.

Fig. 2. Spikes generated in the output layer

Fig. 3. Synaptic connections from the input layer to the hidden layer with a probability of 0.6.

Fig. 4. Synaptic connections from hidden layer to output layer with a probability of 0.4.

Fig. 5. Implementation of backpropagation using python and Brian Simulator.

Fig. 6. Accuracy results after changing hidden layer values.

5 Conclusion

The most complex neural network uses up to several Megawatts of energy, but the human brain uses only 20 Watts despite its complexity. Since the Spiking Neural Network mimics the human brain in terms of functionality, it can perform fundamentally powerful computations and consume less energy than a traditional neural network. The human brain does not do backpropagation, so the concept of backpropagation in Spiking Neural

Networks is over the fence. Therefore, we have proposed a reverse spike-based backpropagation algorithm, which sends spikes to hidden layer neurons from the corresponding output neuron index that the neurons are connected to.

Presently, the model is designed to achieve the desired result with the desired accuracy by doing backpropagation using the reverse spiking mechanism for static images. In the future, the same approach can be used to implement moving object detection. Since Spiking Neural Networks use less time and resources than a conventional neural networks, they would be much more suitable to carry out complex tasks. Spiking neural networks have been shown to work well with dynamic data, so the same approach might be used to train speech recognition models or action detection algorithms, in the future.

References

1. Lee, J.H., Delbruck, T., Pfeiffer, M.: Training deep spiking neural networks using backpropagation. Front. Neurosci. **10**, 508 (2016)
2. Wunderlich, T.C., Pehle, C.: Event-based backpropagation can compute exact gradients for spiking neural networks. Sci. Rep. **11**(1), 1–17 (2021)
3. Kheradpisheh, S.R., Masquelier, T.: Temporal backpropagation for spiking neural networks with one spike per neuron. Int. J. Neural Syst. **30**(06), 2050027 (2020)
4. Patel, K., Hunsberger, E., Batir, S., Eliasmith, C.: A spiking neural network for image segmentation. arXiv preprint arXiv:2106.08921 (2021)
5. Dutta, S., Kumar, V., Shukla, A., Mohapatra, N.R., Ganguly, U.: Leaky integrate and fire neuron by charge-discharge dynamics in floating-body mosfet. Sci. Rep. **7**(1), 1–7 (2017)
6. Zhang, J.: Basic neural units of the brain: neurons, synapses and action potential. arXiv preprint arXiv:1906.01703 (2019)
7. Eshraghian, J.K., et al.: Training spiking neural networks using lessons from deep learning. arXiv preprint arXiv:2109.12894 (2021)
8. Stimberg, M., Brette, R., Goodman, D.F.: Brian 2, an intuitive and efficient neural simulator. Elife **8**, e47314 (2019)
9. Goodman, D.F., Brette, R.: Brian: a simulator for spiking neural networks in python. Front. Neuroinform. **2**, 5 (2008)
10. Ilakymchuk, T., Rosado-Muñoz, A., Guerrero-Martínez, J.F., Bataller-Mompeán, M., Francés-Víllora, J.V.: Simplified spiking neural network architecture and STDP learning algorithm applied to image classification. EURASIP J. Image Video Process. **2015**(1), 1–11 (2015)
11. Wikipedia contributors. Spiking Neural Network. In Wikipedia, The Free Encyclopedia (2022). Accessed 13 Mar 2022, https://en.wikipedia.org/w/index.php?title=Spiking_neural_network&oldid=1085978426
12. Lyashenko, B.: Basic guide to spiking neural networks for deep learning cnvrg.io. In: cnvrg (2022). https://cnvrg.io/spiking-neural-networks
13. Snider, G.S.: Spike-timing-dependent learning in memristive nanodevices. In 2008 IEEE International Symposium on Nanoscale Architectures, pp. 85–92. IEEE (2008)
14. Han, K.S.: A cellular automata approach to biological neurons using matlab. Anal. Appl. Math. **7**, 58–134 (2016)
15. Wikipedia contributors: Backpropagation. In Wikipedia, The Free Encyclopedia (2022). Accessed 13 Mar 2022, https://en.wikipedia.org/w/index.php?title=Backpropagation&oldid=1076560327

16. Lendave, V.: A tutorial on spiking neural networks for beginners. In: Analytics India Magazine (2021). Accessed 13 Mar 2022, https://analyticsindiamag.com/a-tutorial-on-spiking-neural-networks-for-beginners/
17. Lee, C., Sarwar, S.S., Panda, P., Srinivasan, G., Roy, K.: Enabling spike-based backpropagation for training deep neural network architectures. Front. Neurosci. **119**, 1–22 (2020)
18. Comşa, I.M., Versari, L., Fischbacher, T., Alakuijala, J.: Spiking autoencoders with temporal coding. Front. Neurosci. **15**, 936 (2021)
19. Bashar, A.: Survey on evolving deep learning neural network architectures. J. Artif. Intell. **1**(02), 73–82 (2019)
20. Manoharan, J.S.: Study of variants of extreme learning machine (ELM) brands and its performance measure on classification algorithm. J Soft Comput. Paradigm (JSCP) **3**(02), 83–95 (2021)
21. Diehl, P.U., Neil, D., Binas, J., Cook, M., Liu, S.C., Pfeiffer, M.: Fast-classifying, high-accuracy spiking deep networks through weight and threshold balancing. In: International Joint Conference on Neural Networks (IJCNN), pp. 1–8. IEEE (2015)
22. Molka, D., Hackenberg, D., Schöne, R., Müller, M.S.: Characterizing the energy consumption of data transfers and arithmetic operations on X86− 64 processors. In: International Conference on Green Computing, pp. 123–133. IEEE (2010)
23. Alawad, M., Yoon, H.J., Tourassi, G.: Energy efficient stochastic-based deep spiking neural networks for sparse datasets. In: 2017 IEEE International Conference on Big Data (Big Data), pp. 311–318. IEEE (2017)

Smart City Image Processing as a Reflection of Contemporary Information Codes

Valerii Leonidovich Muzykant[1(✉)], Shlykova Olga Vladimirovna[2], Barsukov Kirill Pavlovich[3], Kulikov Sergey Vladimirovich[1], and Efimets Marya Aleksandrovna[4]

[1] Department of Mass Communication, Faculty of Filology at the Peoples' Friendship University of Russia (RUDN University), 6, Miklukho-Maklaya Ulitsa, Moscow 117198, Russian Federation
vmouzyka@mail.ru

[2] UNESCO's Department, Faculty of Public Administration and Management, Russian Presidential Academy of National Economy and Public Administration (RANEPA), room 2087, Vernadsky Prospect, 82/1, Moscow 119571, Russian Federation

[3] Autonomous Nonprofit Organization Moscow Directorate of Transport Services «RULI», Nikoliskaya Ulitsa, 11-13, Stroenie 4, Moscow 107012, Russian Federation

[4] The Research Department at the State Academic University for Humanities, Maronovskiy pereulok, 26, Moscow 119049, Russian Federation

Abstract. The article is about the formation of new audiovisual practices, including network ones, which make qualitative changes in the characteristics of the developing smart society, where the individual's freedom, identity and opportunities for self-realization in virtual and mixed reality environments are understood in a different way. Yandex.Taxi, Go-Jek and Ruli media services in the transport complex of the city of Moscow brings some novelty of the work which lies in the fact that the specifics of transport infrastructure in the culture of a smart city have been clarified: the potential of the mobile application Ruli in the creative economy of the capital is revealed, the value base of the information space as the quality of life of various social groups is identified as a comfortable environment.

Keywords: Ecosystem · Smart city · Information codes · Object recognition technology · Media services

1 Introduction

The ecosystem of a «smart city» implies a balance of interests of all parties interested in sustainable socio-economic and socio-cultural development of the territory. The criteria for the quality of management is improving the quality of life of the population and its well-being, improving the natural environment, achieving personal involvement of everyone in overcoming social tension, creating conditions for self-development of the individual and disclosing internal reserves, responsibility for a sustainable future.

J. I.-Z. Chen et al. (Eds.): ICIPCN 2022, LNNS 514, pp. 519–530, 2022.
https://doi.org/10.1007/978-3-031-12413-6_41

Moscow provides city residents with wide access to more than 330 online service: it became possible to manage their city using the potential of such projects as «Active Citizen», «Our City», and the Moscow Government's online platform crowd.mos.ru. At the same time, it should be noted that 18,000 access points to free unlimited Internet have been installed in the capital [1]. The IT infrastructure has been improved, modern information systems have appeared. One of these innovative projects was the *Ruli* (Drive) service, which was announced in 2018 [2].

2 Smart City as a Core of Modern Media Services and Information Solutions

New media or the Internet and communication technology changed human life in the 21st century. A number of scientists characterize digitalization era as a driver of new paradigms of «transforming public administration» [3]. This is important for accessibility and simplification of smart city management.

Smart city concepts are built depending on how "digital culture" and "smart city" are understood. More often, the central place is given to information and telecommunications, more broadly – to digital technologies. A number of researchers focus on a fundamentally new system of communications and interactions, in which government officials, business entrepreneurship, and civil society institutions are equally involved.

Statistics show that more than 600 cities in Europe alone, including Russia, have the status of a smart city, along with the first of the smart cities - Silicon Valley. Experts from the Centre for Strategic Research confirm that the third generation of smart cities is being born, where the course towards the intellectualization of urban space is associated with total digital transformation.

The concept of «smart city» is becoming a global trend relevant to the global community [4]. For a long time, general cross-sectoral goals were at the forefront, but to a lesser extent, the direct change of specific individual industries. The Safe City and Smart City programs provide mechanisms for creating a comfortable urban environment, including conditions for unloading road and transport infrastructure.

A new concept based on industry platform solutions. From 2017–2018 more and more often, the question of the need to build industry-specific platform solutions focused on interaction with other products through the use of uniform standards and protocols began to arise. A more detailed description of the new concept began to receive in the framework of the implementation of national projects.

And, finally, in 2020, within the framework of the national project «Safe and high-quality roads», the digitalization and automation of the road transport complex was specified in the regulatory act of the relevant federal executive body.

Decree of the Ministry of Transport of Russia dated March 25, 2020 No.AK-60-r «On approval of the Methodology for evaluating and ranking local projects in order to implement the event – Introduction of intelligent transport systems that provide automation of traffic management processes in urban agglomerations, including cities with a population of over 300 thousand People» within the framework of the federal project «General system measures for the development of the road sector» (National project «Safe and high-quality roads») [5].

Eight modules and 35 subsystems of the Unified Transport System Management Platform reflect the directions of digitalization and digital transformation of the road transport industry. Its main advantage is the ability to get away from the narrow-profile tasks of building each individual system. The implementation of each additional system is logically correlated with the results of a common platform solution, allowing you to gradually develop a "window" of a single source and destination of the necessary data.

It is the media that today act as the main mechanism for communication between the people and the authorities, an important link in the system for implementing the state information policy, which makes it possible to regulate the processes of information impact in various spheres of the life of society and the state. On the other hand, by providing an information space for public discussion and dissemination of information and opinions, the media play an important role in establishing feedback with society, allowing the authorities to hear the requests of various social groups, analyse and adjust public policy.

Different social conditions during the design development process have the opportunity to produce different final designs. The SCOT (Social Construction of Technology) conceptual framework consists of four related components, namely: Interpretive Flexibility, The Relevant Social, Closure and Stabilization, The Wider Context [6].

Talking about new media, the Internet or digital technology has changed the way humans interact with one another through the media. The final goal of new media is to cover the appearance of digital media services in the late of the last century [7]. The new media's features can be changed, as well as users' generated content cannot be monopolized by anyone.

The term new media was introduced in 1969 by an academic figure named Marshall McLuhan. McLuhan said that new media is a development of communication technology that, in its history, has expanded the reach of human communication. The development of new communication technology produces a robust cultural effect. Hence, the presence of new media produces a new mass communication model where previously one too much communication became many to much communication.

New media has an interactive and free nature. Interactive means direct interaction between the communicant and the audience through the media they consume. The intended free nature is that the public can freely create media content that contains information. Society holds control over the distribution and consumption of content in new media.

New media terminology includes all technical equipment for processing, storing, and delivering information. Information and communication technology are related to various processes, the use of tools, manipulation, and information processing. The presence of digital media can facilitate humans in their work.

Humans create social reality in the social world because humans are creative actors in creating that reality. Reality itself arises from the power of social construction on the environment. Current technological developments are originating from the reality of human construction of the social world around him so that the technology becomes what fits the reality it creates. One example is today's social life, advances in communication technology that makes all communication and information easily accessible, and everyone without space and time limits. Society has contracted that with this technology narrowing the boundaries of humans in communicating and getting information.

New media emerged as a result of social construction in utilizing the rapid development of information technology advancements in the digital age. Modern society needs something instant and practical to support their lives — mobile application technology innovation to overcome existing problems in society. The SCOT perspective focuses on how technology arises due to social processes. Social constructivism is how social drive influences the discovery of new technology. Not only has the discovery of new technology, but social constructivism also forced existing technologies to continue to innovate to meet the demands of users, in this case, the community. For example, the trend of video blogs in the community has led to the continued innovation of applications such as WhatsApp, Facebook, Instagram, illustrating video features in their applications, known as «Insta story» on Instagram, «My Status» on WhatsApp, and «Facebook Stories» on Facebook. The application creator or application content must be able to capture what the user wants or likes so that the application still exists and is not left by the user.

3 E-government in a Smart City. Regulation of Services

In modern conditions, the use of electronic media in public administration should include not only the access of citizens in electronic form to various kinds of information about the activities of state bodies and the provision of various public services on the Internet. The most promising is the organization of communication, which provides citizens with the opportunity to participate (influence) in the process of making certain managerial decisions, that is, in the development of feedback in the interaction of government bodies with the public.

The introduction of e-government is able to partially implement the functions of the media not only in terms of informing citizens, but also in shaping the public agenda and influencing decision-making processes. On the other hand, by implementing feedback functions, various components of e-government make it possible to more efficiently and quickly select the information necessary for making managerial decisions. This increases the efficiency of the entire system of public administration.

«Yandex.Taxi» Online forms of application of transport technologies of social construction in New Media. Technology comes from experience that translates into knowledge. The development of technologies that looks real at the present time is the development of information and communication technologies. Yandex.Taxi is one of the specific examples of it, which is indicated on Fig. 1.

The digital transport platform was born from the need for cheaper, faster, safer, more efficient and efficient transport services. Technology is familiar with the life of modern society in the information age Technology comes from experience transformed into knowledge, which this knowledge is processed into science to create technology. New media technology applies a business based on a digital transport network called Yandex.Taxi. This $14 billion company is already operating in 17 countries around the world. Russia is the main market for companies based in the city of Moscow. Yandex.Taxi is a subsidiary of a Russian transnational company, a subsidiary of the Yandex holding. Just like Yandex.taxi, which is a picture of social construction in new media technology

Fig. 1. Market share of taxi services in Moscow, Russia, 2020 (Source: https://www.statista.com/statistics/1074260/moscow-taxi-market-distribution-by-company)

in modern society, the online transportation platform is arguably one of the greatest revolutions of modern human life today [8].

Yandex.taxi provides services that can meet the public's needs for modes of transportation that can transport from one point to another easily, safely, and inexpensively through websites and mobile applications. The business concept is ridesharing because the company's form connects passengers and drivers using private and non-commercial vehicles — the Yandex.taxi application was first launched in the city of Moscow, Russia, in 2011 by a Russian multinational company called Yandex (NASDAQ: YNDX). In 2021 it has operated in more than 300 major cities throughout Russia, Belarus and others [9]. The presence of this online transportation application is due to the high level of mobility needs of today's society, which requires fast, efficient transportation, constant and affordable prices.

Facebook, Instagram for example, the trend of vlogs in the community has led to continued innovation in apps like WhatsApp, Facebook, Instagram illustrating video features in their apps known as «Insta story» on Instagram, «My Status» on WhatsApp and «Facebook Stories» on Facebook. Given the broad reach of communication through social media channels, crucial changes seem to occur in the distribution of media power [10]. The creator of the application or the content of the application must be able to capture what the user wants or likes so that the application still exists and is not abandoned by the user. The state of Indonesia has an online application-based transportation service, namely Go-Jek, founded by Nadiy Makarim. Go-Jek was founded on the experience of a regular motorcycle taxi customer. The difficulties of motorcycle taxi drivers who find it difficult to deliver passengers is one of the foundations of the creation of this technology start up.

The State of Indonesia has an online application-based transportation service, namely, Go-Jek, which Nadiem Makarim founded. Go-Jek was founded based on the experience of being a loyal customer of motorcycle taxis [11]. The difficulty of motorcycle taxi drivers who have difficulty getting passengers is one of the foundations for establishing this technology-based startup.

Go-Jek has transformed into a Super App, providing more than 20 services. Go-Jek has been downloaded nearly 10 million times on Google Play Store; besides that, it is also available on the App Store (iOS). Go-Jek currently has more than 2 million partners throughout Southeast Asia [12].

The business concept is ride sharing as the company's form connects passengers and drivers using private and non-commercial vehicles. The digital transport platform Yandex.Taxi continues to expand after the signing on July 13, 2017 of a cooperation agreement with one of the competitors of another digital transport company, Uber. In addition to the expansion, Yandex.Taxi also expanded its business services by purchasing a 100% stake in food delivery company Foodfox in December 2017.

In October 2018, Yandex.Taxi again acquired another food technology called Partiya Edi, which is currently operating in Moscow and St. Petersburg. This service is available in 24 cities across Russia. Yandex.Taxi also launched another feature called Self-Driving Car. Yandex.Taxi continues to innovate by launching our autonomous car project in 2017, heavily supported by Yandex's proprietary mapping, navigation, computer vision and object recognition technology using artificial intelligence (AI) technology. Yandex.Taxi is developing a platform for driving vehicles without human intervention to revolutionize the way people travel.

The development of technology is a process of interaction or discourse between technologists and their relationship with social groups. If one uses this perspective, it means that the philosophy of the formation of a technology depends on humans' fundamental values themselves. The service currently offers autonomous driving services in «Innopolis» (Republic of Tatarstan) and «Skolkovo» (Moscow region) innovation ecosystems from mid-2018, where cars can deliver passengers without anyone in the driver's seat because of robots control. Yandex.Taxi is currently testing this new self-driving car feature in two cities - Moscow and Tel Aviv.

The total volume of digital services in the e-commerce segment amounted to almost 809 billion Russian rubles. The 5th largest country in terms of the number of smartphone users in the world, Russia offers a wide range of business opportunities to attract consumers through the Internet [13].

New media have emerged through social construction, using the rapid development of information technology in our digital age. This modern society needs something practical to support its life - the emergence of a digital platform or mobile application, such as Go-Jek and Yandex.Taxi. The community continues to innovate to find the best way to solve transportation problems in our society. Technology does not define people. However, it is people who drive the operation of the technology itself. Yandex.Taxi, an online taxi service, was the leader in the taxi market in the Russian capital, accounting for more than 63% of total taxi trips in Moscow as of April 2020.

The Citymobil mobile application is the second largest market player with approximately 28%. New media have emerged through social construction, using the rapid development of information technology in our digital age. This modern society needs something instant and practical to sustain its life - the emergence of a digital platform or mobile application such as Go-Jek and Yandex.Taxi.

This service is complemented by the *Ruli* service. This service allows a car enthusiast to rent out his car, earn extra money on it, and the city gets a chance to unload roads, to optimize the management of transport infrastructure.

According to statistics, about 1.2 million cars (of the total volume of cars in the city 4.7 million) are used no more than three days a week, and the rest of the time they are parked. In Moscow, over the past 6 years, the number of users has exceeded 1 million people, and the number of trips has amounted to more than 130 million. In 2015, there were only 350 car sharing vehicles, and the number of registered users was about 30,000. Moscow actively supports private car sharing companies and provides subsidies for fleet renewal. One of the key advantages is free parking in the city, for which the user does not pay due to special conditions for car sharing companies from the Moscow Government. There are totally Moscow eight car sharing operators in the city [14].

4 The Ruli Service as Integrated Part of Smart City Image Processing

For the city, the advantage is that one short-term rental vehicle is used by several drivers per day, which reduces the load on the transport network and improves the environmental situation in the city of Moscow, and retains drivers from buying a personal vehicle.

Pre-project study included a survey was conducted and the reasons for the choice of various car sharing services among users and the refusal of a personal vehicle were identified: in some cases it is cheaper than a taxi ride 68.7%; no need to spend money on the maintenance and service of the vehicle 50.9%; saving time (no need to waste time at service stations, go to gas stations, look for parking in the city) 48.1%; no need to worry about the safety of the vehicle at night 31.8%; there is no possibility to buy a personal vehicle 19.6%.

Ruli's lack of competition in the market lies in the fact that this project requires government support, funding and the creation of its own software. The *Ruli* service is similar to peer-to-peer and includes the idea of creating a closed community of car enthusiasts and their friends. In the future, with the recognition and trust of this project, the possibility of a part of the audience moving from among the users of car sharing to *Ruli* is not ruled out.

This *Ruli* service will create a certain growth for car sharing according to a new model for Moscow from other peer-to-peer companies. The results of pilot studies as follows: native development for IOS and Android; a one vehicle is not used by one person; now, closed testing is underway and the IT platform is being finalized for a full-fledged launch in Moscow and St. Petersburg for everyone.

Car sharing operators exist largely due to the availability of parking permits, they allow services to exist as a business for a personal vehicle, this solution does not apply, due to the potential vulnerability, all people will be able to obtain a permit for free parking. Each model that appears in the car sharing operator's fleet undergoes a series of tests and studies before it is equipped with all the necessary modules, which increases the period of the start of vehicle operation for car sharing operators.

In the *Ruli* service, in order to connect a vehicle, the owner sends an application and installs free special equipment that allows not only opening and closing, but also starting the vehicle from the phone.

The purpose of the project: substantiation of effective mechanisms for managing the innovative service *Ruli* by the process of implementing the IT sharing platform. In accordance with the goal in the final qualification work, the following tasks are solved:

1) to identify signs of innovation and innovative management of the *Ruli* service based on the transport complex of the city of Moscow;
2) to reveal the features of an innovative product on the example of the *Ruli* service;
3) summarize Innovative Sharing Practices (peer-to-peer);
4) identify the innovative components of the service *Ruli*;
5) implement a full-fledged launch of *Ruli* and attract more than 1000 users;
6) present the mission, stages and risks of projects, development prospects.

After the Mayor of Moscow announced the launch of the *Ruli* service, active work began on the selection of a team. After the formation of the project management structure, the team began to work out a strategy for the development and implementation of *Ruli* in the city of Moscow. At first, the team planned to create a marketplace where owners would rent their vehicle to everyone for a minute, but the *Ruli* team settled on a scenario where a person shares a car with relatives and friends. After the launch of closed testing, car owners began to actively register in the mobile application and leave applications, and after two months, Ruli received more than 150 applications from potential car owners and 500 applications from potential tenants as shown on Fig. 2 and 3.

Fig. 2. Example of screen service ruli with applications from potential car owners and l tenants (Source: Mos.ru/ruli)

Fig. 3. Screen of Statistic per day from the site (Source: Mos.ru/ruli)

Over the past 6 years the number of users exceeded 1 million, while the number of repetitions is more than 130 million.

The Department of Transport of Moscow with its subordinate organization ANO Moscow Directorate of Transport Services *Ruli* is studying two more models for the development of *Ruli*:

1. Search and rent cars nearby through the application - within a radius of 300–500 m. In this case, the owner of the car decides whether to give it to another person or not and under what conditions.
2. Delivery and rental of cars on a regular basis. The scenario is similar to the one in which car sharing services are currently working. But this requires a large number of users - both car owners and those who are looking for a car.

The main feature of *Ruli* at the stage of closed testing is that it will be completely free. This package includes: installation of special equipment; driving a vehicle without keys using the Steering Wheels mobile application; registration of «Multidrive» for 3 months.

During the testing process, users will receive: special equipment (allowing you to remotely start the engine, open, close and control the vehicle); insurance of your vehicle for 3 months (*Ruli* will insure the vehicle under Car insurance "Multidrive" for the duration of the testing program); participation in the development of a unique project (the user will be the first to try a new service, his feedback will make the service better).

After entering the open launch of the service, it is planned to launch a subscription in which the user will be able to: convenient access to unlocking the vehicle without a key and all controls, via a mobile application, save your time by handing over your vehicle to your loved ones, without even meeting with the help of the *Ruli* mobile application, control a personal vehicle, the user will always be aware of all the actions that occur: travel history including location, rental duration, mileage, fuel consumption and, in the future, driving style.

For the tenant, the interest lies in the following: beneficial use on individual terms (the owner can hand over his vehicle to the tenant «just like that»). These are more flexible and favorable conditions than rental; simply and quickly, the tenant receives a notification about the available vehicles of his friends and chooses the right one. There is no need to pick up the keys and meet, the tenant will get access to the vehicle through the *Ruli* mobile application; the comfort is that your friend's vehicle, which has not been used by many people; interest for the tenant in providing the opportunity to ride on different vehicles and gain access to unusual models.

The *Ruli* service plans to scale up and add a marketplace with additional services from partners: per-minute travel insurance; sink; remote refueling; driving a car.

Due to additional services, *Ruli* will be able to take a commission from partners, but these are long-term plans and prospects. The project is fully funded by the city, which has no goal of making money on the service. A subscription is needed to cover the cost of equipment and insurance, and a marketplace is needed to provide users with useful and convenient services with the help of partners.

The city is solving a different problem - to activate more than 5 million cars, which stand in one place for 22 h a day, and to reduce the number of newly purchased cars. The *Ruli* service may face problems of user distrust due to the possibility of surveillance. To connect the car to the *Ruli*, the owner must install special equipment from partners of the Department. *Ruli* has already received feedback from users who are afraid of surveillance or leakage of personal data.

As practical and theoretical methods, the following are used: analysis of works in the field of a peer-to-peer product, comparative analysis, generalization, design, expert observation, conversations, analysis of documents and results of activities, etc. The research methodology consists of two stages, based on a pilot study in Moscow, which obtain 13% of the total transport in Russia [15].

Service *Ruli* will help all car owners quickly and safely rent their vehicle only to a trusted circle of close people. At the moment, the closed testing is going well and soon it is planned to start the open launch of the Ruli service in Moscow with the prospect of launching this service based on Ruli in other cities.

A full launch was scheduled for autumn 2021. After attracting 100 vehicle owners and their *Ruli* friends, it is planned to connect at least 1,000 vehicles within six months or a year. For two months, *Ruli* received more than 150 applications from car owners and 500 from potential tenants. Attracting an audience is the main problem for the sharing service, as it is necessary to change the attitude towards a personal car, to convince car owners and tenants that Ruli helps save time: a car can be rented and returned at any place and time, no keys are needed for this.

After closed testing and the establishment of all processes, the service moves to an open launch without limiting the number of Users. At this stage, the *Ruli* service is working in closed testing in Moscow and St. Petersburg [16].

5 Conclusion

The digital services industry plays a significant role in the development of e-commerce in Russia. The total of digital services as an e-commerce segment was measured at nearly 809 billion Russian rubles. The fifth largest country by smartphone users worldwide, Russia offers a wide range of opportunities for businesses to reach consumers via the internet. In 2019, the country's digital services, internet finance, and eTravel had a revenue of almost 2.9 trillion Russian rubles.

New media emerged because of social construction in utilizing the rapid development of information technology advancements in this digital age. This modern society needs something instant and practical to support their lives—the emergence of a digital platform or mobile application like Go-Jek and Yandex.taxi. The community continues to innovate to find the best way to solve transportation problems in our society. Still, humans are the ones who control the actions of technology itself.

An online ride hailing company Yandex.Taxi was the leader on the taxi market in the capital of Russia, accounting for over 63% of total taxi rides in Moscow as of April 2020. The second largest market player was a mobile application Citymobil, whose share amounted to approximately 28%. A study of the media service showed that attracting an audience is the main problem for the sharing service, since it is necessary to change the attitude towards a personal car, to convince car owners and tenants that the *Ruli* service helps to save time, since it is possible to rent a car at any time and return it to anywhere.

These could be the following approaches: a decentralized approach, when digital transformation is carried out with the involvement of business (for megacities); centralized, where the digital transition is coordinated at the level of local governments, mobilizing available resources and interested actors (for large and medium-sized cities); a model of local actions (for medium and small cities), when, due to insufficient resources, single, problematic infrastructure sectors undergo digital transformation.

Acknowledgments. This paper has been supported by the Russian Academy of National Economy and Public Administration (RANEPA) and RUDN University Strategic Academic Leadership Program. It also made in the framework of the State Assignment of the Ministry of Science and Higher Education of the Russian Federation (The topic № FZNF-2020–0001 – «Historical and Cultural Traditions and Values in the context of Global History» at the State Academic University for Humanities – GAUGN).

References

1. Lysenko, E.A.: The development of smart services in the capital: the present and the future. In: MMGU Herald, vol. 4, pp. 3–6 (2019). (in Russian)
2. Khabibrahimov, A.: The city needs fewer personal cars: how the "People's Car sharing" from the Moscow City Hall Works and Who Builds it. Accessed 01 July 2021, https://vc.ru/transport/261403-gorodu-nuzhno-chtoby-bylo-menshe-lichnyh-mashin-kak-ustroen-narodnyy-karshering-ot-merii-moskvy-i-kto-ego-stroit. (in Russian)
3. Smart-city. https://www.csr.ru/ru/publications/smart-city-v-rossii-kak-zastavit-goroda-poumnet, Accessed 04 Apr 2022
4. Shlykova, O.V.: "Digitalization" and "Digital Culture" as new trends of the digital era. In: Kirillova, N.B. (ed.) Audiovisual Platform of Modern Culture: Materials of the International Scientific Conference, Yekaterinburg, pp. 22–32 (2020). (in Russian)
5. Decree of the Ministry of Construction of the Russian Federation dated October, 31st, 2018 No. 695/pr «On Approval of the Passport of the Departmental Digitalization Project of the Urban Economy "Smart City"». http://www.minstroyrf.ru/docs/17594/?sphrase_id=650485. (in Russian)

6. Yousefikhah, S.: Sociology of innovation social construction of technology perspective. Electron. Mater. **2**, 31–43 (2017)
7. Muzykant, V.L., Muqsith, M.A., Burdovskaya, E.Y., Palagina, I.Y., Barabash, V.V., Volkova, I.I.: Contemporary transportation applications as new forms of social construction technology. In: 2021 Fifth World Conference on Smart Trends in Systems Security and Sustainability (WorldS4), London, pp. 6–11 (2021)
8. Fulton, L., Mason, J., Meroux, D.: Three revolutions in urban transportation. Joule **2**(4), 575–578 (2018)
9. Bne IntelliNews, «intellinews.com» (2017). https://www.intellinews.com/yandex-taxi-and-uber-join-forces-in-six-cis-countries-125307
10. Muqsith, M.A., Muzykant, V.L., Kuzmenkova, K.E.: Cyber protest: new media and the new social movement in Indonesia. RUDN J. Stud. Literat. J. **24**(4), 765–775 (2019)
11. Kussanti, M.D.P.: Personal Branding Nadiem Anwar Makarim Melalui Pidato hari. Jurnal Trias Politika **4**(1), 51–65 (2020)
12. Stephanie, C.: «Kompas.com». Accesed 12 Nov 2020, https://tekno.kompas.com/read/2020/11/12/18090947/satu-dekade-beroperasi-gojek-punya-2-juta-mitra-pengemudi-di-asia-tenggara?page=all
13. Dobrolyubova, E.I., Yuzhakov, V.I., Efremov, A.A., Klochkova, E.N., Talapina, E.V., Startsev, Y.: The Digital Future of Public Administration by Results, p. 114. RANEPA, Moscow (2019).(in Russian)
14. Figure of the Day: How Many Cars are There in Russia?. https://www.autonews.ru/news/5ca9c5409a79474a2e7d76d9#ws. (in Russian)
15. Moscow «Smart-city – 2030». Strategy Project. Moscow, p. 111 (2018). (in Russian)
16. Barsukov, K.P., Shlykova, O.V.: Culture of digital communications in the smart-city (on the example of the innovative project “Steering” in Moscow). In: Eurasian Union of Scientists. International Research Issue, vol. 12, no. 93, pp. 9–11 (2022)

Clinical Decision Support System Braced with Artificial Intelligence: A Review

Jigna B. Prajapati[1(✉)] and Bhupendra G. Prajapati[2]

[1] Acharya Motibhai Patel Institute of Computer Studies, Ganpat University, Gujarat, India
jigna.prajapati@ganpatuniversity.ac.in
[2] Shree S K Patel College of Pharmaceutical Education and Research, Ganpat University, Gujarat, India
bhupen27@gmail.com

Abstract. The Healthcare sector is one of the most vibrant & crucial sectors in any development toward smart city substantivity. The health sector is enabled techier faster & faster. Artificial Intelligence (AI) is affecting the massive upliftment of all health-related services. AI is used to improve human decision-making. It performs advanced decision-making with Rules-based expert systems (ES) and machine-learning (ML). ES & ML can be combined to assist clinical releasers in their diagnosis operations for more accurate and effective clinical decisions, reduce clinical errors, and improve safety & efficacy. AI in clinical support is effective to save money to increasing the overall system's quality & performance. This paper examines a variety of studies that used Artificial Intelligence techniques in clinical decision support systems in order to define basic criteria for the usage of intelligent techniques. The use of AI in clinical systems raises ethical concerns. We also discuss the ethical, economic, legal, and societal consequences of AI in clinical support systems.

Keywords: CDSS (Clinical decision support system) · Artificial intelligence · ES (Expert System) · Machine learning · Ethics

1 Introduction

Artificial Intelligence (AI) is the latest innovation for human act & thinking replacement in heterogeneous sectors. of industries [1]. AI provides the intelligent platform for developing a smart system to accomplish tasks themselves which are carried out with the help of human intelligence [2]. AI is popular for functioning as human work. Earlier the usage of AL is limited to certain fields only but it is not true for today's scenarios. AI scope is increasing day by day as it brings significant impacts on many industries such as manufacturing, supply chain, marketing, Healthcare, education, Security and Surveillance &, etc. Artificial intelligence covers many inner fields to work heterogeneously in various sectors. AI incorporates majorly Robotics, ES (Expert System, ML-DL (Machine Learning-Deep Learning), NLP (Natural Language processing), & Speech Recognition [3–8]. Broadly AI can be classified As cognitive approach programming, data mining,

J. I.-Z. Chen et al. (Eds.): ICIPCN 2022, LNNS 514, pp. 531–540, 2022.
https://doi.org/10.1007/978-3-031-12413-6_42

genetic algorithms, knowledge base systems, theorem proving, constraint satisfaction, and theory of computation [9, 10].

AI is used in all most all fields to have better performance in their relent section. There are many innovations in the education sector using AL. Some of these innovations as training with Computer & computer-aided (CBT & CAI) [11]. Technological education using AI and its impact on education has been spread widely in the current era [12]. AI is also important for management analytics using AI-driven functions [13]. AI is becoming popular in corporate decision-making in many fields of core management [14]. AI trend and provide effective engineering and artificial intelligence in mechatronic engineering and gives several real-time problem solutions [15]. AI helps in the Energy optimization of mechatronic systems with PMSG [16]. In modern artificial intelligence, AI tools are used for diagnostic systems of electrical-enabled machines [17]. AI has giant potential to enhance the fight against crime and strengthen national security[18]. Artificial intelligence can manage the financial security of an organizer very well [19]. AI is also paly a vital role in Space engineering and space technology [20]. Military operations are evolutionary changes using Artificial intelligence [21]. Artificial Intelligence and its techniques can enhance the assessment level of food products. The results of the assessment can be of influence each objective attribute of the assessed food with better results [22]. The agriculture industry has maximized its yield production including proper prediction using AI [23, 24].

Due to the large complexity and volume of data in healthcare, artificial intelligence (AI) will be used more frequently [25]. AI is used in Diagnoses, treatment, recommendations, patient engagement, administrative duties, genetic codes, drug discovery and development, transcribing medical documents, clinical decision support system (CDSS) &, etc. [26, 27].

2 Clinical Decision Support System

CDSS (Clinical decision support systems) is working on the collection of EHRs (electronic health records) using computer assistance to analyse patient data for health-related updates [28]. EHR includes the various types of health records which are available on electronic platforms. It covers patient medical history from single-point access. During the treatment, such evidence-based clinical standards are kept in concern. CDSS works in two types majorly as Knowledge-based and non-knowledge-based [29]. İn a knowledge base system, rules are created where if-then condition base rules implementing. The source of rules covers patient history, practice history, or relevant literature. A diagnosis decision support system is a knowledge base CDSS. Such a system suggests the concern diagnoses for the predefined patient on basis of information compilation [30, 31]. It can be classified as broadly in three ways. The first one as high adoption CDSS & effective Usage, the second one as Best Knowledge Available on Need, and the third one as Continuous Improvement of Knowledge [32].

CDSS provides a better health care environment for health staff, patients, clinical assistants, and clinical researchers concerned with health knowledge. CDSSs are used for patients, clinicians, doctors, patients, or related staff for many roles. There is one of the roles of advice to concern with best practices for pre & medical history. CDSS can assist the patients to have alternate treatment on base on previous records of treatment using various tools and technology [33]. Machine Learning is used to process high-complexity clinical data such as text, images, and various types of connected biological data & apply the learned knowledge to derive desired outcomes[34].

The clinical Decision Support system needs to follow the CDSS regulations. FDA's regulatory authority where licensed CDSS associated professionals will work under the direction of predefined regulations. The vendors and users keep concerned about using the software under the supervision of licensed practitioners [35].

Clinical decision support systems assist the range of healthcare providers in many ways. The healthcare provider as pharmacists, technicians, analysts, doctors, podiatrists, chiropractors, psychologists, or any other medical-related practitioner can take the assistance of CDSS in their routine practice for better decision making. CDSS can assist the healthcare provider by sending alerts on specific medical rules. It facilitates the healthcare provider to study, analyse, design, and diagnose many medical situations with its predictive risks & solutions. There are much computational data from CDSS that can assist health care providers with better clinical decision making [36, 37].

The CDSS functions for better healthcare service providing. The important function as sending alerts on problems, suggestions for complex medical-related issues, patient advice, healthcare provider assistance, remote health monitoring, an automated diagnosis from the historical patient data, better documentation for adverse drug effects another workflow.

Fig. 1. CDSS environment

The potential benefits of CDSS are counted as patient safety, cost-effectiveness, administrative automation, being aware of medical errors in advance, and centralization of data which can help to decide at glance. The CDSS can give more accurate medical-related information with quick access. CDSS also benefited from choosing the proper drug, drug dose, and drug reaction history.

3 Early-Stage Working of CDSS Using AI

Friedman and Frank (1983) have discussed rule-based systems as generalizability, structure module, ease of rule acquisition, expandability, and value for patient care with artificial intelligence (AI) [38]. Kunz, J C et al. (1984) focused on a different aspect as clinical algorithms, numerical analysis, decision analysis & symbolic reasoning for patient data, [39]. Clarke, J R et al. (1988) is positive about the possibility of a computerized decision support system that can assist surgeons in developing early definite management strategies for serious trauma patients [40]. Molino, G et al. (1990) focused on AI Techniques for well-defined clinical problems [41] Furlong, J W Et al. (1991) has discussed analysis of neural networks for cardiac enzyme records using AI [42]. Forsstrom and Dalton (1995) have discussed about decision-making in clinical medicine using ANN. They concluded that there is a vital role of ANN in image process examination and signal processing [43]. De Graaf, P M et al.(1997) has a decision support system (DSS) in anesthesia where data is converted into relevant patient information [44]. Buhler and Miller (1999) have discussed distributing knowledge maintenance for CDSS. [45] C Hanson and Marshall (2001) have discussed past and present AI applications for medical care units using MEDLINE records. They used a variety of AI applications to look up patient follow-ups directly. Eta S Berner (2002) has elaborated on a variety of issues of AI in concern with CDSS as legal & ethical [46]. Figure 1 presents the basic CDSS structure. This structure presents the collection of EHRs, Clinical data classification, Labeling of the Same Data & converting it to a training Dataset. Once we have training data set AI mechanism applying with computation advancement for complete CDSS.

4 Current Trends in CDSS Using AI

Tanguay Et al (2022) used AI in depression treatment. They have implemented the tasks with the simulation center clinical decision support system. Novel artificial intelligence models assisted with said treatment selection [47]. Patni Et al (2021) has surveyed the diagnosis of COVID-19 patients using AI for CDSS. They have gone through a multi-dimensional deep learning system. They have worked on the system for the detection of COVID-19 by chest CT images [48]. Giuseppe Citerio has given the light on artificial intelligence (AI) in neurocritical care. He has shown the prospect of a better offer for working doctors using current trends in technology [49]. Mosavi & Santos (2021) have highlighted how to reduce error in medical assisted systems, and how to enhance the treatment performance with the objective to reduce overall cost using an Intelligent Decision Support System. They have used Simon's model as well for further processes [50]. Dongyoung Kim Et al (2022) have discussed automated sleep stage classification using deep learning. They have combined CDSS with CNN & transform supervised learning of three classes of sleep stages only with single-channel EEG data. they derived accuracy in normal = 94.3%, mild = 91.9%, moderate = 91.9%, and severe cases = 90.6% [51]. Fangfang Liu et al. (2022) have discussed the high incidence and low diagnostic accuracy which is the primary headache with their concern. They worked with a few important features such as migraine and tension-type for primary headaches. They have tried to build up CDSS with the help of ML for primary headaches [52]. Damien K. Ming Et al (2022) has worked on the identification of dengue shock syndrome patients using ML models. These models are trained with cloud data of clinical records for better decision-making processes. Their study is carried out on 4,131 patients [53]. Stefan Vogel Et al (2022) have discussed the development of the CDSS interface by "Campbell's Five Rights" workflow. CDSS is complex and requires user-centered planning of assistive interventions [54]. Sira Kim et al. (2022) have presented that EMRs (electronic medical records) are gradually used in CDSS. They evaluate CDSS in such a way that the medical staff also understands the core functions of such systems [55]. Levivien Et al (2022) have shown a hybrid decision support tool for precise to perceive drug-related problems in routine exercise. They found the best possible solution to improve patients' safety & care with certain limitations [56].

5 Challenges for AI-Enabled CDSS

Over the past period, only human was taking healthcare decisions directly, later to this phase. The health care decisions were assisted by some intermediate intervals. As technology grows such intermediate intervals' interference becomes wider. Nowadays smart computers & many tools technology make health care decisions. These process elevations queries about answerability, transparency, consent, and confidentiality. There are many ethical issues and concerns about CDSS implementation. Figure 2 Shows challenges for AI enable CDSS.

5.1 Transparency

Transparency is one of the main causes of trustworthy CDSS. There are many types of EHRs. Among these some images based on which bases the diagnosis happens to lead the treatment. Such records need to be well maintained before applying further knowledge.

5.2 Accountability

The misjudgment toward image processing using AI may lead to unwanted errors in patient diagnosis. who is accountable for such issues which reach up to erroneous treatment in CDSS. The accountability factor is very crucial in predicting a higher risk of disease & action because it may affect the disease-related judgments. Once the error is led by the doctor or a nonmedical person makes a big sense to the further action. The accountable person is a reasonable medical decision-maker for CDSS using AI implementation. The higher risk for the disease can lead to the accountable person will make a proper judgment about the risk & remedies.

Fig. 2. Challenges for AI-enabled CDSS

5.3 Consent to Use

The patient should be well aware of his/her image data which going to use by CDSS applications for clinician interaction. Their concern about the limited study desired study must be part of the patient. The dataset on a particular study should not be used for any other study without the consent of patients [57] The clinicians are responsible educate patients on the usage of AI & ML systems' inputs & outcomes.

5.4 Safety

Safety is always the core important factor & significant concern for CDSS. One of the case studies about the cancer patients where their nature of study claimed as the improper process as they have state-certain labels as wrongly towards "unsafe or inaccurate" state for treatment suggestion [58–60] The algorithms are always used for producing reliable results. But in case of any misleading data sequence or misleading treatment could not reserve the wrong thing happened which is really a big concern in the viewpoint of "safety".

5.5 Privacy

The pre-processed Dataset is prepared for certain clinical trials. The same dataset may merge with other patients' processed data for diagnosis and find acute results [58]. The data of multiple Patients may work on the same background but keep concerned of each one fully informed about the test's data processing. The partial loss of data or sharing of data is a loss of fundamental privacy rights.

5.6 Liability

Current liability frameworks are being challenged by new AI-based technologies. It will be critical in determining obligations and building an effective liability design.

5.7 Regulatory Responsibility

There is a monitoring system by governmental and regulatory agencies to implement and usage of CDSS using consequential technologies. But still, it needs more attention to improve policy processes to get rid of errors and their effect on human societies.

6 Discussion and Conclusion

The challenges of AI-enabled CDSS are a long assessment tradition. The healthcare sector includes a heterogeneous range of sub-sector to be developed with AI. Health care is very dynamic in nature and needs constant attention for new changes. Healthcare needs the support of techy day-by-day path, process & paradigm changes as per the need requirements. The Artificial intelligence-enabled CDSS is joining hands for better healthcare sustainability. The users or practitioners learn & improve from past practice. The frameworks, methodologies, tools, and implementation capacity is increasing as time passes. The EHRs and their types have spread tactically for clinical trials, diagnosis & treatments. The EHRs are used widely and smartly for many medicated decision-making. Especially for clinical trials need historical data which can be easily available from EHRs. On the basis of various types of EHRs, various clinical trials use historical data and confirm certain trial directions. The Capacity enhancements are done using the various AI techniques from primary to ultimate stages of decision. The Decision making becomes faster, more accurate, and knowledge base by specific design, development, selection of trail dataset & deployment of techniques. Clinical Decision support system improves competence, speed, control, and capability for clinical decision-making of healthcare organizations by incorporating, data quantification, cost-benefit, and model manipulation.

Besides this, some of the questions remain unchanged regarding the safety and effectiveness of AI in CDSS as it becomes more challenging over the year and years. The way AI changes its technical support it faces many issues to keep CDSS updated as per the latest technology trends of AI. The effectiveness of AI in CDSS also concerned with patient class, medical treatment, and societal Environment. changeThe medication errors may cause hard impact clinical decisions with patient outcomes. Technical updating is always a major concern for any intelligent system. The CDSS performance matters in human roles. The changing role of the particular segment may affect accuracy and transparency. Some of the common factors which really need potential attention while implementing CDSS is well Computer Learning, Technical Availability, handy IT Support, Clinician Autonomy, & Patient Attitudes.

References

1. Buchanan, B.G.: A (very) brief history of artificial intelligence. AI Mag. **26**(4), 53 (2005)
2. Bhbosale, S., Pujari, V., Multani, Z.: Advantages and disadvantages of artificial ıntelle-gence. Aayushi Int. Interdisciplinary Res. J. 227–230 (2020)
3. Stone, P., Littman, M.L., Singh, S., Kearns, M.: ATTac-2000: an adaptive autonomous bidding agent. J. Artif. Intell. Res. **15**, 189–206 (2001)
4. Al-Ani, A., Deriche, M.: A new technique for combining multiple classifiers using the Dempster-Shafer theory of evidence. J. Artif. Intell. Res. **17**, 333–361 (2002)
5. Becker, A., Bar-Yehuda, R., Geiger, D.: Randomized algorithms for the loop cutset problem. J. Artif. Intell. Res. **12**, 219–234 (2000)
6. Singer, J., Gent, I.P., Smaill, A.: Backbone fragility and the local search cost peak. J. Artif. Intell. Res. **12**, 235–270 (2000)
7. Chan, H., Darwiche, A.: When do numbers really matter? J. Artif. Intell. Res. **17**, 265–287 (2002)
8. Poole, D., Zhang, N.L.: Exploiting contextual independence in probabilistic inference. J. Artif. Intell. Res. **18**, 263–313 (2003)
9. Peng, Y., Zhang, X.: Guest editorial: Integrative data mining in systems biology: from text to network mining. Artif. Intell. Med. **41**(2), 83–86 (2007)
10. Wang, S., et al.: A multi-approaches-guided genetic algorithm with application to operon prediction. Artif. Intell. Med. **41**(2), 151–159 (2007)
11. Beck, J., Stern, M., Haugsjaa, E.: Applications of AI in education. XRDS: Crossroads, The ACM Magazine for Students **3**(1), 11–15 (1996)
12. Guilherme, A.: AI and education: the importance of teacher and student relations. AI Soc. **34**(1), 47–54 (2017). https://doi.org/10.1007/s00146-017-0693-8
13. Haenlein, M., Kaplan, A., Tan, C.-W., Zhang, P.: Artificial intelligence (AI) and management analytics. J. Manage. Anal. **6**(4), 341–343 (2019)
14. Petrin, M.: Corporate management in the age of AI. Colum. Bus. L. Rev. 965 (2019)
15. Xiong, Q.: Overview of the relationship between mechatronic engineering and artificial intelligence. In: 2021 International Conference on Wireless Communications and Smart Grid (ICWCSG), pp. 532–535. IEEE (2021)
16. Tarnapowicz, D., German–Galkin, S.: Energy optimization of mechatronic systems with PMSG. In: E3S Web of Conferences, vol. 46, p. 00016. EDP Sciences (2018)
17. Awadallah, M.A., Morcos, M.M.: Application of AI tools in fault diagnosis of electrical machines and drives-an overview. IEEE Trans. Energy Convers. **18**(2), 245–251 (2003)
18. Sundhar, K.A.: Artıfıcıal ıntellıgence and future of humanıty. Keep your dreams alive. Understand to achieve anything requires faith and belief in yourself, vision, hard work, determination, and dedication. Remember all things are possible for those who believe, 165
19. Melnychenko, O.: Is artificial intelligence ready to assess an enterprise's financial security? J. Risk Finan. Manage. **13**(9), 191 (2020)
20. Girimonte, D., Izzo, D.: Artificial intelligence for space applications. In: Intelligent Computing Everywhere, pp. 235–253. Springer, London (2007)
21. Lewis, L.: Insights for the Third Offset: Addressing challenges of autonomy and artificial intelligence in military operations. Center for Naval Analyses Arlington United States (2017)
22. Goyache, F., et al.: The usefulness of artificial intelligence techniques to assess subjective quality of products in the food industry. Trends Food Sci. Technol. **12**(10), 370–381 (2001)
23. Eli-Chukwu, N.C.: Applications of artificial intelligence in agriculture: a review. Eng. Technol. Appl. Sci. Res. **9**(4), 4377–4383 (2019)
24. Ben Ayed, R., Hanana, M.: Artificial intelligence to improve the food and agriculture sector. J. Food Q. 2021 (2021)

25. Yu, K.-H., Beam, A.L., Kohane, I.S.: Artificial intelligence in healthcare. Nature Biomed. Eng. **2**(10), 719–731 (2018)
26. Sathesh, A.: Computer vision on IOT based patient preference management system. J. Trends Comput. Sci. Smart Technol. **2**(2), 68–77 (2020)
27. Sungheetha, A., Sharma, R.: Real time monitoring and fire detection using internet of things and cloud based drones. J. Soft Comput. Paradigm (JSCP) **2**(03), 168–174 (2020)
28. Mills, S.: Electronic health records and use of clinical decision support. Critical Care Nursing Clinics **31**(2), 125–131 (2019)
29. Ho, K.-F., Chou, P.-H., Chao, J.C.J., Hsu, C.-Y., Chung, M.-H.: Design and evaluation of a knowledge-based clinical decision support system for the psychiatric nursing process. Comput. Methods Programs Biomed. **207**, 106128 (2021)
30. Ancker, J.S., Edwards, A., Nosal, S., Hauser, D., Mauer, E., Kaushal, R.: Effects of workload, work complexity, and repeated alerts on alert fatigue in a clinical decision support system. BMC Med. Inform. Decis. Mak. **17**(1), 1–9 (2017)
31. Kwok, R., Dinh, M., Dinh, D., Chu, M.: Improving adherence to asthma clinical guidelines and discharge documentation from emergency departments: implementation of a dynamic and integrated electronic decision support system. Emerg. Med. Australas. **21**(1), 31–37 (2009)
32. Ortiz, D.R., Maia, F.O.M., Ortiz, D.C.F., Peres, H.H.C., de Sousa, P.A.F.: Computerized clinical decision support system utilization in nursing: a scoping review protocol. JBI Evidence Synthesis **15**(11), 2638–2644 (2017)
33. Rubins, D., et al.: Importance of clinical decision support system response time monitoring: a case report. J. Am. Med. Inform. Assoc. **26**(11), 1375–1378 (2019)
34. Athenikos, S.J., Han, H.: Biomedical question answering: a survey. Comput. Methods Programs Biomed. **99**(1), 1–24 (2010)
35. Sloane, E.B., Silva, R.J.: Artificial intelligence in medical devices and clinical decision support systems. In: Clinical Engineering Handbook, pp. 556–568. Academic Press (2020)
36. Van Der Veen, W., et al.: Association between workarounds and medication administration errors in bar-code-assisted medication administration in hospitals. J. Am. Med. Inform. Assoc. **25**(4), 385–392 (2018)
37. Peris-Lopez, P., Orfila, A., Mitrokotsa, A., Van der Lubbe, J.C.A.: A comprehensive RFID solution to enhance inpatient medication safety. Int. J. Med. Inform. **80**(1), 13–24 (2011)
38. Friedman, R.H., Frank, A.D.: Use of conditional rule structure to automate clinical decision support: a comparison of artificial intelligence and deterministic programming techniques. Comput. Biomed. Res. **16**(4), 378–394 (1983)
39. Kunz, J.C., Shortliffe, E.H., Buchanan, B.G., Feigenbaum, E.A.: Computer-assisted decision making in medicine. J. Med. Philosophy Forum Bioethics Philosophy Med. **9**(2), 135–160 (1984)
40. Clarke, J.R., Cebula, D.P., Webber, B.L.: Artificial intelligence: a computerized decision aid for trauma. J. Trauma **28**(8), 1250–1254 (1988)
41. Molino, G., Ballaré, M., Aurucci, P.E., Meana, V.R.D.: Application of artificial intelligence techniques to a well defined clinical problem: jaundice diagnosis. Int. J. Bio-med. Comput. **26**(3), 189–202 (1990)
42. Furlong, J.W., Dupuy, M.E., Heinsimer, J.A.: Neural network analysis of serial cardiac enzyme data a clinical application of artificial machine intelligence. Am. J. Clin. Pathol. **96**(1), 134–141 (1991)
43. Forsström, J.J., Dalton, K.J.: Artificial neural networks for decision support in clinical medicine. Ann. Med. **27**(5), 509–517 (1995)
44. De Graaf, P.M.A., Van den Eijkel, G.C., Vullings, H.J.L.M., De Mol, B.A.J.M.: A decision-driven design of a decision support system in anesthesia. Artif. Intell. Med. **11**(2), 141–153 (1997)

45. Geissbuhler, A., Miller, R.A.: Distributing knowledge maintenance for clinical decision-support systems: the" knowledge library" model. In: Proceedings of the AMIA symposium, p. 770. American Medical Informatics Association (1999)
46. Berner, E.S.: Ethical and legal issues in the use of clinical decision support systems. J. Healthcare Inf. Manage. JHIM **16**(4), 34–37 (2002)
47. Tanguay-Sela, M., et al.: Evaluating the perceived utility of an artificial intelligence-powered clinical decision support system for depression treatment using a simulation center. Psychiatry Res. **308**, 114336 (2022)
48. Patni, J.C., et al.: COVID-19 pandemic diagnosis and analysis using clinical decision support systems. In: Cyber Intelligence and Information Retrieval, pp. 267–277. Springer, Singapore (2022)
49. Citerio, G.: Big data and artificial ıntelligence for precision medicine in the neuro-ICU: Bla, Bla, Bla. Neurocritical Care, 1–3 (2022)
50. Mosavi, N.S., Santos, M.F.: Characteristics of the intelligent decision support system for precision medicine (IDSS4PM). In: Proceedings of Sixth International Congress on Information and Communication Technology, pp. 675–683. Springer, Singapore (2022)
51. Kim, D., Lee, J., Woo, Y., Jeong, J., Kim, C., Kim, D.-K.: Deep learning application to clinical decision support system in sleep stage classification. J. Personalized Med. **12**(2), 136 (2022)
52. Liu, F., Bao, G., Yan, M., Lin, G.: A decision support system for primary headache developed through machine learning. Peer J. **10**, e12743 (2022)
53. Ming, D.K., et al.: Applied machine learning for the risk-stratification and clinical decision support of hospitalised patients with dengue in Vietnam. PLOS Digital Health **1**(1), e0000005 (2022)
54. Vogel, S., et al.: Development of a clinical decision support system for smart algorithms in emergency medicine. Stud. Health Technol. Inform. **289**, 224–227 (2022)
55. Kim, S., Kim, E.-H., Kim, H.-S.: Physician knowledge base: clinical decision support systems. Yonsei Med. J. **63**(1), 8 (2022)
56. Levivien, C., et al.: Assessment of a hybrid decision support system using machine learning with artificial intelligence to safely rule out prescriptions from medication review in daily practice. Int. J. Clin. Pharmacy, 1–7 (2022).https://doi.org/10.1007/s11096-021-01366-4
57. Cohen, I.G., Amarasingham, R., Shah, A., Xie, B., Lo, B.: The legal and ethical concerns that arise from using complex predictive analytics in health care. Health Affairs **33**(7), 1139–1147 (2014)
58. Gerke, S., Minssen, T., Cohen, G.: Ethical and legal challenges of artificial intelligence-driven healthcare. In: Artificial intelligence in healthcare, pp. 295–336. Academic Press (2020)
59. Brown, J.: IBM Watson reportedly recommended cancer treatments that were 'unsafe and incorrect'. Gizmodo, 25 (2018)
60. Ross, C., Swetlitz, I.: IBM's Watson supercomputer recommended 'unsafe and incorrect'cancer treatments, internal documents show. Stat **25** (2018)

An Extensive Study on Machine Learning Paradigms Towards Medicinal Plant Classification on Potential of Medicinal Properties

R. Sapna[1,2(✉)] and S. N. Sheshappa[1,2]

[1] Sir M Visvesvaraya Institute Of Technology, affiliated to Visveshwaraya Technological University, Bengaluru, India
sapna.aradhya@gmail.com, sheshappa_is@sirmvit.in

[2] Department of Computer Science and Engineering, Presidency University, Bengaluru, India

Abstract. The automatic classification of medicinal plants requires more exploration as it is considered as major issue for conservation, authentication, and manufacturing of medicines. Generally, medicinal plants have been classified by features of the leaf with respect to color, shape and texture. Leaf is a main parameter on analyzing its plant nutrition, plant contentions, plant soil-water association, plant preservation measures, crop ecosystems, plant respiration rate, plant transpiration rate and plant photosynthesis. Classification of the plant species is a primary and highly essential procedure for plant conservation. An object recognition system is required to classify the various species of the plant species and to protect them from various diseases. In this article, a detailed survey on machine learning models has been carried out to identify and classify medicinal plants by considering the texture and shape features of a plant leaf using linear and non linear feature descriptors. However the extracted features from the plant leaf image will be huge containing high redundancy information's. On employment of feature selection techniques through weighted average strategies through metaheuristic techniques, those techniques reduces the redundancy on feature extracted and minimizes the equal error rate to obtain the optimum weighted features. Further numerous classification techniques on supervised and unsupervised types has been employed to classify the optimal feature on various dataset has been experimented and validated using cross fold validation using confusion matrix. It is vital and essential task for providing detailed insight on that classification model for medicinal plant with respect to its medicinal properties. The efficacy of each model has been demonstrated on single plant and multiple plants on basis of classifier and dataset employed. Finally outline of the proposed methodology as framework to classify the medicinal plant has been provided. Evaluation of models has been carried out on the processing of the dataset.

Keywords: Medicinal plant classification · Machine learning · Feature extraction · Feature selection · Feature normalization

J. I.-Z. Chen et al. (Eds.): ICIPCN 2022, LNNS 514, pp. 541–555, 2022.
https://doi.org/10.1007/978-3-031-12413-6_43

1 Introduction

World's human population depends on plant as it is important source of the oxygen. Plant plays an important role in maintaining the earth diversity [1]. Plant is classified into different types. Medicinal plant is one of the popular plant types in the world which utilized to treat various kind of disease causing serious illness in the human body and animal body. In medicinal plant, there numerous herbal remedies which is very crucial to ensure disease free human health and to increase the economic growth of the various countries which has excellent plants growth with medicinal properties. Especially different parts of single plant are efficient for multiple disease treatment. In order to classify the plant on basis of the medicinal properties has been become vital and challenging issues for conservation, authentication, and production of medicines. Generally, strong analysis is required on plant nutrition, plant contentions, plant soil-water associations, plant protection, crop ecosystems, plant respiration rate, plant transpiration rate and plant photosynthesis. Those analyses help to classify the plant and to identify various species using computer vision techniques [2].

Medicinal Plant has been determined on basis of the color spot and patterns in the leaves and stem of the plant. Traditionally manual observation is carried out to identify the types and it is becoming obsolete on considering the large field as it is found in deep forest. Conventional manual identification is inefficient, time-consuming and costly. In some cases, misinterpretation may leads to loss of human life or severe health problem. In order to simplify and enhance the accuracy of the plant detection, image processing technique based machine learning paradigms [3] has been employed in large extent. In identification of the medicinal properties of plant species, Image processing plays a vital role to obtain optimal results and minimizes the outcomes from human attempts. Further Image processing techniques supports to detect plant species with more accuracy and efficiency. In order to achieve best outcomes, discrete image processing techniques have been modelled to identify and classify plant using traditional supervised and unsupervised based machine learning paradigms.

Plant type based on leaves is represented as Tulsi, Peppermint, Bael, Stevia, Lemon Balm and catnip is characterised by texture analysis, color analysis and the shape analysis of the image matrix. At present, the detection of plant through machine learning is capable of identifying plant type on basis of feature extracted [4]. Several image features of plant is represented in shape, texture, and color were determined and those values were utilized for classification of the plant [5]. Further Machine learning architectures incorporates the pathological and morphological changes of the plant for and it results provided with high accuracy. The state of art machine learning based detection technique leads to misclassification error on various types of plants. Considering the large environmental impact of the plant and weeds, slow detection speed of learning model, and its low accuracy.

Machine learning model becomes less adoptable and it cannot be widely applied to prediction of plants due to above mentioned results. Due to above mentioned complexity; it has become an important to make rapid and accurate prediction on plant classification through projection of machine learning architectures with Meta heuristic techniques [6]. Image classification are processed using projected metaheuristic architectures [7], obtains the optimal feature of the plant image. On applying the metaheuristic optimization techniques, plant has been classified with high accuracy for multiple species in the specified region. Feature selection produces the higher performance for images with constraints have been employed as fitness criteria to prevent the model from over-fitting and improving its generalization performance.

The rest of this article is sectioned as follows: A detailed review of literature in terms of machine learning architectures to plant classification is represented Sect. 2 and advantage of image processing techniques represented as machine learning technique for plant classification models has been analyzed on various strategies and constraints in Sect. 3. Section 4 provides research objective of the work and its proposed methodology to identify the plant. Section 5 concludes survey of the work.

2 Related Work

In this section, various state of art approaches applied to classification of the various medicinal plant disease is analyzed by utilizing machine learning model along the feature extraction techniques and feature selection techniques has been detailed as follows

2.1 Analysis of the Feature Extraction Techniques

The affiliated institutions are to be listed directly below the names of the authors. Multiple affiliatons should be marked with superscript arabic numbers, and they should each start on a new line as shown in this document. In addition to the name of your affiliation, we would ask you to give the town and the country in which it is situated. Do not include the entire postal address. E-mail addresses should start on a new line and should be grouped per affiliation.

2.2 Analysis of the Feature Extraction Techniques

Analysis of feature extraction technique on the medicinal plant is carried out on basis of various feature representation like shape, texture and color. Differentiation of plant leaves has been identified with respect to the features extracted. The feature of the leaves extracted are shape, texture, color, edge, tip, base, surface, trichomes, venation and arrangement on stem divisions of the main blade.

2.2.1 Texture based Features Descriptors

Texture based feature descriptors captures the color, shape and texture of the plant through linear and non linear model to achieve effective classification. Various texture based feature descriptors employed for plant leaf classification is as follows.

- **Local Binary Pattern:**

Leaf feature selection is vital part for identification of the plant on basis of the medicinal properties. Local Binary Patterns (LBP) is considered as one of the texture

Fig. 1. Feature extracted on computation of LBP values

based feature extractor model which produces high efficiency and robustness on plant specifies identification and classification (Fig. 1).

In local binary pattern, good size of sampling point has to be defined initially. Local Binary Patterns is employed for rotation invariant texture extraction by obtaining pattern value of local structures on analysis of the plant image. LBP value is obtained using thresholding of the neighborhood rotation pixels from the central pixel and then multiplying the threshold by binary weighting of the local pattern. LBP value is capable of identify the texture properties such as line, edge, point and corner on the leafs image of the dataset [8].

VAR is a feature descriptor to a measure a local contrast property in a texture of the image and it is kept unchanged on employing monotonic transformation of images in gray scale format. Joint LBP with joint distribution of local contrast pattern also considered as a LBPV texture descriptor. LBPV also represented as texture descriptor to gather the local patterns of texture and contrast of the images. LBP method is considered to be appropriate to the image having a difficult texture structures.

- **Modified Local Gradient Pattern**

It is a another type of texture based feature descriptor which employs various channels of color images for extracting more important features to enhance the performance of image classification. It uses multiple color channels to obtain discriminative information of medicinal plants. Further Harmonic Mean of the neighboring absolute gradient (gi) values uses adaptive threshold to reduce outlier's effect. Color channels are capable of the detecting even small variance of the feature on combination of all the information of features.

Initially Harmonic Mean (μh) is computed to get the threshold on utilizing the N × N neighborhood of each pixel of the image [9]. Threshold is used to set the binary value on compare the gradient value (gi) of each local gradient with the Harmonic Mean. If gradient value is greater than or equal to harmonic mean then binary value is set to '1' otherwise binary value is set to '0. Then total neighboring pixels and its distance from center pixel has to be computed (Fig. 2).

Fig. 2. Feature generation process for multiple color channels.

On obtaining the coded pattern of the two color channels, binary code is created and converted into decimal value with respect to the horizontal and the vertical codes of those two channels. Horizontal codes and vertical codes of two channels generate the histograms and it is concatenated. Six combination of feature is generated on its extraction procedure with 4 color channels.

Properties of Multichannel Modified Local Gradient Pattern

- Multiple Color Channel: Various types of plant images has different shade in colors, Multichannel Modified Local Gradient Pattern get more information on the plants to enhance classification accuracy
- Outlier's Effect: Harmonic mean of the image produces lower value when it is affected by outlier pixel for extraction of features of MLGP.
- Discriminative Power: MLGP feature extraction technique computes small changes of images accurately to lower threshold value.

- **Principle Component Analysis**

Principal Component Analysis (PCA) is considered as texture based feature extraction technique to generate the most discriminative features of the images. PCA is implemented to plant image to analyze the various regions and objects to generate patterns in terms of variance and it is formed into classes. Classes are achieved on basis of correlation and covariance of the image represented as matrix. Principle component between classes describes the considerable amount of variance. Pattern is highly complex to identify in the images of high dimentions. PCA can compute the patterns in the image by reducing the number of dimension without loss of the information.

- **Linear Discriminant analysis**

Linear Discriminant Analysis is employed as linear feature extraction model to extract an optimal feature vector in the image by reduction of the non optimal features. It uses the fisher criterion function for the data projection direction to discriminate between the homogenous and heterogeneous patterns in the vector. LDA model [11] determines the optimal feature vector of the image on basis of the scatter matrix. Scatter matrix is considered as transformation matrix. Especially it generates the effective features for medicinal plants. The optimal solution vector is obtained on reduction of the feature space to enhance the recognition accuracy for large number of image samples containing various plant classes.

2.2.2 Shape Features

Shape feature of the medicinal plant image as it less depend on the conditions of the leaf images and on the quality of the plant images. It is computed as follows

- Aspect ratio: It defined as ratio of the width of the leaf to length of the leaf. Sometimes it can be defined as ratio of maximum axial length of leaf to minimum axial length of the leaf. It is considered as feature for determination of slimness.
- Compactness: It is defined as the ratio of the leaf area with 4π to the square of perimeter of the leaf area. It is considered as feature for determination of roundness.
- Dispersion: It is a defined as ratio of the radius of the leaf on its maximum circle enclosing its entire region to minimum circles that contained the same repetitive part in the region.
- It is defined as the ratio between the radius of the maximum circle enclosing the region and minimum circle that can be contained in the region. It is considered as feature for determination of the region spreads. However, Dispersion is insensitive to slight discontinuity in the shape such as crack in the leaf.
- Centroids: It is defined as center part of the leaf or region of interest. The centroids co-ordinates of the leaf are obtained and labelled as centroid x and centroids y. It is considered as feature for determination of center of the leaf region.
- Eccentricity: Eccentricity is a defined as characteristic feature of any conic part of the leaf.
- HU invariant moments: It is considered as Seven moment of the plant images which is captured from the shape. The moments captured as feature are invariant to rotation, translation and scaling. The moment of the image is the addition of moments of individual pixels in the leaf image.

- **Histogram of Oriented Gradient (HOG)**

HOG is another popular feature extraction technique. It extracts the local region of the image under analysis. It is considered as feature descriptors containing local regions which are invariant to geometry structure and image transformation or rotation. Process flow of HOG based feature descriptor is described in the Fig. 3. Initially, input image of plant is divided into N*N small connected regions of the image where each region of the image has equal sizes with representation of dimension X*X. HOG feature extraction determines the gradient orientation and magnitude of the regions.

The histogram of the each gradient direction or edge orientation is computed for all pixels available in the each region. On basis of gradient orientation, every region is discretized into angular bins. Weighted gradient for the each region pixel is calculated to its corresponding angular bin. The adjacent regions of the image are grouped into block to normalize the histogram of the adjacent pixels. Once the histograms of the adjacent regions are normalized, feature descriptors considered on the block histogram.

Fig. 3. Feature extraction using HOG

Color Features

Color moments of the plant image are to represent color features to characterize the image. Features extracted from the color features can be considered as Mean, Standard deviation, skewness and kurtosis.

2.2.3 Leaf Margins

Leaf margin is considered as feature descriptor on various aspects of the leaf which is used for the classification of the several plants. Feature Descriptor of the leaf considered are entire, serrated, wavy pinnatifid, toothed, pinnatisect, lobed etc. Further morphological and zero crossing of the curvature scale space of the leaf region can be considered as feature descriptor..

2.2.4 Vein Structure

The leaf vein structure of the medicinal plant is considered as feature descriptor for classification of the leaves with similar shapes. Feature extraction process uses the morphological operations of the plants to extract the veins and subsequent operations of the veins to measure its distribution in the entire region of the leafs.

2.3 Feature Fusion Technique

Feature Fusion Technique is carried out to combine the feature extracted from the feature descriptors to generate effective feature set on considered large aspect of the image for selection and classification process. Various feature fusion techniques employed to medicinal plant images is described below.

2.3.1 Feature Fusion – Concatenation of LBP Features

Local binary patterns employed for feature extraction is limited to only spatial area which is used to extract the texture of large structures. On varying the sampling points P and radius R of the descriptors, it can increase the extraction of the feature for the small structures in the spatial areas by combining the small structure to form large structures on concatenation operation. It is achieved by computing the histogram of the small structure and concatenation of the histogram of small structures using fusion operators. Further it is enhanced linearly on the increase of the different sizes of sampling points and radius. Fusion of the local binary pattern is achieved in two methods, such that.

- Initially, Histogram of the multiple LBP features on various size of sampling points and radius is computed separately and it is concatenated with multiple histograms.
- Concatenation of multiple histograms has been classified on basis of the structure of the features and the feature fusion is accomplished on histogram of the similar features combinations.

2.3.2 Discriminant Correlation Analysis

It considered as fusion technique for concatenation of the feature vectors from the various feature extraction techniques in addition to the computation of the distance vector and according to combination rule to maximize the pair wise correlation of the feature on combinations.

2.4 Analysis of the Feature Selection Techniques

Feature selection is considered as significant process in the machine learning architecture as it reduces the irrelevant and redundant features in the feature vector. Feature selection generates the optimal feature set to increase the computation accuracy and reduce the computation cost of the classification on elimination of deceptive data. Further it helps to understand and learn the characteristics of the features for efficient classification of the plants.

2.4.1 Differential Evolution Model (DE)

Differential Evolution is a considered as population-based optimization algorithm to establish the efficient search capabilities on the feature extracted. Model composed of four processing step such as population Initialization, Mutation, Recombination and Selection. In the population initialization step, a set of population vectors containing the feature extracted. The feature vector is initialization step is three time of data dimension chosen by other Meta heuristics optimization model for further processing of the model to generate the optimal features with dimensions [10]. Processing of those parameter vectors encounters mutation, recombination and selection process to generate an optimal feature vector. In mutation stage, three parameter vectors have been selected on given parameter vector which is distinct to each other. The fitness function of the DE is given by (1).

$$F_{i,j}(n) = \frac{(n-1)}{M} \sum_{k=0}^{n} X\, h_{i,j}(k) \quad n = 1, 2 \ldots N-1 \tag{1}$$

where M is initial population of the feature vector.

h_{ij} is the mutation.

2.5 Analysis of the Classification Techniques- Feature Classification

The optimal feature subset determined by using feature selection process for plant images has been employed for classification using using supervised or unsupervised classifiers. The methods are considered to be the instance based as they are capable to predict the class label for the feature sample by determining the closest points of the features using distance measures. The closest points of the features samples are predicted by the distance weighting computation to model the classes and label it.

2.5.1 Probabilistic Neural Network:

Probabilistic Neural Network (PNN) is considered as a back propagation neural network for plant image classification. Classification is employed to classify LBP features extracted from the plant images. PNN classifier has diverse advantageous on processing of the features with less no of iteration to obtain the plant classes with support of Bayesian approach. The Network classifies the feature vector into specific class on processing it using four layers such as input layer, pattern layer, summation layer and output layer.PNN structure for classification is represented in Fig. 4

Layer of PNN

- Input Layer: Input layer composed of the feature vector with k value which is to be processed and classified into n classes
- Pattern Layer: Pattern layer of the PNN evaluate the vector distance between the input vector and row weight vectors in weight matrix. The distance of the features is scaled nonlinearly.
- Summation Layer: In this layer, each feature determines the shortest distance among other features to combine the features for each classes
- Output Layer: In this layer, the feature are be classified into class on basis of the distance value, if the value of feature is larger than other feature, classify it into other class.

Fig. 4. Representation of the PNN classifier on medicinal plant classification

2.5.2 Support Vector Machine

SVM model is employed to classify the features, the spatial points is separated by hyper-plane construction. Feature represented as spatial points is transformed as dimensional vector. Vector is processed by computing the condition for separating the features in the dimensional hyper plane. A hyper plane is considered as classes with large separation of features on its categories. This hyper plane is defined as the maximum margin of the hyper plane containing features. Hyper plane of the data points between two classes is represented as (2).

$$G_{gc}=(P,H) + (1 - \Delta)\ .\ O_{reg}(U) \tag{2}$$

where P and H considered as two classes and $O_{reg}(u)$ is considered as kernel function.

It is possible to achieve the good separation of the feature on the hyper plane which has largest distance to the data points of any classes. Finally classifier results in low generalization error on matching the input feature between the classes with respect to the kernel function. It is best suited to the plant leaf images.

2.5.3 Decision Tree

Decision Tree is a classifier for approximating feature vector in the discrete-valued target functions. The decision process is achieved by partitioning the feature points into the discrete classes. The class containing features is represented as decision node, chance node and leaf node. Decision rules for classifying the features on basis of association rules of target function. Target function generates temporal and causal relationship on the features in the hierarchical structure represented as a tree with decision nodes and leaf nodes. Target function is expressed as.

$$G_{func}(u) = S.O_{data}(U) + (1 - \Delta) \, . \, C_{reg}(U) \tag{3}$$

where $SO_{data}(u)$ considered as Support 1 and $C_{reg}(u)$ is considered as Confidence n.

The splitting process of the features is employed using association rules. The rules are represented in form of support and confidence using a greedy approach to identify the feature to be classified. It will be considered to partition the features with iteration for classification of the plant on its medicinal properties. It is suitable to any kind of the plant classes.

2.5.4 K-Nearest Neighbour

K- Nearest Neighbor is a classification process of unlabeled samples into classes containing feature on basis of its distance measures between the feature data points. The classes are considered as the majority class among the k patterns in the feature space. It is non parametric method. Feature is classified into classes among its k nearest neighbor feature computed using distance measures. Class should contain the features with smallest distance among all data points Multi class generation function of the KNN classifier is given by following equation.

$$C_{m,n}(I, j) = \sum_{x} \sum_{y}^{n} x \begin{cases} 1 & m(x, y) = i(x, y) = j \\ 0 & otherwise \end{cases} \tag{4}$$

The important parameters of KNN classifier is k which represent the similar value of the features through the distance metric. The detailed texture feature classification procedure based on CCM is represented in the dependence matrix.

2.5.5 Random Forest Classifier

Random Forest Classifier is an ensemble learning model which combines the many decision trees models using a majority voting mechanism among them. The Random Forest Classifier generates a set of decision trees using gini index for splitting the features for classifying it in tree structure of each classifier. Final class of the each classifier is aggregated and voted by weight value to construct a final medicinal plant. It is best applied to the medicinal plant classification. Computation of Gini index is given by.

$$\text{Gini}(t) = 1 - \sum \left(p_j\right)^2 \tag{5}$$

where p is the frequency of the class j.

2.5.6 Multi-layer Perception (MLP)

Multilayer Perception is a feed forward ANN classifier which is composed of input layer, hidden layer, and an output layer along its computational nodes (neurons) for classifying the feature extracted on the images. MLPs are contains the activation function to map the feature spaces in terms of weighting features on basis of the frequency of occurrence to output layer containing classes which is linearly separable to each other. Activation function of the MLP is given by equation as.

$$L(f) = \frac{1}{n}\sum_{k=0}^{L-1}\left\{\sum_{x}[y \log(p(i,j)) + (1-y)\log(1-P(I,j)\right. \tag{6}$$

MLP is fully connected network as nodes in each layer are connected with features on basis of weight to following layer. It is the back propagation technique as it incorporates the gradient decent to compute the change of the weight among features set. Class is represented as target by perceptron through minimum error.

3 Tabular Representation of the Machine Learning Models for Medicinal Plant Classification

In this part, machine learning model employed for plant classification has been employed using feature extraction and classification model on various illumination effects of the image has been represented in the Table 1.

Table 1. Tabular representation of the plant classification

Title	Feature extraction	Feature classification	Type of medicinal plant	Advantage of the method	Disadvantage of the method	Performance
Identification of Ayurvedic medicinal plants [9]	Texture	K-mean clustering and Gray level co-occurrence matrix (GLCM) method has been employed through Back propagation algorithm	Coleus ambonicus Mentha arvensis Adidirachta indica Solanum trilobatum	It is extended to other type of plant	restricted to selected features	98.75% of accuracy
Classification of selected Medicinal plant [10]	Shape, Texture color	Laplacian filtering method, ANN	Solanum trilobatum	It is capable to identifying 50 varieties of plants	It is Complicated towards interpreting the target variable of the disease	97.75% of accuracy

(*continued*)

Table 1. (*continued*)

Title	Feature extraction	Feature classification	Type of medicinal plant	Advantage of the method	Disadvantage of the method	Performance
Kohonen maps based Plant Classification on data fusion[11]	Morphological	Fluorescence induction and spectral reflection method, Principle Component analysis	Coleus ambonicus Mentha arvensis Adidirachta indica	It is suitable to large canopy	shape and texture relationships is complex	96.75% of accuracy
Semantic Segmentation for Plant Leaf[12]	Texture	Semantic Segmentation	Rosette plant	It is extend to images with color information	discrimination of the disease is complex	95.75% of accuracy
Unsupervised entropy based segmentation [13]	Texture shape color	linear discriminate analysis, support vector machine	Ambonicus Mentha arvensis Adidirachta indica Solanum trilobatum	It capable identifying the different variation in surrounding conditions	Setting model parameters was one of the major complication	97.75% of accuracy

4 Outline of the Proposed Research Framework for Medicinal Plant Classification Through Mobile Application

Mobile application has been designed using optimized framework containing Feed Forward Neural Network to identify and classify plant image from the test images which taken a from particular plant image. Application is fulfilled with abstract steps to process and classify the leaf image through preprocessing, feature extraction, Feature selection and classification [14]. Initially preprocessing of the input image is carried out to remove non leafy regions and empty spaces. Further resizing and contrast of the image is established. Preprocessed image extracts the features of the image on basis of texture characteristics. Extracted features are processed with feature selection technique to yield less no of features to reduce the classification burden [15, 16]. The optimal features are classified using proposed methodologies and finally leaf image is recognized on basis of plant type. The semantic web concepts can be deployed to make ontology for the image data [17].

5 Conclusion

The accuracy of classification of medicinal plants has been analyzed against complex texture structures. This analysis helps to model a mobile application composed of classifier to identify the type of medicinal plant. In this study, machine learning architectures along texture based feature extraction technique and Metaheuristics based feature selection technique has been analyzed in multiple aspects towards effective recognition of different leaf features. These analyses provide knowledge to model a novel methodology for medicinal plant classification against imbalance problem and over fitting issues to enhance performance.

References

1. Turkoglu, M., Hanbay, D.: Recognition of plant leaves: an approach with hybrid features produced by dividing leaf images into two and four parts. Appl. Math. Comput. **352**, 1–14 (2019)
2. Ghasab, M.A.J., Khamis, S., Mohammad, F., Fariman, H.J.: Feature decision-making ant colony optimization system for an automated recognition of plant species. Expert Syst. Appl. **42**(5), 2361–2370 (2015)
3. Zhao, C., Chan, S.S., Cham, W.K., Chu, L.M.: Plant identification using leaf shapes—a pattern counting approach. Pattern Recogn. **48**(10), 3203–3215 (2015)
4. Turkoglu, M., Hanbay, D.: Leaf-based plant species recognition based on improved local binary pattern and extreme learning machine. Phys. A Stat. Mech. Appl. **527**, 121297 (2019)
5. Bantan, R.A., Ali, A., Naeem, S., Jamal, F., Elgarhy, M., Chesneau, C.: Discrimination of sunflower seeds using multispectral and texture dataset in combination with region selection and supervised classification methods. Chaos Interdiscip. J. Nonlinear Sci. **30**, 113142 (2020)
6. Abbas, Z., Rehman, M., Najam, S., Rizvi, S.M.D.: An e_cient gray-level co-occurrence matrix (GLCM) based approach towards classification of skin lesion. In: Proceedings of the 2019 Amity International Conference on Artificial Intelligence (AICAI), Dubai, UAE, pp. 317–320, 4–6 February 2019
7. Cardinali, F., Bracciale, M.P., Santarelli, M.L., Marrocchi, A.: Principal Component Analysis (PCA) combined with naturally occurring crystallization inhibitors: an integrated strategy for a more sustainable control of salt decay in built heritage. Heritage **4**, 13 (2021)
8. Dahigaonkar, T.D., Kalyane, R.: Identification of ayurvedic medicinal plants by image processing of leaf samples. Int. Res. J. Eng. Technol. (Irjet) **5**, 351–355 (2018)
9. Manoj Kumar, P., Surya, C.M.: Department of Electronics and Communication Engineering, Government engineering College Wayanad, Kerala, India. In: Gopi, V.P. (ed) Identification of ayurvedic medicinal plants by image processing of leaf samples Third International Conference on Research in Computational Intelligence and Communication Networks (ICRCICN) (2017)
10. Gopal, A., Prudhveeswar Reddy, S., Gayatri, V.: Classification of selected medicinal plants leaf using image processing. In: International Conference on Machine Vision and Image Processing (MVIP) (2012)
11. Moshoua, D., et al.: Plant disease detection based on data fusion of hyper-spectral and multi-spectral fluorescence imaging using Kohonen maps. Real-Time Imaging, **11**(2), 75–83 (2005)
12. Praveen Kumar, J., Domnic, S.: Image based leaf segmentation and counting in rosette plants. Inf. Process. Agric. **6**(2), 233–246 (2019)
13. Navid, N., Baleghia, Y., Agahib, H.: Maximum mutual information and Tsallis entropy for unsupervised segmentation of tree leaves in natural scenes. Comput. Electron. Agric. **162**, 440–449 (2019)
14. Guo, Z., Zhang, L., Zhang, D.: A completed modeling of local binary pattern operator for texture classification. IEEE Trans. Image Processing **19**(6), 1657–1663 (2010)
15. Bloice, M.D., Stocker, C., Holzinger, A.: Augmentor: an image augmentation library for machine learning, arXiv preprint arXiv:1708.04680 (2017)

16. Pavithra, N., Sivaranjani, K.: Content based image retrieval system data mining using classification technique. Int. J. Comput. Sci. Mob. Comput. **5**(7), 519–522 (2016)
17. Sapna, R., Monikarani, H.G., Mishra, S.: Linked data through the lens of machine learning: an Enterprise view. In: 2019 IEEE International Conference on Electrical, Computer and Communication Technologies (ICECCT), pp. 1–6 (2019). https://doi.org/10.1109/ICECCT.2019.8869283

Medical Imaging a Transfer Learning Process with Multimodal CNN: Dermis-Disorder

Sumaia Shimu[1(✉)], Lingkon Chandra Debnath[1], Md. Mahadi Hasan Sany[1], Mumenunnessa Keya[1], Sharun Akter Khushbu[1], Sheak Rashed Haider Noori[1], and Muntaser Mansur Syed[2]

[1] Department of Computer Science and Engineering, Daffodil International University, Dhaka, Bangladesh
{sumaia15-12806,lingkon15-12781,mahadi15-11173, mumenunnesa15-10100,sharun.cse}@diu.edu.bd

[2] Department of Computer Engineering and Science, College of Science and Engineering, Florida Institute of Technology, Florida, USA
msyed2011@my.fit.edu

Abstract. The skin is the largest and fastest progressive part in the body. The skin's immune cells interact with keratinocytes in a number of ways to trigger a mass of dermatitis. Both living immune cells and skin-penetrating cells can coordinate with keratinocytes to promote pathogenesis of the disease. The activated cells make chemokines that attach to the immune cells in the skin. According to the latest data released by WHO in 2018, the number of deaths due to skin cancer in Bangladesh has reached about 0.04%. The mortality rate is 0.27% per 100,000 populations by age. The author has proposed an exploration using Transfer Learning on six types of skin disease as- Peeling, Acne, Eczema, Heat-rash, Melanoma, and Cold sore. The classification of these skin conditions was done using a Convolutional Neural Network. For the comparison with CNN, four state-of-art Transfer Learning models have been applied such as NASNetLarge, InceptResNetV2, EfficientNetB1 and DenseNet169, in which NASNetLarge (Accuracy 90% & Validation 80%) has given the highest accuracy. And our state of model NASNetLarge can perfectly recognise the disease types than the others. Through images processing, skin experts can initial treatment of skin diseases through observation of images of affected areas. As a result, the type of disease can be ensured and consequently can be ensured to reduce the complexity and disorders of skin diseases.

Keywords: Skin disease · Augmentation · Convolutional neural network · Transfer learning · Image processing

J. I.-Z. Chen et al. (Eds.): ICIPCN 2022, LNNS 514, pp. 556–573, 2022.
https://doi.org/10.1007/978-3-031-12413-6_44

1 Introduction

The current era is the era of computer technology. Now-a-days, everyone cares about their health and is concerned about their skin beauty. However, due to climatic and genetic reasons, most of have different skin problems [5]. Fungal infections, germs, allergens, and viruses, among other things, can cause skin problems. The texture and color of your skin can be affected by a skin disorder. Some skin disorders do not manifest symptoms for several months, allowing the condition to progress and worsen. This is owing to the public's lack of medical knowledge, but the cost of such a diagnostic remains limited and prohibitively expensive, which is why we suggest an image processing approach to diagnose skin illnesses [3]. Medical image processing technology has advanced rapidly in recent years. Some digital imaging devices, such as computed tomography (CT), digital subtraction angiography (DSA), and magnetic resonance imaging (MRI), are commonly employed in people's daily lives [6]. Current imaging methods used to detect skin diseases also have certain disadvantages. The median filter's biggest flaw is its high computational complexity. Furthermore, the median filter's software implementation does not produce accurate results. The issue with the sharpen filter is that if you apply a high pass mask to the image, the resultant image will have negative pixel values [10]. Smart health apps have been more easily incorporated into our lives than ever before, thanks to the increasing availability of powerful IoT devices. These apps currently focus on health monitoring, such as heart rate, body temperature, and ambient atmosphere, as well as an online diagnostic platform. Remote diagnosis has become more crucial than ever during the current COVID19 pandemic, and it requires a greater understanding of how to employ powerful artificial intelligence technologies.

The goal of this work is to discover and classify skin diseases in the human body using image processing techniques. In this work, author have extracted images of various skin diseases from different sources and classified the diseases using the DenseNet169, EfficientNetB1, InceptionResNetV2, NASNetLarge, and the CNN model. Major skin conditions such as acne, cold sore, melanoma, heat rash, eczema and peeling skin, and this study offers an in-depth study of the autoimmune subset of these conditions. The remainder of the proposed work is structured as follows: Sect. 2 discusses recent work/literature review, and Sect. 3 on description of the dataset, Sect. 4 describes the dataset used to test the proposed system, Subsect. 5 presents the results and finally the conclusions presented in this subsection [12].

In Table 1 represented the sample of skin diseases for each class.

Table 1. Samples of each classes disease.

Name of Skin Disease	Sample 1	Sample 2	Sample 3
Peeling			
Acne			
Eczema			
Heat rash			
Melanoma			
Cold sore			

For this work, the data set is divided into train, validation, and test which are shown in the table.

2 Literature Review

Tanvi Goswami et al. [11] proposed an automatic computer-based approach for skin condition detection and categorization based on skin disease images, which was needed to improve diagnosis accuracy while also addressing the lack of human specialists. Millions of data were used to enhance the model with CNN by applying threshold, c-means clustering and watershed algorithm & GMM pre-processing method for the better result where threshold squares 90–98% accuracy and C-means Clustering and watershed algorithm gain 96–98% accuracy. Shuchi Bhadula et al. [12] used 5 various machine learning algorithms that were chosen and dead on skin infection information collection. The applied models were random forest, naive bayes, logistic regression, kernel SVM, and CNN, with CNN providing the best training and testing accuracy of 99.05% and 96%, respectively, with the lowest error rate of 0.04. In [17], the author used 938 Alexnet as the model where CNN gave 71.5% accuracy (30 Epochs) and Alexnet gave 76.1% (30 Epochs). [2] used 9144 images of five classes for the images of three classes for the classification of Skin diseases which are Melanoma(439), Nevus(551), Seborrheic

Keratosis(413). For the better result CNN and classification which are healthy (3014), acne (913), eczema (967), benign (3015) and malignant lesions (1235) are collected from different sources. For the better result Genetic Algorithms, SVM, ANN, CNN are applied as the model where Genetic algorithm gave 76.17% accuracy, SVM gave 83%–90%, ANN gave 81.34%–85.71%, CNN gave 86.21% accuracy. [5] A technique is being developed for dissecting skin problems and utilizing color images without the need for doctor participation. The system was evaluated on three types of skin illnesses (Eczema, Melanoma, and Psoriasis) with a 100% accuracy using the CNN model. In [10], the author used 1619 pictures for the categorization of skin diseases. For the better result VGG16, Inception, ResNet-50, MobileNet, DenseNet, Xception are applied as the model where Inception gave 86.2%; 85.7% accuracy, DenseNet gave 88.6%; 90.1%, ResNet gave 92.5%; 81.8%, Mobilenet gave 97% accuracy. Manoharan et al. [22] used multiple ELM algorithms to solve problems. This technique can be made better and more efficient if using a unique feed-forward algorithm and a neural network. The feed-forward neural network has recently been running at a greater gain and with a longer calculation time. For simple generalized operations, the weight vector and biases of the neural network can be fine-tuned using intelligent assignment and get more accuracy (94.1%) [24, 25]. Used to solve the inverse problems, a revolutionary deep convolutional neural network was used. The traditional technique to coping with this challenge over time has been observed to be regularized iterative procedures. Despite the fact that these methods generate excellent results, they do have disadvantages, such as the difficulty in picking hyper parameters and the high cost of implementation [26]. Used two modifications to improve the CNN's accuracy. The first enhancement entails using Euler methods to change feature vectors and mixing normalized and raw features. Based on the findings, author conducted a comparative analysis using comparable methodologies, and it was determined that the suggested CNN outperforms the others. The proposed methodology can be applied to video and image classification in the future, which will increase the robustness and accuracy of deep networks [27] (Table 2).

Table 2. Related work.

Reference	Approach	Topic	Dataset	Future work	Limitations
[11]	CNN, Inception v3, resnet, VGG16, VGG19, Alexnet	Skin Disease Classification-A Survey	Derm Net NZ Image library, Dermofit Image Library, ISBI-201, ISBI-2017, Ham10000, Stanford Hospital, Peking Union Medical college clinical Database, IRMA Dataset, PH2, MEDNO, DermQuest,Hospital Pedro Hispano, Matosinhos, SD-198	–	Types of disease are less than others research paper

(*continued*)

Table 2. (*continued*)

Reference	Approach	Topic	Dataset	Future work	Limitations
[12]	Random forest, naïve bayes, logistic regression, kernel SVM and CNN	Skin Disease Detection	Many	AI and the advantages of AI-assisted diagnostics be examined	If any disease is absent, the model doesn't predict the disease
[17]	CNN, AlexNet	Skin Disease Detection and Classification	938 images of Three Classes of Skin diseases	Apart from CNN and AlexNet, different design be enforced to enhance the accuracy of the classification	Overall performance is yet to be improved
[2]	Genetic Algorithm, SVN, ANN, CNN	Multi-Class Skin Diseases	9144 images of five classes	Exploring the impact and smartphone based multi-class skin lesion classification to create the intelligent expert system accessible for folks living in remote areas and with restricted resources	Need to some multi class skin lesion classification for get more accuracy
[5]	CNN, SVM, Alexnet	Skin Disease Detection	80 images of every disease (20 Normal images, 20 Melanoma images, 20 Eczema images and 20 Psoriasis images)	First, the strategy of find skin disease should air the mobile application developed, detection the skin lesion in derma layer of the skin	
[10]	VGG16, Inception, ResNet-50, MobileNet, DenseNet, Xception	IoT Enabled Skin Disease Detection	Totally 1619 pictures	Evaluating the proposed two-phase classification, skin-related Continuing the integration of image-based disease, Enhancing the capabilities these four directions are included in our next research	Need new classification to produce better Classification results in skin disease detection

3 Methodology

The aim of the proposed work is to detect various skin diseases of the human body using image classification technology of machine learning. It has several levels, such as data collection (*4.1*), pre-processing (*4.2*), applying the models (*4.3*) and classification (*4.4*), and then checking how accurately the model is working. In this part, all of the sections will discuss elaborately the mentioned steps (Fig. 1).

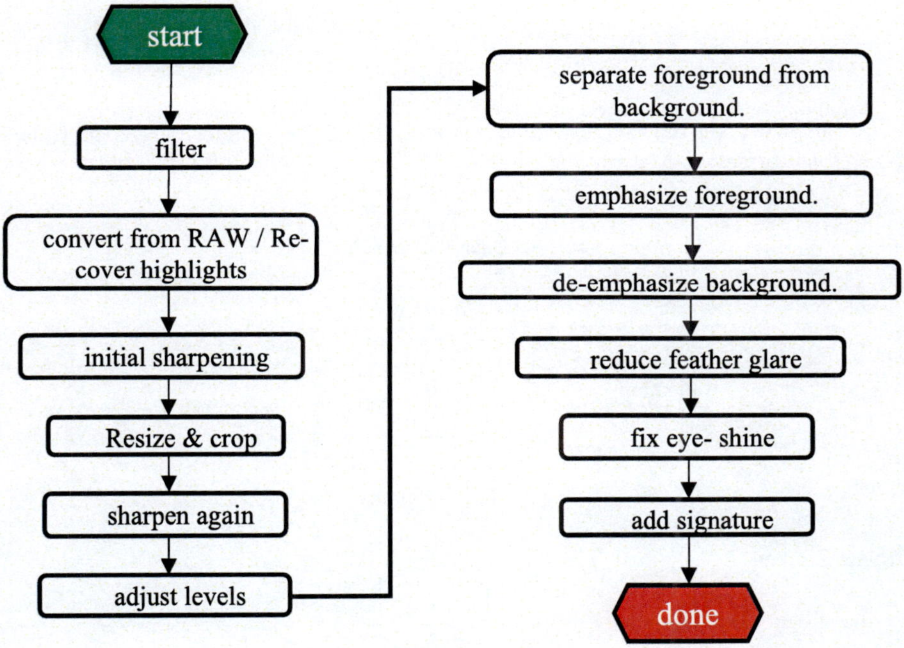

Fig. 1. Flow chart

A. The dataset's Description
We gathered images from various skin disease websites (Google) to create our dataset. The database contains some images of each type of disease which is represented in Table (Table 3).

Table 3. Data collection of each classes.

Name of diseases	The amount of images
Peeling skin disease	300
Acne skin disease	300
Eczema skin disease	300
Heat rash skin disease	300
Melanoma skin disease	300
Cold sore skin disease	300

For this work, the data set is divided into train, validation, and test which are shown in the table (Table 4).

Table 4. Dataset distribution.

Name	Total raw data
Train	1645
Validation	350
Test	105

Dataset Source

Eczema source acne diseases source peeling diseases source
heat rash source melanoma skin disease coldsore skin disease

B. Preprocessing

Preprocessing images is a very important step before applying any kind of algorithms or feeding any neural network model. For this work, author do some preprocessing steps before feeding those model. Here actually collect a lot of data for each class. Then select those photos manually in consideration of the noise data, good resolution, clear portion of disease area, crop the disease area that means in some data there are a lot of objects such as leap, nose, and fingers, etc. Then removed by cropping because the variation of this kind of objects can be effect the model during training (Fig. 2).

C. Augmentation

Image augmentation is one valuable strategy in building convolutional neural networks that can increment the estimate of the preparing set without obtaining unused pictures. The main concept is making some duplicate images with the variety according to the

Fig. 2. Block diagram

kind of augmentation. Tables x1 and x2, are demonstrating the variation of augmentation and the quantity of data after augmentation for train, test, and validation which will be fed the models (Tables 5 and 6).

Table 5. Augmentation with parameters and quantity of data after generating for training.

Name of augmentation	Parameters	Quantity of data after augmentation
rescale	1./255	3290
shear_range	0.2	4935
zoom_range	0.4	6580
horizontal_flip	True	8225
vertical_flip	True	9870
channel_shift_range	0.2	11515
fill_mode	nearest	13160

Table 6. Augmentation with parameters and quantity of data after generating for both validation and testing.

Name of augmentation	Parameters	Quantity of data after augmentation
Rescale	1./255	3290

D. Building and Applying the Models

In this work, CNN [18] and transfer learning [19] are used to classify skin disease classes. Keras [20, 21] framework has been used to build neural network architecture because it provides a python interface for building ANN which is quite comfortable and easy to interact with (Table 6).

E. Convolutional Neural Network

In this proposed CNN model created two convolution layers, where the first layer is extracting features from images and then creates feature maps and then use max-pooling to reduce the image size and then input shape of the image is set (225, 225, 3) which actually resizing the image cause here data set containing the images in various shape. The second convolution layer is like the first convolutional layer with some difference where the filter size for extracting features is more than the previous layer.

Then use flatten the feature maps to change over the output of the convolutional portion of the CNN into a 1D feature vector as like every classifier. The last stage of CNN is classifiers which are called dense layers and it is a fully connected dense layer. Table x32 demonstrates all the parameters which we used in this model (Table 7).

Table 7. All the parameters of our CNN model.

Layers	conv2d_1	maxpool_1	conv2d_2	maxpool_2	FC1
Kernel	3 * 3	2 * 2	3 * 3	2 * 2	–
Channel	32	32	64	64	6

F. Transfer Learning

Transfer learning is a machine learning method in which a model created for one task is reused as the starting point for another model. It actually refers to improving learning in an unused task by transferring information from a previously learned related task. Transfer learning is an optimization that permits quickly progressed execution when modelling the second task [19].

In this work, several state-of-art models are used to classify our skin disease classes. Before applying the models with our data set we did some changes in the pre-processing part. We used the particular pre-processing technique, which is actually used before the particular state-of-art model. In table x34 is showing the models which are used and the pre-processing method name (Table 8).

G. Classification

During the period of building the models are fit for training. After completing the training, checked what actually the models can predict on the testing data and then evaluated prediction and the confusion matrix based on the results.

Let us understand the process of convolution using a simple example. Consider that we have an image of size 3×3 and a filter of size 2×2 (Fig. 3).

Table 8. Name of the models for transfer learning with preprocessing Methods name

Name of the pre-trained models	Preprocessing method's name
NASNetLarge	nasnet
InceptionResNetV2	inception_resnet_v2
EfficientNetB1	efficient net
DenseNet169	densenet

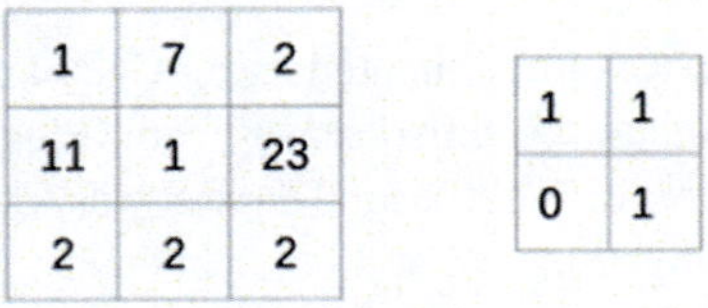

Fig. 3. Confusion matrix

The filter goes through the patches of images, performs an element-wise multiplication, and the values are summed up:

$$\mathbf{(1 \times 1 + 7 \times 1 + 11 \times 0 + 1 \times 1) = 9}$$
$$\mathbf{(7 \times 1 + 2 \times 1 + 1 \times 0 + 23 \times 1) = 32}$$
$$\mathbf{(11 \times 1 + 1 \times 1 + 2 \times 0 + 2 \times 1) = 14}$$
$$\mathbf{(1 \times 1 + 23 \times 1 + 2 \times 0 + 2 \times 1) = 26}$$

The output from the convolution layer was a 2D matrix. Ideally, we would want each row to represent a single input image. In fact, the fully connected layer can only work with 1D data. Hence, the values generated from the previous operation are first converted into a 1D format] (Fig. 4).

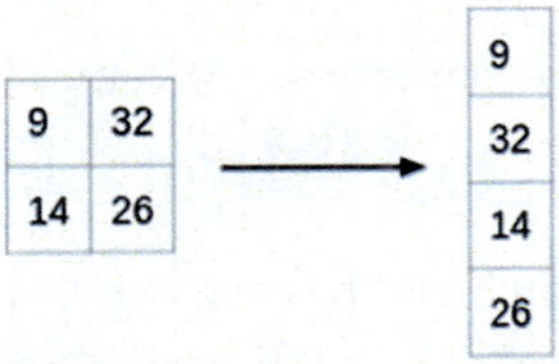

Fig. 4. 2D matrix

Once the data is converted into a 1D array, it is sent to the fully connected layer. All of these individual values are treated as separate features that represent the image.

$$\mathbf{Z = W^T.X + b}$$

Here, X is the input, W is weight, and b (called bias) is a constant. Note that the W in this case will be a matrix of (randomly initialized) numbers (Fig. 5).

Fig. 5. CNN and transfer learning model

4 Result Discussion

The proposed ML algorithm was implemented using CNN and Transfer Learning. This solution makes use of 1645 images of skin diseases from Google that were gathered from various websites. Furthermore, 15% of the total dataset was set aside for testing purposes. The images in the dataset ranged in size and color intensity. The images obtained were most likely (150 × 150) resolutions.

After applying CNN model, four types of Transfer Learning model have been used for the model performance comparison. The graph of CNN model accuracy and model loss is presented in Fig. 6 (a), (b).

(a) (b)

Fig. 6. (a) Training and validation accuracy of our CNN model (b) Training and validation loss of our CNN model.

Because of the over fitting problem from the Fig. 6 and 7, here can see in the 10th time running period the training accuracy is 1.0 when the validation accuracy is 0.5 in epoch no 6. That means a high variance model doesn't accurately modernize new data points. That's why the difference of the model accuracy and validation is too and also the loss is very high. Here, the accuracy of CNN is very low. That's why here use transfer learning methods (DenseNet169, EfficientNetB1, InceptionResNetV2, NASNetLarge).

The graph of DenseNet169 model accuracy and model loss is presented in Fig. 7 (a), (b).

(a) (b)

Fig. 7. (a) Training and validation accuracy of DenseNet169 model, (b) Training and validation loss of DenseNet169 model.

In the DenseNet169 model the model trained 10 times in this figure, it is seen that in epoch 4 when training accuracy 0.8 up then validation accuracy is 0.7. That means the accuracy was 80% when training but validity accuracy 70% after model run. The model loss is lower than the CNN model. The graph of EfficientNetB1 model accuracy and model loss is presented in Fig. 8 (a), (b).

(a) (b)

Fig. 8. (a) Training and validation accuracy of EfficientNetB1 model, (b) Training and validation accuracy of EfficientNetB1 model.

In the EfficientNetB1 model the model trained 10 times in this figure, it is seen that in epoch 4 when training accuracy 0.8 then validation accuracy is 0.7. That means the accuracy was 80% when training but validity accuracy 70% after model run.

The graph of InceptionNetV2 model accuracy and model loss is presented in Fig. 9 (a), (b).

(a) (b)

Fig. 9. (a) Training and validation accuracy of InceptionNetV2 model, (b) Training and validation accuracy of InceptionNetV2 model.

The InceptionNetV2 model in this figure, it is seen that in epoch 4 when training accuracy is 0.9 then validation accuracy is 0.7. That means the accuracy was 90% when training but validity accuracy 70% after model run. In The graph of NASNetLarge model accuracy and model loss is presented in Fig. 10 (a), (b).

(a) (b)

Fig. 10. (a) Training and validation accuracy of NASNetLarge model, (b) Training and validation accuracy of NASNetLarge model.

In the NASNetLage model in this figure, it is seen that in epoch 4 when training accuracy is 0.9 then validation accuracy is 0.8. That means the accuracy was 90% when training but validity accuracy 80% after model run (Fig. 11).

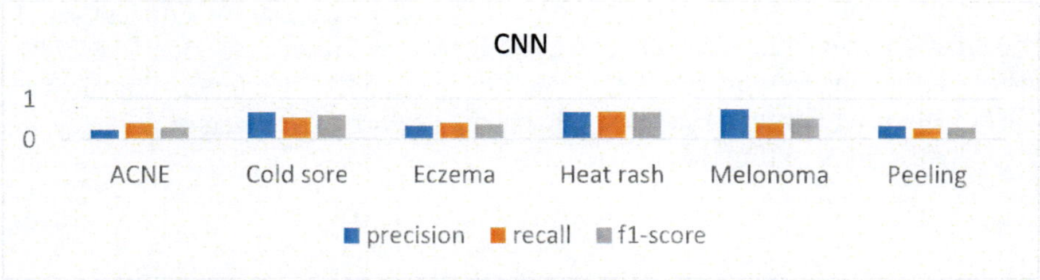

Fig. 11. Precision, recall and f1-score result of our CNN Model.

The bar graph shows the six types of skin diseases accuracy for CNN. Here precision rate 24%, 67%, 33%, 67%, 75%, 33% and recall rate 40%, 53%, 40%, 67%, 40%, 27% and f1-score rate 33%, 59%, 36%, 67%, 52%, 33% for the skin diseases ACNE, Cold sore, Eczema, Heat rash, Melonoma Peeling. Here Average accuracy are 50%, 44%, 47 (Fig. 12).

Fig. 12. Precision, recall and f1-score result of our DenseNet169.

The bar graph shows the six types of skin diseases accuracy for DenseNet169. Here precision rate 71%, 69%, 56%, 81%, 99%, 69% and recall rate 80%, 73%, 60%, 87%, 60%, 73% and f1-score rate 75%, 71%, 58%, 84%, 75%, 71% for the skin diseases ACNE, Cold sore, Eczema, Heat rash, Melonoma Peeling. Here Average accuracy are 74%, 72%, 72.33% (Fig. 13).

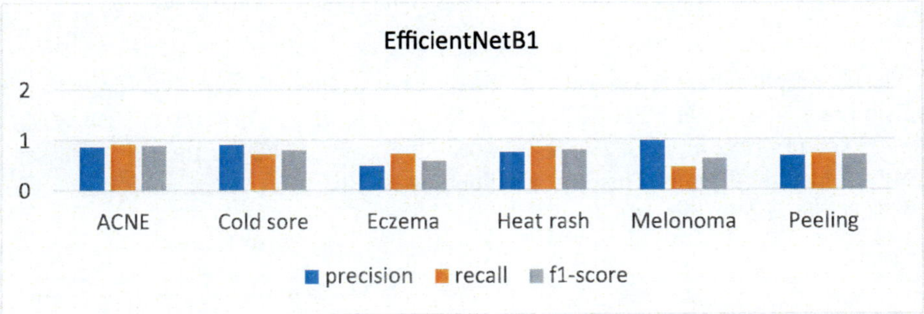

Fig. 13. Precision, recall and f1-score result of our EfficientNetB1.

The bar graph shows the six types of skin diseases accuracy for EfficientNetB1. Here precision rate 88%, 92%, 50%, 76%, 99%, 69% and recall rate 93%, 73%, 73%, 87%, 47%,73% and f1-score rate 90%, 81%, 59%, 81%, 64%, 71% for the skin diseases ACNE, Cold sore, Eczema, Heat rash, Melonoma Peeling. Here Average accuracy are 79%, 74.33%, 74.33% (Fig. 14).

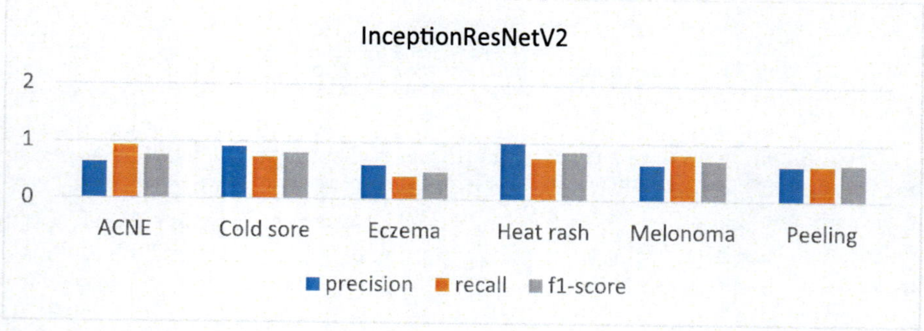

Fig. 14. Precision, recall and f1-score result of our InceptionResNetV2.

The bar graph shows the six types of skin diseases accuracy for InceptionResNetV2. Here precision rate 64%, 92%, 60%, 99%, 63%, 62% and recall rate 93%, 73%, 40%, 73%, 80%, 62% and f1-score rate 76%, 81%, 48%, 85%, 71%, 65% for the skin diseases ACNE, Cold sore, Eczema, Heat rash, Melonoma Peeling.Here Average accuracy are 73,33%, 70.16%,71% (Fig. 15).

Fig. 15. Precision, recall and f1-score result of our NASNetLarge.

The bar graph shows the six types of skin diseases accuracy for NASNetLarge. Here precision rate 74%, 85%, 50%, 80%, 60%, 85% and recall rate 93%, 73%, 33%, 80%, 80%, 73% and f1-score rate 82%, 79%, 40%, 80%, 69%, 79% for the skin diseases ACNE, Cold sore, Eczema, Heat rash, Melonoma Peeling. Here Average accuracy are 72.33%, 72%, 71.5% (Fig. 16).

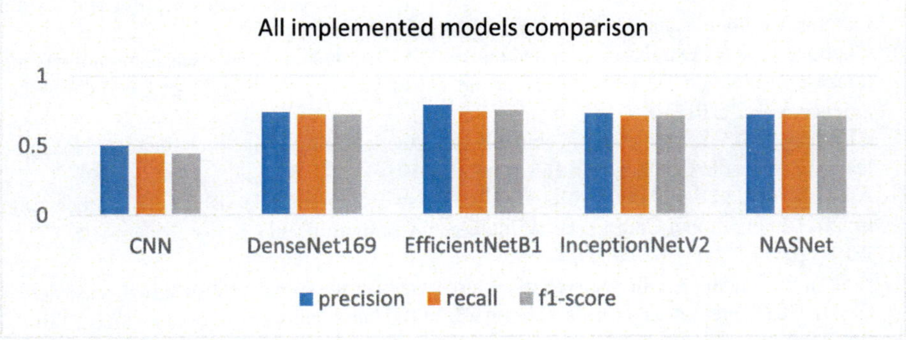

Fig. 16. Comparison of all the implemented models.

Here the comparison bar graph show five deep learning models. Here precision, recall, fi-score rate (50%, 44%, 44%), (74%, 72%, 72%), (79%, 74%, 75%), (73%, 71%, 71%), (72%, 72%, 71%) for models CNN, DenseNet169, EfficientNetB1, InceptionNetV2, NASNet. Here EfficientB1 gives higher accuracy for precision, recall & f1-score.

5 Conclusion and Limitation

In this research paper, an image processing system was introduced to automatically diagnose six different types of dermatitis. Images of inflammatory skin conditions such as acne, cold sore, eczema, heat rash, melanoma and peeling from the dataset were reviewed to test the effectiveness of the system. The main feature of the proposed strategy is a huge feature of the CNN & Transfer learning model used for classification when diseases are detected. The accuracy rates of CNN model is 44% & the Transfer Learning Model are (DenseNet169, EfficientNet, InceptionResNetV2, NASNetLarge, SVM_in_CNN) 73%, 75%, 71% and 72%, respectively. This work certainly helps dermatologists improve their working skills. The main limitation of the research is high variance which occurs for overfitting problems. The image conduct for training with the models, contains noise which we actually completely can't remove from the dataset as it's skin disease. The main reason is the dataset is on skin disease because the noise reduction is too complex. There are various approaches to improve the model described in this study. For example-the data's size is being expanded, properly cleaning and noise reduction will lead to better results. In future we'll try it with GAN for better results.

References

1. Chung, Y.-M., et al.: Topological approaches to skin disease image analysis. In: 2018 IEEE International Conference on Big Data (Big Data). IEEE (2018)
2. Hameed, N., Shabut, A.M., Alamgir Hossain, M.: Multi-class skin diseases classification using deep convolutional neural networks and support vector machines. In: 2018 12th International Conference on Software, Knowledge, Information Management & Applications (SKIMA). IEEE (2018)

3. Wei, L., Gan, Q., Ji, T.: Skin disease recognition method based on image color and texture features. Comput. Math. Methods Med. **2018** (2018)
4. Okuboyejo, D.A., Olugbara, O.O., Odunaike, S.A.: Automating skin disease diagnosis using image classification. In: Proceedings of the World Congress on Engineering and Computer Science, vol. 2 (2013)
5. ALEnezi, N.S.A.K.: A method of skin disease detection using image processing and machine learning. Procedia Comput. Sci. **163**, 85–92 (2019)
6. Ajith, A., et al.: Digital dermatology: skin disease detection model using image processing. In: 2017 International Conference on Intelligent Computing and Control Systems (ICICCS). IEEE (2017)
7. Kolkur, S., Kalbande, D.R.: Survey of texture based feature extraction for skin disease detection. In: 2016 International Conference on ICT in Business Industry & Government (ICTBIG). IEEE (2016)
8. Kumar, M., Kumar, R.: An intelligent system to diagnose skin disease. ARPN JEAS **11**(19), 11368–11373 (2016)
9. Haddad, A., Hameed, S.A.: Image analysis model for skin disease detection: framework. In: 2018 7th International Conference on Computer and Communication Engineering (ICCCE). IEEE (2018)
10. Yu, H.Q., Reiff-Marganiec, S.: Targeted ensemble machine classification approach for supporting IoT enabled skin disease detection. IEEE Access **9**, 50244–50252 (2021)
11. Goswami, T., Dabhi, V.K., Prajapati, H.B.: Skin disease classification from image-a survey. In: 2020 6th International Conference on Advanced Computing and Communication Systems (ICACCS). IEEE (2020)
12. Bhadula, S., et al.: Machine learning algorithms based on skin disease detection. Int. J. Innov. Technol. Explor. Eng. (IJITEE) **9**(2), 4044–4049 (2019)
13. Arivazhagan, S., et al.: Skin disease classification by extracting independent components. J. Emerg. Trends Comput. Inf. Sci. **3**(10), 1379–1382 (2012)
14. UdriŞtoiu, A.-L., et al.: Skin diseases classification using deep learning methods. Curr. Health Sci. J. **46**(2), 136 (2020)
15. Shanthi, T., Sabeenian, R.S., Anand, R.: Automatic diagnosis of skin diseases using convolution neural networks. Microprocess. Microsyst. **76**, 103074 (2020)
16. Srinivasu, P.N., et al.: Classification of skin disease using deep learning neural networks with MobileNet V2 and LSTM. Sensors **21**(8), 2852 (2021)
17. Malliga, I., Sindoora, Y.: Skin disease detection and classification using deep learning algorithms. Int. J. Adv. Sci. Technol. **29**(3s), 255–260 (2020)
18. Kalchbrenner, N., Grefenstette, E., Blunsom, P.: A convolutional neural network for modeling sentences. arXiv preprint arXiv:1404.2188 (2014)
19. Torrey, L., Shavlik, J.: Transfer learning. In: Handbook of Research on Machine Learning Applications And Trends: Algorithms, Methods, and Techniques, pp. 242–264. IGI Global (2010)
20. Ketkar, N.: Introduction to Keras. In: Deep Learning with Python, pp. 97–111. Apress, Berkeley (2017)
21. Manaswi, N.K.: Understanding and working with Keras. In: Deep Learning with Applications Using Python, pp. 31–43. Apress, Berkeley (2018)
22. Manoharan, J.S.: Study of variants of extreme learning machine (ELM) brands and its performance measure on classification algorithm. J. Soft Comput. Paradigm (JSCP) **3**(02), 83–95 (2021)
23. Tripathi, M.: Analysis of convolutional neural network based image classification techniques. J. Innov. Image Process. (JIIP) **3**(02), 100–117 (2021)
24. Vijayakumar, T.: Comparative study of capsule neural network in various applications. J. Artif. Intell. **1**(01), 19–27 (2019)

25. Bashar, A.: Survey on evolving deep learning neural network architectures. J. Artif. Intell. **1**(02), 73–82 (2019)
26. Vijayakumar, T.: Posed inverse problem rectification using novel deep convolutional neural network. J. Innov. Image Process. (JIIP) **2**(03), 121–127 (2020)
27. Karuppusamy, P.: Building detection using two-layered novel convolutional neural networks. J. Soft Comput. Paradigm (JSCP) **3**(01), 29–37 (2021)

Medchain for Securing Data in Decentralized Healthcare System Using Dynamic Smart Contracts

R. Priyadarshini[1] (✉), Mukil Alagirisamy[2], and N. Rajendran[3]

[1] Computer Science Engineering, Faculty of Engineering, Lincoln University College, Malaysia, Malaysia
rspdarshini@gmail.com

[2] Department of Electrical and Electronics Engineering, Faculty of Engineering, Lincoln University College, Malaysia, Malaysia
mukil.a@lincoln.edu.my

[3] Department of Information Technology, School of Computer Information and Mathematical Sciences, Chennai, India

Abstract. In today's era the current medical application accepts and analyze the patient medical data. Developing a system to acquire data that can "Efficient Data Security System for Storing Medical system data". The medical data are obtained by a user, Lab technician (for X-ray, blood test), and hospital. In the proposed research, security is one of the component in optimization of services using Cyber Physical System. The digital records are accessible only to the authorized persons. This paper consists of a medical application that accepts and analyze a patient's medical data to give a pre-diagnosis to check if he/she is having specific disease, ensuring to efficient data security system to secure the patient's medical data stored and accessed in the cloud. The paper aims to design and develop a Block Chain based Security Model for smart healthcare systems to enable Confidentiality, Integrity, Authentication and Authorization for the medical data.To propose the Dynamic Smart Contract Ledgers (DSMCL) for Confidentiality and Integrity handling Sensitive data of patient and DSMCL for Authentication and Authorization handling medical Images(scan, X-ray). The outcome of this paper is to secure the medical data and image using blockchain and smart contract technology. Only the authorized users will be allowed to access patient detail after validation and verification during emergency situation using Key generation which holds individual patient data inside a block. Finally, the accuracy of the execution is compared with the algorithms not involving blockchain and the proposed system is much better than the conventional ones.

Keywords: Blockchain · Medical data · Sensitive data · Proof of work · Dynamic smart contracts · Machine learning · Conseus algorithm · Pre-diagonisis of medical data

J. I.-Z. Chen et al. (Eds.): ICIPCN 2022, LNNS 514, pp. 574–586, 2022.
https://doi.org/10.1007/978-3-031-12413-6_45

1 Introduction

The security is very much essential for transactions that are done in medical and health care domain. The ledger which is not changebale and facillitates the transaction in an commercial network is called dynamic ledgers in Block chain. The smart contracts are service level agreements between two people and it is executed as the code in the blockchain network [1]. Mostly these agreements are implemented based on the business requirements. The significance of the blockchain is that it tis decentralized [2]. These smart agreements disallow the multiple user to tamper or use the data without any changes. Healthcare data security is an important feature of "Health Insurance Portability and Accountability Act of 1996 Rules" (HIPAA). In recent days personal records of medical data has been hacked, fraud or theft is been identified (more than 750 data breaches occurred in 2021). The data stored in the block chain is The block chain consists of blocks which are defined as follows.. The Every block contains several transactions, every transaction is recorded in the ledger. The multiple records are stored as block in a chain and it is monitored by several authorized members. The data in blockchain is decentralized and they have fixed security based identificationa and it is called as block.

Data insecurity leads to loss of confidential data [3] as the key is easily hackable because of a single algorithm applied.This paper focus on Efficient Data Security System where the security is ensured with the help of the algorithms in the smart contracts. When the conditions are staisfied in the smart contracts then the authenticated contacts are verified with the digital services that provide secured implementation, monitoring and control. The transactions of the smart contracts are executed in the block chain platform. The blockchain network consist of many mediators and middle level people who can access the transactions and they are not part of the execution. The significance of the block chain is that any authenticated and authorized person can implement the transactions in the block chain platform. The user data are stored in the cloud by integrating: Confidentiality, Integrity, Authentication, and Authorization. The user data are kept under the control of the user in the cloud. In case of a medical emergency, the records of the user can be accessed by using the "Authentication Key" generated from the user where the records are securely maintained using Smart contracts in blockchain technology.

The data security is performed by using Triple level encryption algorithm [4] and stored via cloud. The data relating to a patient are maintained in terms of confidentiality and integrity. This triple level encryption algorithm (3 DES) it increases the cost of execution time for securing the data, vulnerable to collision attacks and has been instructed by NIST to deprecate its usage by 2023. This application is designed to accept only specific disease data where us other data are not managed and maintained. The proposed work is based on the dynamic smart contract ledgers where the machine learning algorithm is used to choose any of the conseus algorithm.

2 Related Work

In the paper proposed by Omolara [5], discussed about the health sectors and sensitive medical data upload and accessibility. The security is ensured in terms of blocks in the healthcare sector. The blockchain platform uses Dapps which are used in ethereum platform. The data in the blockchain platform is not changeable and it can also not be deleted. There are many D-Apps available to access the data online, the clinical managment and the common man is using this D-App method to access the data. The limitations include that the brute force attack can be used to crack the application.

In the previous leiterature proposed by Junqin Huang et al. [6] the Smart contracts and its usecases are automated based on the transactions. The sectors used for the smart contracts includes textile industry. The activities and operations done in the smart contracts are automatic. Blockchain is used in many fields like communication, health science, malicious attack based applications. The digital transcations used in the fields are automatic and machine learning algorithm can be used to improversie the same. The limitations are the privacy and trust are not incorporated along with CIAA.

In the work of Kai Fan [7], The Light weight RFID protocol in IOT is discussed and the privacy using that protocol is emphazized.

In the work of Massoud Masoumi [8], he discussed about the medical image encryption and decryption with deep-key gen and deep learning based stream cipher. The benefits of the implementation includes the generation of the private key which is used for providing security.

Mauro Mangia et al. [9], discussed about security in IoT sensing and a blockchain inspired approach was given to it.

In the paper published by Xiaoning Liu et al. [10], they have proposed the privacy preserving analytics on the medical time series. The paper focuses on the medical encoding of health-related data which is given to the distributor. The paper focus on the medical aid for encoding and decoding the health records. The deployment of medical analytics is due to privacy corcenrs.

In the work discussed by Yang Yang, Robert H. [11], has privacy-preserving health records based on the finite automation. In the model of encoding the information with the P-Med, the confidentiality, integrity and authentication is used for the full security of the medical records to avoid privacy leakage to the cloud server.

The medical privacy protection in IoT is used is discussed and implemented in this paper by Kai Fan, Light wieght RFID protocol is used for the privacy protection [12]. The lightweighted security and privacy of collected data via secure authentication is done using the current messages.

In the above literature survey, the following strategeis were discussed:

1. Privacy protection of the data with the help of encryption and decryption.
2. Maintenence of confidentiality, integrity and authentication.
3. The encoding model also helps to avoid leakage.
4. The medical images are encoded and decoded with the help of the algorithms.

The limitations of the previous literature includes, there is a need for the security framework which comprises of the customized encryotion and decryption or SHA techniques that are incorporated in the smart contracts itself. There is also need for triple level security which is embedded inside the blockchain algorithms. The encryption techniques are also done using generative adversarial networks and Deep Key Gen algorithms as shown below.

3 Existing System

The issues in the exsiting system are marked in the system which has triple level security and the GAN which used Deep Key Gen algorithm. The encryption and decryption of the keys and the security of the medical images are implemented using the Deep Key Genration and Triple level security with SHA. Even though the advance tecniques are used, the security is breached in terms of brute force attacks. The disadvantage of the approximate error finding method consumes more time and generate less accurate results [8]. In the case of the Deep gen the plot x, y in the image is secured and it is iterated several times to predict the error and correct the same. The estimation of DeepKeyGen using three data sets, i.e., Montgomery County chest X-ray, Ultrasonic Brachial Plexus, and BraTS18. The estimation of results and security analysis exhibit the suggested key epoch grid can conduct high-level protection in producing the private key. The exisiting system data and images are taken for the comparison in the proposed paper. The novelty in the propised system is choosing the conseus algorithm dynamically using Random selection technique and the storage of text, data and image are.

4 Medchain Based Secured Framework

Healthcare data protection is a vital element of HIPAA Rules. Additionally, there were 750 data violations occurred in 2021, the maximum of seven of which extended over 193 million confidential documents to forgery and originality stealing. Data security is the method of safeguarding data from prohibited permits and information crime through its life process [9]. An Efficient Blockchain based Security Model Enabling Confidentiality, Integrity, Authentication, and authorization is designed. The sensitive information of the users is stored secretly which enables privacy in turn confidentiality [10]. The data cannot be tampered with by the hacker by any means which provides integrity which is done by the inbuilt blockchain framework based on CIAA.

The access level security and availability of data only for the authorized and authenticated individuals in the framework are provided. The above-mentioned CIAA-based security model is developed and implemented using smart contracts and enhanced by predictive analysis [11]. No encryption or decryption key is required for accessing data which eliminates the fear of data being hacked or tampered with by the hacker [12]. Thus, this system provides an end to end data security to the medical records of a hospital.Interpolation and Storage of data using Web application is done and in this module, the data has been entered, stored and retrieved from the patient using Web application. These data's are entered using web 3.0 and truffle framework. The data's are validated using Ethereum service provider. Those validated data set are stored has EHR (Electronic health record). The health record storage and security is shown with the help of the truffle framework (Fig. 1).

The authorised user can access and alter the data of their health status in their medical data [13]. Data cannot be tampered by any third party nor data modification can be performed except the authorised user. In this access level of data security is essentially provided to the blockchain database. The ethereum virtual machine is accessed using the truffle framework [14]. The hyperledger of the ehtereum and supported platforms are

Fig. 1. Interpolation and storage of data using Web application

Fig. 2. Interpolation and storage of data using Web application

used to connect the d-app with the web applications. The web front end user interface are integrated via truffle framework. The user interface is integrated with the truffe framework sources which are both transformative and informative [15]. The web 3.0 front end is used for precise connectivity. The framework is semantic and intelligent in nature and it accerelates the modbile devices. It converts mere informative source to interactive source. The overall architecture is shown in the Fig. 2.

Consensus is the crux of the blockchain and its primary goal is to acquire contracts on approved commerce among a distributed method [16]. Consensus algorithm is a protocol via which all the forces of the blockchain network come to a standard deal (consensus) on the current details state of the register and be able to trust unfamiliar peers in a distributed computing environment. The proof of work algorithm is customized to check the validity of the crypto rewards and validate the customized configuration [17]. It secures the image and text data to a greater extent. The web3.0 front end development user interface is given in the Fig. 3a.

Fig. 3a. Web 3.0 frontend development

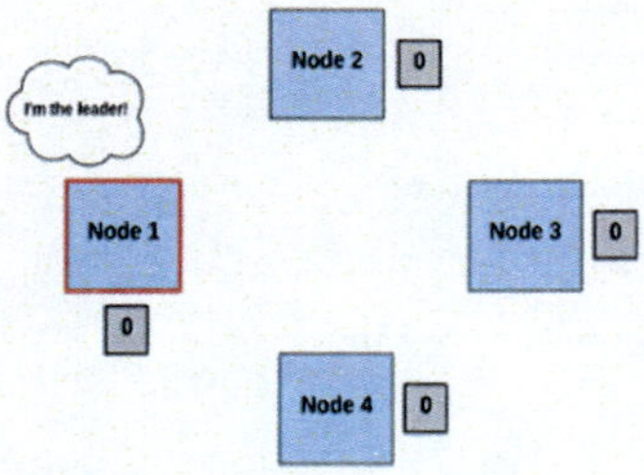

Fig. 3b. Consensus algorithm

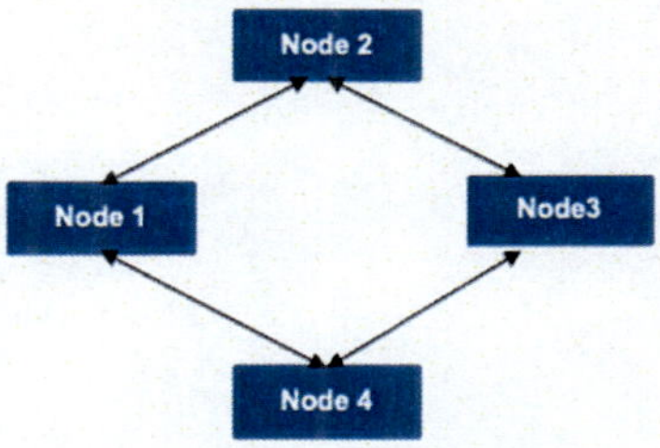

Fig. 3c. Consensus algorithm

4.1 Smart Contract Key Generation

The private key generator generates the random key automatically. It also generates private key that is "Login Key" for accessing medical records of patients. Therefore, the patient record is been authenticated and validated by the web application and then the data can be stored inside the blockchain [18]. The smart key generation and the implementation of conseus algorithm is shown in the Figs. 3b and 3c (Fig. 4).

Fig. 4. Smart contract key generation.

4.2 Metamask Integration

MetaMask is a Chrome extension that lets in customers to apply Ethereum-enabled programs and web sites [19]. Metamask acts as a token pockets to soundly control identities and to save and ship a person's Ether. It additionally lets in web sites and programs to connect with the Ethereum blockchain. It handles account control and connecting the person to the blockchain. The metamask integration is shown in Fig. 5.

Fig. 5. Metamask integration

The Metamask integration is done with the verification of the signature and authentication of signatures.

4.3 Creation of Smart Contracts with Heuristics Rule Based Conditions in the Web Application

The contracts rules are written inside the web application code. When the authorized user access or store the data, it is considered as a transaction and updated in the blockchain network. The records are secured by the blockchain network (Fig. 6).

Fig. 6. Creation of smart contracts with heuristics rule based conditions

4.4 Smart Contract Development

A smart contract is a contract between two or more groups that are accumulated on a blockchain, such as Ethereum. Every such agreement has a predefined group of laws and conditions and is automatically run when those conditions are met [20]. These agreements are secured by the consensus of the entire blockchain web. This makes smart contracts one of the securest and most powerful tools for making agreements between different parties. In this paper, Solidity language is used in smart contract development. The overall outline is shown in the Fig. 7.

Fig. 7. Smart contract development

4.5 Development of Intelligence Service for Information Sharing Using Service

The patient data are entered through web application and accessed using frame work [21]. The credential are required for access is Login Key generated by user, it will be authenticated and verified by the Service Provider (Ethereum). The data is maintained as EHR (Electronic Health Record) in the cloud (Fig. 8).

Fig. 8. Development of intelligence service for information sharing using service provider

5 Performance Analysis

The need to secure the data and image is much needed in the todays health care industry, in the case the medical X-ray reports and medical imaging, there are lot of chances that name of the patient and report id may be interchanged or there may be chances of leakage of data. In this cases there are many conventional methods used to preserve the security, the cipher text image and histogram of the cipher text is processed. Even though the error in the encryption and decryption is identified using GAN, the high level security is achieved with the help of the meta masks and the DSMCL algorithms.

It is necessary to secure the sensitive text data and the image, the text data in medical health is leaked and it leads to unnecessary marketing calls and messages. At the same time, the sensitive information leakage leads to many risky situations for the patient and the hospitals [22]. This Fig. 9 represents a) Montgomery County chest X-ray data set, (b) Ultrasonic Brachial data set., and (c) BraTS18. The training is done for different epochs and the data related to the different sizes and the rate of machine learning are notes and keys are generated.

Fig. 9. Histogram analysis of plaintext images and corresponding cipher text images

Table 1. Study of network hyper parameters on validation data set

Epoch	LR = 0.02			LR = 0.002			LR = 0.0002		
	BS = 1	BS = 6	BS = 10	BS = 1	BS = 6	BS = 10	BS = 1	BS = 6	BS = 10
10000	0.5222	3.7878	3.8766	5.1698	5.9714	6.1303	7.9380	7.7889	7.9004
15000	NaN	3.7878	3.8766	4.6602	2.6587	5.7981	7.9405	7.7093	7.1990
20000	NaN	3.7878	3.8766	5.4456	2.7983	6.1169	**7.9798**	7.9533	7.0686
25000	NaN	3.7878	3.8766	5.1133	2.4782	5.6927	7.9490	7.1992	7.6847
30000	NaN	3.7878	3.8766	7.9003	3.1811	6.6174	7.9555	6.7871	7.4846

The best performance in the above Table 1 is maintained based on the batch size of 10 and the private key generation is implemented with the metamask and the DSMCL algorithms. The customized proof of work concepts helps the DSMCL to outperform GAN method. In the case of GAN, every time the epochs are executed in the particular period of time. The data stored as blocks showcases high rate of accuracy compared to any of the algorithms. However the triple level security gives a encrypted private key for the images. The medical report considered is referred to as (Mont-gomery County chest X-ray data set, Ultrasonic Brachial data set., and BraTS18), it consumes excess time for execution.

Fig. 10. Comparison of GAN technology with blockchain technology

In the Fig. 10 given above, the accuracy in securing the data is executed and showcased. Since the AES algorithm with SHA and the triple level security along with GAN was a conventional method and it dint give security as the proposed one. The customization of the security is configured in the Dynamic Smart Contract Ledger (DSMCL). The algorithms in the proof of works is modfied for retrieval of data and images in batches. In this case the new transactions were updated to the other mediators in the network. In the graph as the time increases the accuracy level as increases.

6 Conclusion

The implemnetation in this paper has been successfully created a Medchain system with which the data of the patient can be securely viewed using any hospital and labs with high security with the permission of the patient where Ethereum blocks play a major role. An Efficient Data Security System for where Smart Contracts and WEB 3.0 are used to secure the data and it provides complete data security where the data cannot be tampered with by the hacker. A framework called WEB 3.0 is developed which has an inbuilt blockchain framework to secure the data at the backend.

The proposed framework comprises of security model and enhanced predictive analyses to store the current status disease in the smart contracts. The proposed system ensures building of smart contract system for secure healthcare system using Blockchain technology.

Thus, this paper successfully provides an end to end data security to the medical records of a hospital. The proposed method is better in terms of accuracy than the conventional methods as shown in the graph in Fig. 10. The image storage is intiated with the blockchain and the proposed deep learning algoirithm and segmentation will be done as the future work.

References

1. Davi, C., Pastor, A., Oliveira, T., de Lima Neto, F.B., Braga-Neto, U., Bigham, A.W.: Severe dengue prognosis using human genome data and machine learning. IEEE Trans. Biomed. Eng. **66**, 2861–2868 (2019)
2. Zhou, C.: Comments on "light-weight and robust security-aware D2D-assist data transmission protocol for mobile-health systems". IEEE Trans. Inf. Forensics Secur. **13**, 1869–1870 (2018)
3. Abbasinezhad-Mood, D., Nikooghadam, M.: Efficient design of a novel ECC-based public key scheme for medical data protection by utilization of NanoPi Fire. IEEE Trans. Reliab. **67**, 1328–1339 (2018)
4. Huang, H., Gong, T., Ye, N., Wang, R., Dou, Y.: Private and secured medical data transmission and analysis for wireless sensing healthcare system. IEEE Trans. Ind. Inform. **13,** 1227–1237 (2017)
5. Omolara, O., Jantan, A., Abiodun, O., Arshad, H., Dada, K., Emmanuel, E.: HoneyDetails: a prototype for ensuring patient's information privacy and thwarting electronic health record threats based on decoys. Health Inform. J. **26**, 146045821989447 (2020). https://doi.org/10.1177/1460458219894479
6. Huang, J., Kong, L., Chen, G., Wu, M.-Y., Liu, X., Zeng, P.: Towards secure industrial IoT: blockchain system with credit-based consensus mechanism. IEEE Trans. Ind. Inform. **15**, 3680–3689 (2019)
7. Fan, K., Jiang, W., Li, H., Yang, Y.: Lightweight RFID protocol for medical privacy protection in IoT. IEEE Trans. Ind. Inform. **14,** 1656–1665 (2018)
8. Masoumi, M.: Novel hybrid CMOS/memristor implementation of the AES algorithm robust against differential power analysis attack. IEEE Trans. Circuits Syst. II Express Briefs **67,** 1314–1318 (2020)
9. Mangia, M., Marchioni, A., Pareschi, F., Rovatti, R., Setti, G.: Chained compressed sensing: a blockchain-inspired approach for low-cost security in IoT sensing. IEEE Internet Things J. **6**, 6465–6475 (2019)
10. Wang, S., Ouyang, L., Yuan, Y., Ni, X., Han, X., Wang, F.-Y.: Blockchain-enabled smart contracts: architecture, applications, and future trends. IEEE Trans. Syst. Man Cybern. Syst., **49,** 2266–2277 (2019)
11. Rao, V.S.H., Kumar, M.N.: A new intelligence-based approach for computer-aided diagnosis of dengue fever. IEEE Trans. Inf. Technol. Biomed. **16**, 112–118 (2012)
12. Yang, W., Dai, X., Xiao, J., Jin, H.: LDV: a lightweight DAG-based blockchain for vehicular social networks. IEEE Trans. Veh. Technol. **69,** 5749–5759 (2020)
13. Liu, X., Zheng, Y., Yi, X., Nepal, S.: Privacy-preserving collaborative analytics on medical time series data. IEEE Trans. Dependable Secure Comput.
14. Yang, Y., et al.: Privacy-preserving medical treatment system through nondeterministic finite automata. IEEE Trans. Cloud Comput.
15. Kamil, Y.M., Bakar, M.H.A., Yaacob, M.H., Syahir, A., Lim, H.N.: Dengue E protein detection using a graphene oxide integrated tapered optical fiber sensor. IEEE J. Sel. Top. Quantum Electron. **25,** 1–8 (2019)
16. Smith, T.F., Waterman, M.S.: Identification of common molecular subsequences. J. Mol. Biol. **147**, 195–197 (1981)
17. May, P., Ehrlich, H.-C., Steinke, T.: ZIB structure prediction pipeline: composing a complex biological workflow through web services. In: Nagel, W.E., Walter, W.V., Lehner, W. (eds.) Euro-Par 2006. LNCS, vol. 4128, pp. 1148–1158. Springer, Heidelberg (2006). https://doi.org/10.1007/11823285_121
18. Foster, I., Kesselman, C.: The Grid: Blueprint for a New Computing Infrastructure. Morgan Kaufmann, San Francisco (1999)

19. Czajkowski, K., Fitzgerald, S., Foster, I., Kesselman, C.: Grid information services for distributed resource sharing. In: 10th IEEE International Symposium on High Performance Distributed Computing, pp. 181–184. IEEE Press, New York (2001)
20. Foster, I., Kesselman, C., Nick, J., Tuecke, S.: The physiology of the grid: an open grid services architecture for distributed systems integration. Technical report, Global Grid Forum (2002)
21. National Center for Biotechnology Information. http://www.ncbi.nlm.nih.gov
22. Ding, Y., Tan, F., Qin, Z., Cao, M., Choo, K.-K.R., Qin, Z.: DeepKeyGen: a deep learning-based stream cipher generator for medical image encryption and decryption. IEEE Trans. Neural Netw. Learn. Syst.
23. Smys, S., Wang, H.: Security enhancement in smart vehicle using blockchain-based architectural framework. J. Artif. Intell. **3**(02), 90–100 (2021)
24. Sivaganesan, D.: Performance estimation of sustainable smart farming with blockchain technology. IRO J. Sustain. Wirel. Syst. **3**(2), 97–106 (2021)
25. Kruthik, J.T., Ramakrishnan, K., Sunitha, R., Prasad Honnavalli, B.: Security model for Internet of Things based on blockchain. In: Raj, J.S., Iliyasu, A.M., Bestak, R., Baig, Z.A. (eds.) Innovative Data Communication Technologies and Application. LNDECT, vol. 59, pp. 543–557. Springer, Singapore (2021). https://doi.org/10.1007/978-981-15-9651-3_45

Multipurpose Linux Tool for Wi-Fi Based Attack, Information Gathering and Web Vulnerability Scanning Automations

Ceronmani Sharmila, J. Gopalakrishnan(✉), P. Shanmuga Prasath, and Y. Daniel

Department of Information Technology, Hindustan Institute of Technology and Science, Chennai, India
csharmila@hindustanuniv.ac.in, gopalakrishnan567890@gmail.com

Abstract. Manual processes of various attacks or scanning are huge time killers for penetration testers and students. Penetration testers, Bug bounty hunters have their own way of doing reconnaissance and penetration testing, for these phases the crowd has to surf through lots and lots of tools and applications in order to deliver a fully-fledged bug report or to identify a bug. Providing a tool for automating those processes would be a great thing for the users. The uniqueness of this tool is accessibility, Accessibility in the sense even though CLI has no attractive element to make the user more engaged to the interface; This Linux Command Line Interface (CLI) tool would satisfy the users by providing automation techniques used for Wi-Fi based attacks and Web application vulnerability scanning. Also, this tool has Open Source Intelligence (OSINT) for information gathering.

Keywords: Scripting · Command-Line-Interface · OSINT · Penetration testing

1 Introduction

Linux is widely known for the most used Operating System (OS) it bundles various Program, tools and services. However operating Linux via CLI is the most used operation among the programmers and security researchers. Penetration testers, Bug bounty hunters have their own way of doing reconnaissance and penetration testing, for these phases the crowd has to surf through lots and lots of tools and applications in order to deliver a fully-fledged bug report or to identify a bug. This tool (PENTROSINT) will bring the most possible techniques and processes required for Information gathering and pen testing by integrating various automations of enormous manual processes. The uniqueness of this tool is accessibility, Accessibility in the sense even though CLI has no attractive element to make the user more engaged to the interface; PENTROSINT brings the user-friendly interface which can be accessible to a person with a basic amount of programming knowledge. There are three modules in PENTROSINT which are reconnaissance, web application scanning and Wi-Fi Based attacks. In reconnaissance module

J. I.-Z. Chen et al. (Eds.): ICIPCN 2022, LNNS 514, pp. 587–598, 2022.
https://doi.org/10.1007/978-3-031-12413-6_46

information gathering techniques like Social media hunting using image, Tracing single IP, IP heat map, URL redirection checker, PDF metadata analysis, URL lookup in Web Pages, Information gathering using name, Phone number verifier and Open Source Intelligence for Instagram. For Web Application scanning techniques like SQL Injection, Clickjacking, Host header injection, Subdomain Enumeration, Reverse IP. And finally Wi-Fi attacks modules embed DE authentication attack and Evil Twin Attack.

2 Related Work

In our search for relevant works, we came across a few OSINT tools, which are divided into two categories based on whether they are available through command line interface (CLI) or graphical user interface (GUI). Further investigation of these CLI and GUI applications revealed that they are designed to automate the search and collecting of publicly available data, with some offering both manual and automated capabilities [1]. Photon is a Python-based crawler that surfs the Web for related OSINT information. Fit for investigating existing sites or those chronicled through administrations, for example, the Way back Machine (https://web.archive.org), Photon can recuperate information, for example, JavaScript documents, email addresses, online media information, passwords, and API keys connected with an objective. The short-sighted, adjustable nature of Photon makes it a strong robotized device for OSINT investigators.

SpiderFoot, a Python programme, provides a Web interface for querying multiple public databases for target information, such as IP addresses, area names, email addresses, and usernames [2]. Despite the fact that SpiderFoot is mostly accessed through software, it also has an order line interface. SpiderFoot improves the visual presentation of obtained data by using several modules to execute port sweeps, language ID, and screenshotting. SpiderFoot, like Maltego, includes both a restricted local area release and a strong, commercial form. Arun S, Bijimol TK's Research work [3] has concluded that NMAP is the most effective and powerful for the Penetration testers. Even though the utility of NMAP is wide, getting to know the processes and techniques seems to be a bit complicated.

2.1 SQL Injection

ART4SQLi [4] presented a new syntax for interested SQL injection attack payloads, and it was aimed to speed up the testing method for managing assault payloads to find SQLi flaws. ART4SQLi decomposed each payload string into tokens and represented each payload as an element vector at first. Following that, ART4SQLi identified a promising payload for evaluation by randomly creating a size-fixed applicant set from the payload collection and selecting the one that was the furthest away from each of the assessed payloads. When a payload discovered a SQLi vulnerability problem, we marked the payload as successful and completed the interaction; in any event, the assessed set of payloads was updated.. These techniques for payload control methodologies are motivated from this paper [2] to assemble a strong Pen testing apparatus for actually looking at SQL infusion on a Web Application.

2.2 Wi-Fi Module

This Module has two goals, one is to get users getting excited about hacking and getting their hands on hacking and let them play with routers and secondly it teaches the users about security. Knowing these possibilities of Wi-Fi based attacks will let the users be aware of risk in using public Wi-Fi. The attack which this module is going to use is an evil twin attack which is as scary as it sounds.

[5] Dr. Dwivedi's paper conveys the method to perform a Wi-Fi DE authentication attack manually using kali Linux using aircrack-ng. The only drawback in these methods is transmitting packets using aircrack-ng may sometimes won't work for routers with multiple transmitting cords.

3 Methodology

3.1 Adaptive Random Testing (ART)

ART stands for "Adaptive random testing". It's because it's widely assumed that trials revealing programming flaws will tend to cluster together in the experiment space. It so proposes that by exploiting the area, haphazardly chosen experiments be distributed more consistently throughout the data region. This paper has been accepted for publication in a subsequent issue of this journal. Except for pagination and statistics from tests that have been completed but have not shown disappointments [6] the content is last as introduced. In this strategy, we take such a stance in order to focus on the SQLi vulnerability discovery issue. Craftsmanship makes use of the space between experiments to allow the test Case contestant to quickly examine the main experiment that leads the question. To register the contrast between experiments, a variety of distance metrics have been proposed. The Euclidean distance is based on the spatial distance between the examples rather than the differences in the qualities of the examples. Instead of focusing on the spatial distance between the instances, the cosine distance focuses on the differences in example qualities. Because payloads are addressed by vectors in a symbolic space, and there is no logical premise to govern diverse aspects of such a space, we chose the last alternative in our study. For object-situated programming (OOS) experiment priority, J. Chen et al. [6] presented a versatile arranging strategy. Because of the ART presentation's distance computation error, J. Chen et al. [6] concentrated. The way inclusion data for ART is collected has been dissected.

3.2 Docker

We can construct an image and execute the application as a container once Docker is installed.

Build: docker build –t osintgram

Docker run –rm –it –v "$PWD/output:/home/osintgram/output" osintgram

We'd like to use the Instagram account target> as our recon target.

The -i flag allows an interactive terminal to use container commands.

On completion, the -v flag mounts a volume between your local file system and prevents cruft buildup. To avoid cruft build-up, the—rm command destroys the container file system after completion.

The -t flag creates a pseudo-TTY for colored output.

Docker Head Start: The osintgram module has an additional feature called Docker quick start. All we need is to install docker and access the module through docker no need to install each and every requirement that is required to run the tool.

Before we access the tool docker needs to be installed and configured properly. These 3 following prerequisites need to be completed.

1. Docker must be installed.
2. Install Docker-composed (if using Docker-compose).
3. Configure credentials—this can be done manually or by using the "make setup" command from the repo's root.

Important: If you don't complete step 3 and configure the credentials, your container will fail.

4 Framework

Fig. 1. Framework of pentrosint.

The above framework shows the architectural flow of PENTROSINT completely. The tool starts with the python generated welcome audio, listing the 3 available modules in the tool to the user. The user has to navigate through the features of the tool just by entering the numerals. Rather memorizing the commands and processes (Fig. 1).

Pseudocode:

```
Main Page:
Start
print (
            Available Modules

        1. Information gathering,
        2. Web vulnerability scanning,
        3. WiFi-Deauthentication
   )
   Print (Note: In Information gathering type 'tools' to find tools.)
   print (Note : In Web vulnerability scanning type 'help' to find tools.)
   a= User input
   Main(a)
          |
          |
definition Main(a):
   if(a=1):
     To  reconinput
   elif(a=2):
     To  Webvuln
   elif(a=3):
          Execute WiJAMMER.sh
End
```

Fig. 2. Main interface

Figure 2 is the main interface of the tool which lists 1. Information gathering, 2. Web vulnerability scanning, 3. Wi-Fi Deauth.

```
Recon input:

definition reconinput:
   inp=User input
   if(inp = 1):
     Execute recon
   elif (inp=2):
      Execute iplocate
   elif(inp=3):
      Execute read_multiple_ip
   elif(inp =4):
      Execute urlinfo
   elif (inp=5):
      Execute pdfinfo
   elif(inp=6):
      Execute Links
   elif (inp=7):
      Execute Nameinfo
   elif (inp=8):
      Execute number
   elif (inp=9):
           Run ('python3main.py [TARGET'S USERNAME]
   Elif (inp=exit):
      exit
   elif(inp=tools):
```

Fig. 3. Reconnaissance module

Figure 3 shows the available reconnaissance module on PENTROSINT. Each modules are independent and has its own purpose.

5 Analysis

This analysis section lists and describes the effectiveness of the tool with respect to the manual process of each tool by its time taken to complete that specific task manually and with PENTROSINT.

Information Gathering:

Fig. 4. Manual reconnaissance time graph

The above displayed time graph Fig. 4 pictures the time taken for a few steps in the reconnaissance process like Social media hunting using image, IP tracing, URL redirection checker, Phone number verifier, and Instagram OSINT operations manually.

Fig. 5. Automated reconnaissance time graph

The above displayed time graph Fig. 5. pictures the time taken for a few steps in a reconnaissance process like Social media hunting using image, IP tracing, URL redirection checker, Phone number verifier, and Instagram OSINT operations with the help of PENTROSINT.

As the graph Fig. 6 shows the red streak depicts manual process and blue streak depicts automated process from PENTROSINT. From the graph itself we can find the huge time difference between the two processes. Therefore manual processes for the above mentioned tasks are the heavy time killers whereas automated ones barely reached 7 s as an average.

Fig. 6. Reconnaissance PENTROSINT versus. Manual time graph

Web Vulnerability Scanning

Fig. 7. Pentesting auto versus manual time graph

The above time graph depicts the huge time taken difference to perform each task in the pen testing module (Fig. 7).

Wi-Fi DE Authentication

See Fig. 8.

Fig. 8. Wi-Fi module auto versus manual time graph

6 Conclusion and Future Works

This part brings the article to a conclusion by listing some upcoming projects that will be implemented on PENTROSINT. We have made a significant contribution, including reconnaissance, Web vulnerability scanning, and Wi-Fi-DE authentication, in light of the work's goal. As a result of examining the work's outcome, we can draw certain conclusions.

- The average time required to perform each and every listed task varies with a huge time gap.
- Future work for this paper will be implementing these tools in various different platforms mainly focused in Windows because of the vast usage.

References

1. Sangwan, S.: Photon GitHub project page. Photon, 06 December 2019. Photon(GitHub). Accessed 20 Apr 2018, 2020

2. SM7 Software OÜ: SpiderFoot Homepage. SpiderFoot (2020). https://www.spiderfoot.net/. Accessed 18 Apr 2020
3. Arun, S., Bijimol, T.K.: A research work on information gathering tools. In: National Conference on Emerging Computer Applications (NCECA), vol. 3, no. 1 (2021)
4. Zhang, L., Zhang, D., Wang, C., Zhao, J., Zhang, Z.: ART4SQLi: the ART of SQL injection vulnerability discovery. IEEE Trans. Reliab. **68**(4), 1470–1489 (2019)
5. Joshi, D., Dwivedi, V.V., Pattani K.M.: Deauthentication attack on wireless network 802.11i using kali Linux. IRJET **04**(01), 1666–1669 (2017)
6. Chen, J., et al.: An adaptive sequence approach for OOS test case prioritization. In: Proceedings of the IEEE International Symposium on Software Reliability Engineering Workshops, pp. 205–212 (2016)
7. Ibarra-Fiallos, S., Higuera, J.B., Intriago-Pazmiño, M., Higuera, J.R.B., Montalvo, J.A.S., Cubo, J.: Effective filter for common injection attacks in online Web applications. IEEE Access **9**, 10378–10391 (2021)
8. Pingle, B., Mairaj, A., Javaid, A.Y.: Real-world man-in-the-middle (MITM) attack implementation using open source tools for instructional use. In: 2018 IEEE International Conference on Electro/Information Technology (EIT) (2018)
9. Sagar, D., Kukreja, S., Brahma, J., Tyagi, S.: Studying open source vulnerability scanners for vulnerabilities in Web applications. IIOAB J. **9**, 43–49 (2018)
10. Wright, T., Whitfield, S., Cahill, S., Duffy, J.: Project umbra. In: IEEE International Conference of Advances in Social Networks Analysis and Mining (ASONAM) (2020)
11. Santhi, V., Raja Kumar, K., Vinay Kumar, B.L.V.: Penetration testing using Linux tools: attacks and defense strategies. Int. J. Eng. Res. Technol. **5**(12) (2016)

Customer Engagement Through Social Media and Big Data Pipeline

Rubeena Rustum[1](✉), J. Kavitha[2], P. V. R. D. Prasada Rao[3], Jajjara Bhargav[4], and G. Charles Babu[1]

[1] Gokaraju Rangaraju Institute of Engineering and Technology(Autonomous), Bachupally, Hyderabad, Telangana, India
rubeena.rustum@gmail.com

[2] BVRIT Hyderabad College of Engineering for Women, Hyderabad, Telangana, India

[3] Koneru Lakshmaiah Education Foundation, Vaddeswaram, Guntur, Andhra Pradesh, India

[4] Chalapathi Institute of Engineering and Technology (Autonomous), Lam, Guntur, Andhra Pradesh, India

Abstract. Engagement of customers through social media has gained considerable popularity in recent years in the field of digital marketing. Especially with the rise of technological revolution in business operations, utilizing sophisticated technology for strategic development of businesses has been seen. In this regard, data pipeline can be considered as an efficient, automated and sophisticated technology that uses a systematic data management process for voluminous data. The paper thus aims to investigate the beneficial scope of aligning social media with data pipeline technology for enhancing customer engagement. Through an empirical analysis of existing secondary resources as a viable and beneficial method for research, the study has developed a comprehensive understanding of data pipelines and social media platforms contributing to the enhancement of customer engagement.

The main findings of the study indicates that through the ETL pipeline (Extraction–Transformation–Loading), large columns data is managed sequentially and swiftly. The automated system can be used with both flexibility and control to manage the data flow. The businesses are able to control the data flow to their advantage and increase visibility and interaction. Simultaneously the analytics process of the big data aids the decision-making process that ensures customer behavior and market demands are being considered accurately. The study also considers certain challenges such as lack of increased storage capacity, high volume data, and consistency on that can be addressed through further development of advanced architecture.

Keywords: Data pipeline · ETL pipeline · Big data management · Big data analytics · Social media · Social media marketing · Digital marketing · Customer engagement

1 Introduction

Social media platforms have been recognized as one of the most effective and engaging platforms through which various industries promote products and services or simply

J. I.-Z. Chen et al. (Eds.): ICIPCN 2022, LNNS 514, pp. 599–608, 2022.
https://doi.org/10.1007/978-3-031-12413-6_47

establish relationships with their customers. Recent inclination to utilize sophisticated technology for enhancing marketing policies among target demographic has been seen throughout various global industries. Organizations are seen to utilize various data analytics technologies for processing vast quantities of customer data and perform tasks such as online engagement rates, response to new and old products, sharing feedback and so on. These are only surface level operations that are aided through sophisticated data analytics technology such as data pipeline.

Data pipeline is an advanced data processing and analytics technology that is able to retrieve and process data and store it in a data warehouse. The three prime elements in data pipelines consist of a source, processing of data and destination [1]. The apparent simplicity of its process is far more critical in application as it is used by organizations to organize real-time data, process it and store it. Organizations are highly interested in processing large-scale digital information and analyze it for assessing market demands, leading to creation of personalized digital marketing policies. On a global scale, the expenditure for social media marketing is expected to increase 10.93% by 2025 [2]. Especially, with the increased rate of using various mobile applications, it has become a necessity to process and store data of each customer. Hence, the paper will address the issues of large-scale data processing in general and aim at investigating the beneficial use of data pipelines in increasing the rate of customer engagement through social media (Fig. 1).

Fig. 1. Big data pipeline architecture [1]

1.1 Analysis of Data Pipeline Technology

The overall usage of digital information gathered from various social media platforms have grown in variety, volume and velocity. Big data or the enormous volume of data gathered through social media interaction can be processed and stored through dada pipelines [3]. Such a sophisticated technology is able to support big data and has evolved to ensure that data processing is conducted in a clear and logical sequence. The interaction of customers throughout various social media platforms are processed and assessed through predictive analytics [4]. It leads to the enhancement of engagement through proper marketing. In other words, as the organizations monitor the rate of engagement, they are able to understand the direction of market demands and cater towards the personal desires of customers (Fig. 2).

Fig. 2. Three vs of big data [3]

Extraction of big data from its current source location, processing or transforming it into a reliable format. A secured loading or storage process is conducted at the third stage of the data pipeline where the stored data is analyzed through sophisticated machine learning technology [5]. The most beneficial aspect of the data pipeline that is used by business is its ability to extract data at any point of data processing [6]. Hence, the data pipeline technology can be used to provide organizations with the required data for analytics. It not only contributes to saving time of conducting data extraction from the source point alone but also enables companies to save the extracted data. Huge volume of data is then analyzed automatically in order to perceive the market demands and customer attitudes towards a certain product or service.

Data ingestion pipeline is used for the purpose of big data analysis. AI has been used in recent times for the improvement of ingestion practices [7]. It also helps in avoiding redundant loading of the processed data that uses an automated system. In essence, data pipeline technologies are able to provide optimal scope for extracting and processing big data in a systematic manner.

1.2 Benefits of Data Pipeline for Customer Engagement

Data pipelines can be used to optimize the process of big data extraction, transformation and loading. Customer data can be extracted from various applications used by them [8]. As companies grow more and more aware about the behavior and demands of customers, they are able to employ policies for raising social media engagement through effective marketing. It also helps in providing the organizations with ample knowledge regarding their retail sales rates and post engagement, aiding business-related decision-making processes. This information is highly beneficial to the decision-making process as it assists in generating strategies for attaining substantial competitive advantage.

The elements of data pipeline technology help in managing the e-commerce segment of various businesses and increases customer engagement through the accurate data analytics. Using social media analytics tools for analyzing customer behavior helps in the decision-making process [9]. Customers express their feedback and attitudes towards a certain product or service through social media posting, sharing, commenting or liking. Such data is then analyzed to generate personalized, suggestive and attractive marketing policies to further enhance customer engagement. On the other hand, through social media platforms, various organizations enhance their interaction with individual customers [10]. It not only establishes a sustainable and loyal relationship among a brand's targeted customer base but it also ensures that through such direct interactions, market demands are being recognized. In essence, the big data extracted from social media platforms through data pipelines contributes to critical business operations that have significant influence on an organization's manufacturing or production policies and marketing policies. Thus, the use of data pipeline technology for big data processing leading to proper analytics systems helps organizations to engage with their customers in a personalized manner that naturally leads to competitive advantage.

2 Methods and Materials

The paper has conducted an empirical analysis on the topic of social media and data pipeline optimization in customer engagement. Construction of systematic and sequential methodology for the conduction of research enhances scope for logical and valid outcomes [11]. Various secondary resources on the topic had been gathered to analyze in order to meet the aim of the paper. In this regard, a descriptive design to the analysis of gathered resources had been conducted. Descriptive research design helps in the development of a research that is based on the development of knowledge through analyzing existing data or information [12]. Furthermore, a deductive research design had been adopted in order to ensure in-depth analysis. The overall qualitative approach undertaken for this particular paper had helped it to generate new ideas and hypotheses based on multidimensional existing literature on the topic.

The materials used for the study had been gathered from a targeted and systematic search for materials in certain electronic databases such as ProQuest and Google Scholar. Peer-reviewed and published journals, newspaper articles and official reports had been gathered. A systematic sampling method had been adopted as well which consisted of inclusion–exclusion criteria. Setting inclusion–exclusion arteritis assists in discovering only those materials which are beneficial for the course of a given research [13]. The criteria set for this particular study pertaining to the selection of beneficial and informative journal articles had included, peer-reviewed articles, publications within the last 5 years, publications in English language and publications containing certain keywords such as data pipeline, big data, big data analytics, marketing, digital marketing, customer engagement and social media. Thus, adoption of these specific methods has aided the process of data collection and data analysis, leading to the findings of the research that is valid, relevant and evidence-based.

3 Results

Advantages of data pipeline technology for enhancing social media engagement
There are various advantages that are attained by a business through a data pipeline. Businesses use data pipelines to enhance their capacity for predictive analysis and measure the rate of activities or engagement in social media. Social media is a thriving platform that allows individuals to interact freely with a brand [14]. Businesses use this opportunity to retrieve the data of users to get accurate information regarding the brand's ability to meet the key performance indicators (KPI) [15]. The activities of a potential customer on social media are assessed to analyze their demands, timing of purchase, and inclination for purchasing within a price range and so on. In essence, as a key advantage of such data streaming and analysis helps businesses to perceive the buying behavior of customers and offer products and services accordingly.

As businesses continue to align their social media platforms with data pipelines, they are essentially in control of regulating the flow of data. This helps businesses to gain relevancy and thus further engagement. Social media is being used as a significant platform for promoting and interacting with customers [16]. The current demographic is more prone to using social media than engaging in any other forms of engagement platforms. With the use of data pipeline technology, a brand is able to manage the data in a flexible manner [17]. Such flexibility leads to the beneficial alignment of data to the customers, leading to enhancement in social media interaction among customers (Fig. 3).

Fig. 3. Social media engagement strategies [16]

Management of such vast amounts of data and using it to meet the specific needs of a business is an advantageous side of data pipeline technology. The ETL pipeline (Extraction–Transformation–Loading) helps in the formation of an effective network [18]. Utilization of Thai networks in social media aids the process of customer engagement (Fig. 4).

Fig. 4. ETL pipeline process [18]

Challenges in big data analytics with data pipeline

There are several challenges that are faced in big data management without proper data pipeline architecture. The large volume of data remains at risk of being lost in the processing phase or may not reach the data warehouse [19]. Additionally, the major challenges identified in the field of big data analytics is proper management and integration process. In many cases, such processes are reliable in human understanding [20]. However, in recent times, development of AI-based systems for such operations has reduced this particular challenge. Regardless, lack of proper knowledge regarding the system limits the ability of a business to use it and optimize information to its own benefits.

There is a critical challenge that is faced by business in the utilization of data pipeline architecture as SMEs fail to integrate it due to financial lack. Both technological and financial infrastructure is required for the development and integration of sophisticated technology [21]. The technology itself may be faced with certain challenges as there is a possibility of a gap between real-time data entry and ETL process (Extraction–Transformation–Loading) [22]. The recent integration of AI-based systems in data processing and analytics has however decreased the time lag. Another challenge faced in recent years is the rising velocity, variety and most importantly volume of data. Each stream of data is to be processed through a data pipeline after which proper analytics can be conducted [23]. Hence, gap in processing capacity due lack of proper digital architecture leads to challenges in implementation as well as lack of consistency, security and so on (Fig. 5).

Optimization of data pipeline for enhancing social media engagement and organizational benefits

The organizations are able to benefit exponentially through data processing and analysis optimization that is aided by data pipeline technology. The organizations are able to gain profit and obtain substantial competitive advantage [24]. Enhancing customer engagement through social media is attained through analysis and predictive assessment [25]. As KPIs are also maintained with the use of data pipelines, it brings about significant opportunities for profitability.

On the other hand, the flexibility provided to the organizations as data pipelines are being used assists in their processes of active engagement. Social media analytics has been considered as an emerging tool for aiding the progress of business [26]. Data pipelines are also able to provide a clear view of the data flow based on which decision-making process is conducted. In essence, it can be stated that linkage of social media

Fig. 5. Challenges in big data analytics [23]

and data pipelines helps in increasing customer engagement and help businesses to formulate strategies for further marketing growth. The data flow of social media is tracked and analyzed in order to provide the organizations with ample insight into the needs of necessary modifications.

Social media marketing has been acknowledged as an effective segment for promoting products and services effectively. As the worldwide wealth of digital information rises, businesses use this opportunity to place their products or services in a strategic manner [27]. It essentially helps increase product visibility and ensure customers are effectively engaged with the brand through social media platforms. Hence, the most significant benefit that data pipelines are able to provide the organizations is the ability to analyze voluminous data and use it to retain relevance and interaction. Keeping track of the customer's opinions and latest trends are also analyze to strategically place product suggestions through various social media platforms, enhancing scope for sustained profitability.

4 Discussion

Analysis of the relevant literature has indicated the positive scope of using data pipeline in connection with social media to ensure a positive growth in customer interaction. Most optimal method of data processing and analysis has been identified through the implementation of a data pipeline system [15]. As a data pipeline provides the optimal scope for arranging data in a sequential and synchronized format, it aids the development of a reproducible system in which businesses are able to gain insight into the gathered data and use it to aid their decision-making process. Additionally, it has also been indicated that the automated ETL pipelines enable a timely analysis of valuable data.

The flexibility offered in data pipelines can be used for social media platforms to ensure that the data flow of raw data travels uninterruptedly to its destination. The point of data flow within a pipeline system ensures sequential flow which can be controlled by a business [22]. The main benefit of aligning data pipelines and social media is the enhancement in capacity to process large volumes of data, generated every second of the day. Being in control of the data processing procedures, it has been indicated that the development of an effective and interactive relationship between the brands and customers is possible by virtue of relevant data flow.

Data pipelines are also able to extract big data and distribute them in various relevant sets. Such a simplified structure for voluminous data helps in backing up data and redistributing in case of a crashed server [23]. Analysis of the data gathered from various social media platforms, a business is also able to assess the behavior and attitudes of customers. Gaining the trust of customers is an important part of effective marketing. However, customer engagement enhancement is not merely limited to the marketing procedures. It is extended towards a controlled flow of data to ensure that product visibility is optimal (Fig. 6).

Fig. 6. Big data analytics for social media [4]

It has also been gathered that certain challenges are faced by organizations in the process of data pipeline integration for management and processing. A comprehensive awareness of the entire system hinders the process of proper utilization [19]. Along with that, the lack of financial support has also been identified as one of the challenging factors pertaining to the implementation of sophisticated technology. Moreover, as the wealth of digital data on a global scale continues to increase, feasible options for adequate management, processing and analysis become highly dependable on the capacity of a system's data storage. Real-time data gathered from social media contributes to the generation of voluminous data every day. Hence, the overall capacity of the data pipeline to stream sequential data poses a challenge. However, despite these challenges, constant modification and innovative discoveries made in the field enhances its overall capacity and enables it to be used efficiently for managing big data and providing scope to the businesses to use it as a digital tool for increasing customer engagement.

5 Conclusion

Social media is used as a platform for active interaction between various businesses and customers. Increased rate of online interaction between a business and its target customers leads to the establishment of loyal and beneficial relationships. Increase in social media engagement is thus considered to be a significant part of increasing profitability and competitive advantage. Fulfillment of such business agenda may be possible with the use of data pipeline for social media engagement enhancement.

The data pipeline uses a systematic process of data extraction, transformation and loading (ETL) for sequential management of big data. With the growth of digital data, processing such voluminous data through a controlled yet flexible technology is required which is provided by a data pipeline. However, the processing and storage capacity is required to be in accordance with the growing volume, velocity and variety of data. It has been gathered that with the alignment of social media and data pipelines, businesses are able to control the data flow and direct the raw data to a beneficial destination that ultimately leads to the enhancement of customer engagement. In conclusion, it can be used as a potent tool for strategic placement, increased interaction and decision-making process.

References

1. del Rio Astorga, D., Dolz, M.F., Fernández, J., García, J.D.: A generic parallel pattern interface for stream and data processing. Concurr. Comput. Pract. Exp. **29**(24), e4175 (2017)
2. Statista.com: Social Media Advertising (2021). https://www.statista.com/outlook/dmo/digital-advertising/social-media-advertising/worldwide. Accessed 2 Nov 2021
3. Prim, J., Uhlemann, T., Gumpfer, N., Gruen, D., Wegener, S., Krug, S., Hannig, J., Keller, T., Guckert, M.: A data-pipeline processing electrocardiogram recordings for use in artificial intelligence algorithms. Eur. Heart J. **42**(Supplement_1), ehab724-3041 (2021)
4. Sebei, H., Taieb, M.A.H., Aouicha, M.B.: Review of social media analytics process and big data pipeline. Soc. Netw. Anal. Min. **8**(1), 1–28 (2018)
5. Helu, M., Sprock, T., Hartenstine, D., Venketesh, R., Sobel, W.: Scalable data pipeline architecture to support the industrial internet of things. CIRP Ann. **69**(1), 385–388 (2020)
6. Therrien, J.D., Nicolaï, N., Vanrolleghem, P.A.: A critical review of the data pipeline: how wastewater system operation flows from data to intelligence. Water Sci. Technol. **82**(12), 2613–2634 (2020)
7. Akanbi, A., Masinde, M.: A distributed stream processing middleware framework for real-time analysis of heterogeneous data on big data platform: case of environmental monitoring. Sensors **20**(11), 3166 (2020)
8. de Oliveira Santini, F., Ladeira, W.J., Pinto, D.C., Herter, M.M., Sampaio, C.H., Babin, B.J.: Customer engagement in social media: a framework and meta-analysis. J. Acad. Mark. Sci. **48**, 1211–1228 (2020)
9. Pääkkönen, P., Jokitulppo, J.: Quality management architecture for social media data. J. Big Data **4**(1), 1–26 (2017). https://doi.org/10.1186/s40537-017-0066-7
10. Li, M.W., Teng, H.Y., Chen, C.Y.: Unlocking the customer engagement-brand loyalty relationship in tourism social media: the roles of brand attachment and customer trust. J. Hosp. Tour. Manag. **44**, 184–192 (2020)
11. Xanthopoulou, D.: Capturing within-person changes in flow at work: Theoretical importance and research methodologies. In: Flow at Work, pp. 50–65. Routledge (2017)

12. Bloomfield, J., Fisher, M.J.: Quantitative research design. J. Australas. Rehabil. Nurses Assoc. **22**(2), 27–30 (2019)
13. Patino, C.M., Ferreira, J.C.: Inclusion and exclusion criteria in research studies: definitions and why they matter. J. Bras. Pneumol. **44**, 84 (2018)
14. Lin, H.C., Swarna, H., Bruning, P.F.: Taking a global view on brand post popularity: six social media brand post practices for global markets. Bus. Horiz. **60**(5), 621–633 (2017)
15. Hunt, K., Gruszczynski, M.: The influence of new and traditional media coverage on public attention to social movements: the case of the Dakota Access Pipeline protests. Inf. Commun. Soc. **24**(7), 1024–1040 (2021)
16. Wang, X., Baesens, B., Zhu, Z.: On the optimal marketing aggressiveness level of C2C sellers in social media: evidence from China. Omega **85**, 83–93 (2019)
17. Baljak, V., Ljubovic, A., Michel, J., Montgomery, M., Salaway, R.: A scalable realtime analytics pipeline and storage architecture for physiological monitoring big data. Smart Health **9**, 275–286 (2018)
18. Bala, M., Boussaid, O., Alimazighi, Z.: A fine-grained distribution approach for ETL processes in big data environments. Data Knowl. Eng. **111**, 114–136 (2017)
19. Elragal, A., Klischewski, R.: Theory-driven or process-driven prediction? Epistemological challenges of big data analytics. J. Big Data **4**(1), 1–20 (2017). https://doi.org/10.1186/s40537-017-0079-2
20. Arunachalam, D., Kumar, N., Kawalek, J.P.: Understanding big data analytics capabilities in supply chain management: unravelling the issues, challenges and implications for practice. Transp. Res. Part E **114**, 416–436 (2018)
21. Wang, L., Alexander, C.A.: Big data analytics in medical engineering and healthcare: methods, advances and challenges. J. Med. Eng. Technol. **44**(6), 267–283 (2020)
22. Moly, M., Roy, O., Hossain, A.: An advanced ETL technique for error-free data in data warehousing environment. Int. J. Sci. Res. Eng. Trends **5**, 554–558 (2019)
23. Ardagna, C.A., Bellandi, V., Bezzi, M., Ceravolo, P., Damiani, E., Hebert, C.: Model-based big data analytics-as-a-service: take big data to the next level. IEEE Trans. Serv. Comput. **14**(2), 516–529 (2018)
24. Liu, Y., Jiang, C., Zhao, H.: Assessing product competitive advantages from the perspective of customers by mining user-generated content on social media. Decis. Support Syst. **123**, 113079 (2019)
25. Li, F., Larimo, J., Leonidou, L.C.: Social media marketing strategy: definition, conceptualization, taxonomy, validation, and future agenda. J. Acad. Mark. Sci. **49**(1), 51–70 (2020). https://doi.org/10.1007/s11747-020-00733-3
26. Lee, I.: Social media analytics for enterprises: typology, methods, and processes. Bus. Horiz. **61**(2), 199–210 (2018)
27. Hajirahimova, M.S., Aliyeva, A.S.: About big data measurement methodologies and indicators. Int. J. Mod. Educ. Comput. Sci. **9**(10), 1 (2017)

Performance Analysis of CNN Models Using MR Images of Pituitary Tumour

Ashwitha Kulal(✉)

Department of MCA, Nitte Meenakshi Institute of Technology, Bangalore, Karnataka, India
ashwitha.gk@nmit.ac.in

Abstract. Pituitary tumours are abnormal growths that form over time in the pituitary gland. An excess of hormones that control key physiological functions is caused by some pituitary tumours. Pituitary tumours are mostly benign (non-cancerous) growths (adenomas). Adenomas are benign tumours that remain in the pituitary gland or surrounding tissues and do not spread to other organs. Cancer cells are the most frequent and intense cancer, with a life expectancy of only few months in the most mature stage. As nothing more than a result, therapy planning is a vital stage in improving the quality of life of victims [2]. In this analysis, MRI imaging is used to diagnose a pituitary tumour in the brain. The MR images were first rescaled and augmented. The results are compared on the basis of accuracy between the different CNN models VGG-16, Inception_v3, and ResNet50, which provided 99%, 86%, and 94% accuracy, respectively. Precision, recall, F-score, and accuracy are used to evaluate CNN model performance, and the results are compared on the basis of accuracy between the different CNN models.

Keywords: Convolutional neural network · Visual geometry group-16 · ResNet50 · Dataset · Pituitary

1 Introduction

The pituitary gland is a tiny, pea-sized small gland behind the bridge of the nose, near the brain. It's a component of the endocrine system that modulates hormones in the body. The pituitary gland is known as the "master endocrine gland" since it expels hormones that affect the functioning of other glands in the body. The pituitary gland is regulated by the hypothalamus, a minor brain area.

Healthy cells mutate and proliferate out of control, developing in a tumour, which would be the cause of cancer. A pituitary tumour can be either dangerous or harmless. A cancerous tumour is one that is capable of developing and transmitting to other areas of the body. The terminology "cancerous lump" describes a tumour that can proliferate but it does not spread.

J. I.-Z. Chen et al. (Eds.): ICIPCN 2022, LNNS 514, pp. 609–624, 2022.
https://doi.org/10.1007/978-3-031-12413-6_48

Pituitary adenomas are the most prominent harmless growths in the pituitary gland. A pituitary gland tumour, on the other side, can infrequently function like a cancerous cell, spreading to other parts of the body or disrupting adjacent tissues and organs.

Pituitary cancers are not the same as brain tumours. The pituitary gland is independent from the brain and is placed beneath it. Endocrine tumours are the medical term for tumours of the pituitary gland.

By interfering with the pituitary gland's normal endocrine activity, both benign and malignant tumours in this organ can cause major medical complications. This develops in some circumstances because the tumour originates in cells that synthesize hormones, leading the tumour to produce excess hormones. Pituitary tumours that produce hormones are categorized as "functional tumours."

Early diagnosis of a brain tumour is critical, and it is now achievable because of breakthroughs in medical research and technology. Software that can detect and classify cancers into different categories can be developed using Artificial Intelligence and Machine Learning [1]. Through different CNN models, a brain tumour classification method is suggested using image processing and machine learning.

Traditional methods are particularly successful for the initial cluster size and cluster centres, according to the research findings. When these clusters change with various beginning inputs, it becomes difficult to categorise pixels. The cluster median value is chosen at random in the popular fuzzy cluster mean approach. This will lengthen the time it takes to obtain the desired result. Radiologists must manually segment and evaluate MRI brain pictures, which is time-consuming; segmentation is done using machine learning approaches, which have lower accuracy and computing speed [12]. For the categorization and recognition of tumours with low accuracy, many neural network techniques have been utilised. The segmentation and detection techniques utilised determine the detection accuracy.

2 Related Work

Kumar et al. [13] suggested a deep wavelet auto encoder-based compression strategy that combines the auto encoder's fundamental filter feature selection property and the wavelet transform's image decomposition property. These procedures have a substantial influence on the growth of the feature set when doing another classification problem with DNN. The suggested DWA-DNN image classifier was tested on a brain image database. They compared the DWA-DNN classifier to other classifiers like DNN and auto encoder and found that it performed better. Deep Nayaka et al. introduced a deep neural network technique for identifying distinct classes of brain disorders using a stacked random vector functional link (RVFL) based auto encoder (SRVFL-AE). The RVFL autoencoders form the foundation for their planned SRVFL-AE. When compared to deep learning approaches based on the autoencoder, the major goal of picking RVFL as a significant component of the SRVFL-AE is to improve learning speed and generalizability. Furthermore, they introduced a ReLU (Rectified Linear Unit) activation function into the deep network, which they proposed for better-hidden representation of training dataset and higher response. They used two typical datasets of MD-1 and MD-2 MRI data to evaluate the efficiency of their technique. On the MD-1 and MD-2 datasets, their proposed technique had an accuracy of 96.67% and 95%, respectively [14].

The suggested Brain Cancer Detection and Classification System by P Gokila et al. has been built using ANN. They adopted Hough voting-based technology, which enables for automated manufacturing identification and fragmentation of the anatomical features of interest. It also uses a learning-based segmentation system that is reliable, inter, versatile, and scalable to many modalities. In order to predict the outcomes, significant ratios of training data and dimensions (2D, 2.5D, and 3D) are used. The picture is analyzed using Convolutional neural networks, Hough voting with CNN, Voxel-wise classification, and Efficient patch-wise assessment with CNN.In the recognition of meningioma, glioma, and pituitary tumor, their model had an overall accuracy of 91.3% and a recall of 88%, 81%, and 99%, respectively [11].

In Detection of Brain Abnormality by a Novel Lu-Net Deep Neural CNN Model from MR Images, Hari Mohan Roy et al. published a Novel CNN model (LU-Net) for the detection of brain tumors in two categories with tumor and without tumor [9]. The suggested technique for segmenting and classifying MR images for brain tumor identification not only had the highest accuracy rate, but it also outperformed existing deep neural models. The experiment revealed that the Lu-Net Model trumps previous CNN models in every way, with an overall accuracy of 98%.

3 Proposed Research

The proposed study compares the performance of the VGG-16, Inception_v3, and ResNet 50 CNN models in diagnosing pituitary tumors employing MRI. The dataset is imported and a CNN model is created. Then, in less time, all three models with datasets are considered and simulated. The accuracy, as well as other characteristics like as precision, recall, and F1 score, are then computed, with VGG-16 achieving the highest accuracy.

4 Methodology

The author of this paper employed brain MRI to detect pituitary tumours and compared numerous CNN models based on assessment criteria after analysing a significant number of research articles on various aspects of image processing.

In this proposed system, CNN is preferred for analysing big data because it employs complicated algorithms and artificial neural networks to instruct systems to learn from experience, categorise, and identify data/images in the same way that a human brain does.

4.1 MRI Dataset

The open source dataset for pituitary tumour MRI is freely accessible on kaggle, and the few positive images are from my personal MRI, which is split into Yes and No files, indicating tumour existence and absence, respectively. The collection contains a total of 1222 MR pictures of pituitary tumours of different shape and dimensions. These datasets are divided into two categories: tumour and non-tumour. There are 827 photos in the tumour category and 395 in the non-tumour category, i.e. normal tissue. The MR Dataset is depicted in Fig. 1 according to its labelling, with tumours in one category labelled "NO" and tumours in the other labelled "YES."

4.2 MRI Data-Pre-processing

The most significant issue of image analytics is information pre-processing. Images of varied widths and heights are included in our dataset because it is heterogeneous. As a result, the initial step is to visualise the distribution's ratio by dividing the width by the height. The width and height of the bar graph are used to visualise the fact distribution ratio. Figure 2 shows that the width-to-height ratio is heterogeneous, with the highest images having a height-to-width ratio of 1. As a result, before detecting a tumour, the pictures must be normalised.

One of most notable gain of Deep Learning is that it eliminates the need to manually extract characteristics from images [3]. During training, the network learns to automatically extract the features. The image is basically sent to the network (pixel values).

Fig. 1. Labelled MRI data set

4.3 Data Distribution

In the classification of MRIs, data distribution is quite significant. The dataset is classified into three categories: training set, validation set, and testing sets. The model is trained using training data. The model improves itself by using the training photos. Validation information can be utilized to examine and demonstrate the training process's validity. The accuracy of the model is determined using the test data. In this work, about 70% of the total data is reserved for training (607), 10 MRI scans have been set aside for testing, and 153 images have been set aside for validation. The percentage of the testing dataset that reflects the majority of the variation in the dataset is low, depending on the size of the dataset [15].

The dataset distribution is depicted in Fig. 3 as a bar graph.

Fig. 2. Full dataset distribution with a width-to-height ratio

4.4 Cropping of MRI Images

Cropping photos ensuring that they only contain needed data by removing undesired portions and spaces are required [6]. To discover the extreme points in this work, I used a technique provided in the pyimagesearch blog. The following steps are followed to crop the images employing extreme point computation:

Step 1: Get your hands on the original image.
Step 2: Find the contour that is the largest.
Step 3: Find the polar opposites with OpenCV.
Step 4: Crop the image.

The cropping procedure is presented in Fig. 4 at each stage. Figure 5 shows the MR images following the cropping operation.

Fig. 3. Data distribution by data set (train, test, and validation)

Fig. 4. The technique of cropping MR pictures

Fig. 5. After cropping-MR dataset with labels

4.5 Image Resizing and Augmentation

Image augmentation is a method of enhancing the size of a data set artificially. This is useful when dealing with a data set that has only a few data samples. In the case of Deep Learning, this is a negative circumstance because the model tends to over-fit when only a few data samples are used to train it [9].

To boost the data sample count, zoom, shear, rotation, pre-processing function, and other picture augmentation factors are typically used [7]. Images with these attributes are generated when these parameters are utilised during the training of a Deep Learning model. In most cases, image augmentation results in a $3\times$ to $4\times$ increase in the size of the existing data sample collection.

The original image and the augmented image are demonstrated in Fig. 6.

Fig. 6. The original image and the augmented MR images.

5 CNN Model Architecture

A convolutional neural network (CNN) is a multi-layer neural network. It retrieves key attributes from data that is organised in a grid-like format. One of the most notable advantages of employing CNNs is that little image pre-processing is required. CNNs are capable of determining the most significant filter properties. We save a lot of time and trial-and-error labour by not requiring as many parameters. CNNs apply filters to data as it is processed. The ability of CNNs to modify the filters during training sets those apart [5]. This method can be used to fine-tune findings when working with large data sets in real time, such as photographs. Filters created by hand are no longer necessary because they may be updated to improve CNN training. This increases the amount of filters we may apply to a data set as well as their relevance [4].

Hand-crafted feature extraction approaches, such as texture analysis, are used in most contemporary MRI investigations, followed by traditional machine learning classifiers, such as random forests and support vector machines [10]. There are a few distinctions to be made between such tactics and CNN. To begin with, CNN does not demand feature extraction by manually. Second, CNN topologies do not always require expert knowledge to segment tumours or organs. Third, because there are millions of learnable parameters to estimate, CNN is significantly more data hungry and computationally costly, necessitating the use of graphics processing units (GPUs) for model training.

There are four layers in a convolutional neural network [8]:

i) The convolutional layer
ii) The pooling layer
iii) The ReLU correction layer
iv) The fully-connected layer

5.1 VGG-16

VGG16 is a key innovation that set the way for a number of subsequent innovations in this sector. VGG16 was found to be the most reliable technique of all the configurations on the ImageNet dataset. In their research "Very Deep Convolutional Networks for Large-Scale Image Recognition," K. Simonan and A. Zisserman of the University of Oxford described VGG16 as a CNN model.

In this research, I adopted VGG-15 because it has a high number of weight layers and utilizes comparatively small receptive fields (3 × 3 with a stride of 1).

ImageNet was employed in this model, which contains over 14 million photos divided into over 1000 classes and 22,000 categories. Nearly 1.2 million trained images, 50,000 validated images, and 150,000 test images constitute the ImageNet database [4]. The Architecture of VGG-16 is shown in Fig. 7.

Fig. 7. The architecture of VGG-16

5.2 Inception_v3

By updating prior Inception topologies, Inception_v3 emphasizes on reducing excess computational capabilities. The first study of this paradigm was reported in 2015, with Rethinking the Inception Architecture for Computer Vision. Christian Szegedy, Vincent Vanhoucke, Sergey Ioffe, and Jonathon Shlens partnered on the script.

The third Deep Learning Convolutional Architecture from Google, Inception_v3, is the third in a series of Deep Learning Convolutional Architectures. The original ImageNet dataset comprises over 1 million training photos, however the TensorFlow counterpart has 1001 classes because of the inclusion of a "background" class that was not representative of the original ImageNet. Inception_v3 was constructed for ImageNet's Large Visual Recognition Challenge, where it really took second place.

5.3 ResNet50

A Microsoft team developed a residual learning framework to train networks that are significantly deeper than those actually employed at a quicker speed, because deep neural networks take a long time to train and are prone to over fitting. The findings of this research were published in the journal Deep Residual Learning for Image Recognition in 2015. The well-known ResNet50 (short for "Residual Network") was born as a result.

When training deep neural networks, there comes a point where the accuracy hits a saturation point and subsequently rapidly decreases. This is referred to as the "degradation problem." This demonstrates that not all neural network topologies are created equal.

ResNet50 uses a technique called "residual mapping" to solve this problem. Rather than trusting that every few layered levels will satisfy a chosen underlying mapping, the Residual Network allows these layers to specifically fulfil a residual mapping.

6 Performance Metrics

I created a confusion matrix to define numerous evaluation criteria in order to verify the findings. To make comparisons of all three types of models, qualitative components of performance evaluation criteria have been used. The confusion matrix is being used to calculate the result, which will then be documented in terms of recall, precision, F-score, specificity, and accuracy rate.

7 Results and Discussions

On the discovery of the tumour, the simulation is run on a pituitary tumour MRI dataset of 1222 images. The experiment was carried out on many models, and the results were examined using performance indicators. The simulation is carried out in a Keras and TensorFlow environment using Python. Prior to training, the model images are cropped and resized to a specific width and height. Training, validation, and testing are the three categories in which the data is organised. Each of the three models, the input information is combined into 607 training data, 10 test data, and 153 validation data.

Fig. 8. VGG-16 variations in accuracy and loss of cropped MR images

The same assumptions were used to analyse all of the CNN models. With a learning rate of 0.0003, a decay of 1e-4, a batch size of 32, and an epoch number of 120, the Adam optimizer is implemented for training. The resolution of the MR image input is 224*224*3.

The VGG-16 model was used on pre-processed images in the first trial, yielding nearly 100% accuracy on training data and 99% accuracy on validation data. In terms of accuracy and loss, the VGG-16 model performed brilliantly on both training and validation data. Figure 8 exhibits the effectiveness of the VGG-16 framework in order of training and validation accuracy, as well as training and validation loss.

The experiment's following phase was carried out on Inception_v3. The training accuracy of this model is around 98%, and the validation accuracy is around 82%. Figure 9 indicates the performances of the Inception_v3 model in the context of training and validation accuracy, as well as training and validation loss.

Fig. 9. Inception_v3 variations in accuracy and loss of cropped MR images

Fig. 10. ResNet50 variations in accuracy and loss of cropped MR images

The ResNet50 model was used in the last experiment, which had a training accuracy of 100% and a validation accuracy of 95%. Figure 10 measures the performance of the ResNet50 model on the basis of training and validation accuracy, along with training and validation loss.

The validation dataset's confusion matrix shows the validation performances of all three approaches. In the validation set, there are 153 high-resolution MR images which were not trained on. The correct interpretation of tumours using VGG-16, Inception_v3, and ResNet50 was 137,126, and 137, respectively, according to the confusion matrix, whereas only 15, 0 and 8 non-tumour scans were reviewed thoroughly out of 15 MR data. As a result, the VGG-16, Inception_v3, and ResNet50 validation accuracy rates reached 99, 92, and 95, respectively. Figure 11 demonstrates the confusion matrix for each of the three models.

Based on the confusion matrix, Table 1 illustrates the True Positive, True Negative, False Positive, False Negative, and cumulative validation findings for all three models. When compared to Inception_v3 and ResNet50, VGG-16 and ResNet50 have the highest and identical True Positive images, however VGG-16 has the highest True Negative images. For every CNN Model, the indicators Recall, Precision, F-Score, Specificity, and Total Accuracy were estimated and tabulated in Table 2.

According to the tabulated results, VGG-16 has great accuracy of 99%, ResNet50 has a second-best accuracy of 95%, and the Inception_v3 model has the lowest accuracy of 92%. Recall, Precision, F-score, Specificity, and Accuracy of the VGG-16, Inception_v3, and ResNet50 models are (0.99,1.00,1.00,1.00,0.99),(0.91,0.89,0.90,0.00,0.82), respectively (0.99,0.95,0.97,0.53,0.95). VGG-16 has the best performance across all evaluation criteria, followed by ResNet50. The performance of all three models is depicted in Fig. 12 in terms of assessment evaluation metrics. VGG-16 is efficient and accurate, based on the classification accuracy attained using the state-of-the-art model used in this paper.

(a)

(b)

(c)

Fig. 11. Confusion matrix of validation data of (a) VGG-16, (b) Inception_v3 and (c) ResNet50 models

Fig. 12. The performance of VGG-16, Inception_v3 and ResNet50 models

Table 1. Confusion matrix assessment values for VGG-16, Inception_v3 and ResNet50 models

CNN model	TP	TN	FP	FN	Total
VGG-16	137	15	0	1	153
Inception_v3	126	0	15	12	153
ResNet50	137	8	7	1	153

Table 2. Metrics for analysing the VGG-16, Inception_v3, and ResNet50 models' efficiency

CNN-model	Recall	Precision	F score	Specificity	Accuracy
VGG-16	0.99	1.00	1.00	1.00	0.99
Inception_v3	0.91	0.89	0.90	0.00	0.82
ResNet50	0.99	0.95	0.97	0.53	0.95

8 Conclusion

In this work, three CNN models are being used to analyze the diagnosis of pituitary tumours using brain MRI: VGG-16, Inception_v3, and ResNet50. Before being loaded into the CNN model, the clinical MR images are pre-processed, shrunk, compressed, and optimized. After training and validation on 1222 slightly elevated MR image datasets with 21 augmentations, the findings of CNN models VGG-16, Inception_v3, and ResNet50 are compared. The VGG16 model outperformed the other two models with a 99% accuracy, as well as high precision, recall, and F-score. The accuracy of the system can be improved by adding more dataset and increasing epochs.

References

1. Chong, C., Coukos, G., Bassani-Sternberg, M.: Identification of tumor antigens with immunopeptidomics. Nat. Biotechnol. **40**, 175–188 (2022). https://doi.org/10.1038/s41587-021-01038-8
2. Seetha, J., Raja, S.S.: Brain tumor classification using convolutional neural networks. Biomed. Pharmacol. J. **11**(3), 1457–1461 (2018)
3. Shahriar Sazzad, T.M., Tanzibul Ahmmed, K.M., Hoque, M.U., Rahman, M.: Development of automated brain tumor identification using MRI images. In: 2019 International Conference on Electrical, Computer and Communication Engineering (ECCE), pp. 1–4 (2019). https://doi.org/10.1109/ECACE.2019.8679240
4. Umri, B.K., Wafa Akhyari, M., Kusrini, K.: Detection of Covid-19 in chest X-ray image using CLAHE and convolutional neural network. In: 2020 2nd International Conference on Cybernetics and Intelligent System (ICORIS), pp. 1–5 (2020). https://doi.org/10.1109/ICORIS50180.2020.9320806
5. Soda, P., D'Amico, N.C., Tessadori, J., Valbusa, G., Guarrasi, V., Bortolotto, C., et al.: AIforCOVID: predicting the clinical outcomes in patients with COVID-19 applying AI to chest-X-rays

6. Murugavel, M., Sullivan, J.M., Jr.: Automatic cropping of MRI rat brain volumes using pulse coupled neural networks. Neuroimage **45**(3), 845–854 (2009). https://doi.org/10.1016/j.neuroimage.2008.12.021
7. Hamamci, A., Kucuk, N., Karaman, K., Engin, K., Unal, G.: Tumor-Cut: segmentation of brain tumors on contrast enhanced MR images for radiosurgery applications. IEEE Trans. Med. Imaging **31**(3), 790–804 (2012). https://doi.org/10.1109/TMI.2011.2181857. Epub 2011 Dec 26 PMID: 22207638
8. Mohsen, H., et al.: Classification using deep learning neural networks for brain tumors. Future Comput. Inform. **3**, 68–71 (2018)
9. Rai, H.M., Chatterjee, K.: Detection of brain abnormality by a novel Lu-Net deep neural CNN model from MR images. Mach. Learn. Appl. **2** (2020)
10. Yamashita, R., Nishio, M., Do, R.K.G., Togashi, K.: Convolutional neural networks: an overview and application in radiology. Insights Imaging **9**(4), 611–629 (2018). https://doi.org/10.1007/s13244-018-0639-9
11. Gokila Brindha, P., Kavinraj, M., Manivasakam, P., Prasanth, P.: Brain tumor detection from MRI images using deep learning techniques. IOP Conf. Ser.: Mater. Sci. Eng. **1055**, 012115 (2021)
12. Chen, T., Lin, L., Zuo, W., Luo, X., Zhang, L.: Learning a wavelet-like auto-encoder to accelerate deep neural networks. In: Proceedings of the AAAI Conference on Artificial Intelligence, New Orleans, LA, USA, 2–7 February 2018, pp. 74–93
13. Mallick, P.K., Ryu, S.H., Satapathy, S.K., Mishra, S., Nguyen, G.N., Tiwari, P.: Brain MRI image classification for cancer detection using deep wavelet autoencoder-based deep neural network. IEEE Access. **7**, 46278–46287 (2019). https://doi.org/10.1109/ACCESS.2019.2902252
14. Abd El Kader, I., et al.: Brain tumor detection and classification on MR images by a deep wavelet auto-encoder model. Diagnostics (Basel, Switzerland) **11**(9), 1589 (2021). https://doi.org/10.3390/diagnostics11091589
15. Chattopadhyay, A., Maitra, M.: MRI-based brain tumour image detection using CNN based deep learning method. Neurosci. Inform. **2**(4), 100060 (2022). ISSN 27725286. https://doi.org/10.1016/j.neuri.2022.100060

Detection of Facebook Addiction Using Machine Learning

Md. Zahirul Islam[1](✉), Ziniatul Jannat[1], Md. Tarek Habib[1], Md. Sadekur Rahman[1], and Gazi Zahirul Islam[2]

[1] Department of Computer Science and Engineering, Daffodil International University, Dhaka, Bangladesh
{zahirul15-7822,ziniatul15-7679}@diu.edu.bd, sadekur.cse@daffodilvarsity.edu.bd

[2] Department of Computer Science and Engineering, University of Information Technology and Sciences, Dhaka, Bangladesh

Abstract. A popular social media platform today is Facebook. Facebook addiction is ill-defined. As responsible citizens, we must help society avoid this addiction. Using machine learning techniques, we can forecast the danger of being hooked to Facebook. First, we look into the elements that affect Facebook. This article is great since it informs readers about the hazards of addiction and the elements that lead to it. More than 1,000 people of different ages and backgrounds, addicted or not, have provided data. This article examines people's Facebook addiction and everyday routine. We use SVM, *k*-nearest neighbors, decision trees, Gaussian naive Bayes, logistic regression, and random forest to predict Facebook addiction. SVM is the most commonly used algorithm, followed by *k*-nearest neighbors and decision tree. We evaluate their overall performance using a variety of metrics. We utilize PCA to mathematically reduce data. Our results show that SVM outperforms the other algorithms by 85.00%.

Keywords: Facebook addiction · Prediction system · Machine learning classifier · SVM · PCA · Decision tree

1 Introduction

Facebook is both a social network and a news source. We can use Facebook data to increase our ROI and marketing success. Online in 2019: 2.50 billion. With 1.66 billion daily users [1].

Globally, more individuals use Facebook. Twitter and Facebook have more users in Bangladesh as 96.3 percent of Bangladeshis use Facebook, with YouTube having a lower user base [2]. In Bangladesh, 73.8 percent of Facebook users are men, while 46.3 percent are women. Athenian Plague, Tudor English Sweating Sickness, and the Black Death have all plagued mankind at various times in history [3]. Now a decade-long societal ill appears to be fading. In 2017, according to a Princeton research, Facebook would lose 80% of its members. Nobody likes toothpaste marketers on Facebook except Nick Clegg

J. I.-Z. Chen et al. (Eds.): ICIPCN 2022, LNNS 514, pp. 625–638, 2022.
https://doi.org/10.1007/978-3-031-12413-6_49

[4]. Many of them are on Facebook, raising questions about disease dynamics in social networks. The Princeton researchers make their case using epidemiological modeling, acronyms, and formulae. To be honest, this means F/U + C*K all to me [4].

Recent pandemics have forced governments to lockdown, forcing citizens to stay home for their safety. Bangladesh is in the same boat. During the lockdown, people can only use social media to express their views. We are addicted to social media. This study uses machine learning to categorize Bangladeshi Facebook addiction.

These methods retrieve important data from the dataset with little client input. After algorithms, we utilize PCA. Performance is determined by $F1$-score and accuracy. Machine Learning uses simple statistics to learn [5]. A machine learning model is either predictive or descriptive. The predictive model is a full model that employs prescient displaying technologies to anticipate a certain outcome [5]. At last, compare the outcomes with other research works.

We created an SVM machine learning model to estimate the risk of Facebook addiction. The forecasts help patients grasp their addiction.

To illustrate this paper: Introduction: Sect. 1. Section 2 summarizes relevant works. Section 3 discusses the Facebook addiction prediction system architecture. Section 4 describes the research process. Section 5 summarizes feature and data descriptions. An experimental evaluation is in Sect. 6. The findings are compared in Sect. 7. Section 8 concludes and proposes new work.

2 Literature Review

Associating the performances of multiple machine learning algorithms in diverse circumstances is difficult.

Habib et al. [7] investigate the recognition of papaya disease using the machine learning classification method. They use bad papaya analogies. They make all images 300 × 300. It uses bicubic interpolation and histogram equalization. Their approach uses 126 photos of faulty and defect-free objects. Their dataset is split into two parts: training and testing. They employed a variety of machine learning classification methods. Many strategies are used to reduce errors, including SVMs, C4.5, *k*-NN, naive Bayes, random forest, backpropagation neural networks (BPN and CPN), logistic regression, and backpropagation neural networks (BPN and CPN). They've dealt with five common ailments. SVM outperforms the others. Among all classifiers, SVM achieved 95.2% accuracy.

Mamun and Griffiths [8] propose classification techniques like logistic regression (95%) that are considered for this investigation. Unnecessary data is removed from records before mining. These two researchers used three data mining methods. Extraction of hidden data from a database, model building, and validation against a test dataset. The Classification Matrix represents Adequacy. The best model to predict problems for users appears to be Logistic Regression, followed by Neural Networks and Decision Trees.

Mahmood and Farooq [9] perform the prediction of students' performances using classification algorithms (standard deviation and linear regression). For this, an examination is performed for the expectation of Facebook addiction in connection with students' performances.

Ainin et al. [10] check how Facebook affects socialization and performance. They discuss how Facebook usage affects students' grades. The more time spent on Facebook, the better students perceive and perform. On antecedents, process (Facebook users), and output. Regression analysis shows 79 percent Facebook usage. They correlate it very precisely. The procedure is easy to use and accurate to 90%.

An enormous amount of data was mined by Abdulahi Abdulahi et al. [11] to predict the likelihood of becoming addicted to Facebook using various data mining techniques such as *k*-Nearest Neighbor, Logistic Regression, Support Vector Machine, Decision Trees, Naive Bayes and Linear Discriminant Analysis. We have discovered that the existing systems have demonstrated fewer features than our proposed system. We use both public and some private datasets for the algorithms and find the best accuracy.

Facebook has caused some addictive behavior among its users, according to Jafarkarimi et al. [12]. It was based on the Bergen Facebook Addiction Scale. Facebook was found to be addictive to 47% of Universiti Teknologi(UT) Malaysia students interviewed. This ratio is nearly the same for Malaysian and non-Malaysian postgraduate students. The study included 459 UT Malaysia students. The sample was chosen for its SNS users. To create a balanced sample of men and women aged 17 to 48, we removed incomplete or incorrect questionnaires. 25 years, SD 5.41. 58 percent were undergrads and 42% were grads.

Eshleman et al. [13] use obsession-related social media content to plan a computational epidemiological approach for predicting a user's propensity for seeking drug revitalization interventions. Effective treatment, identifying recoverable groups, and resource allocation require solving this problem. Our technology analyzes the relations that impact a drug user's behavior along with predicts their likelihood to join addiction treatment groups using machine learning techniques. Using real-world data from Reddit and Twitter, the suggested technique may accurately identify people in need of addiction treatment. We utilized user activity to predict addiction. Whether or not a person has posted in drug rehabilitation forums determines their ground truth designation. Out of 24,551 members, 8,697 were posted in revitalization forums. Table 3 summarizes the results. The *k*-NN classifier with $k = 11$ had the greatest overall performance, scoring 0.848 on the $F1$ scale. However, Random Forests had the greatest accuracy (0.914) and recall (0.843) (*k*-NN with $k = 3$). $k = 5, 7, 9$, and 13 were also attempted.

3 System Architecture

The Facebook addiction prediction system. This design is simple enough for anyone. The user can inspect the UI and query it. Data on daily Facebook usage, anxiety over status updates, photo sharing, and the need to use Facebook are gathered. Form data sent to system expert by a user. Data integration and standardization are pre-processing steps for analysis. Two-set data division. The training set is half the test set. SVM-based automated learning classifier addresses the findings. The online application will show the result. It revealed addiction in the end. The architectural model can produce a definite result, as shown in Fig. 1.

Fig. 1. Detection of Facebook addiction system architecture

4 Research Methodology

Facebook addiction is predicted using *k*-NN, SVM, RF, decision tree, naive Bayes, and logistic regression. It's important to know where Facebook users live. Recent research shows the six machine learning methods work well. Preprocessing data makes it usable. It involves removing duplicates and fixing errors. So data efficiency improves. Transforming data involves moving it. A source configuration change is common. Randomizing numerical or alphabetic data. Making sense of big data necessitates. This method is also known as variable subgroup selection (variable indicators). Figure 2: Study plan.

"We execute six machine-learning algorithms on 27-feature datasets. Then we use PCA, which collects data variance in orthogonal linear projections. PCA reduces dimensionality. A model's dimensionality is an independent variable. For the next objective, PCA selects only the essential variables [17]".

Fig. 2. Steps of our proposed methodology for detection of Facebook addiction

The first algorithm we use in our machine learning work is k-NN. The k-NN is the simplest and one of the machine learning algorithms which is based on supervised learning. The k-NN technique can be used to explain classification and regression difficulties, as well as a variety of other problems. For identifying the concealed test data, the k-NN algorithm memorizes the training observation. The technique of k-NN grabs interconnected things that exist in a close quarter [6] using Minkowski distance which is determined by (1) [16].

$$\left(\sum_{i=1}^{k} |x_i - y_i|^p\right)^{\frac{1}{p}} \tag{1}$$

SVM is a machine learning method for regression and classification which is supervised. It is unlike any other machine learning implementation. This method is also used in issue classification and regression. They are given in the selected coordinate and arranged in n-dimensional space. It's termed a hyper plane because it creates the most homogenous points in each subdivision [6]. By using (2), we can find out the divider.

$$W.X + b = 0 \tag{2}$$

A logistic function is used in the process of logistic regression, also called a sigmoid. Real values 0 and 1 are converted by an S-shaped curve [6]. The logistic function is:

$$f(x) = \frac{1}{1 + e^{-t}} \tag{3}$$

A probabilistic machine learning model called naive Bayes is used to classify data. Sentiment analysis, spam filtering, and recommendation systems can all benefit from this data type. Analytical methods such as probability and statistics are employed in this case. It has class and conditional probability. It enhances features by the use of Gaussian distributions [6]. The Gaussian distribution is discussed in detail with its mean and standard deviation (4).

$$P(x = v|C_k) = \frac{1}{\sqrt{2\pi {\sigma_k}^2}} e^{-\frac{(v-\mu_k)^2}{2{\sigma_k}^2}} \tag{4}$$

On may also use decision trees to do supervised machine learning. By separating category features into smaller portions with equal answer values, preprocessing is reduced. Using the divide-and-conquer method, create a tree diagram [6]. With the use of the Eqs. (5) and (6), the user may decide which feature to split on.

$$H(X) = -\sum_{i=1}^{n} p(x_i) - \log p(x_i) \tag{5}$$

$$IG(T, a) = H(T) - \sum_{v \in vals(a)} p_a(v) H(S_a(v)) \tag{6}$$

Random forest is an excellent alternative for rapidly generating any model; to evaluate its performance, training is provided early in the model construction process. During the learning process, the algorithm creates a large number of decision trees. It outputs the class that is the mode of the classes (classification) or mean guess (regression) of the individual trees [6], as shown in Fig. 3.

Fig. 3. Working principle of the random forest classifier

These algorithms were deployed on test data after training with training data. In this regard, it is worth noting that classifier performance might vary greatly depending on the dataset balance [18]. Thus, researchers always use the confusion matrix, which is a 2×2 matrix for binary classification and an $n \times n$ matrix for multiclass classification. True positives (TP) and true negatives (TN) are shown in the confusion matrix (TN). This data is used to determine the precision, recall, and $F1$-score. The ratio of genuine positive to anticipated positive. The recall is the ratio of exactly predicted explanation to all observations in the class. The $F1$-score is the summed accuracy and recall. It calculates both positive and negative values [9]. However, the confusion matrix predicts accuracy, precision, recall, and $F1$-score as follows:

$$Accuracy = \frac{TP + TN}{TP + TN + FP + FN} \times 100\% \tag{7}$$

$$Precision = \frac{TP}{TP + FP} \times 100\% \tag{8}$$

$$Recall = \frac{TP}{TP + FN} \times 100\% \tag{9}$$

$$F_1 score = \frac{2 * precision * recall}{precision + recall} \times 100\% \tag{10}$$

The holdout method [7, 18] along with the evaluation metrics from (7) to (10) were used to evaluate the performance of our classifiers.

5 Feature and Data Description

5.1 Feature Selection

The main causes of Facebook addiction are studied to develop a feature set that can identify a Facebook addict.

Table 1. Detailed descriptions of the algorithms that were used.

Algorithm	Specifications
k-NN	No. of neighbors = 5
	Weight function, $f(w) = c$, where c is a constant
	Size of leaf = 30
	Distance metric = Minkowski
SVM	$\text{MinkowskiDistance} = \left(\sum_{i=1}^{k} \lvert x_i - y_i \rvert^p\right)^{\frac{1}{p}}$
	C = 1.0
	Size of cache = 200
	Weight function, $f(w) = c$, where c is a constant
	Gamma = auto deprecated
	Kernel = linear
	Random state = 0
	Weight function, $f(w) = c$, where c is a constant
Random forest	Criterion:
	$\text{Gini}(D) = 1 - \sum_{i=1}^{c} (p_i)^2$
	Maximum depth = 2
	Randomstate = 0
	C = 1.0
	No. of penalty = 12
	Random state = 0
	Normalization flag = False
Logistic regression	Distribution:
	Gaussian distribution, $\int (x, \mu, \sigma) = \frac{1}{\sigma\sqrt{2\pi}} e^{\frac{-(x-\mu)^2}{2\sigma^2}}$
	Mean:
	$Mean,\ \mu_x = \frac{1}{N}\sum_i (x)^i$
Naïve Bayes	Variance:
	$Variance,\ \sigma_x = \frac{\sum_{i=1}^{n} (x_i - \overline{x})^2}{n}$

Six algorithms are utilized here. Each algorithm has parameters whose values vary. Table 1 shows the parameter values for all techniques used to train the model. We would choose the optimal algorithm for our issue area based on its performance. The list of features of Facebook addiction is shown in Table 2.

Table 2. Features for Facebook addiction prediction.

Sl. no.	Feature title	Based on	Sl. no.	Feature title	Based on
1	Age	[1, 8]	20	The Easiest way to contact	[20]
2	Gender	[8]	21	Feel part of an expansive exciting world	[14]
3	Opening first id	[19]	22	Play games by using Facebook	[20]
4	Spending time	[19]			
5	Status update anxiety	[20]	23	Take selfies and post them on Facebook	[20]
6	Make new friends	[20]	24	Visit food-related pages, and groups on Facebook	[14]
7	Excessive Time	[20]	25	Share political and religious views	[20]
8	Compare Facebook life and real life	[20]	26	Feel the urge to use Facebook	[20]
9	Facebook design	[20]	27	Forget personal problems	[20]
10	Sharing personal life	[20]			
11	Checking Facebook	[14]			
12	Share images	[14]			
13	Spending hours browsing through Facebook	[20]			
14	Influenced to add more friends	[20]			
15	Attractive online social life	[14]			
16	Virtually date	[20]			
17	One or more fake id	[20]			
18	Addiction to Facebook	[20]			
19	Facebook posts change the mood	[20]			

Note: Class value: *Yes/No*

We must consider each of these aspects to determine the likelihood of being hooked to Facebook. We learned about these characteristics from a variety of physicians, as well as a variety of papers and writings located on pertinent websites such as [8, 10, 15, 20].

5.2 Data Collection and Preprocessing

Sets are vital in categorization studies. Our dataset was prepared both online and off. A questionnaire was first created. Then we traveled to Dhaka's schools, universities, and venues to collect data from various age groups and occupations. Using a Google Form, we circulated the questionnaire to various target groups and individuals through the internet. We have 1001 data points from 27 factors. Then we categorized each record with the aid of educational psychologists and student counselors.

After collecting data, we gather missing, category, numerical, and text data. Then we determine how to process the data so that it may be used in algorithms. Data preparation is the capacity to prepare data after collection. Processing data in a given format facilitates output generation.

Figure 4 shows our data preparation effort. The first is data cleansing. We search for erroneous incentives. We encoded the text to a numeric level. In this case, we used imputer and median to deal with missing values. In the following step, we looked at the correlation matrix. This matrix shows the proportion of each piece of information. A positive value implies a strong connection, whereas a negative number suggests a poor connection. Outlier quartile detection cleans up noise. Then it was dropped. Each element has its histogram to better display data. Standardization finished the information modification. So we have the informative index.

Fig. 4. Steps of data preprocessing

6 Experimental Evaluation

Our data come from a variety of sources, both offline and online. On-and-offline polling yielded 1001 data training and test datasets (67 and 33 percent) divided using the handover approaches. They include 671 and 330 data respectively. The feature's connectivity is named using a correlation matrix.

The data had some unacceptably noisy values. We tried to reduce jitter. A box plot was used to show this. Figure 6 depicts the causes of noisy readings. We used an outlier quartile to clean up the figure's initial id and time spent characteristics. To clean up the opening initial id and time spent characteristics (Fig. 7).

As shown in Table 3, the PCA improves the accuracy, precision, recall, and $F1$-score. It says that before PCA, k-NN was 70% accurate, SVM was 83%, logistic regression was 77%, naive Bayes was 76%, decision tree was 73%, and random forest was 80%. When compared to the other six algorithms, SVM outperformed PCA in terms of accuracy (8) - (10). The optimum values for SVM are accuracy (83%), recall (86%), and $F1$-score (85%). The best recall value is generated by random forest (88 percent).

Table 3. Classifier performance evaluation before and after applying PCA

Before					After			
Algorithms	Accuracy (%)	Precision (%)	Recall (%)	*F1*-score (%)	Accuracy (%)	Precision (%)	Recall (%)	*F1*-score (%)
k-NN	70	72	77	74	72	90	64	85
SVM	83	84	86	85	85	67	100	80
Decision Tree	73	77	74	76	72	21	25	23
Logistic Regression	77	79	80	79	54	65	31	42
Random Forest	80	81	88	84	78	82	75	80
Naïve Bayes	76	82	73	77	62	34	44	28

Table 4. Correlation with outcome features.

Features	Correlation Values	Features	Correlation Values
Age	0.16	Status update anxiety	0.10
Spending hours browsing through Facebook	0.15	Influenced to add more friends	-0.11
Share Images	0.15	Take selfies and post them on Facebook	-0.11
Feel part of an expansive exciting world	0.14	Spending time	-0.11
Excessive Time	0.14	Checking Facebook	-0.12
Make new friends	0.14	Play games by using Facebook	-0.13
Virtually date	0.13	Facebook design	-0.13
Compare Facebook life and real life	0.13	Feel the urge to use Facebook	-0.13
Share political and religious views	0.12	Forgot Personal Problem	-0.14
The Easiest way to contact	0.12	Visit food-related pages, and groups on Facebook	-0.15
Attractive online social life	0.11	One or more fake id	-0.15
Sharing personal life	0.11	Addiction to Facebook	-0.16
Status update anxiety	0.10	Gender	-0.16
		Facebook posts change the mood	-0.17
		Opening first id	-0.19

After PCA, we can see that some approaches are more accurate than others. SVM, and k-NN were improved by PCA. Table 3 shows precision, recall, and $F1$-score after PCA.

Figure 5 shows the difference before and after PCA. SVM had 83 percent accuracy before PCA and 85 percent accuracy after PCA.

Table 4 shows the correlation between the features. It also shows the features' correlation with the outcome.

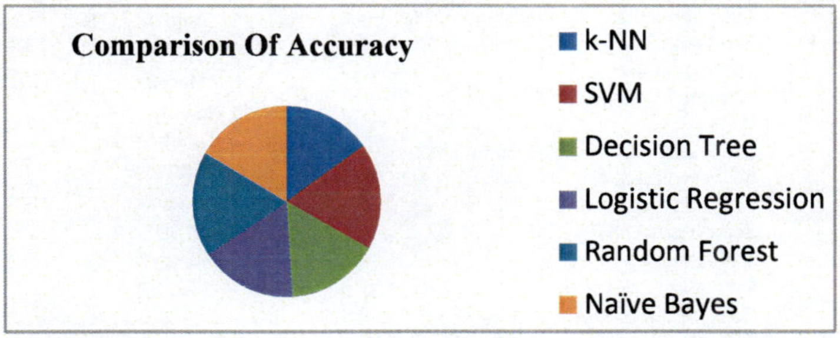

Fig. 5. Comparison of accuracy

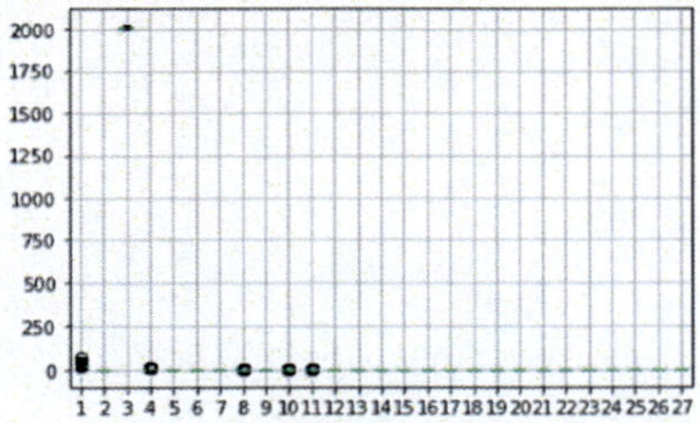

Fig. 6. In a box plot, the values are noisy in Opening first id and 'Spending time'

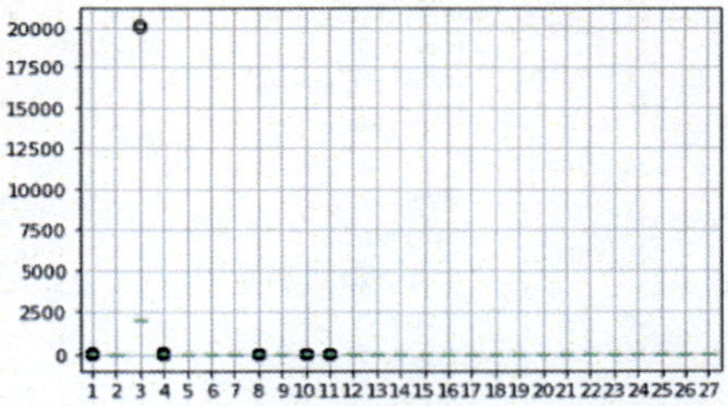

Fig. 7. In a box plot, the values are not noisy in Opening first id and Spend time

7 Comparative Analysis of Results

Recent research allows us to assess our proposed system's efficiency. Widespread but unrelated prediction research We did some research on Facebook addiction. It's difficult to compare our work to others. Comparing our efforts to others:

Table 5 compares our work to other works. Mamun and Griffiths [8] work on the relationship between Facebook addiction as well as depression: A pilot survey study surrounded by Bangladeshi students. Works with 300 data and 13 features. Calculated 95% accuracy when applying some machine learning algorithms. Ainin et al. [10] work on Facebook usage, socialization, and academic performance. They use regression analysis for their works. Jafarkarimi et al. [12] work on Facebook Addiction among Malaysian Students. Among 441 data 47% were addicted to Facebook. Eshleman et al. [13] work for Identifying those Amenable to Drug revitalization Interventions through Computational Analysis of Addiction Content in Social Media. The data set contains 24,551 data and the best performing algorithm is k-NN.

So, as you can see in the table above, we work in Bangladesh. All ages and social classes were interviewed. Our data set is small, but we used various machine learning techniques to maximize accuracy.

Table 5. Results of comparing our work with others.

Method / Work Done	Prediction dealt with	Problem Domain	Sample Size	Size of feature set	Algorithm	Accuracy
This Work	Detection of Facebook addiction using machine learning	Prediction	1001	27	Support Vector Machine (SVM)	85%
Mamun and Griffiths [8]	The association between Facebook addiction and depression: A pilot survey study among Bangladeshi students	Addiction *Addiction*	300	13	Logistic Regression	95
Ainin et al.[10]	Facebook usage, socialization, and academic performance	Addiction	1165	16	Regression Analysis	NM
Jafarkarimi et al.[12]	Facebook Addiction among Malaysian Students	Addiction	441	10	NM	47% Addicted
Eshleman et al.[13]	Identifying Individuals Amenable to Drug Recovery Interventions through Computational Analysis of Addiction Content in Social Media.	Prediction	24,551	NA	*K*-NN	NM

- NM: Not Mentioned
- NA: Not Assigned

8 Conclusion and Future Plans

We gathered data on Facebook addiction in Bangladesh. We used multiple machine learning algorithms to forecast Facebook's addiction level. People are unaware of their cardiac problems. Some are overeating issues. Facebook distracts some of them from their schoolwork. Facebook robs people of their time and energy. Some are going through an identity crisis. Facebook addiction lacks feeling due to improper use. After all, there are too many posts to read.

Our study uses machine learning to predict Facebook addiction. Our work and model are limited. We used a small data collection. More data would have been better. There are many complex data processing methods, and the model can be presented in many ways. Our method can detect Facebook addiction. This study shows that SVM works well with our data. They also work well with raw data. Our approach may help future research on preventing Facebook addiction and educating people on how to manage their current situation. This issue area holds great promise for handling massive data sets.

References

1. Zephoria. Top 15 Facebook Statistics for 2020 – The Year in Review. Zephoria Inc. (2019). https://zephoria.com/top-15-valuable-facebook-statistics/
2. Statcounter: Social Media Stats Bangladesh. StatCounter Global Stats (2020). https://gs.statcounter.com/social-media-stats/all/bangladesh
3. Napoleoncat: Facebook Users in Bangladesh – January 2019 (2019). https://napoleoncat.com/stats/facebook-users-in-bangladesh/2019/01
4. Mahdawi, A.: If Facebook is an infectious disease, here's a guide to the symptoms. The Guardian (2014)
5. Shin, T.: All Machine Learning Models Explained in 6 Minutes | by Terence Shin | Towards Data Science (2020). https://towardsdatascience.com/all-machine-learning-models-explained-in-6-minutes-9fe30ff6776a
6. Ray, S.: Commonly Used Machine Learning Algorithms | Data Science. Analytics Vidhya (2017). https://www.analyticsvidhya.com/blog/2017/09/common-machine-learning-algorithms/
7. Habib, Md. T., Mia, Md. J., Uddin, M.S., Ahmed, F.: An in-depth exploration of automated jackfruit disease recognition. J. King Saud Univ. Comput. Inf. Sci. S1319157820303530 (2020) https://doi.org/10.1016/j.jksuci.2020.04.018
8. Mamun, M.A.A., Griffiths, M.D.: The association between Facebook addiction and depression: a pilot survey study among Bangladeshi students. Psychiatry Res. **271**, 628–633 (2019)
9. Mahmood, S., Farooq, U.: Facebook Addiction: A Study of Big-Five Factors and Academic Performance amongst Students of IUB, p. 18 (2014)
10. Ainin, S., Naqshbandi, M.M., Moghavvemi, S., Jaafar, N.I.: Facebook usage, socialization and academic performance. Comput. Educ. **83**, 64–73 (2015)
11. Abdulahi, A., Samadi, B., Gharleghi, B.: A study on the negative effects of social networking sites such as Facebook among Asia Pacific University scholars in Malaysia. Int. J. Bus. Soc. Sci. **5**, 133–145 (2014)
12. Univeristi Teknologi Malaysia: He is also with Department of Computer, Damavand Branch, Islamic Azad University, Damavand, Iran et al. Facebook addiction among Malaysian students. IJIET **6**, 465–469 (2016)
13. Eshleman, R., Jha, D., Singh, R.: Identifying individuals amenable to drug recovery interventions through computational analysis of addiction content in social media. In: 2017 IEEE International Conference on Bioinformatics and Biomedicine (BIBM), pp. 849–854. IEEE (2017). https://doi.org/10.1109/BIBM.2017.8217766
14. Mouri, D., Ali Arshad, C.: Social networking in Bangladesh: boon or curse for academic engagement? Manage. Market. **11**, 380–393 (2016)
15. Chakraborty, A.: Facebook addiction: an emerging problem. Am. J. Psych. Res. J. **11**, 7–9 (2016)
16. NPMJS: Compute-Minkowski-Distance. NPM (2020). https://www.npmjs.com/package/compute-minkowski-distance

17. Jolliffe, I.T., Cadima, J.: Principal component analysis: a review and recent developments. Phil. Trans. R. Soc. A. **374**, 20150202 (2016)
18. Tan, P.-N., Steinbach, M., Kumar, V.: Introduction to Data Mining, p. 169 (2006)
19. Aslam, S.: 63 Facebook Statistics You Need to Know in 2020 – Omnicore (2020). https://www.omnicoreagency.com/facebook-statistics/
20. Ryan, T., Chester, A., Reece, J., Xenos, S.: The uses and abuses of Facebook: A review of Facebook addiction. J. Behav. Addict. **3**, 133–148 (2014)

Mask R-CNN based Object Detection in Overhead Transmission Line from UAV Images

D. Satheeswari[1(✉)], Leninisha Shanmugam[2], N. M. Jothi Swaroopan[3], and Nirmala Venkatachalam[4]

[1] Department of Electronics and Communication Engineering, Meenakshi College of Engineering, Chennai, India
satheeswari.ykp@gmail.com
[2] School of Computer Science and Engineering, VIT University, Chennai, India
leninisha.shanmugam@vit.ac.in
[3] Department of Electrical and Electronics Engineering, RMK Engineering College, Chennai, India
jothi.eee@rmkec.ac.in
[4] Department of Computer Science and Engineering, St Joseph's College of Engineering, OMR, Chennai, India

Abstract. The transmission of electricity through Overhead Transmission Line (OTL) is a difficult task. The defects in OTL are caused by the presence of a bird's nest in the transmission tower, which frequently affects the entire surrounding area. Therefore in this article proposed two methods as preliminary research. In the first method, birds' nests and transmission towers are located using template matching, a conventional technique. As a second method, introduced a convolutional neural network (CNN) with instance segmentation, which recognizes birds' nests and transmission towers from images captured by an unmanned aerial vehicle (UAV). The second method is carried out through preprocessing, manual annotation, and feature extraction using the backend architecture Resnet-50 and trained into Region of Interest (ROI)Align, which produces output with two classes, as well as the bounding box and mask for the respective classes. When compared to traditional method, the proposed Mask Region based Convolutional Neural Network (Mask R-CNN) method achieves 98.7% accuracy.

Keywords: Annotation · Mask R-CNN · Instance segmentation · UAV images · ResNet-50

1 Introduction

Distribution of electricity, far and near power stations through OTL is most essential. Liability induced by bird's nest in transmission tower are shooted up and causes great hazards to the living environment. Bird's nest is composed of sheep wool, grasses, moss, leaves, sticks, twigs and acts as conductors that lead to stumbling of the transmission

J. I.-Z. Chen et al. (Eds.): ICIPCN 2022, LNNS 514, pp. 639–653, 2022.
https://doi.org/10.1007/978-3-031-12413-6_50

line and leads to high risk. To legislate the hazards, it is necessary to monitor and inspect OTL. Traditional manual inspection was performed with more labour in the early years, but the cost was high and the process took a long time. Later, robots are used to inspect OTL, which also takes more time but does not involve humans. Unmanned Aerial Vehicle (UAV) has recently been used, which automatically inspects and monitors the OTL in less time, and the analysis performed on UAV images is evaluated using Deep Convolutional Neural Network (DCNN).

- Images with chaotic backgrounds were examined from various lighting angles.
- The pixel values of each target object differ.
- The structure, composition, and appearance of each target object vary.
- The characteristics of an images are accomplished by extracting multilevel features and selecting areas of various sizes and aspect ratios at each location.
- The coordinate position and category score of each candidate area are used to evaluate it.
- The co-ordinate position is balanced and the loss function is calculated based on the candidate region.

2 Related Work

Many segmentation methods are investigated to segment the images based on spatial clustering method, region growing method, hybrid linkage of region growing method, split and merge method [1]. Machine learning techniques are very good at evaluating feature selection and extraction. Feature selection states that reduction occurs while selecting a subset of features, whereas dimensionality reduction occurs during feature extraction [2]. The Deep Learning (DL) architecture is proposed for image segmentation. Manual instance segmentation is done by labelling the samples, which is a difficult task. Once the samples have been manually annotated, they are trained into various pre-trained CNN architectures, which improve accuracy and performance while maintaining robustness[3]. The author proposed the detection and location of power line, is done by continuous monitoring and to find the rotating angle of insulator from image, rotation invariance is proposed and SVM is used to extract feature through sliding window through which achieved good performance [4]. Mask R-CNN is a deep learning technique used to implement the ship target detection algorithm, and soft non maximum suppression (soft-NMS) is used to detect the ship in Google Earth images [5]. A network structure with invariant edges is investigated. Edge detection of objects is critical, and the expansion of the scale based on background has improved their connection [6]. Fast-RCNN is used to detect foreign obstacles via visual saliency and collaborates with regions to determine the relationship between areas and environment [7]. R-CNN detects small target objects at multiple scales. As a result, to identify the broken insulators and bird's nest in a high voltage line, a DCNN method is supported by a Faster R-CNN method [8]. The Faster R-CNN solves the target detection and recognition problem [9]. CNN accurately detects power lines and obtains the necessary information from each layer. Feature maps are extracted from the model based on the available valid information and provide the exact detected output [10]. To indicate the presence of power lines on aerial images, the DCNN architectures VGG and ResNet were designed to identify objects in the ImageNet dataset [11]. Faster R-CNN is an improved version of R-CNN, which recognizes and detects objects [12, 13]. The Region Proposal Network (RPN) is used by the Faster R-CNN to select a random region of the image as the proposal region and trains it to locate the component in the image, because RPN generates regions for identification and thereby the faster R-CNN solves the computation problem [14–16]. With the help of Gabor features, SVM is used to identify insulators from UAV images [17]. The automatic detection of bird nests is proposed using DL technique-based ROI and RCNN. The K-means algorithm is used to calculate the dimensions, which improves accuracy and the focal loss function is assessed using foreground and background images. Each class's losses are reduced by RCNN [18]. The author proposes using Faster RCNN with ResNet-50 to identify a bird's nest on a transmission tower automatically. The experimental results show that this method can detect bird's nest targets with an accuracy of 95.38% [19]. In this paper, a DL algorithm based on Mask RCNN is proposed for power-line inspection using an unmanned aerial vehicle (UAV). In this case, the UAV platform was the Dragan fly XP-4. The Resnet-50 architecture was used for the backbone network,the Feature Pyramid Network (FPN) and Region Proposal Network (RPN) were used to extract features [20]. According to the reviewed paper, here proposed the fault detection in OTL exists due to birds' nests and towers and is located using (i)template matching and (ii) Mask R-CNN. The first method, template matching, requires a significant amount of time to detect the target classes. This method finds the similar templates by comparing a source image to a template (birds nest and tower). During the template matching process, each pixel value of the source image is compared to the template image one at a time. The OpenCV library includes a function called cv. matchTemplate(). It compares the template to the patch of the input image beneath the template image (as in 2D convolution), and its main drawback is that it cannot detect bird's nests and towers exactly due to poor occlusion or a cluttered background. As a result, the Mask R-CNN, which creates a mask over the segmented target objects in the second method, aids in precisely detecting and locating bird's nest and tower.

3 The Proposed Model

Monitoring and detection of OTL is required for healthy environmental conditions. The detection and location of a bird's nest and a tower in OTL is based on a poor background image collected from publicly available UAV images [19].

3.1 CNN in Object Detection

CNN makes use of a technique known as Convolution. The network uses a convolutional network to generate feature map from the provided input image. CNN can be used to use a deep network to locate objects and train a high-capacity model with little data. It achieves excellent object detection accuracy with DCNN.

3.2 Region Based Convolutional Neural Network (RCNN)

This process utilizes selective search, method to generate regions. As a result, convolutional networks are evaluated independently for each Region of Interest (ROI) to classify more than one image region into the proposed class. The RCNN architecture was developed for image detection. The R-CNN architecture is also the basis of Mask R-CNN. We are using a Python script to perform inference or prediction by applying selective search to an input image, thereby determining which regions of the input image should be classified by selective search, and finally showing the R-CNN results.

3.3 Target Detection Using Template Matching

For all images, the traditional method known as template matching [20] is used. Based on the template, this method searches for and locates the bird's nest and transmission tower. The images used are 8-bit in size. The detection methods available are
'cv2.TM_CCOEFF',
'cv2.TM_CCOEFF_NORMED',
'cv2.TM_CCORR',
'cv2.TM_CCORR_NORMED',
'cv2.TM_SQDIFF',
'cv2.TM_SQDIFF_NORMED'.

"TM_SQDIFF" is the best method for our dataset. The size of the template image should not be larger than the size of the original image. For example, if the original image has X and Y, the template image should have x and y. As a result, Eq. (1) gives the following expression:

$$(X - x + 1)\,(Y - y + 1) \tag{1}$$

In this case, template matching is analogous to a sliding window (rectangular box) with a fixed width and height, through which the template image is slid over the input image (similar to convolution) and the overlapping patch is compared by sliding the template over the input images. The Similarity score is the threshold value which decides to create the box for the birds' nest and tower. The threshold valve for template matching is set to 0.1. When the similarity score matches, a correlation for the pixel value in the image is established, and a sliding window is generated for the target objects. When the similarity score does not match, no correlation exists, and a false box is created for the target objects.

3.4 A Mask R-CNN-Based Target Detection Method

Mask R-CNN with instance segmentation groups each object separately and displays those using bounding boxes of varying colours in the proposed work. The proposed methodology's schematic diagram is revealed in Fig. 1.

3.4.1 Datasets of OTL

The proposed research methodology begins with the collection of data, which consists of 400 UAV images with a cluttered background, with the birds' nest and transmission tower being considered as target objects to be segmented and detected. The image 8688 × 5792 was preprocessed and resized to 768 × 512 pixels.

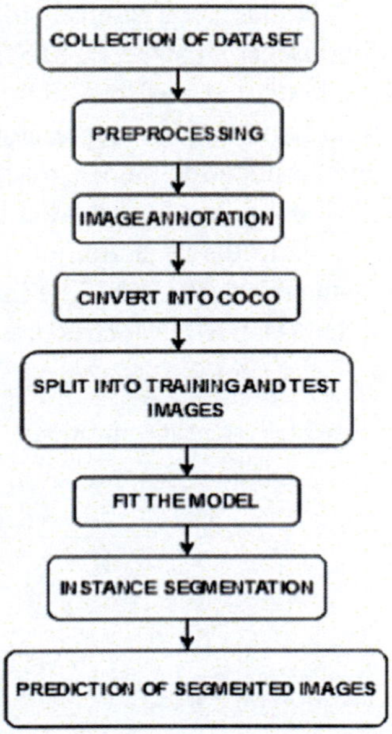

Fig. 1. Schematic diagram of proposed research methodology

3.4.2 Annotation of Images

VGG image annotator [21] is used on image datasets. It is easy to annotate images and has proven to be more efficient for this type of algorithm. The collected images are annotated here by labelling the classes with the tool polygons. When the annotation process is finished, the annotated images are saved in json format. The dataset was then divided into training (80%) and testing images (20%).

3.5 Mask R-CNN Combined with Instance Segmentation

The target objects are separated based on the anchor box. The annotated UAV images were converted into the coco (Common Objects in Context) format, which supports the Mask R-CNN [22, 23]. Coco is frequently used for target image recognition when comparing the performance of real-time object detection algorithms. Based on the anchor box, instance segmentation can easily detect foreground and background objects separately. This method generates a pixel-by-pixel mask for each object in the image. Instance segmentation divides objects into labels and assigns colours to each object based on its label name. All of the target objects in the images with different pixels are classified and grouped. The proposed work is carried out with the system details and is expressed in Table 1. Object detection and segmentation is a CV task that involves the process of locating objects in images. It is a DCNN technique that is primarily used to solve segmentation problems. It entails the recognition, localization, and classification processes. Mask R-CNN, in particular, precisely locates objects using an open-source library, namely Tensor flow and Python. Mask R-CNN is a more advanced version of Faster RCNN [24–26]. However, in addition to Faster RCNN, it also creates a mask for the segmented image. Mask R-CNN shares the similarly identified feature map in stage I, and it tunes and solves pipeline issues in stage II. ROI generates a binary mask for losses such as classification loss, bounding box loss, and binary cross entropy loss, in addition to class classification and bounding box regression. The key concept in many recent models when performing classification, and regression is the classification that is based on mask predictions. The bounding box created by Mask R-CNN in parallel for classification and regression aids in the reduction of multi-stage R-CNN pipeline. The data can be easily trained using Mask R-CNN. The intersection over union (IOU) bounding box is generated, and the Mask R-CNN structure is displayed in Fig. 2.

Table 1. System specifications

Python	3.10.4
Tensorflow	1.13.1
Keras	2.0.8
TPU	P100-PCIE-16GB
Google Colaboratory PRO Ram size	16 GB
Processor	I5-7200U
OS	Windows 10

Fig. 2. Structure of mask R-CCN

3.5.1 Deep Learning Model ResNet-50 in Mask R-CNN

Mask R-CNN model [27, 28] can be used along with many architectures such as DenseNet [29, 30], VGG16 [31] ResNet-50 [32–34], SqueezeNet [35] and Inception-V4 [36]. In our proposed work, ResNet-50 architecture which is best suited architecture for the specified dataset and is basically a good DL model which consists of subsequent residual modules which forms the building block of ResNet-50. The advantage of ResNet-50 architecture is, it can create even thousands of residual layers and train them appropriately to obtain the output. It is a residual network, helps in extracting all level of features and this stacked layer are enhanced with many features map classifiers in an end-to-end classification method, when the network converges there occurs a problem called degradation which increases the depth of network and the accuracy gets saturated and suddenly degrades rapidly. Table 2 shows the layers in ResNet-50 architecture. Instead of hopping, short connections are used (skipping) for identity mapping and are added along with the stacked layers to produce the output. As the depth of layer increases in ResNet-50, training error gets increased thereby increasing the gain of the network. ResNet-50 gives very good accuracy than the existing other networks, as the depth of the model is improved. From the input layers it learns the residual function rather than the unreferenced function. When the identity mapping is optimal then it pushes the residual to zero and fits the identity mapping by stacking non-linear layers. If the previous layer dimension does not match the next layer, 3 × 3 convolution layer will change the spatial dimension of an image from the 32 × 32 to outside of spatial dimension 30 × 30, so proposed up sampling, the presence of previous input layers identifies the skip connections using up sampling. Hence there exist two ways, zero padding and no extra parameters. Expanding the dimension with 1 × 1 convolution, the residents look like in contrast to 34 layers plane network which is a convolutional layer followed by activation functions followed by batch normalization (BN). The residual when tested with 50 layers on ImageNet and this gets their state of art results. The ensemble event resonance is able to achieve 3.75% error on ImageNet test data. ResNet- 50 extract the

Table 2. Layers in ResNet-50

Convolutional Layer	Size of the output	ResNet-50 Layers
Convol_1	112 × 112	7 × 7, 64, S = 2 (S = Stride)
Convol_2	56 × 56	3 × 3 Max Pooling, S = 2
		1 × 1, 64 × 3
		3 × 3, 64 × 3
		1 × 1, 256 × 3
Convol_3	28 × 28	1 × 1, 128 × 4
		3 × 3, 128 × 4
		1 × 1, 512 × 4
Convol_4	14 × 14	1 × 1, 256 × 6
		3 × 3, 256 × 6
		1 × 1, 1024 × 6
Convol_5	7 × 7	1 × 1, 512 × 3
		3 × 3, 512 × 3
		1 × 1, 2048 × 3
Avg_Pool	1 × 1	1000, SoftMax
FLOPS		3.8×10^9

features from the image dataset and classifies different bounding box based on same region proposed algorithm. Similarly, it uses a BN after each convolution and before activation uses the initializers. The batch size, learning rate, weight decay, no dropout is used because they do not predict on test images, as layers are increased the error occurs. When skipping the layers, the dimension doesn't match the zero pad, 1×1 convolution known as projection that they do not get a significant performance and also have extra parameter, i.e., when the 50 layers are trained, the skip connection is extended and it skips a head of two layers, rather than the residual building block. This is done to save the training time. ResNet-50 has more than 3 blocks. Even when the depth increases, it has less complexity when compared to other architecture.

3.5.2 Feature Pyramid Network (FPN) in Mask R-CNN

FPN is a feature extractor which takes the arbitrary size as input. It is used to generate feature maps which will be in a fully convolutional fashion method. It will be the independent process of backbone convolutional architecture. It is used to build feature pyramids inside DCNN which is used for object detection. Multiple spatial resolutions are taken for ResNet-50 architecture. Segmentation process involves identification of objects which is based on classes. Class requires a configuration object as a parameter which is used during training or inference. Based on the number of classes the process must be carried out to obtain the output. Any number of classes can be identified that is based on our decision. Each class will be represented by different pixels and colours. The pixel values will be grouped together to form classes.

3.5.3 Region Proposal Network in Mask R-CNN

RPN proposes candidate object bounding boxes. It is used to generate feature maps of various sizes. The generated feature maps will have various convolutional layers with a lot of information about the images embedded in them. The RPN output is essentially a region proposal like areas where objects can be easily found. Each proposal is unique in its own way. IOU is used to create the anchor box. Here the foreground classes are captured if the IOU is greater than 50% and the background classes are captured if the IOU is less than 50% finally, there exist the layer is called the fully connected layer (FC).

3.5.4 Region of Interest Align in Mask R-CNN

The ROI portrays the feature map by means of bilinear interpolation. The ROI Align [37, 38] extracts the features from RPN layer and pretrained architecture ResNet-50. The size of the generated feature maps varies. Using ROI align, these are converted into fixed size feature maps. The received feature maps of the birds' nest and tower in ROI are from RPN and ResNet-50, and they differ in scale and aspect ratio. As a result, all features must be converted to fixed dimensions, which is accomplished through ROI pooling, which yields 7×7 kernels with 512 channels, followed by two FC that flatten all features, and finally analysed using a classifier, regressor, and mask generator. Bird's nest and tower are classified as present or not by the SoftMax classifier. The regressor aids in the plotting of bounding boxes for the classified birds' nest and transmission tower, and a mask is generated for the predicted classes.

4 Results and Discussions

Two methods are used to collect, annotate, and implement UAV images of bird nests and towers. First, the traditional template matching method is used to detect birds' nests and towers based on the template, and the recognition error of this framework is that the rectangular box is not exactly positioned for the target classes, and it takes a long time to detect for the entire dataset. Figure 3 depicts the location of target objects using template matching. Figure 3a shows the original image with the target objects. Figure 3b is the detection of target objects (birds' nest and towers). This figure clearly shows that there is poor detection of target object, i.e., birds' nest and tower are not precisely (bounding box not in exact location) located, for our dataset and accuracy is poor. As a result, to improve the detection process a second method called MASK R-CNN was proposed, which aids in the resolution of problems encountered when segmenting images using vision computing. The experimental results are examined in conjunction with the extracted features from the sample images. The model employs with the parameters such as 0.001 learning rate, 145 epochs, and 32 batch size. It separates the various objects in the image. ResNet-50 is the backend network used in Mask R-CNN instance segmentation. There are two options. The bottom-up pathway, which employs ResNet-50, that aids in the extraction of features from the original images. The second pathway is the bottom-up one, which generates feature maps of the same size as the previous one. It performs the convolution and addition between the two. The performances of FPN is extremely good in extracting features at various resolution levels. Initially RPN checks all feature maps generated by FPN and identifies the regions for the target objects present in the images. Generating feature map is an efficient way, Once the features match the original images, generates the feature maps and it is necessary to create the anchor box for the corresponding images, and assigned bounding box which is based on the IOU value. Once the anchors box created are matched with the feature maps, RPN identifies and decides the size of the bounding box. Here the convolution with up sampling and down sampling is evolved to keep the features received should be same as the original images. Later the other neural network utilizes the proposed regions generated by initial level, which locates the respective areas and generates the classes, bounding boxes and masks for the corresponding birds' nest and tower which is present in the images. Figure 4 illustrates the images taken from different angles, including a bird's nest and a tower of the target object. Figure 4a be the original images, Fig. 4b is the resized images, Fig. 4c are the manually annotated images and Fig. 4d defines the masked output images. The performance of Mask R-CNN is displayed in Table 3, based on the implementation process.

Fig. 3. Detection using template matching

Fig. 4. Images from various angles. (a) Original images. (b) Resized images. (c) Manually annotated images. (d) Masked output images

Table 3. Performance of Mask R-CNN

Models	Accuracy of bird's nest	Accuracy of tower	mAP (%)
Mask RCNN	99.5	99.4	98.7 (m_{13})
	99.2	99.1	98.6 (m_{18})
	99.5	99.3	99.3 (m_{23})
	99.8	99.9	99.7 (m_{42})
	99.8	99.6	99.5 (m_{68})
	99.5	99.7	99.4 (m_{94})
Faster RCNN (VGG 16)	96.83	NIL	NIL
Faster RCNN (ZFNET)	91.8	NIL	NIL
Faster RCNN (RESNET 50)	98.41	NIL	NIL

Fig. 5. Comparison chart of proposed model with existing model

4.1 Validation and Visualization of mAP

Once the process of training and testing (inference) is completed. The mean average precision (mAP) is calculated for the validation images. Once the training and testing (inference) process is completed. For the validation images, the mAP is evaluated. When the 13th image in the validation folder is selected for testing, it predicts the segmented output with the mask and mAP. It calculates the actual length of the predicted images as 3 because there are three bounding boxes, and the AP is 99.5% for the birds' nest, 99.7% for the tower, and 99.4% for the tower. Finally, the precision of the mAP for the entire image in the dataset is 98.7%. Similarly, several images are tested for validation datasets, including the 18th, 23rd, 42nd, 68th, and 94th. Table 3 shows the mAP for the verified images. Figure 5 explains the comparison chart of proposed deep learning Mask R-CNN model and existing model. From the figure it is understood that proposed model gives very good accuracy of 99.7% in detection than the surveyed paper [19].

4.2 Loss Function in Mask R-CNN

Mask Rcnn's losses are evaluated after each epoch and visualised with Tensorflow. This procedure aids in the monitoring of the training process and parameters. After the training process is completed, the model's losses for the training and inference dataset are recorded after 145 epochs. With the LR reset at epochs 50, 74 and 86 the loss of the first epoch is less than the loss of the last epoch, especially for training loss. Figure 6 depicts various losses plotted against epochs and loss values. Figure 6a explains the RPN losses for RPN bounding box and RPN class losses which gradually decreases and maintains the constant level. It performs the task of different loss function. i.e., it does the process of classification, location and creating mask for the target objects in OTL and is shown in Fig. 6b. The class loss and the bounding box loss tries to learn and create mask for every class is based on ResNet-50. The mask loss automatically creates mask based upon the losses in bounding box. The overall losses are the loss that occur during training and testing process. The losses are RPN losses and Mask R-CNN losses when it is trained for the epochs to 145 is summarised in Fig. 6c and its loss function is obtained by Eq. (2).

$$\begin{aligned} \text{Losses} = {} & \text{RPN Class}_{\text{loss}} + \text{RPN Bbox}_{\text{loss}} + \text{MRCNN Class}_{\text{loss}} + \\ & \text{MRCNN Bbox}_{\text{loss}} + \text{MRCNN Mask}_{\text{loss}} \end{aligned} \tag{2}$$

Fig. 6. Displays different losses. (a) RPN losses. (b) Mask R-CNN losses. (c) Overall losses

5 Conclusion

OTL faults are caused by birds' nests in transmission towers. To solve this problem, we proposed mask R-CNN with instance segmentation as a result of our research. As a result, the proposed CNN method solves the problem that occurs when using template matching in the prediction of target objects from blurred backgrounds and achieves a higher accuracy of 98.7% than previously reviewed methods. This method precisely locates target objects with very fast image prediction, which improves the training process in a short period of time. Based on this approach, automatic detection of other common equipment in OTL images will be investigated further.

Acknowledgements. This research received no specific grant from any funding agency in the public, commercial, or not-for-profit sectors.

Conflict of Interest. The authors declare that there is no conflict of interest.

References

1. Haralick, R., Shapiro, L.: Image segmentation techniques. Comput. Vis. Graph. Image Process. **27**(3), 389 (1984). https://doi.org/10.1016/S0734-189X(85)90153-7
2. Khalid, S., Khalil, T., Nasreen, S.: A survey of feature selection and feature extraction techniques in machine learning. In: Science and Information Conference (2014). https://doi.org/10.1109/SAI.2014.6918213
3. Minaee, S., Boykov, Y.Y., Porikli, F., Plaza, A.J., Kehtarnavaz, N., Terzopoulos, D.: Image segmentation using deep learning: a survey. In: IEEE Transactions on Pattern Analysis and Machine Intelligence (2021). https://doi.org/10.1109/TPAMI.2021.3059968
4. Jabid, T., Udin, M.Z.: Rotation invariant power line insulator detection using local directional pattern and support vector machine. In: International Conference on Innovations in Science, Engineering and Technology (ICISET), pp. 1–4 (2016). https://doi.org/10.1109/ICISET.2016.7856522
5. Nie, S., Jiang, Z., Zhang, H., Cai, B., Yao, Y.: Inshore ship detection based on mask R-CNN. In: IGARSS 2018 – 2018 IEEE International Geoscience and Remote Sensing Symposium, pp. 693–696. IEEE (2018). https://doi.org/10.1109/IGARSS.2018.8519123
6. Wang, X., Ma, H.M., Chen, X., You, S.: Edge preserving and multi-scale contextual neural network for salient object detection. IEEE Trans. Image Process. **27**(1), 121–134 (2018). PMID:28952942. https://doi.org/10.1109/TIP.2017.2756825
7. Huang, L., Xie, R., Xu, Y.: Invasion detection on transmission lines using saliency computation. In: IEEE International Symposium on Signal Processing and Information Technology (ISSPIT), pp. 320–325 (2015). https://doi.org/10.1109/ISSPIT.2015.7394352
8. Huang, J., Shi, Y., Gao, Y.: Multi-scale faster-RCNN algorithm for small object detection. J. Comput. Res. Dev. **56**(2), 319 (2019). https://doi.org/10.7544/issn1000-1239.2019.20170749
9. Lei, X., Sui, Z.: Intelligent fault detection of high voltage line based on the Faster R-CNN. Measurement **138**, 379–385 (2019). https://doi.org/10.1016/j.measurement.20z19.01.072
10. Zhang, H., Yang, W., Yu, H., Zhang, H., Xia, G.S.: Detecting power lines in UAV images with convolutional features and structured constraints. Remote Sens. **11**(11), 1342 (2019). https://doi.org/10.3390/rs11111342

11. Yetgin, Ö.E., Benligiray, B., Gerek, Ö.N.: Power line recognition from aerial images with deep learning. IEEE Trans. Aerosp. Electron. Syst. **55**(5), 2241–2252 (2018). https://doi.org/10.1109/TAES.2018.2883879
12. Ren, S.Q., He, K.M., Girshick, R., Sun, J.: Faster R-CNN: towards real-time object detection with region proposal networks. İn: International Conference on Neural Information Processing Systems, pp. 91–99 (2018). https://doi.org/10.1109/TPAMI.2016.2577031
13. Girshick, R., et al.: Rich feature hierarchies for accurate object detection and semantic segmentation. İn: IEEE Conference on Computer Vision and Pattern Recognition, pp. 580–587. IEEE Computer Society (2014). http://arxiv.org/abs/1311.2524v1
14. Uijlings, J.R., Van De Sande, K.E., Gevers, T., Smeulders, A.W.: Selective search for object recognition. Int. J. Comput. Vision **104**(2),154–171 (2013). https://doi.org/10.1007/s11263-013-0620-5
15. Girshick, R.: Fast R-CNN[C]. In: IEEE International Conference on Computer Vision, pp. 1440–1448 (2015)
16. Yu, S., Wu, Y., Li, W., Zeng, W.: A model for fine-grained vehicle classification based on deep learning. Neurocomput. **257**, 97–103 (2017). https://doi.org/10.1016/j.neucom.2016.09.116
17. Wang, X., Zhang, Y.: Insulator identification from aerial images using support vector machine with background suppression. In: International Conference on Unmanned Aircraft Systems (ICUAS), pp. 892–897 (2016). https://doi.org/10.1109/ICUAS.2016.7502544
18. Li, F., et al.: An automatic detection method of bird's nest on transmission line tower based on Faster_RCNN. IEEE Access **8**, 164214–164221 (2020). https://doi.org/10.1109/ACCESS.2020.3022419
19. Li, J., Yan, D., Luan, K., Li, Z., Liang, H.: Deep learning-based bird's nest detection on transmission lines using UAV imagery. Appl. Sci. **10**(18), 6147 (2020). https://doi.org/10.3390/app10186147
20. Xiu, C., Pan, X.: Tracking algorithm based on the improved template matching. In: 29th Chinese Control And Decision Conference (CCDC), pp. 483–486. IEEE (2017). https://doi.org/10.1109/CCDC.2017.7978142
21. Dutta, A., Gupta, A., Zissermann, A.: VGG İmage Annotator (VIA) (2016).
22. Leninisha, S., Vani, K, Agasta Adline, A.L., Vani, V.: Damaged road detection in rural areas for ımproving agricultural marketing. In: Technological Innovation in ICT for Agriculture and Rural Development (TIAR), pp. 90–95. IEEE (2015). https://doi.org/10.1109/TIAR.2015.7358537
23. He, K., Gkioxari, G., Dollár, P., Girshick, R.: Mask r-cnn. In: IEEE international conference on computer vision, pp. 2961–2969 (2017)
24. Satheeswari, D., Shanmugam, L., Swaroopan, N.M.J.: Recognition of bird's nest in high voltage power line using SSD. In: First International Conference on Electrical, Electronics, Information and Communication Technologies (ICEEICT), pp. 1–7 (2022). https://doi.org/10.1109/ICEEICT53079.2022.9768651
25. Naveed, S.: Early diabetes discovery from tongue images. Comput. J. **65**(2), 237–250 (2022). https://doi.org/10.1093/comjnl/bxaa022
26. Venkatachalam, N., Shanmugam, L., Heltin, G.C., Govindarajan, G., Sasipriya, P.: Enhanced segmentation of ınflamed ROI to ımprove the accuracy of ıdentifying benign and malignant cases in breast thermogram. J. Oncol. **2021**, 17 (2021). Article ID 5566853. https://doi.org/10.1155/2021/5566853
27. Shanmugam, L., Gunasekaran, K., Natarajan, A., Kaliaperumal, V.: Quantitative growth analysis of pulp necrotic tooth (post-op) using modified region growing active contour model. IET Image Process. **11**(11), 1015–1019 (2017)
28. Liu, Y., Huo, H., Fang, J., Mai, J., Zhang, S.: UAV transmission line inspection object recognition based on mask R-CNN. J. Phys. **1345**(6), 062043 (2019). https://doi.org/10.1088/1742-6596/1345/6/062043

29. Shanmugam, L., Kaliaperumal, V.: Water flow based geometric active deformable model for road network. ISPRS J. Photo. Remote Sens. **102**, 140–147 (2015). https://doi.org/10.1016/j.isprsjprs.2015.01.013
30. Zhang, K., Guo, Y., Wang, X., Yuan, J., Ding, Q.: Multiple feature reweight DenseNet for image classification. IEEE Access **7**, 9872–9880 (2019). https://doi.org/10.1109/DASC50938.2020.9256456
31. Tammina, S.: Transfer learning using vgg-16 with deep convolutional neural network for classifying images. Int. J. Sci. Res. Publ. (IJSRP) **9**(10), 143–150 (2019). https://doi.org/10.29322/IJSRP.9.10.2019.p9420
32. Targ, S., Almeida, D., Lyman, K.: Resnet in resnet: Generalizing residual architectures. arXiv preprint arXiv:1603.08029 (2016). https://doi.org/10.48550/arXiv.1603.08029
33. Shanmugam, L., Kaliaperumal, V.: A junction aware water flow approach for urban road network extraction. J. IET Image Proc. **11**, 227–234 (2016) https://doi.org/10.1049/iet-ipr.2015.0263. ISSN: 1751-9659, IF: 0.67
34. Wen, L., Li, X., Gao, L.: A transfer convolutional neural network for fault diagnosis based on ResNet-50. Neural Comput. Appl. **32**(10), 6111–6124 (2019). https://doi.org/10.1007/s00521-019-04097-w
35. Wang, A., Wang, M., Jiang, K., Cao, M., Iwahori, Y.: A dual neural architecture combined SqueezeNet with OctConv for LiDAR data classification. Sensors **19**(22) 4927 (2019). https://doi.org/10.3390/s19224927
36. Szegedy, C., Ioffe, S., Vanhoucke, V., Alemi, A.A.: Inception-v4, inception-resnet and the impact of residual connections on learning. In Thirty-first AAAI conference on artificial intelligencze (2017)
37. Gong, T., et al.: Temporal ROI align for video object recognition. Proc. AAAI Conf. Artif. Intell. **35**(2), 1442–1450 (2021)
38. Shanmugam, L., Adline, A.A., Aishwarya, N., Krithika, G.: Disease detection in crops using remote sensing images. In: IEEE Technological Innovations in ICT for Agriculture and Rural Development (TIAR) (2017). https://doi.org/10.1109/TIAR.2017.8273696

Insights into Fundus Images to Identify Glaucoma Using Convolutional Neural Network

Digvijay J. Pawar[1](✉), Yuvraj K. Kanse[2], and Suhas S. Patil[2]

[1] Electronics Engineering, Shivaji University, Kolhapur, MS, India
djp@amgoi.edu.in

[2] Department of Electronics and Telecommunication Engineering, K.B.P. College of Engineering, Satara, MS, India
{yuvraj.kanase,suhas.patil}@kbpcoes.edu.in

Abstract. Glaucoma, an eye disease, is a multi-factorial neuro-degenerative disease that vitiates vision over time and which may cause permanent vision impairment. In recent years, machine learning has been used with the idea of using algorithms to find patterns and/or mark extrapolations based on a collection of data. The detection of glaucoma has been achieved with various deep learning (DL) models so far. This paper presents the Convolutional Neural Network (CNN) approach for the diagnosis of glaucoma with remarkable performance. In this approach, glaucoma and healthy images can the differentiated because it forms patterns that can be detected with the CNN. The fundus images are used as image modality which includes publically available retinal image datasets as IEEE DataPort, Drishti-GS and Kaggle Dataset. The analysis is performed for selected datasets, it is observed that, the IEEE DataPort dataset gives better results than others and obtained values of accuracy, sensitivity and specificity are 95.63%, 100% and 91.25% respectively.

Keywords: Early detection · Glaucoma · Convolutional neural network · Fully connected layer · Softmax classifier

1 Introduction

By the year 2050, it is anticipated that around 61 million individuals will be blind, 474 million will have moderate and extreme vision impairment, 360 million will have gentle vision impairment, and 866 million will have uncorrected presbyopia [4]. This means the need for eye care is set to surge in the coming years.

Glaucoma disease is one of the essential apprehension for the human eye. Glaucoma symptoms can be seen in the human eye which in-turns results in loss of vision. Glaucoma is a huge defect that could result in blindness if early detection is not achieved [1, 2].

Glaucoma is generally alluded to as "quieted robber of vision" since the manifestations at a beginning phase of glaucoma are not expressly characterized and are difficult to measure. In the event that the progression of glaucoma is not halted at the starting stages it brings about the extreme damage of the optic nerve and as an outcome, it

J. I.-Z. Chen et al. (Eds.): ICIPCN 2022, LNNS 514, pp. 654–663, 2022.
https://doi.org/10.1007/978-3-031-12413-6_51

will prompts serious visual deficiency [3, 5]. Glaucoma is an ocular disorder observed because of increase in intra-ocular pressure (IOP) which in-turns damage at optic nerves and astrocytes. Also, there is another important indicator for glaucoma is Cup-to-Disc Ratio (CDR). Generally, if CDR is increased more than 0.4 then it is concluded that it is glaucoma infected or glaucomatous image otherwise it is normal or healthy image [16] as depicted in Fig. 1.

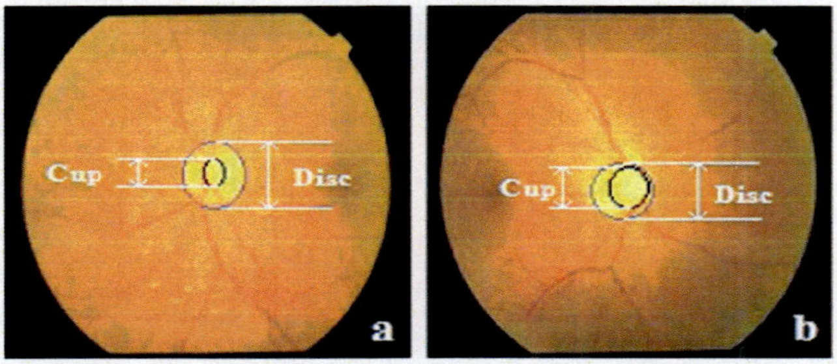

Fig. 1. Retinal fundus images - (a) healthy image (b) glaucomatous image

It is important to recognize glaucoma at an early phase because of following realities:

- There are no recognizable signs in its primer stages.
- It is an extreme issue as the harm it causes is irremediable.
- It prompts unending loss of sight on the off chance that not relieved quickly.
- There is no prophylactic treatment for glaucoma, yet it is feasible to stay away from visual deficiency by detecting, treating and overseeing glaucoma at a very primary stage [6].

If the early recognition of glaucoma can forestall the visual deficiency and can helps to prevent blindness in the human eye. So there is a need of appropriate detecting model to detect this disease. Many endeavors are taken to expand this kind of framework. The present approach is to detect the glaucomatous image pattern from suspected images using Deep Convolutional Neural Network (D-CNN). The presented system will apply the CNN technique for the classification of fundus images. The fundus image is the most used imaging technique for capturing and evaluating the human eye for any retinal diseases. The CNN model will be utilized to differentiate the trends within the founded information for glaucoma detection [8, 12, 17].

In Artificial Intelligence (AI), Convolutional Neural Network (CNN) is a specific kind of feed-forward neural network. It is popular tool for image recognition. The input data is represented by CNN as multi-dimensional arrays. It functions admirably for an enormous number of marked data. The receptive field, is what CNN uses to extract each and every portion of the input image. It assigns weights to each neuron based on the receptive field's importance. So that it can distinguish between the neurons from each other. CNN is also computationally efficient [9].

Benefits of CNN

- Among all image prediction algorithms, it has the highest accuracy.
- Low reliance on pre-processing, reducing the amount of human effort required to develop its functionalities.
- It is effective for both supervised and unsupervised learning.
- It is simple to comprehend and implement.

To classify different images through machine learning, the generalized structure of CNN is represented in Fig. 2.

Fig. 2. Generalized structure of convolutional neural network

Three types of layers of CNN architecture are as –

- **Convolution Layer –** This layer contains input vectors, such as an image, filters as a feature identifier, and output vectors as a feature map. Feature or activation maps are generated with the convolution filters in this layer. The weights are subject to their size. Here, the actual masking of an image is carried out, where certain types of connexions can be featured, which is recognized as the convolution operation. It filters out input data and find information. The differences between the convolution, the filter, and the output feature map is shown in Fig. 3.

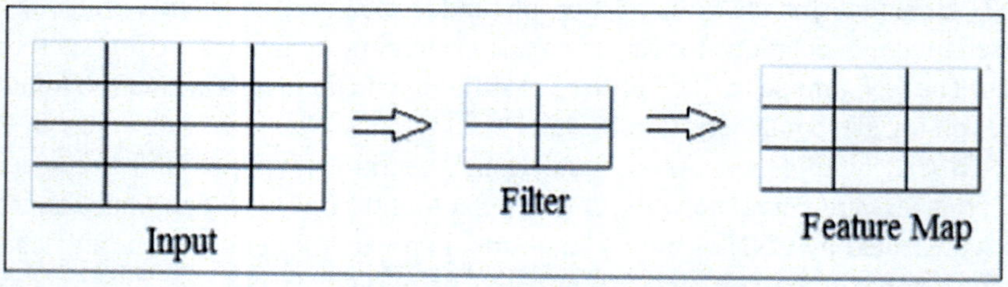

Fig. 3. Difference between convolution, filter and feature map

Padding and stride are featured with this convolution process, which adds more accuracy to the image analysis.

Generally, after the convolution action, the feature map will have reduced in dimensions than the input function. To keep the original size of the image, to maintain the dimensions of the output map and to prevent it from shrinking, the addition of zeroes round the input signal can be done which is termed as padding. The stride is nothing but the amount by which the filter function moves at each step over input function.

- **Pooling Layer -** Each feature map is considered separately at this pooling layer. Based on how elements are chosen from the feature map, it is decided to use the max-pooling or average pooling method. This layer helps to minimize the parameter count and computation in the network.

This two layers can be repeated several times, depending on the database intricacy.

- **Fully Connected Layers –** Finally, the parameters extracted are transformed into a vector as an input of the multi-layer perceptron. The Convolutional Neural network has then trained with the usage of these parameters and weights using functions like softmax or sigmoid.

Weights and biases are the two types of parameters in a CNN. The sum of all weights and biases equals the total number of parameters. The number of convolution layer weights equals the number of convolution layer biases.

2 Proposed Methodology

The layer-based CNN architecture will distinguishes the image samples for glaucoma detection from the fundus images. The proposed system is shown in Fig. 4 as –

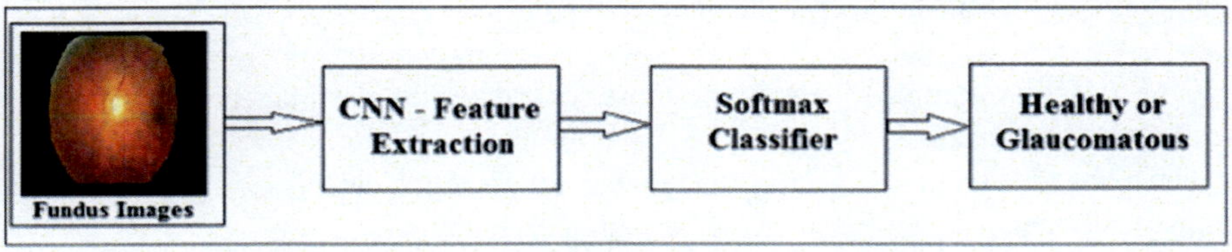

Fig. 4. Block diagram of proposed system

2.1 Insights to CNN Architecture

The related values of parameters and hyper-parameters must first be determined before performing data transformation and classification. The determination of the parameter values for each hidden layer, as well as hyper-parameter values for convolution layer, number of hidden layers, and number of nodes in each hidden layer, using CNN [7]. The architectural details and the parameter values are summarized below-

- Image size = 30 × 45
- Convolution Layer = 1
- Number of Kernels = 20

Parameter Sets in Advance

- Loss Function: Cross-entropy
- Padding: Zero padding to input borders as default.
- Weight Initializers: Glorot
- Solver Name: Stochastic Gradient Descent with Momentum (SGDM)
- Optimizer: Adam
- Hidden Layer Activation Function: Rectified Linear Unit (ReLU)
- At Output Layer Function: Sigmoid

Hyper-parameters

- Kernel matrix = 15 × 15
- Number of blocks = 20
- Number of Epochs = 69
- Number of batches = 10
- Batch size = 10
- Initial Learning Rate = 0.0001
- No. of Epochs = 69
- Filter Size = 15
- No. of Filters = 20
- Pooling Size = 2
- Output Size = 2 (positive or negative)

CNN based system can identify the most critical features of glaucoma disease by use of above parameters.

The main advantage of CNN over its predecessors is that it detects important features automatically and without the need for human intervention.

2.2 Dataset Description

There are various image modalities available for retinal image capture like fundus, OCT, HRT, ultrasound etc. The fundus camera is a type of modern imaging technique. Examining the internal structure of the eye is carried out by fundus images which are captured from the fundus camera. The several public and private databases of retinal images can be used for research related to retinal diseases like glaucoma, diabetic retinopathy, cataract, myopia etc.

By evaluating several image patterns from numerous open-access datasets, the image databases namely IEEE DataPort, Drishti-GS and Kaggle has been chosen for the experimentation.

- **IEEE DataPort -** The various fundus images are publicly available at IEEE DataPort which is submitted by Wheyming Tina Song [11]. This dataset comprises a total of 1450 images captured by a fundus camera, in which glaucomatous images are 899 and healthy or non-glaucomatous images are 551.
- **Drishti-GS -** Drishti-GS is a publicly available dataset with 101 images. There are 50 training images and 51 testing images in this set. This database was developed at Aravind eye hospital, Madurai by manifesting all of the images by four eye experts with varying clinical experience [13, 14].
- **Kaggle Dataset –** Similar to the above dataset, it is a publically available dataset. It consists of 520 trained images and 130 images are availed for validation purposes [15].

The experimentation is carried out with the particular images from these three datasets for training and testing purpose with the CNN structure which is mentioned below in Table 1.

Table 1. Details of dataset used

Dataset used	No. of trained images		No. of testing images	
	Positive	Negative	Positive	Negative
IEEE DataPort	140	140	80	80
Drishti-GS	35	15	15	12
Kaggle	100	100	80	80
Total images	**530**		**347**	

2.3 Experimental Set-Up

- Hardware Used: Intel core i3, Intel motherboard, 4GB RAM, 140GB SSD Hard Disk.
- Software Used: windows 10 (64-bit), Matlab 2017.
- Datasets: IEEE DataPort, Drishti-GS and Kaggle Database.

3 Results and Discussions

For an experiment, from the mentioned three datasets, the 530 fundus images have been trained and 347 images are tested with the deep neural network.

In this experimental work, we have initiated with learning rate of 0.0001 and used an Adam optimization algorithm for optimization as it is efficient because it is a combination of stochastic gradient descent with momentum (SGDM) and robust learning. So, it gives better results than rmsprop and other algorithms. The softmax classifier has been used to automatically distinguish the inputted images as normal or glaucomatous. The configured model is trained for 69 epochs with batch size of 10.

The performance analysis for the different datasets have done with accuracy, sensitivity, specificity, ROC curve, True Positive (TP) value, and False Positive (FP) value. The performance parameters are measured by using equations as mentioned below:

$$Accuracy = (TP + TN) / (TP + FP + TN + FN) \tag{1}$$

$$Sensitivity\ or\ True\ Positive\ Value = TP/(TP + FN) \tag{2}$$

$$Specificity\ or\ True\ Negative\ Value = TN/(TN + FP) \tag{3}$$

$$Precision\ or\ True\ Predicative\ Value = \mathrm{TP}/(\mathrm{TP} + \mathrm{FP}) \tag{4}$$

where,

True Positive (TP): Glaucomatous images are accurately identified as Glaucomatous.
False Positive (FP): Healthy images are inaccurately identified as Glaucomatous.
True Negative (TN): Healthy images are accurately identified as Healthy.
False Negative (FN): Glaucomatous images are inaccurately identified as Healthy.

The proposed system works with an accuracy of 95.63%, 83.78%, and 90.63% for IEEE DataPort, Drishti-GS, and Kaggle databases respectively, which is depicted in Fig. 5 as a comparison of performance for different datasets.

After a comparison of the system's output for different three datasets, the dataset of IEEE DataPort works better with an accuracy of 95.63%, sensitivity of 100%, and specificity of 91.25% as shown in Table 2.

Table 2. Performance analysis for different datasets

Dataset used	Accuracy	Sensitivity	Specificity	TP rate	FP rate
IEEE DataPort	95.63%	100.00%	91.25%	91.95%	8.75%
Drishti-GS	83.78%	88.00%	75.00%	88.00%	25.00%
Kaggle	90.63%	92.50%	88.75%	89.16%	11.25%

Figure 6 represents the confusion matrix giving the classification as healthy class 0 image and the glaucoma image class 1 for the IEEE DataPort dataset.

The ROC is plotted as a curve that indicates the trade-off between TPR (True Positive Rate) and TNR (True Negative Rate). The FPR (False Positive Rate) = (1 – TNR), which is mentioned with TPR to show the ROC curve as shown in Fig. 7, which is obtained for 69 epochs.

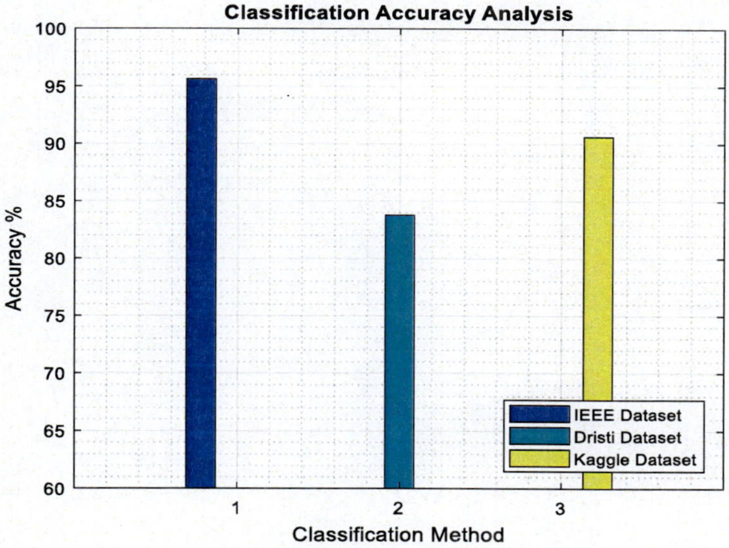

Fig. 5. Accuracy comparison of the proposed system's performance for different datasets

Fig. 6. Confusion matrix for IEEE DataPort dataset

Fig. 7. The ROC curves obtained after training for 69 epochs for different datasets

4 Conclusion

The proposed system matured with the use of features from convolutional neural networks and it helps to categorize normal and glaucomatous fundus images with remarkable accuracy. The raw fundus images are directly fed to the CNN structure. The key features of the glaucoma disease are automatically extracted from multi-layers of the filters convolved along with the raw input image.

The work can be expanded in the future to include implementing the proposed system with a much larger sample of datasets and various convolutional neural network model architectures to gain a more precise identification of glaucoma disease.

References

1. Olver, J., Cassidy, L.: Ophthalmology at a Glance. Blackwell Science Ltd., Hoboken (2005)
2. Jogi, R.: Basic Ophthalmology, 4th edn. Jaypee Brothers Medical Publishers (P) Ltd., New Delhi (2009)
3. Choplin, N.T., Lundy, D.C.: Atlas of Glaucoma, 2nd edn. Informa UK Ltd., London (2007)
4. Trends in prevalence of blindness and distance and near vision impairment over 30 years: an analysis for the Global Burden of Disease Study, vol. 9. Elsevier Ltd., February 2021, https://doi.org/10.1016/S2214-109X(20)30425-3. www.thelancet.com/lancetgh
5. Noronha, K.P., Rajendra Acharya, U., Prabhakar Nayak, K., Martis, R.J., Bhandary, S.V.: Automated classification of glaucoma stages using higher order cumulant features. Biomed. Signal Process. Control **10**, 174–183 (2014)
6. Mohamed, N.A., Zulkifley, M.A., Zaki, W.M.D.W., Hussain, A.: An automated glaucoma screening system using cup-to-disc ratio via Simple Linear Iterative Clustering superpixel approach. Biomed. Signal Process. Control **53**, 101454 (2019)

7. Song, W.T., Lai, I.-C., Su, Y.-Z.: A statistical robust glaucoma detection framework combining Retinex, CNN, and DOE Using fundus images. IEEE Access (2021). https://doi.org/10.1109/ACCESS.2021.3098032
8. Deepa, N., Esakkirajan, S., Keerthiveena, B., Bala Dhanalakshmi, S.: Automatic diagnosis of glaucoma using ensemble based deep learning model. In: 7th International Conference on Advanced Computing and Communication Systems (ICACCS). IEEE (2021). ISBN: 978-1-6654-0521-8/21
9. Saxena, A., Vyas, A., Parashar, L., Singh, U.: A glaucoma detection using convolutional neural network. In: International Conference on Electronics and Sustainable Communication Systems (ICESC 2020). IEEE Xplore (2020). ISBN: 978-1-7281-4108-4
10. Vaghjiani, D., Saha, S., Connan, Y., Frost, S., Kanagasingam, Y.: Visualizing and understanding inherent image features in CNN-based glaucoma detection. In: 2020 Digital Image Computing: Techniques and Applications (DICTA). IEEE (2020). ISBN: 978-1-7281-9108–9/20. https://doi.org/10.1109/DICTA51227.2020.9363369
11. https://ieeedataport.org/documents/1450-fundus-images-899-glaucoma-data-and-551-normal-data
12. Gua, J., Wangb, Z., Kuenb, J., Mab, L., et al.: Recent advances in convolutional neural networks. arXiv:1512.07108v6 [cs.CV], 19 October 2017
13. Sivaswamy, J., Krishnadas, S.R., Chakravarty, A., Joshi, G.D., Syed, T.A.: A comprehensive retinal image dataset for the assessment of glaucoma from the optic nerve head analysis. JSM Biomed. Imaging Data Pap. **2**(1), 1004 (2015)
14. Sivaswamy, J., Krishnadas, S.R., Joshi, G.D., Jain, M., Tabish, A.U.S.: Drishti-gs: retinal image dataset for optic nerve head (ONH) segmentation. IEEE. 978-1-4673-1961-4/14/$31.00 © (2014)
15. https://www.kaggle.com/datasets/sshikamaru/glaucoma-detection?resource=download
16. Pawar, D.J., Kanse, Y.K., Patil, S.S.: Classification based automated glaucoma detection using retinal fundus images. GIS Sci. J. **8**(5), 636–641 (2021). ISSN: 1869-9391
17. Aloudat, M., Faezipour, M., El-Sayed, A.: Automated vision-based high intraocular pressure detection using frontal eye images. IEEE J. Transl. Eng. Health Med. **7** (2019). https://doi.org/10.1109/JTEHM.2019.2915534

An Implementation Perspective on Electronic Invoice Presentment and Payments

B. Barath Kumar[1], C. N. S. Vinoth Kumar[1](✉), R. Suguna[2], M. Vasim Babu[3], and M. Madhusudhan Reddy[4]

[1] Department of Networking and Communications, College of Engineering and Technology (CET), SRM Institute of Science and Technology, Kattankulathur, India
{bb7318,vinothks1}@srmist.edu.in

[2] Department of Computer Science and Engineering, Bannari Amman Institute of Technology, Sathyamangalam, India
sugunar@bitsathy.ac.in

[3] Department of Electronics and Communication Engineering, KPR Institute of Engineering and Technology, Coimbatore, India

[4] Department of Electronics and Communication Engineering, K.S.R.M. College of Engineering, Kadapa, Andhra Pradesh, India
msreddy@ksrmce.ac.in

Abstract. A company's goods or services are sold to clients on a daily basis in Business to Business (B2B) situations. Companies like Tata Motors, Amazon to name a few, use COMPANY ABC (CLIENT) to purchase desktop and personal computers (PCs) for use in the workplace. There is a giant corporation, COMPANY ABC (CLIENT), which everyday sell a significant number of its personal computers (PCs) to a variety of other corporations. As a result, these corporations must pay COMPANY ABC (CLIENT) for all of the items they have acquired. As a result, COMPANY ABC has a wide range of clients (CLIENT). Accounting software such as Net Suite or SAP is used by COMPANY ABC (CLIENT) to keep track of all of its clients' unpaid bills. Using EIPP, companies all around the globe may exchange documents like invoices, purchase orders, and credit notes digitally instead of on paper. Electronic Invoice Presentment and Payment Users can rapidly monitor, query, authorize, manage, and pay for all of their payable and receivable transactions online with the touch of a mouse. With the help of the solution, businesses may replace expensive paper-based operations with more efficient technology, cut operational expenses, and make faster settlements and faster payments.

Keywords: Payment · Invoice · Cash flow · Presentment

1 Introduction

Daily transactions on a wide scale are done in Business to Business (B2B) situations for the items or services that a business delivers to its clients. Each buyer is billed immediately upon purchase. This is accomplished by creating unique invoices for each

J. I.-Z. Chen et al. (Eds.): ICIPCN 2022, LNNS 514, pp. 664–671, 2022.
https://doi.org/10.1007/978-3-031-12413-6_52

purchase made by a single client, including the anticipated payment amount and the due date by which it must be paid, among other things. Customers make payments against open invoices by sending checks or electronic funds transfers (EFTs) to the company's bank account. The bank then delivers payment files including the payment details for all payments made by each client. A single payment made by a client may cover the amount of any number of open invoices. As a result, the client must additionally transmit remittance information to the firm, including all bills paid in a given payment as well as any deductions (if any). Payments and remittances originate from distinct sources - the bank and the customers - and must be reconciled in order to cancel open invoices in the customers' General Ledger (GL). Because these transactions are B2B, they entail not just big payments sent across the world, but also a massive volume of transactions that any human system would struggle to keep track of. In these situations, it is critical to have a safe and secure payment mechanism in place, as well as a handy way to see all invoicing history, a list of all open and closed invoices, a record of all communication delivered to date, and a variety of other extra functions. The introduction discusses the challenges associated with manual payments and communication today, as well as the solution provided by automating the process via the use of automation software. This is followed by an outline of the implementation stages and a review of prior research in this topic. Following that, the terms and technology are described in depth. A consultant's important contribution is in the application of rules that justify and execute business processes.

2 Related Work

After the post international financial crisis, the cash flow management has become the primary focus for Multinational enterprises. With the technological advancements in today's era, it is necessary for the modern enterprises to re-examine the cash flow management from the internal perspective. Attempts have been made under the guidance of the application of internal control to build the scientific system, risk control system and value creation system of cash flow management along with the financial activities application guidance to create its new ideas, new concepts and new models [1]. But analyzing the growing complexity of the Financial Supply Chain Management (FSCM), we not only need systems and procedures to handle this complexity but also automated processes such as automation software to enable the cash flow management electronically with the least possible manual intervention [2, 3].

Electronic Invoice Presentment and Payment (EIPP), is one such attempt in this regard which duly addresses the problems of the Accounts Receivables section of the FSCM by automating the entire process of Payments and Presentment for B2B scenario [4, 5]. The EIPP software follows an iterative model of the Software Development Lifecycle (SDLC) wherein the entire lifecycle of development of the software is carried out based on the initial master framework developed in the first cycle of SDLC of EIPP.

Since the cash flow information can be applied well in the evaluation of an enterprise's financial situation, Li *et al.* in their paper [6] have regarded the cash profit, which is the information connotation of cash flow, as a basis of financial evaluation index system. Two early warning models based on cash profit and traditional accounting profit have been analyzed here to give an early warning before a financial distress arises. Financial institutions are motivated by the need to meet increased regulatory requirements for risk measurement and capital reserves. Wu *et al.* in their paper [7, 8] describe and demonstrate a model to support risk management of accounts receivable. They have presented a credit scoring model to assess account credit worthiness. However, a continuous monitoring of the Credit extended by a Business to its Customer Businesses in large scale Business to Business (B2B) operations can prevent or mitigate any such financial crisis. This raises a demand for more automation software that can not only manage the cash flow in B2B operations but also handle large volume of transactions with ease on a day to day basis keeping a check on the Credit Management of the company [9–11].

3 The Process of Order to Cash Cycle

The order to cash cycle, or O2C cycle, refers to the whole cash flow cycle, from the time an order is made to the time payment is received. This cycle is sometimes abbreviated as bill to cash or quote to cash. The O2C cycle may be better comprehended by thinking of it as a sequence of stages. Prior to making an order, the buyer inquires about the product's specifications and confirms his or her intention to acquire the item or service. Immediately after the acceptance of an order for a product/service, the credit risk of the client is assessed using the customer's portfolio. The Fig. 1 demonstrates that the consumer is financially capable of purchasing the items. The next stage is when the order is fulfilled, which means that the correct quantity of goods must be sent to the customer in accordance with the order placement. If there are any discrepancies, the customer must be notified immediately to avoid unnecessary complications during the payment process.

Once the order is packed, logistics takes over. The order is prepared to be turned over to logistics, and once the order is delivered [12–15], the O2C team gets evidence of invoicing and proof of landing from the carrier service. If there is a disagreement over an order, these papers are critical in settling it. Post-delivery invoicing is critical; this is completed by the billing and invoicing team and is provided to the client through mail, fax, or regular mail. The organization should maintain records of every credit extended and guarantee that payments are made on time [16–18]. Once payment has been received, it must be matched to the appropriate one-off bill and recorded in the ledger.

Fig. 1. Stages in O2C

4 The Proposed Electronic Invoice Presentment and Payment

Electronic Invoice Presentment and Payment (EIPP) is a very efficient billing and invoicing software solution for managing accounts receivables. EIPP streamlines and simplifies the customer experience by combining paper-based and electronic invoicing into a single automated end-to-end procedure. Additionally, this has a payment interface that allows businesses to collect payments on invoices from customers. This is a cloud-based, high-precision solution that smoothly interfaces with existing enterprise resource planning (ERP) systems [19]. Additionally, this provides insight into the status of invoicing and dispute resolution in real time. EIPPs use regional print hubs to give paper invoices to customers who need them, hence reducing postal costs. From the Fig. 2 Certain EIPPs provide tax compliance, record keeping, and electronic signatures, making them very adaptable to today's large, growing business.

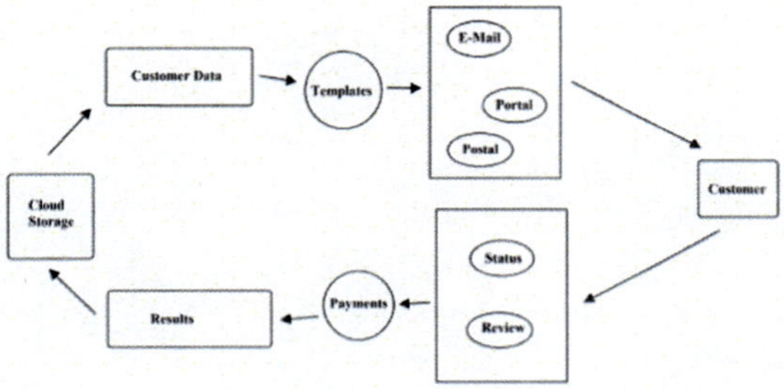

Fig. 2. Block diagram of the proposed model

The Proposed system consists of the following modules like Data Pre-processing, Data loading and Preparation, Blueprinting as per client requirements, Implementation of EIPP, Testing and in cycles, Deployment Using Production. Invoicing is crucial in creating a firm's customer relationship since it is the main form of communication between the business and the client. The conventional method of invoicing includes considerable human processing, which is not only time demanding but also prone to error. The time necessary for invoice delivery was excessively lengthy since it was solely dependent on the company's postal services, and foreign invoices were delayed; however, with the present technique of invoicing, invoices may be delivered quickly by using the wide connections of printing hubs. Due to the fact that the bulk of bills were on paper, invoices were housed in warehouses, and in the event of a dispute or other disturbance, the company was necessary to physically identify the relevant invoice in a sea of invoices. Because the invoices are printed on paper, they are susceptible to wear and tear from environmental conditions, making them non reliable. Electronic invoice presentment and payment, also abbreviated as EIPP, is a process through which a client may electronically receive and pay an invoice from you. The word "Presentment" is used instead of "presentation," since invoices resemble a formal request for payment.

EIPP operates by providing an internet connection, or gateway, to your customers' invoices, which are normally sent through email. The EIPP system, once verified the customer's identification, they will be able to see their current invoice as well as any previous ones. The consumer may then pay their invoice electronically via online banking, or directly by a debiting link on the EIPP system or an online card payment. This then results in a real-time money transfer, which means EIPP saves you both time and money!

The use of EIPP in the invoicing workflow intrinsically digitizes formerly paper-based operations, significantly increasing the efficiency and speed of the accounts receivable process and providing several additional advantages to both senders and receivers of electronic invoices.

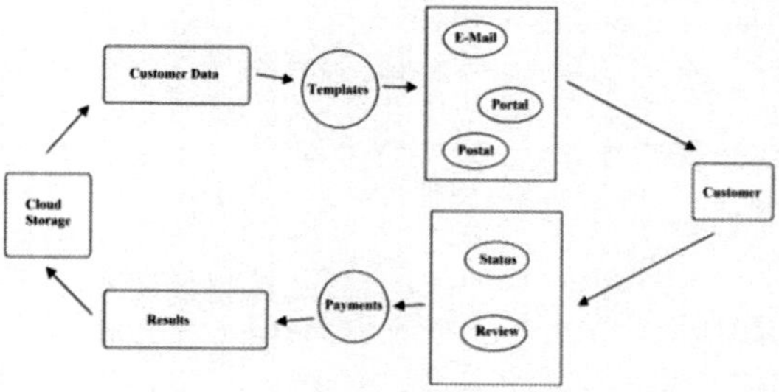

The Maximised automatic process can be achieved through B2B Practising with the proposed payments method. The EIPP Sample Application from Fig. 4 presents the complete information includes Invoice data and sale of report. From the available preset data-set Fig. 5 shows the improvised transactions using proposed EIPP. The graph represents the gradual increase in the Days sale outstanding and acknowledges the automation process.

Fig. 5. Shows the progress of improvement in days sale outstanding through the automation

5 Result and Future Work Discussion

EIPP allows users to edit, add, delete, or manage invoices using a straightforward user interface, enabling them to avoid mistakes, increase productivity, and so improve the overall efficiency of the business and its workers. Additionally, it helps us to preserve data, recognize patterns within the data, and make real-time modifications to trends that may assist in future decision-making, With this system, efficiency of payment cycle can increased by 10% and also the manual billing cost can be cut down by 50% also integrated with client ERP this enables accurate and efficient presentment of invoices. The advantages of an Electronic Invoice Presentment and Payment Application obviously outweigh the downsides of traditional invoicing and given the application's potential for future advancements, small and medium-sized enterprises should convert to EIPP, as it will be more convenient for them. While it may be time-intensive for larger businesses, the rewards are far higher. EIPP is a project that is always being enhanced. This cloud-based initiative is smashing all prior records for payment processing and collection, enabling the business to develop much faster. Joining an EIPP provides firms with a variety of advantages, including faster processing and a decrease in DSO. When cloud and the machine learning are fully integrated, the procedure becomes exceedingly simple. Leveraging deep learning and robotics for management and deeply understanding trends in invoicing will make things exceedingly simple.

References

1. Liu, F.: Cash flow management of management modern enterprises under the guidance of application of internal control. In: 2011 2nd International Conference on Artificial Intelligence, Management Science and Electronic commerce (AIMSEC), Deng Leng, China, pp. 6176–6178 (2011)
2. Dhaya, R., Kanthavel, R.: Edge computing through virtual force for detecting trustworthy values. IRO J. Sustain. Wirel. Syst. **2**(2), 84–91 (2020)
3. Vivekanandam, B.: Evaluation of activity monitoring algorithm based on smart approaches. J. Electron. **2**(03), 175–181 (2020)
4. He, M., et al.: Financial supply chain management. In: Proceedings of 2010 IEEE International Conference on Service Operations and Logistics, and Informatics, Qingdao, Shandong, pp. 70–75 (2010)
5. Li, T., Wang, W.: The study on application value of cash flow information connotation in financial early-warning. In: 2010 International Conference on Management and Service Science, Wuhan, pp. 1–4 (2010)
6. Chen, J.H., Chen, W.H.: Factoring account receivables towards mitigating cash flow fluctuation for construction projects. In: 2008 IEEE International Conference on Communications, Beijing, pp. 5538–5542 (2008)
7. Wu, D.D., Olson, D.L., Luo, C.: A decision support approach for accounts receivable risk management. IEEE Trans. Syst. Man Cybern. Syst. **44**(12), 1624–1632 (2014)
8. Zhang, L., Allam, A., Gonzales, C.A.: Service-oriented order-to-cash solution with business RSS information exchange framework. In: 2006 IEEE International Conference on Web Services (ICWS 2006), pp. 841–848 (2006). https://doi.org/10.1109/ICWS.2006.121
9. Španić, D., Ristić, D., Vrdoljak, B.: An electronic invoicing system. In: Proceedings of the 11th International Conference on Telecommunications, pp. 149–156 (2011)
10. Cedillo, P., García, A., Cárdenas, J.D., Bermeo, A.: A systematic literature review of electronic invoicing, platforms and notification systems. In: 2018 International Conference on eDemocracy & eGovernment (ICEDEG), pp. 150-157 (2018). https://doi.org/10.1109/ICEDEG.2018.83
11. Saranya, A., Naresh, R.: Efficient mobile security for E health care application in cloud for secure payment using key distribution. Neural Process. Lett., 1–12 (2021). https://doi.org/10.1007/s11063-021-10482-1
12. Saranya, A., Naresh, R.: Cloud based efficient authentication for mobile payments using key distribution method. J. Ambient Intell. Humaniz. Comput., 1–8 (2021). https://doi.org/10.1007/s12652-020-02765-7
13. Naresh, R., Vijayakumar, P., Jegatha Deborah, L., Sivakumar, R.: A novel trust model for secure group communication in distributed computing. J. Organ. End User Comput. **32**(3), 1–14 (2020). https://doi.org/10.4018/JOEUC.2020070101. Special Issue for Security and Privacy in Cloud Computing
14. Naresh, R., Sayeekumar, M., Karthick, G.M., Supraja, P.: Attribute-based hierarchical file encryption for efficient retrieval of files by DV index tree from cloud using crossover genetic algorithm. Soft Comput. **23**(8), 2561–2574 (2019). https://doi.org/10.1007/s00500-019-03790-1

15. Sakthipriya, S., Naresh, R.: Effective energy estimation technique to classify the nitrogen and temperature for crop yield based green house application. Sustain. Comput. Inform. Syst. (2022). https://doi.org/10.1016/j.suscom.2022.100687
16. Srivastava, G., Vinoth Kumar, C.N.S., Kavitha, V., Parthiban, N., Venkataraman, R.: Two-stage data encryption using chaotic neural networks. J. Intell. Fuzzy Syst. **38**(3), 2561–2568 (2020)
17. Vinoth Kumar, C.N.S., Suhasini, A.: Secured three-tier architecture for wireless sensor networks using chaotic neural network. In: Satapathy, S., Prasad, V., Rani, B., Udgata, S., Raju, K. (eds.) Proceedings of the First International Conference on Computational Intelligence and Informatics. AISC, vol. 507, pp. 129–136. Springer, Singapore (2017). https://doi.org/10.1007/978-981-10-2471-9_13. ISSN 2194-5357
18. Vinoth Kumar, C.N.S., Suhasini, A., IEEE Explorer Digital Library entitled: Improved secure three-tier architecture for WSN using hop-field chaotic neural network with two stage encryption, 15 August 2017. https://doi.org/10.1109/ICCECE.2016.8009540. ISBN 978-1-5090-4432-0
19. Sarma, P., Kumar, U., Vinoth Kumar, C.N.S., Vasim Babu, M.: Accident detection and prevention using IoT & Python Opencv. Int. J. Sci. Technol. Res. (IJSTR) **9**(04), 2677–2681 (2020). ISSN: 2277-8616

Multi-model DeepFake Detection Using Deep and Temporal Features

Jerry John(✉) and Bismin V. Sherif

Muthoot Institute of Technology and Science, Kochi, India
mtech20ai07@mgits.ac.in

Abstract. Deepfakes are one of the most advanced technological frauds that can be seen in the outer world, and it has been classified as one of the significant adverse impacts of deep learning. Deepfakes are synthetic media created by superimposing a targeted person's visual characteristics into a source video. This results in a video which contains content that the targeted person has never done. This kind of digital fraud can cause many social relevant problems like damaging the image and dignity of famous public figures, hate campaigns, blackmailing etc. Because of these reasons, it is high time to find some methods to detect these deep fakes even before they are published. For that, a deepfake detection method is proposed using a deep neural network. A combination of a temporal model-based and a deep model-based deepfake detection is used. For the temporal based model, a combination of ResNext and LSTM architectures and for the deep model based deepfake detection, a triplet model architecture is used. The datasets used for training this model are DFDC, Celeb-DF, and Faceforensics++, composed of different deepfake creation techniques. The extensive experiments show that the temporal model obtained the highest testing accuracy of 92.42% accuracy, at a frame rate of 100, and the triplet model obtained an accuracy of 91.88%. The final pipeline of these models obtain a testing accuracy of 94.31%.

Keywords: DeepFake detection · LSTM Resnext model · Triplet CNN · Multi model deepfake detection · Deepfake · Temporal model

1 Introduction

The advancement in technologies, the internet and communication systems have brought many changes in how people see the world. Even though deep learning and the machine learning were invented long back, they were not used to their full potential. The main reason behind is the lack of computational power and insufficient data. But as the internet boom occurred, both of these limitations were surpassed. From the beginning of the twentieth century, the computation power compared to their cost has been reduced by a large extent, and due to the wide use of the internet, millions of data are created and stored each second. This data can be used effectively to train different deep learning and machine

J. I.-Z. Chen et al. (Eds.): ICIPCN 2022, LNNS 514, pp. 672–684, 2022.
https://doi.org/10.1007/978-3-031-12413-6_53

learning algorithms with high-end computers to solve many real-life problems starting from the educational level to aerospace.

Deep Neural Networks works by learning from previous data and taking appropriate decisions based on that pattern. Most of the times this Neural Network models performs better than humans. Deep learning and artificial intelligence have a wide variety of positive applications. They are used in all kinds of areas to improve performances. The positive applications range from smart farming, analyzing and predicting different diseases even before they come, areas in auto-driving cars, giving customized product suggestions, and the list goes on. But deep learning also has its own negative sides. One of the most advanced and recent among them is deepfake creation.

Deepfakes are digitally tampered with or altered video and image data, where a target person's face is used to superimpose on a source so that it can be used for different purposes, starting from fun to fraud. Simply, the deepfake creates photos or gestures that a person doesn't do. This type of visual deepfakes can be made using many deep learning teachings like autoencoders, GANs etc. In auto encoder techniques, there will be an encoder and decoder. Both of them are trained separately using the target person's and the real person's face. After the training process, the decoders are interchanged to generate the deepfake (Fig. 1).

Fig. 1. DFDC dataset sample

Along with the visual part, the deepfakes can also be created with the audio, where the target person's voice is used to create a fake speech, which that person never said. It works by creating source audio by the deepfake creator. Then they use the target person's different audios to train a deep learning model, and then the source audio created initially is given to the deep learning model. This model converts the source audio into the voice of the targeted person. This audio deepfake combined with the visual deepfake can create very harmful videos because the voice with correct head and lip movement will be so accurate that we can't distinguish them with human eyes. This paper mainly concentrates on visual-based deep face detection.

This broad use of deepfakes can create many socially relevant issues. False messages from reputed persons can be easily created using deepfakes. This can lead to many personal and communal problems. Once these deepfakes are uploaded on the internet, they cannot be easily controlled. Because of these issues, large organisations, countries and commercial companies are conducting relevant research and developments in this area so that these deepfakes can be identified even before uploading to the internet.

2 Literature Survey

The deepfake detection works by extracting features from the face, which are then used for classification purposes. Basically, for deepfake detection, most of the algorithms initially detect the whole face, or by using the face landmarks, they detect the face. Then all the processing is done in this facial area. This deepfake detection is a binary classification task (detecting as real or fake). Based on the feature selection, the deepfake detection methods are classified into three: visual feature-based, deep feature-based, and temporal feature-based. The visual feature is based on deep fake detection, mainly concentrating on features that are seen or noticed visually. That includes eye movement, head pose, eye blinking, miss matching of facial regions, lip movement etc. The deep feature-based deep fake detection uses deep models to detect the digital manipulation on a pixel level. This method uses deep learning architectures for detection. Temporal feature-based deepfake detection uses temporal information to classify as real or fake; temporal information means continuous information. It is mainly used when a video input is used, where there will be a correlation between the consequent frames. This correlation can be used to detect deep fakes (Fig. 2).

In [1] Li *et al.* introduce a method that mainly focuses on the lack of eye blinking to detect the digital tampered video. This method is based on the conclusion that this type of physiological signs are not well captured in the digital tampered video. Convolutional Neural Network and LSTM architectures are combined. Where the CNN is used to detect whether the eye is closed or open, and the LSTM is used to identify the temporal information, as the eye blinking has a good correlation between the nearby frames. A cross-entropy loss is used as the loss function.

In [2] X. Yang *et al.* proposes an approach that can be used to detect deep fakes, which are created by placing the synthesized face in the source image. This method is based on the conclusion that the difference in head pose estimated between the central region and the whole face will be minimal for the real images and high for the fake images. The face detection is done initially, and then the

Fig. 2. Data preprocessing work flow

face landmark points are detected. Using these face landmark points, the head pose differences are estimated. These differences are then given to a Support Vector Machine(SVM) for classification.

In [3] Li *et al.* proposes an image representation method known as the face X-ray for deepfake detection. The face X-ray is basically a grayscale image, which can be used to distinguish whether an image is real or fake by checking whether the image is made by blending two images. If the image is fake, there will be a blending boundary, and there will be no blending boundary for the actual images. This method can generate negative examples of its own from the original images. Where initially, they collect positive images. Then randomly select one from them and detect the face landmarks from that. Another similar image is taken from the positive image set based on the face landmark similarity. Both of these images are blended to get the fake image. At the same time, after detecting the face landmarks, based on the landmark, an initial mask is generated, and it is finally converted into a face X-ray. This generated training set is used to train the fully connected convolutional neural network (Fig. 3).

In [4] Lyu *et al.* proposes a method that is used to detect the deepfakes which are created using affine transform. They are using a Convolutional Neural Network(CNN) architecture. This method is based on the observation that there will be a resolution inconsistency between the facial and surrounding regions for the deepfake contents. This resolution inconsistency is caused because of the compression step in the deepfake creation. The affine transform works by generating a transformation matrix using the face landmark points detected. This transformation matrix is used to create the affine transform to the targeted image to create the final deepfake. In this method, negative samples can also be generated from the positive samples, using a face alignment and gaussian blur technique.

Fig. 3. Face detection

Afchar *et al.* [5] uses a neural network called Meso-4 to detect the digital tampering of the video. This architecture concentrates on detecting digital tampering mainly based on deepfake and face2face methods. The Meso-4 primarily consists of four convolutional and pooling layers, followed by a dense layer network, including an input layer, one hidden layer and one output layer. To improve the generalization for each convolution layer, a ReLU activation function is used. This model concentrates on a fixed-sized Region Of Interest (ROI) of size $256 \times 256 \times 3$. That means a coloured input is used in this method.

In [6] Chintha *et al.* proposes a method using a bidirectional recurrent LSTM network, which can get both forward and backward temporal information. They initially convert the video into frames, and from each frame, the faces are extracted using the DLib Face Extractor. This face is then given to an XceptionNet which will extract the featural information(a vector representation is obtained). This is then passed through the bidirectional LSTM to get the temporal information, followed by a fully connected network for the final classification. They use a combination of both FaceForensics and Celeb datasets (Fig. 4).

Guera *et al.* [7] uses a system composed of a convolutional LSTM structure. This architecture mainly consists of two parts, CNN is used for the facial feature extraction, and LSTM is used to extract the temporal information. A fully connected network follows these networks for the classification.

In [8] Ekramm Sabir *et al.* proposes a technique that consists of a combination of ResNext and LSTM. A transfer learning technique is used to train the ResNext, and this ResNext is used for the feature extraction part. A feature vector is given as the output. In this method, the videos that need to be trained are converted into a preprocessed separate dataset. Only the face parts are cropped out and combined to generate this new video dataset. Which is then trained

Fig. 4. Preprocessing: detecting and cropping of face

using ResNext and LSTM, and this LSTM is used for the temporal information extraction.

In [9] Daniel Mas Montserrat *et al.* propose a method for detecting the deepfakes based on a combination of three different neural structures like MTCNN, CNN and RNN. This method can be divided into three parts; where in the first part, the face detection and face landmark points are extracted using MTCNN. This MTCNN is a multitask cascaded model that can provide both of these features. Then the face feature extraction is done using a CNN model. Then comes the automatic face weighting system. This automatic face weighting is used because a CNN only provides predictions for a single image. But in this process, a projection for an entire video is required. In ordinary cases, we choose the average, but this approach has many drawbacks. So we use an automatic face weighting mechanism that gets the most reliable regions where faces have been detected and discard the non-valuable frames. These features obtained are then given to a Gated Recurrent Unit (GRU) for the learning process (Fig. 5).

3 Purpose and Practical Implications

The deepfake is the most recent technological fraud, which pops up in the late 2017s. The development in computation and vast internet availability worldwide boost this process. The problems caused by deepfakes cant be scaled; it affects individuals to organisations. Once the data is published on the internet, it is challenging to control them. So it is crucial to evaluate them before publishing them on the internet. These deepfake detection models can be used on social media platforms (where most public videos are released). Before uploading to the public world, they can detect any sides of deepfakes on that videos. If found, they can prevent them from being public and take appropriate actions against them.

Fig. 5. Deep model work flow

4 Methodology

4.1 Data Collection

A combination of three datasets is used as the primary dataset for this model. The first one is the Deepfake Detection Challenge (DFDC) [11] dataset, which was created by Facebook using paid actors and consists of more than 10,000 videos. The main advantage of this dataset is that it has a wide diversity in its dataset like age, colour, tone, gender, etc. This dataset not only concentrates on any particular media manipulating technique, but it also contains deepfakes created using a wide variety of methods.

The second dataset is the FaceForensics++ [10] dataset, consisting of about 1000 videos using various deepfake methods like Face2Face, FaceSwap, Neural-Textures etc. The third dataset used is the celeb-DF [12] dataset, consisting of more than 5000 synthesis videos. For the proposed models a dataset consist of more than 10,000 videos are used, which is more than 10,00,000 frames.

All this dataset consists of videos, which are created using different deep fake creation methods. Some datasets like the celeb-DF dataset consist of original and corresponding deep videos. The dataset videos containing multiple faces are ignored for this work for the easiness of the work.

4.2 Data Preprocessing

The next step is to convert the data from various sources into a single format so that it can be easily given to various neural network architectures. The Deepfake Detection Challenge (DFDC) dataset consists of data blocks of almost 10 GB, consisting of videos and their corresponding details in a JSON file. These blocks of video data are combined to get all the data in one place. The JSON file is also converted into CSV format. The Celeb-DF and FaceForensics++ do not have any labelled CSV or JSON files, the deepfake and real videos are stored in separate folders for these two datasets. So a new CSV file is created using a separate program, which contains the file names and their corresponding label. This data is combined with the DFDC dataset to get the final dataset to be processed. Before the preprocessing step, a final sorting is done to remove the corrupted videos and the videos with a low frame rate (Fig. 6).

Fig. 6. Model workflow

Considering deepfake detection, the main region that needs to concentrate on is the face part. So, for the training process's easiness, all other surrounding regions can be avoided. Two libraries are mainly used for accurately detecting the face, one is the mediapipe, and the second is the face recognition. Each video from the dataset is consider separately, which is then converted into frames; then the faces are detected from each frame. This face region is cropped out, and all these cropped images are combined to create a preprocessed version of that original video dataset. This preprocessed set of video datasets and its corresponding labelled file is considered as the final dataset which is used for the model training.

4.3 Model Creation

For this deepfake detection model, a dual neural architecture is used. One is for getting the temporal information of the video, and the other one is used to extract the pixel-wise in-depth deep information from the video (Fig. 7).

Temporal Based Model. In most cases, there will be a correlation between the pixels. So when it comes to deepfake detection, it is essential to capture this information. This temporal information plays an essential role in improving the detection accuracy. This is achieved by using a combination of resnext and LSTM architectures. The neural network layers used in this model include ResNext CNN layer, where a pre-trained model of RCNN is used, consisting of 50 layers of 32×4 dimensions. After that, a sequential layer is used, which stores the feature vector coming from the ResNext model and which is then passed to

Fig. 7. Training vs validation graph

the LSTM layer. A transfer learning approach is used for the ResNext layer for better performance.

The LSTM layer is used to get the sequential data and spot the temporal changes among the video frames. In LSTM, the frames are processed sequentially by comparing the frame at 't' second with the frame a 't-s' seconds, and 's' is the number of frames. For this temporal model, a non-linear activation function called ReLU is used, which returns zero for all inputs less than 0, and if the input is greater than zero, it will provide the same input as the output. A dropout layer with a value of 0.4 is used to avoid overfitting (Fig. 8).

During the train test split, 30% of the data is given for testing, and the rest 70% is given for training. A balance between the real and fake data is also ensured for the better performance of the model. The data loader is used to load the video with their corresponding labels. The training is done for 50 epochs, and adam optimizer is used to enable the adaptive learning rate.

As deepfake detection is a binary classification problem, the cross-entropy approach is used to calculate the loss function. At the final layer, a softmax layer is used to classify as real or fake. This neural architecture is trained using the preprocessed data.

Deep Model. In the deep model, the dataset is converted into a collection of triplets. Triplets are a combination of three images, a positive image (real image), a negative image (fake image) and an anchor image (real image). These triplet images are passed separately through three different Convolutional Neural Networks. A triplet loss is calculated based on the difference between the negative and anchor image and the similarity between the positive and the anchor image.

Fig. 8. Model workflow

The model is trained to reduce this triplet loss by updating the neural network weights. The output of the Convolutional Neural Networks will be a vector. After the training, the average image vector for the fake and real images is calculated (Table 1).

The new video is first converted into frames on the prediction part, and faces are detected on each frame. These face images are then given to pretrained anchor CNN architecture to obtain the vector value. The Euclidean distance between the vector value obtained from the new image and the mean vector value obtained during the training process is compared. Based on this value, the images are concluded as fake or real. Suppose the Euclidean distance calculated between the new image and the mean real image vector is smaller. Then the video is categorised as real video and vice versa.

Table 1. Temporal models accuracy

Model name	Number of frames	Training accuracy	Testing accuracy
Model 1	10	82.55%	81.97%
Model 2	20	85.73%	83.77%
Model 3	50	89.18%	89.12%
Model 4	75	93.47%	90.06%
Model 5	100	95.68%	92.42%

5 Major Research Findings and Observations

A preprocessed dataset of more than 10,000 videos is used to train the temporal and the deep model separately. For the temporal model, input videos of different frame sizes are used to train multiple models for comparison purposes, and it is observed that the model with the highest frame rate of 100 obtains the highest training accuracy of 95.68% and testing accuracy of 92.42%. After training, the triplet based deep model obtained the highest testing accuracy of 91.88%. After extensive experiments based on different test datasets, it is observed that the temporal based model is more accurate in detecting deepfakes which are having a temporal correlation between the frames than the frames with complex pixel features. On the other hand, the triplet-based deep model performs well with deepfakes with complex pixel-level informations. Based on all the observations and comparisons, both of these models are combined to create a pipeline of models so that the system's overall accuracy increases. The combined final model obtained a better accuracy of 94.31% on testing, which outperforms most of the winning models of the DeepFake Detection Challenge [12] (Table 2).

Table 2. Models and performance

Model name	Training accuracy	Testing accuracy
Temporal Model	95.68%	92.42%
Triplet Deep Model	93.10%	91.88%
Final Model	96.52%	94.31%

6 Research Limitations and Future Works

The model proposed is limited to a frame rate of 100 frames per second. With a better dataset and good computational power, the model accuracy can be improved. However, in this step, the model will be able to capture more temporal information. There is a need to research other non-neural methods that can be combined with neural models so that the purpose of deepfake detection can be performed in a better way. Instead of concentrating only on the face region, research can be done on full-body deepfake detection. The proposed method only concentrates on the video-based deepfake; audio-based deepfake detection can also collaborate with the video-based deep fake.

7 Conclusion

In recent years deepfake technologies are growing at an unpredictable rate, resulting in many social relevant issues like hate campaigns, damaging the dignity of famous public figures, blackmailing etc. The spreading of fake information can affect the well being of society and individuals. To control these deepfakes, so that these deepfakes are detected and removed from the internet even before they are published, large organizations, governments and commercial companies are conducting relevant research and developments in this area. In this paper, a novel multi-model deepfake detection algorithm is proposed. An extensive dataset is used to improve the performance, along with some state of the art face detection libraries. The performance of different models based on different frame rates is analysed and compared. The temporal and deep models are combined to create a new pipeline to get better accuracy.

References

1. Li, Y., Chang, M.C., Lyu, S.: Exposing AI created fake videos by detecting eye blinking. In: 2018 IEEE International Workshop on Information Forensics and Security (WIFS), pp. 1–7. IEEE, December 2018
2. Yang, X., Li, Y., Lyu, S.: Exposing deep fakes using inconsistent head poses. In: ICASSP 2019 IEEE International Conference on Acoustics, Speech and Signal Processing (ICASSP), Brighton, United Kingdom, pp. 8261–8265 (2019)
3. Li, L., et al.: Face X-ray for more general face forgery detection. In: Proceedings of the IEEE/CVF Conference on Computer Vision and Pattern Recognition, pp. 5001–5010 (2020)
4. Li, Y., Lyu, S.: Exposing deepfake videos by detecting face warping artifacts. In: Proceedings of the IEEE Conference on Computer Vision and Pattern Recognition Workshops, pp. 46–52 (2019)
5. Afchar, D., Nozick, V., Yamagishi, J., Echizen, I.: MesoNet: a compact facial video forgery detection network. In: 2018 IEEE International Workshop on Information Forensics and Security (WIFS), pp. 1–7. IEEE, December 2018
6. Chintha, A., et al.: Recurrent convolutional structures for audio spoof and video deepfake detection. IEEE J. Sel. Top. Signal Process. **14**(5), 1024–1037 (2020)
7. Guera, D., Delp, E.J.: Deepfake video detection using recurrent neural networks. In: 2018 15th IEEE International Conference on Advanced Video and Signal Based Surveillance (AVSS), pp. 1–6. IEEE, November 2018
8. Sabir, E., Cheng, J., Jaiswal, A., AbdAlmageed, W., Masi, I., Natarajan, P.: Recurrent convolutional strategies for face manipulation detection in videos. In: Applications of Computer Vision and Pattern Recognition to Media Forensics at CVPR 2019 (2019)
9. Montserrat, D.M., et al.: Deepfakes detection with automatic face weighting. In: 2020 IEEE/CVF Conference on Computer Vision and Pattern Recognition Workshops (CVPRW), pp. 2851–2859 (2020)

10. Rossler, A., Cozzolino, D., Verdoliva, L., Riess, C., Thies, J., Nieaner, M.: FaceForensics++: learning to detect manipulated facial images. In: ICCV (2019)
11. Li, Y., Yang, X., Sun, P., Qi, H., Lyu, S.: Celeb-DF: a large-scale challenging dataset for DeepFake forensics. In: IEEE Conference on Computer Vision and Patten Recognition (CVPR) (2020)
12. Dolhansky, B., Howes, R., Pflaum, B., Baram, N., Ferrer, C.C.: The deepfake detection challenge (DFDC) preview dataset. arXiv preprint arXiv:1910.08854 (2019)
13. Singh, A., Saimbhi, A.S., Singh, N., Mittal, M.: DeepFake video detection: a time-distributed approach. SN Comput. Sci. **1**(4), 1–8 (2020). https://doi.org/10.1007/s42979-020-00225-9
14. Lyu, S.: DeepFake detection. In: Sencar, H.T., Verdoliva, L., Memon, N. (eds.) Multimedia Forensics. Advances in Computer Vision and Pattern Recognition, Springer, Singapore (2022). https://doi.org/10.1007/978-981-16-7621-5_12
15. Hao, H., et al.: Deepfake detection using multiple data modalities. In: Rathgeb, C., Tolosana, R., Vera-Rodriguez, R., Busch, C. (eds.) Handbook of Digital Face Manipulation and Detection. Advances in Computer Vision and Pattern Recognition, Springer, Cham (2022). https://doi.org/10.1007/978-3-030-87664-7_11
16. Jiang, L., Wu, W., Qian, C., Loy, C.C.: DeepFakes detection: the DeeperForensics dataset and challenge. In: Rathgeb, C., Tolosana, R., Vera-Rodriguez, R., Busch, C. (eds.) Handbook of Digital Face Manipulation and Detection. Advances in Computer Vision and Pattern Recognition, Springer, Cham (2022). https://doi.org/10.1007/978-3-030-87664-7_14
17. Suratkar, S., Sharma, P.: A simple and effective way to detect DeepFakes: using 2D and 3D CNN. In: Rao, U.P., Patel, S.J., Raj, P., Visconti, A. (eds.) Security, Privacy and Data Analytics. Lecture Notes in Electrical Engineering, vol. 848, pp. 227–238. Springer, Singapore (2022). https://doi.org/10.1007/978-981-16-9089-1_19
18. Kaliyar, R.K., Goswami, A., Narang, P.: DeepFakE: improving fake news detection using tensor decomposition-based deep neural network. J. Supercomput. **77**(2), 1015–1037 (2020). https://doi.org/10.1007/s11227-020-03294-y
19. Chen, J.I.Z., Smys, S.: Social multimedia security and suspicious activity detection in SDN using hybrid deep learning technique. J. Inf. Technol. **2**(02), 108–115 (2020)
20. Kumar, T.S.: Construction of hybrid deep learning model for predicting children behavior based on their emotional reaction. J. Inf. Technol. **3**(01), 29–43 (2021)

Real-Time Video Processing for Ship Detection Using Transfer Learning

V. Ganesh(✉), Johnson Kolluri, Amith Reddy Maada, Mohammed Hamid Ali, Rakesh Thota, and Shashidhar Nyalakonda

KITS Warangal, Warangal, Telangana, India
ganeshveluru08@gmail.com

Abstract. Automatic ship classification and detection is an interesting research field concerning maritime security. Automatic ship detection systems are important for maritime security and surveillance. These systems can be used to monitor marine traffic, illegal fishing and illegal activities which deal with the prospects of maritime security. This research has gained interest because many ships that are sailing on the ocean or sea do not install transponders which are used for tracking the ships. It will be a serious threat to the nation, mankind, and sea-life if we do not keep an eye on these kinds of ships. Therefore, in this we presented a novel Deep Learning method that will be used to detect ships by using satellite Images. This approach uses TensorFlow object detection API to detect objects in the images as we are concerned with object detection. The dataset used for this purpose is Maritime Satellite Imagery (MASATI)-v2 dataset which consists of various satellite images that are captured under different weather and dynamic conditions. As the real-time satellite monitoring is a kind of video thing, we proposed an approach to perform video processing to detect ships by using the model that is trained using MASATI-v2 dataset. For training the model we are using a Transfer Learning technique by utilizing (Single Shot Detector) SSD MobileNetV2. This algorithm is different from normal convolution neural networks, and it uses depth wise and pointwise separable filters to perform the task.

Keywords: Ship detection · Transfer learning · Object detection · Single shot detector · MobileNetV2

1 Introduction

The remote sensing technology is being used in many applications such as security and surveillance, illegal activities and pollution control, spills, or oil slicks. Leveraging the use of rapidly growing remote sensing technologies can help in maritime security. This technology uses many advanced sensors and also uses Automated Identification System (AIS), the Vessel Monitoring System (VMS), and the Synthetic Aperture Radar (SAR), images in a visible spectrum, as well as hyperspectral imaging acquired by Earth Observation Satellites (EOS) [11, 15]. Among these, Automated Identification System (AIS) and Vessel Monitoring System (VMS) uses Very High Frequency (VHF) and

J. I.-Z. Chen et al. (Eds.): ICIPCN 2022, LNNS 514, pp. 685–703, 2022.
https://doi.org/10.1007/978-3-031-12413-6_54

Global Positioning System (GPS) to wirelessly transmit the identity and present location of a ship. But here the main concerning thing is that some ships can go undetected or can be invisible because all the ships may not carry/install transponders, and in some cases, they can be turned off with wrong intentions to avoid radar detection so that they can carry out illegal activities. To overcome these problems Remote sensing technologies are being used.

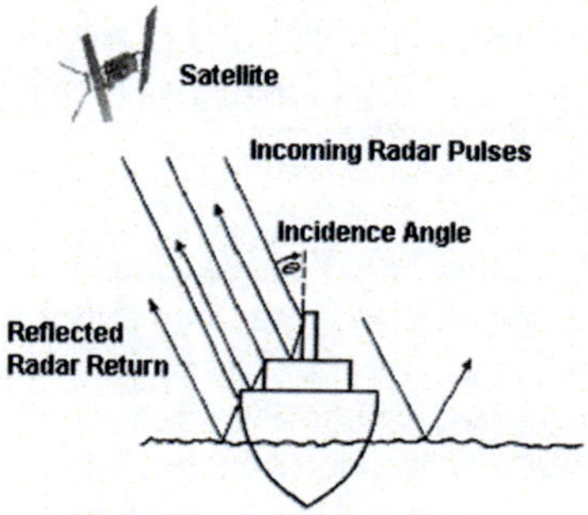

Fig. 1. Synthetic aperture radar system

Traditionally the Synthetic Aperture Radar (SAR) system as shown in Fig. 1. was used to obtain remote sensing imagery. It uses Radio wave High frequency radar signals to obtain remote sensing imagery. This Technology is neither affected by changes in light i.e., day or night nor by adverse weather conditions. Many satellites are equipped with SAR. But SAR images have some limitations because images have low radiometric and spectral resolution. To overcome these limitations, we have optical remote sensing technology which gives higher spatial resolution optical images, this improves object detection and recognition. The ship detected using optical satellite images can be identified by AIS or VMS and if it is not identified by these systems then this alerts us if either ship is performing illegal activities or is an Unidentified Floating Object (UFO). The advancement in optical technology has made it possible to obtain high spatial resolution satellite images. But limitation with optical imagery is that they cannot be used at night (i.e., in low lighting) or under bad weather conditions (unclear or cloudy). To accomplish the requirements of real-time ship monitoring, we have focused on researching what are the possible ways to detect ships in videos using optical satellite images. We propose the effective method of classifying and detecting ships from optical aerial imagery acquired in the visible spectrum by using Convolutional Neural Networks (CNN), a Deep Learning Object Recognition and Classification Technique.

In this project we develop a model which detects ships in satellite images using transfer learning techniques and it involves the reuse of an already trained model for a specific task. The first and middle layers are transferred in the process of transfer learning, and the other remaining layers are retrained to produce a new model [14]. The model uses labeled data to get retrained. This technique improves accuracy and leads to a good Deep Learning model.

TensorFlow object detection API which is easy for constructing, training, and deploying the objects detection model is open-source framework. TensorFlow model garden is a repository provided by TensorFlow which contains different kinds of implementations of state-of-the-art (SOTA) models, these models are raised for good performance and easy to read. TensorFlow object detection API uses protobufs library which is used for configuring the models and training parameters. protobufs it is basically the method of translating the image files to a format that can be stored in a file. This library should be installed and compiled before using it.

2 Literature Survey

In this section we discuss the research and related work on ship detection in optical aerial images.

Gallego, A.Pertusa and Gil [1] proposed Convolutional Neural Networks (CNN) based approach for ship recognition in optical aerial images. It also combines the neural codes which are extracted from CNN with the use of K- Nearest Neighbor method to improve performance. In this work, they have configured and evaluated many CNN models to obtain best fit hyper parameters to get best results. Nur Jati Lantang Marfu'ah and Arrie Kurniawardhani [2] have used ML Based and computer vision algorithms for detection of ships in satellite imagery. They have compared Convolutional neural Networks (CNN), a deep learning technique and Support Vector Machine (SVM), a ML code on the satellite imagery data set. By this comparison it is concluded that CNN has higher accuracy to that of the SVM and CNN is better in detecting ships in satellite images. Here CNN takes more time than SVM because CNN has more, and complex computing steps involved.

Xiaoshi, Yifan, Q. Guo, H. Zhang [3] in their work, they compared different ML algorithms such as Random Forests, Regression, K-nearest neighbor, Convolutional Neural Networks (CNN), Support Vector Machine (SVM). HOG feature extraction was used to improve the accuracy. In this comparison, the CNN model, which was built in this work, outperformed with a 99% accuracy which proves that CNN is better in image classification tasks. The sliding windows detection technique was used to detect all ships in satellite images with bounding boxes.

Guo-Qing,Y. Liu, Z. Kuang, HY Cui and [4] presented a ship recognition and categorization method on remote sensing optical images. In this method they extracted CDF 9/7 wavelet coefficients from the raw input data, and they also did candidate extraction with the use of LL subband in ships to reduce the complexity and processing time. They used techniques such as image enhancement, target-background segmentation to extract candidate ships in satellite images. Then they trained the CNN model to detect original ships from all candidate ships extracted. The model was trained on Google Earth which are remote sensing. This model achieved 95% classification accuracy and 99% recognition accuracy. This proves that CNN is best for object detection and classification.

Mohd Stofa, M.Asyraf and Siti Zaki [5] proposed a method based on DenseNet to classify and detect ships in remote sensing images. DenseNet is a complex, refinement algorithm which gives more than 90% accuracy in object detection. In this experiment the best results i.e. hyperparameter is obtained when a batch size was considered sixteen and a learning rate of 0.0001. zur des, I.doktor Agrar and U. Fakultät der [6] have developed a process for VHR and MR optical satellite images. They have presented a complete real time application for maritime security which integrates the AIS and ship recognition method based on CNN. Here they dealt with two types of datasets. In this project the VHR training dataset was acquired from the set of Worldview and GE-1 images which has nearly forty thousand of annotated images divided into fourteen categories. And the MR training dataset from the set of Landsat-8 images which has nearly fourteen thousand of annotated images of 7 categories.

Tian, X. Bai, Z. Feng, W. Fan and Tao Mingliang [7] proposed a DCNN based ship detection algorithm for PolSAR images. They focused to mitigate the problem of detection of ships along the coastline, for this they have used a three-class classifier based on DCNN, to extract images or samples which contain ships. Furthermore, they compared the accuracy of the model to that of the conventional Faster R-CNN and modified Faster R-CNN. This resulted in concluding that Faster R-CNN is conventionally faster and accurate in detection of ships which are of varying sizes. H. Liu, S. Han, F. Chen, and Y. Dong [8] have proposed a Faster-R-CNN method to detect the ships in High resolution optical remote sensing images. The active rotating filter technique has been used for feature extraction which uses oriented response networks, channel attention, and spatial attention. They used active rotating filters and neural networks to improve the performance of the model. They compared their proposed model to that of other detection methods based on CNN. The mean average precision of their proposed model has been improved by 5.49% to that of the other conventional CNN Algorithms.

3 Preparing the Dataset

3.1 Dataset

The dataset used here is MASATI-v2 dataset which is an acronym of Maritime Satellite Imagery dataset, and this dataset consists of nearly 7300 images. These images are colored and captured under different typical dynamic environments. These images are real maritime scenes and mainly used for our evaluation in various ship recognition methodologies. The number of ships in a particular image is one or multiple at different weather conditions. The dataset is primarily divided into 7 classes namely ship, detail, land, multi, coast, coast-ship, and sea. Here the main classes we are concentrated on are ships and no ship as our main concern is ship detection. The different classes in the dataset are given in Table 1. The images in the dataset are captured from various continents and several oceans. All Images should be in the same size for training, and all images are resized to 320 $\times$ 320. The sample images from our dataset are shown in Fig. 2.

3.2 Annotation

We humans can understand how a ship will look like and what object is what based on learning in our life, but computers may not understand the perceptual information like us and to make our model to act similarly humans we must express to model what is

Fig. 2. Sample ımages from dataset

the main goal and what it must interpret to make decisions. For this so much of data is required to get better understanding about its goal and to make those objects recognizable Annotation plays a major role in the field of object recognition, and we can simply define Annotation as a sequence of labeling the data which is mainly used in case of supervised learning and without the intervention of data Annotation the model is not capable of learning the input patterns. So, to get a good model it requires high quality of data which is annotated by manpower and annotation is used to improve accuracy and for correct prediction.

Table 1. Dataset

Main class	Sub class	Description	Number of ımages
ship	Detail	Elaborate details of ship	1789
	Ship with coast	Images of ship near coast	1037
	ship	Ship in ocean	1027
	Multiple ship	Many ships on sea	304
No ship	Land	Normal area on earth without ships	1078
	sea	Sea area with no ships	1022
	coast	Coast without ships	1132

3.3 LabelImg

The Image annotation tool that we have used for Annotating Images in our Dataset is LabelImg as it is free open source, and it is created using python and QT. We can draw a bounding box around the object by dragging the mouse. After drawing a box, we must assign a class for that object for the sake of learning. The file that is created for each image is a xml file that consists of coordinates where our interested object is located. If there are multiple objects in a single image, then coordinates for each object will be generated in the same file. For images which have no objects there will be no coordinates in it. The respective class for each object will also be there in the.xml itself. The Annotated Image and its respective.xml file are shown in Fig. 3.

3.4 Partition the Dataset

After the images in the dataset are annotated the next task is to divide the Dataset into two parts. One part is used to make our model Learn the features of the ship and another part is used to check whether our model learnt well or not. The subset of data that is used to make our model learn is called Training dataset and the dataset used to evaluate our model is called Testing dataset. The ratio of Training and Testing dataset that are used is 80, 20 respectively in this project. In our case there are separate folders for each subclass. Therefore, from each folder we must divide into our training and testing ratio. Using the Python program, we divided each subclass folder into 2 parts and placed each subclass Training data under the train folder and testing data into test folder.

3.5 Label Map

Our task of detecting ships requires a label map which serves the purpose of mapping a class to an integer value. This Label map is used by both testing and training processes. Since we are only concerned with ships our Label map will looks like

Fig. 3. Annotated image

item {
id:1
name: "ship"
}

Here the name should match our class that we have used while annotating our dataset.

3.6 TFRecords

Images and their respective.xml files which have bounding box information are there. After this, create a.csv file which consists of all information about images. This.csv file will serve as input to develop the TFRecord format. TFRecord is a data format and used in TensorFlow. We need to generate a TfRecord format for both train and test data. The problem with normal data is there will be many images and they may be scattered over disk, to overcome this problem TfRecord format is used. It will store data into sequential format as shown in Fig. 4.

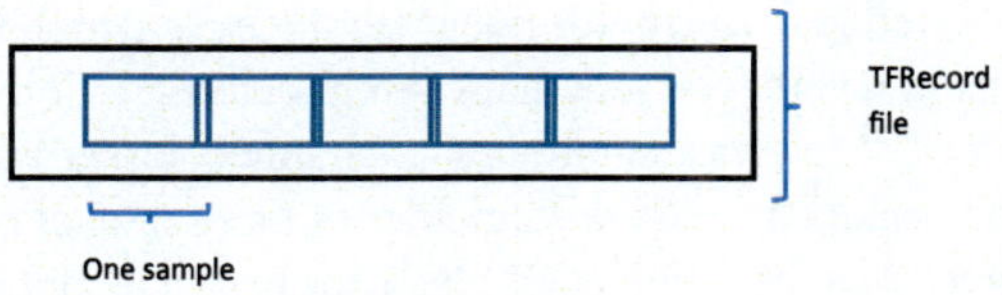

Fig. 4. TFRecord

Here every single data is considered as one Example and it is generally a dictionary that stores key and our data. The Example has many inner components which are called as features. We can think of why we cannot store our data in the form of numpy arrays. The main advantage is TFRecord enables quick access as it is stored in sequential manner. There is also a possibility native integration into TensorFlow API's and the data in TFRecord format can be shuffled dynamically by itself which will be helpful in training and testing phases. As it is storing data in binary format the performance will Improve and the time to train our model will decrease and that binary data consumes less storage on disk.

In TFRecord the data will be stored in the form of binary strings, and we must predefine some structure before writing it into file. First, we can preserve our data and after that we can serialize our data using tf.python.io.TFRecordWriter to place it on disk. Here the protocol buffers are used for serialization and deserialization purposes and stores our data in the form of bytes. The building blocks for TFRecords are protobufs and they are like json and xml formats but these TFRecord formats are even simpler than xml and useful in faster transmission. These protobufs are developed by google and they are not much readable like xml formats. The complete process of preparing data for training is shown in Fig. 5.

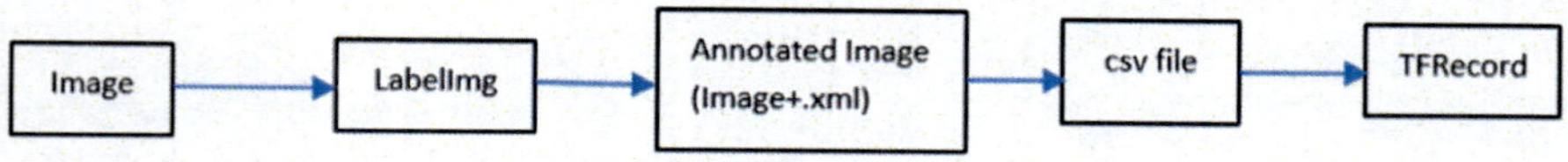

Fig. 5. Flow of data preparation

4 Architecture

To perform the training, we need to download a pre-trained model from the tensor flow model garden and the model we used here is ssd_mobilenet_v2.

4.1 MobileNet-v2

MobileNetV2 is the better improvement of MobileNetV1 [9, 10] uses state of the art (SOTA) for visual recognition which includes object detection, classification, and segmentation. Mobile Net v2 is a new layer model or block which uses depth wise separable convolutions called inverted residuals with linear bottlenecks which is to filter the features in between expansion layers. The schematic Architecture of MobileNetV2 is shown in Fig. 6. In MobileNetV2 it makes the number of channels lesser which is also called projection layer which projects the data with greater number of channels to lower number of channels or dimensions. MobileNetV2 also uses residual connections which is a new thing in V2 which helps in the flow of gradients through the network.

Fig. 6. MobileNetV2 architecture

4.2 Single Shot Detector (SSD)

SSD is used for object detection; it will predict which object is present in image with their location. SSD makes many predictions for every class and scores will be there for respective detections. The input is given to MobileNetV2 Architecture which is basic Architecture in SSD which is mainly used for Image Classification for extracting feature maps. To get Predictions there are Convolution feature layers after MobileNetV2 Architecture [13]. Each Layer will generate Predictions using convolution filters. Finally non-maximum suppression is used to get final Detections based on confidence scores. The architecture of the SSD is shown in Fig. 7.

Fig. 7. Single shot detector architecture

5 Configuring a Training Pipeline

Evaluation and Training processes are configured using protobuf files. The pipeline.config consists of five parts. First one is related to the model which has parameters like model used, activation function and Regularizer etc. Second one in train_config which has information about parameters that are used to train the model, for example augmentation, optimizer etc. Third one is eval_config contains the metrics that will be used for evaluation purposes and the metrics used here is coco_detection_metrics. Fourth part is train_input_config which has Information about Labelmap and input TfRecord file for training. The final part is eval_input_config which has a path to test tfrecord file and Label map. The Parameters used in the config file are given in Table 2 and Table 3.

Table 2. Parameter values for model

Model	
Parameter	Value
max_detections_per_class	100
height	320
width	320
initializer	truncated_normal_initializer
depth_multiplier	1.0
regularizer	L2 regularizer
box_coder	faster_rcnn_box_coder
classification_loss	weighted_sigmoid_focal
similarity_calculator	iou_similarity
localization_loss	weighted_smooth_l1
matcher	Argmax_matcher
num_layers_before_predictor	4
localization_weight	1.0
score_converter	sigmoid
depth	256
classification_weight	1.0
kernel_size	3
score_threshold	0.3
iou_threshold	0.6
num_classes	1
max_total_detections	100
feature_extractor- type	ssd_mobilenet_v2_fpn_keras
activation	RELU_6

Table 3. Parameter values for train config

train_config	
Parameter	Value
data_augmentation_options	random_horizontal_flip, random_crop_image
Batch_size	8
total_steps	50000
learning_rate	cosine_decay_learning_rate
Fine_tune_checkpoint_type	detection
optimizer	momentum_optimizer
max_number_of_boxes	100
momentum_optimizer_value	0.8999999
warmup_steps	2000
warmup_learning_rate	0.013333
learning_rate_base	0.039999

6 Object Detector

6.1 Training the Model

After configuring the training pipeline, the next phase is to train the model so that the model learns the training data to detect ships in satellite images. The training job can be started by using model_main_tf2.py.

6.2 Evaluating the Model

After the training process is done the next task is to evaluate our model on a testing dataset to check whether our model is doing good or not on new data of satellite images. During the training process it will generate various checkpoints and by using these checkpoints now our model is going to be evaluated on our testing images data. By using the metrics defined below we are going to see the performance of the model on the testing dataset. These metrics we must download and install, the metrics we are using in this instance is COCO metrics and we include these metrics in the.config file.

Metrics

a) *True positive:* It says that our model predicted as positive, and it is true means the correct decision is taken by our model in detecting objects.
b) *False Positive:* The object is not there but our model is drawing bounding boxes there.

c) *False Negative:* Our model missed in detecting the object although it is there.
d) *True Negative:* It refers to our model detected as Negative and it is true, but here in object detection we do not deal with detecting backgrounds instead of objects. While at the time of annotating we did not do any explicit annotation for backgrounds which are not our interest. Therefore, our model has no need to detect background regions correctly.
e) *Intersection over Union (IOU):* IOU is one of the object detection metrics that tells the difference between ground-truth objects and detecting boxes as shown in Fig. 9. means it finds the degree of overlap between actual and predicted boxes.

$$IOU = \frac{area(gt \cap pd)}{area(gt \cup pd)} \tag{1}$$

where gt means ground truth and pd means predicted.

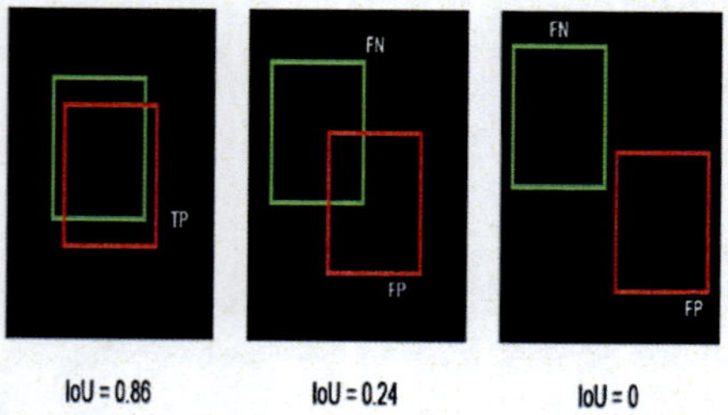

Fig. 8. TP, TN, FP, FN

Fig. 9. Intersection over union

In our task of ship detection, the model may predict multiple boxes for each object and this IOU will act as a metric to decide which detection is correct or not. IOU will act as a threshold(alpha). And it can remove unnecessary bounding boxes.

The Fig. 8 shown above has TP, FP, FN's and if we change the threshold IOU = 0.2 then the second image will have no FN, FP and it becomes TP. Therefore TP, FP, FN are dependent on threshold value.

f) *Precision:* It is the measure of correctness of our model

$$Precision = \frac{(TP)}{(TP + FP)} = \frac{TP}{all\ detections} \tag{2}$$

g) *Recall:* it is the measure of true predictions from all correctly detected data.

$$Recall = \frac{(TP)}{(TP + FN)} = \frac{TP}{all\ ground - truth\ images} \tag{3}$$

If a model has high precision and high recall, then we can say that it is a good model.

h) *Average Precision:* In an instance to compare any precision-recall curves is difficult as they are in a kind of zig-zag manner. Therefore, to represent this precision-recall curve as a single value, average Precision is used. The formula for AP is given as

$$AP = \sum_{k=0}^{k=n-1} [Recalls(k) - Recalls(k+1)] * Precisions(k) \tag{4}$$

where n = no of thresholds, Recall(n) = 0, Precision(n) = 1

i) *F1-Score:* Another way to see balance between Precision and recall is to use F1-score. If both are good and balanced then F1-score will be better.

$$F1\,Score = \frac{2 * Precision * Recall}{Precision + Recall} \tag{5}$$

j) *Mean average Precision:* Generally average precision is calculated separately for each class and there will be as many average precisions as classes. Now the average of these average Precision for each class gives mean average Precision.

$$mAP@ \propto = \frac{1}{n} \sum_{i=1}^{n} AP_i \; for\, n\, classes \tag{6}$$

AP_i = AP of class i

k) *mAP@0.5IOU* - It gives us the mean average precision with IOU threshold with 0.5. Similarly, mAP@0.75IOU means IOU = 0.75.
l) *mAP(small)* - Here small indicates we are calculating mAP for boxes that have small size and the classification of small based on area sizes is given in the Table 4. Similarly, for (large, medium).

Area = all indicates the metric is calculated for all the given sizes that are small, medium, large. If we have not there that much area bounding boxes it will simply give us −1 as result.

Table 4. Bounding box area vs size

Size	Area
Small	$<32^2$
Medium	$32^2 <$ area $< 96^2$
Large	$>96^2$

xiii) AP@IOU = 0.5:0.95 – it gives the average Precision value for thresholds from 0.5 to 0.95 with 0.05 increment each time.

xiv) AR@1 – here ‘1’ is maxdet, it gives average recall for images with at most one detection. Similarly for 10 and 100.

xv) Localization Loss: The Localization is nothing but regression of coordinates, the loss related to offset prediction of bounding boxes is Localization loss.

xvi) Classification Loss: the loss related to conditional class probabilities is called classification Loss.

xvii) Regularization Loss: the extra loss that is getting generated by the regularization function.

xviii) Total Loss: it is here described as sum of Localization loss, classification Loss and Regularization Loss.

6.3 Exporting Trained Model

After the training job is completed, we must extract that model so that it can be used to detect objects in real-time and we can use this saved model anywhere when it is required. The script exporterv2_main.py in the training demo can be used to save the model using the last checkpoint by executing the command. Then we will get a model file which will be used for performing object detection for images as well as videos.

6.4 Object Detection with Saved Model

Image

To get bounding boxes on a particular image if a ship is detected we are using visualization_utils from object_detection. utils. First, we must load our saved model and convert the input image to tensor as our model accepts input image as tensor. After changing to tensor, pass that tensor to the loaded model. Make a separate copy of the image of input to draw boxes based on coordinates returned by model. The bounding boxes for image after using trained model is shown in Fig. 10.

Fig. 10. Results on test data

Make a separate copy of the image of input to draw boxes based on coordinates returned by model. The bounding boxes for image after using trained model is shown in Fig. 10. The below code in Fig. 11 shows the procedure of drawing a box for detected ships.

Fig. 11. Sample code for drawing boxes in image

Video

The real time satellite tracking system is monitoring continuously the oceans/ maritime environment means it is working on the real time videos. Therefore, we have to use our model on videos to verify whether it is detecting objects(ships) in videos. We can define video as a set of images [12]. We can simply make a video by merging images in the dataset so that we can use that video for our model. Now for the model we have to give this video as input, but our model is trained on images, and it accepts only images as input. For video processing what we have done is by reading the video file we have to capture images from them and send this image as input for our loaded model. By using the above procedure, the ship in the image is shown by drawing bounding boxes and after this again we must write this image in video so that the final output we receive is video. The sample code for video processing is shown below in Fig. 12.

Fig. 12. Sample code for drawing boxes for video

Deployment
After Building and testing the model, the next main phase is deployment. We need to develop an interface so that users can upload images or videos into it and get the output with detected ships. For that we are using flask which is a micro web framework in python and the template engine used by this framework is jinja. The workflow of our application is shown in Fig. 13.

7 Evaluation Results

The evaluation results for the testing data set are (Tables 5, 6, 7 and 8):

Table 5. Average precision values

Evaluation results-average precision variations			
IOU	Area	maxDets	result
AP@[IOU = 0.50:0.95]	All	100	0.392
AP@IOU = 0.50	All	100	0.796
AP@IOU = 0.75	All	100	0.323
AP@[IOU = 0.50:0.95]	Small	100	0.356
AP@[IOU = 0.50:0.95]	Medium	100	0.861
AP@[IOU = 0.50:0.95]	Large	100	−1.000

Table 6. Average recall values

Evaluation results-average recall variations			
IOU	Area	maxDets	result
AR@[IOU = 0.50:0.95]	All	1	0.325
AR@[IOU = 0.50:0.95]	All	10	0.503
AR@[IOU = 0.50:0.95]	All	100	0.556
AR@[IOU = 0.50:0.95]	Small	100	0.539
AR@[IOU = 0.50:0.95]	Medium	100	0.878
AR@[IOU = 0.50:0.95]	Large	100	−1.000

Fig. 13. Flow of Application

Table 7. Mean average precision values

Evaluation metrics at step 50000-mean average precision	
mAP	0.392215
mAP@0.50IOU	0.795573
mAP@0.75IOU	0.322740
mAP(small)	0.355678
mAP(medium)	0.860700
mAP(large)	−1.0000
mAP	0.392215

Table 8. Loss values

Loss	value
Localization loss	0.072644
Classification loss	0.113660
Regularization loss	0.100630
Total loss	0.286934

The Information about Losses are discussed Previously in Sect. 4.2 and the following graphs that are shown in Fig. 14 captured using Tensorboard while Training the model.

Fig. 14. Loss graphs

8 Conclusion and Future Scope

Real time video processing of ship detection is proposed here as illegal activities are going on across the ocean like illegal fishing, women trafficking and oil spilling etc. In this Transfer learning approach is used because the pre-trained model will get trained faster compared to normal one and satellite monitoring is like video processing therefore, the main concentration is on processing the video using the model which is trained on images. MobileNet-v2 algorithm is used for this task as it is an extension of MobileNet-v1 and achieved a loss of 0.28.

In the future we can use many other pre-trained models like R-CNN, ResNet50, Faster R-CNN, Retinanet to compare with the present one and by using all these models we can even perform Ensemble Object Detection to get more accurate results for detecting ships in real-time.

References

1. Gallego, A.-J., Pertusa, A., Gil, P.: Automatic ship classification from optical aerial images with convolutional neural networks. Remote Sens. **10**, 511 (2018)
2. Marfu'ah, N.J.L., Kurniawardhani, A.: Comparison of CNN and SVM for ship detection in satellite ımagery. Department of Informatics, Faculty of Industrial Technology,Yogyakarta ,Islamic University of Indonesia. Naskah publikasi (2020)
3. Li, Y., Zhang, H., Guo, Q., Li, X.: Machine learning methods for ship detection in satellite ımages (2003)
4. Liu, Y., Cui, H.-Y., Kuang, Z., Li, G.-Q.: Ship detection and classification on optical remote sensing ımages using deep learning. ITM Web Conf **12**, 05012 (2017)
5. Stofa, M.M., Zulkifley, M.A., Zaki, S.Z.M.: A deep learning approach to ship detection using satellite imagery. IOP Conf. Ser. Earth Environ. Sci. **540**, 012049 (2020)
6. Deep Learning-based Vessel Detection from Very High and Medium Resolution Optical Satellite Images as Component of Maritime Surveillance Systems. zur Erlangung des akademischen Grades,Doktor der Ingenieurwissenschaften (Dr.-Ing.) an der Agrar and Umweltwissenschaftlichen Fakultät der Universität Rostock (2020). https://doi.org/10.18453/rosdok_id00002876
7. Fan, W., Zhou, F., Bai, X., Tao, M., Tian, T.: Ship detection using deep convolutional neural networks for PolSAR images. Remote Sens. **11**, 2862 (2019)
8. Dong, Y., Chen, F., Han, S., Liu, H.: Ship object detection of remote sensing image based on visual attention. Remote Sens. **13**, 3192 (2021)
9. Sinha, D., El-Sharkawy, M.: Thin MobileNet: an enhanced mobilenet architecture. In: 2019 IEEE 10th Annual Ubiquitous Computing, Electronics & Mobile Communication Conference (UEMCON), pp. 0280–0285 (2019). https://doi.org/10.1109/UEMCON47517.2019.8993089
10. Dong, K., Zhou, C., Ruan, Y., Li, Y.: MobileNetV2 model for ımage classification. IN: 2020 2nd International Conference on Information Technology and Computer Application (ITCA), pp. 476–480 (2020). https://doi.org/10.1109/ITCA52113.2020.00106.Ssd
11. Shetty, N.: Investigation of operational efficiency using stochastic models for electric propulsion in ships. J. Electr. Eng. Autom. **2**(2), 84–91 (2020)
12. Ginimav, I.: Live streaming architectures for video data-a review. J. IoT Social Mob. Anal. Cloud **2**(4), 207–215 (2020)
13. Chen, S., Hong, J.,X Zhang, J., Li, J., Guan, Y.: Object detection using deep learning: single shot detector with a refined feature-fusion structure. In: 2019 IEEE International Conference on Real-time Computing and Robotics (RCAR), pp. 219–224 (2019). https://doi.org/10.1109/RCAR47638.2019.9044027
14. Zhuang, F., et al.: A comprehensive survey on transfer learning. Proc. IEEE **109**(1), 43–76 (2021). https://doi.org/10.1109/JPROC.2020.3004555
15. Schwehr, K.D., McGillivary, P.A.: marine ship automatic identification system (AIS) for enhanced coastal security capabilities: an oil spill tracking application. Oceans **2007**, 1–9 (2007). https://doi.org/10.1109/OCEANS.2007.4449285

Smart Shopping Using Embedded Based Autocart and Android App

V. Sherlin Solomi, C. Srujana Reddy, and S. Naga Tripura(✉)

Department of ECE, Hindustan Institute of Technology and Science, Chennai, India
vsherlins@hindustanuniv.ac.in, {18121092, 18121118}@student.hindustanuniv.ac.in

Abstract. In the present situation, shopping in malls became daily activity to the human kind. After completion of collecting the products, the customers need to wait in a long queue for billing. To show the difference in ordinary shopping and the smart shopping this system is going to propose. Instead of searching for the products in the mart and wasting lots of time at the billing counters by standing in a long queue, we are proposing a system to save the customer's time. It also helps the customers to follow the social norms, as it is very important for the human kind in the current pandemic situation. At times people were busy and, in a hurry, people may forget to maintain social distance. In such situation an alert will be given as an Ultrasonic sensor is interfaced with in the frame work. In our proposed system, each smart cart is equipped with RFID reader, a Arduino, an LCD display, a Bluetooth module and GSM module.

Keywords: RFID reader · RFID tags · GSM module · LCD display · Arduino UNO · Bluetooth module · Ultrasonic sensor

1 Introduction

At the present time, from the past two years we are listening a word "corona" where ever we go. Because of this more than 180 countries were affected and until now the countries were facing many problems, many countries lost their economy also. And the least countries are having poverty because of this epidemic. So, to avoid those pandemics many are afraid of going out and are unable to get the necessary requirements. If any wants go out then definitely need to follow some social-norms like wearing mask, sanitization and the main thing to maintain social distance. To create awareness regarding the social norms to the people, an ultrasonic sensor has been inserted to the framework which will alert the people when they over rule the social norms.

Now a days people were too busy in their works. Making a machine human friendly this system is going to be proposed. The human kind was addicted to smart work. For instance, while people were going to super markets, they are wasting lot of time in buying the products and at the billing counter. This system may help the customers to save their time and makes their work smartly [2]. Each and every time the customers in the super market are wasting their time in searching of products and also near the billing counters.

J. I.-Z. Chen et al. (Eds.): ICIPCN 2022, LNNS 514, pp. 704–712, 2022.
https://doi.org/10.1007/978-3-031-12413-6_55

To save the customer time, in the proposed system an app is developed to search the for the product in the mart and for the final payment also. The customers can automatically pay the bill by themselves after buying the products. So, this project will helpful for customers and will saves their time.

2 Literature Survey

In [4] during last 2 years the government of every state announced lock down due to the increasing of covid cases rapidly. All the shops were closed and people felt very difficult to go and get the necessary items and daily needs. So, the government gave permission to open shops in a particular time so that the public can go and get their required things. In that particular limited time, it is very difficult for all to buy the things by following social norms.

A BBC report in July 2020 reported that since social distance needs to be maintained, many sectors must innovate especially the super markets and the grocery stores [5]. There are many customers standing in long queues and maintaining social distance of two meters apart from one another following the protocols by wearing masks and frequent sanitization. So, it takes a long time for the customers and many of them may not be able to buy their requirements.

Majority of the articles [6] reported that the spread of the virus is mainly due to the grocery stores. It came to know that the virus is spreading through contact. So, the government-imposed lockdown to reduce the crowd and this mainly affected the people to buy the essentials. Since all the shops were closed and the shops that are opened need to follow the covid norms by maintaining a social distance of at least two meters from one person to another to avoid the spreading of virus to other person.

Since the pandemic situation around the world has forced us to remain home and the government has suggested to open shops with social distancing. This is the main reason for us to help the people in pandemic by smart and easy shopping [1].

In a smart shopping cart [9] is implemented for the clients to examine the thing and naturally refreshes the bill hence forestalling long lines at the checkout. Likewise, one more fascinating element of this brilliant shopping basket is the truck to-truck correspondence that assists the clients with shopping corresponding with loved ones.

In this [10] framework is intended for explicit execution subtleties of every part, zeroing in on the different RFID label advancements that will make incorporation costs reasonable. Shopping spaces can then roll out sufficient improvements to fit the clients' assumptions.

3 Existing System

In addition to Android and RFID, [3, 11] the current framework uses the Internet of Things innovation. With this framework, customers may come across a prepaid shopping framework. With this framework, customers need to charge trucks as much as they need. For each item added to the cart, the total will be deducted from the reactivated total. In addition, if additional items are not removed from the truck, the amount will be credited to the customer's prepaid shopping account. The entire framework is delivered via IOT and is used to handle communications to recognize the subtleties of factors such as cost. All work is done through the Node MCU, which is best suited for working with IOT Internet applications. The remote camera takes into consideration live web based of video transmission with a safe association with a cell phone or any PC through cloud. The main drawbacks of existing systems are:

1. There is a structure created using IOT. This framework requires you to save your data before you go shopping. In addition, all you have to do is preload your shopping cart [7].
2. On some trolleys, the case is made of a Raspberry Pi, which is more expensive than the Arduino. The edges of this set are a lot of work.
3. In the current framework, the framework is planned with great effort, and the correspondence between the frameworks is not ensured by the increased responsibility.
4. It is used to improve RFID reads that are performed with low iterations.

4 Proposed Work

The innovative technique flows through the following five phases.

1. Pairing mobile with the cart through the Bluetooth. That is when the customers enter the mart he/she needs to turn on the bluetooth app in their mobile and need to search for the paired devices and should connect to the HC-05 Bluetooth and can start shopping.
2. Scanning the product with the help of RFID tags and RFID reader.
3. Displays the scanned products on the Mobile and LCD. As the customer is connected through the Bluetooth app the scanned products will get displayed on LCD and also on the mobile bluetooth app.
4. After giving the hint code on the below of the app we will receive an SMS. When the shopping gets completed the customer need to give the code to the cart as shopping completed so that the cart can start the process of billing and sends the total bill along with the individual product price to the registered mobile number through SMS.
5. We will get the total amount of the shopped products also.

So, In the system we are using Arduino as it is affordable when compared to other microcontrollers like Raspberry pi. As the framework should be interfaced with every cart in the mart, it will become more expensive for the shop owners to use any other microcontrollers (Fig. 1).

RFID is a specific built-in wireless card in the suggested system. A loop antenna is incorporated in a chip. Implicit A 12-digit card number is addressed by the coordinated chip. RFID The reader is a circuit that creates a 125 kHz signal magnetically. A loop antenna attached to this circuit transmits the magnetic signal that helps to grasp the RFID card number. This system is provided with the power supply of 12V [13].

The RFID card is utilized as a access card for security in this project. As a result, each product has its own distinct personality. The product name is represented via an RFID card. RFID reader is attached to the microcontroller. The Microcontroller here is

Fig. 1. Block diagram

a glimmer type reprogrammable microcontroller, the card number is now programmed. Or The microcontroller is connected to the keyboard [15].

RFID reader is used to read the RFID labels and the stores the information of products like price etc.,

In [12] the system the Bluetooth module is inserted because it is used to connect the mobile with the whole system to start the shopping. To take a next step in the super mart first the system needs to be connected to the Bluetooth and then the customer can proceed to shop. To connect the system to the Bluetooth an application named blue serial need to be installed in the mobile.

In this system GSM module is used to send the SMS to the registered mobile number that lets the customer to know the total amount of the shopped products along with the individual product price. So that the customer can directly proceed to the payment process.

Here two types of switches are used in the system. One is utilized to eliminate the item from the truck while checking and other is utilized to end the shopping. After adding the products to the basket purchaser can remove the selected product by using the decrement switch. The LCD is used to show up the products that are added to the cart and also displays the total amount of the purchased products.

In this system, the ultrasonic sensor is used to sense the distance between the customers. Whenever the customer violates the protocol, this sensor will detect the situation and gives an alert through buzzer and it also get displayed on the LCD like maintain distance (Fig. 2).

Fig. 2. Circuit diagram

5 Flow Chart

See Fig. 3.

Fig. 3. Flowchart of proposed system

1. Start the program.
2. Initialize the system and connect through Bluetooth. First, connect mobile with the system through the Bluetooth HC-05 which was designed.
3. Scan the required products. Next, Read the RFID tags with the help of RFID reader.
4. If the tags got scanned then the reader will get the information and will display it on LCD and the mobile. Then products which was scanned or read is displayed on the LCD Displayed and also in the mobile app.
5. If it is not scanned properly then the product should be scanned again. Between scanning of products, some seconds of time should be given.
6. Then it will be shown as items added to the cart. If it is scanned properly, then the item is displayed on LCD showing that item is added to the cart.
7. If any item is removed from the cart, then the total bill will be displayed on LCD and the mobile. If the customer needs to remove the product from the cart then they need to scan a product with pressing decrement button.
8. If not, then the total amount will remain same.
9. If the customer feels the shopping is completed then they need to give the keyword Shopping Completed in their mobile. Then its starts billing, the final bill is generated and the SMS is sent to the registered mobile number. And also, the total amount also displayed on the LCD.
10. If any customer violates the social distance protocol an alert will be given.
11. End the program.

6 Working Methadology

When the customer enters the super market first the customer needs to connect his/her mobile to the cart through Bluetooth and then the customer can start shopping. Here, RFID label goes about as an item and RFID pursuer (EM18) fills in as a scanner [8]. At the point when the cards arrive at the pursuer, the pursuer will figure out mark and LCD show item subtleties like item name and cost. Assuming that the client chooses to eliminate the added items can utilize press button to eliminate items in the bin. After each product, cost will be added.

After purchasing all the required products, the customer will get the total amount to their mobile through SMS and also the total amount will be displayed on LCD. The customer can pay the bill by using any payment card or by online payments. An app is existed for the customers to connect with the super market via Bluetooth, which shows the price of the individual product and the total amount so that customers can directly proceed to the payment. Also, the social distance feature is interfaced in the system to make the customers follow the social protocols. connections are given as follows (Fig. 4):

7 Result

Hence the below figure shows the output of the proposed hardware system. When the customer enters the mart and connects his/her mobile with the framework inserted in to the cart through Bluetooth then it will display as Welcome to the Smart Shopping and the customer can start buying their needs (Fig. 5).

Fig. 4. Framework of the system

Fig. 5. Output of the system

In a hurry sometimes the customers may fails to follow the social protocols. So, whenever the customer fails to maintain the social distance an alert will be given and also gets displayed on the LCD as maintain distance so that the customer can be aware (Fig. 6).

Fig. 6. Output display when violates protocol

8 Conclusion

The principal goal of our system is to stay away from mass individuals at the shopping counters and furthermore to save the customers time in a smart manner. This system will be very helpful for the customers to shop in a smart way. Also helps the customers to follow the covid protocols without any fail.

9 Future Scope

Based on the study of the Smart Shopping using Embedded based Auto cart, it aids in the future and reduces the pressure of searching for a product in the Supermarket. Then, based on the successful creation of that application, we can proceed with our shopping. As a result, we can now shop in grocery marts with ease.

References

1. Anandakumar, H., Umamaheswari, K.: A bio-inspired swarm intelligence technique for social aware cognitive radio handovers. Comput. Electr. Eng. **71**, 925–937 (2018). https://doi.org/10.1016/j.compeleceng.2017.09.016
2. Wang, Y.C., Yang, C.C.: 3S-cart: light weight interactive Sensor based cart for smart shopping in super market (2016)
3. Shahroz, M., Mushtaq, M.F., Ahmad, M.: IoT based smart shopping using radio frequency identification (2020)
4. Yatra, A.: Coronavirus: Haryana Govt orders shops and offices to remain shut on weekends except the shops selling essential goods published on TIMESNOWNEWS, August 2020
5. Rahmanan, A.: How Covid-19 impacts Shopping in day-to-day life – BBC, July 2020

6. Suryaprasad, J., Praveen Kumar, B.O., Roopa, D., Arjun, A.K.: A novel low-cost intelligent shopping cart. In: 2011 IEEE 2nd International Conference on Networked Embedded Systems for Enterprise Applications (2011)
7. Dhavale Shraddha, D., Dhokane Trupti, J., Shinde Priyanka, S.: IOT based intelligent trolley for shopping mall. Int. J. Eng. Dev. Res. **4**, 1283–1285 (2016)
8. Karmouch, A., Salih-Alj, Y.: Aisle-level scanning for pervasive RFID-based shopping applications. Int. J. Eng. Dev. Res. (2013)
9. Saravanakumar, S., Ravichandran, K., Jeshwanthraj, R.: IOT based smart card with automatic billing for futuristic shopping experience. Int. J. Comput. Sci. Eng. (2019)
10. Kamble, S., Meshram, S., Thokal, R., Gakre, R.: Developing a multitasking shopping trolley based on RFID technology. Int. J. Soft Comput. Eng. (2014)
11. Li, R., Song, T., Capurso, N., Yu, J., Couture, J., Cheng, X.: IoT application on secure smart shopping system (2017)
12. Kaur, A., Garg, A., Verma, A., Bansal, A., Singh, A.: Arduino based smart cart. Int. J. Adv. Res. Comput. Eng. Technol. (IJARCET) **2**(12) (2013)
13. Yathisha, L., Abhishek, A., Harshith, R., Darshan Koundinya, S.R., Srinidhi, K.: Automation of shopping cart to ease queue in malls by using RFID (2015)
14. EktaMaini, J.: Wireless intelligent billing trolley for malls. Int. J. Sci. Eng. Technol. **3**(9), 1175–1178 (2014)
15. Vinutha, M.L.: Shopping and automated using RFID Technology. Int. J. Electron. Commun. Eng. Technol. **5**(8), 132–138 (2014)

Gastric Cancer Diagnosis Using MIFNet Algorithm and Deep Learning Technique

Mawa Chouhan[1], D. Corinne Veril[1], P. Prerana[1], and Kumaresan Angappan[2](✉)

[1] Department of Information Technology, Hindustan Institute of Technology and Science, Chennai, India

[2] School of Computer Science and Engineering, Vellore Institute of Technology, Vellore, India
kummaresan@gmail.com

Abstract. Gastric cancer (GC) is one among many common forms of cancer with poor prognosis. Accompanied with such a poor prognosis, Gastric Cancer [GC] appears as one of the deadliest diseases across the globe. To eliminate this drawback pathologists are now adopting Artificial Intelligence [AI] techniques to enhance the efficiency and diagnostic accuracy. A sort of biological inconsistency either microsatellite inconsistency or chromosomal instability is shown in the majority of cancer tissues and has been characterized as an initial phase in tumor development. The existing cancer cell standard based upon histological parameters, genetics, including molecular variants assist in providing a deeper understanding on the overall characteristics of each subtype to enhance the process of early diagnosis, screening, and chemotherapy. This research study presents MIFNet as an alternative for the existing semantic segmentation techniques that were unable to resolve the challenges observed in precise segmentation and efficient inference. Moreover, this research study introduces a novel deep learning driven methodology to support mostly during the pathological diagnosis of gastric cancer, which generally involves a series of tests.

Keywords: MIFNet algorithm · Gastric cancer · Image processing · Deep learning · Pathologist

1 Introduction

With the population of 1.3 billion the number of existing pathologist are far less than what is required. When a patient is suffering through any kind of disease the most important thing is to get treated as soon as possible, but in the case of cancer the patient has to go through multiple tests before the doctors come to a conclusion. It is usually diagnosed using a variety of procedures such as biopsies, endoscopies and blood tests [2]. Not only time consuming these tests can cost a fortune to many. The earlier individuals were detected, the higher their possibilities of rehabilitation as well as surviving. So, it has been proposed as an opportunity for us to make a contribution to the medical field.

Gastric cancer (GC) is among the most commonly diagnosed diseases that has a pretty awful prediction. It maintains to be among the deadliest severe diseases with such

J. I.-Z. Chen et al. (Eds.): ICIPCN 2022, LNNS 514, pp. 713–724, 2022.
https://doi.org/10.1007/978-3-031-12413-6_56

a poor prediction. Gastroscopy is a frequently used technique for identifying gastric lesions and providing early (GC) screening and detection. Conventional GC screening procedures, on the other hand, are only as good as the gastroscopy specialist's medical expertise [3]. The primary goal of this research study is to make this procedure less time consuming, more precise, and less expensive. The proposed research study has developed a novel deep learning algorithm to successfully assist in the pathological diagnosis of gastric cancer, which requires a number of tests to obtain a result. In this perspective, this research study has developed the advanced MIFNet algorithm to more precisely detect the presence of cancer.

1.1 Problem Statement

Pathologists' image analysis appears to be less efficient and accurate in standard pathological visual diagnosis. It can lead to a health concern where multiple investigations are required to diagnose gastric cancer.

MIFNet

Multi-scale edges and multi-scale segmentation data are part of the Multi-Information Fusion Network (MIFNet). It also provides global context information and analyses network training to fuse various scales and datasets. Analyses on a set of natural-coloured photos from the given dataset; MIFNet beats [4] state-of-the-art approaches, as per the conclusions. After the collected dataset and preprocessing, it will be considered for training with the Deep Learning algorithms. MIFNET, another advanced algorithm, is being used to help diagnose better accurately. In this study, the multi-input MIFNet model is used to segment the lesions in the pathological picture and increase the dice

Fig. 1. System architecture of MIFNet algorithm

coefficient in the segmentation of gastric cancer case images to 81.87%, which is a significant improvement over current segmentation algorithms [5]. The combination of the three different algorithms (Multi-task Net, Fusion Net and Global Net) will provide an accurate prediction of gastric cancer without any additional diagnosis. This algorithm is presented as an alternative to previous semantic segmentation techniques that were unable to resolve the issues of precise segmentation and efficient inference [6] (Fig. 1).

2 Literation Review

Seungwoo Song et al. implemented a CMOS capacitive biosensor [7] for detecting the cancer cells in human blood. A peptide receptor is used, and an electrochemical reaction with VEGF changes the capacitance between the two microneedles. VEGF is observed with 15 fMRMs resolution in the range of 0.1 to 1000 pM, indicating evident selectivity. Using peptide based cancer biomarker detection with functionalized micro needle cancer is detected at an early stage.

Chih-Hung Chan et al. proposed a texture-map-based [8] network based on the texture feature images and also a branch-collaborative network based on a texture-map-based network is presented in this article for detecting and marking cancer in the buccal region of the oral cavity. In order to extract the texture feature images, wavelet transformation and Gabor filter are used. The standard deviation values have then been computed using a sliding window. Using the standard deviation values, a feature map is constructed and divided into multiple patches and then they will be used to build a Deep Convolutional Network [DCN]. This network model incorporates a branch for detecting oral cancer as well as a branch for identifying ROIs. Two different architectures have been used for the detection of oral cancer. According to the experimental results, Gabor filter has many useful properties than FCN. The outcomes are as follows: The ROI segmentation and labelling process are performed by using a Feature Pyramid Network (FPN) framework and Deep Convolution Network (DCN) framework. Experimental results show that the Gabor filter provides more useful features than DCN. In the results, it has been confirmed that the model can accurately identify high-risk regions for oral cancer and serve as an important screening tool.

Seong Ryeol Moon et al. DNA-binding, examined individual mutation positions, [9] which can be analyzed with the following five aspects, which includes the whole body gene, known mutation hotspots. Out of 289 samples, which included a mutation and matched expression dataset from the TCGA, 138 samples had mutations in TP53, while 151 samples had TP53 wild type. Among the findings, it was concluded that.

(1) Gastric cancer did not impact overall TP53 status-wild-type vs. mutated
(2) Mutations in the DNA-binding domain (DBD) did not much affect survival when compared to DBD domain
(3) Secondary structure of the TP53 helix or turn was much more detrimental than disruptions to a beta-strand
(4) R248 patient mutations showed inferior survival to those with mutations at other positions. The approach here is used for studying the consequences of phenotypes of specific genetic order and in gaining a deeper understanding of physiologic re-responses involving master transcription regulators like TP53.

Jean-Sebastien et al. introduced the nanorobots [10] to assist in the recognition of multifocal disease from a multimodal optimization (MMO) perspective. Then proposed another system of processing roused cancer detection procedure (MCDP). Under the notion of MCDP, the growth of foci has to be identified and viewed as the arrangements of the goal work, where the tissue locale around the disease regions addresses the boundary space, and the nanorobots stacked with contrast medium atoms for disease recognition when compared to the improvement specialists. The interaction in which the nanorobots detect tumours by swimming in the high-risk tissue region may be considered as the cycle in which professionals strive for the solutions of a target task in the boundary space for certain limits. This research work has additionally fostered an m-NGA model, which can be utilized to work on the exhibition of MCDP. Gaining from the enhancement strategy of NGA, the NGA-roused MCDP is proposed to find the cancer targets effectively while considering a reasonable in-vivo spread and controlling of nanorobots, which is unique by concerning the utilization situation of the standard NGA. Mathematical models have shown the adequacy of the proposed procedure for the bloodstream speed profile, which is incited by cancer angiogenesis. Future work might incorporate a working on the presentation of the calculation to identify all the malignant growth regions with many nanorobots.

Shaolong Shi et al. utilized the non-thermal plasmas (NTP) [11] to provide a disease treatment, which has turned into a potential field during the recent years. The theories about the potential use of NTP as another malignant growth therapy usually range from in vitro research to early clinical trials. Regardless, with a diverse array of NTP sources and natural tests examining various disease models, it is difficult to build a complete vision on the influence of NTP on malignant tumor cells, even in vitro. Natural limits should be set and regulated as much as possible to analyse the disease treatment capability of various NTPs while understating the changeability. In segment 2, the convertible plasma fly gadget is portrayed with few fundamental plasmas. Further examinations should be performed to explain the idea of the cell's conduct and passing after plasma openness. At last, as the cytotoxicity of direct plasma treatment was uniquely viewed by concerning one of the plasma-treated mediums, where the system prompting cell demise depend on the gas stage or fluid stage brief of RONS.

3 Proposed Methodology

This research study develops a method using deep learning algorithms to assist the pathological diagnosis of gastric cancer as it involves numerous tests to conclude. An advanced algorithm such as MIFNet is used to diagnose the presence of cancer more accurately. MIFNet is a combination of three different algorithms such as Multi-task Net, Fusion Net and Global Net, which gives accurate prediction of gastric cancer without any additional diagnosis. Thus, this project helps in effectively diagnosing gastric cancer with higher accuracy than the existing models [12].

3.1 Gastric Cancer Dataset Collection

The datasets are made up of many different types of information, which can be used to create a training model to uncover predictable outcomes within the original data. The initial step is to generate a collection. After generating a dataset, the most important aspect to consider is the primary data source. For this project, datasets are collected from Kaggle and UCI. Then the dataset is divided into two parts, Training dataset is used to train the proposed algorithm. Testing dataset is used to evaluate how well the proposed algorithm was trained (Fig. 2).

Fig. 2. Dataset collection

3.2 Dataset Preprocessing

Real-world data is usually composed of noise, entry errors with an incomprehensible structure which could not be used immediately in deep learning methods. Data preprocessing is responsible for cleaning raw reports and processing it in an acceptable manner for the deep learning methods, which further improves the model's effectiveness and precision. The network architecture and also the intermediate data type should always be thoroughly considered while developing an appropriate machine learning algorithm [13]. The set of images, image length, image breadth, number of connections, and number of levels per pixel are the most frequently considered image input parameters (Fig. 3).

Fig. 3. Dataset preprocessing

3.3 Image Annotation

Image annotation is the process of labelling or classifying an image by using either text, annotation tools. At the pixel level, the segmentation approach is used to recognize and grasp the information present in the image [14] (Fig. 4).

Fig. 4. Annotation

Semantic segmentation divides the images into different segments depending on the pixels. Each pixel in an image will be labeled [15]. MakeSense is a free and open-source annotation tool utilized in this project (Fig. 5).

Fig. 5. Uploading images for annotation

3.4 MIFNet Algorithm Training

After the dataset is collected, the pre-processing will be carried out for training the Deep Learning [DL] algorithms. An advanced algorithm such as MIFNet is used to accurately diagnose the presence of cancer. MIFNet is a combination of three different algorithms such as Multi-task Net, Fusion Net, and Global Net. The combination of the algorithm gives an accurate prediction of gastric cancer without any additional diagnosis.

3.5 Validation and Evaluation

After applying the MIFNet algorithm and training, validation is carried out. After training the data, model file will be generated and the testing data will be given to the trained model file. The features have been extracted as a model file. When an input image is given for the disease prediction process, it will predict the presence of the disease. If the condition is presented yes or no, the output result will be compared with actual data. If the data is true, it will be accurate and if the data is false, it will not be accurate.

At last, it will determine the presence of disease with higher accuracy. Epoch means the training process done by the algorithm inside the images folder for one time is 1 epoch and so the x-axis represents the number of times it has been trained by using the proposed algorithm. Loss is been given in percentage and also it has been reduced from 100 to 0. The red line indicates the training loss whereas the blue line indicates the validation loss. The percentage of loss has been decreased with the increasing number of epochs (Figs. 6 and 7).

Fig. 6. Validation and evaluation

Fig. 7. Loss over training and validation

Epoch represents the training process done by the algorithm inside the images folder for one time which is represented as 1 epoch and so the x-axis represents the number of times it has been trained using the algorithm. The red line indicates training accuracy whereas the blue line indicates validation accuracy (Fig. 8).

Fig. 8. Accuracy over training and validation

4 Algorithm

Stage 1. Applying the combination of three algorithms (MIFNet) to increase the efficiency and accuracy [16].
Stage 2. Getting the dataset from Kaggle to train the Algorithm.
Stage 3. Preprocessing (Using Multi-task Net) the data for segmentation so that the image will reduce the noise and all the images will be equal in size.
Stage 4. Augmentation (Global Net) increase the dataset amount to train the application to increase the accuracy of the results.
Stage 5. Annotation (with the help of "make sense") spotting and marking the area where the tumors are present.
Stage 6. Combining both the algorithm and providing the result with the help of Fusion Net (Fig. 9).

Fig. 9. MIFNet algorithm

5 Result and Output

MIFNet played the major role in the working of our project and coding has been done in python and code has been run on Google colab (Figs. 10, 11 and 12).

Fig. 10. Prediction of normal cancer

Fig. 11. Prediction of Stage 1 cancer

Fig. 12. Prediction of Stage 2 cancer

6 Conclusion

This research work has successful estimated the prediction efficiency for those patients who are suffering from gastric cancer and the proposed model has the potential to reduce the health risk by providing the results in less time. The very first stage of the disease can usually be cured with right medication. As a result, it becomes critical to detect the disease in the earlier stages in order to help sufferers. The major goal of this study is to find the best prediction model, i.e. the best Deep Learning [DL] technique for distinguishing patients with gastric cancer and healthy people. The final trained model file has indeed been generated after verification and evaluation. It has successfully predicted the ailment and evaluated if the information is accurate by providing an input image. Finally the proposed model has easily predicted the presence of disease with higher accuracy.

References

1. Chen, Y., et al.: A machine learning model for predicting a major response to neoadjuvant chemotherapy in advanced gastric cancer. Front. Oncology **11** (2021). ISSN: 2234-943X. https://doi.org/10.3389/fonc.2021.675458. https://www.frontiersin.org/article/10.3389/fonc.2021.675458
2. Cao, R., Gong, L., Dong, D.: Pathological diagnosis and prognosis of Gastric cancer through a multi-instance learning method. EBioMedicine **73**, 103671 (2021). https://doi.org/10.1016/j.ebiom.2021.103671
3. Ding, S., Hu, S., Li, X., Zhang, Y., Wu, D.D.: Leveraging multimodal semantic fusion for gastric cancer screening via hierarchical attention mechanism. IEEE Trans. Syst. Man Cybern. Syst., 1–14 (2021). https://doi.org/10.1109/TSMC.2021.3096974
4. Pan, Z., Dou, H., Mao, J., Dai, M., Tian, J.: MIFNet: multi-information fusion network for sea-land segmentation. In: Proceedings of the 2nd International Conference on Advances in Image Processing, ICAIP 2018, Chengdu, China, pp. 24–29. Association for Computing Machinery (2018). ISBN: 9781450364607. https://doi.org/10.1145/3239576.3239578
5. Cai, L., Gao, J., Zhao, D.: A review of the application of deep learning in medical im age classification and segmentation. Ann. Transl. Med. **8**(11) (2020). ISSN: 2305-5847. https://atm.amegroups.com/article/view/36944
6. Cheng, J., Peng, X., Tang, X., Tu, W., Xu, W.: MIFNet: a lightweight multiscale information fusion network. Int. J. Intell. Syst. https://doi.org/10.1002/int.22804. eprint: https://onlinelibrary.wiley.com/doi/pdf/10.1002/int.22804. https://onlinelibrary.wiley.com/doi/abs/10.1002/int.22804
7. Song, S., et al.: A CMOS VEGF sensor for cancer diagnosis using a peptide aptamer-based functionalized microneedle. IEEE Trans. Biomed. Circuits Syst. **13**(6), 1288–1299 (2019). https://doi.org/10.1109/TBCAS.2019.2954846
8. Chan, C.-H., Huang, T.-T., Chen, C.-Y., Lee, C.-C., Chan, M.-Y., Chung, P.-C.: Texture-map-based branch-collaborative network for oral cancer detection. IEEE Trans. Biomed. Circuits Syst. **13**(4), 766–780 (2019). https://doi.org/10.1109/TBCAS.2019.2918244
9. Moon, S., Balch, C., Park, S., Lee, J., Sung, J., Nam, S.: Systematic inspection of the clinical relevance of TP53 missense mutations in gastric cancer. IEEE/ACM Trans. Comput. Biol. Bioinform. **6**, 1693–1701 (2018). https://doi.org/10.1109/TCBB.2018.2814049
10. Boisvert, J.-S., Lafontaine, J., Glory, A., Coulombe, S., Wong, P.: Comparison of three radio-frequency discharge modes on the treatment of breast cancer cells *in Vitro*. IEEE Trans. Radiat. Plasma Med. Sci. **4**(5), 644–654 (2020). https://doi.org/10.1109/TRPMS.2020.2994870
11. Shi, S., Chen, Y., Yao, X.: NGA-inspired nanorobots-assisted detection of multi-focal cancer. IEEE Trans. Cybern., 1–11 (2020). https://doi.org/10.1109/TCYB.2020.3024868
12. Zhang, Y., Kong, J., Qi, M., Liu, Y., Wang, J., Lu, Y: Object detection based on multiple information fusion net. Appl. Sci. **10**(1) (2020). ISSN: 2076-3417. https://www.mdpi.com/2076-3417/10/1/418
13. Joshi, R.C., Singh, D., Tiwari, V., Dutta, M.K.: An efficient deep neural network based abnormality detection and multi-class breast tumor classification. Multimedia Tools Appl. **81**(10), 13691–13711 (2022). ISSN: 1380-7501. https://doi.org/10.1007/s11042-021-11240-0
14. Lundervold, A. S., Lundervold, A.: An overview of deep learning in medical imaging focusing on MRI. Zeitschrift für Medizinische Physik **29**(2), 102–127 (2019). Special Issue: Deep Learning in Medical Physics. ISSN: 0939-3889. https://doi.org/10.1016/j.zemedi.2018.11.002. https://www.sciencedirect.com/science/article/pii/S0939388918301181

15. Yang, R., Yu, Y.: Artificial convolutional neural network in object detection and semantic segmentation for medical imaging analysis. Front. Oncol. **11**, 638182 (2021). https://doi.org/10.3389/fonc.2021.638182
16. Akram, N., et al.: Exploiting the multiscale information fusion capabilities for aiding the leukemia diagnosis through white blood cells segmentation. IEEE Access **10**, 48747–48760 (2022). https://doi.org/10.1109/ACCESS.2022.3171916

Cold Chain Logistics Method Using to Identify Optimal Path in Secured Network Model with Machine Learning

Vijaykumar Janga[1], Desalegn Awoke[1], Assefa Senbato Genale[1], B. Barani Sundaram[1], Amit Pandey[1], and P. Karthika[2(✉)]

[1] College of Informatics, Bule Hora University, Bule Hora, Ethiopia
[2] Kalasalingam Academy of Research and Education, Krishnankoil, India
karthikasivamr@gmail.com

Abstract. A secured networking model is essential to handle the logistics networks. In this article are going to see an intelligent secured networking model to identify the optimal path for cold chain logistics to hospitals. The optimal path finder is used to find the path between point A to point B, which is short & best. It also considers the road traffic and cost of transport. The cold chain logistics to the hospitals, includes medicines & vaccines which are to be stored in a particular temperature. Thus path optimization is more essential in cold chain logistics to hospitals than other type of logistics. In this research Bee-Ant Optimization Algorithm (BAOA) is proposed to perform the intelligent transportation to the hospitals. The proposed algorithm is compared with the existing Ant Colony Optimization (ACO), Bee Colony Optimization (BCO), and Neural Network Model. From the results it can be observed that the proposed algorithm shows 98.83% for accurate delivery of logistics to the hospitals.

Keywords: Hospitals · Logistics network · Secured networking model · Optimal path · Bee-Ant Optimization Algorithm (BAOA) · Artificial Intelligence (AI)

1 Introduction

Cold Chain Logistics seems to be the technology or method that enables for the safe transportation of thermal commodities and products throughout the supply chain. To assess and account for the relationship between temperature with perishability. Cold Chain Management is used in hospitals to preserve vaccines at specific temperatures along the supply chain. The general level of technology with in cold chain logistics business is still very modest [1–5]. The major goal of this research [6–8] is to develop an intelligent algorithm for cold chain logistics redistribution optimization based using secured networking. Cloud technology can be utilised to link the cold chain logistics database with external clients utilising cold chain logistics. As either a result, any data type terminal can monitor and update data [9, 10].

J. I.-Z. Chen et al. (Eds.): ICIPCN 2022, LNNS 514, pp. 725–735, 2022.
https://doi.org/10.1007/978-3-031-12413-6_57

2 Literature Survey

The origins of augmented reality technology can be traced back to the 1950s, some generations earlier. With the rapid advancement of computer graphics processors, computing power, and other technologies throughout recent years, interactive virtual technology-related equipment has begun to evolve. The commercialization of mobile intelligence, the growth of the mobile internet, and more growth of mobile augmented reality technology should gradually mature with the advancement of diverse skill sets and technologies. First, hardware manufacturing will eventually achieve industrialization and scale [11–13]. Quality, square meter refresh rate, latency, computation, and rendering abilities are essential indications of virtual reality-related technologies. Second, as technological resources have improved, the interoperability of software systems and hardware has also advanced rapidly [14–18]. Whenever inadequate temperature as well as relative high humidity exist during shipment and storage, the quality of the goods can rapidly deteriorate. Temperature fluctuations can be observed in warehousing, handling, as well as transportation. This same industry's principal concern is the study the analytics monitoring temperature differences inside buildings with complete refrigerators, containers, especially trucks [19]. In addition, cold chain logistics seems to be a subset of logistics that refers to the systematic effort of ensuring that chilled goods are always kept at a low temperature throughout the entire process of manufacture, shipping, storage, consumption, including marketing. This is done to guarantee the quality of something like the items is not compromised [20]. The proposed path finding methodology addressed the lack of information about the nearby environment, including control-feasibility of such investigated paths, such as the steepest possible turns a vehicle should make, and also the computational requirements of such a time-dependent surroundings [21–23]. A new diverting rule has already been formed as a result of the cascading dips in the procedure there under edge failure situation, and therefore the distribution expense which has been related with the transport speed. Its cold chain logistics distributing route optimization model is built with both the goal of lowering distribution costs in mind [24–26]. Cloud computing's information and technology (IT) resources can also be viewed as applications. Cloud computing already has had a significant impact on technical progress as a result of the new IT installation. Cloud computing can deliver crucial software for corporate administration while lowering Something and hardware & system maintenance expenses. Cloud computing will provide small and medium-sized businesses with access to proper IT solutions while requiring less IT investment. The cloud-based cold chain logistics solution can also be used to integrate the database among cold chain logistics with external stakeholders. As a result, each user defined terminal can monitor and update the information [27, 28]. Here, a cold chain logistics technology provided by cloud computing has indeed been developed. This system is made up of several functional components, including configuration items, business development, knowledge updating as well as transfer, control strategy, evidence gathering, including data calculation. The interaction among cold chain logistics but also their clients may be strengthened, which allows for co-control and product sales and inventory, boosts the efficiency the cold chain logistics, therefore benefits all involved [29].

3 Proposed Work

The optimization model appears to take some many influencing effects into consideration to represent the real situation, including such time-varying transportation online traffic, traffic levels, customer time consumed, sleekness of innovative products, but it also pathway going to charge delay time. The Bee-Ant Optimization Algorithm was created to help with the Electric Vehicle Routing for Logistics Vehicles (EVRP). According to simulation results, the proposed scheme can effectively prevent congestion problems during the delivery process, lower total distribution costs, and improve the efficiency of a cold chain logistic distribution system with new items.

The set of all logistics nodes in the network is denoted by $K = P \cup M\{0\}$. The letter P stands for the collection of customer touch points. The processing facility is denoted by the number 0; M denotes a collection of fast chargers, whereas K denotes a collection of all endpoints in a logistics system. H is the set of time periods that occur throughout the day, $H = \{H_1, H_2, ..., H_n\}$; n is the overall number of time intervals G denotes the collection of road segments in the road infrastructure, $G = \{G_1, G_2, ..., G_n$; G denotes the number of road section types. $D = \{D_1, D_2, ..., D_n\}$; D is the number of area types in the area set.

Let ρ_i signify the required to charge stations able to charge equipment utilisation ratio $\rho_i = (\lambda_i/x_i\mu_i)$ According to the conventional slight Equation, the waiting period for such $g^{th}SK$ and trying to charge point i is represented by Eq. (1).

$$h_{ig}^{G} = \sum_{i=1}^{x} \frac{(x_i\rho_i)^{s_i}\rho_i}{\lambda_i x_i!(1-\rho_i)^2\varphi_{D_i}} \cdot \left(\sum_{n=0}^{x_i-1} \frac{(x_i\rho_i)^n}{n!} + \frac{(x_i\rho_i)^{x_i}}{s_i!(1-\rho_i)}\right)^{-1} H_{ig} \times g^{th}SK \quad (1)$$

In the charging method, the load voltage for the $g^{th}SK$ at end point i is as shown in Eq. (2).

$$h_{ig}^{G} = \sum_{i=1}^{g} \frac{S_{max} - S_{ig}^{H}}{n_h} \quad (2)$$

Electric energy consumption is influenced not just by vehicle's disposition, but by its loads and speeds. Whenever an SK with a charge travels at A_k transportation distance η km/h k on a flat road, the able to run power $D(A_k, k)$ is represented by Eq. (3).

$$D(A_k, k) = \frac{((y + A_k).g.\int k + (R_n.Z_i.k^3/22.56))}{3700\eta km/hk} \quad (3)$$

The genuine speed of such a vehicle in a sufficiently short span of time that is used to speed proportional with the determined period of time. The Eq. (4) is updated to explain variation in the existing roads through which driver has planned to travel.

$$k_{ij}(t) = \sum_{i=1}^{j} \begin{cases} k_{ij}^{1}, \ t \in H_1, \\ k_{ij}^{2}, \ t \in H_2, \ (i,j) \in G \\ k_{ij}^{n}, \ t \in H_3, \end{cases} \quad (4)$$

In x is the allocation of velocity fee on the street does now no longer alter in identical timeframe. $(i, j).M$ range of things are considered, together with avenue type, delivery time period, however additionally $M_1, M_2, ..., M_n$ time-various velocity of the car together with such location and additionally transition as with inside the following Eq. (5).

$$M = \sum\{M_1, M_2, \ldots\ldots M_n\} \tag{5}$$

The transmitter and g_{ij} the receiver need to be one of a kind from the alternative vehicles, and the operator can certainly be clearly connected (WSN with AI) to the desired tool with out dispute. Transition pace is $n_i - \overline{n}$ is classed into: on the spot transition and time change. Space time transformation, inclusive of the preceding MH, takes a while to comply with the series is represented within the Eq. (6).

$$MH = k_{ij}(t) \sum_{i=1}^{x} \frac{x\sum_{i=1}^{x}\sum_{j=1}^{x} + \sum_{g_{ij}}(n_i - \overline{n})(n_j - \overline{n})}{\sum_{i=1}^{x}\sum_{j=1}^{x} + g_{ij}(n_i - \overline{n})^2} \tag{6}$$

The following Eq. (7) represents the Cold Chain Logistics Vehicles primarily based totally extrude within the transition time.

$$MH = k_{ij}(t) \sum \frac{x\sum_{i=1}^{x}\sum_{j\neq 1}^{x} + g_{ij}(n_i - \overline{n})(n_j - \overline{n})}{K^2 \sum_{i=1}^{x} \times \sum_{j=1}^{x} g_{ij}} \tag{7}$$

Equation (8) then deals with the calculation of the number of storage media to modify the transaction depending on the time of the cold chain logistics.

$$k_{ij}(t) = \sum_{i=1}^{j} \frac{\sum_{j=1}^{q}\sum_{d=1}^{q}\sum_{u=1}^{q}\sum_{y=1}^{x_d}\left|s_{ij} - S_{dy}\right|}{2x^2 t} \tag{8}$$

Multiple sentences related to various A vehicle types and data add-ons can indeed be described by vehicle position and conversion, and the number of $s_{ji} - s_{dy}$ data can indeed be represented by $t_j + t_d$ by the focus arc vehicle speed and duration, according to the following Eq. (9), WSN network and AI technology work together as presented in the Eq. (9).

$$petri_{jd} = \frac{\sum_{A=1}^{d_j}\sum_{y=1}^{x_d}\left|S_{ji} - S_{dy}\right|}{x_j x_d (t_j + t_d)} \tag{9}$$

A WSN net with code operating procedures based on AI technology is described by the Eq. (10) provides the ϑ direction of the vehicle $\vartheta_n^2 + \vartheta_x^2 + B_1$. B_1 Transformations such as $2\vartheta_n\vartheta_x + B_1$ correlation between different τ objects of a given statement, and $2\tau_{nx} + B_2$ oriented action sequences have very different types $\tau_n^2 + \tau_x^2 + B_2$ of velocities and times in the declaration.

$$M_{xn} = 1 - \Sigma \frac{(2\vartheta_n\vartheta_x + B_1)(2\tau_{nx} + B_2)}{(\vartheta_n^2 + \vartheta_x^2 + B_1)(\tau_n^2 + \tau_x^2 + B_2)} t \in h_i \tag{10}$$

Equation (11) clarifies the modeling method of the influence of the code based on the M_{xn} network on the optimization performance of the code behavior cold chain modeling based on the h_i network; the v_q correlation between the y network component and the code component is also in accordance with the Eq. (12).

$$K(h_i, g_i) = \sum K(h_i)K(g_j/h_i); \sum_{i=1}^{j} K(g_j/h_i) = \sum_{q=1}^{q} K(g_j/v_q)K(v_q/h_i) \quad (11)$$

$$k_y = \sum_{y=1}^{y} \frac{2y}{y+1} + \left[\frac{1}{2} + \frac{1}{2y}\right] \sum_{x=1}^{x} \left[\frac{x_2 - x_1}{3}\right]^2 + \sum_{x=1}^{x} \frac{2(x_2 - x_1)}{3} \quad (12)$$

In order to complete the evolution of $x_2 - x_1$ from rules to networks, static path analysis techniques are used to analyze and process executable files, as shown in the following Eq. (13)

$$P_{jn} = \int_0^{\infty} hs_j(x) \int_0^{s} (x - n) \times hS_n(x) \quad (13)$$

One program document $x - n$ provides an associated process M_{ih} with a certain input object set that has been processed, and the digital output object P_{jn} that is produced is presented as a result in the Eq. (14) is implemented for the process.

$$\ln\left(\frac{M_{ih}}{M_{ih} - 1}\right) = \sum_{i=0}^{h} \alpha + \beta \ln M_{ih} - 1 + (2\vartheta_n v_s + B_1)(2\tau_{ns} + B_2) \quad (14)$$

$$\sum_{n=1}^{\theta} Wn * l = \frac{\sum_{n=1}^{\theta} \left(\frac{\left(T + \frac{Y_i}{\sum_1^x D_s}\right)}{M} + Y + YK \right)}{\beta + \alpha \ln M_{in} + \frac{K}{Y}} \quad (15)$$

Equation (15) represents the sum of the unit time of $\sum_{n=1}^{\theta} W_n * l$ and the total execution time. Execution time includes $\beta + \alpha \ln M_{in} + \frac{K}{Y}$ time spent mainly on roads and client points, and time spent not only on queuing but also on charging points.

4 Results and Discussion

Kindly In x the availability of speed fee on the road no longer changes in the same timeframe. (i, j). M a number of factors are taken into account, including avenue type and delivery time frame, but also $M_1, M_2, \ldots, M_n$ time-varying velocity of the car in conjunction with so much location and additionally transformation as with within the following Eq. (5) to show in Fig. 1 depicts data optimization analysis performed on cold chain logistics vehicles. The optimization method employs the use of intelligent wireless sensor networks for online ordering of materials, which necessitates the preservation of temperature during resource transportation. This is accomplished through the use of intelligent sensor devices attached to the transporting vehicle to continuously monitor the status of the resources. Table 1 shows the minimum distribution, travel distance, and energy consumption for transportation.

Fig. 1. Analysis of cold chain logistics vehicles in different data sets using WSN and AI techniques

Table 1. Effectiveness conclusion analysis for cold chain logistics hospitals vehicles in different data sets using WSN and AI techniques.

EX	Minimum distribution cost (TC)			Minimum travel distance (km) (VD)			Minimum energy consumption (kWh) (ECC)		
	TC	VD	ECC	TC	VD	ECC	TC	VD	ECC
C103	12783.4	978.5	345.1	13756.3	987.6	298.1	12234.9	980.2	256.7
R202	6245.67	1587.9	487.5	6234.2	1034.2	502.3	6567.3	1034.7	321.3
RC203	75 78.54	1 67.3	49 4.3	75 68.3	1 3 25.7	47 4.1	76 45.1	15 67.2	3 99.1

Fig. 2. The Bee-Ant Optimized Algorithm was used to analyze cold chain logistics hospitals vehicles in various data sets

One program document $x - n$ associates a process M_{ih} with a specific input item set which has been filtered, and also the digital signal object P_{jn} which is produced has been proffered as a direct consequence in the process's above Eq. (14). The performance analysis represented in Fig. 2 is carried out on various and varying sizes. Table 2 shows the quantity of data and the status of the analysis. When the size of the data increases, the corresponding analysis metrics reach their peak values faster than when the data size is small.

Table 2. Cold chain logistics hospitals vehicle performance analysis in different data sets using the Bee-Ant Optimized Algorithm in WSN with AI

Various data	TC	EC	VL	VN	CPUT
10	6734.1	4356.7	752.5	11	5
9	6345.9	4675.2	689.4	9	0
7	5136.1	5258.6	623.2	13	1
5	6567.9	5753.8	356.0	16	1

Fig. 3. Using the Bee-Ant Optimized Algorithm, we analyzed existing systems and cold chain logistics hospitals vehicles in various data sets

The quantity of the unit of time of $\sum_{n=1}^{\theta} W_n * l$ and also the maximum completion time is given by Eq. (15). The execution time includes $\beta + \alpha \ln M_{in} + \frac{K}{Y}$ time spent primarily on roads but also client points, as well as effort consumed not only having to queue but also charging stations charging points is represented (Refer Fig. 3). The vehicles used for transportation in cold chain logistics are equipped with temperature monitoring and many other comparative sensors for constant monitoring of a transporting vehicle. According to Table 3, the number of hospitals vehicles considered in the Virtual Network (VN) and the considered qualify for evaluation is 1000. CPUT represents the time it takes to generate all possible paths and is measured in seconds. Following the selection of the optimal path using Artificial Intelligence, the system will be able to train in that path but also test the remaining transportation. It has been discovered that the implementation of the ABC algorithm results in an accuracy of 98.83% in optimal path selection but also transportation of cold logistics along that path.

Table 3. Using the Bee-Ant Optimized Algorithm, we compared the results of existing systems and cold chain logistics hospitals vehicles in various data sets

Algorithm	VN - number of vehicle	CPUT - running time (s)	Training/Testing (%)	Accuracy (%)
Bee-Ant Optimized Algorithm	1000	453.3	94.89	98.83
Existing method: Ant Colony Optimization	1000	1427.9	1200.9	97.67
Existing method: Bee Colony Optimization	1000	762.87	637.1	96.98
Existing method: Neural Network Method	1000	563.6	89.23	95.54

Cold Chain Logistics refers to the safe storage, logistics, and management of temperature-sensitive goods from the point of manufacture to the point of consumption. Vehicles were not used in this process, but rather warehouses outfitted with a cooling system. Cold chain logistics ensures that items such as processed foods, medicine, blood, eyes, and kidneys are safely transported to consumers. Researchers will investigate the path optimization strategy with logistics and transport trucks obtained with the proposed evolutionary algorithm in this scientific article. The logistics and transport vehicle plan is essential for reaching consumers in less time, distance, and expense. Data from road transportation is used to develop routes. The Bee-Ant Optimized Algorithm is compared with the results of existing systems (95.54%) and Cold Chain Logistics hospitals Vehicles (98.83%) in various data sets. It compared to the existing method is better result provide the in our methods.

5 Conclusion

An intelligent transport system has become mandatory in the updated world with the technological evolution. The role of an intelligent system have to focus on all the desired users or clients. The products to be transports should be handled with care with optimal path routing for timely delivery. In recent years, the transportation of logistics have to be concentrated as it involves different medicine that needs certain temperature to be maintained during the transportation. Hence, in this research Cold Chain Logistics concepts is proposed to monitor the transportation of medicine with the specified temperature and Bee-Ant colony Optimization Algorithm to transport in the optimal path. From the results, it can be observed that the proposed model has acquired an accuracy performance in optimal path transportation as 98.83%. which is higher when compared to the existing Bee Colony Optimization, Ant colony Optimization and the neural network model.

References

1. Nedumaran, A., Ganesh Babu, R., Kass, M.M., Karthika, P.: Machine level classification using support vector machine. In: AIP Conference Proceedings of International Conference on Sustainable Manufacturing, Materials and Technologies (ICSMMT 2019), Coimbatore, Tamil Nadu, India, pp. 020013-1–020013-10, 25–26 October 2019
2. Karthika, P., Vidhya Saraswathi, P.: A survey of content based video copy detection using big data. Int. J. Sci. Res. Sci. Tech. **3**, 114–118 (2017)
3. Barani Sundaram, B., Pattanaik, B., Kannaiya Raja, N., Elangovan, B., Masthan, K., Kumar, B.S.: Performance suitability interms of fault tolerane of MANET's on AOMDV and MDART in combination with ECDS. J. Crit. Rev. **7**(14) (2020). ISSN: 2394-5125
4. Barani Sundaram, B., Elemo, T.K.: Node isolation attack on OLSR, reputation relied mitigation. Palarch's J. Archaeol. Egypt/Egyptol. **17**(9) (2020). ISSN 1567-214x
5. Barani Sundaram, B., Kedir, T., Sorsa, T.T., Geleta, R., Srinivas, N., Genale, A.H.: An Approach for rushing attack resolution in AOMDV using arbitrary ID in MANET. Palarch's J. Archaeol. Egypt/Egyptol. (2020)
6. Barani Sundaram, B., Kannaiya Raja, N., Sreenivas, N., Mishra, M.K., Pattanaik, B., Karthika, P.: RSA algorithm using performance analysis of steganography techniques in network security. In: International Conference on Communication, Computing and Electronics Systems (ICCCES 2020), pp. 21–22, October 2020
7. Barani Sundaram, B., Kedir, T., Mishra, M.K., Yesuf, S.H., Tiwari, S.M., Karthika, P.: Security analysis for Sybil attack in sensor network using compare and match-position verification method. In: Second International Conference on Mobile Computing and Sustainable Informatics (ICMCSI 2021) (2021)
8. Barani Sundaram, B., Pandey, A., Abiko, A.T., Vijaykumar, J., Genale, A.H., Karthika, P.: Wireless sensor network to connect isolated nodes using link assessment technique. In: 3rd International Conference on Intelligent Communication Technologies and Virtual Mobile Networks (ICICV 2021) (2021)
9. Barani Sundaram, B., et al.: Steganalysis for images security classification in machine learning using SVM. In: 4th International Conference on Computational Vision and Bio Inspired Computing (ICCVBIC 2020) (2020)
10. Barani Sundaram, B., Maurya, S., Karthika, P., Vidhya Saraswathi, P.: Enhanced the data hiding in geometrical image using stego-crypto techniques with machine laerning. In: 6th International Conference on Inventive Computation Technologies (ICICT 2021) (2021)
11. Barani Sundaram, B., Rajkumar, P., Ananthi, M., Sravan Kumar, V., Vijaykuma, J., Karthika, P.: Network security analysis for signal strength based packet filitering. In: 3rd International Conference on Intelligent Sustainable Systems (ICISS 2020) (2020)
12. Barani Sundaram, B., Srinivas, N., Elemo, T.K., Mishra, M.K., Thirumoorthy, D., Sors, T.T.: Renewable energy sources efficient detection in triangulation for wireless sensor networks. In: International Virtual conference on Robotics, Automation, Intelligent Systems and Energy (IVC-RAISE 2020) (2020)

13. Sowjanya, S., Sundaram, B.: Overview on E-Voting System and Security Challenges (2018). http://www.ijaerd.com/index.php
14. Sowjanya, S., Sundaram, B.: Discovering IP Spoofers Locations from Path Backscatter (2018). http://ijsart.com/Home/IssueDetail?id=21325
15. Barani Sundaram, B., Sowjanya, S., Andavar, V., Reddy, N.R.: Opportunities and challenges of E-commerce in the case of Ethiopia. Int. J. Res. Technol. Stud. **5**(4) (2018). ISSN (online): 2348–1439
16. Barani Sundaram, B., Sowjanya, S., Andavar, V., Reddy, N.R.: Effectiveness of geographic information system and remote sensing technology as a decision support tool in land administration the case of Yeka sub city. Addis Ababa Int. J. Innov. Res. Sci. Eng. Technol. **7**(3) (2018). ISSN(Online): 2319-8753, ISSN (Print): 2347-6710
17. Sowjanya, S., Sundaram, B.: Review on stylization analysis using stylometric approach. Int. J. Sci. Adv. Res. Technol. (Int. Peer Rev. Open Access J.) (2018). ISSN [Online]: 2395-1052
18. Sowjanya, S., Sundaram, B., Deresa, S.: Humans emotions extraction from image analysis. Int. J. Sci. Adv. Res. Technol. (Int. Peer Rev. Open Access J.) (2018). ISSN [Online]: 2395-1052
19. Masthan, K., Kumar, B.S., Barani Sundaram, B., Pattanaik, B., Elangovan B., Kannaiya Raja, N.: Approach for active event correlation for detecting advanced multi-level cyber-attacks. Solid State Technol. **63**(2s) (2020)
20. Elangovan, B., Kannaiya Raja, N., Masthan, K., Kumar, B.S., Barani Sundaram, B., Pattanaik, B.: The image steganographic method to hide the secret text in an image using spiral pattern and cryptographic technique to increase the complexity for eavesdropper. J. Adv. Res. Dyn. Control Syste. **12**(08), 614–619 (2020)
21. Barani, B., et al.: Analysis of machine learning data security in the Internet of Things (IoT) circumstance. In: Jeena Jacob, I., Gonzalez-Longatt, F.M., Kolandapalayam Shanmugam, S., Izonin, I. (eds.) Expert Clouds and Applications. LNNS, vol. 209, pp. 227–236. Springer, Singapore (2022). https://doi.org/10.1007/978-981-16-2126-0_20
22. Barani Sundaram, B., Kedir, T., Mishra, M.K., Yesuf, S.H., Tiwari, S.M., Karthika, P.: Security analysis for Sybil attack in sensor network using compare and match-position verification method. In: Shakya, S., Bestak, R., Palanisamy, R., Kamel, K.A. (eds.) Mobile Computing and Sustainable Informatics, vol. 68, pp. 55–64. Springer, Singapore (2022). https://doi.org/10.1007/978-981-16-1866-6_4
23. Thirumoorthy, D., Rastogi, U., Sundaram, B.B., Mishra, M.K., Pattanaik, B., Karthika, P.: An IoT implementation to ATM safety system. In: Proceedings of the 3rd International Conference on Inventive Research in Computing Applications, ICIRCA 2021, pp. 744–749 (2021)
24. Andavara, V., Sundaram, B., Bacha, D., Dadi, T., Karthika, P.: The impact of perceived ease of use on intention to use mobile payment services for data security applications. In: Proceedings of the 2nd International Conference on Electronics and Sustainable Communication Systems, ICESC 2021, pp. 1875–1880 (2021)
25. Pattanaik, B., Barani Sundaram, B., Mishra, M.K., Thirumoorthy, D., Rastogi, U.: Industrial speed control of im based model predictive controller using Zeta converter. J. Phys. Conf. Ser. **1964**(6), 062075 (2021)
26. Rastogi, U., Pattanaik, B., Barani Sundaram, B., Mishra, M.K., Thirumoorthy, D.: Investigation of DSR protocol performance in wireless MANET using correlation method. J. Phys. Conf. Ser. **1964**(4), 042042 (2021)
27. Barani Sundaram, B., Mishra, M.K., Thirumoorthy, D., Rastogi, U., Pattanaik, B.: ZHLS: security Enhancement by integrating SHA256, AES, DH in MANETS. J. Phys. Conf. Ser. **1964**(4), 042003 (2021)

28. Pattanaik, B., Hussain, S.M., Kumar, R.S., Sundaram, B.B.: Design of smart inverter for distribution system using PV-STATCOM. In: International Conference on Intelligent Technologies, CONIT 2021 (2021)
29. Sundaram, B.B., Pandey, A., Abiko, A.T., Vijaykumar, J., Genale, A.H., Karthika, P.: Wireless sensor network to connect isolated nodes using link assessment technique. In: Third International Conference on Intelligent Communication Technologies and Virtual Mobile Networks (ICICV), pp. 39–42. IEEE (2021)

Brain Tumor Image Enhancement Using Blending of Contrast Enhancement Techniques

Deepa Abin(✉), Sudeep Thepade, Yash Vibhute, Sphurti Pargaonkar, Vaishnavi Kolase, and Priya Chougule

Department of Computer Engineering, Pimpri Chinchwad College of Engineering, Pune, Maharashtra, India
{deepa.abin,sudeep.thepade}@pccoepune.org

Abstract. A tumor is an abnormal tissue in the brain that damages the cell's ability to operate. Hence, detecting a brain tumor is a difficult undertaking. Manual tumor identification is dangerous since it necessitates the insertion of a needle into the brain. As a result, automated brain tumor detection technologies are required. Medical image processing gives fundamental information about brain abnormalities and assists doctors in making the best treatment decisions. Early detection and treatment reduce the odds of cancer worsening, enhances the survival rate, and improves the chances of a healthy life. In this work, enhancement of brain tumor MRI ('Magnetic Resonance Imaging') and therefore, better detection of the tumor present, if any. This paper proposes blending of existing algorithms like BBHE ('Bi-histogram Equalisation'), CLAHE ('Contrast Limited Adaptive Histogram Equalisation'), RESIHE ('Recursive Exposure-based Sub-Image Histogram Equalisation'), MSRCR ('Multi Scale Retinex with Colour Restoration') and more. Out of the ones that were experimented, CLAHE + MSRCR performed better; it's BRISQUE ('Blind/Reference less Image Spatial Quality Evaluator') value was found to be 29.805718 which shows the tumor is better visible.

Keywords: Brain tumor · Image enhancement · BBHE · CLAHE · BRISQUE · MSRCR

1 Introduction

Brain tumors can be fatal, have a severe impact on quality of life, and completely transform a patient's and their family's lives. One cause for the rising cases of brain tumors in the younger generation is the widespread use of cell phones [34]. A brain tumor is caused by abnormal cell growth in the brain. malignant and benign tumors are the two most common forms of tumors. Primary brain tumors have an effect on the tissues around them. Secondary brain cancers spread from the brain to other regions of the body. The location and kind of tissue used to detect brain tumors are used to classify them [31]. The World Health Organization (WHO) has recognised and classified approximately 120 different types of tumors. A person's life expectancy is likely to be extended if a tumor is detected early. Raymond V Damadian conducted research on the human body

J. I.-Z. Chen et al. (Eds.): ICIPCN 2022, LNNS 514, pp. 736–747, 2022.
https://doi.org/10.1007/978-3-031-12413-6_58

and developed MRI in 1969. MRI is a commonly utilised tool in hospitals because it can readily show soft abnormalities in the brain and does not require ionising radiation [30]. Brain tumors develop as a result of unregulated and excessive cell proliferation. Primary and secondary tumors of the brain are distinguished. With the use of MRI, tumor heterogeneity can be recognised with high resolution. Traditional imaging technology like computerized tomography (CT) or X-ray provide better views of the brain, muscles, heart, and malignant tissues than MRI. Noise and image clarity are restrictions in MRI imaging [23]. Unwanted information in photographs is referred to as noise. Noise can sometimes affect the edges and small details, limiting the contrast resolution. It is difficult to detect the exact boundaries and classification of the tumor due to Noise, making it an emerging topic of image processing research. However, if the cancer is identified early and a precise diagnosis is made, the person who is suffering from it may have a better chance of surviving. As technology advances at a rapid pace, the opportunities for data recording expand, resulting in a plethora of new channels for information flow.

The dataset used is a 'Kaggle' dataset: "Brain MRI images for Brain Tumor Detection." [2]

The paper has been put forth as: '2. Literature Survey', '3. Materials and methods', '4. Results and Discussions', '5. Conclusions'.

2 Literature Survey

The purpose of this work is to give a literature review that demonstrates how various papers have led to a better understanding of these algorithms and their current structure.

2.1 Image Enhancement

Image enhancement is an essential technique for such work, as it makes the detection of tumors easier. Such algorithms enhance the borders of tumor regions and make it easier to identify them, which is why we need to explore this domain for enhancing medical images and aiding in early detection of the same. Before detection of the above mentioned tumors, we need to preprocess the images. Preprocessing requires the noise to be removed, which is also achieved by applying these algorithms for enhancement of the MRIs.

Some of the algorithms that were experimented on, are:

'Bi-histogram Equalisation (BBHE)': This method is an extension of the standard histogram measurement, which is able to keep the image's luminosity in the end definition of bi-histogram equity. In this way we can overcome the problem of normal histogram measurement [5].

In this way we can split the image into two parts and take it as a threshold. One part has a set of sample values greater than the definition value and one component with a set of lower values. Then BHE measures a decaying image into two smaller images independently on the basis of their histogram [4].

'Dualistic Sub-Image Histogram Equalization (DSIHE)': The process for achieving this approach is similar to that of the BBHE algorithm, other than the image divergence of histograms is generated using the median value of an original image [1].

'Contrast Limited Adaptive Histogram Equalisation (CLAHE)': It is a variation of adaptive histogram equalisation (AHE) that corrects for contrast overamplification. CLAHE works on tiles, which are small areas of an image rather than the complete image. The contrast of photographs can be improved with this approach [9].

'Multi Scale Retinex with Colour Restoration (MSRCR)': This method improves photographs shot in a variety of non-linear lighting circumstances to the point where the user can see them in real time. The acquired image is used to compare the expansion and adjustment of global Gamma to enhance the image [24].

'Recursive Exposure-based Sub-Image Histogram Equalisation (RESIHE)': R-ESIHE iteratively executes the ESIHE technique till the exposure leftover is less than a predefined threshold in each cycle [26].

'Recursively Separated Exposure-based Sub-Image Histogram Equalisation (RSESIHE)': It performs recursive histogram decomposition and is a recursion variation of ESIHE. It iteratively breaks down the input histogram using exposure limits from particular sub histograms [25, 26].

2.2 Comparative Study of Existing System

The purpose of this work is to give a literature review that demonstrates how various papers have led to a better understanding of these algorithms and their current structure.

As discussed in [5], when compared to basic BBHE, DSIHE, advanced approaches produce superior outcomes with maximum entropy and improved contrast. Sakshi Patel et al. [7] it was discovered that the BBHE algorithm preserves the image's brightness, while the modified versions DSIHE and RSIHE, yield better outcomes. As mentioned in [1] DSIHE, MMBEBHE are modifications of the BBHE approach that preserves maximum brightness. For MRI Brain Images, it can execute effective contrast enhancement procedures. The input image and the CLAHE contrast-enhanced image are dissected for three layers, as indicated in [9]. At each stage of decomposition, approximation coefficients are merged with detailed coefficients via an averaging technique. Here the CLAHE enhances the image's local features more effectively and also gives more accurate results. Most preceding contrast enhancement approaches, such as Brightness preserving BBHE, DSIHE, and RSIHE, are utilised for normal greyscale images, as noted in [6]. BBHE enhances greyscale images and improves image contrast. According to [8], the HE approach is a global process; hence the image luminance is not preserved. Local-HE and luminance-conserving HE approaches have been suggested to fix this problem. Best performance brightness image enhancement is the outcome of the specified task. As discussed in [9] The BBHE method was established premised on the notion that Histogram Equalization does not account the image's mean luminosity. As a result, BBHE utilizes the image's average luminance and here there is no amplification of noise signal. Simulation results showed that RMSHE, instead of HE, BBHE, DSIHE, and MMBEBHE, is the superior equalisation strategy [13] by Varsha, Manju Mathur. It has already been discovered that MMBEBHE is much effective than BBHE and DSIHE for conserving an image's new luminosity when used in the context of bi-histogram equalisation and this study shows us high accuracy. A bright and specific-conserving HE approaches with better contrast enhancement effect has been an aim of much current HE researches, as mentioned in [14] by Parmeshwar Kumar, Manoj Yadav, Kailash Patidar and here BBHE solves mean shift problem of traditional histogram equalization technique. As shown in [15], on the actual image, enhancement techniques such as BBHE were initially used to create a new image with a histogram indicating three zones which. This work results in usage of computer aided diagnostic tools for doctors for tumor specification. The BBHE equalizes the sub-images independently, which has high accuracy. We apply filters on improved images for image classification of brain tumors, as discussed in [16]. The technique manages the image histogram in such a manner that no rebinding of the histogram maxima occurs and this study is useful to overcome the problem that occurred due to intensity inhomogeneity. The paper suggests a unique extension of BBHE called minimum mean brightness error bi-histogram equalisation (MMBEBHE) as stated in [17] and here MMBEBHE provides maximum brightness preservation, BBHE preserves the original brightness. In this work when it comes to utilising the sample photos, MMBEBHE performs in a similar way to BBHE and DSIHE. BBHE has been suggested and theoretically studied to recover the natural luminance to a certain extent, as mentioned in [19] and also it shows the scalable brightness preservation with RMSHE. The DSIHE technique utilizes the entropy value for histogram segmentation, as indicated in [18]. MMBEBHE is an expansion of the BBHE technique that gives better contrast enhancement and here the techniques perform contrast enhancement very well. Previous histogram equalisation approaches have been plagued by amplified contrast, artefacts, and a surprise abnormal appearance of the processed images, as stated in [7] and it also results a high level of visibility quality combined with a lot of useful knowledge. This study offers an iterative, mean, and multi-threshold selection criterion with plateau limits, which consists of histogram fragmentation, to resolve these shortcomings and this study results in very high accuracy. As stated in [20], histogram equalisation is the most common method used to improve contrast in the feature domain. Regarding MRI glioblastoma brain tumors, a unique efficient and simple pre-processing framework based on combination of noise removal and contrast enhancement approaches has indeed been presented in this study and this gives best accuracy in terms of accuracy.

The comparative study of existing system done as seen in Table 1.

Table 1. Related work summarization

Sr. no	Author	Technique	Strength	Limitation	Result	Accuracy
1.	M. Agarwal, R. Mahajan	BBHE, DSIHE	Accuracy of Diagnosis is High	Time consuming	Effectively used in video processing, image processing	91%
2.	S. Patel, Bharath K. P.	BBHE, DSIHE	BBHE works better	More complexity, takes more time to give the output	Gives best results for better diagnosis	91.78%

(*continued*)

Table 1. (*continued*)

Sr. no	Author	Technique	Strength	Limitation	Result	Accuracy
3.	M. Pawar, S. Talbar	BBHE, CLAHE	CLAHE improves local details of image more efficiently	Risk of overfitting	Gives more accurate results	97%
4.	Z. Y. Lim, K. S. Sim	BBHE, DSIHE	BBHE enhances greyscale images	Comparatively low accuracy	Improves image contrast	90%
5.	Ashwini Sachin Zadbuke	BBHE, DSIHE	Best performance brightness image enhancement	Time consuming	Gives the best result	96%
6.	R. Sunita, Kaur Amandeep	BBHE, CLAHE	No amplification of noise signal	Limits the amplification by clipping histogram	CLAHE gives good results, high accuracy	91%
7.	Varsha, Manju Mathur	BBHE, DSIHE	High accuracy	Depends upon high dimensional data	Should not get revealed to unauthorised channel	95%
8.	P. Kumar, Manoj Yadav, K. Patidar	BBHE, DSIHE, Contrast Enhancement	BBHE solves mean shift problem of traditional histogram equalization technique	Contrast Enhancement isn't very effective	Gives accurate results using BBHE, DSIHE	92%
9.	Pratik Vinayak Oak	BBHE, MMBEBHE	BBHE equalizes the sub-images independently	Occurrence of noise	Computer aided diagnostic tools for doctors for tumor specification. High Accuracy	90%
10.	Mr. A. K. Mittal, Dr. Sukwinder Singh	Histogram Equalization	Solves the problems due to intensity inhomogeneity	Comparatively low accuracy	Segments the tumor from the background of the brain MRI	89%
11.	Soong-Der Chen, Abd. Rahman Ramli	BBHE, MMBEBHE, DSIHE	MMBEBHE provides maximum brightness preservation	Time consuming	When utilising photos, MMBEBHE performs in a similar way to the other two	96.2%

(*continued*)

Table 1. (*continued*)

Sr. no	Author	Technique	Strength	Limitation	Result	Accuracy
12.	P. A. Mohrut, Dr. D. Shrimankar	DSIHE, BBHE, MMBEBHE	Performs contrast Enhancement well	Brightness of an image is not preserved	Improves image contrast without over enhancement	91%
13.	M. Raghavendra, G. B. Reddy	MMBEBHE, DSIHE	All approaches preserve maximum luminosity	Low accuracy	Performs contrast enhancement for MRI well	85%
14.	S. D. Chen, A. R. Ramli	BBHE, DSIHE, RMSHE, MMBEBHE	Scalable brightness preservation with RMSHE	Comparatively low accuracy	High Accuracy	90%
15.	Md Q. Ali, Z. Yan, Hua Li	BBHE, DSIHE	High accuracy	Time consuming	A high quality images	92%
16.	H. Mzoughi, B. Slima	BBHE, CLAHE,	Give the best result	Complicated process	Highly Accurate	93%

3 Materials and Methods

3.1 Image Enhancement

The need to study and experiment on such algorithms arises from the increasing number of patients inflicted with such tumors, that can sometimes turn fatal. To prevent such fatalities, this project can help better image quality, which will, in turn, help in early detection of tumors. Detection, although, an important step, without preprocessing, it's a very difficult task to undertake. For better preprocessing of these images, these algorithms provide better noise reduction, edges highlights, etc., which help in further making the detection easy. The algorithms that were used for experimentation are:

'Bi-histogram Equalisation (BBHE)': It is a mathematically evaluated technique that can retain natural luminosity to some extent. It does the segmentation of the histogram of the input image first, into two parts, then equalises them separately. Another histogram is created using the pixel values. The final image is calculated by measuring each histogram independently using the histogram equalisation approach, and then adding them together to produce the final image. In order to retain the original luminosity to some extent, the concept of BBHE [5] was already developed and theoretically researched. Depending on the information mean, it separates the histogram of the input image into equal halves and equalises them individually.

'Dualistic Sub-Image Histogram Equalization (DSIHE)': It works a lot like BBHE, which splits the histogram of the first image into two sub histograms based on exact value of the median. It decomposes an image depending on the grey matter that has accumulated with the goal of maximizing the resulting image's Shannon's entropy. The input image is separated into two distinct images, one black and one light, and the lower decay images are then applied with HE [7].

'Contrast Limited Adaptive Histogram Equalisation (CLAHE)': It can overcome over magnification of the sound problem in the same image area with standard histogram measurement. The CLAHE [9] algorithm is different from the standard HE in that regard CLAHE works in small regions in the picture, called tiles, and counts a few histograms, each associated with a different category of picture and use them to redisperse the light values of picture.

'Multi Scale Retinex with Colour Restoration (MSRCR)': This colour fog imaging algorithm based on adaptive scaling is proposed. The colour restoration coefficient is first tested in the RGB colour space. The local weight correction function is then applied to each channel's pixel values, and the appropriate scale's Gaussian kernel is calculated. Finally, the image is enhanced by contrast stretching and global Gamma correction using the resulting image [24].

'Recursive Exposure-based Sub-Image Histogram Equalisation (RESIHE'): The difference in exposure between iterations determines the number of recursions. Histogram subdivision and equalisation are the most important steps in the technique [26].

'Recursively Separated Exposure-based Sub-Image Histogram Equalisation (RSESIHE)': It decomposes the input histogram recursively, whereas ESIHE decomposes it just once based on exposure thresholds of specific sub histograms. Individually, the decomposed sub histograms are equalised [25, 26].

3.2 Performance Metrics

'Nature Image Quality Evaluator (NIQE)': It is a blind quality analyser (IQA) that uses only measurable deviations from data sets seen in original images, without training in distorted images that are man-made, and, among other things. exposure to these images. However, all of the current standard non-reference IQA, in the form of training models, algorithms require information regarding predicted distortions and the effects of linked human perceptions [22, 28].

'Perception based Image Quality Evaluator (PIQE)': The scoring image is distorted due to artifact blocking and Gaussian noise. It produces a quality local mask featuring high-performance block blocks, visual art blocks, and audio blocks in a photo [22].

'Blind/ Reference less Image Spatial Quality Evaluator (BRISQUE)': BRISQUE is a spatial-domain distortion-generic blind/no-reference (NR) image quality assessment (IQA) model based on natural scene statistics (NSS). It instead uses scene statistics of locally normalised brightness coefficients to assess probable losses of 'naturalness' in the image due to the presence of distortions, resulting in a holistic measure of quality [27].

Entropy: Entropy is a measure of impurity or uncertainty in a set of data used in information theory. It determines how data is split by a decision tree. In a closed system, entropy is a measure of disorder and randomness. In other words, a high entropy number indicates that your system's randomness is high, making it impossible to anticipate the state of its atoms or molecules. If the entropy is low, however, forecasting that state is significantly easier [29].

4 Results and Discussion

As a result, all of the algorithms were executed, including BBHE, DSIHE, CLAHE, MSRCR, RESIHE, RSESIHE, and it can be seen that CLAHE has the best results.

Initial results shown in Fig. 1.

Fig. 1. Results of algorithms applied on dataset images.

The adoption of combining/blending of algorithms like CLAHE+BBHE, CLAHE+MSRCR, CLAHE+RESIHE, and CLAHE+RSESIHE results in a clearer visualisation of the tumor, implying that the tumor is more illuminated and easily detected. The outcomes of this experiment are as shown below in Figs. 2 and 3:

In both these examples, it can be seen that the combination CLAHE+MSRCR showed the best results and the tumor is easily seen and detected.

After this, performance metrics mentioned above were calculated and tabulated. Values of these metrics help in understanding the quality of these images, like the BRISQUE values of the combination algorithms were found to be lesser in comparison to the original value which shows that the image has been enhanced and as it can be seen, the tumor is better visible. The BRISQUE value for the input image is 38.618412, the BRISQUE values of blended algorithms should be less than this to indicate better quality, enhanced images. The table and graph for the BRISQUE values are as shown below in Table 2 and Fig. 3:

Fig. 2. Results of combination of the algorithms on dataset images

Fig. 3. Results of blending of the algorithms on dataset

From the graph, Fig. 4, it can be concluded that images with the lowest BRISQUE value, the image quality is very good. Therefore, proving that CLAHE+MSRCR has generated better results.

Table 2. BRISQUE values tabulated according to algorithms tested.

ALGORITHM	BRISQUE values
BBHE [5]	40.064878
DSIHE [1]	40.560203
CLAHE [9]	39.18355
RESIHE [26]	39.315245
CLAHE+BBHE	37.946967
CLAHE+MSRCR	29.805718
CLAHE+RESIHE	36.827191
CLAHE+RSESIHE	38.366146

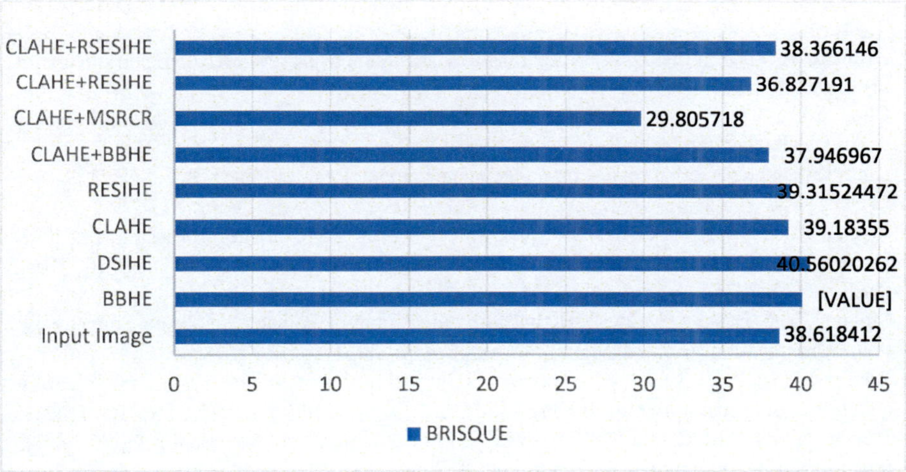

Fig. 4. Graph for BRISQUE values of tested algorithms.

5 Conclusion

Combining the other algorithms with CLAHE proved to be a successful attempt to enhance such images for easy and early tumor detection. These images, obtained by CLAHE+MSRCR, show the edges and shape of the tumor very clearly, as it's BRISQUE value was 29.805718 which was less than the input image's value and hence, proves that CLAHE+MSRCR was a successful blend. This will greatly help in early detection of any tumors that might go unnoticed in lower quality images. With these results at hand and some more information, this method can be used in helping to classify such tumors in the near future, which will, in turn, be very beneficial to treat any patient and help them live a tumor-free, healthy and long life.

References

1. Raghavendra, M., Vikramsimhareddy, M., Reddy, G.: Preprocessing MRI images of colorectal cancer. Int. J. Comput. Sci. Issues (IJCSI) **14**(1), 48 (2017)
2. https://www.kaggle.com/navoneel/brain-mri-images-for-brain-tumor-detection
3. Logeswari, T., Karnan, M.: An improved implementation of brain tumor detection using segmentation based on soft computing. J. Cancer Res. Exp. Oncol. **2**(1), 006–014 (2009)
4. Agarwal, M., Mahajan, R.: Medical images contrast enhancement using quad weighted histogram equalization with adaptive gamma correction and homomorphic filtering. Procedia Comput. Sci. **115**, 509–517 (2017)
5. Sim, K.S., Chung, S.E., Zheng, Y.L.: Contrast enhancement brain infarction images using sigmoidal eliminating extreme level weight distributed histogram equalization. Int. J. Innov. Comput. Inf. Control (IJICIC) **14**(3), 1043–1056 (2018)
6. Sakshi Patel, K.P., Bharath, S.B., Muthu, R.K.: Comparative study on histogram equalization techniques for medical image enhancement. In: Das, K.N., Bansal, J.C., Deep, K., Nagar, A.K., Pathipooranam, P., Naidu, R.C. (eds.) Soft Computing for Problem Solving. AISC, vol. 1048, pp. 657–669. Springer, Singapore (2020). https://doi.org/10.1007/978-981-15-0035-0_54
7. Zadbuke, A.S.: Brightness preserving image enhancement using modified dualistic sub image histogram equalization. Int. J. Sci. Eng. Res. **3**(2), 1 (2012)
8. Chen, S.D.: A new image quality measure for assessment of histogram equalization-based contrast enhancement techniques. Digit. Signal Process. **22**(4), 640–647 (2012)
9. Kaur, K., Bathla, A.K.: A review on image enhancement. Res. Cell Int. J. Eng. Sci. **22**, 31–36 (2016)
10. Methil, A.S.: Brain tumor detection using deep learning and image processing. In: 2021 International Conference on Artificial Intelligence and Smart Systems (ICAIS), pp. 100–108. IEEE, March 2021
11. Mohan, G., Subashini, M.M.: MRI based medical image analysis: survey on brain tumor grade classification. Biomed. Signal Process. Control **39**, 139–161 (2018)
12. Pawar, M., Talbar, S.: Local entropy maximization based image fusion for contrast enhancement of mammogram. J. King Saud Univ.-Comput. Inf. Sci. **33**(2), 150–160 (2021)
13. Jabeen, A., Riaz, M.M., Iltaf, N., Ghafoor, A.: Image contrast enhancement using weighted transformation function. IEEE Sens. J. **16**(20), 7534–7536 (2016)
14. Oak, P.V.: Automatic tumor detection and area calculation from brain MR image using image processing
15. Mittal, A.K., Singh, S.: Brain tumor detection with histogram equalization and morphological image processing techniques. Int. J. Innov. Res. Sci. Technol. **1**(3), 28–31 (2014)
16. Chen, S.D., Ramli, A.R.: Preserving brightness in histogram equalization-based contrast enhancement techniques. Digit. Signal Process. **14**(5), 413–428 (2004)
17. Mohrut, P.A., Shrimankar, D.: Image contrast enhancement techniques: a report. Int. Res. J. Eng. Technol. (IRJET) **2**(2) (2015)
18. Chen, S.D., Ramli, A.R.: Minimum mean brightness error bi-histogram equalization in contrast enhancement. IEEE Trans. Consum. Electron. **49**(4), 1310–1319 (2003)
19. Ali, Q.M., Yan, Z., Li, H.: Iterative thresholded bi-histogram equalization for medical image enhancement. Int. J. Comput. Appl. **114**(8) (2015)
20. Mustafa, W.A., Kader, M.M.M.A.: A review of histogram equalization techniques in image enhancement application. In: Journal of Physics: Conference Series, vol. 1019, no. 1, p. 012026. IOP Publishing, June 2018
21. Schlett, T., Rathgeb, C., Henniger, O., Galbally, J., Fierrez, J., Busch, C.: Face image quality assessment: a literature survey. ACM Comput. Surv. (CSUR) (2021)

22. Jijja, A., Rai, D.: Efficient MRI segmentation and detection of brain tumor using convolutional neural network. Int. J. Adv. Comput. Sci. Appl. **10**(4), 536–541 (2019)
23. Parthasarathy, S., Sankaran, P.: An automated multi scale retinex with color restoration for image enhancement. In: 2012 National Conference on Communications (NCC), pp. 1–5. IEEE, February 2012
24. Singh, K., Kapoor, R., Sinha, S.K.: Enhancement of low exposure images via recursive histogram equalization algorithms. Optik **126**(20), 2619–2625 (2015)
25. Singh, N., Kaur, L., Singh, K.: Histogram equalization techniques for enhancement of low radiance retinal images for early detection of diabetic retinopathy. Eng. Sci. Technol. Int. J. **22**(3), 736–745 (2019)
26. Mittal, A., Moorthy, A.K., Bovik, A.C.: Blind/referenceless image spatial quality evaluator. In: 2011 Conference Record of the Forty Fifth Asilomar Conference on Signals, Systems and Computers (ASILOMAR), pp. 723–727. IEEE, November 2011
27. Liu, Y.H., Yang, K.F., Yan, H.M.: No-reference image quality assessment method based on visual parameters. J. Electron. Sci. Technol. **17**(2), 171–184 (2019)
28. Zhuang, L., Guan, Y.: Adaptive image enhancement using entropy-based subhistogram equalization. Comput. Intell. Neurosci. (2018)
29. Nazir, M., Wahid, F., Ali Khan, S.: A simple and intelligent approach for brain MRI classification. J. Intell. Fuzzy Syst. **28**(3), 1127–1135 (2015)
30. Sirven, J.I., Wingerchuk, D.M., Drazkowski, J.F., Lyons, M.K., Zimmerman, R.S.: Seizure prophylaxis in patients with brain tumors: a meta-analysis. In: Mayo Clinic Proceedings, vol. 79, no. 12, pp. 1489–1494. Elsevier, December 2004
31. Sharma, P., Diwakar, M., Choudhary, S.: Application of edge detection for brain tumor detection. Int. J. Comput. Appl. **58**(16) (2012)

Flower Recognition Using VGG16

Md. Ashikur Rahman[1], Md. Saif Laskar[1], Samir Asif[1], Omar Tawhid Imam[2], Ahmed Wasif Reza[1], and Mohammad Shamsul Arefin[3,4](✉)

[1] Department of Computer Science and Engineering, East West University, Aftabnagar, Dhaka 1212, Bangladesh
wasif@ewubd.edu
[2] Department of Computer Science and Engineering, Bangladesh University of Engineering and Technology, Dhaka, Bangladesh
[3] Department of CSE, Daffodil International University, Dhaka 1341, Bangladesh
[4] Department of Computer Science and Engineering, Chittagong University of Engineering and Technology, Chattogram 4349, Bangladesh
sarefin@cuet.ac.bd

Abstract. The purpose of our model is to classify five types of flowers from input images. The flowers are Sunflower, Rose, Tulip, Daisy, and Lavender. We have also built our own CNN model for the task and compared it with the modified VGG16 network. Our modified VGG16 model gives better accuracy than the existing works. We have achieved a test accuracy of 96.64% by using the proposed model. As the accuracy is quite good, we were able to recognize the flowers accurately. Agriculture institutes and flower nurseries can be benefitted by using this model.

Keywords: Flower recognition · VGG16 · Model · Accuracy

1 Introduction

Flower recognition is a very important task for students, researchers, and other people in agriculture, plantation, and flower nursing including other related sectors. Recently, a lot of image processing research has been done. In image processing and computer vision, the neural network is an important architecture. A convolutional neural network (CNN) is a high-performance Machine Learning architecture. It has so many successes in image processing and the computer vision field.

A lot of work has been done for automatic plant recognition based on plant images. Among the plants, flowers can be distinguishable easily because different flowers contain different types of features such as different shapes, different colors, or textures. Each flower has its unique feature. These features are stable. If the weather condition changes or the age of flowers increases the features remain the same. Therefore, flower images can be a good task for identification or recognition. But the identification is not so easy. There are some challenges like object deformations, intra-cluster similarity, inter-cluster similarity, and occlusion showed by [2], Nilsback, Maria-Elena, and Andrew

J. I.-Z. Chen et al. (Eds.): ICIPCN 2022, LNNS 514, pp. 748–760, 2022.
https://doi.org/10.1007/978-3-031-12413-6_59

Zisserman. There are many approaches to flower recognition. All of the approaches consist of four steps. These steps are classification, segmentation, pre-processing, and feature extraction. Usually, a flower image has a complex background which makes these steps very hard. Complex background can affect the accuracy. Because of the complex background sometimes the obtained accuracy is decreased.

Transfer learning is getting popular. Recently it has shown several successes in image classification, segmentation, and object recognition. Transfer learning can also be used for clustering type problems and regression. Reusing a pre-trained model for another task is known as transfer learning. VGG16 is a very well-known pre-trained model. 'VGG' stands for Visual Geometry Group and the value '16' means it has 16 layers. This paper uses VGG16 pre-trained model to classify the flower images.

2 Related Work

Plant identification based on images of plant organs may be accomplished in two ways. They are hand-crafted features with a focus on deep learning. Kernel Descriptors and other hand-crafted (or hand-designed) characteristics have been used for flower identification (KDES). For flower recognition, rank 1 accuracy of 11.38% and rank 2 accuracies of 38.05% have been achieved in [1] by Thi-Lan Le, Nam-Duong Duong, Hai Vu, and Thanh-Nhan Nguyen.

Now, [2], Maria-Elena, and Nilsback proposed their research about shape, color, and texture. On a dataset of 17 flower categories, the authors retrieved numerous types of features such as HSV values, MR8 filter, SIFT, and Histogram of Oriented Gradients (HOG) in [2]. They then blend several linear weighted kernels with a Support Vector Machine (SVM) classifier. They analyze and pick the best characteristics, which they then apply to a dataset of 102 categories and species. A high rate of recognition is attained (up to 80%).

In [3], Rodrigo, Ranga, Kalani Samarawickrame, and Sheron Mindya used the shape and color features to extract images of the flowers. Principal Component Analysis (PCA) from several types of flowers to distinguish species have been used. From PCA, they have achieved 77.5% accuracy. And using SVM they got accuracy up to 82.5%.

In [4], HONG An-Xiang, CHEN Gang, LI Jun-li, CHI Zhe-RU, ZHANG Dan did flower region extraction operation and shape features were used in recognition. As a database, a dataset containing 885 flower images was used in this paper. In color, features recall was up to 0.67. In shape feature, the recall was up to 0.56, and combining color and shape feature up to 0.8 recall was achieved.

In [5], Hervé Goëau, Pierre Bonnet, Alexis Joly did plant identification research. The authors have used image data of 1000 different species instead of 500. The total number of observations was 41,794. 123 research groups registered worldwide in 2015 in this research. The maximum score of 0.667 was achieved by SNUMED INFO run4 (5-fold GoogLeNet Borda+). In terms of feature learning methodologies, some research teams in the PlantCLEF 2015 [6] competition used CNN for the plant identification challenge based on multi-image plant observation queries. A complete plant, branch, fruit, leaf, flower, stem, or leaf scan is one of seven view kinds for a query observation. To the best of our knowledge, only a few studies have used CNN to focus on floral images. In this work, we show how CNN may be used to identify plants based on their flowers. To demonstrate CNN's robustness, we compare its performance to that of a hand-designed feature approach.

In [7], S Christian et al. have researched to improve the performance of the neural networks. They tried to go into the depth of the neural network and improve it. Architectural details of the neural network were studied. Their implementation was CPU-based. 1.2 million images were used for training, 50,000 for validation, and 100,000 for testing. They achieved the lowest error rate of 6.67% only.

In [6], Liefeng Bo, Xiaofeng Ren, Dieter Fox proposed KDES is a robust feature extraction approach that allows you to develop hierarchical models starting at the pixel level and working your way up to the patch and/or entire flower picture level. After generating KDES, the authors use an SVM classifier to classify the data. For leaf-based plant identification, KDES produced extremely encouraging results. When KDES is used for flower-based identification, however, the recognition rate remains unacceptably low. The Images, used in the experiment, were not more than the size of 300 $\times$ 300. Laplacian kernel SVMs have been found as over efficient than linear SVMs.

In [8], Alex Krizhevsky, Ilya Sutskever, Geoffrey E. Hinton used CNN in their paper. The training was done on multiple GPUs. ImageNet dataset has been used. ImageNet has over 15 million high-resolution images. The nonlinearity, used in the experiment, is ReLU. Their architecture has 60 million parameters. ILSVRC's 1000 classes impose a 10-bit limit on the mapping from image to label for each training sample, however, this is inadequate to learn that many parameters without significant overfitting. Data augmentation was done, 224 $\times$ 224 patches were extracted from 256 $\times$ 256 images. Top-1 and Top-5 error rates on this dataset were 67.4% and 40.9%.

In [9], Thi Thanh Nhan Nguyen, Van Tuan Le, Thi Lan Le, Hai Vu, Natapon Pantuwong, Yasushi Yagi have used deep convolutional neural networks to identify the flower species. As a dataset, they extracted the flower images from the dataset of PlantCLEF 2015. Their initial experiment was comparing the performance of multiple CNNs on a pre-processed flower database. They have used three neural network architectures-Caffenet, Alexnet, Googlenet. From the experiment, they have found that the best accuracy among the architecture was from the Googlenet, which is 66.60%. The Alexnet accuracy was 50.60% and the Caffenet accuracy was 54.84%. The Googlenet performs better because it has more layers than Alexnet and Caffenet.

In [10], G. Ranganathan analysed many literatures works which contains different techniques and implemented with pre-processing step and without pre-processing step. Author compared the accuracy of literature works contains pre-processing step with literature works that do not contain pre-processing step. Author finds that pre-processing technique improves the accuracy on deep learning applications though the deep learning algorithms can be trained without applying any pre-processing technique. Pre-processing removes noise from data. It gives better result for signal classification algorithms, data, and image.

In [11], Akey Sungheetha and Rajesh Sharma R researched the remote sensing images classification. The images are actually the representation of some of the parts of earth captured from space. For that, they have used two CNNs parallelly on the same image for double feature extraction. After taking the input image, the image was pre-processed with feature extraction by CNN, and it was done twice separately. After normalizing, the SVM classifier was used. As an augmentation part, the images were rotated at 90, 180, and 270°. Accuracy for pre-trained CNN, Single classifier SVM, and Probability-based CNN is respectively 69.83%, 80% 84%. But accuracy of their proposed hybrid CNN framework is 97%. Unlike the existing methods, the proposed hybrid algorithm gives more accurate accuracy.

In [12], flower recognition has been performed based on color and GIST Features. In preprocessing part, RGB images were converted to HSV. Statistical color features like mean and standard deviation have been calculated for those HSV images and combined to obtain a threshold to create binary images. Binary images were created and used as masks on the RGB images. In this work, the highest accuracy obtained on the test dataset is 85.93% using the Support Vector Machine. As a test dataset, 16 images were taken.

In [13], a review of flower recognition works has made. Authors have reviewed 15 papers and among those papers, the highest rank-1 accuracy is 93.41 using Inception-v3 on the FLOWERS102 dataset [17]. FLOWERS102 dataset has 8189 images in 102 categories and each category has a different number of images.

In [14], the Oxfordflower-102 public dataset has been used. VGG-16, Improved AlexNet, and CNN with On/Off ReLU were used. VGG-16 performs best with an accuracy of 91.9%.

This [15] work is a multimodal Biometric recognition system. Biometric features like iris, face, fingerprint, finger vain, palm print, and finger vein pattern can be used to recognition of a person on an online signature. The system, which has been built is a multimodal system. It performs the recognition based on the feature it gets. The proposed biometric model is a multimodal CNN approach that performs n Iris, face, fingerprint, vein, and palm print and identification accuracy is 94%.

3 System Architecture and Design

Figure 1 describes the proposed method. The framework contains (1) Data initialization, (2) Model building, (3) Model Testing.

Fig. 1. System architecture of the proposed framework

3.1 Dataset Description

We have worked with 5 categories of flowers. The categories are Rose, Daisy, Lavender, Sunflower, and Tulip. Rose, Daisy, Lavender, Sunflower, Tulip contains the number of images respectively 155, 159, 116, 202, and 113. So, we are processing a total of 745 images. Figure 1 shows some samples of each category of flowers. Most of the images contain a single piece of flower, but some of the images also contain multiple flowers of the same category. For example, we can see that, in Table 1, a rose and daisy contain only a single flower, but tulips, lavender, and sunflowers have multiple flowers in a single frame. No image contains multiple flowers with different categories in our dataset.

Table 1. Sample flowers of the dataset

Serial	Image			Name	Total number of images
1				Sunflower	202
2				Tulip	113
3				Rose	155
4				Daisy	159
5				Lavender	116

3.2 Data Preprocessing

We have selected five flowers for our dataset, and they are Rose, Daisy, Lavender, Sunflower, Tulip. First, we have resized all the images and the size of each image is 224 $\times$ 224. We read images with cv2.imread and used cv2. IMREAD_COLOR which read images as BGR color format. We have divided all the pixel values by 255.0 to range from 0 to 1. For the own build CNN model, we have used the data augmentation, but no accuracy was updated and there was no benefit.

Here the flower recognition is done through three steps-Dataset Initialization, Model Building, Model Testing.

3.3 Dataset Initializing

Algorithm1: Dataset Initializing

Input: Folders Directory and Category

1. **begin**
2. **for** each image in the directory **do**:
   initialize path;
   read image in an array using the path;
   resize the image;
   append image;
   append image category
4. **end for;**
5. **end;**

As a dataset, we took the images from the folders. A loop is considered, which runs to get all the images in the directory. At first, the path is initialized by the directory and category. Then, images are read into an array using the path. Then resizing is done and then, the resized images and image category/flower category is appended to the dataset.

3.4 Model Building

Algorithm2: Model Building

Input: dataset
1. **begin**
2. **İmport** the VGG16 architecture as vgg;
3. İnitalize architecture's parameters(weight and include_top);
4. **İmport** the Model and Dense;
5. Initialize the model with parameters.
   [inputs= vgg.input and Outputs=(5 and activation_function= 'softmax')];
6. train model;
7. Save model;
8. **end;**

We have used VGG16 for flower recognition. Using the training dataset, we have had to build the Flower Recognition model. So, as training data of the model, we have considered our flower dataset. In this section, the VGG19 was imported, then, the parameters of the imported architecture were defined. Then the model for Flower Recognition was built with parameters. After that, the model was trained by the training dataset. After Building the model, it was saved, so that it will not need to rebuild the model.

3.5 Model Testing

Algorithm3: Model Testing

Input: image given by the user

1. **begin**
2. import saved model;
3. resize input image;
4. predict the category using the model.
5. print the flower Category
6. **end;**

Input is the user given here. The saved model would be imported first. Then, the user input image will be resized. Then the model would predict the input flower category and tell the user which kind of flower the input image has.

4 Proposed Method

CNN

Convolutional neural networks are a sort of artificial neural network that uses multiple hidden neurons to analyze picture inputs and have learnable weights and bases to several portions of images that can separate themselves. Convolutional Neural Networks have the advantage of leveraging the utilization of local spatial coherence in the input images, allowing them to have fewer weights because some parameters are shared. In terms of memory and complexity, this procedure was efficiently proposed by [16], Fei Liu, Yong Wang, Fan-Chuan Wang, Yong-Zheng Zhang, Jie Lin.

In our own built CNN, there are two 2 types of layers:

Convolutional 2D Layer
For the next layer, a feature map is required, which is built by passing a kernel matrix over the input matrix in the convolutional layer. When the kernel matrix is slid over the input matrix, we perform convolution. At each of the points, matrix multiplication is done element-wise and the resulting total is plotted on the feature map. It is a linear operation that is frequently utilized in a range of disciplines such as image processing, statistics, and physics. It can be used on more than one axis. If we have a 2-Dimensional image input, I, and a 2-D kernel filter, K, we may generate the convoluted picture as follows:

$$S(x, y) = \sum_{n=1}^{p} \sum_{n=1}^{q} I(p, q)k(x - m, y - n)$$

Pooling Layer
The disadvantage of the convolutional layer's convolutional feature output is that it records the accurate location of features in the input. This indicates that any cropping, rotation, or other tiny adjustments to the input image will result in a new feature map. To address this issue, we propose downsampling of convolutional layers. Applying a pooling layer after the nonlinearity layer allows for downsampling. Pooling aids in making the representation roughly robust to tiny translations of the input. Translation variability states that if we translate the input by a little bit, the values of the majority of the pooling outputs do not change. In our own CNN, we have customized the **maxpooling** layers.

VGG-16 Model
VGG16 consists of five layers and there is a total of 13 convolutional layers. The input layer takes images and sends them to the kernel layers. In the first layer, there are two 224 × 224 kernels with depth 64 and one 112 × 112 maxpooling layer. In the second layer, there are two 112 × 112 kernels with depth 128 and one 56 × 56 maxpooling layer. In the third layer, there are three 56 × 56 kernels with depth 256 and one 28 × 28 maxpooling layer. In the fourth layer, there are three 28 × 28 kernels with depth 512 and one 14 × 14 maxpooling layer. In the fifth layer, there are three 14 × 14 kernels with depth 512 and one 7 × 7 maxpooling layer. "Softmax" has been considered as the activation function. include_top = true means it connects the top three layers of the VGG16 architecture. Softmax classifier activation is used to load the pretrained weights.

The modified VGG16 can recognize flowers far better than the existing related works and our model. The accuracy is 96.64% of VGG16, CNN is 91%. We can conclude that VGG16 is more robust for flower recognition.

VGG-16 Architecture
See Fig. 2.

VGG-16 Architechture:

Fig. 2. Architecture of VGG-16

5 Implementation and Experimental Setup

5.1 Experimental Setup

Dataset of Rose, Daisy, Sunflower, and tulip has been collected from Kaggle. And another category lavender is collected from google. We have implemented the system on Kaggle with a machine of 8GB RAM. We have added a 4GB GPU with Kaggle to make the epochs faster.

5.2 Implementation

The dataset contains flower images with a single category. As all the images are different from each other in size, that's why we have converted all the images to the same size. Then the resized images have been sent to the main procedure. Then each of the categories of flowers were stored in a list and the corresponding name of that category in another list. Then split the dataset where 80% for training and the rest of the images of testing. While reading images with cv2.imread cv2.IMREAD_COLOR has been set which converts images from RGB to BGR color images. Figure 3(a) is the actual image and Fig. 3(b) is the BGR color image after converting. Then we have moved to the model creation, where the weights were taken from the ImageNet and the include_top is set to be true. We have used a checkpoint of the model so that it can be saved for later work. The number of epochs we have run in our system is 30. After 30 epochs the validation_accuracy becomes 0.9664. Finally, the user can give any flower image to test that in which category the given flower exists.

(a) (b)

Fig. 3. (a) Actual image (Sunflower), (b) BGR color image (Sunflower)

5.3 Performance Evaluation

As mentioned in the section on implementation, we got a validation accuracy of 0.9664. We have 745 images in total, and we have taken 20% for testing from the total images. So, the number of images for testing = (20/100) * 745 = 149 As we have taken 20% for testing from total images, we have 149 images for testing. So, we can calculate the accuracy for each category by using Eq. 1 (Table 2).

$$\text{Accuracy for each category} = \frac{\textit{Number of correctly detected flowers}}{\textit{No. of Images}} \times 100\% \quad (1)$$

Table 2. Performance analysis for proposed system (VGG16)

Category name	No. of images (for testing)	Number of correctly detected flowers	Accuracy for each category
Sunflower	40	38	95.00%
Tulip	19	17	89.8947%
Rose	35	32	91.43%
Daisy	34	32	94.12%
Lavender	21	21	100%

The main challenge was to detect images for similar types of two-category flowers. Similar types of categories conflict with each other because they have almost the same type of color and shape. Color and shape are one of the most important factors for differentiating flowers from each other. Sunflower and Daisy have almost similar types of shapes, that's why accuracy for sunflower and daisy is almost the same. It is hard to increase the accuracy of the training or testing images containing multiple flowers of the same category, though our model can recognize it. Usually, lavender exists in the form of groups. The number of images is also too small than the rest of the images except tulip. Like the shape, color, and other properties of lavender is very different from other flowers, that's why it's having the maximum accuracy, because of its uniqueness.

5.4 Comparison with Other Existing Frameworks

The purpose of the proposed method is to recognize flowers. The proposed method in [2] is SVM (Support Vector Machine). They used 102 categories and species to train SVM. Shape and color feature extraction was used with PCA (Principal Component Analysis). In [9], the author used 3 different neural networks. These are GoogLeNet, Alexnet, Caffenet. Among them, GoogLeNet gives the best accuracy. We also built our own CNN model, where we did augmentation before processing. After that we made a model with 5 layers, containing max-pooling and conv2D. This model gives around 91% validation accuracy. The accuracy based on methods is given in Table 3. From Table 3, we can see that our proposed method (VGG16) is far better than the existing methods.

Table 3. Comparison with other existing frameworks

Method	Val_Accuracy
SVM [2]	80%
PCA [3]	77.5%
GoogLeNet [9]	66.60%
Alexnet [9]	50.60%
Caffenet [9]	54.84%
CNN (own build model)	91%
Proposed method	96.64%

6 Conclusion

In this paper, we have proposed a method that gives quite better accuracy than previous related works. This method is significant to recognize flowers. Our system will be able to recognize only five types of different flowers. In the future, we can increase the category. It will be helpful for the robots to train. Robots will be now able to recognize flowers. It can be used in agriculture institutions and for other purposes also. The difficulty we have faced is selecting good images for training and differentiating similar types of two flowers. But in the end, we have achieved a good accuracy.

References

1. Le, T.L., Duong, N.D., Vu, H., Nguyen, T.N.: MICA at LifeCLEF 2015: multi-organ plant identification. In: CEUR Workshop Proceedings, vol. 1391 (2015)
2. Nilsback, M.: An automatic visual Flora – segmentation and classification of flower images. Thesis, p. 20 (2009)
3. Rodrigo, R., Samarawickrame, K., Mindya, S.: An intelligent flower analyzing system for medicinal plants. In: 21st International Conference on Central Europe on Computer Graphics, Visualization and Computer Vision in Co-operation with EUROGRAPHICS Association, WSCG 2013 - Poster Proceedings, pp. 41–44 (2013)

4. Hong, A.-X., Chen, G., Li, J., Chi, Z., Zhang, D.: A flower image retrieval method based on ROI feature. J. Zhejiang Univ. Science A **5**(7), 764–772 (2004). https://doi.org/10.1631/jzus.2004.0764
5. Goëau, H., Bonnet, P., Joly, A.: LifeCLEF plant identification task 2015. In: CEUR Workshop Proceedings, vol. 1391 (2015)
6. Bo, L., Ren, X., Fox, D.: Kernel descriptors for visual recognition. In: Advances in Neural Information Processing Systems, 24th Annual Conference on Neural Information Processing Systems, NIPS 2010, vol. 23, June 2010
7. Szegedy, C., et al.: Going deeper with convolutions. In: Proceedings of IEEE Conference on Computer Vision and Pattern Recognition, 7–12 June 2015, pp. 1–9, October 2015. https://doi.org/10.1109/CVPR.2015.7298594
8. Krizhevsky, A., Sutskever, I., Hinton, G.E.: ImageNet classification with deep convolutional neural networks. Adv. Neural Inform. Proc. Syst. **25** (NIPS 2012)
9. Thanh, T., et al.: Flower species identification using deep convolutional neural networks. In: AUN/SEED-Net Regional Conference for Computer and Information Engineering 2016 (RCCIE 2016), p. 6 (2016)
10. Ranganathan, G.: A study to find facts behind preprocessing on deep learning algorithms. J. Innov. Image Process. **3**(1), 66–74 (2021). https://doi.org/10.36548/jiip.2021.1.006
11. Sungheetha, A., Sharma R.R.: Classification of remote sensing image scenes using double feature extraction hybrid deep learning approach. J. Inf. Technol. Digit. World **3**(2), 133–149 (2021). https://doi.org/10.36548/jitdw.2021.2.006
12. Lodh, A., Parekh, R.: Flower recognition system based on color and GIST features. In: Proceedings of 2nd International Conference on Devices for Integrated Circuit (DevIC) 2017, pp. 790–794 (2017). https://doi.org/10.1109/DEVIC.2017.8074061
13. Patel, R., Panda, C.S.: A review on flower image recognition. Int. J. Comput. Sci. Eng. **7**(10), 206–216, (2019). https://doi.org/10.26438/ijcse/v7i10.206216
14. Lv, R., Li, Z., Zuo, J., Liu, J.: Flower classification and recognition based on significance test and transfer learning. In: 2021 IEEE International Conference on Consumer Electronics and Computer Engineering (ICCECE), pp. 649–652 (2021). https://doi.org/10.1109/ICCECE51280.2021.9342468
15. Vijayakumar, T.: Synthesis of palm print in feature fusion techniques for multimodal biometric recognition system online signature. J. Innov. Image Process. **3**(2), 131–143 (2021). https://doi.org/10.36548/jiip.2021.2.005
16. Liu, F., Wang, Y., Wang, F.C., Zhang, Y.Z., Lin, J.: Intelligent and secure content-based image retrieval for mobile users. IEEE Access **7**, 119209–119222 (2019). https://doi.org/10.1109/ACCESS.2019.2935222
17. Gogul, I., Kumar, V.S.: Flower species recognition system using convolution neural networks and transfer learning. In: 2017 4th International Conference on Signal Processing, Communication and Networking, ICSCN 2017, pp. 1–6 (2017). https://doi.org/10.1109/ICSCN.2017.8085675

A Smart Garbage System for Smart Cities Using Digital Image Processing

P. Sivaranjani(✉), P. Gowri, Bharathi Mani Rajah Murugan, Ezhilarasan Suresh, and Arun Janarthanan

Department of Electronics and Communication Engineering, Kongu Engineering College, Erode, India
{sivaranjani,gowri.ece}@kongu.ac.in

Abstract. Garbage management is one of the most obvious challenges that humanity will face in the near future. The primary need is to produce a dust-free atmosphere. It is difficult to have a clean waste system in large cities. An automated system is put up beside the trashcan to keep track of the individual's rubbish disposal habits. The majority of people, out of laziness, throw the dust out of the bin. The camera positioned over the trashcan records whether or not dust is deposited in the bin. When the dust is removed from the bin, the camera catches the defaulter's face. The defaulter's facial data is compared to an existing data collection and identified. The kit's warning system will issue a warning signal to the appropriate user. We describe a smart waste management system that requires the user to handle dust with caution. The user database is hosted in the cloud and may be viewed from anywhere in the city.

Keywords: Garbage management · Face recognition · Deep stack model · Automated warning system · Raspberry Pi camera

1 Introduction

Garbage management in developing nations like India is a huge undertaking that takes a lot of time, money, and effort. People are less conscious of rubbish management and its future impacts on the environment and climate change. Garbage has a negative impact on air quality because it emits poisonous gases. When non-biodegradable garbage is left outside the bin for an extended length of time, it pollutes the soil. By just throwing the dust in the bin, these consequences can be greatly decreased. According to a statistical survey, 6 out of 10 people in India litter dust outside the bin owing to laziness and a lack of understanding about the need of a clean rubbish management system. Littering not only harms the environment, but it also places a significant load on local municipal staff who must deal with the debris.

J. I.-Z. Chen et al. (Eds.): ICIPCN 2022, LNNS 514, pp. 761–770, 2022.
https://doi.org/10.1007/978-3-031-12413-6_60

Human involvement and local government supervisors have always been used to keep track of waste management systems. However, it is not a comprehensive solution. The necessity of the hour is for an automated system that tracks users and penalizes defaulters. Checking each and every user at an apartment's garbage terminal and insisting that they place the dust in the bin is quite tough. Thus, in smart cities, we suggest an automatic approach of monitoring and recognizing each and every user at the apartment level, where the user is fined immediately if discovered to be polluting the dust. Labor and time are saved by automating the waste collection system. The system includes a face recognition system that employs a deep stack modelling approach for increased accuracy and reduced calculation time. The literature review, full explanation of several steps in the proposed model, findings and conclusions, and scope for improvement are the sections that follow in this study.

2 Literature Review

The first attempts to deploy a face recognition system were made in the 1960s, when facial data was used to recognize corporate personnel. In the 1990s, trash management systems began to automate. The United Nations Organization emphasized the importance of an efficient waste management system.

In smart city initiatives, an Internet of Things (IoT)-based waste management system has been suggested. The ultrasonic sensors played a critical part in these projects, since the amount of waste in the dust bin is continually monitored by the ultrasonic sensors, and an alarm is given to the local garbage management authority when the bin reaches its maximum level. The bin contents are not categorized into degradable and non-degradable in this manner [1].

The author of [2] suggests a trash management system that includes a gas detecting panel. The technology identifies toxic gas created by biodegradable and non-biodegradable garbage in the dustbin. The ultrasonic sensor installed inside the dustbin monitors the level of waste in the bin and sends rubbish data to the cloud. The data for the smart city rubbish management sector is available via the cloud. When the bin fills up, they respond quickly. An alarm-based warning mechanism is built into the poisonous gas indicator. When harmful gas is released from the trash, the alarm sounds.

The reference [3] describes a facial recognition-based attendance tracking system. The facial data of the company's personnel is saved as a dataset in the local system. It recognizes the user's face using the Eigen Face Algorithm. Linear Discriminant Analysis and Faster – Neural Network Analysis are used in the procedure. In the training phase, face data is received as input and transformed from RGB photos to grayscale images. The facial data is taken from the image as layers and saved in the database. During the recognition phase, the input picture is processed to generate a grayscale image, which is then converted into layers using the DCFNN algorithm and compared to the given dataset to identify the individual.

The author of [4] proposes the Principal Component Analysis (PCA) and Linear Discriminant Analysis (LDA) methodologies. The PCA technique is commonly used to discover patterns in data and to turn data into eigen vectors. PCA enhances accuracy to some amount, and accuracy improves with illumination. This system uses the Haar – Cascades frontal face detection method.

The authors of [5] suggested a system based on real-time facial recognition that is reliable, secure, and rapid but might be improved in varied lighting circumstances.

The study on numerous approaches, components, and their operation was derived from the above-mentioned literature review. In paper [1] we discuss in detail how an automation system is required for the garbage management and collection system, as well as the need for better garbage management system optimization in relation to combining the Smart City project and the Swachh Bharat project, both of which were initiated by our Central Government. The investigation for the requirement of automating the waste system that may be done using microcontrollers and gas sensor panel setup that tend to display the harmful content in the bin is presented in paper [2].

The relevant data is communicated to the local authorities using the ZigBee technique. The study on face detection system utilizing Eigen face method and its implementation in real-time system is covered in article [3]. The authors present a full investigation on PCA and LDA algorithms and their roles in face recognition in article [4]. Research on real-time implementation of a face recognition system utilizing a deep stack method is presented in article [5]. It outperforms other facial recognition algorithms in terms of efficiency.

Fig. 1. Block diagram of the existing method

In Fig. 1, the block diagram depicts the training and recognition phases of the existing approach (Eigen Values Algorithm for face detection). The person's face datasets are pre-processed before entering the training phase and being stored in the database. During the recognition phase, the face data is gathered again and pre-processed in order to compare it to the pre-stored face datasets in the database. If there is a match in the face databases, the face is recognized by the system. Otherwise, it reports that it is an unknown face (Table 1).

Table 1. A tabular representation of all classifiers used in the literature review and their performances are discussed in detail.

Performance evaluation conditions	PCA + distance classifier	LDA + distance classifier	PCA + SVM	PCA + Bayes	LBPH + distance classifier
False positive rate	55%	53%	51%	52%	25%
Distance of object for correct recognition	7 feet	7 feet	7 feet	7 feet	4 feet
Training time	1081 ms	1234 ms	24570 ms	29798 ms	563 ms
Recognition rate (static images)	93%	91%	95%	94%	95%
Recognition rate (real time video	61%	58%	68%	65%	78%
Occluded faces	2.5%	2%	2.8%	2%	2.3%

3 Proposed Model

Figure 2 illustrates the suggested architecture. The deep stack model is used in conjunction with the Faster-RCNN algorithm in this technique to provide an effective and cost-effective waste management system, particularly for smart apartments in smart city projects. When a resident of a specific apartment has to toss dust in the bin, he goes to the floor's garbage terminal and throws the dust into or near the bin. The Raspberry Pi camera will keep an eye on the area where the dust is being flung. The Pi camera records the defaulter's face when the dust flung falls out of the Region Of Interest (ROI). The input picture is converted to grayscale using the deep stack model. There are two parts to the face recognition process: pre-processing or training and face recognition. Pre-processing the data, such as face alignment, feature extraction, feature matching, and lastly face recognition from the dataset, allows the model to recognise the face from the image.

PCA – Principal Component Analysis, LDA – Linear Discriminant Analysis, CNN – Convolutional Neural Network.

Face data is captured at the camera in the proposed model, then pre-processed into positive, negative, and part samples. The Haar-cascades facial recognition method is then applied to the picture. The saved face dataset is compared to the already stored face datasets supplied into the training module during the recognition phase. The face is initially identified, aligned, and face data is extracted as an.xml file using annotations in the recognition step. If the user is discovered to be the defaulter, his information is retrieved from the database, and he is sent a warning message via the GSM module.

Fig. 2. Block diagram of the proposed method

4 Result

The images below show the example face inputs and their corresponding face identified outputs. The warning is issued to the user based on these identified faces. When the input is presented to the camera as an image for saving the data in the dataset, the pictures are

Fig. 3. Training the face data in Roboflow platform

taken as an RGB file and then pre-processed to a grey scale image for the development of the input's feature images. The face data are saved in the database, which establishes a match between the input and the pre-available dataset during the recognition face.

The face of the users is trained in the training phase of the proposed method, using the Roboflow platform as shown in the Fig. 3. The user details are also collected as a database by the community authorities.

Fig. 4. Training the object detection in Roboflow platform

Using the Roboflow platform, the system is taught to detect dust outside the bin and bin overflow, as demonstrated in Fig. 4. The picture is separated into two sections, with the dustbin (object) detection occurring in the lower region of the image.

Fig. 5. Training output of face data in YOLO

The users' facial data is gathered as a 10-s movie, with each second segmented into 20 frames per second. As a result, each user now has a training set of 200 photos. This is followed by augmentation, in which the face data for the training set is vastly increased to over 2000 photographs for each user, as seen in Fig. 5.

Fig. 6. Testing the dataset using Convolution Neural Network

The dataset produced for model testing is fed into the R-CNN architecture, which performs layer convolution based on a region-based technique. The input picture is separated into multiple parts for simpler identification of the face and faster production of feature face, as shown in Fig. 6.

Fig. 7. Littering detected and warning message sent to user

When the system detects and recognizes a specific user littering using the facial recognition algorithm, it retrieves the user's contact information from the apartment's community database and sends a warning message to the user, as illustrated in Fig. 7. If the defaulter continues to litter on a regular basis, the system automatically penalizes the user by sending a warning message to the user as well as the authorities of the society or community of the apartment in question.

Fig. 8. RGB image converted into Gray Scale

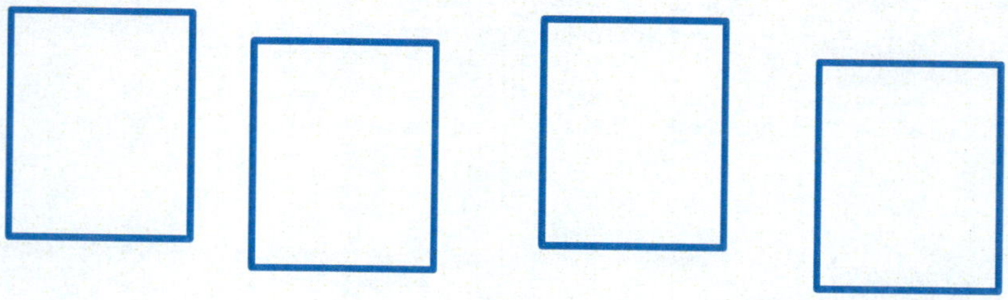

Fig. 9. Face Recognition from the pre-stored data

Figure 8 shows that the input RGB picture is used as the system's input. The OpenCV module then converts the supplied RGB picture to grayscale. The grayscale image is now being pre-processed in preparation for the training and recognition phases.

During the training phase, the deep stack model classifier looks for faces in the picture. Based on the face structure, an.xml annotation file is generated and saved in the database as face data. Different perspectives, saturation, emotions, and texture of the face are collected and taught in a similar manner.

If the user throws the dust outside the bin (litters the dust), i.e., dust thrown inside the area of interest (ROI), the system attempts to detect the user's face as illustrated in Fig. 9. In the recognition phase, the feature face data is extracted and the annotations are compared with the pre-stored data base; if both the annotations are equal, the contact details are retrieved from the database and a warning message is sent via the GSM module.

Table 2. Efficiency parameters of augmentation matrix

Augmentation matrix parameters	Training values
Number of Input images	150 images per person
Number of users	5
Number of images in Training dataset	128580 images
Number of images in Validation dataset	359808 images
Number of layers used in convolution network	21 layers
Number of augmentation methods used	8
Total params	4,714,688
Trainable params	4,714,688
Non-trainable params	0

Fig. 10. Training metrics for the face recognition in Roboflow platform

The metrics of the augmentation matrix in Yolo V4 are represented in Table 2. After the training and testing phases are done, a set of new inputs from the same user is presented at the validation step to assess the accuracy and precision of the user identification. As shown in Fig. 10, the accuracy of a model grows as the number of pictures in the training dataset increases.

5 Conclusion

This waste management system based on facial recognition delivers precise information about defaulters in a simple manner and warns them in real time. This technology is user-friendly and provides increased security and a cleaner environment. The technique mentioned in this article is extremely successful and may be implemented with minimum resources and cheap effort. The system can identify and recognize the user in 0.02 s, which is far quicker than any other facial recognition algorithm currently available. Under ideal lighting circumstances, it can achieve a 96% accuracy.

References

1. Singh, G., Goel, A.K.: Face detection and recognition system using digital image processing. In: 2nd International Conference on Innovative Mechanisms for Industry Applications (ICIMIA), pp. 348–352 (2020). https://doi.org/10.1109/ICIMIA48430.2020.9074838
2. Bah, S.M., Ming, F.: An improved face recognition algorithm and its application in attendance management system. Array **5**, 100014 (2020). https://doi.org/10.1016/j.array.2019.100014. (ISSN: 2590-0056)
3. Vinod, V.M., Tamilselvan, K.S., et.al.: An IoT enabled smart garbage management system for smart cities – Indian scenario. Int. J. Innov. Technol. Explor. Eng. (IJITEE) **9**(4) (2020). ISSN: 2278–3075
4. Dipak, S., Aithal, S.: Smart city waste management through ICT and IoT driven solution. Int. J. Appl. Eng. Manag. Lett. (IJAEML) **5**(1), 51–65 (2021). (ISSN: 2581-7000)
5. Gaikwad, A.T.: LBP and PCA based on face recognition system, pp. 368–373, November 2018. ISSN 2348-8034
6. Bhangale, K.B., Jadhav, K.M., Shirke, Y.R.: Robust pose invariant face recognition using DCP and LB. Int. J. Manag. Technol. Eng. **8**(9) 1026–1034 (2018). ISSN NO: 2249–7455
7. Kalaiselvi, P., Nithya, S.: Face recognition system under varying lighting conditions. IOSR J. Comput. Eng. **14**(3), 79–88 (2013). (e-ISSN: 2278-0661, p-ISSN: 2278-8727)
8. Cai, Y., Xu, G., Li, A., Wang, X.: A novel improved local binary pattern and its application to the fault diagnosis of diesel engine. Shock Vib. **2020**, 15 (2020). https://doi.org/10.1155/2020/9830162
9. Ding, C., Choi, J., Tao, D., Davis, L.S.: Multi-directional multi-level dual-cross patterns for robust face recognition. IEEE Trans. Pattern Anal. Mach. Intell. **38**(3), 518–531 (2016). PDF: Robust Pose Invariant Face Recognition using DCP and LBP
10. Pedregosa, F., et al.: Scikit-learn: machine learning in Python. JMLR **12**, 2825–2830 (2011)

A Productive On-device Face Authentication Architecture for Embedded Systems

G. Renjith(✉) and S. Aji

Department of Computer Science,University of Kerala,
Thiruvanathapuram, Kerala, India
{renjithg,aji}@keralauniversity.ac.in

Abstract. Face authentication is one of the effective cyber security confirmation techniques used nowadays, and it is designed and implemented using embedded systems. Face authentication systems need cloud or server-based platforms to manage the face database and to perform image processing and deep learning. The world is moving towards a situation where data privacy has a severe concern compared to any other era. Sharing private data to remote servers for face authentication is highly sensitive. The existing works focus on machine learning and deep learning aspects rather than the practical or implementation aspects of embedded systems. In this paper, we focus the major benchmark inventions on the embedded implementation perspective. We have practically deployed and validated the existing architecture on various embedded platforms such as smartphones, development kits and tablets etc. The architecture was evaluated on different versions of the Android operating system as well. We have analysed and compared different practical aspects such as model size, latency, power consumption, efficiency, and memory usage, along with the machine learning aspects such as prediction accuracy, precision etc.

Keywords: On-device · Face authentication · Embedded systems · Contactless authentication · Masked face recognition

1 Introduction

Face authentication is one of the emerging technologies which can potentially replace the existing cyber security authentication methods. With respect to the practical requirements, capable hardware is required for implementing efficient image processing techniques such as biometric iris, fingerprints and face authentication etc. Considering the scenarios of Covid-19 situations, where the world is switching towards a contactless approach in all aspects of the territory. When it comes to face recognition systems, the various methods have evolved within a short range of period and with improved performance. The face authentication techniques needs an efficient hardware processing component to handle consistent performance. The unique features of the human face are the iris of the eye, the shape of the eyes, nose, lips, the distance between the face key points, and

J. I.-Z. Chen et al. (Eds.): ICIPCN 2022, LNNS 514, pp. 771–786, 2022.
https://doi.org/10.1007/978-3-031-12413-6_61

the curvature of the face. These unique features of the human faces are called face features, which help to distinguish one person from another. Existing the literature primarily discusses the system performances, mainly from the machine learning perspective, such as accuracy, recall, and precision and gives less focus to the embedded aspects, such as GPU, CPU, power, and latency usage etc. Face authentication mostly consists of face detection, the liveness of the face, face feature extraction, face classification, and prediction of the identity. Face recognition systems are achieving major transitions for both architecture concepts in machine learning and the implementation aspects for the embedded devices. Most of the recent works consider the architectural changes from the machine learning perspective, but a comparison of the face authentication system on an embedded devices is not given importance. This paper analyses both the architectural and the embedded perspective of the face authentication system, such as model size, latency, consumption, efficiency, memory usage etc. The practical features of existing systems are analysed by implementing the respective method on embedded devices such as smartphones and tablets with different versions of the Android operating system. Here we discuss the overall system design and the implementation, which explains the procedural, theoretical, and experimental aspects.

This paper is organized as follows. Section 2 describes various studies published on face authentication, face antispoofing, and embedded deployment. Section 3 discusses the overview and the design aspects. Section 4 discusses the verification and the performance analysis of the system in a real-world environment, the results of the experimental parts are correlated and summarised in the result section. Section 5 describes the summary of the works with conclusions and future scope.

2 Literature Review

Most of the previous work portrays a clear idea of the machine learning aspects, such as the architectural changes of the model, the theoretical enhancements, and the efficiency of face recognition systems. In [1], they considered the 3D and 2D face recognition that generally describes face recollection without considering the authentication.

The convolution neural networks (CNN) are used to take out the face features in FaceID [2]. The Principal Component Analysis (PCA) [3], Linear Discriminant Analysis (LDA) [4], and Deep Learning [5–7] are also used for face feature extraction. PCA and LDA are feature reduction techniques and produce linear combination face features. LDA method was able to identify a group with similar features, and this method is simple and computationally efficient. In the case of convolutional networks, the training time grasping is high. To avoid retraining the whole dataset on the addition of each person, we freeze all layers except the top layers used for the classification. Cloud-based face recognitions like Azure Face API of Microsoft have an accuracy of 90–95%. In Fig. 1, the blocks are considered the leading design for the face authentication systems. Face recognition is a technology capable of identifying a face through an image by detecting

Fig. 1. Block diagram of face authentication system

and extracting the facial feature, where security plays a major role due to a kind of risks like attacks and spoofing that may turn into fraud and other types of crimes. Therefore, in order to design a stable face recognition system in a real framework, antispoofing techniques should be a top priority from the initial design of the system, and some of the antispoofing techniques have been mentioned in Fig. 1. Deploying a model is also very important for the real-time implementation.

Some of the works in [10,11] deal with blurred and low-resolution images, meanwhile [8,9] focuses on high-quality images. Different types of systems are developed to deceive the face recognition techniques such as mask attack, print attack, and replay attack. Most of the works related to face antispoofing are working with texture features - LBP [12,13], HOG [14], LBP-TOP [15], DoG [16]. In [17], face antispoofing is classified into four groups that are texture-based, motion-based, 3D-shape-based, and multispectral reflectance-based. In [18], the analysis is done on the Fourier spectra, which depict the photographs that contain a higher frequency spectrum compared to the original. In [19], the above method is changed by applying the combination of the Linear regression, which improves the result and makes it suitable for the antispoofing of the face in high light illumination conditions. The paper [20] deals with the hardware architecture consideration for the always-on systems, which are systems that work for 24×7. For event-driven face verification, researchers achieved insufficient accuracy due to the use of hand-crafted features [21]. Some of the works adapted from CNN and proposed dedicated hardware with dynamic voltage, accuracy, and frequency scaling for energy efficiency [22]. By considering covid-19 in [23], they investigated face mask identification systems and monitored people to avoid the spreading of the coronavirus. To detect the face mask, they proposed an efficient two-stage identification facemask for identification and proved as a very efficient and good accurate algorithm. But it fails to identify when the mask is not positioned properly. To improve the manipulating of light utilization and extraordinary operations they [24], have proposed the infrared flat metalens inbuilts in attaining the AR imaging with the enhanced attributes sidesteps that prevailing obstacles that are caused by the existing systems and reports a narrow vision and largeness. The metalens gives a wider view in the absence of changing the colour of the objects even on the heavy closer to, eye spectacle.

3 Methodology

Face authentication differs from face recognition, and the features added to the system will increase its reliability. The liveness detection feature against the print attacks and the replay attack is an example. The section is divided into face recognition and the authentication.

3.1 Design of Face Recognition

The main aspect incorporated into the proposed system is to provide an on-device training for faces which is not seen much in the literature. The face authentication system embedded on a device has a capacity for training and the prediction that will go through different stages, as mentioned below.

Collecting Images. In this stage, the continuous capturing of images from the real-time camera at a fixed time interval will happen. The captured image will be converted into a colour model, say RGB, to use in the next phase.

Face Detection. The faces in the images are detected in this phase, and we used the BlazeFace model [25] for face detection.

Face Security. The detected face is conformed to be an alive face through various techniques such as eyeblink detection, eye aspect ratio checking, face movements and flash reflection techniques on the face. The detected face is given to the eye blink detection model with a CNN, the eye blink is detected from the bitmaps of the face reaching the model. After the confirmation of the liveness of a face then, it is given for a face feature extraction. Liveness detection [17] also checks the print and the replay attacks.

Face Feature Extraction. The recognition of a face can be completed only when the features of a person's face are identified uniquely. Face features are extracted in Euclidean space where every person's face has a unique space. Here we have tested with different models as a face feature extractor as arcface, Facenet, VargFaceNet, probabilistic face embedding, sphere face, and CosFace to compare their efficiency in terms of the reliability of prediction and the efficiency of using embedded hardware.

Face Classification. The identification of a person will happen in this stage. The database created by the face extractor, where every person is considered as a separate class, is used to train the classification model.

3.2 Deep Face Recognition

Deep learning models extract the features from different layers, and the models move in-depth as the layer increases. FaceID is mainly used as a recognition of one from others. The revolutionization in deep face recognition starts to be trending as the different architectures using the convolution layers, such as AlexNet, InceptionNet and MobileNet gain popularity in image classification [26]. In [27], the deep face is one of such benchmarks, which achieved 97.35% in the LFW dataset. Here are some of the architectures used in face recognition with the different loss functions. In [2], the loss concept in face features maps between the best Euclidean distance that make the different faces apart and the similar faces close. Angular margin is another parameter used along with Euclidean distance. Contrastive and triplet losses are the most used losses, which have drawbacks of extending the embedding space region. The region selected and discriminative features are highly affected by the sample mining. Sphere face uses a full mini-batch and is able to bypass the step of finding the triplet/pairs [28]. SoftMax loss gives a good interpretation of the embedding space by considering the geometrical analysis and the angular distance in the embedding space.

The cosface [29] is the loss function which was proposed in large margin cosine loss (LMCL), which changes the SoftMax function into an approximate cosine function. The radial variation that existed in the SoftMax loss is avoided by adjusting the feature and weight vectors by L2 normalization. SoftMax loss was introduced to strengthen the separation of the classes by the introduction of the cosine margin. Cosine margin and the L2 normalization help to provide a minimum intra-class variance and maximum inter-class variance. Arcface [30] is considered the benchmark of face recognition architecture to train an efficient architecture of neural networks. The inter and intra class margins are introduced, and the center loss creates the separation between the classes by considering the created Euclidean distance between the class centres, which doesn't believe the angular margin. The sphere face representation of the class centres in the annular space is the last dense neural network that performs the linear transformation of the features obtained from the pool layers of the CNN, which lead to the penalize weights, features obtained from the previous layer in a multiplicative way.

Capturing high-resolution images is difficult in the embedded environment for face recognition. In Fig. 2, have specified both the training and production processes. During the training process, the different image will be given as input to the system, once the pre-processing is done, the facial features are extracted from photos or video to identify and validate the person's face, is done based on training process model will be generated and deployed in real-time systems. During the real-time production process, the camera captures the live face of a human and extract the feature from the face and detect it. Once the face is predicted, it sends to a security operation to perform the action.

Fig. 2. Training and production pipeline flow

3.3 Dataset

VGGface2 dataset [34] is the most commonly used dataset for training the face recognition systems. The dataset is created by extracting the images from google image search, and that makes it a diversified dataset without a bias in age, illumination, pose, ethnicity and profession. The masked face authentication became a major challenge with the Covid-19 pandemic, so the dataset was modified to compatible with the recognition of both faces with masks and without a mask. Based on that, we have developed a masked face dataset for the training model to identify masked faces. VGGFace2 dataset was modified into a masked face dataset by incorporating the digital mask to cover the face with the help of face key point identification in face detection. All the dataset is converted into a mask, which makes the model difficult to identify unmasked faces. Dataset is created as having a fixed proportion of masks and unmasked faces of each person. We have chosen a ratio of 1:3 for masked and unmasked faces, respectively.

The popular face datasets are LFW [31], mega face [32], IIJB-C [33], vggface2 [34] or ms-movie star-1m [35] are imbalanced each in gender and skin colouration [36]. The [37] offers a positive dataset which is manipulated to have an independent dataset. IIJB-C [38] consist of 12,549 public domain photos, and there are 152,917 pics from 6,139 identities. Annotations were performed by evaluating attributes like gender, pores, and skin colour and valid negative attributes age group, head pose, photograph supply, sporting glasses and abounding field size.

The face recognition model developed is all affected by the bias of the dataset towards the white face. These reasons are mostly considered due to the dataset used for the training of a white face, and the development of the photographic instruments focusing on the white faces. By considering this bias of face recognition as a major drawback of the system. The scarcity in getting the black face dataset leads to the usage of GANS in the generation of faces. Bias to the dataset will not create a wrong prediction, but the probability of the prediction decreases. Another effect of the bias in facial expression is the wrong prediction of the gender, the model developed from the biased dataset gives wrong predictions in the identification of the women.

3.4 Face Security Techniques

Face antispoofing consists of the security-related features that are added to increase the reliability of the prediction in the face recognition. Motion-based antispoofing considers the motions in the face elements such as eyelids, lips and mouth. Blinking of the eyes is considered the most popular one among these types of methods as, in [26], which describes the blinking of eyes at various stages by making use of a specific set of conditions. Lip movements and their kinematics were used by [39]. Apart from merely checking the movements of the lip, it focuses on the lip reading by giving a specific set of words to read and tracking the activities. Optical flow analysis was used to track the plane of the face to know whether it is a 3D or 2D planar, which allows for segregating out the 2D planar photographs as in [40], which is based on the difference in the motion of the central regions of the face with that of the out regions. In [41], the distinguishing features in the foreground and background images were considered [42,43] using a combination of the above methods 3D shape-based antispoofing, the consideration of the 3D perspective in the real face as the method suggested in [44], but this method has its drawback as it can't defend against the wrapped images as they are not limited by the coplanar restriction as of other 2D photographs. To improve the above drawbacks [45], suggest a method for the recreation of the 3D shapes by analysing in-depth, but the method fails with the dataset, that is the, 3D Mask Attack database (3DMAD) [46]. Multispectral reflectance-based anti-spoofing differentiates the real and fake from the change in the illumination spectrum of both on reflection and absorption from the spectrum in a high light source which is depicted in the paper [47]. A Method that focuses on the reflection intensities, a gradient-based multispectral method, was described in [48]. The method is not widely used due to the extra devices for the creation of the reflectance variance and the capturing images on the invisible spectrum. Methods such as Remote Photoplethysmography (PPG) signals [49] were also considered in recent times.

3.5 Deployment Techniques

The camera-based face authentication has its merit over the conventional fingerprint authentication, which requires a contact surface for initiating the authentication process. In the era of wearables, face authentication was used as the security method. Authentication systems are always-on systems that try to detect an event, and after the detection, the event-driven process will be processed.

The authentication system, probably working in 24×7 mode, should be capable of identifying a person when the camera captures the face. The power consumption cycles of the system should be managed efficiently. Feature extracted by the FaceNet represents a partial feature set and is given to the model that has already trained with the data of the known person. When the model predicts an unknown person, the proceeding steps are skipped. Hence the power consumed by the model for these layers is efficiently conserved. The authors in [50], has proposed a low-power CNN-based face recognition system for user authentication in smart devices. The system comprises an always-on functional CMOS image sensor (CIS) for imaging and face detection and a low-power CNN processor (CNNP) for face verification. They were implemented in 65-nm CMOS technology, where the system consumes 0.62 mW to evaluate one face at one fps and achieves 97% accuracy.

The machine learning models have a large number of parameters and require complex computational resources, which should be limited in embedded systems. The accuracy and improved performance are attained by the deep learning models with these high amounts of computation and the memory space taken by the model during the prediction. So, the whole process and memory usage can't be reduced, but the changes can be brought in the execution and the memory mapping. More generally used techniques are pruning, quantization, model distillation, network design strategies, and low-rank factorization [25]. The process of pruning is to remove the unimportant and inefficient parameters by evaluating the performance of the model on each parameter. Quantization is a method, and the memory mapping is reduced by decreasing the resolution of the single-step. Here the number of bits is required to represent the parameter. The quantization is also done on conversion to the TensorFlow Lite model. Model distillation is the technique where a large network is represented as a small network by mapping the computationally expensive process into a compatible process with the less computational infrastructure. The low-rank factorization technique has been used to accelerate and compress neural networks.

4 Experimental Evaluation

The existing studies which are considered for the benchmarking techniques in face recognition were ported to the embedded environment. Models specified were converted into tflite to get compatible with the Android operating system. Major models are taken from [7,30,51]. The pre-processing and image classification algorithms in [19,26,50,52] were also incorporated. The usage of the memory, fps and latency were analysed for comparison.

4.1 Setup

The models are deployed on smartphones and tablets. The operating system used is Android. Python and Java are mainly used languages for the training and deployment of models. Python modules that are used for training and preprocessing are TensorFlow, Keras, Tensorflowlite, ONXX, PyTorch, Caffe, and Scikit-learn.

4.2 Implementation

The efficiency of the model in real-time is tested by training a person and predicting the person on different conditions such as varying sunlight and different artificial lights. The person with similarities in the facial features such as beard, moustache, eyebrows, face curves, glasses etc., the similar person with a beard, mask and without beard and mask were tested. Seven people with 30 photos were trained for each person, and the prediction was tested on the 1104 images. The prediction labels are stored each time when the prediction happens with a bitmap image in the Android application. That application was set up to have the bitmap image from the frame of a camera at an interval of 3 s. Testing of the face extractor dataset is divided into testing and validation, with 30% of the dataset used as a validation set. This validation set is created by 10-fold cross-validation so that the validation set has distributed data of every class. To test an SVM, the classifier used a similar 30% dataset.

The reliability of the system is evaluated on the similar procedure by deploying the package on mobile devices in Android versions like Oreo, Pie, Android 10 etc. Graphics processing unit (GPU), Digital signal processor (DSP), Neural processing unit are also configured. All the packages are deployed in smartphones and tablets thus evaluating the model dependency with GPU, CPU. For analysis, the impact on camera resolution, frame rates of camera and frame per second(fps) are changed in a range of values and evaluated the prediction delays and accuracy.

We have implemented the face recognition models to work on the embedded devices and are tested real-time in embedded boards with different android versions, which gives a proper study of the system in the embedded environment. Researchers are focusing on increasing reliability with reduced latency and power consumption. The testing dataset is a custom build by including the considerations of the dataset bias towards race, and gender [37].

4.3 Results

The different architecture performances are measured with fps and memory usage. Loss functions performances are measured by testing with different datasets and observing the accuracy measure. The observations are noted in Tables 1, 2, and 3. In Table 1 comparison is made between ArcFace, SphereFace and CosFace, all 3 have different kinds of margin penalty. The ArcFace has a high accuracy compared to SphereFace and CosFace because ArcFace has a constant linear angular margin throughout the whole interval and also exact correspondence to the geodesic distance. Where SphereFace and CosFace have a nonlinear angular margin.

Table 1. Comparison of accuracy of different Loss functions on different dataset

Loss function	LFW	CFP-FP	AgeDB-30
ArcFace	99.46	95.47	94.93
SphereFace	99.11	94.38	91.70
CosFace	99.51	95.44	94.56

4.4 Discussion

Different models are shown during the training and prediction in Table 2, and Table 3. Every model has its own focused area of advantage for the embedded environment. Memory usage, fps and latency are considered the essential measurements for the embedded environment. Arcface performs best in memory usage, which consumes very little memory compared to others. Vargfacenet performs well with high fps use cases.

4.5 Comparison with Different Architectures and Parameters

The different attributes used in the research work are compared in Table 4. Attributes considered are different datasets used, power usage, memory usage, whether considering masked face recognition, deployment in an embedded environment, mode of training, and low-quality image aspects.

The Tables 2 and 3 provides an overview of the training and prediction scenario on embedded system by performing in different models like ArcFace, VarGFacenet, and FaceNet. ArcFace can be applicable in large scale as it has a capability of identifying millions of faces by confronting the GPU memory and computational cost. VarGFaceNet is also able to reach better performance in face recognition and validation. It has more parameter networks, and embedding settings have essential features for extracting more information. FaceNet recognizes the face network, which is trained as per the distance in the embedding space, and tries to separate the positive pair from the negative by distance margin. Figure 3 provide the computational cost with the accuracy tradeoff, and Fig. 4 provides the decrease in training speed with an increase in the training samples.

Fig. 3. The usage of the multi add flop and the accuracy tradeoff

Table 2. Performance of different models during training on embedded devices

Models	Frame per second (FPS)	GPU memoryread per bytes (M)	GPU memory write per bytes (M)	Render time
ArcFace	28.8	191.3 M	333.14 M	44.8 min
Vargfacenet	31.4	216.81 M	188.58 M	37 min
Facenet	32	671.33 M	677.62	121.36 min

Fig. 4. The training speed with the number of samples

5 Future Scope and Conclusion

The usage of the dataset in [37] to avoid the bias in the model, and help to create a more reliable system. Improvements in the face recognition with the masked faces should be given with importance because of the evolving scenarios of the COVID-19. As per recent studies, the various places in the world are discussing towards the ban of the face recognition system as an authentication method because of the reliability issues. The unreliability of the system on the bias scenarios arising with the skin colour, facial expression, gender, age etc. is very challenging issues. Though face recognition is getting more attention, it should attain high reliability while keeping lower computational complexity and cost.

Table 3. Performance of different models during prediction on embedded device

Models	Frame per second (FPS)	GPU memory read per bytes (M)	GPU memory write per bytes (M)	Render time
ArcFace	29.4	192.24 M	174.70 M	29.8 min
Vargfacenet	31.2	211.1 M	364.63 M	52.96 min
Facenet	29.9	597.20	361.48	77.57 min

Table 4. Performance of different architectures

Dataset	Eigen faces [51] PCA Algorithm	PCA + SVM [3]	FaceNet VGGFace2 [7]	Face authentication on mobile [52]	Face authentication IR liveness [55]	Face security system [54]	Vargfacenet [50]	Support guided [53]
Yale database	88.26	93	100	–	–	90	–	–
JAFFE	71.2	–	100	88	–	–	–	–
AT&T	89	98.75	100	–	99.99	–	100	–
Georgia Tech	–	76	100	76	–	–	–	–
Essex faces94	70	–	99.37	69	–	–	94	
Essex faces95	70	70	100	–	100	89	–	
Essex faces96	70	–	77.67	–	–	–	79	
Essex Grimace	70	–	100	–	–	–	–	
LFW	–	–	–	98	99	–	99	99
Peak memory usage on feature extraction	–	758MB	1078.14MB	887MB	1209MB	–	–	–
Masked Face recognition	No	No	No	No	No	No	No	No
GPU memory	–	–	1017MB	–	1056MB	–	–	–
Deployment on edge device	No	No	No	No	No	No	Yes	Yes
Feature extraction model size	–	–	91.3MB	–	91.3MB	–	20MB	–
Liveness detection	No	No	–	No	Yes	No	No	–
Retraining needed with whole dataset	Yes	Yes	Yes	Yes	Yes	Yes	–	–
Low light aspect	No	No	No	No	No	No	No	
Low power aspect	No	No	No	No	No	No	No	

References

1. Abate, M.N., Riccio, D., Sabatino, G.: 2D and 3D face recognition: a survey, pattern recognition letters. In: International Conference on Pattern Recognition (ICPR) (2007)
2. Liu, W., Wen, Y., Yu, Z., Yang, M.: Large-margin Softmax loss for convolutional neural networks. In: ICML (2016)
3. Chen, X., Song, L., Qiu, C.: Face recognition by feature extraction and classification. In: 2018 12th IEEE International Conference on Anti-counterfeiting, Security, and Identification (ASID), pp. 43–46 (2018)
4. Rane, M.E., Pande, A.J.: Multi-modal biometric recognition of face and palm-print using matching score level fusion. In: 2018 Fourth International Conference on Computing Communication Control and Automation (ICCUBEA), pp. 1–6. IEEE (2018)
5. Korkmaz, M., Yilmaz, N.: Face recognition by using back propagation artificial neural network and windowing method. J. Image Graph. 15–19 (2016)
6. Parkhi, O.M., Vedaldi, A., Zisserman, A.: Deep face recognition (2015)
7. Schroff, F., Kalenichenko, D., Philbin, J.: Facenet: a unified embedding for face recognition and clustering. In: Proceedings of the IEEE Conference on Computer Vision and Pattern Recognition, pp. 815–823 (2015)
8. Bowyer, K.W., Chang, K., Flynn, P.: A survey of approaches and challenges in 3D and multi-modal 3D+ 2D face recognition. In: Computer Vision and Image Understanding, pp. 1–15 (2006)

9. Ding, C., Tao, D.: A comprehensive survey on pose-invariant face recognition. In: ACM Transactions on Intelligent Systems and Technology (TIST), pp. 1–42 (2016)
10. Li, P., Prieto, L., Mery, D., Flynn, P.J.: On low-resolution face recognition in the wild: comparisons and new techniques. In: IEEE Transactions on Information Forensics and Security, pp. 2000–2012 (2019)
11. Li, P., Prieto, M.L., Flynn, P.J., Mery, D.: Learning face similarity for re-identification from real surveillance video: a deep metric solution. In: 2017 IEEE International Joint Conference on Biometrics (IJCB), pp. 243–252. IEEE (2017)
12. Abdenour Hadid, M., Pietikinen, M.: Face spoofing detection from single images using micro-texture analysis. In: 2011 International Joint Conference on Biometrics (IJCB), pp. 1–7. IEEE (2011)
13. Yang, J., Lei, Z., Liao, S., Li, S.Z.: Face liveness detection with component dependent descriptor. In: 2013 International Conference on Biometrics (ICB), pp. 1–6. IEEE (2013)
14. de Freitas Pereira, T., Anjos, A., De Martino, J.M., Marcel, S.: LBP-TOP based countermeasure against face spoofing attacks. In: Park, J.-I., Kim, J. (eds.) ACCV 2012. LNCS, vol. 7728, pp. 121–132. Springer, Heidelberg (2013). https://doi.org/10.1007/978-3-642-37410-4_11
15. Tan, X., Li, Y., Liu, J., Jiang, L.: Face liveness detection from a single image with sparse low rank bilinear discriminative model. In: Daniilidis, K., Maragos, P., Paragios, N. (eds.) ECCV 2010. LNCS, vol. 6316, pp. 504–517. Springer, Heidelberg (2010). https://doi.org/10.1007/978-3-642-15567-3_37
16. Peixoto, B., Michelassi, C., Rocha, A.: Face liveness detection under bad illumination conditions. In: 2011 18th IEEE International Conference on Image Processing, pp. 3557–3560. IEEE (2011)
17. Lin, H.-Y.S., Su, Y.-W.: Convolutional neural networks for face anti-spoofing and liveness detection. In: 2019 6th International Conference on Systems and Informatics (ICSAI), pp. 1233–1237. IEEE (2019)
18. Li, J., Wang, Y., Tan, T., Jain, A.K.: Live face detection based on the analysis of Fourier spectra. In: Biometric Technology for Human Identification, vol. 5404, pp. 296–303 (2004)
19. Peixoto, B., Michelassi, C., Rocha, A.: Face liveness detection under bad illumination conditions. In: 2011 18th IEEE International Conference on Image Processing, pp. 3557–3560. IEEE (2011)
20. Bong, K., Choi, S., Kim, C., Yoo, H.-J.: Low-power convolutional neural network processor for a face-recognition system. pp. 30–38 (2017)
21. Choi, J., Shin, J., Kang, D., Park, D.-S.: Always-on CMOS image sensor for mobile and wearable devices. pp. 130–140 (2015)
22. Li, H., He, P., Wang, S., Rocha, A., Jiang, X., Kot, A.C.: Learning generalized deep feature representation for face anti-spoofing. In: IEEE Transactions on Information Forensics and Security, pp. 2639–2652 (2018)
23. Dhaya, R.: Efficient two stage identification for face mask detection using multiclass deep learning approach. J. Ubiquit. Comput. Commun. Technol. (2021)
24. Raj, J.S.: Vision intensification using augmented reality with metasurface application. J. Inf. Technol. Digit. World (2019)
25. Bazarevsky, V., Kartynnik, Y., Vakunov, A., Raveendran, K., Grundmann, M.: BlazeFace: sub-millisecond neural face detection on mobile GPUs. In: Google Research 1600 Amphitheatre Pkwy, Mountain View
26. Chen, Y., Zheng, B., Zhang, Z., Wang, Q., Shen, C., Zhang, Q.: Deep learning on mobile and embedded devices: state-of-the-art, challenges, and future directions. In: ACM Computing Surveys (CSUR) (2020)

27. Taigman, Y., Yang, M., Ranzato, M., Wolf, L.: Deepface: closing the gap to human-level performance in face verification. In: Proceedings of the IEEE Conference on Computer Vision and Pattern Recognition, pp. 1701–1708 (2014)
28. Liu, W., Yu, Z., Li, M., Raj, B.: SphereFcae: deep hypersphere embedding for face recognition. In: IEEE Conference on Computer Vision and Pattern Recognition (2017)
29. Wang, H., et al.: Cosface: large margin cosine loss for deep face recognition. In: Proceedings of the IEEE Conference on Computer Vision and Pattern Recognition, pp. 5265–5274 (2018)
30. Deng, J., Guo, J., Xue, N., Zafeiriou, S..: Arcface: additive angular margin loss for deep face recognition. In: Proceedings of the IEEE/CVF Conference on Computer Vision and Pattern Recognition, pp. 4690–4699 (2019)
31. Learned-Miller, E., Huang, G.B., Chowdhury, A.R., Li, H., Hua, G.: Labeled faces in the wild: a survey. In: Advances in Face Detection and Facial Image Analysis, pp. 189–248 (2016)
32. Kemelmacher-Shlizerman, I., Seitz, S.M., Miller, D., Brossard, E.: The megaface benchmark: 1 million faces for recognition at scale. In: Proceedings of the IEEE Conference on Computer Vision and Pattern Recognition, pp. 4873–4882 (2016)
33. Maze, B., et al.: Iarpa Janus benchmark-c: face dataset and protocol. In: 2018 International Conference on Biometrics (ICB), pp. 158–165 (2018)
34. Cao, Q., Shen, L., Xie, W., Omkar, M.P., Zisserman, A.: Vggface2: a dataset for recognising faces across pose and age. In: 2018 13th IEEE international conference on automatic face and gesture recognition (FG 2018), pp. 67–74 (2018)
35. Guo, Y., Zhang, L., Hu, Y., He, X., Gao, J.: MS-Celeb-1M: a dataset and benchmark for large-scale face recognition. In: Leibe, B., Matas, J., Sebe, N., Welling, M. (eds.) ECCV 2016. LNCS, vol. 9907, pp. 87–102. Springer, Cham (2016). https://doi.org/10.1007/978-3-319-46487-9_6
36. Yu, J., Hao, X., Xie, H., Yu, Y.: Fair face recognition using data balancing, enhancement and fusion. In: Bartoli, A., Fusiello, A. (eds.) ECCV 2020. LNCS, vol. 12540, pp. 492–505. Springer, Cham (2020). https://doi.org/10.1007/978-3-030-65414-6_34
37. Kortylewski, A., Egger, B., Schneider, A., Gerig, T., Morel-Forster, A., Vetter, T.: Analyzing and reducing the damage of dataset bias to face recognition with synthetic data. In: proceedings of the IEEE/CVF Conference on Computer Vision and Pattern Recognition Workshops (2019)
38. Maze, B., et al.: Iarpa Janus benchmark-c: face dataset and protocol. In: 2018 International Conference on Biometrics (ICB), pp. 158–165 (2018)
39. Pan, G., Sun, L., Wu, Z., Lao, S..: Eyeblink-based anti-spoofing in face recognition from a generic webcamera. In: 2007 IEEE 11th International Conference on Computer Vision, pp. 1–8 (2007)
40. Kollreider, K., Fronthaler, H., Bigun, J.: Evaluating liveness by face images and the structure tensor. In: Fourth IEEE Workshop on Automatic Identification Advanced Technologies (AutoID'05), pp. 75–80 (2005)
41. Anjos, A., Chakka, M.M., Marcel, S.: Motion-based counter-measures to photo attacks in face recognition. In: IET Biometrics, pp. 147–158 (2014)
42. Chingovska, I., et al.: The 2nd competition on counter measures to 2D face spoofing attacks. In: 2013 International Conference on Biometrics (ICB), pp. 1–6 (2013)
43. Kollreider, K., Fronthaler, H., Bigun, J.: Verifying liveness by multiple experts in face biometrics. In: 2008 IEEE Computer Society Conference on Computer Vision and Pattern Recognition Workshops, pp. 1–6 (2008)

44. Chetty, G., Wagner, M.: Audio-visual multimodal fusion for biometric person authentication and liveness verification. In: ACM International Conference Proceeding Series, vol. 163, pp. 17–24 (2006)
45. Joshi, T., Dey, S., Samanta, D..: Multimodal biometrics: state of the art in fusion techniques. Int. J. Biometrics, 393–417 (2009)
46. De Marsico, M., Nappi, M., Riccio, D., Dugelay, J.-L.: Moving face spoofing detection via 3D projective invariants. In: 2012 5th IAPR International Conference on Biometrics (ICB), pp. 73–78 (2012)
47. Pavlidis, I., Symosek, P.: The imaging issue in an automatic face/disguise detection system. In: Proceedings IEEE Workshop on Computer Vision Beyond the Visible Spectrum: Methods and Applications (Cat. No. PR00640), pp. 15–24 (2000)
48. Hou, Y.-L., Hao, X., Wang, Y., Guo, C.: Multispectral face liveness detection method based on gradient features. In: Optical Engineering (2013)
49. Liu, Y., Jourabloo, A., Liu, X.: Learning deep models for face anti-spoofing: bbinary or auxiliary supervision. In: Proceedings of the IEEE Conference on Computer Vision and Pattern Recognition, pp. 389–398 (2018)
50. Yan, M., Zhao, M., Xu, Z., Zhang, Q., Wang, G., Su, Z.: Vargfacenet: an efficient variable group convolutional neural network for lightweight face recognition. In: Proceedings of the IEEE/CVF International Conference on Computer Vision Workshops (2019)
51. Dadi, H.S., Krishna Mohan, P.G.: Performance metrics for Eigen and Fisher feature based face recognition algorithms. In: International Journal of Computer Science and Network Security (2016)
52. Kremi, E., Subai, A.: The implementation of face security for authentication implemented on mobile phone. Int. Arab J. Inf. Technol. (2011)
53. Wang, X., Wang, S., Zhang, S., Fu, T., Shi, H., Mei, T.: Support Vector Guided Softmax Loss for Face Recognition (2018)
54. Michel, O., Dergham, A., Haber, G., Fakih, N., Hamoush, A., Abdo, E.: Face recognition security system. In: Conference Paper on Computer, Information, and System Science (2013)
55. Liu, S., Song, Y., Zhang, M., Zhao, J., Yang, S., Hou, K.: An Identity Authentication Method Combining Liveness Detection and Face Recognition (2019)

An Analysis on Compute Express Link with Rich Protocols and Use Cases for Data Centers

Eslavath Lakshmi Bai and Shital A. Raut(✉)

Department of Computer Science and Engineering, Visvesvaraya National Institute of Technology, Nagpur, Mahrashtra, India
saraut@cse.vnit.ac.in

Abstract. The proliferation of data has led the semiconductor industry to make breakthrough architectural shifts that will radically transform data center performance, efficiency, and cost. Machine learning applications produce millions of megabytes of data which forces server designs to take a quantum leap forward to address a problem that has persisted for decades. The idea of decomposition and the universal interfaces has existed. But eventually, the industry is firmly settling on the Compute Express Link (CXL) with respect to cache coherence interconnection of components. CXL is meant to support diversified processing and storage systems along with applications in AI, ML, communication, and high performance computing to meet the expanding needs of high performance computational workloads. It brings many benefits to the data center, including improved performance, increased efficiency, high bandwidth, and low latency by offering protocols. Processing the data in such new applications demands a wide combination of scalars, vectors, matrices, and spatial architectures so the coherency and memory semantics in a heterogeneous environment is becoming increasingly crucial. So the paper explains about the CXL and it's role in computing.

Keywords: Cache coherence · Protocols · Latency · Performance · Devices

1 Introduction

Compute Express Link, a new open interconnect standard [1], is aimed at intense workloads for processors and custom-built accelerators that require access to memory between the Host and a Device that is efficient as well as coherent [2, 3]. Compute Express Link technology preserves the coherency of memory between the central processing unit memory area and the associated device storage, allowing shared resource features to achieve improved performance, decreased complexity of software stack, and system with less cost. Now users can concentrate on their intense workloads rather than focusing on the accelerator's memory management hardware which is not essential [4–6]. As accelerators are increasingly used to complement CPUs to support the development of applications such as artificial intelligence and machine learning, the CXL is designed as an industry-standard high-speed communication interface. The Consortium of CXL identified three major types of devices that will add benefit from the new connections [7, 8].

J. I.-Z. Chen et al. (Eds.): ICIPCN 2022, LNNS 514, pp. 787–800, 2022.
https://doi.org/10.1007/978-3-031-12413-6_62

Before moving to the new interconnect standard, Peripheral component interconnect express (PCIe) is an interface for connecting host processors and devices which has been around for a long time with various generations like PCIe 1.0, 2.0, 3.0, 4.0 and the newly finished version 5.0 of the PCIe basic specification now allows CPUs and peripherals to communicate at speeds of up to 32 GT/s. PCIe, on the other hand, has some drawbacks in a system with big shared memory pools and various devices which require high bandwidth and low latency [9–11]. Because each PCIe hierarchy uses a single 64-bit address space, as a result, there is no support for coherency and cannot manage separate memory pools properly. Furthermore, Peripheral component interconnect express connectivity can have too much delay to effectively manage memory across several devices which are shared in the system.

The CXL tackles a few of these issues by offering an interface that uses the PCIe 5.0 physical layer and electrical to provide memory access with extremely low latency and cache coherence between host processors and memory-sharing devices such as accelerators and memory expanders. The standard modes provided by CXL revolve around a 32GT/s PCIe 5.0 PHY in an x16 lane arrangement (Table 3). For allowing bifurcation, x8 and x4 lane configurations also support 32 GT/s. Degradation mode is defined as being smaller than x4 lanes or much slower than 32 GT/s. This is not expected to be a clear advantage in the target application. While compute express link can provide considerable performance benefits for numerous applications, certain devices only require signal task submission and completion events, which is commonly the case when working with big data items or contiguous streams. PCIe works well as an accelerator interface for such devices, whereas CXL offers no meaningful advantages. Before CXL 2.0, we have CXL 1.0 and CXL 2.0 generations available. CXL 1.0 is built on the top of PCIe 5.0 layer.It came up with few features like, cache coherency, single device connection and resource sharing then CXL 1.1 came into picture with few enhancements by supporting CXL 1.0 features. CXL 1.1 gave support for persistent memory idea. These two generations gave high performance over PCIe interconnect. Currently CXL 2.0 is introduced with so many features like memory pooling and memory expansion and so on (see Table 1).

In this paper analysis of a high-performance interconnect, Compute Express Link has been done. The following are the important contributions of this research: Sect. 2 describes various protocols which are introduced by CXL, Sect. 3 includes an explanation of CXL devices and their benefits, Sect. 4 elaborates on the security that is provided by CXL, Sect. 5 discusses the performance of previous PCIe generations and CXL specification, Sects. 6 and 7 includes future work of CXL and conclusion of paper respectively.

2 Protocols

Before being transferred at 32 GT/s via a conventional PCIe 5.0 physical layer, the CXL standard enables three dynamically multiplexed protocols [1, 2, 9, 12].

- CXL.io
- CXL.cache
- CXL.mem

2.1 CXL.io

A protocol used for Initialization, detection, and enumeration of devices, linking, and access of registers. Moreover, it is an upgraded genre of the PCIe gen-5 IO protocol that provides a load or store command interface for the Input/Output devices which are non-coherent [13].

2.2 CXL.cache

Provides Coherent access from the device to the CPU host using a request and answer technique. This protocol describes interactions between a host and a device, allowing CXL devices to cache host memory efficiently and with very low latency.

2.3 CXL.mem

Allows host processor to use load and store commands to access the memory of an associated device, with CPU host and CXL device serving like owner and slave respectively (master, subordinate). This approach can support both volatile and permanent memory architectures.

The CXL.cache and CXL.mem protocols have been merged (see Fig. 1) and share a single link layer and transaction layer, whereas the CXL.io protocol has a link layer and transaction layer of its own.

Fig. 1. CXL device design with PHY, controller and application

Before being forwarded to the PCIe 5.0 PHY for transmission at 32GT/s, the data generated by each of these three protocols are dynamically multiplexed with the arbitration and multiplexing block. The requests from the CXL link layers (CXL.io and CXL.cache/mem) are arbitrated by ARB/MUX and then multiplexed based on the arbitration results, which are created by a host-maintained weighted round-robin arbitration process [14, 15]. The ARB/MUX also handles power-state transition requests from the link layers, creating a single request to the physical layer to power down in a regulated way.

CXL has fixed-width flits with a 528-bit width, made up of a two-byte CRC and four 16-byte slots. Slots can be dedicated either to the CXL.cache or CXL.mem protocols, and they are described in various ways. The slot's format is specified in the flit header. The transaction layer routes data to the appropriate protocols based on the information available on this header.

3 Devices

The section describes CXL devices which are classified into three distinct types based on different combinations of the protocols (see Fig. 2) [1, 9].

Fig. 2. CXL Device types with different protocol combinations

As the CXL.io protocol is the foundational protocol needed for initialization, device discovery, and link-up, all CXL devices should support it, and if the protocol fails, the connection will be lost.

3.1 Type – 1

CXL type 1 device is a combination of CXL.io and CXL.cache protocols. Type 1 CXL devices have special requirements that require a fully coherent cache where devices become worthy. For such types of devices, standard producer-consumer ordering models won't work well. Examples of such devices are Accelerators, Smart NICs with a coherent cache that implement a completely coherent cache but does not support host manageable memory devices and also enhances PCIe protocol capability which performs complex atomic operations. These devices support two types of transactions namely Device to Host (D2H) and Host to Device (H2D) snoop transactions. The cache size depends on the filtering capacity of the host's snoop. CXL supports this kind of device with the help of an optional cache link over which accelerators can use the CXL.cache protocol for cache-coherent transactions (see Fig. 3).

Fig. 3. H2D snoop transaction exchange flow

3.2 Type – 2

These devices support all the three protocols offered by CXL namely CXL.io, CXL.cache, and CXL.mem. Type 2 devices include all the CXL.mem/cache transactions [16] and implement an optional coherent cache and device memory that is managed by the host. Typical applications are devices that have attached memories with high bandwidth [17, 18].

- CXL.cache Host-to-Device Snoop Transaction Exchange:
  The host sends a request to the device for data (see Fig. 3). Upon receiving the request from the host, the device will send a response to the host as RSPI_FWDM. Afterward, data is sent to the host from the device. Upon receiving a response from the device to the host, the transaction is complete.

  CXL.cache contains three channels: Request, Response, and Data in each direction, which defines the interaction between host and device as well independent to achieve both decoupling and higher effective throughput per wire. (see Fig. 4) shows the transaction description of the CXL.cache protocol. Compute Express Link has provided two types of coherency biases for type 2 devices. Bias modes include device-bias and host-bias mode, which will determine how the coherence data between host-attached memory and device is processed by CXL [1, 4].

(A)

(B)

Fig. 4. (A) CXL.cache protocol Read flow and (B) CXL.cache protocol Write flow

3.3 Type – 3

This type of CXL device supports two protocols CXL.io and CXL. mem. A memory expander for the host is a good example of such a kind of CXL device. With a coherent interface and low latency, the CXl.mem protocol can enable power-efficient performance with developing applications such as AI, DL, and HPC [19, 20], and communications. The device works through CXL.mem to process the requests which are sent by the host. For this type of device two types of flows are supported: read flow and write flow [2, 12]. These are basic memory flows for how the CXL type 3 devices interface with CPUs and some of the flows involved.

These are memory buffers and memory expansion devices that allow us to provide memory bandwidth, memory capacity expansion, or access to other types of memories such as persistent media and so on. So CXL provides us basically with a common standardized interface that allows us to put any number of different types of media or memory behind that interface. This provides the flexibility to support a variety of different memory devices could be as DDR3/DDR4/DDR5, LPDDR3/4/5/persistent memory, and so on. The media controller enables flexibility for different media characteristics (persistent, latency, BW, endurance, and so on) [21].

- Write Flow
  Host initiated the write transaction using memory write (MemWr) semantic. The device responded with the CMP response (see Fig. 5).

Fig. 5. Memory write flow

The host or the CPU will send a MemWr command which is a request with the data and in some cases, it includes some Metadata as well which can be used for tracking a coherency state. Once the memory device receives that request then the memory controller will send that request to the media as a write operation typically with some ECC and include the metadata as well if it's available and supported. The media device will send back the completion to the memory controller and then the CXL memory device will send the completion with no data response back to the host saying that the write has been completed.

- Read Flow

  The host requested the data using memory read (MemRd) semantic to the CXL device with no operation assumption and this request sent to memory media. The device responded with the data using a Meta value response (see Fig. 6).

Read flow is similar to write flow, the host sends a request as a NoOp assuming it's supported. CXL device memory controller will send a read to the media. Media will respond to the device with the data, ECC if applicable, and in addition to the metadata and return it responds to the CPU by sending a response with data and metadata (it's just stated as it is not applicable for a read operation).

Fig. 6. Memory read flow

- Memory Validation Flow
Similarly, there will be a case where a memory will be available and validate it and have to issue. This might be useful in a couple of different cases, relevant example is it could be used to tell the device to flush its caches out to persistent media. So the way this works is the CPU will send a memory to validate command generally that includes some metadata in this context uses the value of zero (see Fig. 7). The memory controller goes out and reads the media itself to determine what the current state is. The media will respond with the data + ECC and the current value of that metadata.

This device includes use cases such as an expansion of memory bandwidth and capacity as well storage class memory.

Emerging memories that are often transactional might have asymmetric and non-deterministic timing in reading or writing. This makes them not suitable for sharing the DDR bus with Dynamic Random Access Memory (DRAM). CXL 1.1 specification does not provide support for multiple Host-to-Device connections however it supports a single Host-to-Device connection [22]. CXL 2.0 comes up with some enhancements over CXL 1.1 specification by introducing three main features: Switching, persistent memory support [21], and security with full backward compatibility and memory can be shared across multiple hosts based on particular application needs and support coherency with switching feature by providing benefits such as pooling and expansion (see Table 1) [12].

- Memory Pooling
CXL attached memory is a combination of hardware (Memory Devices, Platforms, Switches), software (Fabric Manager, Operating system), and protocol (Compute Express Link) upgrades that enable dynamic management and allocation of hardware resources for effective usage of hardware resources. Memory pooling allows for more efficient use of memory resources inside a system (Rack), as well as allocating or deallocating memory resources dynamically and lowers the ownership total cost (TCO). Memory pooling allows most efficient use of memory resources by attaching multiple CXL devices to the host which are memory expanders. So when ever any read or write request comes to the host then it will distribute to multiple memory devices in parallel hence in this we will achieve more performance with effective utilization of devices. Internally device can divide into multi logical devices and allows 16 hosts at a time to access the memory resources.

Memory pooling [11, 12], is a fundamental use case for the CXL specification, and it is supported by a variety of topologies, including single logical devices pooling, pooling without the use of a switch, and pooling inside multi-logical devices. The CXL specification 2.0 defines an application-specific interface called fabric manager (FM) that supports pooling applications and managing platforms by providing setup and control features.

Fig. 7. Memory ınvalidate flow – used for read and writes meta value

Table 1. Characteristics of CXL 1.0 and CXL 2.0 specification.

CXL 1.1	CXL 2.0
Resource sharing	Memory pooling and switching
Coherency interface	Persistent memory
Single level device connection	Integrity and data encryption
Memory protocol	Memory expansion
CXL I/O protocol	Cache coherency
High bandwidth	Hot plug
Low latency	Supports previous CXL generations

4 Security

The last, but feasibly more important feature which has always been a concern is security over the protocols (see Fig. 8).

Fig. 8. Security enhancements with CXL 2.0 specification

CXL provides security with the addition of two main security components namely authentication & key management, Integrity, and Data Encryption (IDE) [23]. Compute Express Link uses an AES-GCM cryptographic scheme with 256 bit key size for confidentiality, integrity, and replay protection.

4.1 RAS

RAS is for error detection, handling, and error reporting.

Reliability (R). The ability of a device or any equipment to perform particular proposed functionality without failure for a given interval of time.

Availability (A). Should be available for the users whenever there is a need.

Serviceability (S). The quality of being able to provide good service. Usability, Usefulness, Utility – the quality of being of practical use. Table 2 includes RAS features that are supported by CXL specifications [14, 15].

RAS provides one feature called LINK CRC and Retry. Cyclic Redundancy check (CRC) is an error-detecting code used in the data link layer – Link CRC and transaction layer – ECRC (End to End CRC) for error checking. Retry happens due to Timeout or NACK (No acknowledgment) and if the number of retries is more than three times, the physical layer will be notified to link retraining and recovery which will be covered by the LTSSM state machine (11 states, including detect, hot reset, recovery, polling, loopback or uplink, configuration, L0, L1, L0s, L2, disable or deactivate). When an uncorrectable error is detected, enhanced downstream port containment (eDPC) defines an error containment mechanism that can automatically disable a link to prevent the spread of corrupted data. At last, all CXL protocol errors have reflected the OS via PCIe Advanced Error Reporting (AER) mechanisms and handled.

Table 2. Supportive features of CXL 1.1 versus CXL 2.0 specification.

Feature	CXL 1.1	CXL 2.0
PCIe RAS mechanisms - link and protocol errors	Yes	Yes
Viral	Yes	Yes
Error injection - compliance testing	Yes	Yes
Data poisoning	Yes	Yes
Function Level Reset - CXL.io	Yes	Yes
CXL Reset - CXL.mem, CXL.cache	No	Yes
Global Persistent Flush and Dirty Shutdown tracking	No	Yes
Detailed memory error logging	No	Yes
Support for viral propagation through switches	No	Yes
Scan Media/Internal Poison List Retrieval	No	Yes
Poison injection	No	Yes
Hot-plug support	No	Yes

5 Discussion

For so many years, Peripheral component interconnect express (PCIe) has been used as an interface for the connection of peripheral components with various PCIe versions which includes some enhancements in each stage. PCI Express version 1.0 had introduced in the year 2005 with some features. Later PCIe 2.0 came up with doubled transfer rate compared to version-1 and some additions like the motherboard being backward compatible with the presence of PCIe-v1, an increase of Per-lane output (250 MBps to 500 MBps), point-to-point data transfer protocol along with the software architecture. In 2007, PCIe 3.0 was announced with more bit rate (see Table 2) and as well more compatible with the updated encoding scheme. In 2017, PCIe 4.0 came up with more performance compared to previous versions of PCIe. In 2019, PCI 5.0 was announced to support CPU lanes and as well with high bandwidth, and recently PCIe 6.0 with a bit rate of 64 GT/s [9]. Each generation of PCI Express has come up with doubled speed and with some more add-ons. However, PCIe interconnects also have a few limitations such as less bandwidth, high latency, and not providing coherency. As the data is growing day by day, applications of machine learning, artificial intelligence, cloud computing, and so on need high performance, low latency and a high bandwidth interconnect [24]. To overcome this, compute express link the new open standard coherent interconnect came into the picture with many benefits which relies on top of PCIe gen-5 physical layer [10, 25, 26].

Table 3. Speed table of PCIe and CXL.

Protocol specification	Lane/ Bandwidth/Way	Link bandwidth	Raw bit rate	Total bandwidth
PCIe 1. 0	250 MB/s	2 Gb/s	2.5 GT/s	8 GB/s
PCIe 2. 0	500 MB/s	4 Gb/s	5 GT/s	16 GB/s
PCIe 3. 0	1 GB/s	8 Gb/s	8 GT/s	32 GB/s
PCIe 4. 0	2 GB/s	16 Gb/s	16 GT/s	64 GB/s
PCIe 5. 0 (CXL 1.1)	4 GB/s	32 Gb/s	32 GT/s	128 GB/s
CXL 2.0	8 GB/s	64 Gb/s	64 GT/s	256 GB/s

6 Future Work

The CXL Consortium is moving beyond CXL 2.0, having issued two revisions of the standards in less than a year. Looking forward to the next iteration of CXL, CXL 3.0, which will include more useful scenarios and give even higher performance, based on feedback from the computer industry and the end-user community.

7 Conclusion

In addition to one platform of accelerators and storage expansion devices, the new CXL specification supports new applications such as switching, pooling of resources in racks, persistent memory flow, and enhanced security with full backward compatibility. In multiple working groups of CXL, reliability, availability, and serviceability in future intercepts of the CXL standard continue to be a hot topic, with efforts to define additional mechanisms which will set up on the principles and capabilities outlined in the paper.

The CXL Consortium is working on the co-development of this open standard and welcomes the active participation of a wide range of industries. The journey is imminent with many exciting products and innovations.

References

1. Sharma, D.D., Ward, G., Bowman, K.: An introduction to compute express LinkTM (CXL) technology. CXL Consortium Webinar, 12 December 2019
2. Tavallaei, S., Blankenship, R., Lender, K.: Exploring coherent memory and innovative use cases. Webinar, CXL Consortium, 12 March 2020
3. Sharma, D.D.: Understanding compute express link: a cache-coherent interconnect. In: Proceedings of Storage Developers Conference, September 2020
4. Coughlin, T.: Digital storage and memory. Computer **55**(1), 20–29 (2022)
5. Lynch., M.: Bank of America Merrill Lynch Global Semiconductors Report, 2 October 2016. investor.bankofamerica.com
6. Shenoy, N.: Intel news: a milestone in Moving data, 11 March 2019. https://newsroom.intel.com/editorials/milestone-moving-data/#gs.vy0a60

7. Sharma, D.D.: CXL: Coherency, memory, and I/O semantics on PCIe infrastructure. Electronic Design, 28 April 2021
8. Sharma, D.D.: Compute Express Link. Compute Express Link Consortium, White Paper, March 2019. https://docs.wixstatic.com/ugd/0c1418_d9878707bbb7427786b70c3c91d5fbd1.pdf
9. Sharma, D.D.: A low latency approach to delivering alternate protocols with coherency and memory semantics using PCI Express® 6.0 PHY at 64.0 GT/s. In: 2021 IEEE Symposium on High-Performance Interconnects (HOTI), pp. 35–42. IEEE (2021)
10. CXL Use-cases Driving the Need For Low Latency Performance Retimers (2021). https://www.microchip.com/en-us/about/blog/learning-center/cxl--use-cases-driving-the-need-for low-latency-performance-reti
11. Webinar, Compute Express Link™ 2.0 Specification: Memory Pooling, March 2021
12. Petersen, C., Chauhan, P.: Memory Challenges and CXL Solutions. Webinar, CXL Consortium, 6 August 2020
13. Sharma, D.D.: Innovations in loadstore I/O causing profound changes in memory, storage, and compute landscape. In: Keynote at Storage Developers (SDC) Conference, 28 September 2021
14. Gurumurthi, S., Advanced Micro Devices, Inc., Branover, A.J., Advanced Micro Devices, Inc., Hornung, B., Micron Technology, Inc., Michna, V., Hewlett Packard Enterprise Company Mahesh Natu, Intel Corporation Chris Petersen, Facebook, Inc.: An overview of reliability, availability, and serviceability (RAS). In: Compute Express Link™ 2.0
15. Academia.edu. https://www.academia.edu/45413748/Compute_Express_Link_RAS
16. VIP Experts. Verification Central, "An Introduction to the CXL Device Types", 4 March 2020. https://blogs.synopsys.com/vip-central/2020/03/04/an-introduction-to-the-cxl-device-types/
17. Sharma, D.D., Tavallaei, S.: Compute Express LinkTM 2.0. White Paper, CXL Consortium, November 2020
18. Taubenblatt, M., Maniotis, P., Tantawi, A.: Optics enabled networks and architectures for data center cost and power efficiency. J. Opt. Commun. Netw. **14**(1), A41–A49 (2022)
19. Taubenblatt, M.A.: Optical interconnects for large scale computing: how do we get beyond the cost & power wall? In: Optical Fiber Communication Conference, pp. Th4I-4. Optical Society of America (2021)
20. Sharma, D.D.: Keynote 1: Compute Express Link (CXL) changing the game for Cloud Computing. In: 2021 IEEE Symposium on High-Performance Interconnects (HOTI), pp. xii-xii. IEEE (2021)
21. A Computer Weekly buyer's guide to "computational storage and persistent memory." Computer Weekly (2021). https://www.computerweekly.com/ehandbook/A-Computer-Weekly-buyers-guide-to-computational-storage-and-persistent-memory
22. Sharma, D.D., Tavallaei, S.: "Compute Express Link 1.1 specification blog" at CXL website, March 2020. https://www.computeexpresslink.org/post/compute-express-link-1-1specificationnowavailable-to-members
23. Webinar, Compute Express Link™ (CXL™) Link-level Integrity and Data Encryption (CXL IDE), September 2021
24. Li, H., et al.: First-generation memory disaggregation for cloud platforms. arXiv preprint arXiv:2203.00241 (2022)
25. Van Doren, S.: HOTI 2019: compute express link. In: 2019 IEEE Symposium on High-Performance Interconnects (HOTI), p. 18. IEEE (2019)
26. Compute Express Link: The Breakthrough CPU-to-Device Interconnect (2020). https://www.computeexpresslink.org

Stability Investigation of Ensemble Feature Selection for High Dimensional Data Analytics

Archana Shivdas Sumant[1(✉)] and Dipak Patil[2]

[1] MET's Institute of Engineering, Nashik Affiliated to Savitribai Phule Pune University, Pune, Maharashtra, India
archana.vaidya@ges-coengg.org

[2] Gokhale Education Society's R. H. Sapat College of Engineering, Management Studies and Research, P T A Kulkarni Vidyanagar, Nashik 422005, Maharashtra, India
dipak.patil@ges-coengg.org

Abstract. In the selection of feature subsets, stability is an important factor. In literature, however, stability receives less emphasis. Stability analysis of an algorithm is used to determine the reproducibility of algorithm findings. Ensemble approaches are becoming increasingly prominent in predictive analytics due to their accuracy and stability. The accuracy and stability dilemma for high-dimensional data is a significant research topic. The purpose of this research is to investigate the stability of ensemble feature selection and utilize that information to improve system accuracy in high-dimensional datasets. We conducted a stability analysis of the ensemble feature selection approaches ChS-R and SU-R using the jaccard similarity index. Ensemble approaches have been found to be more stable than previous feature selection methods such as SU and ChS for high-dimensional datasets. The average stability of the SU-R and ChS-R ensemble approaches is 56.03 and 50.71%, respectively. Accuracy improvement achieved is 4 to 5%.

Keywords: High-dimensional data · Ensemble learning · Stability analysis · Feature selection · Jaccard index

1 Introduction

Due to the rising availability of data, data analysis is becoming more prevalent in all industries. Feature subset selection is used in data analysis to improve predictive accuracy and speed of prediction. To improve system accuracy, ensemble approaches are used at various phases of feature subset selection. Stability is a crucial consideration when choosing a feature subset for examining the reproducibility of the same results with the same subset on other data in the future. HyunJi Kim et al. in [1] discuss the stability of the selected feature subset in addition to the prediction accuracy. They propose Q-statistics as a new measure to boost the algorithm. In feature subset selection, a forward selection is applied for high dimensional data. However, change in the decision of the initial feature may lead to a completely different feature subset and the stability of the selected feature

J. I.-Z. Chen et al. (Eds.): ICIPCN 2022, LNNS 514, pp. 801–815, 2022.
https://doi.org/10.1007/978-3-031-12413-6_63

set will be very low. S. Nogueira and G. Brown in [2] states that Pearson's correlation co-efficient has all necessary properties for stability measure.

It has been established that in high-dimensional problems feature reduction could improve classification accuracy as discussed by Das [3] and Xing, Jordan and Karp [4]. As a result, it has long been associated with the concept of feature selection stability. However, there is no way to distinguish between the behavior of the feature selection algorithm and the behavior of the classifier when the input changes. A low stability score in that situation provides no information about whether the feature selection technique produces a non-optimal feature subset or the classifier's sensitivity to the data is too high.

The rest of the paper is laid out as follows. Section 2 delves into the fundamentals of feature selection as well as the issue of stability for ensemble methods. We analyzed our proposed ensemble methods stability with jaccard index based on instance learning and we compare the performance of our methods to current feature selection techniques. The experiments employed with four classification algorithms and seven datasets in Sect. 3. Section 4 proposes the use of stability analysis to improve system accuracy. Section 5 brings this paper conclusion.

2 Related Work

One of the first researchers to look into feature selection stability individually was Kalousis et al. [5]. He suggested a number of similarity metrics that may be used independently of any learning model and account for the various outputs that feature selection methods produce.

These algorithms do one of three things: give features a weight or a score, rank them, or pick a restricted subset. Kalousis presented three independent metrics as a result: Pearson's correlation coefficient for weight-scoring outputs, Spearman's rank correlation coefficient for ranking, and Tanimoto distance for determining set similarity. Each of the $\mathrm{S_x}$ similarity measures discussed below only works on two sets of data. The sum of the similarity scores for each of the two subgroups is averaged.. This can be used to measure the total similarity S of m sequences $A = \{A_1, A_2, A_3, \ldots A_m\}$ by averaging the sum of the similarity scores for every two subsets.

As a result, regardless of the similarity measure used, a generalized approach for computing total stability is developed is defined as in Eq. 1.

The sum of each individual set's similarity score to all other subsets is computed and averaged, where m is the number of subsets.

$$\mathrm{S(A)} = \mathrm{S(A)} = \frac{\sum_{i=1}^{m-1}\sum_{j=i+1}^{m} S_x(A_i, A_j)}{\frac{m(m-1)}{2}} \tag{1}$$

2.1 Stability Measure Tanimoto Distance

An adaptation of Tanimoto distance has been proposed by Kalousis et al. [5] as a possible measure for set similarity. It relies on computing the number of elements in the intersection of two sets *s* and *s'* divided by the number of elements in the union and can be expressed as follows:

$$S_t(s, s') = \frac{|s \cap s'|}{|s \cup s'|} \tag{2}$$

The values are in the range of [0, 1], where 0 characterizes two completely different subsets, and 1, two subsets which are identical. This measure has advantages, such as computational efficiency and simplicity of implementation. However, upon further investigation one apparent disadvantage emerges, as discussed by Kuncheva [6] and Lustgarten et al. [7] discusses inability of this measure to handle 'chance'. As the number of selected features approaches the total number of features, the Tanimoto distance produces values close to 1.

2.2 Stability Measure Hamming Distance

When evaluating the stability of different feature selection algorithms, Dunne et al. [8] proposed another similarity measure based on the Hamming distance. It employs the masks generated by the feature selection algorithm, 1 denoting a feature has been selected and 0 that it has not. Later Kuncheva [6] expressed the Hamming distance in terms of two sets in the following manner, where '\' denotes the set difference operation.

$$S_h\left(s, s'\right) = 1 - \frac{\left|s \backslash s'\right| + \left|s' \backslash s\right|}{n} \tag{3}$$

Much like Tanimoto distance this similarity measure does not factor for chance.

2.3 Stability Measure Kuncheva's Consistency Index

In order to resolve this problem Kuncheva [6] introduced several desirable properties of a similarity measure, which are monotonicity, limits and correction of chance. Her research empirically proved that the previously used measures do not meet these requirements, thus emphasizing the need for a new one. She introduced the 'consistency index' which can be formulated as follows.

$$S_c\left(s, s'\right) = \frac{rn - k^2}{k(n-k)} \tag{4}$$

where r is $|s \cap s'|$, k is number of selected features, n is total number of features, s and s' are subsets of features with the same size. The values range between [−1, 1], where 1 can be obtained for two identical sets ($r = k$). The minimal value is produced when $= 0$. As noted by Kuncheva the value of the index is not defined for $k = 0$ and $k = n$. Nevertheless, for completeness these can be implemented as 0. By taking into account the total number of features the consistency index satisfies all requirements introduced above.

2.4 Stability Measure Pearson's Correlation Coefficient

Pearson's correlation coefficient has been widely used in statistics and economics as a measure for correlation between two variables. In the context of feature selection stability Kalousis [5] proposed its use for measuring similarity of two separate weighting-scoring outputs produced by feature selection algorithms. It can be formulated as follows, where μw and $\mu w'$ are the sample means of w and w' respectively.

$$S_p(w, w') = \frac{\sum_i (w_i - \mu_w)(w_i' - \mu_{w'})}{\sqrt{\sum_i (w_i - \mu_w)^2 \sum_i (w' - \mu_{w'})^2}} \tag{5}$$

Its values are in the range of $[-1, 1]$, where values less than 0 denote low similarity. Pearson's correlation coefficient was reintroduced in this research, and its utility in stability estimation was demonstrated.

2.5 Stability Measure Jaccard Index

To evaluate the similarity between two selected subsets of features, Jaccard Index is used. Equation (6) calculates the proportion of intersection of the selected features to the total number of selected features in the two subsets combined.

Jaccard has several advantages as a stability measure. First, it is symmetric, thus, $J(F_1, F_2) = J(F_2, F_1)$. Also, it is monotonic. In other words, the larger the intersection between F_1 and F_2, the larger J will be. Besides that, Jaccard is always between 0 and 1.

$$J(F_1, F_2) = \frac{|F_1 \cap F_2|}{|F_1 \cup F_2|} \tag{6}$$

Assume there are two feature subsets, F1 and F2. Suppose size of each set is 20. When these sets are compared, say 12 features are the same. Then $|F1 \cap F2|$ becomes 12 and $|F1 \cup F2|$ becomes 28. Jaccard index will be calculated as 12/28 as 0.43.

Salem Alelyani [9] states that variance in data is the main cause of instability of feature selected. Bagging ensemble method is used with bootstrap sampling to reduce data variance. By majority vote method aggregated feature subset is selected. Selection stability is evaluated with Jaccard coefficient. They have achieved 20 to 50% stability improvement with their method. The 80% overlap is taken for generating sample subspace.

Afef Ben Brahim [10] proposes filter method to improve feature selection stability and to deal with accuracy stability tradeoff. The proposed method uses instance learning to select relevant features initially. Candidate features are selected by aggregating ranking, occurrence frequency, weighted mean aggregation and redundancy elimination. The proposed approaches achieves better stability and accuracy over existing relief, mRMR [11], t-test and entropy. When assessing the stability of gene subsets, Haury et al. [12] evaluated the role of overlap. When comparing feature lists derived from subsamples of the original data with either 80% or 0% overlap, the researchers analyzed the fraction of instances in common, in addition to additional analyses of their datasets. They also looked at feature lists from four other (but related) datasets. They discovered that the similarity measurements for the 0% overlap case resembled the between-datasets case more closely than the results from the 80% overlap case. It's challenging to generalize the method for creating datasets with randomly selected overlaps.

Let's say we have n samples and p features. Here features are attributes, independent variables, explanatory variables. High dimensional data is data having $n \ll p$ and p are usually high in thousands or ten thousand. So dealing with this number of dimensions with obtaining high predictive accuracy is the challenge. The solution is dimensionality reduction that selects the subset of relevant features. So our aim in this study is to analyze stability of our systems on high dimensional datasets and improve system accuracy.

3 Experimental Setup and Datasets

In this study, we investigated the stability of our proposed ensemble methods ChS-R and SU-R [13]. These ensemble ranking methods are used for dimensionality reduction. As discussed earlier the high dimensional dataset feature dimension is larger than number of samples.

The Fig. 1 gives detail flow of stability analysis. Stratified random sampling is used to split the dataset into ℓ folds. As the datasets are imbalanced datasets stratified random sampling gives balanced representation of classes in each fold. For each fold ensemble ranking methods described in our paper [13] are used to rank the features. The first method, ChS-R, is an ensemble feature selection method that ranks using the Chi-squared (ChS) and ReliefF measures. For ranking, the second method, SU-R, employs symmetric uncertainty (SU) and reliefF. Please refer to our published paper [13] for more details.

The top m rated features are then chosen for further analysis. The value of m is fixed to five for performance comparison. Random Forest (RF), Support Vector Machine (SVM), K-nearest Neighbor (KNN), and Multi Layer Perceptron (MLP) classifiers are used to assess the performance of specified features. The RF, KNN, SVM, and MLP classifiers were chosen after extensive testing with various classifiers.

The languages Python and R are used in system development. Our aim is to test stability of each fold for that purpose jaccard similarity index score is calculated as per Eq. 6. The Jaccard similarity index is a prominent stability metric which is used in [9]. Here, a feature subset from each ℓ fold is compared to a subset selected from the entire dataset. In Sect. 2.5 details of this index has discussed with example.

Fig. 1. Stability assessment processing flow followed in the experiment

Table 1. Dataset details used in experiment

DN	Dataset name	Ck	n	p	Ratio n/p
1	COLON	2	62	2000	0.031
2	Lung	2	203	12600	0.016
3	Leukemia	2	72	7129	0.01
4	Prostate	2	102	12600	0.008
5	MLL	3	72	12582	0.006
6	Lymphoma	3	45	4026	0.011
7	SRBCT	4	83	2308	0.036

The Table 1 gives dataset [14, 15] details used in experiment. Here n are number of samples, p is number of features and C_k denotes number of classes. DN 1 to 7 are high dimensional microarray cancer datasets. On chromosome 11q23, the MLL (mixed-lineage leukemia) gene is found. SRBCT stands for Small Round Blue Cell Tumors and is a cancer dataset. Here n is substantially lower than p, and the n/p ratio is less than 1, ranging from 0.005 to 0.031. DN 1 to 4 are binary datasets while DN 5 to 7 are multiclass datasets.

Accuracy is used to measure system performance and calculated using Eq. 7. To calculate accuracy of classifier fivefold cross validation technique is used. Accuracy is calculated as total correctly predicted instances N_{cc} divided by total number of instances N_T multiplied by 100 for percentage accuracy.

$$\%Accuracy = \frac{N_{CC}}{N_T} * 100 \tag{7}$$

Table 2 shows the stability and accuracy of the SU-R technique in five folds. As these datasets are having less number of samples and also class imbalances fivefold cross validation technique is used. For the COLON dataset, the MLP classifier performed best, whereas the RF classifier performed worst. For this dataset, the KNN classifier has consistently performed well. For the LUNG dataset, RF had the worst average performance, whereas SVM had the best average performance. For the LUNG dataset, the MLP classifier has consistently performed well. For the leukemia dataset, there has been observed variation in performance by KNN and RF. SVM and MLP, on the other hand, have consistently performed. For the Prostate dataset, all folds and classifiers had consistent accuracy.

The MLP classifier has the highest average accuracy of 82.80, while the SVM classifier has the lowest average accuracy of 78.40.

Fold 3 has the lowest accuracy in the MLL dataset, while Fold 2 has the greatest. For the MLL dataset, the lowest accuracy is 64.7 in fold 5.In addition, the MLP accuracy for all samples was found to be as low as 68.The average accuracy of five folds for the MLP classifier is 80.38, which is higher than the accuracy of all samples.

Table 2. Accuracy and stability measured with 5 FOLDS for SU-R method on seven high dimensional datasets

DN	Name		RF	SVM	KNN	MLP	SU-R	SU	ChS	Δ_1	Δ_2
		FOLD	78	69	95	95	0.43	0.29	0.25	0.14	0.18
		FOLD	76	64	93	90	0.33	0.18	0.14	0.16	0.19
		FOLD	95	65	77	99	0.43	0.14	0.21	0.29	0.22
1	COLON	FOLD	99	100	97	100	0.25	0.11	0.14	0.14	0.11
		FOLD	90	96	91	100	0.43	0.18	0.18	0.25	0.25
		AVG	88.60	78.8	90.60	96.80	**0.37**	**0.18**	**0.18**	**0.19**	**0.19**
		Acc*	87	90	85	84					
		FOLD	89	100	100	100	0.52	0.47	0.39	0.04	0.13
		FOLD	89	100	99	99	0.69	0.64	0.69	0.06	0.00
		FOLD	91	100	99	99	0.85	0.67	0.67	0.19	0.19
2	Lung	FOLD	87	100	94	100	0.72	0.72	0.69	0.00	0.03
		FOLD	91	100	98	100	0.67	0.52	0.69	0.15	-0.03
		AVG	89.40	100	98.00	99.60	**0.69**	**0.60**	**0.63**	**0.09**	**0.06**
		Acc*	98	100	94	100					
		FOLD	98	98	95	99	0.59	0.37	0.33	0.22	0.25
		FOLD	99	98	97	100	0.64	0.45	0.43	0.19	0.21
		FOLD	89	100	96	100	0.67	0.54	0.43	0.13	0.24
3	Leukemia	FOLD	96	100	99	100	0.59	0.37	0.30	0.22	0.29
		FOLD	98	97	87	95	0.64	0.43	0.32	0.21	0.32
		AVG	96.00	98.60	94.80	98.80	**0.62**	**0.43**	**0.36**	**0.19**	**0.26**
		Acc*	99	100	99	89					
		FOLD	80	78	81	84	0.67	0.25	0.54	0.42	0.13
		FOLD	81	77	81	81	0.48	0.22	0.48	0.26	0.00
		FOLD	80	79	82	82	0.67	0.34	0.43	0.33	0.24
4	Prostate	FOLD	82	80	82	84	0.82	0.25	0.43	0.57	0.39
		FOLD	80	78	79	83	0.67	0.49	0.48	0.18	0.19
		AVG	80.60	78.40	81.00	82.80	**0.66**	**0.31**	**0.47**	**0.35**	**0.19**
		Acc*	82	79	80	85					
		FOLD	94.11	88.23	88.23	70.58	0.45	0.25	0.05	0.20	0.40
		FOLD	94.11	97.05	100	97.05	0.48	0.34	0.05	0.14	0.43
		FOLD	88.88	88.88	77.77	77.77	0.30	0.25	0.08	0.05	0.22
5	MLL	FOLD	88.88	83.33	89.74	91.79	0.32	0.18	0.05	0.14	0.27
		FOLD	94.11	94.11	94.11	64.7	0.37	0.22	0.08	0.15	0.29
		AVG	92.02	90.32	89.97	80.38	**0.39**	**0.25**	**0.06**	**0.14**	**0.32**
		Acc*	91.2	82.8	92.5	68					
		FOLD	92	96	97	99	0.59	0.21	0.25	0.38	0.34
		FOLD	93	95	96	100	0.67	0.25	0.21	0.42	0.45
		FOLD	93	95	95	99	0.59	0.18	0.25	0.41	0.34
6	Lymphoma	FOLD	94	94	95	100	0.67	0.33	0.25	0.33	0.42
		FOLD	93	96	96	100	0.75	0.33	0.29	0.42	0.46
		AVG	93.00	95.20	95.80	99.60	**0.65**	**0.26**	**0.25**	**0.39**	**0.40**
		Acc*	94.73	96.37	96.83	100					
		FOLD	90	83	86	94	0.56	0.48	0.54	0.08	0.03
		FOLD	91	82	87	96	0.54	0.54	0.54	0.00	0.00
		FOLD	91	83	88	96	0.54	0.43	0.43	0.11	0.11
7	SRBCT	FOLD	92	84	87	96	0.47	0.48	0.43	-0.01	0.04
		FOLD	91	83	88	95	0.56	0.43	0.48	0.14	0.08
		AVG	91.00	83.00	87.20	95.40	**0.54**	**0.47**	**0.48**	**0.06**	**0.05**
		Acc*	92	84	88	96					
							Average improvement			**20%**	**21%**

*Accuracy with all samples

Table 3. Accuracy and stability measured with 5 FOLDS for ChS-R method on seven high dimensional

DN	Name		RF	SVM	KNN	MLP	ChS-R	SU	ChS	Δ_3	Δ_4
		FOLD	83	54	91	90	0.54	0.29	0.25	0.25	0.29
		FOLD	87	59	90	90	0.29	0.18	0.14	0.11	0.15
		FOLD	100	100	100	100	0.43	0.14	0.21	0.29	0.22
1	COLON	FOLD	90	35	83	80	0.25	0.11	0.14	0.14	0.11
		FOLD	58	89	86	67	0.38	0.18	0.18	0.20	0.20
		AVG	83.60	67.40	90.00	85.40	**0.38**	**0.18**	**0.18**	**0.20**	**0.19**
		Acc*	91	90	88	77					
		FOLD	85	100	98	100	0.47	0.47	0.39	0.00	0.08
		FOLD	93	100	99	100	0.79	0.64	0.69	0.15	0.10
		FOLD	83	100	93	100	0.96	0.67	0.67	0.29	0.29
2	Lung	FOLD	93	99	99	99	0.92	0.72	0.69	0.20	0.23
		FOLD	92	100	98	100	0.82	0.52	0.69	0.30	0.13
		AVG	89.20	99.80	97.40	99.80	**0.79**	**0.60**	**0.63**	**0.19**	**0.17**
		Acc*	89	100	96	100					
		FOLD	83	93	85	94	0.47	0.37	0.33	0.10	0.14
		FOLD	93	95	91	98	0.61	0.45	0.43	0.16	0.18
		FOLD	95	98	97	98	0.49	0.54	0.43	-0.05	0.06
3	Leukemia	FOLD	97	98	95	98	0.61	0.37	0.3	0.24	0.31
		FOLD	88	91	84	91	0.56	0.43	0.32	0.13	0.24
		AVG	91.20	95.00	90.40	95.80	**0.55**	**0.43**	**0.36**	**0.12**	**0.19**
		Acc*	97	97	97	95					
		FOLD	90	88	82	85	0.6	0.25	0.54	0.35	0.06
		FOLD	89	87	81	82	0.48	0.22	0.48	0.26	0.00
		FOLD	88	89	80	80	0.48	0.34	0.43	0.14	0.05
4	Prostate	FOLD	90	91	83	86	0.6	0.25	0.43	0.35	0.17
		FOLD	91	91	83	85	0.6	0.49	0.48	0.11	0.12
		AVG	89.60	89.20	81.80	83.60	**0.55**	**0.31**	**0.47**	**0.24**	**0.08**
		Acc*	90.9	90.9	83.87	85.15					
		FOLD	88.23	94.11	88.23	58.82	0.21	0.25	0.05	-0.04	0.16
		FOLD	100	94.11	94.11	82.35	0.54	0.34	0.05	0.20	0.49
		FOLD	94.44	88.88	94.44	77.77	0.25	0.25	0.08	0.00	0.17
5	MLL	FOLD	88.88	88.88	88.88	38.88	0.48	0.18	0.05	0.31	0.43
		FOLD	88.23	94.11	86.84	76.47	0.21	0.22	0.08	-0.01	0.13
		AVG	91.96	92.02	90.50	66.86	**0.34**	**0.25**	**0.06**	**0.09**	**0.27**
		Acc*	95	90.2	92.4	63					
		FOLD	80	78	81	84	0.59	0.21	0.25	0.38	0.34
		FOLD	81	77	81	81	0.37	0.25	0.21	0.12	0.16
		FOLD	80	79	82	82	0.27	0.18	0.25	0.10	0.02
6	Lymphoma	FOLD	82	80	82	84	0.32	0.33	0.25	-0.01	0.07
		FOLD	80	78	79	83	0.75	0.33	0.29	0.42	0.46
		AVG	80.60	78.40	81.00	82.80	**0.46**	**0.26**	**0.25**	**0.20**	**0.21**
		Acc*	94.74	100	91.57	95.78					
		FOLD	91	95	92	87	0.39	0.48	0.54	-0.09	-0.14
		FOLD	90	96	91	88	0.52	0.54	0.54	-0.02	-0.02
		FOLD	91	96	92	85	0.45	0.43	0.43	0.02	0.02
7	SRBCT	FOLD	92	95	92	88	0.47	0.48	0.43	-0.01	0.04
		FOLD	92	95	92	87	0.56	0.43	0.48	0.14	0.08
		AVG	91.20	95.40	91.80	87.00	**0.48**	**0.47**	**0.48**	**0.01**	**0.00**
		Acc*	92	96	92	88					
							Average improvement			**15%**	**16%**

*Accuracy with all samples

For the Lymphoma dataset, RF classifier performed the worst, whereas MLP performed the best. For this dataset, excellent accuracy has been seen in all folds. SVM has the lowest performance of 83 on the SRBCT dataset, while MLP has the greatest score of 95.40. The performance of all folds has been consistently good.

The stability and accuracy of the ChS-R approach are shown in Table 3 in five folds. Here we are discussing about accuracy and later we will discuss stability. For all datasets, all classifiers performed consistently. For the COLON dataset, SVM had the lowest average performance, while KNN had the highest. For the LUNG dataset, all classifiers performed consistently. The best results were achieved by SVM and MLP, while the worst results were achieved by RF. All classifiers performed poorly on the Leukemia dataset in folds ℓ and 5. The performance of the KNN classifier is the worst of all, whereas MLP is the best.

The lowest performance for the Prostate dataset was observed 81.80 by the KNN classifier. RF and SVM have continuously demonstrated great performance, with scores.

MLP has the lowest accuracy in fold 4 and fold ℓ for the MLL dataset. The average fold performance of MLP and RF classifiers is superior to the average performance with all samples. In all folds, there is diversity in performance for all classifiers.

The all-fold average performance for all classifiers in the Lymphoma dataset is much lower than the performance with all samples. In all folds, the performance of all classifiers is consistent.

All classifiers perform consistently well on the SRBCT dataset. MLP classifier has once again showed the lowest average performance, whereas SVM has shown the highest average performance.

We compare the proposed methods to existing methods to assess their stability. Existing feature selection methods, such as SU and ChS stability, are compared to our proposed SU-R and ChS-R stability. To calculate stability improvement $\mathbf{\Delta_i}$ is computed as in Eq. 8. The average stability improvement for each dataset is estimated as the sum of improvements in 5 folds divided by 5.

$$\Delta_i = \text{Proposed method Stability} - \text{Existing method Stability} \tag{8}$$

$\mathbf{\Delta_1}$ and $\mathbf{\Delta_2}$ are computed to show improvement of proposed SU-R with existing SU and ChS feature selection methods respectively in Table 2. The average improvement $\mathbf{\Delta_1}$ and $\mathbf{\Delta_2}$ have been found to be 20 and 21% correspondingly. The SRBCT dataset has the lowest stability improvement, whereas the lymphoma dataset has the most.

$\mathbf{\Delta_3}$ and $\mathbf{\Delta_4}$ are computed to show improvement of proposed ChS-R with existing SU and ChS feature selection methods respectively in Table 3.

The average improvement $\mathbf{\Delta_3}$ and $\mathbf{\Delta_4}$ have been found to be 15 and 16% correspondingly. Less stability improvement is observed for prostate dataset with ChS-R when compared with ChS. For MLL dataset less improvement is observed when compared to SU. Again for SRBCT dataset lowest improvement is observed when compared to existing method. Similarly for lymphoma dataset highest improvement is observed. Our purpose is to check stability improvement for high-dimensional datasets. From this analysis it has been observed that SU-R is more stable than ChS-R.

Fig. 2. Stability comparison of existing SU, ChS methods with proposed ensemble ChS-R and SU-R method

Figure 2 compares the proposed SU-R and ChS-R to SU and ChS in terms of stability. The proposed ensemble feature selection algorithms outperformed conventional SU and ChS for high-dimensional datasets, with an average stability gain of 20%. Existing approaches work well on the lung dataset. As a result, we've discovered that our suggested ensemble feature selection method is more stable than existing methods for high-dimensional datasets.

Table 4. Five fold average accuracy and stability calculated for SU-R and ChS-R method on seven datasets

Dataset	Average SU-R Stability	Average Fold SU-R Accuracy	Average ChS-R Stability	Average Fold ChS-R Accuracy
COLON	37.38	88.45	37.73	81.6
Lung	69.05	96.75	79.17	96.55
Leukemia	62.4	97.05	55.03	93.1
Prostate	65.99	80.7	55.03	86.05
MLL	38.51	88.17	34	85.33
Lymphoma	65.3	95.9	46	80.7
SRBCT	53.56	89.15	48	91.35
Average	**56.03%**	90.88%	**50.71%**	87.81%

Table 4 illustrates the average accuracy and percentage stability achieved by the SU-R and ChS-R techniques in five folds, as calculated from Table 2 and Table 3 data. The average SU-R stability is 56.03 and accuracy of 90.88 percent.

The average stability of the ChS-R is 50.71%, while the average accuracy is 87.81%.

Fig. 3. Stability and Average fold accuracy plotted for SU-R and ChS-R method

We plotted average accuracy and stability for each dataset obtained by both approaches in Fig. 3 to investigate the accuracy-stability tradeoff. This graph depicts the link between accuracy and stability. When the accuracy is lower, the stability is worse, as observed in the MLL dataset. When stability is considerable, as in the Lung, Lymphoma, and Leukemia datasets, accuracy is likewise high. As a result, we discover that for these high-dimensional datasets, there is a linear connection between accuracy and stability.

4 Proposed Systems SA-SU-R and SA-ChS-R

Individual rankings provided in ℓ folds to each feature have been observed in stability analysis to be varied. So we proposed methods SA-SU-R (Stability Analysis-Symmetric Uncertainty-ReliefF) and SA-ChS-R (Stability Analysis-Chi-Squared-ReliefF).

The working of these systems is shown in Fig. 4. The system computes feature ranking in each fold with SU-R and ChS-R ranking methods which is ensemble ranking method devised in [13]. The ranked features are then sorted in descending order as per score computed by ranking method. The system generates feature subset by combining the highest-ranking features from each fold. If the same feature is at highest rank, then second rank feature is added from that fold. In the next step feature subset is validated with RF, SVM, KNN and MLP classifiers.

Table 5 depicts the results of the new devised approaches on the seven high-dimensional datasets listed in Table 1. Here, Δ_5 and Δ_6 are computed to compare how well SA-SU-R and SA-ChS-R work.

The COLON dataset demonstrates no improvement with SVM and KNN classifiers when using the SA-ChS-R approach. Both SVM and MLP had no improvement on the Lung dataset. SVM has showed no improvement using the SA-SU-R approach for the leukemia dataset. In the case of prostate cancer, every classifier has improved. The

Fig. 4. Proposed SA-SU-R and SA-ChS-R system with stability analysis

MLL dataset showed the most improvement, and the MLP classifier showed the most improvement. With the SA-SU-R technique, MLP classifier showed no improvement in lymphoma and SRBCT datasets. When it comes to classifier performance, MLP has demonstrated the most progress. After MLP, SVM, KNN, and RF are ranked.

Table 5. Proposed methods with stability analysis SA-SU-R and SA-ChS-R shows accuracy measured on seven high dimensional datasets

DN	Name	Classifier	Acc. of SU-R with all samples	SA-SU-R	Δ_5	Acc. of ChS-R with all Samples	SA-ChS-R	Δ_6
1	COLON	RF	87	90	3	91	92	1
		SVM	90	92	2	90	90	0
		KNN	85	87	2	88	88	0
		MLP	84	87	3	77	90	13
2	Lung	RF	98	100	2	89	92	3
		SVM	100	100	0	100	100	0
		KNN	94	98	4	96	98	2
		MLP	100	100	0	100	100	0
3	Leukemia	RF	99	100	1	97	98	1
		SVM	100	100	0	97	100	3

(*continued*)

Table 5. (*continued*)

DN	Name	Classifier	Acc. of SU-R with all samples	SA-SU-R	Δ_5	Acc. of ChS-R with all Samples	SA-ChS-R	Δ_6
		KNN	99	100	1	97	98	1
		MLP	89	95	6	95	100	5
4	Prostate	RF	82	87	5	90.9	93	2.1
		SVM	79	89	10	90.9	94	3.1
		KNN	80	87	7	83.87	88	4.13
		MLP	85	90	5	85.15	95	9.85
5	MLL	RF	91.2	100	8.8	95	98.36	3.36
		SVM	82.8	100	17.2	90.2	99.09	8.89
		KNN	92.5	95.45	2.95	92.4	95.45	3.05
		MLP	68	90.9	22.9	63	86.36	23.36
6	Lymphoma	RF	94.73	100	5.27	94.74	98.84	4.1
		SVM	96.37	100	3.63	100	100	0
		KNN	96.83	100	3.17	91.57	100	8.43
		MLP	100	100	0	95.78	97.89	2.11
7	SRBCT	RF	92	96	4	92	93	1
		SVM	84	97.6	13.6	96	97	1
		KNN	88	95.2	7.2	92	94	2
		MLP	96	96	0	88	92	4
		Average	90.48	95.47	**4.99**	91.38	94.35	**3.91**

The total average accuracy reached with the SA-SU-R approach is 95.47%, while the SA-ChS-R method is 94.35%. On average, the new approach SA-SU-R improved by 4.99%, whereas SA-ChS-R improved by 3.91%. As a result, stability analysis has enhanced system correctness and consequently plays a crucial role in dealing with high-dimensional data. Both systems succeed in their goals of stability and accuracy.

5 Conclusions

Stability and accuracy are two sides of the same coin when it comes to data analytics. Furthermore, both are equally important. When the data pattern is not consistent, a stable model is more important. We used the Jaccard similarity measure to assess the stability of our proposed ensemble feature selection strategies. The data is separated into five folds, and the proposed methods are used to pick out features from each fold. The amount of identical features selected by each fold is determined by feature selection stability. Our proposed techniques, SU-R and ChS-R, yield 56.03 and 50.71% stability, respectively. Ensemble approaches have the advantage of combining the benefits of various underlying methodologies. When compared to previous SU and ChS approaches, SU-R achieves an overall improvement of 20 to 21% for high dimensional datasets. ChS-R, on the other hand, improves by 15 to 16%. Stability analysis is used to improve system accuracy. The percentage improvement for the SA-SU-R and SA-ChS-R is 4.99 and 3.91, respectively. In the future, an ensemble classifier will be used during the prediction stage to improve the accuracy of the system.

References

1. Kim, H., Choi, B.S., Huh, M.Y.: Booster in high dimensional data classification. IEEE Trans. Knowl. Data Eng. **28**(1), 29–40 (2016). https://doi.org/10.1109/TKDE.2015.2458867
2. Nogueira, S., Brown, G.: Measuring the stability of feature selection. In: Frasconi, P., Landwehr, N., Manco, G., Vreeken, J. (eds.) ECML PKDD 2016. LNCS (LNAI), vol. 9852, pp. 442–457. Springer, Cham (2016). https://doi.org/10.1007/978-3-319-46227-1_28
3. Das, S.: Filters, wrappers and a boosting-based hybrid for feature selection. In: ICML, vol. 1 (2001)
4. Xing, E.P., Jordan, M.I., Karp, R.M.: Feature selection for high-dimensional genomic microarray data. In: ICML, vol. 1 (2001)
5. Kalousis, A., Prados, J., Hilario, M.: Stability of feature selection algorithms: a study on high-dimensional spaces. Knowl. Inf. Syst. **12**, 95–116 (2007). https://doi.org/10.1007/s10115-006-0040-8
6. Kuncheva, L.I.: A stability index for feature selection. In: Artificial Intelligence and Applications, pp. 421–427 (2007)
7. Lustgarten, J.L., Gopalakrishnan, V., Visweswaran, S.: Measuring stability of feature selection in biomedical datasets. In: AMIA, pp. 406–410 (2009)
8. Dunne, K., Cunningham, P., Azuaje, F.: Solutions to instability problems with sequential wrapper-based approaches to feature selection. J. Mach. Learn. Res. **1**, 22 (2002)
9. Alelyani, S.: Stable bagging feature selection on medical data. J. Big Data **8**(1), 1–18 (2021). https://doi.org/10.1186/s40537-020-00385-8
10. Ben Brahim, A.: Stable feature selection based on instance learning, redundancy elimination and efficient subsets fusion. Neural Comput. Appl. **33**(4), 1221–1232 (2020). https://doi.org/10.1007/s00521-020-04971-y
11. Peng, H., Long, F., Ding, C.: Feature selection based on mutual information: criteria of max-dependency, max-relevance, and min-redundancy. IEEE Trans. Pattern Anal. Mach. Intell. **27**, 1226–1238 (2005)
12. Haury, A.C., Gestraud, P., Vert, J.P.: The influence of feature selection methods on accuracy, stability and interpretability of molecular signatures. PLoS ONE **6**(12), e28210 (2011). https://doi.org/10.1371/journal.pone.0028210
13. Sumant, A.S., Patil, D.: Ensemble feature subset selection: integration of symmetric uncertainty and Chi-square techniques with RReliefF. J. Inst. Eng. (India) Ser. B **103**, 831–844 (2021). https://doi.org/10.1007/s40031-021-00684-5
14. https://archive.ics.uci.edu/ml/datasets.php
15. https://csse.szu.edu.cn/staff/zhuzx/Datase

Pneumonia Prediction on X-Ray Images Using CNN with Transfer Learning

N. Krishnaraj(✉), R. Vidhya, M. Vigneshwar, K. Gayathri, K. Haseena Begam, and R. M. Kavi Sindhuja

Department of Computer Science and Engineering, Sri Krishna College of Technology, Coimbatore, Tamilnadu, India
{krishnaraj.n,vidhya.r,18tucs249,18tucs045,18tucs036, 18tucs055}@skct.edu.in

Abstract. Pneumonia is a lung infection that fills your air sacs with fluid or pus. Pneumonia can range from mild to life threatening. Countries like Morocco are very concerned since this disease kills several hundreds of children every day. So, being able to diagnose pneumonia can greatly benefit both health care and patients. This work proposes a new Convolutional Neural Network architecture model based on ResNet50 with the help of transfer learning. Using this model on the x-ray dataset of paitents made a phenomenal performance of 94.3% testing accuracy.

Keywords: Adaptive Moment Estimation (ADAM) · Binary Cross Entropy (BCE) · Root Mean Square Propagation (RMSP)

1 Introduction

A. *Deep Learning*

Deep learning is a subset of machine learning. It is based on the structure of the human brain. Deep learning makes use of a multi-layered model of algorithms known as multi-layered perceptron's or neural networks. Neural Networks are mainly used to do regression or classification. Long before Deep learning got famous; ML algorithms like Support Vector Machine, Logistic Regression, and Decision Trees were used all the time. These are called 7 Flat algorithms. Another important thing to note here is that ML algorithms need an additional step called feature extraction whereas DL algorithms don't need this additional step [1].

Deep Learning models tend to improve in accuracy as the amount of training data increases, but typical machine learning models such as Decision Tree and Logistic regression plateau after a certain point

J. I.-Z. Chen et al. (Eds.): ICIPCN 2022, LNNS 514, pp. 816–825, 2022.
https://doi.org/10.1007/978-3-031-12413-6_64

B. *Artificial Neural Networks*

A neural network is made up of interconnected units or nodes. These nodes are referred to as neurons. These are loosely patterned after the real neurons in the brain. A typical neural network design is made up of many layers. The input layer is the initial layer. The input layer, as the name implies, accepts input. The last layer is the output layer, which produces a vector containing the result. The values in the output layer are expressed by the elements in this vector. Anything in between these two levels is simply referred to as a hidden layer (Fig. 1).

Fig. 1. Structure of feed forward neural network

C. *Convolutional Neural Networks*

Although ANNs are good, they perform very poorly if the dataset is of images. For image classification, CNNs are usually preferred. This is primarily because, with ANNs, a 2D image is first transformed into a 1-D vector before model training [2]. This results in more parameters needed for the Neural Network (Fig. 2).

Fig. 2. Architecture of CNN

CNN is made up of three basic layers: 1) Convolution layer, 2) Pooling layer, and 3) Fully linked layer by sliding the filter over the input picture, the first layer of the convolutional network extracts features. For each sliding motion, the result is the element-wise aggregate of the picture's filters and their sum. The principal aim of the pooling layer is to lower the number of learnable parameters and, as a result, the computational burden [3]. Fully linked layers are the last layers that decide the output. The arriving input is reduced into a one-dimensional vector and then passed into the fully linked layer.

D. *Transfer Learning*

Transfer learning is not a novel idea that is exclusive to deep learning. Transfer learning makes use of previously learned models' weights and other parameters to train newer models. Knowing what, when and how to transfer aids in the development of the deep learning model [4]. Amongst the most important prerequisites for transfer learning is the availability of existing models that perform effectively. In this study, the VGG-19 and ResNet models are evaluated and utilized to build the final model. In transfer learning, we usually de-couple the output layer of the existing model and introduce our own layers and then train this new model (Fig. 3).

Fig. 3. Architecture of a typical transfer learning model

The rest of the article is structured as follows: Sect. 2 discusses the dataset as well as pre-trained models. Section 3 demonstrates how to fine-tune the model.

Finally, Sect. 4 highlights the findings of this research.

2 Materials and Methods

A. *Pneumonia Dataset*

The collection of X-ray pictures was obtained from Kaggle. The dataset is divided into three folders: train, test, and val, with subfolders for each picture class (Pneumonia/Normal). There are 5,863 JPEG X-Ray pictures. It's composed of images of about 2.3 GB, decomposed for testing, training and validated parts (Figs. 4, 5 and 6).

Fig. 4. X-ray images showing no signs of pneumonia

B. *Data Augmentation*

It is well known that Neural Networks are very dataset hungry. The more input data there is, the higher the accuracy. So, augmenting the data to create more inputs can help us achieve a better performance as long as the model doesn't over-fit [5]. As we'll be using regularization, which is explained later in this section, we don't have to worry about our model over fitting the training data.

In the x-ray dataset, the data augmentation is done with respect to rescale, shear, zoom, brightness, width shift and tilt. Other parameters like Horizontal and vertical flip are not really useful as it doesn't make sense medically and it should be counted as human error if we ever get an x-ray that is flipped or enlarged abnormally (Fig. 7).

Fig. 5. X-ray images showing signs of pneumonia

Fig. 6. Graph depicting the number of photos that exhibit and do not indicate pneumonia.

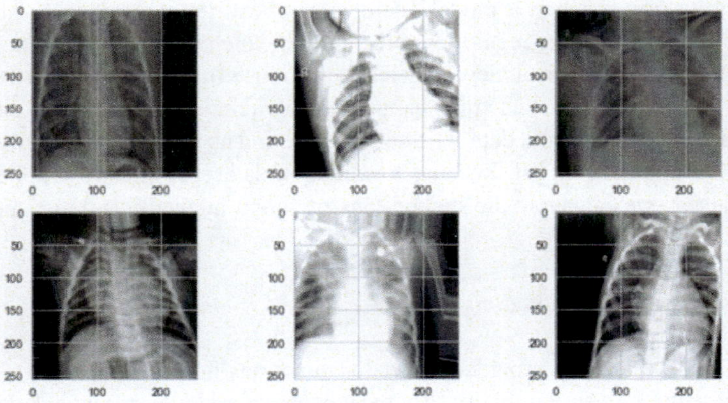

Fig. 7. X-ray images produced with data augmentation

C. ***Optimization***

The Gradient Descent method is perhaps one of the most significant algorithms in a Neural Network. Gradient Descent is employed in the loss function to aid in the discovery of local minima. It simply works by determining the slope of a curve at a given point and then calculating the difference between the weight and the product of slope and learning rate [6].

$$weight_{(t+1)} = weight_{(t)} - \alpha\ m_{(t)} \tag{1}$$

where, m = slope (or) gradient, α *= learning rat, t = time.*

This works well in general, but it may be improved by combining it with optimization methods. The Adam optimizer, which includes both moment and RMSP, is applied. The benefit of adopting this optimization algorithm is that it allows us to have an adaptable learning rate, which greatly aids in identifying a function's local/global minima [7].

D. ***Regularization***

Regression is a frequent method for dealing with model over fitting. Regularization via early halting will be utilized in this case (Fig. 8).

Fig. 8. Comparsion of underfit justfit and overfit with dataset

Over-fitting occurs when a model attempts to tweak its parameters to match the training data. It's almost like memorizing the model, which is undesirable since

Over-fitting can impair accuracy when the model is used to forecast against a different set of data [8]. To prevent this from happening, we create a new dataset called the validation data set from the old data set, and when we find no gain in validation accuracy, we simply halt the training [9]. Because a validation dataset is already supplied in the provided dataset, we can avoid the step of constructing it manually and just apply it on our model.

E. *Activation and Loss Functions*

Aside from the already provided activation functions in ResNet50 layers, we also have to choose the activation functions for the custom dense layers we added along with it. We'll be using Relu followed by sigmoid in the two dense layers (Fig. 9).

Fig. 9. Comparsion of Sigmoid and ReLU

As we are dealing with binary class prediction using sigmoid before the output layer is very preferred as it neatly allows you to have the value between 0 to 1 which is exactly what we want. Loss function which is also called cost function is an algorithm that allows us to define a cost for our function in an epoch and thereby allowing us to reduce the loss function which in turn also increases our accuracy. Choosing the loss function is pretty straight-forward as realistically the only two options are MSE and BCE. BCE is used here as the class type we are predicting is binary anyway.

3 Experimentation

A. *Vanilla CNN*

A simple CNN model with following layers shown in Table 1 is created and then tested against the dataset:

The above model was created using the ADAM optimiser and the BCE loss function. This model has a testing accuracy of 73.08% and a validation accuracy of 62.50%. Although CNNs outperformed ANNs in terms of accuracy, this is insufficient.

Table 1. Represents the layers of vanilla CNN

Layer	Output shape	Parameters
Conv2d_1 (Conv2D)	(None, 62, 62, 32)	896
Max_pooling2d_1	(None, 31, 31)	0
Conv2d_1 (Conv2D)	(None, 29, 29, 32)	9248
Max_pooling2d_2	(None, 14, 14, 32)	0
Flatten_1	(None, 6272)	0
Dense_1 (Dense)	(None, 128)	802944
Dense_2 (Dense)	(None, 1)	129

B. *Pre-trained Models*

VGG-19 and Resnet50 models were used for experimentation. Replacing the output layer with below layers in both the models, and making the other layers non-trainable, transfer learning is achieved. Although VGG-19 usually achieves better results in almost any image classification tasks, ResNet performs exceptionally well and almost always outclasses VGG-19 when it comes to medical image classification tasks as seen in the Table 2 below.

Table 2. Showing the remaining layers coupled with the existing pre-trained model

Layer	Output shape	Parameters
global_average_ pooling2d	(None, 62, 62, 32)	896
dense_1 (Dense)	(None, 31, 31, 32)	0
dense_2 (Dense)	(None, 29, 29, 32)	9248

The above two models were compiled with ADAM optimizer and BCE as its loss function. The VGG-19 model and the ResNet50 model were able to 84.45% and 91.98% testing accuracy respectively.

C. *Fine Tuning*

Clearly, these two transfer learning models were able to achieve a far better accuracy compared to a vanilla CNN. However, out of these two, ResNet50 was able to outperform VGG-19 by a significant margin. Although the accuracy is fine, we can do better with fine tuning [10]. By making the last 90 layers of the ResNet50 model trainable, we should be able to obtain improved accuracy, and it did, with a testing accuracy of 94.39%.

4 Results

The following result was obtained after experimenting with vanilla CNN, ResNet50, VGG-19, and fine-tuned ResNet50-based models (Table 3).

Table 3. Indicates the accuracy of the tested models

Model	Accuracy (%)
Vanilla CNN	73.08
VGG-19	84.45
ResNet50	91.98
ResNet50 (after fine-tuning)	94.39

This ResNet50 based model after fine tuning was able to achieve phenomenal results of 94.39% testing accuracy, and 0.1630 testing loss. Using the history of the model, plots for validation loss and accuracy were plotted for both before and after fine tuning, the ResNet50 based model (Figs. 10 and 11).

Fig. 10. Displays the Resnet50 model's loss and accuracy prior to fine tuning

Fig. 11. Displays the Resnet50 model's loss and accuracy after fine tuning

5 Conclusion

Tens of millions of people are killed by pneumonia each year. In addition, it is one of the primary causes of newborn death. Deep learning has advanced, and image classification is performing better than we could have hoped, making CNNs more dependable than ever. Transfer learning has been a significant advance in Deep Learning and, to a greater extent in image classification. As a result, suggested transfer learning model would inspire others to create new transfer learning models for medical healthcare challenges. In this work, a CNN model based on ResNet50 is suggested, with an accuracy of over 94%, which would hopefully aid medical specialists in the diagnosis of pneumonia.

References

1. Zhang, J., et al.: Viral pneumonia screening on chest X-rays using confidence-aware anomaly detection. IEEE Trans. Med. Imaging **40**(3), 879–890 (2021). https://doi.org/10.1109/TMI.2020.3040950
2. Kanakaprabha, S., Radha, D.: Analysis of COVID-19 and pneumonia detection in chest X-ray images using deep learning. In: 2021 International Conference on Communication, Control and Information Sciences (ICCISc), pp. 1–6 (2021). https://doi.org/10.1109/ICCISc52257.2021.9484888
3. Wan, S., Hsu, C.-Y., Li, J., Zhao, M.: Depth-wise convolution with attention neural network (DWA) for pneumonia detection. In: 2020 International Conference on Intelligent Computing, Automation and Systems (ICICAS), pp.136–140 (2020). https://doi.org/10.1109/ICICAS51530.2020.00035
4. Singh, A., Shalini, S., Garg, R.: Classification of pediatric pneumonia prediction approaches. In: 2021 11th International Conference on Cloud Computing, Data Science & Engineering (Confluence), pp. 709–712 (2021). https://doi.org/10.1109/Confluence51648.2021.9376884
5. More, K., Jawale, P., Bhattad, S., Upadhyay, J.: Pneumonia detection using deep learning. In: 2021 International Conference on Smart Generation Computing, Communication and Networking (SMART GENCON), pp. 1–5 (2021). https://doi.org/10.1109/SMARTGENCON51891.2021.9645844
6. Abubakar, M.M., Adamu, B.Z., Abubakar, M.Z.: Pneumonia classification using hybrid CNN architecture. In: 2021 International Conference on Data Analytics for Business and Industry (ICDABI), pp. 520–522 (2021). https://doi.org/10.1109/ICDABI53623.2021.9655918
7. Ayan, E., Ünver, H.M.: Diagnosis of pneumonia from chest X-ray ımages using deep learning. In: 2019 Scientific Meeting on Electrical-Electronics & Biomedical Engineering and Computer Science (EBBT), pp. 1–5 (2019). https://doi.org/10.1109/EBBT.2019.8741582
8. Swetha, K.R., Niranjanamurthy, M., Amulya, M.P., Manu, Y.M.: Prediction of pneumonia using big data, deep learning and machine learning techniques. In: 2021 6th International Conference on Communication and Electronics Systems (ICCES), pp. 1697–1700 (2021). https://doi.org/10.1109/ICCES51350.2021.9489188
9. Pant, T.R., Aryal, R.K., Panthi, T., Maharjan, M., Joshi, B.: Disease classification of chest X-ray using CNN. In: 2021 IEEE 6th International Conference on Computing, Communication and Automation (ICCCA), pp. 467–471 (2021). https://doi.org/10.1109/ICCCA52192.2021.9666246
10. Hasan, M.M., Jahangir Kabir, M.M., Haque, M.R., Ahmed, M.: A combined approach using ımage processing and deep learning to detect pneumonia from chest X-ray ımage. In: 2019 3rd International Conference on Electrical, Computer & Telecommunication Engineering (ICECTE), pp. 89–92 (2019). https://doi.org/10.1109/ICECTE48615.2019.9303543

Big Data Distributed Storage and Processing Case Studies

Tariqul Islam(✉) and Mehedi Hasan Abid

Daffodil International University, Dhaka, Bangladesh
{tariqul15-2250,mehedi15-226}@diu.edu.bd

Abstract. This research paper summarizes the evolution of big data concept and the techniques that are closely connected to big data concept, including CAP theorem, which happens to be one of the most important rules in modern distributed computing. Moreover, this paper focuses on the open-source software built from the Apache big data platform. It looks under the domain of NoSQL database Cassandra, which is the primary software considered in this study, and real use cases of Cassandra and related technologies in real-world scenarios. The proposed study also describes the development of certain use cases.

Keywords: Big data · MapReduce · NoSQL · Cassandra · Sparse column database

1 Introduction

Big data is a new innovation and technology that is actively reshaping the IT sectors present across the globe. However, it should be noted that this concept is extensively defiled and frequently used in the advertising certain products. The primary goal is to provide a comprehensive and objective perspective of what big data is, as well as to clearly distinguish what it is not, so that the reader can demonstrate after reading the work under this notion that certain technologies, methods, and alternatives were implemented for social good. Another motivation for this research study is the need to study something novel and intriguing in the current scenario. As big data technology is highly dynamic and constantly evolving, describing the term big data is not always an easy task and also there is no precise definition for this term. The fact that it is widely used in marketing, it also contributes to a more difficult definition of the term communication. The term "big data" refers to the manipulation of large scale datasets where typical tools and databases fail to process them. Data collection, organization, storage, search, sharing, analysis, and visualization are all the examples of dataset manipulation. Finding this limit or its exact definition is the most difficult challenge. Many additional works on this subject are already published. Big data preference for large datasets over several smaller ones that have the same volume and carry the same amount of information has been initiated due to searching and discovering the non-existent correlations, manifestations of business trends, and data evaluation in real-time or near real-time environment. Big data is not only about the data, it is actually describes about a more complex categorization, where other

J. I.-Z. Chen et al. (Eds.): ICIPCN 2022, LNNS 514, pp. 826–837, 2022.
https://doi.org/10.1007/978-3-031-12413-6_65

characteristics that are marked in the literature will also play a role in the abbreviation of 3 V derived from the initial letters of the three categories: Volume, Velocity and Variety as mentioned below.

Fig. 1. Scheme describing 3 V using set diagrams [20]

1.1 Volume

Today, data is far from being only in text form as it can also be stored in the form of music, image, or video. Due to this fact, there is an exponential increase in the amount of stored data and it is not exceptional for enterprise systems to store terabytes or petabytes of data. The data will generally constitute information, which is often evaluated from different angles and then saved and re-evaluated, although the original data remain unchanged when their number grows in an unprecedented manner. At this point, volume can be viewed as one of the characteristics of big data.

1.2 Velocity

The speed can be analyzed from two different perspectives. The first perspective expresses the speed at which the data is increasing and its appropriateness. For example, the theory of exchange rate development depends on the information whose value is unspoken and changes each minute. For instance: The speed of the information conveyed from a newspaper has also changed and television stations also obtain information through social networks. The volume of data is growing quickly and the timeliness of information has been rapidly shortened. The second perspective deals with the speed at which a particular user need to process data. Always, there will be different pieces of information that flow to us very often but their evaluation makes sense. For example: only once every 24 h. However, they apply to the data that we must handle in real time as it comes in. A good example of such data can be the live streaming information from the meteorological stations. The speed at which data is processed today is highly increased and therefore not only the volume of data but also the speed of processing represents big data.

1.3 Variety

Figure 1 shows the diversity of the data as structured/unstructured. It has already been mentioned that the data can take many forms. Moreover, the data present in the same form (for example, text) can be differently structured. This fact needs to be adapted and further the data should be stored and processed in other formats.

2 Background

From the beginning of computer age, data has to be analyzed [13]. With the increasing availability of modern technologies and their general acceptance in society, the boundaries of this need have shifted from government organizations to private where a huge amount of information and its analysis is required even in a small business environment. In the period of 1950s and 1960s, computers (and data processing and analysis) are only used in large corporations and research laboratories. For example, ENIAC computer generated the first model for weather forecasting. Analysts have also solved the shortest path challenge and many more [13]. In the period of 1970s to 1990s, the analytical activity has been extended to include medium-sized enterprises and startups. A good example of such cases is the design and development of first prediction model for decline and growth shares. It is also worth mentioning the first commercial tool for model-driven decision-making. Another important milestone is the emergence of companies such as eBay and Amazon. The increasing need to personalize online shopping has been initiated and Google is the first platform to implement the first search algorithm to increase the relevance of the results. In the period of 2000 to present, analytics has expanded to the area of small businesses and experts (individuals). It is starting to have a huge impact on human lives [2]. Dynamic change in commodity prices, and recommendations are becoming a matter of course for products, music, films, or traffic management. Fields such as analysis and processing of natural language from newspapers, e-mails, or social networks require an effective technology. Nothing stands in the way of big data, owing to the lack of computer power and data processing. It is anticipated that in the future, analytical work would govern day-to-day activities of every individual. In everyday life, the benefits of data analysis include predictive analysis and attack detection, research in or entirely personalized consumer engagement, even for small business chains [19].

Retail customer care and increased profits are the main motives for data processing and analysis. Based on the big data resources (web activity, customer card, anonymous customers) the user behavior can be predicted in each stage of the purchase. This behavior can also be linked to corporate data and search for correlations by using Map reduce mechanisms. For example, The biggest pioneer merger between big data and retail is undoubtedly the Tesco chain with its own Club Card loyalty card. They compile based on the customer's purchase history product rankings to recommend or even estimate the period of pregnancy of their customers [5]. Researchers are not surprised that big data is being used in measurement results or for searching the correlations in measured values. Nobel laureate Peter Higgs used the NoSQL database system and Cassandra, to process the data and proved the existence of the so-called Higgs boson [21]. There are several uses in the financial sector. For example, the above mentioned product recommendations according to the history of transactions and collection of personal data. Banks and other financial institutions offer customers with suitable financial products such as mortgage. Much more interesting is the most common use of big data is fraud detection, where banks are based on the analysis of all transactions to look for the patterns of fraudulent behavior and report certain transactions as suspicious and thus protect their clients or themselves. Based on the storage and subsequent processing of all user behaviors on this page, companies can optimize the website and its content may be restructured. After analyzing specific users' activity, it is possible for them to automatically customize the page's content and the interesting items for them without having to click on them in a complicated way. Big data is used in bioinformatics to map genomes and analyze sequences, for example. This data aids in the understanding DNA as well as the prevention and treatment for genetic abnormalities and congenital illnesses [16]. On the Internet, we can discover various organizations involved in data analysis and visualization, with all analysis and visualization taking place in third-party software. Among the most well-known is Good Data [1]. These organizations, on the other hand, specialize in the processing of company data and visualization in their BI solutions. Customized corporate solutions will serve as an alternative to select a comprehensive solution from a big data company, which will supply tools for data storage, processing, and visualization. Component programming or configuration is the customer's responsibility, and these firms provide licensing, training, and technical support. For example, this option is provided by IBM. Open Source and solutions based on the last option is to use open-source tools to save, analyze and visualize the data. This is the path selected for this research study. A small extension of this solution may be the companies offering commercial packages of these Open-Source solutions. It is also acting as a very popular approach in cases where it is necessary to combine several tools. Their configuration is very difficult and therefore the commercial packages offer ready-made solutions mostly with a new superstructure, which allows to perform some above-standard processes and activities.

3 Big Data Approach and Techniques

It is highly required to first define the basic technological frameworks of big data. Big data depends on the technologies and paradigms developed by technology giants, who were the first to hit technological frontiers and push them further. The order of these smart technologies and paradigms derive from the logical sequence as they relate and follow each other.

3.1 Distributed Systems

Distributed systems is a concept that extends far beyond the realm of big data. Following that a division is introduced to simplify the following phrase: Distributed systems are classified based on the processing power, distributed storage, or its combination. Distributed computing power is a system where the calculation of one will spread to the tasks over multiple computers. Parallel calculations are known in information technology for many years and are most often used for the scientific parallel compilation of source code or for other operations that would consume more time on one computer.

- Backup - We also know from the perspective of personal computers the trend where it depends on multiple physical devices to prevent their loss in the event of a technical failure on the equipment.
- Lack of capacity - From the realm of personal computers, we know users less needed files are stored on external peripherals because the capacity of disks in personal computers is usually in the order of hundreds of GB.
- Availability - If the user wants to access one file from both the computer and the work computer, this file must be physically located on both computers or must use some file-sharing software. The combined approach is obvious. It is used both the computing power of individual computers in the system, as well as their storage space.

Big data uses all these approaches. The huge amount of data discussed in the introduction is better handled on multiple computers. It is also logical that we store a large amount of data on multiple computers, and this is true even if we want to back up the data. However, we do not always use a combined approach. Commonly, companies are building huge computer farms that serve only as data warehouses and data centers. It always depends on the specific situation and method of use.

3.2 CAP Theorem

At the advent of distributed systems in 2000, scientist Eric Brewer published an article describing the so-called CAP Theorem [4], which is that distributed systems are directly related. This theorem states that distributed systems have these 3 main Properties:

- Consistency - A property that determines whether each request returns the correct result to the server. This means that the answer is equivalent to the specification of the required service. The exact meaning of consistency depends on the type of service. In the case of data, we define it so that each server has current and the same data.
- Availability - A property that respond for each request will provide an answer. A faster response is mostly preferred than a smaller one but in the context of the theorem, the answer must arrive at all possible situations. However, it widely known from the practice that a very late answer is as bad as the answer. Therefore this feature will be simplified and say that must always be available.
- Partition tolerance - This feature (as the only one) determines the behavior of the support system on which the service is running, and the conduct of the service itself. This property indicates whether or not the system can continue to function in the event of a system outage. According to Brewer, each distributed system may fulfil a maximum of two of these three properties.

In 2012, Brewer wrote another article [3] describing the state of his edema after 12 years. He explains that from the beginning, the label was "only 2 of 3" misleading and vague because it simplified a lot of things too much. For example, in a system with high granularity, several choices are made between the choice of C and A levels and all properties have values over time rather than binary and also that it depends on the construction of the system and its small nuances. At the same time, however, writes that, as a result, the theorem served its purpose and opened up systemic throwers' eyes when designing distributed systems and made them think over the advantages and disadvantages of individual system features. In the same year, another interesting article was published describing the current state of the CAP theorem [15]. It mainly describes the relationship of systems to CAP property choices.

3.3 Best Possible Segmented Consistency and Availability

The most common choice is guaranteed consistency with the maximum degrees. This is a natural choice for most systems. So, the server for each price returns the correct answer and then tries to optimize the highest possible degree and the shortest possible response time concerning network conditions. This approach makes the most sense if the computers are in the same data center and run the same service. A typical representative is "lock service" and service managing metadata for a distributed system with low granularity. The second most common group are systems for which there is a loss of availability unthinkable and therefore guaranteed and strives for the highest possible level of consistency. This procedure is best suited in situations where we have computers distributed across several data centers. In this case, availability can be rapid decrease with any error and therefore needs to be guaranteed. In these cases, therefore, designers sacrifice consistency to guarantee a fast enough answer, although it may not always be entirely correct. They are an ideal example of web caches and image servers. This is the most interesting option and is crucial for this work. Some systems do not have uniform requirements for all aspects of the service. Some require strong consistency and some high availability. To comply with the CAP theorem, the most natural possibility seems to be to divide the system into several components that will be specifically set. Thus, the whole system does not guarantee consistency or availability, but every part of the system provides it with the qualities it needs.

3.4 MapReduce

It is a programming model and framework first introduced by Google [10] for processing a large dataset. The user specifies a mapping function that processes pairs (key, value) and generates transient pairs (key, value), which are then passed to a reduction function that merges all the inter-values with the same inter-key. Many real-world examples can be translated into this paradigm. The advantage of these functionally written programs is that they are automatically well parallelizable. The system takes care of the details of data distribution and scheduling individual tasks on individual computers and handles errors and expected states. It allows even programmers with very little knowledge of parallel programming to write highly parallel and efficient programs. This paradigm has become the most important building block for big data. Thanks to this mechanism, we can process terabytes of data quickly and efficiently and save the calculated result to the database again. All big data processing procedures are directly or indirectly based on MapReduce (Fig. 2).

Fig. 2. MapReduce schema [10]

3.5 Cassandra

Although NoSQL databases are suitable for storing and processing big data there is a lot, we think cassandra is the most interested in golden in terms of architecture. Achieves great results in comparisons tests [8], used in production and large clusters of well-known companies, has the largest community and more widely used database in its category [9]. We has encountered this database before and we wanted to find out where the database is for here time moved. Due to the ease of integration with other software, war from the Apache big data Stack, Cassandra becomes the ultimate big data tool. Graduates may also benefit from the fact that Cassandra is industrial the most used database of its kind. Cassandra's architecture consists of several key points, each of which fulfills its role. I will briefly explain them in the following subsections. Cassandra is a fully symmetrical system where there is no single point of failure. Its architecture was designed with the premise of supporting hardware and software may fall. That is why Casandra offers symmetrical architecture where they are all nodes are equal and all data is stored across all nodes in the cluster. Commit logs on each node capture all entries and data are as well cached. When it is full, the data is written to disk and automatically replicate and then distribute throughout the cluster.

Virtual nodes Cassandra from version 1.2 brings major improvements in the form of virtual nodes, where each node splits into several virtual ones (each gets its random slice), rapidly reducing the number of tokens that belong to the node. Virtual nodes downplay and improve several tasks in Cassandra:

- Tokens for new nodes are assigned randomly themselves and for new nodes already it is not necessary to count and assign tokens manually.
- Repairing a dead node is much faster because everyone is involved do their nodes in the cluster and changes are incremental.
- After adding a new node, we don't have to do a complicated cluster predominance, local nodes are created evenly between the other nodes and take so much smaller amount of data from each node.
- Improves the use of heterogeneous machines in a cluster. In machines with different capacities, we can set a different number of virtual nodes (Fig. 3).

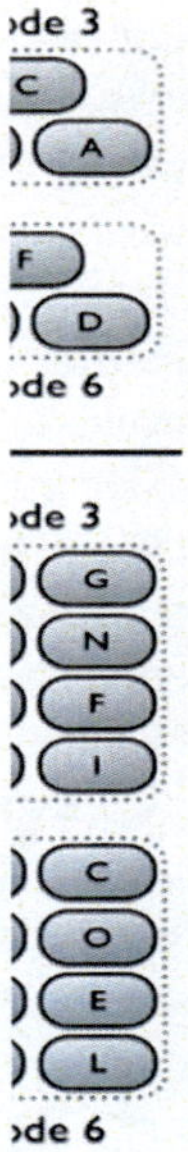

Fig. 3. Comparison of data storage in a circle with and without virtual nodes

When each (virtual) node has its own set of tokens assigned and we have determined how we will obtain tokens, data replication needs to be addressed. Number replica is given by the replication factor that is set for each database. A replication factor of 1 means that each row will exist only once. A replication factor of 2 means that each row will be saved twice and each time on another machine. In Cassandra, all nodes are equal, and the same is true for data replicas. So, there is no main replica, they are all of the same importance. The replication factor should not exceed the maximum number of nodes in the cluster, the data would then be redundant and lose power.

4 Implementation

Implementation of selected use cases generate online reports based on various criteria. This seemingly simple problem involves several non-trivial tasks. Above all, it is the amount of data that needs to be processed. If we were processing data sequentially, so we would have to go through approximately 1860 GB of system logs, which we certainly could not do in any relational database in a reasonable way time. We receive the data incrementally every day (always for the previous day) in several batches from various servers around the world. We decided to store the data in a table in Cassandra and all we will write is the data with the TTL attribute so that it will be automatically in half a year deleted. In here database table severity column shows the "danger" of the query as it was evaluated proxy server. We chose the design pattern as the primary key for time-lapse data, the user and the time stamp. To simplify this, in this example, we assume that the user will not access more than one-second website. In the real world, it would be good to add a web key page or refine the timestamp to milliseconds. The first solution to get data for reports is to query directly over the table a bunch of logs. The idea might seem completely wrong, but the opposite is true. If it were a specific date, access to the database is very fast despite the huge number of records and it turns out how powerful Cassandra is. The time required to access all data is for this amount of records also excellent, but for larger records, we would already do not reach the times suitable for an online report. Moreover, we can easily sort the data, and all the sorting and evaluation we would have to subsequently execute in code, which would be very time-consuming and inefficient. Not to mention a situation where we would ask about different departments and have to children questions a few. The data in the table does not correspond to real operation, all of the nodes are located on a one not very powerful computer with a single disk, so implementation of access log analyzer operations is very limited and subsequent communication with the test pro-Wednesday takes place over a local area network. Fair values would be many times lower, the aim is to point out the difference in values in different situations. The partition key is followed by other parts of the primary key, such as the date because according to the results will be sorted inside the partition, which is exactly what we want, plus date and webpage switched, we would have a query a specific website so we can ask at all date, which is meaningless given the features of the application. As can be seen, the order of the columns forming the primary key is very important. The mapping function is very simple in this case, we need to create a verbal schema of the key so that we can add up the visits on a given day. Therefore, we "normalize" the date. In the case of a date, we only need to trim it for the year, month and day. We are not interested in the time of the request. From information standardized date, address, user, and severity level, we will create the key to which we rank 1 because it is a single visit to the website in one day by a specific user. By reducing for each key we make the sum of values the number of visits and we write the result in the above table with the primary key described above, which is equal to the same key we created in the mapper. We before saw how to write a MapReduce job using the algorithm, defining the mapping and reduction function. In this case, It was relatively easy and it didn't give us much work. Nevertheless, we need to write a few classes and compile everything correctly and then run it. In the second we will describe how to use the Hive query language, which translates the queries themselves, puts in MapReduce code that runs and returns the result. In storage table we will keep the results of MapReduce jobs the same as in the first case, the results blanket is the same.

4.1 Implementation of Access Log Analyzer

This query looks complicated, but it is a simple SQL query that is well understood by programmers using SQL. The only more complicated thing is a function that trims our date. The result of this query is calculated for daily site visits for all users. We will run this query classically via the Hive JDBC connector and its results via the Cassandra JDBCwe write vector. The result is the same as in the previous case with a minimum writing code, at the cost of a little longer runtime, which is at the expense of generating a MapReduce work. Hive also experimentally supports embedding in Cassandra via external bulk, insert the INSERT statement into the table, where the table is the name of our external table. However, this procedure is really experimental and does not work for yet all cases and therefore its use is not generally recommended. When connected via an external table would therefore suffice for a single short query and data would be imposed on us as in the first case. The disadvantage, is that thus we cannot use the TTL function when writing to Cassandra. One possible optimization is to limit the entry to predefined units of time. For example, one day back, last week, or last month. So that choice the data will not be arbitrary, which would reduce the number of entries in the report table and the selection of this data would be faster.

4.2 Implementation of Selected Use Cases

The second and somewhat more fundamental optimization is that in the previous examples we always sort from the whole data set. We can prevent this by doing this only on today's data. Optimization is suitable for the incremental increase of data because we will not work unnecessarily with data that we have already calculated. Department Groups Part of the assignment is that the user may want to view a summary report for the whole department. In the current variant, we would have each user in the department make a separate query and calculate the results. We will expand the table with logs by a column with department information and create a second MapReduce or Hive query that will not take into account the column with the user, but the one about the department and the table with logs, we will also have records of specific departments. This will lead to duplication of data, which we do not mind because it is the least regard to other solutions disadvantageous. In case users would like to see summary statistics, such as the number of pages visited yesterday (overall) or the most popular pages in general, just create a Hive query for each of these statistics. Exactly for such purposes, Hive is the ideal choice. Evaluation of the use case I consider this case to be very useful from a practical and illustrative point of view and complex. This is a large case pointing to several technologies. From the point of view of a newly prepared subject, this example can be used for exercises dedicated to Cassandra, MapReduce, and Hive. It is necessary to understand the knowledge of mechanisms from the lecture. Taking into account the scope and complexity, we agreed with the supervisor to use it as the main one and detailed and other cases will be described and not implemented. This use case points to a modern approach to image storage and at the same time, the use of an approach where the database schema is not fixed. Although we model a table that has a fixed structure, internally it is translated into the format without a fixed schema. Again, this is a real use case, which is used and currently stores 7 GB of data.

The current imaging services store a significant number of images and require them to be shown. Sending a universal image to all inquiries is inappropriate, as image reduction depend on quantity. There are various drawbacks in saving images directly to the file system. This requires a solution to resolve the backup and distribution of the entire FS, control of individual files, and, most importantly the associated limitations. Cassandra is more than able to handle this task. The whole imaging system will consist of the following parts: image cache layer, image editing layer, layer for saving images and reading them. The first two levels are of interest in theoretical aspect and based on the specific conditions, the solution will be used in the system. Image caching is very useful because we want frequently requested images to never run out from disk. The image editing layer again depends on the specific choice. In our scenario, any image-capable software or library will be sufficient to downsize. The layer that saves and reads the images is far more crucial. When a request is received for an image, we will check to see whether we have an image in this size saved in this database. If yes, we will return the image; if not return but the image is still accessible in its original size (i.e., we store the image in the database), we will reduce it to the needed size and deliver it to the user. The image is then saved back to the database with these dimensions and prepared for future usage. If we also have a caching layer, it is appropriate to paint the image straight into cache. The primary key is the identifier, including the size of the image. The data is therefore they will partitioned according to this identifier (the first part of the primary key). The data column will contain a serialized image and the metadata column can contain specific metadata for this image. If we wanted all images with the same id to maintain one metadata would be better.

5 Conclusion and Future Work

The process of data segmentation during the initial data collection process face some significant challenges related to processing the big data resources as it has different interpretations. The majority of the articles and research proposals concentrated on specific sub-sectors and provided no comprehensive information. With the proposed considerable research study, the big data challenges with basic principles and its background have also been well explored. The most significant issue encountered in the process was the technology's inadequacy. During the explanation of technologies and software in the Apache Big data Stack, which supplies complicated features face a fully functional big data platform, this research study has discovered that the situation is not clear. These software tools are continually developing and have several versions, which are frequently incompatible retrospectively, and some versions are incompatible with other applications present on this platform. However, when designing the architecture of systems that employ more of this tool pile, this poses a severe challenge. It is not uncommon to discover that we must re-register a new function since the new version is incompatible with other applications. We have almost completed action from the selection of acceptable variants. This research study has investigated the most significant impediment that has been added to the usage and deployment of this platform.

References

1. GoodData: [cit. 2013-11-10]. http://www.gooddata.com
2. Boyd, D., Crawford, K.: Critical questions for big data: provocations for a cultural, technological, and scholarly phenomenon. Inf. Commun. Soc. **15**(5), 662–679 (2012)
3. Brewer, E.: CAP twelve years later: how the "rules" have changed. Computer **45**(2), 23–29 (2012)
4. Brewer, E.A.: Towards robust distributed systems. In: PODC, p. 7 (2000)
5. Bughin, J., Chui, M., Manyika, J.: Clouds, big data, and smart assets: ten tech-enabled business trends to watch. McKinsey Q. **56**(1), 75–86 (2010)
6. Chang, F., Dean, J., Ghemawat, S., et al.: Bigtable: a distributed storage system for structured data. ACM Trans. Comput. Syst. (TOCS) **26**(2), 4 (2006)
7. Datastax: When to use an index. [cit. 2014-01-05]. http://www.datastax.com/documentation/cql/3.1/cql/ddl/ddl_when_use_index_c.html
8. Datastax: Independent Benchmark Results Showing That Apache Cassandra Outperforms NoSQL Competitors by a Wide Margin (2013). http://www.datastax.com/2013/02/datastax-releases-independent-benchmark-results-showing-that-apache-Cassandra-outperforms-NoSQL-competitors-by-a-wide-margin
9. DBEngines: DB-Engines Ranking of Wide Column Stores (2014). http://db-engines.com/en/ranking/wide+column+store
10. Dean, J., Ghemawat, S.: MapReduce: simplified data processing on large clusters. Commun. ACM **51**(1), 107–113 (2004)
11. Dolák, O.: Big data, analyza dat. Cit. 2011-07-07. http://www.systemonline.cz/business-intelligence/big-data.htm
12. Ebay:Cassandra at Ebay (2013). http://www.datastax.com/wp-content/uploads/2012/12/DataStax-CS-eBay.pdf
13. Fico.com: The Analytics Big Bang [online]. [cit. 2011-07-07]. http://visual.ly/look-history-and-future-predictive-analytics-and-big-data
14. Ghemawat, S., Gobioff, H., Leung, S.-T.: The Google file system. In: ACM SIGOPS Operating Systems Review, vol. 37, pp. 29–43. ACM (2003)
15. Gilbert, S., Lynch, N.A.: Perspectives on the CAP Theorem. Institute of Electrical and Electronics Engineers (2012)
16. Swiss Big Data User Group: Big Data Across Industriess [online]. [cit. 2012-25-09]. http://www.slideshare.net/SwissHUG/big-data-usecases-across-industries-Georg-Polzer-teralytics
17. Lakshman, A., Malik, P.: Cassandra: a decentralized structured storage system. ACM SIGOPS Oper. Syst. Rev. **44**(2), 35–40 (2010)
18. Lynch, C.: Big data: how do your data grow? Nature **455**(7209), 28–29 (2008)
19. Manyika, J., Chui, M., Brown, B., et al.: Big data: the next frontier for innovation, competition, and productivity (2011). http://www.mckinsey.com/insights/business_technology/big_data_the_next_frontier_for_innovation
20. MCFadin, P.: 3 V of BigData. [cit. 2014-05-10]. http://smartdatacollective.com/yellowfin/75616/why-big-data-and-business-intelligence-one-direction
21. MCFadin, P.: Cassandra and Time Series [online]. [cit. 2013-11-10]. http://www.slideshare.net/patrickmcfadin/cassandra-20-and-timeseries
22. Schneider, R.D.: Hadoop for Dummies, John Wiley and Sons Canada, special edition. ISBN 978-1-118-25051-8
23. White, T.: Hadoop: The Definitive Guide: The Definitive Guide. O'Reilly Media (2009)

Author Index

J. I.-Z. Chen et al. (Eds.): ICIPCN 2022, LNNS 514, pp. 839–841, 2022.
https://doi.org/10.1007/978-3-031-12413-6

R

S

T

U

V

W

Z